Orthopaedic Biomechanics in Sports Medicine

Jason Koh • Stefano Zaffagnini
Ryosuke Kuroda • Umile Giuseppe Longo
Farid Amirouche
Editors

Orthopaedic Biomechanics in Sports Medicine

ISAKOS
International Society of Arthroscopy, Knee Surgery and Orthopaedic Sports Medicine

Springer

Editors
Jason Koh
Department of Orthopaedic Surgery
Orthopaedic & Spine Institute
NorthShore University HealthSystem
Skokie, Illinois
USA

Stefano Zaffagnini
IRCCS – Rizzoli Orthopaedic Institute –
2nd Orthopaedic and Trauma Unit
University of Bologna
Bologna
Italy

Ryosuke Kuroda
Department of Orthopaedic Surgery
Kobe University Graduate School of
Medicine
Kobe
Japan

Umile Giuseppe Longo
Department of Orthopaedic and
Trauma Surgery
Campus Bio-Medico University
Trigoria, Rome
Italy

Farid Amirouche
Department of Orthopaedic Surgery
Orthopaedic & Spine Institute
NorthShore University Health System
Evanston, Illinois
USA

ISBN 978-3-030-81548-6 ISBN 978-3-030-81549-3 (eBook)
https://doi.org/10.1007/978-3-030-81549-3

This Springer imprint is published by the registered company Springer Nature Switzerland AG
The registered company address is: Gewerbestrasse 11, 6330 Cham, Switzerland

Foreword

Biomechanics is the bridge between biology and orthopaedic surgery!

Musculoskeletal biomechanics has a very long history. In the fifteenth century, Leonardo da Vinci studied biomechanics of muscles and bones in the human body. Later, Galileo Galilei used mechanics to describe force, motion, and strength of materials. By the seventeenth century, Sir Isaac Newton gave us the laws of physics.

Since the 1970s, orthopaedic biomechanics has experienced an explosive growth as engineers and surgeons began to work together and used biomechanical principles to study tissue behaviour and joint motion, to evaluate body function as well as to develop mathematical models and computation methods for better implant design. Biomechanics has become a part of the regular curriculum in orthopaedic education. It became clear that a good understanding of biomechanics and quantitative data is essential for successful clinical management of patients.

It is so timely for the International Society of Arthroscopy, Knee Surgery, and Orthopaedic Sports Medicine (ISAKOS) to ask Drs. Jason Koh, Stefano Zaffagnini, Ryosuke Kuroda, Umile Giuseppe Longo, and Farid Amirouche to lead the efforts in preparing a textbook entitled *Orthopaedic Biomechanics in Sports Medicine*.

These editors are to be congratulated for carefully selecting and assembling a strong list of corresponding authors from around the world. The topics covered (33 chapters) are relevant and encompassing. These chapters are logically organized into eight (8) parts that flow seamlessly from Basic Principles of Biomechanics and Properties of Materials to Biomechanics of Shoulder, Elbow, Hip, Knee, Ankle, and Foot, and ending with Tissue Repair Techniques.

Moreover, the materials presented have a strong international flavour. This writer knows almost all of the corresponding authors personally and has a high regard for their knowledge and clinical expertise. As such, each chapter will provide the reader with materials on biomechanics that are comprehensive and clinically relevant as well as translational.

It is without a doubt that this book will be of significant value to the ISAKOS community, especially for our young and up and coming surgeons in orthopaedic sports medicine.

Savio L. -Y. Woo
Musculoskeletal Research Center
Department of Bioengineering
Swanson School of Engineering
University of Pittsburgh
Pittsburgh, PA, USA

Preface

On behalf of the editors, contributing authors, and the International Society of Arthroscopy, Knee Surgery, and Orthopaedic Sports Medicine (ISAKOS), it is with great pleasure that we invite you to enjoy our book *Orthopaedic Biomechanics in Sports Medicine*.

Knowledge of orthopaedic biomechanics is fundamental to understanding normal function, injury, repair, and recovery in sports medicine. Over the past several years, there have been many recent advances, spurred by the growth and development of such areas as image analysis, robotics, and finite element analysis. Researchers have used these techniques to build upon existing information to gain a better understanding of how the body moves and functions.

In this book, an international team of experts has come together to provide a comprehensive introduction to orthopaedic biomechanics. These clinicians and researchers have a profound understanding of biomaterials, tissues, and joint function from the shoulder and elbow to the hip, knee, and ankle. Based on the newest research and analytical techniques, this book is intended to assist clinicians, researchers, and students in gaining a background in orthopaedic biomechanics as it relates to sports medicine.

As editors and authors, it has been a pleasure to create this work, and we hope you find this book interesting and informative. We have all been students and learned from many pioneers before us and hope that this contributes to further understanding and future progress in orthopaedic biomechanics.

Many thanks,

October 2021

Evanston, IL, USA — Jason Koh
Bologna, Italy — Stefano Zaffagnini
Kobe, Japan — Ryosuke Kuroda
Rome, Italy — Umile Giuseppe Longo
Chicago, IL, USA — Farid Amirouche

Contents

Part I

Basic Principles in Orthopaedic Biomechanics

Biomechanics of Human Joints

1

Farid Amirouche and Jason Koh

1.1 Introduction

The mechanics of the shoulder joint has been the subject of numerous studies focusing mostly on three clinical complications such as the glenoid implant loosening [1–3], rotator cuff repair [4, 5], and total shoulder arthroplasty [6, 7]. TSA has consistently provided successful outcomes for patients with severe arthritis [8], and patients where rotator cuff repair was unsuccessful resulting in joint instability and cruciate pain. The GH joint is the most mobile joint in the human body, supported and guided by a structure composed of rotator cuff muscles, ligaments, and an intrinsic cartilaginous glenoid surface that conforms to the humeral head geometry.

F. Amirouche
Department of Orthopaedic Surgery, Orthopaedic & Spine Institute NorthShore University Health System, Evanston, Illinois, USA

Department of Orthopeadic Surgery, College of Medicine, University of Illinois, Chicago, Illinois, USA
e-mail: FAmirouche@northshore.org, amirouch@uic.edu

J. Koh (✉)
Department of Orthopaedic Surgery, Orthopaedic & Spine Institute NorthShore University HealthSystem, Skokie, Illinois, USA

University of Chicago Pritzker School of Medicine, Chicago, Illinois, USA

Northwestern University McCormick School of Engineering, Evanston, Illinois, USA

Related challenges to TSA include edge contact-friction loading and rim stress due to eccentric loading of glenoid implant resulting from migration and translation of the humeral head causing the rocking-horse phenomenon [9], this is an inherent problem due to implant bone fixation. So, bone–material interface and methods of fixation including the bony stock conditions for the glenoid dictate the course of success and become a real challenge to overcome. This is associated with several contributing parameters such as glenoid asymmetric bone loss, glenoid rim defect, bony conditions and depth, and sizing of the onlay or inset and bone removal.

It's obvious that the complexity of the GH joint is one that requires modeling and analysis to gain insight into its kinematic-kinetic function. Modeling of the human joint using a simple 2D or 3D model has usually evolved around the knee, hip and spine but very few models exist for the shoulder joint. This is due in part to the shoulder griddle complexity in terms of articulating surfaces and muscle structures and ligaments that make up the anatomy of the GH joint. Current shoulder models include the one by a group in Sweden [10, 11] related to the model of Hogfors et al. [12, 13], the Newcastle shoulder model [14], the shoulder part of the AnyBody Modeling System [15], the open SIMM Stanford model [16], and the Delft Shoulder Model (DSM) to name few.

These complex musculoskeletal models of the shoulder and upper arm have been developed to

J. Koh et al. (eds.), *Orthopaedic Biomechanics in Sports Medicine*,
https://doi.org/10.1007/978-3-030-81549-3_1

study the shoulder joint and its muscles during abduction-adduction. Biomechanical models are essential tools for simulation and predictions of clinical scenarios not feasible with standard in vivo studies.

Computer modeling and simulation of human joints has risen to new heights in recent years, mainly because of the medical needs and advances in imaging and sensors technology. The growing belief that models can provide more quantitative explanations of how the neuromuscular and musculoskeletal systems interact to produce movement is vital to orthopedic surgeons.

Dynamic models provide knowledge of muscles and joint reaction forces, stress-strain behavioral responses to load, micromotion measurements and potential pitfalls which are invaluable to the clinician. They provide essential parameters to the design, validation of orthopedic implants and patient-specific orthopedic solutions. Recent studies use computed tomography (CT) images to build patient-specific 3D models which are then used to assist the clinicians to visualize the pathology more in depth and propose new orthopedic solutions. Through different simulation studies one can minimize complications by selecting the right size of implants and demonstrate a definitive path for best outcome solution. This is all virtual allowing for interaction and evaluation of clinical benefits to patients.

To create realistic shoulder models cadaveric studies are used to improve on the assumptions of the GH joint under different constraints conditions and address the limit of range of motion associated with translation and rotation. Preserving the patient natural motion is key to understanding the shoulder joint stability and kinematics.

The role of the muscles forces and their corresponding moment arms are crucial in the analysis of less conforming prostheses designed for the replacement of GH joint. Detailed multibody muscle forces provide the only feasible method to quantifying or evaluating the dynamics of the shoulder joint under different loading conditions.

Early work by Duca et al. [17] used a simple 2D model to investigate the relationship of the forces between the deltoid and supraspinatus muscles. The relationship between these forces can be observed by calculating the absolute force of the individual groups of muscle fibers. It was confirmed that a linear relationship between muscle force and cross-sectional area is found in all muscles. Hogfors early work [18], investigated the force vectors for different muscles around the shoulder joint including greater pectoral muscle, trapezius, deltoid, infraspinatus muscle, and subscalpular muscle. Each muscle is defined as a vector with coordinates system defined by the relative insertion point of each muscle on the major bones in the system.

Pitching in baseball has drawn a lot of interest from past and current researchers as the number injuries of professional athletes became crucial in understanding the mechanism of injury due to high forces and moments produced by the muscles. Andrews et al. [19] study was performed to try to understand how the biceps are potentially responsible for the tear of the glenoid labrum.

Karlsson et al. [20] followed with additional work on muscle forces predictions at the shoulder joint as function of abduction angle in the sagittal plane. The goal is to create 3D force contribution to the joint and understand the balance issue as it relates to each muscle. The challenge of validating these forces is to compare the model data to ECG. This is still work in progress but modeling in 3D is evolving rapidly with the help of imaging techniques and multibody dynamics software. As is the case most of the 3D models are redundant systems and require optimization. Predicted forces deviate considerably from EMG results.

The first Delft Shoulder Model [21] consists of a comprehensive 3D inverse-dynamic model of the shoulder used to evaluate the muscle and joint contact forces, muscle lengths, and moment arms. The muscle force data was validated using force–time curves and EMG signals [22]. The study by Debski et al. [23] was used to measure the in situ force distribution inside the glenohumeral joint capsule as well as the compliance of the joint under a load. Ten cadaver shoulders were studied and the in situ force was measured during varying angles of abduction.

Research by Huang et al. [24] focused on the supraspinatus muscle and the supraspinatus tendon under uniaxial loading. This a load to failure test where strains and stresses are recorded until the tendon failed. Additional force modeling and inverse dynamics can be found in work by Holzbaur et al. [25] where the muscle force-generating characteristics are determined using a Hill-type muscle model.

Additional experimental work was needed to validate modeling techniques of muscle forces. The shoulder joint muscles moment arms were evaluated by Ackland et al. [26], where eight cadavers were used and the moments for seven major shoulder muscle groups were recorded for three different motions: scaption, coronal plane abduction, and flexion.

The contact pressure at the glenoid humeral interface is critical in determining the effects of rotator cuff repair, total shoulder arthroplasty on the balancing and kinematics of the joints. In essence, reconstruction of the joint affects the contact and hence several studies looked into this problem. Hammond et al. [27] used seven cadavers to measure the contact area and contact pressure of the glenohumeral joint at different abduction angles. Measurements for contact area and pressure were made using Tekscan sensors.

There is some evidence that the humeral head translation contributes to the instability and constraints resulting from Reverse shoulder construction surgery of the shoulder joint [28], clinically through X-rays and CT scan one can measure how much translation the humeral head actually migrates in the superior direction otherwise we need to use cadaveric shoulders and try to measure the humeral head during abduction-adduction of the upper arm and validate the translation part. Quental et al. [29] assumed an ideal spherical GH joint and validated the assumptions with cadaver and in vivo studies. He simulated the humeral translation and reported that the GH joint translations can't be measured experimentally because the limitations of conventional motion systems.

Finally, it is worth mentioning that the geometry of the glenoid bone loss can be evaluated, and custom glenoid models can be designed and 3D printed to address patient-specific orthopedic problems. Gunther et al. [30] described how the design of onlay and inset can be addressed as function of the bone quality of the patients. Modeling and simulations combined leads to better understanding of the GH shoulder implants design.

1.2 Anatomy of the Shoulder

The shoulder and its structural support constitute the human joint with greatest mobility. Its anatomy is such that it possesses motion capabilities to perform and support extraneous loads and tasks. From pitching in baseball and mastering the speed of ball delivery to heavy lifting, upper extremities strength and stability are essential to successful gymnasts trying to achieve another high level performance, and therefore, with versatility and strength come injuries and risks. The glenohumeral joint (GH) has a bony congruity at its articulating surfaces which render its kinematics and stability a major factor contributing to the shoulder integrity and longevity. The shoulder joint has to rely on its connective tissues, adjacent ligaments, and muscles to provide and maintain stability.

The Rotator Cuff is made up of muscles tendons that merge together to form a capsule around the shoulder joint used to keep the humeral head in the socket joint. Hence, it supports the GH joint in its rotational motion and constraints its movement to maintain stability. Tendons in the rotator cuff can be injured easily because of their limited range of motion and load as they bounded to limited stretch within a bounded joint space.

The four important tendons that form the rotator cuff are Subscapularis, Supraspinatus, Infraspinatus, and Teres Minor (see Fig. 1.1). These muscles are made up of numerous muscle fibers which in turn are made up of myofibrils. These myofibrils contain regions called Z discs where two adjacent Z discs make up the Sarcomeres. The sarcomeres are the smallest functional unit of a muscle and are responsible for the contractile activities of the muscle [31].

The role each of these muscles play has been the subject of a number of investigations, for instance the subscapularis is seen to have the

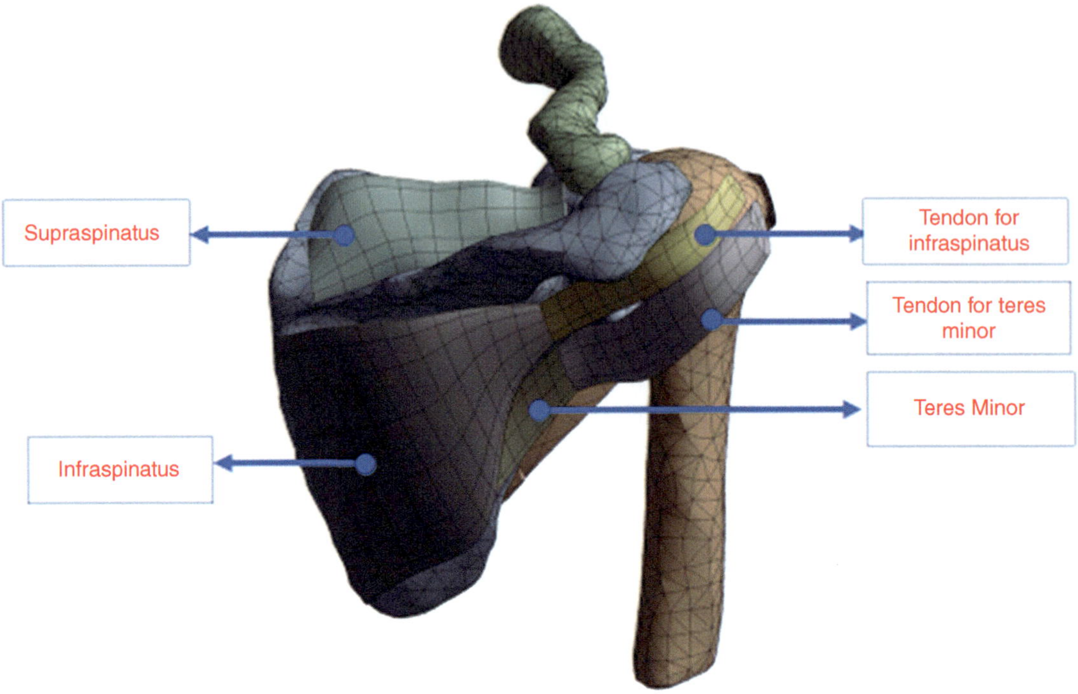

Fig. 1.1 Anterior View of Right Shoulder. A CT based model with Rotator Cuff muscles

highest capacity of force production followed by infraspinatus, supraspinatus and, teres minor [32]. Subscapularis has upper 60% insertion tendinous and lower 40% insertion muscular [33]. In absence of the supraspinatus, Deltoid takes over the function of shoulder elevation completely [34] where otherwise the deltoid is only responsible for a 90° abduction before the supraspinatus perform the rest. In absence of Supraspinatus, it is found that there is 101% increase in middle deltoid force generation to initiate the abduction [35].

In addition to the rotator cuff and its muscles the humeral head sits in a congruent surface called the glenoid which forms a cavity that acts as the articular surface at the lateral side of the scapula. It is concave with a cartilaginous glenoid fossa, retroverted and measures 6–8 cm^2. Inferiorly, the articular cartilage may be thin and fibrous. Superiorly, intra-articularly, the supraglenoid tubercle serves as the attachment site for the long head of biceps tendon.

There are three main shapes of the glenoid surface described in the literature: (1) concave, (2) flat or (3) convex. There are various postero-inferior rim shapes which play a crucial role in fitting and keeping the humeral head to be secure within the glenoid cavity. The rim is still not quite well understood in terms of whether it should be preserved during a resurfacing as it is completely sacrificed during the Onlay resurfacing of the glenoid. In clinical studies, there are three main shapes that have been proposed that best describe the rim geometry: triangular, J, and delta. The J and delta shapes are found to play a key role in posterior shoulder instability [36].

1.3 Kinematics of the Shoulder

The humeral head and the glenoid articular surface are congruent and rotate similar to a ball socket joint within certain range of motion. The humeral head is believed to be more convex in the anterior-posterior direction than in the superior-inferior.

Kinematics is strictly limited to angular displacements, angular speed and angular accelerations for rotational motion. It is also associated with translational motion where displacement, velocity, and acceleration of specific points such as the humeral head center or any other points critical to the analysis. These kinematical parameters are often affected by the joint pathology, geometric properties, kinematical and geometrical constraints resulting from muscles and ligaments. In the case of the GH joint we are mostly concerned by changes in kinematics resulting from rotator cuff tears, arthritic conditions at the joint or simply the arthroplasty effects on the joint and patient joint mobility.

The humeral head looks like a sphere and this was confirmed by Soslowsky et al. [37] in their experiments where they concluded that the articular surface of the humeral head could be approximated by a sphere with small deviations of less than 1% of the radius. This is further supported by Boileau and Walch [38] where they estimated that the difference between the two diameters of the humeral head assumed is less than 1 mm in 88.2% of the tested specimens. They also reported during the arm elevation that the superior-inferior translation of the humeral head is only 0.3–0.35 mm in normal shoulders.

Glenohumeral kinematics is driven by connective forces that have some points of attachments to the joint structure mechanism. Any changes to the muscles forces result in different kinematics of the shoulder joint. Tracking the humeral elevation for a healthy human joint can provide a base reference from which one can assess the results of rotator cuff tears which are associated with superior migration of the humeral head during arm elevation. Motion usually is a result of drivers such as muscles designed to achieve certain tasks. In the presence of different pathologies some of these forces effort diminishes resulting in more work to achieve our goal or simply be limited to certain range of motion.

Here are some clinical evaluations that provide some insight into the changes in kinematics as a result of shoulder instability. For example, in stiff shoulder joints, the humeral head moves upward during the first degrees of arm elevation. Browne et al. observed that the maximal glenohumeral elevation was obtained in a plane 23"anterior to the scapular plane with the arm in 35" of external rotation. Harryman et al. [39] demonstrated that translation of the humeral head is accompanied by passive movements of the glenohumeral joint. The humeral head translates either anteriorly or posteriorly and this can be seen when the arm is in flexion or extension. This forced translation is thought to be induced by the tightening of the capsule-ligamentous structures during motion. Excessive tightness of the anterior capsule after anterior capsulorrhaphy leads to posterior subluxation.

1.4 Modeling and Dynamic Analysis of the GH Joint

1.4.1 FE Models of Glenohumeral Joint Stability

The *shoulder* is the most frequently *dislocated joint*. The dislocation is usually due to a traumatic injury or loosing of the capsular ligaments. Anterior dislocation is defined as when the top of the humerus is displaced forward, toward the front of the body. This is the most common type of shoulder dislocation, accounting for more than 95% of cases. A shoulder subluxation refers to a partial dislocation of the shoulder joint. The 10-year incidence *rate* of *shoulder dislocations* in US Army soldiers is around 3.13%. Complexity of the GH joint is such that one needs more sophisticated tools to examine the shoulder response to shoulder structural integrity, articular joint surfaces, and rotator cuff partial and full tears.

In the past decade there has been several studies conducted to investigate the contact problem and instability of the glenohumeral joint using 3D Finite Element (FE) models (see Fig. 1.2). The contact stress is usually indicative of high peak loads wear and most FE models will provide the stress and strains of the complete meshed surface or volume which includes bones, cartilages and ligaments. Buchler et al. [40] built a FE model which consisted of the major rotator cuff muscles and investigated the contact stresses due to the changes of the geometrical shape of the

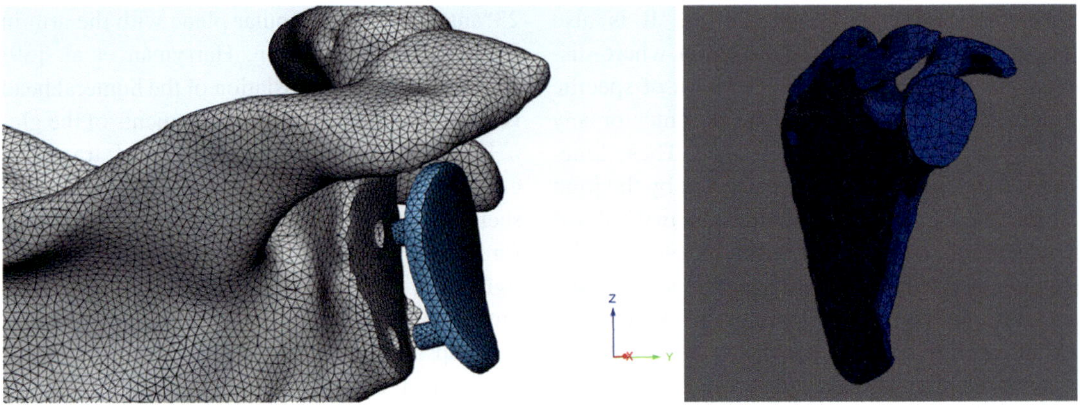

Fig. 1.2 Finite Element shoulder model with onlay prosthesis implant with two pegs

humeral head and the glenoid contact surface in both healthy and pathological conditions. The result of the study suggests that the changes in shape of the humeral head may lead to joint instability. However, one of the drawbacks of this study as is the case with a large number of FE models is lack of validation. FE model once developed and the user has determined that it is reliable it can be used for analyzing the biomechanical effect of the shapes of prosthetic humeral heads after shoulder arthroplasty.

Terrier et al. [41] FE model of the shoulder focused on the supraspinatus deficiency as it relates to osteoarthritis. The result suggested that supraspinatus deficiency increases the upward migration of the humeral head resulting in increased eccentric loading, and thus decreases GH joint stability. An extension of this FE shoulder model was by Walia [42] to investigate the effect of combined defects of the humeral head and of the glenoid. It was found that the glenohumeral joint stability was decreased significantly.

The shoulder joint is a dynamic system that uses forces generated by the muscles to perform certain tasks preserving the stability of the joint and limiting the stress conditions between the articular surfaces. So in reality the FE models alone wouldn't suffice in creating realistic models. As is the case in gait and motion analysis, inverse dynamics is used to evaluate forces and that require the solution of equations of motion with defined constraints and initial conditions. The use of multibody software packages such as OPEN SIM allows creating shoulder models mimicking the natural dynamics of a person. As an option in the analysis an additional FE model can be used to perform the contact problem using the loading conditions resulting from the inverse dynamics solution. Recent FE model of the glenohumeral joint used estimated muscle forces as the loading condition, and the humerus was allowed to move freely with six degrees of freedom [43]. These loading and boundary conditions enable the FE analyses to simulate motions of the shoulder complex closer to its realistic physiological condition than those with predescribed or artificially defined constraints. This study provides a useful framework for future FE studies of the shoulder complex. All models have limitations and one needs to be aware of what is perceived as acceptable conditions under which a model is being simulated including the geometry and structure of muscles and other soft tissues.

1.4.2 FE Models of Rotator Cuff Tears

Rotator cuff tendon tears technically can be associated with any of the muscles/tendons that form that capsule which support the humeral head. One of the most common tendon tear is the supraspinatus tendon often a result of trauma injury.

Tears when not treated can cause chronic shoulder pain, and may lead to secondary degenerative changes in the shoulder. The fundamental mechanism of what initiate a rotator cuff tear remains a problem of concern and certainly it is still unclear what are the shear forces that initiate or produce these tears. A number of studies have been conducted to explore the underlying biomechanical stress-strain responses which might cause rotator cuff tears using FE shoulder models.

Early work in FE addressed the stress-contact relation at the shoulder joint in the presence of rotator cuff tear. For instance, Luo et al. [44] investigated the supraspinatus tear using a 2D FE shoulder model by looking into the tress distribution and how that is related to rotator cuff tear. The stress was evaluated at different abduction angle of 0°, 30° and 60° respectively and also under two different acromial conditions with and without subacromial impingement. It was found that the high stress concentration due to the subacromial impingement could initiate a tear. Another aspect of their findings also showed that this tear may occur on the bursal side, the articular side or within the tendon rather than only on the bursal side as previously reported. This 2D modeling lead to other FE models used to investigate this claim further.

Two further studies based on Luo's model were conducted by first Wakabayashi et al. [45] where they applied histological differences at the tendon insertion in their FE model to analyze the stress distribution of the supraspinatus tendon. Their results found that the maximum principal stress of the tendon occurs at the region in contact with the humeral head rather than at the insertion point. Secondly, Sano et al. [46] in their model they examined the stress distribution in the pathological rotator cuff tendon and revealed potential partial-thickness tears at three different locations: on the articular surface, on the bursal surface and in the mid-substance close to the insertion. It was found that high stress concentration occurs at the articular side of the insertion and the site of tear. The two studies used same modeling method and conditions as Luo's original model, but employed different histological parameters to simulate pathological conditions. With additional information gained about the rotator cuff tears both these models lacked experimental data, a common problem with FEA models.

The first 3D FE model of rotator cuff tears was reported by Seki et al. [47] in 2008 and was used to analyze similarly to previous 2D models the stress distribution in the supraspinatus tendon. It was found that the maximum stress occurs in the anterior portion of the articular side of the tendon insertion rather than at the tendon contact point with the humeral head as suggested by 2D analyses. The analysis provided an important information that explains the frequent occurrence of rotator cuff tears at this site. The 3D model analysis provided an insight that was not possible in 2D, where the anteroposterior direction was investigated showing that the anterior part of the rotator cuff is not in contact with the superior surface of the humeral head. However, this study only analyzed part of the shoulder at 0° abduction without experimental validation.

Adams et al. [48] used a 3D FE model of the glenohumeral joint to investigate the effect of moment arms on muscle forces and rotator cuff tears. An important of this study shows that the moment arms of infraspinatus and teres minor muscles were generally decreased. Consequently, the muscles attached to the torn tendons are required to generate more forces for the same motions, and the overall strength of the shoulder is decreased. This study revealed a potential relationship between shoulder strength reduction and the rotator cuff tears location.

Another 3D FE study of rotator cuff tears was conducted by Inoue et al. [49] which included the rotator cuff muscles as well as the deltoid muscle. This model was used to investigating the rotator cuff tears and stresses at the articular and bursal sides of the supraspinatus tendon were found to cause shearing between the two layers, which is believed to initiate partial-thickness tears. As is the case of other FEA models this 3D model limitation needed further imaging techniques and experimental studies to construct soft tissues and muscles to build additional fidelity into the muscle forces.

1.5 3D Reconstruction and Design of Patient-Specific Shoulder Prosthesis

Patient-specific models are today feasible if CT data is available with a sequential imaging file with a minimum of 0.5 mm interval. High resolutions images will provide better rendering and surface smoothing which allows for a CAD model and mesh that is close to the actual bony structure of the patient. Further modeling of the muscles, tendons, ligaments, and cartilage require MRI images and additional volumetric methodologies to model the muscles associated with the rotator cuff such as the supraspinatus muscle.

The processing of CTs is commonly performed by software that uses Hounsfield units as a contrast to differentiate between soft tissues and bones to isolate the geometry being formed. MIMICS. The data conversion method used to develop the CAD surface model and fill to obtain a solid model of the shoulder can be obtained using MIMICS software [50]. The CAD drawings in its final form is exported to either Mimics own analysis module or other FE software to create their corresponding mesh needed for the analysis. Usually, we perform all the CAD steps in terms of design of implants such as the onlay in CAD environment, perform the resurfacing of the glenoid to accommodate different designs before the assembly. Once the complete stage of sequential imaging, and CAD steps are complete, we move into the final stage of analysis where we can investigate fixation, stability, cycling loading, stress, and any outcome that we perceive will be best to evaluate the shoulder implant surgical success (see Fig. 1.3).

The assignment of material properties for bone and cartilage are usually assumed based assumptions one makes on bone quality, thresholding, DEXA, and mathematical bone density evaluation based on the Hounsfield unit. Isotropic elastic material properties for bones and cartilage can be used, and soft tissues can be modeled as Ogden hyper-elastic material properties [51].

The onlay design with pegs or Keel shape can be addressed by comparing the contact stress as function of loading and different abduction angle, internal or external rotation of the arm (see Fig. 1.4). The effects of muscle moment arms on joint stability can also be simulated to assess the shoulder joint performance.

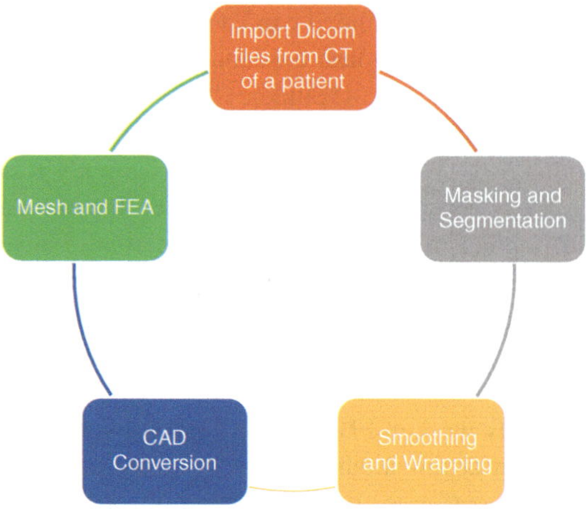

Fig. 1.3 Modeling phases in patient-specific implant design and analysis

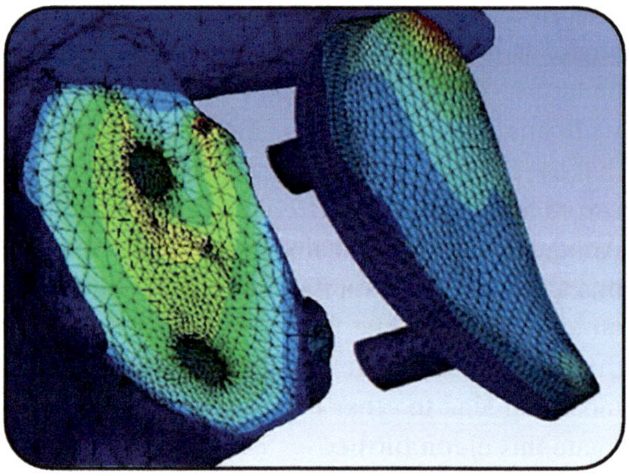

Fig. 1.4 Stress analysis of the implant-scapula assembly

1.6 Future Biomechanics Work and Clinical Challenges

Biomechanics uses mechanics principles to define mathematical equations and functions to explain parameters that are used for modeling of musculoskeletal systems including bones, and soft tissues. It employs imaging techniques, testing, and experimental studies to validate its assumptions of its biomaterials. A powerful tool such as FEA which is founded on the first law of mechanics is becoming a standard tool used to perform mechanical tests accepted by the FDA to obtain approval for implants failure analysis, quality control, stress, and strains and a number of tests that usually are limited to experimental testing. FEA when combined with visual graphics can help translate its output in a form that is understandable and easy to translate. Today we are merging into a multiscale Multiphysics stage integrating several concepts into one framework to be able to investigate the interaction of bone with fluid, investigating bone ingrowth, and connecting our understanding of mechanics from the cell level to the macroscale so one can model and understand how fractures heal and how different fixations affects its healing process. Simulation and 3D printing provide an even more promising world to orthopedics as we become patient specific to address different pathologies and optimize the treatment protocol including implant design process. In fact, simulations performed by use of computers through a display of augmented reality graphics or virtual simulations are referred to as in silico contrary to in vitro and in vivo, where all studies are performed on a computer via modeling and simulation. In silico medicine has been in use in orthopedics where surgical planning of complex cases is studied and potential solutions are examined beforehand giving the surgeon the best option for patient-specific remedy and orthopedic solution. In Silico is work in progress as these models are currently undergoing constant changes to bring these models alive where they can be biologically driven to show the evolution of pathologies and diseases, progression and bone remodeling, and become dynamical so when can examine different stages and progress in real time of bone-implant interaction, fracture healing, spine fusion among and potential soft tissues and nerve regeneration.

What follows are some of the challenges that lie ahead to orthopedic biomechanics. We need to

1. Address how realistic can a patient-specific model be represented using CTs and MRI imaging and realistic connective tissues such as muscles. What are the limits of realistic simulations and can Multiphysics provide better understanding of the nonlinearity of material properties. When modeling a human joint is it possible to collect and store patient data of bone and tissues in healthy conditions so that they can be compared to joints with pathological conditions, including definitions of boundary and loading conditions based on individualized physiological data. Special attention should be paid to the consistency between different FE models and models obtaining from multibody models used for muscle force estimation through inverse dynamics.
2. Define comprehensive models describing the complete joint to capture its complexity including appropriate 3D representations of all major connective tissues and their delicate interactions.
3. Define standard testing protocols to validate these models so that can be developed for clinical use.
4. Allow for more group testing of Dynamic FE simulations so that they are built based on physiologically realistic boundary and loading conditions that are measurable and can be used by others. This allows for better simulation and understanding of human joints.
5. Develop advanced medical imaging, sensing techniques tools to quantify in vivo strain and stress distributions of soft and hard tissues in both normal and pathological conditions so as to validate FE models more rigorously.
6. Develop augmented and virtual reality tools to assist surgeons overcome information redundancy and improve on real time interaction and feedback when needed to improve patient's outcome.
7. Develop artificial intelligence and robotics platform for minimally invasive procedures, complex cases and integrate "savoir faire" and surgical techniques with validated simulation studies.
8. Improve on in silico Multiphysics and multi scale modeling to allow for interaction of fluid and structure interface for better understanding of bone–implants interfaces, and better 3D bioprinting of scaffolding prosthesis.

References

1. Raphael BS, Dines JS, Warren RF, Figgie M, Craig EV, Fealy S, Dines DM. Symptomatic Glenoid loosening complicating total shoulder arthroplasty. HSS J. 2010 Feb;6(1):52–6. https://doi.org/10.1007/s11420-009-9148-1.
2. Jeffrey T. Abildgaard, Jared C. Bentley, Richard J. Hawkins, John M. Tokish "Arthroscopic removal of a loose polyethylene Glenoid component with bone grafting and patch augmentation for Glenoid Osseous defect" Arthrosc Tech. 2017 Jun; 6(3): e529–e535. Published online 2017 May 1. doi: 10.1016.
3. Albert S, Gee O, Angeline ME, Dines JS, Dines DM. Shoulder instability after total shoulder arthroplasty: a case of arthroscopic repair. HSS J. 2014;10(1):88–91. https://doi.org/10.1007/s11420-013-9373-5.
4. Jancuska J, Matthews J, Miller T, Kluczynski MA, Bisson LJ. A systematic summary of systematic reviews on the topic of the rotator cuff. Orthop J Sports Med. 2018;6(9):2325967118797891. https://doi.org/10.1177/2325967118797891.
5. Rossi LA, Rodeo SA, Chahla J, Ranalletta M. Current concepts in rotator cuff repair techniques: biomechanical, functional, and structural outcomes. Orthop J Sports Med. 201;7(9):2325967119868674. https://doi.org/10.1177/2325967119868674.
6. Nitin B. Jain, Ken Yamaguchi "The contribution of reverse shoulder arthroplasty to utilization of primary shoulder arthroplasty "J Shoulder Elbow Surg. Published in final edited form as: J Shoulder Elbow Surg. 2014 Dec; 23(12): 1905–1912. Published online 2014 Oct 7. doi: https://doi.org/10.1016/j.jse.2014.06.055
7. Schoch BS, King JJ, Wright TW, Vigan M, Werthel JD. Defining the tipping point for primary shoulder arthroplasty. JSES Open Access. 2019;3(4):273–7. https://doi.org/10.1016/j.jses.2019.09.009.
8. Pia C ten Voorde, Jeppe V Rasmussen, Bo S Olsen, Stig Brorson Resurfacing shoulder arthroplasty for the treatment of severe rheumatoid arthritis: Outcome in 167 patients from the Danish Shoulder Registry. Acta Orthop. 2015; 86(3): 293–297. Published online 2015 May 13. doi: https://doi.org/10.3109/17453674.2015.1018761
9. Levy DM, Abrams GD, Harris JD, Bach BR Jr, Nicholson GP, Romeo AA. Rotator cuff tears after total shoulder arthroplasty in primary osteoarthritis: A systematic review. Int J Shoulder Surg. 2016;10(2):78–84. https://doi.org/10.4103/0973-6042.180720.
10. Asadi Nikooyan A, Veeger HEJ, Chadwick EKJ, et al. Development of a comprehensive musculoskeletal model of the shoulder and elbow. Med Biol Eng Comput. 2011;49:1425–35. https://doi.org/10.1007/s11517-011-0839-7.
11. Praagman M, Chadwick EKJ, van der Helm FCT, et al. The relationship between two different mechanical cost functions and muscle oxygen consumption. J Biomech. 2006;39(4):758–65.
12. Makhsous M, Hogfors C, Siemien'ski A, et al. Total shoulder and relative muscle strength in the scapular plane. J Biomech. 1999;32(11):1213–20.
13. Hogfors C, Peterson B, Sigholm G, et al. Biomechanical model of the human shoulder joint—II. The shoulder rhythm. J Biomech. 1991;24(8):699–709.
14. Kontaxis A, Johnson GR. The biomechanics of reverse anatomy shoulder replacement-a modelling study. Clin Biomech (Bristol, Avon). 2009;24:254–60.
15. Flores-Hernandez C, Eskinazi I, Hoenecke HR, D'Lima DD. Scapulothoracic rhythm affects glenohumeral joint force. JSES Open Access. 2019;3(2):77–82. https://doi.org/10.1016/j.jses.2019.03.004.
16. Mansouri M, Reinbolt JA. A platform for dynamic simulation and control of movement based on OpenSim and MATLAB. J Biomech. 2012;45(8):1517–21. https://doi.org/10.1016/j.jbiomech.2012.03.016.
17. De Duca CJ, Forrest WJ. Force analysis of individual muscles acting simultaneously on the shoulder joint during isometric abduction. J Biomech. 1973;6(4):385–93. https://doi.org/10.1016/0021-9290(73)90098-5.
18. Högfors C, Sigholm G, Herberts P. Biomechanical model of the human shoulder--I. Elements J Biomech. 1987;20(2):157–66.
19. Andrews JR, Carson WG, Mcleod WD. Glenoid labrum tears related to the long head of the biceps. Am J Sports Med. 1985 Sep-Oct;13(5):337–41.
20. Karlsson D, Peterson B. Towards a model for force predictions in the human shoulder. J Biomech. 1992;25(2):189–99.
21. der Helm V. A finite element musculoskeletal model of the shoulder mechanism. J Biomech. 1994;27(5):551–69. https://doi.org/10.1016/0021-9290(94)90065-5.
22. Nikooyan AA, Veeger HE, Westerhoff P, Graichen F, Bergmann G, van der Helm FC. Validation of the Delft shoulder and elbow model using in-vivo glenohumeral joint contact forces. J Biomech. 2010;43(15):3007–14. https://doi.org/10.1016/j.jbiomech.2010.06.015.
23. Gabriel MT, Wong EK, Woo SL, Yagi M, Debski RE. Distribution of in situ forces in the anterior cruciate ligament in response to rotatory loads. J Orthop Res. 2004;22(1):85–9. https://doi.org/10.1016/S0736-0266(03)00133-5.
24. Huang CY, Wang VM, Pawluk RJ, et al. Inhomogeneous mechanical behavior of the human supraspinatus tendon under uniaxial loading. J Orthop Res. 2005;23(4):924–30. https://doi.org/10.1016/j.orthres.2004.02.016.
25. Holzbaur KR, Delp SL, Gold GE, Murray WM. Moment-generating capacity of upper limb muscles in healthy adults. J Biomech. 2007;40(11):2442–9. https://doi.org/10.1016/j.jbiomech.2006.11.013.
26. Ackland DC, Pandy MG. Moment arms of the shoulder muscles during axial rotation. J Orthop Res. 2011;29(5):658–67. https://doi.org/10.1002/jor.21269.

27. Hammond G, Tibone JE, McGarry MH, et al. Biomechanical comparison of anatomic humeral head resur-facing and hemiarthroplasty in functional glenohumeralpositions. J Bone Joint Surg. 2012;94:68–76.
28. Browe DP, Voycheck CA, McMahon PJ, Debski RE. Changes to the mechanical properties of the glenohumeral capsule during anterior dislocation. J Biomech. 2012;45(11):2028–34. https://doi.org/10.1016/j.jbiomech.2013.10.040.
29. Quental C, Folgado J, Ambrósio J, Monteiro J. A new shoulder model with a biologically inspired glenohumeral joint. Med Eng Phys. 2016; https://doi.org/10.1016/j.medengphy.2016.06.012.
30. Gunther SB, Lynch TL. Total shoulder replacement surgery with custom glenoid implants for severe bone deficiency. J Shoulder Elb Surg. 2012;21(5):675–84.
31. Wilson A, Lichtwark G. The anatomical arrangement of muscle and tendon enhances limb versatility and locomotor performance. Phil Trans R Soc B. 2011;366:1540–53. https://doi.org/10.1098/rstb.2010.0361.
32. McCausland C, Sawyer E, Eovaldi BJ, et al. Anatomy, shoulder and upper limb, shoulder muscles. In: StatPearls [Internet]. Treasure Island, FL: StatPearls Publishing; 2020. Available from: https://www.ncbi.nlm.nih.gov/books/NBK534836/
33. Altintas B, Bradley H, Logan C, Delvecchio B, Anderson N, Millett PJ. Rehabilitation following subscapularis tendon. Repair. Int J Sports Phys Ther. 2019;14(2):318–32.
34. Werthela J-D, Bertelli J, Elhassana BT. Shoulder function in patients with deltoid paralysis and intact rotator cuff. Orthop Traumatol Surg Res. 2017;103:869–73.
35. Escamilla RF, Andrews JR. Shoulder muscle recruitment patterns and related biomechanics during upper extremity sports. Sports Med. 2009;39(7):569–90.
36. Di Giacomo G, Piscitelli L, Pugliese M. The role of bone in glenohumeral stability. EFORT Open Rev. 2018;3(12):632–640. doi: https://doi.org/10.1302/2058-5241.3.180028.
37. Soslowsky LJ, Thomopoulos S, Tun S, Flanagan CL, Keefer CC, Mastaw J, Carpenter JE. Overuse activity injures the supraspinatus tendon in an animal model: A histologic and biomechanical study. J Shoulder Elb Surg. 2000;9:79–84.
38. Szabo I, Boileau P, Walch G. The proximal biceps as a pain generator and results of tenotomy. Sport Med Arthrosc. 2008;16:180–6. https://doi.org/10.1097/JSA.0b013e3181824f1e.
39. Harryman DT 2nd, Sidles JA, Clark JM, McQuade KJ, Gibb TD, Matsen FA 3rd. Translation of the humeral head on the glenoid with passive glenohumeral motion. J Bone Joint Surg Am. 1990;72(9):1334–43.
40. Zheng M, Zou Z, Bartolo PJ, Peach C, Ren L. Finite element models of the human shoulder complex: a review of their clinical implications and modelling techniques. Int J Numer Method Biomed Eng. 2017;33(2):e02777. https://doi.org/10.1002/cnm.2777.
41. Terrier A, Reist A, Vogel A, Farron A. Effect of supraspinatus deficiency on humerus translation and glenohumeral contact force during abduction. Clin Biomech (Bristol, Avon). 2007;22(6):645–51. https://doi.org/10.1016/j.clinbiomech.2007.01.015.
42. Walia P, Miniaci A, Jones MH, Fening SD. Theoretical model of the effect of combined glenohumeral bone defects on anterior shoulder instability: a finite element approach. J Orthop Res. 2013;31(4):601–7. https://doi.org/10.1002/jor.22267.
43. Zheng M, et al. Finite element models of the human shoulder complex: a review of their clinical implications and modelling techniques. Int J Numer Methods Biomed Eng. 2017;33(2):e02777. https://doi.org/10.1002/cnm.2777.
44. Luo ZP, Hsu HC, Grabowski JJ, Morrey BF, An KN. Mechanical environment associated with rotator cuff tears. J Shoulder Elb Surg. 1998;7(6):616–20. https://doi.org/10.1016/S1058-2746(98)90010-6.
45. Wakabayashi I, Itoi E, Sano H, Shibuya Y, Sashi R, Minagawa H, Kobayashi M. Mechanical environment of the supraspinatus tendon: a two-dimensional finite element model analysis. J Shoulder Elb Surg. 2003;12(6):612–7. https://doi.org/10.1016/s1058274603002143.
46. Sano H, Wakabayashi I, Itoi E. Stress distribution in the supraspinatus tendon with partial-thickness tears: an analysis using two-dimensional finite element model. J Shoulder Elb Surg. 2006;15(1):100–5. https://doi.org/10.1016/j.jse.2005.04.003.
47. Seki N, Itoi E, Shibuya Y, Wakabayashi I, Sano H, Sashi R, Minagawa H, Yamamoto N, Abe H, Kikuchi K, Okada K, Shimada Y. Mechanical environment of the supraspinatus tendon: three-dimensional finite element model analysis. J Orthop Sci. 2008;13(4):348–53. https://doi.org/10.1007/s00776-008-1240-8.
48. Adams C, Baldwin M, Laz P, Rullkoetter P, Langenderfer J. Effects of rotator cuff tears on muscle moment arms: a computational study. J Biomech. 2007;40(15):3373–80. https://doi.org/10.1016/j.jbiomech.2007.05.017.
49. Inoue A, Chosa E, Goto K, Tajima N. Nonlinear stress analysis of the supraspinatus tendon using three-dimensional finite element analysis. Knee Surg Sports Traumatol Arthrosc. 2013;21(5):1151–7. https://doi.org/10.1007/s00167-012-2008-4.
50. https://www.materialise.com/
51. Rosen DP, Jiang J. A comparison of hyperelastic constitutive models applicable to shear wave elastography (SWE) data in tissue-mimicking materials. Phys Med Biol. 2019;64(5):055014. https://doi.org/10.1088/1361-6560/ab0137.

2 Shoulder Muscles Moment Arm Contribution to Glenohumerol Joint Motion and Stability

Jason Koh, Lorenzo Chiari, and Farid Amirouche

2.1 Introduction

Surgeons are now performing more arthroscopic rotator cuff repairs than ever before and more of these repairs are being performed in ambulatory surgery settings. The NSAS data from 1996 and 2006 is indicative of such claim where ambulatory surgical procedures increased from 21,236,913 to 34,738,440 and rotator cuff repairs went from 58,846 (0.3%) to 272,148 (0.8%) in respective years [1]. The NHDS data also shows how the rotator cuff impatient hospitalization for the same period decreased by almost half in 1996 and was reduced further to 0.05% in 2006. Similarly, anterior crucial ligaments reconstruction are among the most common sports medicine procedures performed in the United States [1, 2]. The focus of many orthopedic surgeons and of rotator cuff-related research is on the prevention of rotator cuff tears and the development of prevention programs and test procedures that help become the identifiers without the burden of cost. What is more at stake is the challenge to determine when does a rotator cuff patient be able to return to normal activities after rehabilitation [3].

There are several different methods of evaluating shoulder joint ligament injury, symptoms and performance before and after surgery. Most are based on patients' symptoms performance outcome measures in a test, and surgeon clinical findings [4]. In sports there are specific strength assessment tests such as arm power which can discriminate between the injured and non-injured patients and those who had soft tissue reconstruction. Other tests used such as isokinetic tests are not always predictors of the functional stability of the shoulder joint but work well for the knee [5, 6]. To overcome some of these tests limitations additional methods were introduced [7–9].

In the absence of standardized methods to assess joint function and stability after arthroplasty, rotator cuff repair surgeons rely great deal on basic methodologies [10, 11]. Conclusive correlations have yet to be completely formulated between joint function and its kinematics and

J. Koh
Department of Orthopaedic Surgery, Orthopaedic & Spine Institute NorthShore University HealthSystem, Skokie, Illinois, USA

University of Chicago Pritzker School of Medicine, Chicago, Illinois, USA

Northwestern University McCormick School of Engineering, Evanston, Illinois, USA

L. Chiari
IRCCS Istituto Ortopedico Rizzoli, Bologna, Italy

F. Amirouche (✉)
Department of Orthopaedic Surgery, Orthopaedic & Spine Institute NorthShore University Health System, Evanston, Illinois, USA

Department of Orthopeadic Surgery, College of Medicine, University of Illinois, Chicago, Illinois, USA
e-mail: FAmirouche@northshore.org, amirouch@uic.edu

J. Koh et al. (eds.), *Orthopaedic Biomechanics in Sports Medicine*, https://doi.org/10.1007/978-3-030-81549-3_2

muscle–ligament forces. This is partly due to the complexity of the joint and the reliability of manual measurement collected at the clinic or lab.

Moment arm is a tool that is mostly used to assess the torque relation to lever arm and muscle contributing power at the joint. It has a clinical value in measuring torque producing capacity and its effect on kinematics and joint range of motion. Moment arm is usually a measure used when the actual force is not known or difficult to measure but the torque and center of the joint rotation is measurable. Since the torque is indicative of moment which is the product of the force and the distance from the center of rotation, so changes in moment arms provide an insight into the power capacity produced by individual muscles and ligaments after rotator cuff repairs and ACL reconstruction.

The focus of this chapter is to present a study on moment and provide detailed look into its calculation through experimental investigation of rotator repair techniques using cadaver shoulders specimens. The experiment provides an insight how moment arms of the major muscles spanning the glenohumeral joint during abduction, flexion and axial rotation are measured and the benefits we gain from their calculations. Moment arm data for deltoid, supraspinatus, infraspinatus, subscapularis and teres minor will be reported when measured using the definition of force-moment and intricate geometries of tendons insertion points as well using motion capture devices to monitor positions during abduction.

It is important to highlight some of the clinical settings where the moment arms can be advantageous in comparing surgical outcomes of rotator cuff repair tears. Arthroscopic and open rotator cuff repairs are two procedures with distinct advantages and disadvantages. Arthroscopy allows for preservation of the deltoid muscle whereas open repair techniques allow for easier transosseous fixation to better replicate the footprint of the supraspinatus tendon [12].

Surgeons seem to favor impatient surgery more for rotator cuff and for those with high volume. Mini-open repair was the most frequently used method of repair (46.2%) followed by open (36.6%) and arthroscopic (14.5%) [13, 14]. The moment arm will be further explained in the assessment of joint stability, GH range of motion, and muscle forces joint contribution during abduction.

The four muscles joining together to support the humeral head to form the rotator cuff are the subscapularis (supporting the shoulder anteriorly); the supraspinatus (maintains the superior aspect); the infraspinatus and teres minor (located on the posterior shoulder) [15]. Rotator cuff failure usually results from tendinopathy that transforms from partial to full thickness tears involving the supraspinatus tendon and may proceed to entangle the infraspinatus tendon and/or the subscapularis tendon [16].

Surgeons rely on imaging techniques such as ultrasound and MRI to identify the cause of tendon injuries [17, 18]. The decision-making in the treatment of the rotator cuff relies mainly upon the surgeon diagnosis, patient conditions, and the extent of the tear. Other factors such as the muscle quality and patient health conditions including pain play a role in the development of the right protocol for treatment.

Moment arm has been used extensively to evaluate the muscle contribution to joint rotation, kinematics, and moment-torque capacity it produces including the different moment contributions in different planes of rotation. In a vector form the moment is defined by the cross product of the displacement vector measured from the center of rotation to the force line of action. This method is referred in the literature as the geometrical approach used in calculating the moment arm [19–22]. It requires two points to define a vector for the force line of action and the knowledge to accurately measure the center of rotation by which the force is driving the body segment (usually the humerus in case of GH joint). With advances in imaging techniques, geometry reconstruction using CTs and MRIs combined with motion tracking systems this method is currently becoming more popular.

The other method used in estimating the moment arm stems from the concept of virtual work where virtual displacement brings in the excursion rate of change divided by the joint angle differential. Essentially it is a direct rewriting of the virtual displacement used to develop the work performed by a force or a muscle force

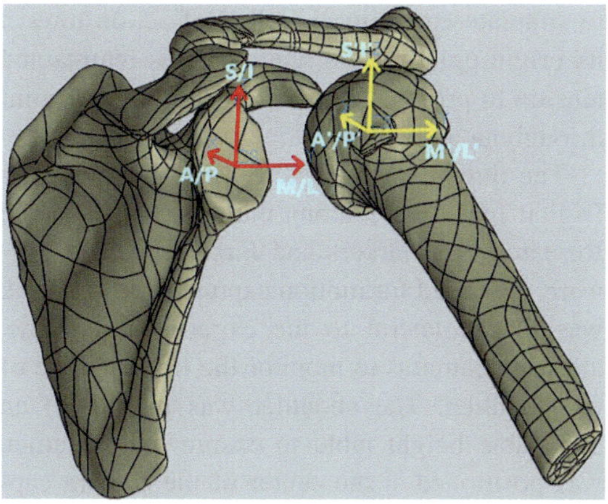

Fig. 2.1 The model was created from CT scans using Materialise Mimics and imported into SolidWorks. The coordinate systems (axes) have been extracted at the approximations of glenoid surface and humeral head. (S/I denote superior inferior, M/L medial lateral, A/P anterior and posterior)

in this case. It has usually been found advantages to the geometric method because you do not need to know the location of the center of rotation point [23–28].

Most kinematic data provide the measurement of the rotation angle and excursion changes associated with the muscle force line of action (insertion points of muscle points). It is unconceivable that an angle can be measured in void. Whether assuming a 2D plane motion of 3D general joint motion a reference axis is usually assumed at the center joint of rotation and hence one needs to be able to position and define the center of rotation (see Fig. 2.1). This is usually performed by measuring the intersection of two lines of one moving body and one being fixed.

2.2 Experimental Methods Used in RC Repairs and Muscles Moment Arm Evaluation

Rotator cuff surgical repair techniques are reported to fail structurally (20–70%), and the belief that they can restore the GHJ normal function has yet to be resolved. Optimum treatment of RCT and satisfactory clinical outcomes to restore and stabilize the joint during shoulder motion vary and experimental validation using imaging techniques and cadaveric specimens are being pursued to provide reputable data to assess the GH repairs.

2.2.1 Imaging Techniques

Previous in vitro and in vivo studies have investigated the GHJ kinematics in the presence of RC tears and repairs [29–31]. Cadaveric studies have shown that RC tears size play a crucial role in the GHJ stability and range of motion [32]. Other cadaveric experiments showed how the humeral position resulted in an inferior translational shift [33]. Additional studies by Yamaguchi et al. [34] using static radiographs in patients with both asymptomatic and symptomatic rotator cuff tears found that both groups of patients demonstrated superior translation of the humeral head with increasing arm elevation. As stated by Bey et al. [35], these studies use conventional radiographs collected under static conditions at specific arm positions and are limited in scope to understanding a three-dimensional problem. Other experiments focused on the correlation between the extent to which patient-reported outcomes after rotator cuff repair are associated with measures of shoulder function. For example, a study by Nho et al. [36] reported that shoulder strength was predictive of patients' American Shoulder and Elbow Surgeons (ASES) score. This design experiment is often controversial as successful RC repairs can lead to poor outcome [37]. In a study by Bey et al. [39], Twenty of 21 rotator cuff repairs appeared intact at 24 months after surgery. Their finding further supports the fact that the humerus of the patients' repaired shoulder was positioned more superiorly on the glenoid than both the patients' contralateral shoulder and the dominant shoulder of control participants. This is another study that differentiate between the GHJ kinematics and clinical outcome associated with shoulder strength.

Kozono et al. [38], used X-ray images and CT derived digitally reconstructed radiographs to evaluate further the kinematics of RCT and measure the humerus translation relative to the scapula. This dynamic analysis of the glenohumeral during scapular plane abduction and axial rotation for 11 RCT patients with large to massive full thickness and 10 healthy control subjects showed the humeral head center with a medial shift at the late phase of scapular plane full abduction, and an anterior shift at the internal rotation position during full axial rotation.

2.2.2 Cadaveric Experiments of RC Tear and Repairs

The experiments conducted in our biomechanics laboratory are designed to evaluate the shoulder kinematics and the changes in muscle forces before and after RC repairs techniques in maintaining a stable joint at different abduction angle. The experiment setup follows Hirata et al. [40] apparatus with the added optotrack and tracking monitoring of muscles insertion and attachment points. This allows for calculation of moment arms using the geometric approach or the excursion method as well. Five fresh frozen human shoulder specimens were used for this study (all female, age range 60–79 years, two right and three left shoulders). Shoulders were thawed and carefully dissected removing unnecessary tissues and muscles were dissected from their respective fossa while keeping their insertion tendons intact. Sutures were secured to the tendinous muscle insertions. One screw was inserted at the lateral end of the scapular spine and another at the humeral neck. These screws served as reference points in the coronal and sagittal planes respectively. The scapula was then placed in a square Plaster of Paris (PoP) mold and PoP was poured to confine the medial two-thirds of the scapula within a block, while the lateral third of the scapula remained exposed.

Muscle-tendon units of all shoulder muscles were maintained in tension by nylon strings, which were attached to muscle-tendon units at one end, passed through multiple pulley systems, and had 3 N free-hanging weights attached to the other end. The nylon strings attached to each shoulder muscle insertion point were positioned to simulate each muscle's line of action toward its origin point on the scapula. This maintained tension to provide support for the shoulder joint throughout abduction.

The three-camera Optotrak Certus (Northern Digital Inc., 2006), along with two Orthopaedic Research Pin markers and First Principles software, was used for motion capture. The Optotrak was placed lateral to the experimental frame, allowing cameras to monitor the frontal plane of the shoulder. The shoulder was placed on an adjustable height table to ensure the specimen was positioned at the center of the camera capture range. Two Optotrak markers were used: one placed on a stationary rod slightly superior to the scapula and one inserted at the insertion site of the deltoid in its musculotendinous. Each Optotrak marker contained three sensors, which had to be placed within the camera frame of the Optotrak. During setup, a digitizing probe was connected and used to assign the digital points for the origin and insertion sites of the RC muscles. This produced digitized points marked on the shoulder girdle that were relative to the two markers previously placed.

Each shoulder underwent four different motion capture experiments with Optotrack recording conducted for each condition: (1) Intact rotator cuff, (2) Complete rotator cuff tear, (3) SCR, and (4) RSA.

A complete tear was created by incising the supraspinatus at its humeral insertion. SCR was conducted to repair the complete tear then followed with an RSA for the same cadaveric model. During RSA, the humeral head was incised, and the glenoid and humeral shaft were prepared for the reverse prosthesis. The prosthesis was fixed with the center of rotation placed infero-medially, following standard surgical procedure.

After the points were digitalized on each model, the humerus was manually abducted in the coronal plane from 0 to 90° at a constant rate, allowing the Optotrak to capture relative motion in a time frame of 20 s. The Optotrak data output featured displacement values in millimeters for the digitized points after each of the experiment. This was later used to calculate moment arm of the four RC muscles. (see Fig. 2.2).

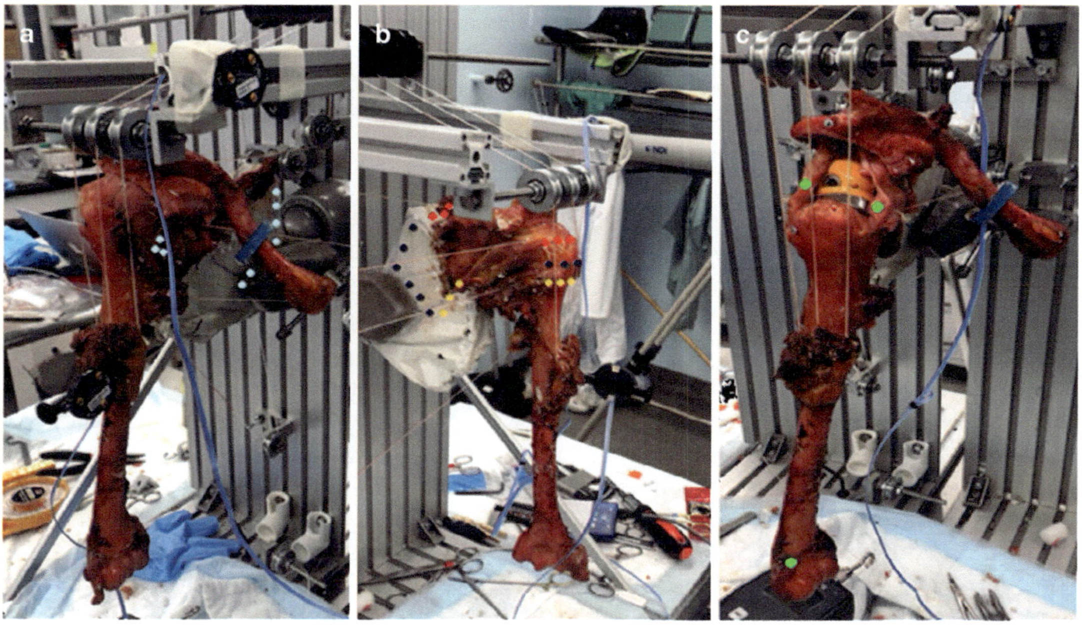

Fig. 2.2 Shoulder experimental apparatus with all four rotator cuff muscles and deltoid. Three views **a**, **b**, and **c** include the anterior, posterior, and lateral, respectively. Optotrack was used to track the digitized points during abduction from 0 to 90°

2.2.3 Moment Arm Calculations

Previous techniques for the calculation of moment arm have been discussed above and their advantages and disadvantages rely great deal on the accuracy of data measured experimentally. The method of geometric measurement requires both the joint center of rotation and the force line of action. In our experiments we digitized the muscles insertion points and so the line of action for the forces can be evaluated. However, it is difficult to make a point to point map for the evaluation of these forces but different combinations of line actions might provide a better insight into the moment arm contribution to the muscle force and torque capacity or moment produced at the joint.

The proposed experimental method is well suited for the tendon-joint excursion method as the different insertion and connecting points of RC muscles are available throughout the abduction angle. Hence, one can compute the relative change in muscle length as function of the joint angle of rotation. While this method was limited to 2D plane motion analysis, the optotrack sensing technology allows for 3D analysis and possibly the calculation of all three joint angles of rotations. The experimental setup in our experiments relies on pulleys and cables to dictate the load force direction which is a balancing act to allow for weight adjustment to stabilize the GH **t.** joint. The pulley system used static weights to balance the shoulder joint throughout abduction. This system allowed the deltoid to be driven by a force sensor dynamometer that allows for measurement of the abduction arc angle and force.

The moment arms (MA) of the rotator cuff muscles were calculated using the origin (O) and insertion (I) digitized points data obtained from Optotrak along with the two additional digitized points on either side of the humeral head used to calculate the center of the shoulder joint.

We define the center of the humerus head as O and as the humerus translate during abduction our optotrack measure of the new center during full range of rotation is reevaluated. The line of action defining the muscle force direction is defined by the points denoting the tendon insertion and attachment points as depicted in Fig. 2.2

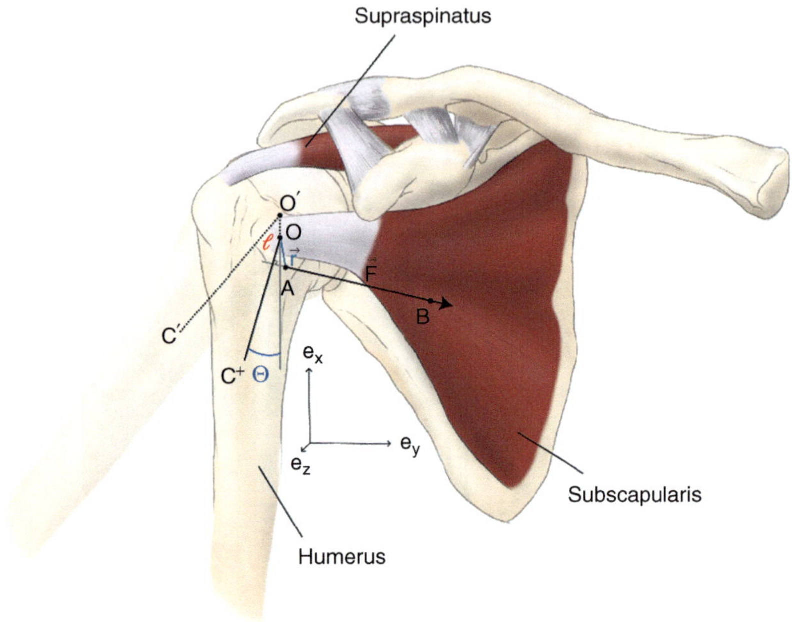

Fig. 2.3 Tracking of the Humerus head center before and after abduction. A and B points along the muscle line of action defining the insertion point and the origin point used for the computation of the moment arm. O and O′ and C and C′ denote the before and after abduction positions. F is the force vector along AB, and r and r' denote the projection from O to AB. The fixed reference frame is given by the unit vectors e_x, e_y, and e_z

during the experiment. Let A be a point on subscapularis as shown in Fig. 2.3 then the moment vector arm $\overrightarrow{OA}$, also referred as $\vec{r}$, is defined as the distance between point O and A. Consider B another point on the force line of action and we define a unit vector $\vec{u}$ along that line. The moment is given by $\vec{M} = \vec{r} \times \vec{F}$.

2.3 Results

The mean value for the intact subscapularis model moment arm starts with a large adducting ability (Fig. 2.4). As the shoulder is abducted, the mean moment arm shows a steady upward trend, reaching a large abducting moment arm at 90° abduction. The mean values for the SCR and RSA models start with approximately 50% of the adducting ability of the intact model. The SCR and RSA models then trended upwards, ending shoulder abduction with a near neutral mean moment arm. Mean moment arm values and percent differences across shoulder abduction were −1.9 mm for intact, 5.5 mm (256.3%) for complete tear, −5.5 mm (336.0%) for SCR, and −4.4 mm (194.4%) for RSA. ANOVA testing revealed that moment arms across the subscapularis of the anatomic shoulder was not statistically significantly different than the three models ($p = 0.148$). There was no significant difference in percent difference testing between model groups ($p = 0.814$).

The mean value for the supraspinatus moment arm started with a large abduction value and features a steady downward trend into a small adducting moment arm over the course of shoulder abduction (Fig. 2.5). All other models also followed a downward trend; however, SCR model was the only one to gain any adduction moment arm at higher shoulder abduction angles. Both the complete tear and RSA models have moment arm values that remained in abduction potential throughout the entire course of shoulder abduction. The RSA model showed the closest alignment with the intact model, with a largely increased percent difference only at 90° abduction. The SCR model showed close alignment with the intact model for the first 30° and then a larger difference at angles above 60°.

The intact supraspinatus showed slightly different behavior between its two muscle fiber groups. Both fiber groups started with large abducting moment arms, but the anterior fibers

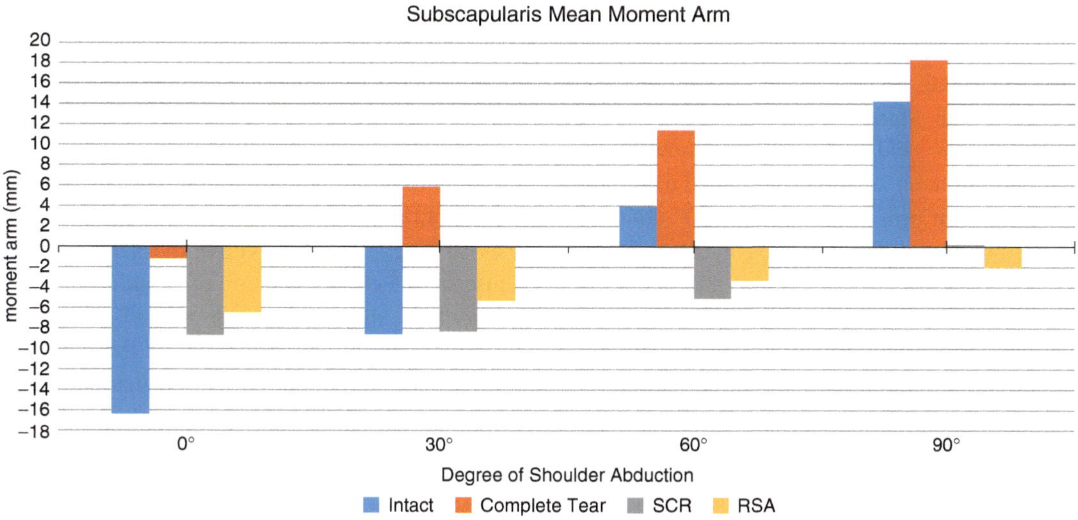

Fig. 2.4 The mean values for moment arm of subscapularis at 0, 30, 60, and 90°

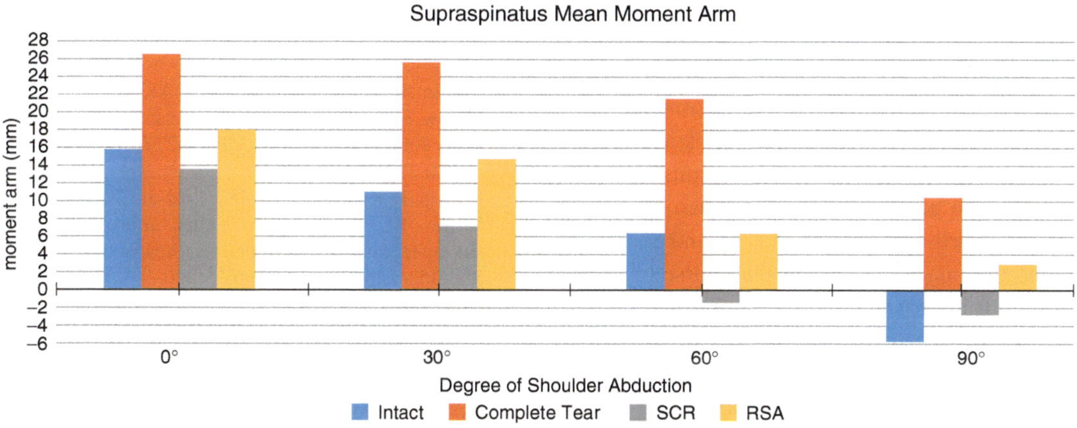

Fig. 2.5 The mean values for moment arm of supraspinatus at 0, 30, 60, and 90°

moved toward a neutral position during abduction while the posterior fibers moved into a large adducting moment toward the end of shoulder abduction. After a complete tear, the supraspinatus fibers showed a similar initial large abducting moment arm; however, although the moment arm values decreased, there was no transition from an abducting to adducting moment arm in the posterior fibers.

The SCR and RSA models maintained similar trends when compared to the intact model, starting with large abducting moment arms that moved into small adducting moments as the shoulder was abducted.

2.4 Discussion

Previous work by Hughes et al. [24] among others have raised the potential or induced from the calculation of moment arms related to the assumption of the humeral head joint center of rotation assumed fixed during abduction. This limitation was clearly resolved in our study as the imaging technique employed allowed us to recompute the humeral head center of rotation.

A study by Ruckstuhl et al. also used the origin-insertion method for calculating moment arm in specific positions using MRI images to construct a 3D anatomical reconstruction [41].

Hamilton et al. used 3D modeling software and found different RSA reconstructions displayed similar trends over abduction when compared to our study results [21, 22]. Both the subscapularis and infraspinatus showed large negative moment arms which trended toward positive moment arm values over 0–90° abduction. Additionally, efficiency testing by Hamilton and percent difference changes from intact models in our study all failed to reach statistical significance in all models, secondary to large standard deviations. However, the large standard deviations in our study were likely secondary to a limited number of cadaveric models and innate differences between shoulder anatomy and strength [42].

An illustration of how the method of discretization of the tendons points at their fixation site provides an insight into the bundled fibers contribution and how its span area of fixation produces different moment arms. A muscle is complex dynamic force producer and a partial or complete tear into these tendons have a direct impact on the overall glenoid-humeral joint stabilization, kinematics and overall kinetics. Our results show that the average percent differences when comparing subscapularis contribution of RSA and SCR models to the intact model over the course of abduction both improved over 20% from the complete tear average percent difference. Complete tears demonstrated a loss of the stabilizing effect typically provided from the fibers producing a negative moment arm at the start of abduction. While this stability was significantly restored in the SCR and RSA models, it did not perfectly mimic the native, intact rotator cuff, especially at higher angles of abduction. The supraspinatus moment arms average percent differences when comparing SCR and RSA models to the intact model were significantly different, with 64% and 28% difference respectively, while a complete tear featured an average difference of 74%.

2.5 Conclusion

The moment arm along with the kinematic data provide valuable information for future studies investigating RC repairs techniques, patient outcome and potentially other pathological conditions of the shoulder. For clinical relevance, quantitative assessment of the dynamic kinematics of shoulders with RCTs will need imaging techniques combined with experimental and 3D can lead to development of indicators for achieving functional restoration. Furthermore, our analysis of fiber divisions within the same muscle illustrate the complex nature of shoulder muscles themselves, and future studies should aim to better explore and model their function. Infraspinatus and subscapularis show the greatest variance when comparing RSA and post-reconstruction to intact. It is in these muscles where there is the most loss of potential abducting or adducting force.

References

1. Colvin AC, Egorova N, Harrison AK, Moskowitz A, Flatow EL. National trends in rotator cuff repair. J Bone Joint Surg Am. 2012;94(3):227–33. https://doi.org/10.2106/JBJS.J.00739.
2. Bishop J, Klepps S, Lo IK, Bird J, Gladstone JN, Flatow EL. Cuff integrity after arthroscopic versus open rotator cuff repair: a prospective study. J Shoulder Elbow Surg. 2006;15:290–9.
3. Gustavsson A, Neeter C, Thomee P, Silbernagel KG, Augustsson J, Thomee R, Karlsson J. A test battery for evaluating hop performance in patients with an ACL injury and patients who have undergone ACL reconstruction. Knee Surg Sports Traumatol Arthrosc. 2006;14:778–88.
4. Morse K, Davis AD, Afra R, Kaye EK, Schepsis A, Voloshin I. Arthroscopic versus mini-open rotator cuff repair: a comprehensive review and meta-analysis. Am J Sports Med. 2008;36:1824–8.
5. Anderson MA, Gieck JH, Perrin DH, Weltman A, Rutt RA, Denegar CR. The relationships among isometric, isotonic, and isokinetic concentric and eccentric quadriceps and hamstring force and three components of athletic performance. J Orthop Sports Phys Ther. 1991;14:114–20.
6. Wilk KE, Romaniello WT, Soscia SM, Arrigo CA, Andrews JR. The relationship between subjective knee scores, isokinetic testing, and functional testing in the ACL-reconstructed knee. J Orthop Sports Phys Ther. 1994;20:60–73.
7. Lephart SM, Perrin DH, Fu FH, Gieck JH, McCue FC, Irrgang JJ. Relationship between selected physical characteristics and functional capacity in the anterior cruciate ligament-insufficient athlete. J Orthop Sports Phys Ther. 1992;16:174–81.

8. Keays SL, Bullock-Saxton J, Keays AC. Strength and function before and after anterior cruciate ligament reconstruction. Clin Orthop Relat Res. 2000:174–83.
9. Tibone JE, Antich TJ, Fanton GS, Moynes DR, Perry J. Functional analysis of anterior cruciate ligament instability. Am J Sports Med. 1986;14:276–84.
10. Giove TP, Miller SJ, Kent BE, Sanford TL, Garrick JG. Non-operative treatment of the torn anterior cruciate ligament. J Bone Joint Surg Am. 1983; 65:184–92.
11. Barber SD, Noyes FR, Mangine RE, McCloskey JW, Hartman W. Quantitative assessment of functional limitations in normal and anterior cruciate ligament-deficient knees. Clin Orthop Relat Res. 1990: 204–14.
12. Park MC, Cadet ER, Levine WN, Bigliani LU, Ahmad CS. Tendon-to-bone pressure distributions at a repaired rotator cuff footprint using transosseous suture and suture anchor fixation techniques. Am J Sports Med. 2005;33:1154–9.
13. Dunn WR, Schackman BR, Walsh C, Lyman S, Jones EC, Warren RF, Marx RG. Variation in orthopaedic surgeons' perceptions about the indications for rotator cuff surgery. J Bone Joint Surg Am. 2005;87: 1978–84.
14. Green LB, Pietrobon R, Paxton E, Higgins LD, Fithian D. Sources of variation in readmission rates, length of stay, and operative time associated with rotator cuff surgery. J Bone Joint Surg Am. 2003;85:1784–9.
15. Bianchi S, Martinoli C. Ultrasound Musculoskeletal Syst. 2007; 246-256.
16. Lewis JS. Rotator cuff tendinopathy. Br J Sports Med. 2009;43(4):236–41.
17. Nogueira-Barbosa MH, Volpon JB, Elias J Jr, Muccillo G. Diagnostic imaging of shoulder rotator cuff lesions. Acta Ortop Bras. 2002;10(4):120.
18. Joseph O. de Jesus, Laurence Parker, Andrea J. Frangos, Levon N. Nazarian accuracy of MRI, MR arthrography, and ultrasound in the diagnosis of rotator cuff tears: a meta-analysis. AJR. 2009; 192:1701–7.
19. Rutten MJ, Jager GJ, Blickman JG. From the RSNA refresher courses: US of the rotator cuff: pitfalls, limitations, and artifacts. Radiographics. 2006;26(2):589–604.
20. Walker DR, Struck AM, Matsuki K, et al. How do deltoid muscle moment arms change after reverse total shoulder arthroplasty? J Shoulder Elb Surg. 2016;25:581–8.
21. Hamilton MA, Roche CP, Diep P, et al. Effect of prosthesis design on muscle length and moment arms in reverse total shoulder arthroplasty. Bull Hosp Jt Dis. 2013;71(Suppl 2):S31–5.
22. Hamilton MA, Diep P, Roche C, et al. Effect of reverse shoulder design philosophy on muscle moment arms. J Orthop Res. 2015;33:605–13.
23. Greiner S, Schmidt C, Konig C, et al. Lateralized reverse shoulder arthroplasty maintains rotational function of the remaining rotator cuff. Clin Orthop Relat Res. 2013;471:940–6.
24. Hughes RE, Niebur G, Liu J, et al. Comparison of two methods for computing abduction moment arms of the rotator cuff. J Biomech. 1998;31:157–60.
25. An KN, Takahashi K, Harrigan TP, et al. Determination of muscle orientations and moment arms. J Biomech Eng. 1984;106:280–2.
26. Ackland DC, Pandy MG. Moment arms of the shoulder muscles during axial rotation. J Orthop Res. 2011;29:658–67.
27. Ackland DC, Pak P, Richardson M, et al. Moment arms of the muscles crossing the anatomical shoulder. J Anat. 2008;213:383–90.
28. Kuechle DK, Newman SR, Itoi E, Niebur GL, Morrey BF, An KN. The relevance of the moment arm of shoulder muscles with respect to axial rotation of the glenohumeral joint in four positions. Clin Biomech. 2000;15(5):322–9.
29. DeFranco MJ, Bershadsky B, Ciccone J, Yum JK, Iannotti JP. Functional outcome of arthroscopic rotator cuff repairs: a correlation of anatomic and clinical results. J Shoulder Elb Surg. 2007;16(6):759–65.
30. Cofield RH, Parvizi J, Hoffmeyer PJ, Lanzer WL, Ilstrup DM, Rowland CM. Surgical repair of chronic rotator cuff tears: a prospective long-term study. J Bone Joint Surg Am. 2001;83(1):71–7.
31. Graichen H, Englmeier KH, Reiser M, et al. An in vivo technique for determining 3D muscular moment arms in different joint positions and during muscular activation – application to the supraspinatus. Clin Biomech. 2001;16:389–94.
32. Su WR, Budoff JE, Luo ZP. The effect of posterosuperior rotator cuff tears and biceps loading on glenohumeral translation. Arthroscopy. 2010;26(5):578–86.
33. Yu J, McGarry MH, Lee YS, Duong LV, Lee TQ. Biomechanical effects of supraspinatus repair on the glenohumeral joint. J Shoulder Elbow Surg. 2005;14(1 Suppl S):65–71.
34. Yamaguchi K, Sher JS, Andersen WK, et al. Glenohumeral motion in patients with rotator cuff tears: a comparison of asymptomatic and symptomatic shoulders. J Shoulder Elb Surg. 2000;9(1):6–11.
35. Bey MJ, Kline SK, Zauel R, Lock TR, Kolowich PA. Measuring dynamic in-vivo glenohumeral joint kinematics: technique and preliminary results. J Biomech. 2008;41(3):711–4. https://doi.org/10.1016/j.jbiomech.2007.09.029.
36. Nho SJ, Brown BS, Lyman S, Adler RS, Altchek DW, MacGillivray JD. Prospective analysis of arthroscopic rotator cuff repair: prognostic factors affecting clinical and ultrasound outcome. J Shoulder Elb Surg. 2009;18(1):13–20.
37. Borgmastars N, Parabola M, Remes V, Lohman M, Vastamaki M. Pain relief, motion, and function after rotator cuff repair or reconstruction may not persist after 16 years. Clin Orthop Relat Res. 2010;468(10):2678–89.
38. Kozono N, Okada T, Takeuchi N, et al. Dynamic kinematics of the glenohumeral joint in shoulders with rotator cuff tears. J Orthop Surg Res. 2018;13(1):9. https://doi.org/10.1186/s13018-017-0709-6.

39. Bey MJ, Peltz CD, Ciarelli K, et al. In vivo shoulder function after surgical repair of a torn rotator cuff: glenohumeral joint mechanics, shoulder strength, clinical outcomes, and their interaction. Am J Sports Med. 2011;39(10):2117–29. https://doi.org/10.1177/0363546511412164.
40. Mihata T, et al. Clinical results of arthroscopic superior capsule reconstruction for irreparable rotator cuff tears. Arthrosc - J Arthrosc Relat Surg. 2013; https://doi.org/10.1016/j.arthro.2012.10.022.
41. Ruckstuhl H, Krzycki J, Petrou N, Favre P, Horn T, Schmid S, Stussi E. Shoulder abduction moment arms in three clinically important positions. J Shoulder Elb Surg. 2009;18(4):632–8. https://doi.org/10.1016/j.jse.2008.10.021.
42. Wu W, Lee PVS, Bryant AL, Galea M, Ackland DC. Subject-specific musculoskeletal modeling in the evaluation of shoulder muscle and joint function. J Biomech. 2016;49(15):3626–34. https://doi.org/10.1016/j.jbiomech.2016.09.025. Epub 2016 Sep 23

Part II

Biomechanical Properties of Orthopaedic Materials

Biomechanics of Ligaments

3

Jonathan D. Hughes, Calvin K. Chan, Sene K. Polamalu, Richard E. Debski, and Volker Musahl

3.1 Introduction

Ligaments are soft tissue structures that are composed of collagen fiber bundles and connect bone to bone. Ligaments play a vital role in the stability of joints throughout the body, including the shoulder and knee. Ligaments act as individual uniaxial restraints on the joint that work together to resist a variety of external loads. Injury to ligaments disrupts the mobility and stability of the joint, potentially leading to significant morbidity and pain for the patient. Understanding the histology, joint kinematics, and biomechanical properties of ligaments will help physicians approach injuries to this structure and formulate an appropriate treatment plan for their patients.

J. D. Hughes · V. Musahl (✉)
Department of Orthopaedic Surgery, UPMC Freddie Fu Sports Medicine Center, University of Pittsburgh Medical Center, Pittsburgh, PA, USA
e-mail: hughesjd3@upmc.edu; musahlv@upmc.edu

C. K. Chan · S. K. Polamalu · R. E. Debski
Department of Bioengineering, University of Pittsburgh, Pittsburgh, PA, USA
e-mail: CAC296@pitt.edu; SKP39@pitt.edu; genesis1@pitt.edu

3.2 Histology

A fundamental understanding of the underlying histology is crucial to comprehend the biomechanical properties of ligaments and will aid in designing biomechanical studies. Ligaments are composed of water, elastin, proteoglycans, glycolipids, and densely packed collagen fibers that run in a longitudinal direction. These fibers are arranged into a series of bundles called fascicles and surrounded by a loose connective tissue called paratenon [80]. Fibroblasts, the predominant cell type, are arranged in rows between subunits of collagen fibers called fibrils. However, in order to withstand multiaxial loads, the fibrils are not uniformly oriented [2]. Ligaments are relatively hypovascular and hypocellular [9, 64].

Approximately 70–80% of the dry weight of a normal ligament is composed of type I collagen, with smaller amounts of types III (3–10% of dry weight), V, IX, X, XI, XII, and XIV collagen [72]. Type V collagen associates with Type I collagen to regulate collagen fibril diameter [6]. Type III collagen has the ability to form intermolecular disulfide bonds essential for wound healing [13]. Type XII collagen provides lubrication between collagen fibers [52]. Types IX, X, and XI are present at the ligament-bone interface in

J. Koh et al. (eds.), *Orthopaedic Biomechanics in Sports Medicine*,
https://doi.org/10.1007/978-3-030-81549-3_3

conjunction with Type II collagen [30]. The water and proteoglycans provide spacing and lubrication to the ligaments, allowing the collagen fibers to glide over each other.

Ligaments are metabolically active, with continued cell and matrix turnover that occurs at a slow rate [1]. Since this occurs at a slow rate, along with the inherent hypovascularity, ligaments heal at a slower rate than other tissues. Extra-articular ligaments, such as the medial collateral ligament, heal in four phases after injury: hemorrhage, inflammation, proliferation, and remodeling [25]. In the initial hemorrhage phase, there is bleeding at the site of injury with hematoma development between the torn edges. After approximately 72 hours, inflammatory cells, including monocytes, leukocytes, and macrophages, appear at the site of injury and secrete cytokines and growth factors. This induces an inflammatory reaction and granulation tissue formation. Fibroblasts synthesize Type III collagen, as well as small amounts of Type I collagen, which forms an immature scar. The cells proliferate, creating an immature vascular neo-ligament. The final phase consists of synthesis of Type I collagen and further organization of the collagen into a parallel arrangement of fibers along the functional axis of the ligament.

Intra-articular ligaments, such as the anterior cruciate ligament (ACL), demonstrate minimal healing potential and do not follow the four phases of healing mentioned above. A thin synovial sheath surrounds the ACL and provides it blood supply. When this is disrupted, blood leaks into the surrounding synovial fluid and is unable to create a hematoma at the site of injury. Additionally, cytokines and growth factors are unable to reach the site of injury. Due to this, there is low healing potential for the ACL after an acute rupture.

Ligament insertions onto bone help dissipate forces from the soft tissues to the bone [70]. These insertions are classified as either direct or indirect. Direct insertions consist of four distinct zones: ligament, fibrocartilage, mineralized fibrocartilage, and bone [16]. The indirect insertions have superficial and deep fibers. The superficial fibers attach directly to the periosteum, while the deep fibers are anchored to the bone via Sharpey's fibers [16]. Various ligaments, including the medial collateral ligament, contain both types of insertions: direct femoral insertion and indirect tibial insertion [72].

The collagen fibrils in ligaments are in various amounts of crimp, which allows for recruitment of more fibrils when a high tensile load is applied to the ligament. The ligament responds to tensile load by elongating, thus allowing normal joint kinematics during movement by allowing the joint to move freely. During high strenuous activities, the stiffness of the ligament increases in order to maintain joint stability and restrict excess motion of the joint. When this applied tensile load exceeds the maximum capacity of the ligament, injury and/or rupture of the ligament can occur [72].

3.3 Joint Kinematics and Function

Evaluation of joint motion is an important method of understanding joint injury. A joint is constrained by its ligaments, supporting soft tissue structures, and bony geometry. The joint constraints allow a combination of motion and stability to a joint along its six degrees of freedom (6-DOF), three translational and three rotational. Traditionally in the knee joint, the three translational axes are proximal-distal, medial-lateral, and anterior-posterior, and the three rotational axes are internal-external, flexion-extension, and varus-valgus [32]. The knee axes are defined by the long axis of the tibial shaft, the line between the femoral insertion points of the lateral and medial collateral ligaments, and the perpendicular line to the long axis of the tibial shaft and the line between the femoral insertions as described above, respectively. By examining the motion of the 6-DOF joint kinematics, researchers and clinicians can determine the role and extent of injury to individual structures.

Measuring kinematics in a laboratory setting have been done in a variety of ways. Early techniques used mechanical loading linkages and weights to measure the motion of the knee in response to specific external loads [7, 41, 55]. Biplanar radiography was also used as a way to visually measure the motion of cadaveric knees based on radiographs from two known angles [56, 63]. Another approach involved using electromagnetic tracking by placing sensors on the bones in order to track the joint motion of cadavers and patients in a three dimensional space [10, 49]. In recent years, robotic technology has become the gold standard, in which joint kinematics are measured by combining a 6-DOF robotic arm with a universal force/moment sensor (UFS) to measure joint motion and forces [26, 27]. All of these methods aimed to compare injured knees to healthy, contralateral knees, in order improve injury diagnosis and treatment options.

The primary passive stabilizers in multiple diarthrodial joints are the ligaments found around and within the joint. Ligaments act as individual uniaxial restraints on the joint that work together to resist a variety of external loads. Prior studies have outlined the role of individual ligaments in providing joint stability against certain external forces. The authors examined 6-DOF joint kinematics in response to external loads before and after resecting the ligament in question in order to see how the ligament provided stability to the joint [46, 60]. A separate study performed a simulated Lachman test on ACL intact and ACL deficient knees. The authors found ACL deficiency increases knee anterior-posterior laxity from its normal 3–5 mm range and could possibly shift loads to other structures [29].

When one or more ligaments is injured, the knee will lose stability and respond to external loads differently. Since ligaments normally carry some passive load in the joint to provide stability, repairing or reconstructing the ligaments will restore that native in situ force. Experimental measurements of ligament in situ force have used contact methods, such as implantable force transducers, buckle transducers, and strain gauges, to directly measure the force in individual ligaments in response to external loads [31, 38, 45]. One study utilizing direct force measurement found that an internal tibial torque of 10 N-m produced a force on the order of 100 N in the ACL at 20° of flexion, while an external tibial torque of 10 N-m produced a smaller force around 50 N by using a load cell attached to the ACL through the tibial insertion [47]. In contrast, noncontact methods of measuring in situ force sought to reduce error from the interface of the ligament to the measuring device. A method using a UFS requires measuring the 6-DOF forces and moments applied to a rigid body connected to the sensor and calculating those forces through the joint based on a series of rigid body transformations [26]. A combined robotic/UFS system can measure kinematics and kinetics of an intact joint in response to external loads and then repeat the intact kinematics after a ligament was transected or repaired in order to record a new set of kinetics. The difference between the two set of kinetic data (intact and resected) is the in situ force of the ligament or graft [33, 59].

3.4 Biomechanical Properties

3.4.1 Structural Properties

Due to the primary function of ligaments to resist tensile loads, several investigations have focused on their biomechanical properties under uniaxial tensile testing. This tensile testing is difficult due to a multitude of factors, including the complex composition of ligaments. In order to apply tensile loads, the ends of the ligament need to be properly secured, but issues with slippage are common. In order to combat slippage, variations of soft tissue clamping have been created utilizing serrated clamps while freezing the secured ends [14, 42, 57, 58]. However, slippage can still happen, and the stress concentrations that now occur near the clamps can cause premature failure and not accurately represent the native tissue.

The structural properties of bone-ligament-bone complexes in response to tensile loads have been investigated. A benefit of testing the entire complex is the inclusion of the insertion sites of the ligaments which have different properties than the midsubstance.

Structural properties that characterize the response of ligaments to loads along the longitudinal axis of the ligament include ultimate load, the greatest load that can be applied to a tissue before failure, ultimate elongation, the deformation of the tissue when failure occurs, and stiffness, the tissue's ability to resist deformation in response to an applied load [70, 75]. These structural properties can be determined from the load-elongation curve (Fig. 3.1a).

3.4.2 Mechanical Properties

The mechanical properties of ligaments refer to the intrinsic behavioral response of the tissue based on its composition and micro-structure and can be obtained from the same uniaxial tensile testing. These mechanical properties are obtained by normalizing the load by the cross-sectional area of the midsubstance of the ligament, and strain, the elongation by the initial length of the midsubstance of the ligament. The stress-strain curve can be used to determine tensile strength, the greatest stress in the tissue before failure, ultimate strain, the greatest strain before failure, modulus, the tissue's resistance to strain in response to stress, and strain energy density, the energy stored in the tissue before failure (Fig. 3.1a,b). This quantitative data can guide clinicians on their graft choice for ligament reconstructions, as the replacing tissue should have a similar mechanical response as the native tissue to best restore intact joint kinematics.

To determine accurate measure of stress, an accurate cross-sectional area (CSA) is needed, as errors in CSA can cause substantial effects on stress calculations. CSA measurement falls into two general categories: contact and noncontact methods. Contact methods include the use of

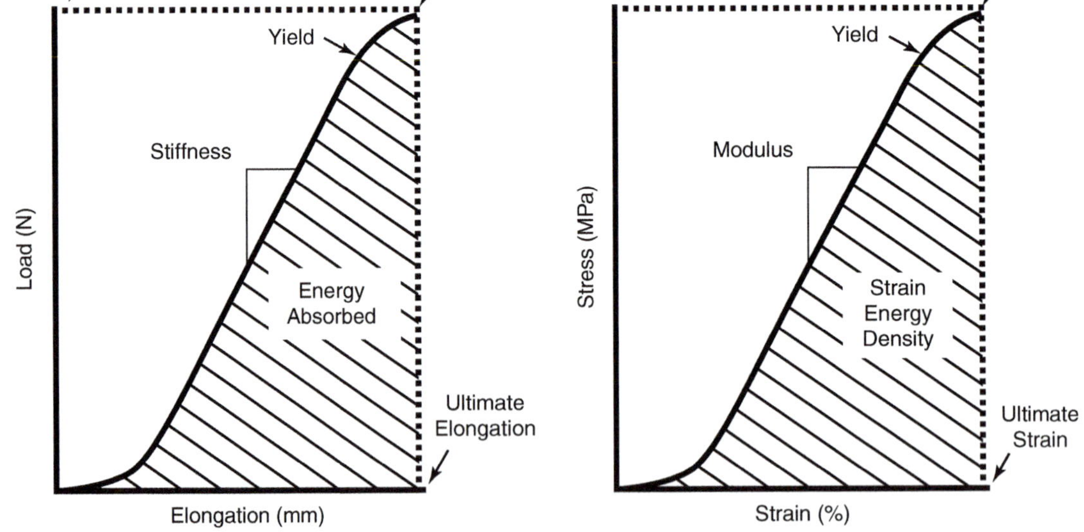

Fig. 3.1 (**a**) Load-elongation response of a bone-ligament-bone complex displaying the structural properties that can be interpreted from the curve. (**b**) Stress-strain response of a ligament displaying the mechanical properties that can be interpreted from the curve

calipers and molding [4, 28, 54, 61, 82]. However, these methods involve touching the soft tissue and thus risk causing deformation or incorrect assumptions of rectangular shape. Noncontact methods have developed from shadow amplitude and light profile methods [20, 39] to the use of laser micrometers, which have been found to be the most accurate and reproducible [67]. However, issues with concavity exist for all of these noncontact measurements. In order to address these issues, a laser reflectance system has been developed [12].

Strain accuracy has similar difficulties as stress, in which an accurate measure of elongation is needed, otherwise the strain measurements are unreliable. Contact and noncontact tissue elongation methods have been developed as well. Contact methods include the use of a differential variable reluctance transducer (DVRT) [5, 22, 48]. A DVRT outputs a voltage that has a linear relationship with the distance and elongation of tissue between the barbs. The limitations of DVRT measurements include not measuring strain throughout the ligament and possibly damaging the tissue by inserting metallic barbs. Noncontact strain measurements involve the use of optical video tracking of either markers or strain, and address both of the limitations of the contact methods [68, 77]. The position of the markers or strain can be tracked, and strain can be calculated by determining the difference in their position before and after applying a load.

3.4.3 Contributing Properties

Ligament compositions vary within different portions of the tissue from the insertion site to the midsubstance and have a complex three-dimensional heterogeneous architecture [53, 66]. This variation has an impact on the mechanical properties and thus the mechanical response. Stress and strain response near the insertion sites are different than that experienced at the midsubstance [11, 43]. Furthermore, a distinct quality of ligaments is their ability to resist tensile loads in their longitudinal direction due to their highly parallel collagen fibers. This fiber alignment causes the mechanical response of ligaments to differ when loads are applied in directions other than along the longitudinal axis of the ligament. Thus, it is pertinent to be mindful of the orientation of the ligament with respect to external loads when measuring mechanical properties.

3.4.4 Viscoelasticity

A distinct property of ligaments and biological tissues is their viscoelasticity, meaning their mechanical response is load and rate dependent. Ligaments demonstrate both viscous and elastic properties during deformation. Elastic response refers to a substance's ability to return to its original shape and size after deformation, while viscous response refers to a substance's ability to resist flow. Viscoelastic substances display different properties from elastic materials in isometric, isotonic, and cyclical loading. In isometric loading, or applying a constant deformation, the tissue displays a decreased stress response over time, known as a stress relaxation (Fig. 3.2a). In isotonic loading, or applying a constant force, the tissue deforms more with time (Fig. 3.2b). When applying cyclical loading to a biological tissue, the unloading curve is below the loading curve, which represents energy dissipating from fluid friction of the water moving in the ligament (Fig. 3.2c). When analyzing the strain response over time during this cyclic loading, the peak stresses to reach the same strain decrease with each cycle as the curve becomes more repeatable (Fig. 3.2d). Viscoelasticity is important clinically as the tissue relaxes during daily activities such as walking or jogging.

3.5 Factors Affecting Mechanical Properties of Ligaments

Numerous biological factors determine the mechanical properties of different ligaments within the body. The same tissue from two

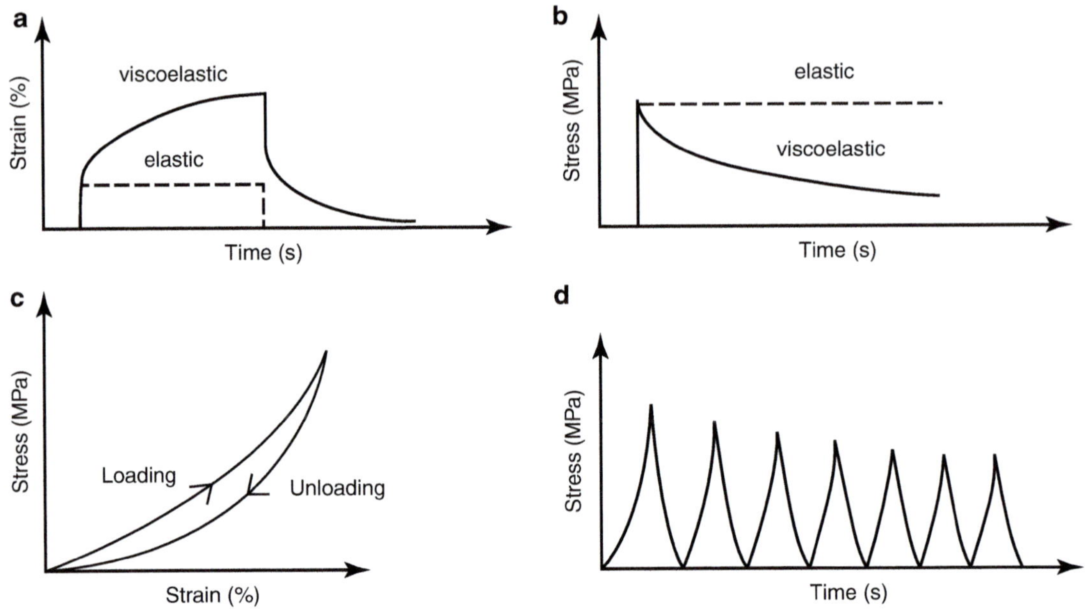

Fig. 3.2 (**a**) Creep of viscoelastic and elastic materials and in response to isotonic loading. (**b**) Stress relaxation of viscoelastic and elastic materials in response to isometric loading. (**c**) Stress-strain curve of a viscoelastic material in response to loading and unloading. (**d**) Stress relaxation of a viscoelastic material in response to cyclic loading demonstrating a decreased peak stress at each cycle

different sources can present two different competing sets of data due to differences in skeletal maturation, age of the source, location of the ligament, and physical activity of the source [65, 66, 76, 78]. When evaluating which mechanical properties to use for individual ligaments, researchers should pay attention to these biological factors to ensure an accurate comparison can be made.

3.5.1 Biological Factors

Age and skeletal maturation play a significant role in determining the mechanical property of a ligament [78, 79]. Younger and skeletally immature specimens still have open physes that allow growth of the bone compared to older and skeletally mature specimens with closed physes. Specimens from younger animals exhibit lower ultimate load, cross-sectional area, and stiffness compared to skeletally mature specimens. Younger specimens tend to fail due to tibial avulsion, whereas mature specimens fail midsubstance [78]. Among skeletally mature specimens, ligaments from younger sources (aged 22–35 years) will demonstrate significantly higher stiffness and ultimate load values (242 ± 28 N/mm and 2160 ± 157 N, respectively) than older specimens [69].

The location of the ligament sample plays an important role in determining the mechanical properties of the tissue being investigated. Different areas of a ligament demonstrate different degrees of fiber organization, mineralization, and protein composition. Initially, authors defined ligaments as homogenous-appearing tissues and used the mechanical properties of an entire tissue complex [23, 77]. However, authors later examined the properties of individual parts of the ligament in order to understand the functional properties of the tissue [11, 24, 81]. Through this later examination, the anteromedial bundle of the anterior cruciate ligament was found to have larger moduli, maximum stresses, and strain energy densities than the posterolateral bundle [11]. Tissue closer to the insertion sites of the ligament will exhibit different strain distribution than tissue at the midsubstance [81].

Physical activity levels of the tissue donor can affect the properties of the tissue in two ways. Some swine exercise models showed femur-MCL-tibia complexes have increases in ultimate load, linear stiffness, tensile strength, and ultimate strain after 12 months of physical exercise when compared to the sedentary group [74]. However, joint immobilization, which usually results from musculoskeletal injuries, has been shown to cause decreases in ultimate loads, elastic moduli, and cross-sectional area in studies done on rabbit ACL midsubstance and canine femur-MCL-tibia complexes [51, 73].

3.5.2 Environmental Factors

When investigating the role of ligaments in an in vitro setting, researchers should keep in mind several external factors that affect the mechanical properties of ligaments. Some of these factors, such as specimen orientation, should be considered long before testing begins [21, 34, 35, 50, 71]. Other properties, such as hydration and surrounding temperature, are important factors that should be monitored during an actual experiment [3, 15, 36, 37, 62, 66].

Before testing begins, specimens will often need to be stored frozen in order to preserve their biomechanical properties as much as possible. Studies have found differing results on the effects of freezing on ligaments. One study that put a bone-patella-tendon-bone complex through multiple freeze-thaw cycles found the tissue can undergo 8 freeze-thaw cycles without affecting any biomechanical properties [40]. Other studies found that fresh harvested rabbit medial collateral ligaments have little to no structural biomechanical difference when compared to ligaments that were properly frozen and then thawed. However, these studies did find that during the first few cycles of cyclical loading, frozen specimens had a significant decrease in the area of hysteresis that became insignificant with more cycles [50, 71]. Researchers should be mindful about limiting the number of freeze-thaw cycles ligament samples undergo in order to accurately measure mechanical properties.

Due to the viscoelastic nature of ligaments, the mechanical behavior of ligaments varies greatly due to their surrounding temperature and hydration. There are established effects of varying degrees of hydration on the mechanical properties of different ligaments [15, 36, 37, 62]. During experimental procedures, specimens should either be immersed in a saline solution or constantly kept moist in order to accurately correlate measurements to in vivo mechanical properties. Studies that examined the relationship between temperature and the mechanical properties of ligaments yielded varying results. One study reported a decline in elastic modulus and stiffness with increasing temperature [3]. Another study showed an inverse relationship between stiffness and temperature, and the ligament relaxed to lower values under cyclic loading performed at higher temperatures [44]. In order to minimize variance between data sets, experimental procedures should also dictate temperature.

When designing an experiment, researchers should pay attention to the strain rate used to apply loads. One study examined rates of 0.003 mm/s, 0.3 mm/s, and 113 mm/s in mature rabbit ACL and patella tendons and found that increasing strain rates increase modulus by 31% in the ACL and 94% in the patella tendon [19]. One study tried to emulate the rates at which injuries occur in vitro with rates as high as 500% per second [18]. However, a different study examining rabbit MCLs found that increasing strain rate from 0.01 to 155% per second increases the ultimate strength of the tissue by 40% [79]. Two studies found loading different ligaments (rabbit MCL and canine LCL) at a faster rate than >1% /s, even up to physiological injury rates (>1000% /s), will not drastically affect mechanical properties from the medium rates of loading [8, 17]. At higher speeds, failure mode usually shifts from midsubstance to the insertion points of the specimen [83].

3.6 Summary

Ligaments are soft tissue structures that govern the stability of joints throughout the body, such as the shoulder, hip, or knee. Understanding the histology, joint kinematics, and biomechanical properties of ligaments helps physicians to precisely approach injuries and formulate an appropriate treatment plan for their patients, i.e., anatomic restoration of the native ligament insertion sites and biomechanical properties to restore native joint function. Many biomechanical studies have been performed to provide scientists and surgeons an in-depth understanding of ligament properties. It is imperative to understand the complex nature of ligaments in response to various external loading conditions, as well as to truly understand their form and function in order to ultimately develop the most appropriate and individualized treatment for patients.

References

1. Amiel D, Frank C, Harwood F, Fronek J, Akeson W. Tendons and ligaments: a morphological and biochemical comparison. J Orthop Res. 1983;1:257–65.
2. Amis A. Biomechanics of bone, tendon and ligament. Sciences Basic to Orthopaedics. 1998:222–39.
3. Apter J. Influence of composition on thermal properties of tissues, vol. 217. Engelwood Cliffs, NJ: Prentice-Hall; 1972.
4. Atkinson TS, Ewers BJ, Haut RC. The tensile and stress relaxation responses of human patellar tendon varies with specimen cross-sectional area. J Biomech. 1999;32:907–14.
5. Beynnon B, Johnson R, Fleming B, Renström P, Nichols C, Pope M, et al. The measurement of elongation of anterior cruciate-ligament grafts in vivo. J Bone Joint Surg Am. 1994;76:520–31.
6. Birk DE, Mayne R. Localization of collagen types I, III and V during tendon development Changes in collagen types I and III are correlated with changes in fibril diameter. European journal of cell biology. 1997;72:352–61.
7. Blankevoort L, Huiskes R, De Lange A. The envelope of passive knee joint motion. J Biomech. 1988;21:705–20.
8. Bonner TJ, Newell N, Karunaratne A, Pullen AD, Amis AA, Bull AM, et al. Strain-rate sensitivity of the lateral collateral ligament of the knee. Journal of the mechanical behavior of biomedical materials. 2015;41:261–70.
9. Bray R, Rangayyan R, Frank C. Normal and healing ligament vascularity: a quantitative histological assessment in the adult rabbit medial collateral ligament. J Anat. 1996;188:87.
10. Bull A, Amis A. Accuracy of an electromagnetic tracking device. J Biomech. 1997;30:857–8.
11. Butler DL, Guan Y, Kay MD, Cummings JF, Feder SM, Levy MS. Location-dependent variations in the material properties of the anterior cruciate ligament. J Biomech. 1992;25:511–8.
12. Chan S. The development of a low-cost laser reflectance transducer system for the measurement of cross-sectional shape and area of soft tissue. ASME Adv Bioeng BED. 1995;31:123–4.
13. Cheung DT, DiCesare P, Benya PD, Libaw E, Nimni ME. The presence of intermolecular disulfide cross-links in type III collagen. J Biol Chem. 1983;258:7774–8.
14. Cheung JT-M, Zhang M. A serrated jaw clamp for tendon gripping. Med Eng Phys. 2006;28:379–82.
15. Chimich D, Shrive N, Frank C, Marchuk L, Bray R. Water content alters viscoelastic behaviour of the normal adolescent rabbit medial collateral ligament. J Biomech. 1992;25:831–7.
16. Cooper RR, Misol S, Stimmel P. Tendon and ligament insertion: a light and electron microscopic study. JBJS. 1970;52:1–170.
17. Crisco JJ, Moore DC, McGovern RD. Strain-rate sensitivity of the rabbit MCL diminishes at traumatic loading rates. J Biomech. 2002;35:1379–85.
18. Crowninshield R, Pope M. The strength and failure characteristics of rat medial collateral ligaments. J Trauma. 1976;16:99–105.
19. Danto MI, Woo SLY. The mechanical properties of skeletally mature rabbit anterior cruciate ligament and patellar tendon over a range of strain rates. J Orthop Res. 1993;11:58–67.
20. Ellis DG. Cross-sectional area measurements for tendon specimens: a comparison of several methods. J Biomech. 1969;2:175–86.
21. Figgie HE III, Bahniuk EH, Heiple KG, Davy DT. The effects of tibial-femoral angle on the failure mechanics of the canine anterior cruciate ligament. J Biomech. 1986;19:89–91.
22. Fleming B, Beynnon B, Nichols C, Renström P, Johnson R, Pope M. An in vivo comparison between intraoperative isometric measurement and local elongation of the graft after reconstruction of the anterior cruciate ligament. J Bone Joint Surg Am. 1994;76:511–9.
23. Frank C, Amiel D, Woo S, Akeson W. Normal ligament properties and ligament healing. Clin Orthop Relat Res. 1985;12:15–25.
24. Frank C, McDonald D, Lieber R, Sabiston P. Biochemical heterogeneity within the maturing rabbit medial collateral ligament. Clinical Orthop Relat Res. 1988;236:279–85.
25. Frank C, Woo SL, Amiel D, Harwood F, Gomez M, Akeson W. Medial collateral ligament healing. A

multidisciplinary assessment in rabbits. Am J Sports Med. 1983;11:379–89.
26. Fujie H, Livesay GA, Woo SL, Kashiwaguchi S, Blomstrom G (1995) The use of a universal force-moment sensor to determine in-situ forces in ligaments: a new methodology.
27. Fujie H, Mabuchi K, Woo SL-Y, Livesay GA, Arai S, Tsukamoto Y (1993) The use of robotics technology to study human joint kinematics: a new methodology.
28. Fujie H, Yamamoto N, Murakami T, Hayashi K. Effects of growth on the response of the rabbit patellar tendon to stress shielding: a biomechanical study. Clin Biomech. 2000;15:370–8.
29. Fukubayashi T, Torzilli P, Sherman M, Warren R. An in vitro biomechanical evaluation of anterior-posterior motion of the knee. Tibial displacement, rotation, and torque. J Bone Joint Surg. 1982;64:258–64.
30. Fukuta S, Oyama M, Kavalkovich K, Fu FH, Niyibizi C. Identification of types II, IX and X collagens at the insertion site of the bovine achilles tendon. Matrix Biol. 1998;17:65–73.
31. Glos D, Butler D, Grood E, Levy M (1993) In vitro evaluation of an implantable force transducer (IFT) in a patellar tendon model.
32. Grood ES, Suntay WJ (1983) A joint coordinate system for the clinical description of three-dimensional motions: application to the knee.
33. Harner C, Livesay G, Choi N, Fujie H, Fu F, Woo S. Evaluation of the sizes and shapes of the human anterior and posterior cruciate ligaments: a comparative study. Trans Orthop Res Soc. 1992;17:123.
34. Haut R (1983) Age-dependent influence of strain rate on the tensile failure of rat-tail tendon.
35. Haut RC, Powlison AC. The effects of test environment and cyclic stretching on the failure properties of human patellar tendons. J Orthop Res. 1990;8:532–40.
36. Haut TL, Haut RC. The state of tissue hydration determines the strain-rate-sensitive stiffness of human patellar tendon. J Biomech. 1997;30:79–81.
37. Hoffman AH, Robichaud DR II, Duquette JJ, Grigg P. Determining the effect of hydration upon the properties of ligaments using pseudo Gaussian stress stimuli. J Biomech. 2005;38:1636–42.
38. Holden DL, James S, Larson R, Slocum D. Proximal tibial osteotomy in patients who are fifty years old or less. A long-term follow-up study. J Bone Joint Surg Am. 1988;70:977–82.
39. Iaconis F, Steindler R, Marinozzi G. Measurements of cross-sectional area of collagen structures (knee ligaments) by means of an optical method. J Biomech. 1987;20:1003–10.
40. Jung HJ, Vangipuram G, Fisher MB, Yang G, Hsu S, Bianchi J, et al. The effects of multiple freeze–thaw cycles on the biomechanical properties of the human bone-patellar tendon-bone allograft. J Orthop Res. 2011;29:1193–8.
41. Kinzel G, Gutkowski L (1983) Joint models, degrees of freedom, and anatomical motion measurement.
42. Kiss M-O, Hagemeister N, Levasseur A, Fernandes J, Lussier B, Petit Y. A low-cost thermoelectrically cooled tissue clamp for in vitro cyclic loading and load-to-failure testing of muscles and tendons. Med Eng Phys. 2009;31:1182–6.
43. Lam T, Shrive N, Frank C (1995) Variations in rupture site and surface strains at failure in the maturing rabbit medial collateral ligament.
44. Lam T, Thomas C, Shrive N, Frank C, Sabiston C (1990) The effects of temperature on the viscoelastic properties of the rabbit medial collateral ligament.
45. Lewis JL, Lew W, Hill J, Hanley P, Ohland K, Kirstukas S, et al. (1989) Knee joint motion and ligament forces before and after ACL reconstruction.
46. Livesay GA, Rudy TW, Woo SLY, Runco TJ, Sakane M, Li G, et al. Evaluation of the effect of joint constraints on the in situ force distribution in the anterior cruciate ligament. J Orthop Res. 1997;15: 278–84.
47. Markolf KL, Gorek JF, Kabo JM, Shapiro MS. Direct measurement of resultant forces in the anterior cruciate ligament. An in vitro study performed with a new experimental technique. JBJS. 1990;72:557–67.
48. Markolf KL, Willems MJ, Jackson SR, Finerman GA. In situ calibration of miniature sensors implanted into the anterior cruciate ligament. Part I: Strain measurements. J Orthop Res. 1998;16:455–63.
49. Milne A, Chess D, Johnson J, King G. Accuracy of an electromagnetic tracking device: a study of the optimal operating range and metal interference. J Biomech. 1996;29:791–3.
50. Moon DK, Woo SL, Takakura Y, Gabriel MT, Abramowitch SD. The effects of refreezing on the viscoelastic and tensile properties of ligaments. J Biomech. 2006;39:1153–7.
51. Newton PO, Woo SLY, MacKenna DA, Akeson WH. Immobilization of the knee joint alters the mechanical and ultrastructural properties of the rabbit anterior cruciate ligament. J Orthop Res. 1995;13:191–200.
52. Niyibizi C, Visconti CS, Kavalkovich K, Woo SL-Y. Collagens in an adult bovine medial collateral ligament: immunofluorescence localization by confocal microscopy reveals that type XIV collagen predominates at the ligament-bone junction. Matrix Biol. 1995;14:743–51.
53. Quapp K, Weiss J (1998) Material characterization of human medial collateral ligament.
54. Race A, Amis AA. Cross-sectional area measurement of soft tissue. A new casting method. J Biomech. 1996;29:1207–12.
55. Reuben JD, Rovick JS, Schrager RJ, Walker PS, Boland AL. Three-dimensional dynamic motion analysis of the anterior cruciate ligament deficient knee joint. Am J Sports Med. 1989;17:463–71.
56. Rhoads DD, Noble PC, Reuben JD, Tullos HS. The effect of femoral component position on the kinematics of total knee arthroplasty. Clin Orthop Relat Res. 1993;12:122–9.
57. Riemersa D, Schamhardt H. The cryo-jaw, a clamp designed for in vitro rheology studies of horse digital flexor tendons. J Biomech. 1982;15:619–20.

58. Rincon L, Schatzmann L, Brunner P, Stäubli H, Ferguson S, Oxland T, et al. Design and evaluation of a cryogenic soft tissue fixation device—load tolerances and thermal aspects. J Biomech. 2001;34:393–7.
59. Rudy T, Livesay G, Woo S-Y, Fu F. A combined robotic/universal force sensor approach to determine in situ forces of knee ligaments. J Biomech. 1996;29:1357–60.
60. Sakane M, Livesay GA, Fox RJ, Rudy TW, Runco TJ, Woo S-Y. Relative contribution of the ACL, MCL, and bony contact to the anterior stability of the knee. Knee Surg Sports Traumatol Arthrosc. 1999;7:93–7.
61. Schmidt KH, Ledoux WR. Quantifying ligament cross-sectional area via molding and casting. J Biomech Eng. 2010;132
62. Thornton G, Shrive N, Frank C. Altering ligament water content affects ligament pre-stress and creep behavior. J Orthop Res. 2001;19:845–51.
63. van Dijk R, Huiskes R, Selvik G. Roentgen stereophotogrammetric methods for the evaluation of the three dimensional kinematic behaviour and cruciate ligament length patterns of the human knee joint. J Biomech. 1979;12:727–31.
64. Woo S. Anatomy, biology, and biomechanics of tendon and ligament. Orthop Basic Sci. 2000:582–616.
65. Woo S, Gomez MA, Sites T, Newton P, Orlando C, Akeson W. The biomechanical and morphological changes in the medial collateral ligament of the rabbit after immobilization and remobilization. J Bone Joint Surg Am. 1987;69:1200–11.
66. Woo SL-Y, Abramowitch SD, Kilger R, Liang R. Biomechanics of knee ligaments: injury, healing, and repair. J Biomech. 2006;39:1–20.
67. Woo SL-Y, Danto MI, Ohland KJ, Lee TQ, Newton PO (1990) The use of a laser micrometer system to determine the cross-sectional shape and area of ligaments: a comparative study with two existing methods.
68. Woo SL-Y, Gomez MA, Woo Y-K, Akeson WH. Mechanical properties of tendons and ligaments. Biorheology. 1982;19:397–408.
69. Woo SL-Y, Hollis JM, Adams DJ, Lyon RM, Takai S. Tensile properties of the human femur-anterior cruciate ligament-tibia complex: the effects of specimen age and orientation. Am J Sports Med. 1991;19:217–25.
70. Woo SL-Y, Lee TQ, Abramowitch SD, Gilbert TW. Structure and function of ligaments and tendons. Basic Orthop Biomech Mechano-Biol. 2005;301
71. Woo SL-Y, Orlando CA, Camp JF, Akeson WH. Effects of postmortem storage by freezing on ligament tensile behavior. J Biomech. 1986;19:399–404.
72. Woo SL, Debski RE, Withrow JD, Janaushek MA. Biomechanics of knee ligaments. Am J Sports Med. 1999;27:533–43.
73. Woo SL, Inoue M, McGurk-Burleson E, Gomez MA. Treatment of the medial collateral ligament injury. II: structure and function of canine knees in response to differing treatment regimens. Am J Sports Med. 1987;15:22–9.
74. Woo SL, Kuei SC, Gomez MA, Winters JM, Amiel D, Akeson W. The effects of immobilization and exercise on the strength characteristics of bone-medial collateral ligament-bone complex. Paper presented at: ASME Biomechanical Symposium1979.
75. Woo SL, Newton PO, MacKenna DA, Lyon RM. A comparative evaluation of the mechanical properties of the rabbit medial collateral and anterior cruciate ligaments. J Biomech. 1992;25:377–86.
76. Woo SL, Ohland KJ, Weiss JA. Aging and sex-related changes in the biomechanical properties of the rabbit medial collateral ligament. Mech Ageing Dev. 1990;56:129–42.
77. Woo SLY, Gomez M, Seguchi Y, Endo C, Akeson W. Measurement of mechanical properties of ligament substance from a bone-ligament-bone preparation. J Orthop Res. 1983;1:22–9.
78. Woo SLY, Orlando CA, Gomez MA, Frank CB, Akeson WH. Tensile properties of the medial collateral ligament as a function of age. J Orthop Res. 1986;4:133–41.
79. Woo SLY, Peterson RH, Ohland KJ, Sites TJ, Danto MI. The effects of strain rate on the properties of the medial collateral ligament in skeletally immature and mature rabbits: a biomechanical and histological study. J Orthop Res. 1990;8:712–21.
80. Yahia LH, Drouin G. Microscopical investigation of canine anterior cruciate ligament and patellar tendon: collagen fascicle morphology and architecture. J Orthop Res. 1989;7:243–51.
81. Yamakawa S, Debski RE, Fujie H. Strain distribution in the anterior cruciate ligament in response to anterior drawer force to the knee. J Biomech Sci Eng. 2017;12:16–20.
82. Yamamoto E, Hayashi K, Yamamoto N. Mechanical properties of collagen fascicles from stress-shielded patellar tendons in the rabbit. Clin Biomech. 1999;14:418–25.
83. Yamamoto S, Saito A, Nagasaka K, Sugimoto S, Mizuno K, Tanaka E, et al. The strain-rate dependence of mechanical properties of rabbit knee ligaments. Paper presented at: Proceedings of the 18th International Technical Conference on the Enhanced Safety of Vehicles (ESV)2003.

4 Biomechanics of Bone Grafts and Bone Substitutes

Daniel R. Lee and James W. Poser

Several comprehensive reviews have been published on bone graft substitutes and provide valuable background information on the vast array of available materials and products [1–3] Rather than provide yet another product or materials review to add to this comprehensive collection, the focus here is to catalog the attributes of these materials in a context that may help guide the surgeon's selection of the bone graft substitutes for particular clinical applications.

Achieving the best possible clinical outcome while satisfying the patient's expectations of return to functionality should be the principal determinants in choosing which of the myriad of bone graft substitutes is the best option for any clinical application. It is known that the structural requirements should be considered in the choice of the appropriate bone graft [4]. Many times the choice of bone graft is relegated to availability in the surgical setting, commercial representation or historic clinical experiences of the clinicians. The choice of a bone graft may also be dependent on a delicate balance between biology and biomechanical stability [5]. Patient age, health status, and activity level, coupled with clinical presentation, compliance, rehabilitation options, and economics are all considerations that should be weighed in the choice to achieve the desired clinical outcome. Most bone grafts and bone substitutes initially provide very little clinically relevant structural stability and ultimately rely on biology to restore structural stability and function.

Deciding which bone graft material to select can be a confusing and daunting process. How does one differentiate between the products? Are product claims supported by reliable science and clinical experience? How should variables such as composition, handling, mechanical and biological properties, patient clinical presentation, intended outcomes and price are all factors that weigh in decision-making?

The objective in this chapter is to provide some of the information that will be useful for the clinician in making that decision. Several publications provide general reviews of the bone grafting options that are available to the clinician [6, 7]. In the current paper, particular emphasis will be placed on the mechanical properties along with material and biological properties of the bone graft with respect to short- versus long-term outcomes and patient satisfaction. With this objective we will discuss bone grafting options from the following clinical perspectives and considerations, see Fig. 4.1.

Bone grafts are used to repair and rebuild missing, damaged, or diseased bones in a human body where the clinical situation where the bone may not heal by itself or healing might be

D. R. Lee (✉)
CellRight Technologies, LLC,
Universal City, TX, USA

J. W. Poser
RegenTX Partners, LLC, San Antonio, TX, USA

J. Koh et al. (eds.), *Orthopaedic Biomechanics in Sports Medicine*,
https://doi.org/10.1007/978-3-030-81549-3_4

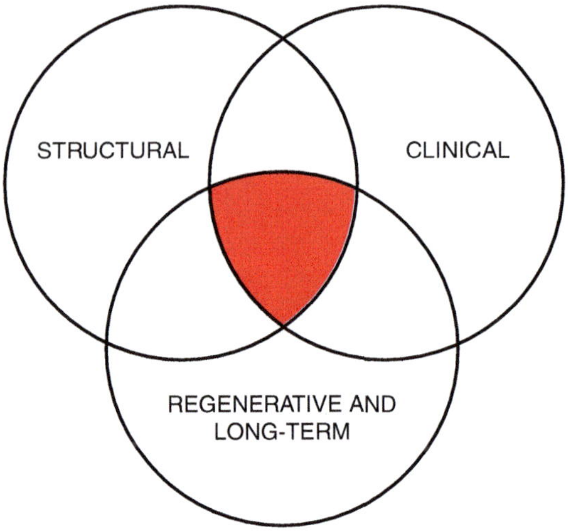

Fig. 4.1 Bone Grafting Considerations

compromised. It has been reported that there are over two million bone grafting procedures performed annually [8]. The considerations for the selection of the optimal bone graft involve biomechanical, biomaterial, and biological considerations. Among the many options available to the surgeon it is safe to say that the ideal bone graft that performs as well as the patient's own bone (i.e., autograft) has yet to be developed, but many of the available bone grafting options do have desirable and often suitable properties for a particular situation.

4.1 Bone Basics: Basic Properties and Concepts

Human bone consists of 80% cortical bone and 20% cancellous bone. Cortical or compact bone is 70% inorganic material (principally hydroxyapatite), 22% organic material (e.g., collagen, non-collagenous proteins, cells, hyaluronic acid), and the rest is water [9]. Cancellous or trabecular bone has the same constituents as cortical bone but has a lower calcium content, tissue density and ash fraction [10]. Cancellous bone also has higher water content (27% compared to 23% for cortical bone) [10]. Cortical bone is dense, strong, and difficult to fracture and thus provides most of the structural support to the body, while cancellous bone is extremely porous as it is more involved in bone remodeling. The turnover rate of trabecular bone is 25% per year versus only 3% for cortical bone [10]. With respect to mechanical properties, the compressive strength of human cortical bone ranges between 90 and 230 MPa and the tensile strength ranges between 90 and 190 MPa [11]. The compressive strength of human cancellous bone ranges from 2 MPa for osteopenic cancellous bone to 45 MPa for dense cancellous bone [12]. The Young's modulus of cancellous bone is 10.4–14.8 GPa, while cortical bone measures 18.4–20.7 GPa [9]. These physical differences also manifest themselves in biological remodeling and the susceptibility to pathologies such as osteoporosis. For cortical bone to be remodeled, osteoclasts and osteoblasts are required so the greater surface area of cancellous bone allows for more rapid revascularization and remodeling.

Bone regeneration has been defined to involve three pillars: osteogenesis, osteoinduction, and osteoconduction [13]. Osteogenesis is the synthesis of new bone by cells derived from either the graft or the host. Osteoinduction is the process where mesenchymal stem cells (MSCs) are recruited and induced to differentiate into functionally competent osteoblasts and chondroblasts. Osteoconduction is the ability process where biological interactions along the graft result in fusion of the graft with the host's bone. The overall functionality of a bone graft is dependent on the graft's ability to perform these three processes.

The timeline for normal bone regeneration and repair involves three distinct phases: inflammatory, proliferative and then remodeling [14]. The inflammatory stage (weeks 1–3) occurs after the insult or injury (hours to days, postoperatively) and involves the recruitment and proliferation of inflammatory cells and growth factors and differentiation into repair cells and the formation of callus. The proliferative phase (weeks to months, post operatively) causes the callus to organize together with a periosteal response which replaces the callus with immature woven bone predominated with vascular ingrowths and collagen matrices. The remodeling phase (months to years, postoperatively) is predominated by the restoration of bone to its baseline strength if mechanical loaded over time. The mineralized callus is replaced with mature mineralized bone and remodeling changes the area to its original size and shape. It is this final phase of bone repair that returns the bone to its previous biomechanical state [14]. Through this complex sequence of healing events bone exhibits one of its unique properties: it is the only tissue in the body that has the ability to undergo perfect, restorative repair.

4.2 Regulation of Bone Graft Substitutes

Allograft bone products are regulated as human tissue if they comply with the jurisdictional requirements. For example, in the United States those requirements are codified in 21 CFR 1271 and USPHSA Section 361, and in the EU similar requirements are codified in a series of Commission Directives (EU), e.g., 2004/23/EC and 2015/565, 566/EC. It is noteworthy that allogeneic cell tissue-based graft substitutes, which contain processed bone and bone marrow cells derived from the same donor, are often referred to as the third generation bone graft products and are approved as stand-alone graft materials under USPHS 361. Bone graft materials like synthetic CaP bone substitutes and xenograft, that do not meet these narrowly defined requirements, are regulated in the United States and in other countries as either medical devices or biologics.

The aforementioned principles and concepts provide the basis for the design and rationale of many bone grafting options and materials (Table 4.1).

Table 4.1 Bone grafting options advantages and disadvantages [6]

Bone graft (BG)	Advantages	Disadvantages
Autologous	• High osteoconductivity	Need of an additional surgery
	• Highest degree of biological safety	
	• No risk of immune reaction	
Xenografts	• Architecture and geometric structure resemble bone	• Possible disease transmission and potential unwanted immune reactions
	• Well documented	• Lacks viable cells and biological components
	• Predictable clinical outcome	• Resorption rate is highly variable
	• Slow bio-absorbability preserves augmented bone volume	• Reduced future availability due European regulatory changes?
Natural biomaterials	Similarity to native extracellular matrix	Mechanical properties poor -biodegradability less controllable
Synthetic polymers	• Tuneable physicochemical properties	• Low cell attachment
	• Tuneable degradability	• Timing of absorption (alteration of mechanical properties)
		• Release of acidic degradation products
Synthetic bioceramics	• High biocompatibility	• High brittleness
	• Osteoinductive properties	• Low ductility
	• Chemical similarity with bone	• Not predictable absorption
	• Stimulation of osteoblast growth	
Composite xenohybrid substitutes	• High similarity with human cancellous bone	• Cleaning and sterilization process partially alters biological performances
	• Higher bioactivity	• Limited clinical data
	• Tailored degradation rates	
	• Incorporation of active biomolecules	

4.3 Autograft: The "Gold Standard" Bone Graft

Autograft, the patient's own bone, has long been considered the standard to which all other graft materials are compared. Bone is one of the tissues with the innate ability to regenerate in adult humans. With respect to bone grafting, autograft or harvesting bone tissue from one anatomical site to another site on a recipient, is considered the "gold standard." Autograft earns this designation since it possesses all the properties essential for bone formation: osteogenicity, osteoinductivity, and osteoconductivity. By containing these three elements, this enables autograft to be replaced more rapidly by host bone wherever it is implanted and provides the advantages of being histocompatible, non-immunogenic and minimizing any infective-related risks [3, 8]. Introducing some combination of these biological properties has been an aspirational goal for allograft, synthetic, and any other bone grafting materials.

Structural considerations—Autogenous bone grafts may be used in structural indications primarily if they have been sourced from an area of cortical bone. Cortical grafts are used for their structural capacity, such as compression and torsion, and are frequently used in conjunction with some form of hardware to provide additional stability or structure (i.e., plates and screws, spinal implant). Cortical bone may be used as an onlay or as an inlay graft. Onlay grafts are used frequently due to the ease of handling and placing of the graft and where there is a structural need to increase the volume of bone at the repair site. The ability to position this three-dimensional graft plays a role in graft success, incorporation and ability to sustain structural forces, i.e., spinal fusion [1]. Inlay grafts are mainly used to fill a bone defect within the anatomical skeleton. Cancellous autograft does not provide immediate structural support but is the most widely utilized form of autogenous bone graft, often harvested from the surgical site and referred to as "local bone," or from the iliac crest Because these grafts lack mechanical strength they may be used to augment bone healing and fill voids in conjunction with a more structural implant, i.e., titanium or PEEK (polyetheretherketone). Mechanical properties of cancellous bone can be enhanced using impaction but the volume available is usually inadequate for most clinical applications [15]. It should also be noted that mechanical properties of autogenous bone may vary widely and are determined by the harvest site and the patient age [16].

Clinical considerations—Autogenous bone that is harvested from a second surgical site generally results in significant harvest site morbidity. Autograft can be taken surgically removed from the iliac crest, distal femur, fibula, proximal or distal tibia, proximal humerus, distal radius, chin, ribs, mandible and some parts of the skull, all resulting in morbidity, higher complications rate, scarring and additional operative time [1, 6]. Many articles have reported the drawbacks of autograft are related to the harvesting process and time, donor site complications including infection and pain, increased blood loss, and limited volume of material [3]. Local bone can be from intramedullary reamings from the femur or tibia, or bone remnants salvaged from decortication and drilling during spinal fusion procedures. The quality of the autograft may also be related to the patient and its source site due to factors such as graft components, volume and complications [5]. It has been reported that major and minor complication rates from bone harvesting are 8.6% and 20.6%, respectively [14]. The limited autograft availability, volume and configuration, may impact utility in larger defects. Autografts may be harvested from cortical or cancellous bone and the successful choice may be related to the survival and proliferation of the osteogenic cells, conditions at the recipient site, type of graft chosen, handling of the graft, and shaping of the graft during the operative procedure [1]. If rapid osteogenesis is desired then cancellous bone is preferred. Cancellous grafts are commonly used in fracture nonunion, dental defects, maxillofacial defects, spinal fusion, and other small bone defects [1].

Regenerative and long-term clinical considerations—The structural and density of cortical bone naturally limits the number of osteoprogenitor cells that reside in the tissue. Remodeling of cortical bone is mainly mediated by osteoclasts and the revascularization and remodeling processes are hampered by the dense architecture of cortical bone [3]. As a result cortical bone grafts will undergo a longer-term incorporation where there is osteoclastic surface resorption and appositional bone growth. This appositional bone growth over a necrotic core is the dominant means of incorporation. As a result the remodeling process for a cortical bone graft may take years depending upon the graft size and volume and the implantation site [3]. Cancellous bone is osteogenic due to its larger surface area and osteoblasts rapidly incorporate new bone and revascularization happens relatively quickly [13]. Cancellous bone's lack of early mechanical strength may be outweighed by its ability to rapidly produce new bone and develop mechanical strength. Therein lies one of the conundrums in bone grafting: which is more important—immediate mechanical strength or long-term restoration of a more natural bone structure?

Additional Autogenous Regenerative Considerations—Bone graft materials are often supplemented with bone marrow aspirates (BMAs) and platelet-rich plasma (PRP). The use of these materials is intended to enhance regenerative properties and thus clinical outcomes.

Platelets are a rich source of endogenous growth factors (GF), and platelet-rich plasma (PRP), also referred to as autologous growth factor (AGF) concentrate, is a readily available source of the GFs from the patient. PRP has been used in sports medicine and orthopedics based upon reports of encouraging tissue healing, and in tissues with low healing potential and treated in other specialties such as dermatology, ophthalmology, plastic and maxillofacial surgery, neurosurgery, urology, and cardiovascular surgery. The interest in this biological procedure as an augment to bone implants with bone and soft tissue has resulted in musculoskeletal treatments using PRP on cartilage, bone, muscle, tendon, and ligament regeneration. In a review article which looked at the role of PRP as an augmentation procedure, the conclusion was that knowledge on this topic was still preliminary and prospective randomized clinical trials were needed to support the potential of this approach to improve outcomes in implant integration [17]. In a review of bone grafts for spine fusion, the studies referenced did not demonstrate improved fusion or fusion rates even when PRP was used to enhance autograft [18].

Bone marrow is a rich source of nucleated cells, including MSCs, and bone marrow aspirate (BMA) has been utilized in orthobiologic repairs as a means of providing the cellular elements for tissue regeneration. Bone marrow can be harvested from various sites in the body including the posterior and anterior iliac crests, distal femur, proximal tibia, and distal humerus. As a matter of common practice, BMA is often concentrated in the operating room using specialized centrifugation and concentrating devices to produce bone marrow aspirate concentrate (BMAC) which can increase the nucleated cell (MSC) concentration by up to ten-fold over BMA. This concentrated cell preparation is then applied onto a scaffold or carrier and used to treat cartilaginous lesions, bone defects, and tendinous injuries [19]. It is estimated that 5–10% of all fractures result in delayed union or nonunions and this complication can result in significant additional cost per patient [19]. One of the focused applications of BMA is in recalcitrant bone nonunions where 94% successful arthrodesis was achieved when BMA was combined with allograft compared to conventional autologous cancellous bone graft alone. In patients with compromised healing capacity, the use of BMAC in ankle fractures in subjects with diabetes demonstrated greater union rates versus patients receiving autologous bone graft [20]. Patients with diabetes treated with BMA reported a union rate of 82.1% with minimal complications compared to the 62.3% union rate with major complications in patients receiving iliac bone graft alone. BMAC has also shown promise in osteochondral lesions due to its potential benefits in healing hyaline cartilage, but also in increasing integrative potential with autologous osteochondral transplantation [21]. Some evidence also indicates that BMAC alone may result in improved defect filling, border repair integration and surface tissue repair [22].

A more recent autologous cell enhancement are adipose-derived stem cells. Adipose-derived stem cell (ASC) treatments focus on isolating the ASCs from adipose tissue. Adipose tissue is harvested percutaneously and mechanically processed to remove lipids and disrupts adipose tissue clusters while maintaining the stromal vascular fractions that are rich in adipose-derived stem cells (ASC). By volume, the number of MSCs may be greater in adipose tissue than in BMAC [23]. The current literature on ASCs is limited regarding preparation, formulation and clinical therapeutic potential as numerous studies are ongoing. ASCs are regulated as a biologic and are not approved for any clinical indication.

4.3.1 Allograft: "the best alternative" to Autograft

Allogeneic bone grafts refer to bony tissue that is harvested from one individual and transplanted to a genetically different individual of the same species [2] [24]. Allografts, which can include tissue from both living human donors and cadavers, represent the second most common bone grafting material used worldwide [6]. Since autogenous bone grafting involves the risk of complications and the amount of graft that can be harvested is limited, allograft is accepted as the next best alternative.

Allografts do have the potential risks of eliciting an immunoreaction, transmitting infection or communicable disease, and higher failure rates over long-term use [3]. The potential for an immune response to allograft bone may be mitigated by narrowing histocompatibility differences [24]. Historically, one of the concerns with allogeneic bone has been the risk of disease transmission due to bacterial and viral contamination. Fresh, unprocessed human allograft is seldom used because of this potential for triggering a clinically significant immune response combined with the risk of disease transmission. Particularly concerning are diseases which are difficult or impossible to detect. For these reasons, most allograft bone used today are processed in established tissue banks where Federal regulations, industry standards and state-of-the-art processing and testing technologies have all but eliminated the risk of infection from allograft tissues. For example, the potential for HBV or HCV transmission has been reduced to one occurrence every 500 years [25, 26]. Processing bone has consequences as, in producing an arguably safer graft, elements essential to bone formation may be removed from the tissue. For example, decellularization removes donor histoincompatibility but removes osteogenic cells required for bone formation. Processing to remove pathogens may also reduce osteoconductive and osteoinductive potential [7]. Allograft safety is further insured by sterilization which can also significantly impact the structure and biologic properties of the allograft. It is intuitive that the goal in processing of allograft is to strike a balance between eliminating the risk of disease transmission while preserving critical properties necessary for the performance of the allograft. Overall, allograft bone is generally considered osteoconductive and weakly osteoinductive [8]. The integration process of allogeneic bone is similar to what is undergone by nonvascularized autogenous bone graft, where the volume of graft influences the time of incorporation [7].

Structural considerations—Allogeneic bone is available in many forms including morselized cortical, cancellous, and corticocancellous, cortical and cancellous grafts, osteochondral, whole bone segments, demineralized bone matrix and more recently introduced cellular allografts. Cancellous and cortical allografts are generally available through musculoskeletal tissue banks. The morselized or cube shaped cancellous bone grafts have little mechanical strength, are not suitable for use in applications requiring load bearing and are primarily used to fill voids. Cortical allografts can provide structural support and confer rigid mechanical properties and are widely utilized in applications where immediate load-bearing resistance is required [27]. An advantage of cortical bone is its natural elasticity and ease of incorporation at the graft-host interface. Furthermore, cortical allografts can be machined and customized to enhance its ease of use and meet the demands of specific applications. For example, machined cortical allografts are commonly used as structural supports in spinal fusion as "biological cages" in forms such as cervical spacers and as femoral ring allografts [28]. There is a note of caution as the mechanical performance of structural allografts may be a disadvantage, impacted by the effects of tissue processing or preservation (i.e., freeze-drying) and the less predictable effects on strength thru fatigue and postoperative remodeling on strength [27]. The freeze-drying process can reduce the mechanical strength of bone tissue by 20% [16]. Graft fractures occur and are often related to the anatomical site of implantation. Another factor that may influence the structural integrity or uniformity in tissue is processing methods.

Clinical considerations—Cancellous bone allografts are one of the most commonly used types of allografts. Since most cancellous grafts possess little mechanical strength, they are mainly used in applications such as graft extenders in spinal fusion procedures and void fillers for partial bone defects, including large depressed articular fractures, rather than segmental bone defects. Commonly freeze-dried cancellous allograft is used to pack defects in revision arthroplasty or after curettage of benign lesions [29]. As cancellous allograft is devitalized bone, it is used primarily as an osteoconductive substrate or autograft bulking agent since it lacks key elements necessary for bone formation. Used alone, cancellous allograft may result in poor clinical outcomes including failed or delayed arthrodesis or fracture. Nonetheless, used properly, allografts have a high success rate and have demonstrated similar fusion rates to autografts in spinal procedures [13].

Cortical allografts, like autograft, can be used as an onlay graft. They have been used to treat pathologies such as fibrous dysplasia, giant cell tumor, surface-based lesions resections, segmental defects after trauma or resections for sarcoma, and replacing bone lost in revision total joint arthroplasty [29]. Cortical allografts are used in spinal augmentation (spacers and wedges in various forms and designs) for filling large skeletal defects where immediate load-bearing resistance is needed. Whether the graft is frozen or freeze-dried, the cortical allograft undergoes incorporation via creeping substitution, a process initiated by osteoclastic resorption followed by sporadic formation of new appositional bone via osteoconduction [3]. The volume of allograft used affects the remodeling and conversion into host bone. Even after 5 years post implantation, large allografts may show only 20% "internal repair" Has occurred [30]. The persistence of an unincorporated and necrotic core can develop microfractures, decrease bone mineral density, and result in reduced mechanical strength and failure many years after implantation [30]. In opening wedge osteotomies, allograft performed well in mean time to union with a low loss of correction, but union time versus autograft was longer with a higher delayed or nonunion rate [31]. Allograft bone in primary arthrodesis and osteotomy procedures in foot and ankle surgery compares favorably with autograft in terms of fusion rates and clinical outcomes with fewer complications [32]. In anterior cervical discectomy and fusion, cervical spacer allografts have the highest fusion rate for the relatively low cost to other bone grafting options with equivalent clinical outcomes [33]. In revision anterior cruciate ligament (ACL) procedures, the bone tunnel frequently requires grafting so allograft unicortical dowels have been employed to provide either a single or two-stage technique [34, 35].

Regenerative and long-term clinical considerations—Compared to autografts, a slower sequence of events happens in the remodeling process. In some cases, the allograft may be delayed by a host inflammatory response which causes fibrous tissue formation around the graft which entraps the allograft and results in incomplete resorption for many years post implantation. Fresh frozen and freeze-dried bone allografts induce more prompt graft vascularization, incorporation and bone regeneration than fresh allograft [1]. The process of freeze-drying bone offers safety advantages but it renders the tissue substantially weaker. Its use as a morselized graft in impaction bone grafting in hip surgery has demonstrated 86–90% graft survival rates 7–8 years post implantation [30].

4.3.2 Demineralized Bone Matrix (DBM)

Demineralized Bone Matrix (DBM) is an osteoconductive and osteoinductive bone graft substitute composed of allograft bone with the inorganic materials removed [36]. Produced from ground human cortical, corticocancellous or cancellous bone this highly processed allograft tissue has 40–100% of the mineral removed from the organic bone matrix by exposure to mild acid, a process that leaves the collagens, non-collagenous proteins, and growth factors naturally present in bone largely intact [3]. Demineralization renders the bone osteoinductive by "unmasking" the inductive proteins present in the bone extracellular matrix. This residual matrix is frequently combined with a carrier to improve handling and performance properties. As a result, the DBMs available for clinical use come in a variety of forms ranging from moldable and injectable putties, pastes, pastes with chips, strips, and sponges.

The collagenous and non-collagenous proteins preserved throughout the demineralization process create an osteoconductive scaffold. Furthermore, the DBMs are osteoinductive by virtue of the remaining growth factors, which are directly correlated with the preparation method and can include bone morphogenetic protein (BMP), fibroblast growth factor, transforming growth factor beta (TGF-β), and vascular endothelial growth factor (VEGF) [13, 36]. Natural donor-to-donor variability, as well as the differences in tissue processing methods, results in an intrinsic range of osteoinductive potential in commercial DBMs as evidenced by testing in validated osteoinductivity assays [37].

Structural considerations—DBMs lack mechanical strength and are used primarily for filling bone defects or as autograft extenders. The carriers used with particulate DBMs minimize DBM migration and can provide resistance to displacement during lavage or motion. In an attempt to provide some structural integrity, incorporate additional osteoconductive scaffold, slow remodeling rate and maintain or increase the overall bone volume, some DBM preparations incorporate cancellous allograft chips or cortical fibers. Demineralized "sponges" produced from blocks and strips of cancellous bone are able to maintain their inherent shape and volume at the site of implantation and are frequently combined with bone marrow aspirate (BMA) to provide a more biologically complete graft.

Clinical considerations—DBM products have a long history of safe and effective clinical use and their benefit to patients are widely recognized. However, the lack of industry standards for the production and performance of DBM products has resulted in vast inconsistencies in these products. There are several attributes to consider when selecting a DBM product. DBMs vary with respect to the donor sources, how they are extracted, processed, formulated, and packaged. This variability has important clinical consequences and offers assorted advantages or disadvantages. Clinicians are encouraged to understand the processing and testing of the DBM products through a review of materials provided by the manufacturer or the comprehensive literature available on the subject (Table 4.2) [37].

Table 4.2 Considerations in the selection of DBM products

Consideration	Issues
Safety	• Clinical experience • Tissue processor • Sterility • Regulatory status
Composition	• DBM concentration • Carrier • Processing methods
Biological properties	• Osteoinductivity testing • Lot-to-lot variation
Forms	• Putty • Paste • Gel • Strips
Handling properties	• Graft extender • Injectable • Moldable • Limited migration

The biologic properties of DBM graft materials that should be considered include: (1) risk of disease transmission or immune-mediated rejection; (2) remodeling and incorporation into the host tissue; (3) promotion of surface-level bone growth or osteoconductivity; (4) osteoinductive activity; and (5) foster controlled new bone growth [37]. DBM physical properties to consider may include: (1) increased total volume of the graft when used as a graft extender; (2) effectiveness as a scaffold with good mechanical strength; and (3) handling characteristics which facilitates graft manipulation and placement and; (4) resistance to migration or displacement by irrigation and movement [37]. The carriers used with DBMs and the DBM:carrier ratios greatly affect the biologic potential, clinical handling and utility in various indications [38]. Some forms of DBM are more ideal for mixing with autogenous bone or BMA, strips or sponges may be used where some support or scaffolding are required, and some may be more amenable for defect filling or injected thru a syringe. Familiarity with the various forms of DBM (putties, pastes, injectable putties, pastes with chips, strips, and sponges) will aid in the proper selection to optimize the desired clinical outcome.

Regenerative and long-term considerations—There are no industry standards for demineralization processes employed by various producers and different processes may result in variable essential components in the DBM depending on the processes used. Inconsistency is compounded by the donor-to-donor variability discussed previously. Some of the processes used may even result in partial or complete inactivation of the BMPs resulting in reduced product efficacy [39]. One of the reasons for the variability of DBM products is many DBMs are regulated in the U.S. as tissue products, where the focus is on tissue safety and traceability versus processing, formulation, and indications for use.

DBMs have also been used in revision ACL procedures to address tunnel defects or tunnel widening in two-stage procedures [40]. However, much of the effort to improve the regenerative properties of DBM has been focused on spinal applications where these products have gained wide acceptance. Numerous clinical reports in spinal applications describe DBM efficacy as an alternative to autograft [13]. There have been reports of graft collapse due to the inferior structural composition in lumbar fusion [13] and a similar report regarding graft collapse in cervical fusion [28] illustrating the need for additional long-term clinical studies.

Animals have proven useful to define best practices in the use of DBMs in spinal fusion. In a posterolateral fusion model in athymic rats, DBMs were able to successfully demonstrate fusion at a higher rate compared to other allograft alternatives but also lot variability was also evident [41]. Animal studies have also demonstrated a range of effectiveness in the different commercial DBM products [39].

The effort to improve DBM performance is proceeding on several fronts. The addition of different materials to DBM, i.e., nanofiber-based collagen scaffolds, are under investigation to enhance the biological or mechanical performance of the DBM [42]. Future products may incorporate carriers that optimize the environment to recruit cells, encourage angiogenesis, facilitate early healing and produce new healthy bone [37].

To reduce the graft rerupture rate and improve tendon-to-bone healing in ACL reconstruction procedures, a recently developed technique combined a proprietary DBM formulation, BMA, and autologous bone collected during tunnel drilling to produce a graft mixture to fill the tunnels prior to the tendon graft passage [43].

4.4 Cellular Allografts: An Autograft Alternative

Recently, cellular allografts or cellular bone matrices (CBMs) have been developed to provide mesenchymal or osteoprogenitor cells for osteogenic grafting without the need for an autograft harvest. These products are designed to provide the scaffold, the signals, and the cells. CBMs are made using proprietary techniques which are aimed to preserve mesenchymal stem cells (MSCs). Because they contain live cells, the source and living status of the donor, the screening process, and the review of donor medical and social history is more stringent than traditional allografts [44]. Screening also includes serological and microbiological testing. The MSCs may be sourced from cancellous bone or adipose tissue. Within 72 h of death, a cadaver's donor bone is harvested and after initial evaluation, processing begins with isolation of cancellous bone chips and the milling and demineralization of cortical bone. The cancellous bone is treated to minimize immunologic issues, cryopreserved and then mixed with demineralized bone or cancellous bone ships. Other processors use cadaveric adipose tissue as the source of the human MSCs which are then mixed with demineralized bone and then cryopreserved. Total nucleated cellular concentration ranges from 250,000 to 3,000,000 cells/cc. where the number of actual mesenchymal stem cells may be a small fraction of the total. In the U.S., most commercially available CBMs fall under the HCT/P guidelines and are regulated as tissue, but some have encountered regulatory challenges due to unsubstantiated claims regarding composition and clinical efficacy.

Structural considerations—CBMs lack mechanical strength and are used in void or implant filling. Handling is similar to that seen with particulate bone grafts combined with BMA.

Clinical considerations—CBMs are regarded as premium products that are challenging and costly to produce, but nonetheless an attempt to provide the triad of components required for the ideal bone healing product. In a comparative test in posterolateral fusion in athymic rats, CBMs did not fuse as well as other grafting options [41]. Concern has been raised whether the athymic model is appropriate for testing a viable cell product. The products in the marketplace have a great variability in cellular concentration and the ideal cell concentration and type of cells may not be present [20]. This cell concentration issue can also be manifested in the donor age at the time of graft harvest as there is an age-related decline in the number of cells. The source of CBM, whether from bone or fat, may affect their ability to undergo osteoblastic differentiation in the human in vivo environment [44]. Other considerations include the number of viable cells that survive storage, thawing, transplantation, presentation at the site and recipient immunogenic response.

Regenerative and long-term considerations—CBMs conceptually have much promise and have captured much interest. In challenging foot and ankle arthrodesis and revision nonunion procedures, a specific CBM formulation demonstrated a high union rate [45]. The key to differentiating its regenerative capacity is the viability of the cell component, concentration, viability, and ability to differentiate. So, donor sourcing and selection can be an important factor. As a tissue product, any additional processing needed to ensure adequate cell concentrations would create regulatory challenges. Additionally, the manufacturing of these products requires some unique logistical challenges for tissue processors so broad availability may be limited. Additional studies will identify the efficacy of this product.

4.5 Xenografts: An Unlimited Biologic Alternative

Xenografts or heterologous grafts are bone grafts sourced from non-human sources. Xenografts are harvested from one individual and transplanted into another individual of a different species [1]. Xenografts are sourced from coral, porcine, or bovine sources. The main advantage of xenografts is the theoretical unlimited supply if they can be processed to be safe for human transplantation. Concerns with immunogenicity and disease transmission, including prions, are the primary objections to the use of xenograft tissue. To denature the proteins responsible for immunogenicity and remove lipids, some use chemical methods to remove these elements prior to terminal sterilization [30]. Other processes use high temperature thermal cycling to deproteinize the bone tissue. These processes destroy the arrangement of collagen fibrils and crosslinking [16]. These processes also effect osteogenic and osteoinductive properties since the remaining tissue is primarily a calcified hydroxyapatite scaffold and collagen. Deproteinization using milder reagents can preserve the inherent collagen architecture and can influence bone regeneration. Other processes just seek to remove cells and lipids which can elicit an immune reaction but maintain the collagen integrity and mechanical properties.

Structural considerations—Most xenografts, like allografts, are available as powders or particulates and some structural forms have also been developed. Though the mechanical properties of the xenograft bone may be similar or superior to allograft bone, the advantages are the almost unlimited availability and a mechanically consistent product. The source of the xenograft tissue is usually controlled and so a higher probability of being reproducible.

Clinical considerations—Xenograft cancellous or cortical bone incorporates in a similar manner to allograft. The clinical use of xenografts has demonstrated success as a graft extender when mixed with autogenous tissue. There have been reports of early graft resorption, loosening, and foreign body reactions, as well as satisfactory clinical reports when the xenograft is mixed with bone marrow [1, 30]. In some studies the xenograft trabecular graft was second best to the autogenous cancellous bone [1].

Regenerative and long-term considerations—To increase the biological activity of xenograft bone, the potential of xenogeneic DBM has been investigated with mixed early results. With respect to processing, multiple physical, chemical, and enzymatic methods have been used to remove antigens, but preserve the extracellular matrix and mechanical and functional characteristics. These acellularization methods may result in reducing or eliminating immunogenicity of xenografts which may enhance graft incorporation [1]. Xenografts are globally regulated as medical devices so commercialization requires demonstration of product safety and clinical efficacy. Additionally, commercial availability of xenograft bone is limited globally.

4.5.1 Synthetic Bone Graft Substitutes

Most synthetic bone graft substitutes are approved as autograft extenders, requiring that they be combined with the patient's own bone or bone marrow principally because they lack viable osteogenic cells which embody one of the triad of essential elements present in autograft.

Approval for commercial sale requires that the supplier indicate the composition and intended use of each product which should serve as a guide to the end user.

4.5.2 Calcium Sulfate

Calcium sulfate (plaster of Paris) is considered a first generation osteoconductive bone graft substitute and is still in use as a bone void filler or autograft extender. It is generally used as a nonsetting pelletized material that lacks the mechanical properties to provide sustained structural support [46]. The primary applications for calcium sulfate bone graft substitutes have been as a nonstructural bone void filler or for antibiotic delivery [47].

4.5.3 Calcium Phosphates

With a long history of safe and effective use, calcium phosphates (CaPs) are the largest category of synthetic bone graft materials. Differences in chemical composition, crystallinity, processing, shape, and porosity produce a spectrum of CaP graft materials with variable physical properties, and dissolution or degradation profiles. CaPs include β-tricalcium phosphate (β-TCP), hydroxyapatite (HA), combinations of β-TCP and HA, CaP collagen composites, ion-substituted CaPs, CaPs combined with other calcium salts, and CaP Cements. Recent reviews provide relevant and comprehensive information on these materials and useful background for the topics presented here [48, 49].

Often referred to as osteoconductive, synthetic bone graft substitutes, including CaPs, are medical devices approved as bone void fillers with the following use restrictions: intended for use as a *"resorbable bone void filler for voids or gaps that ARE NOT* (emphasis added) *intrinsic to the bone structure and…should not be used to treat large defects that in the surgeon's opinion would fail to heal spontaneously."* Other restrictions may include that the bone graft substitute is for use only as a graft extender to be combined with autograft or bone marrow aspirate (Table 4.3) [33, 50].

Table 4.3 Characteristics of bone graft options (Reproduced with permission from [29])

Graft, substitute, or augment	Comments	Advantages and disadvantages	Common applications	Product examples
Cancellous autograft	Gold standard No disease transmission	Limited availability Donor-site morbidity No structural support	Curettage and cancellous grafting	n/a
Cortical autograft (fibula, etc.)	Rapid union, osteogenic	Technically demanding if vascularized Donor-site morbidity	Segmental diaphyseal defects	n/a
Cancellous allograft	Fresh frozen has some growth factors preserved Freeze dried lowest likelihood of disease transmission	10–15% infection rate Limited shelf life (~1 year at −20 ° C)	Curettage and cancellous grafting	MTF™ cancellous chips
Cortical allograft	Structural support Osteoarticular with ligaments and tendons	10–15% graft failure 10% nonunion Immunogenic	Intercalary osteoarticular strut	MTF™ osteoarticular distal femur graft
Calcium Sulfate	"Plaster of Paris"	Rapid resorption (4–12 weeks) Inconsistent setting wound drainage	Can be mixed with antibiotics	Osteoset™ (Wright medical, TN)
Calcium phosphate	High compressive strength (4-10× cancellous bone)	Slow resorption (95% resorbed in 26 to 86 weeks)	Periarticular voids	Norian SRS™ (Synthes, PA) Hydroset™ (Stryker, MA)
Demineralized bone matrix (DBM)	Variable osteoinductivity based on formulation	No structural support possible reaction to carrier (e.g., glycerol)	Commonly mixed with allograft void filler	Grafton DBM™ (Osteotech, NJ)
Bone morphogenetic proteins (BMP)	rhBMP-2 and rhBMP-7 approved for humanitarian device exception	No osteoconductive or structural support Requires collagen or bone mineral substrate osteolysis, ectopic bone	No role in oncologic setting due to risk of increased oncogenesis	Infuse™ (Medtronic, TN) Op-1™ (Stryker, MA)
Polymethyl methacrylate (PMMA)	Unlimited supply, low cost	Excellent in compression Poor mechanical properties in shear/tension	Currettage and till periarticular voids	Simplex™ (Stryker, MA) Palacos™ (Heraeus, Germany)

This section of the review will focus on the characteristics and use of the subset of bone graft substitutes that provide stabilization and mechanical support at the site of application. Biological characteristics and biomechanical properties, material handling and physical form, along with patient requirements, are the primary factors in determining which bone graft substitute is most appropriate for the intended application.

4.6 Items to Consider

4.6.1 Porosity

The degree of material porosity plays a critical role in the biology and clinical performance of a bone graft substitute. Porosity creates the local mechanical and biological environment while providing a necessary route for neovascularization, and migration, proliferation and phenotypic expression of regenerative mesenchymal stem cells, pre-osteoblasts and osteoblasts [11, 51]. Importantly, within comparable classes of biomaterials, porosity generally correlates with mechanical strength and resorption rates [11]. For these reasons, porosity should be one of the most important material characteristics in determining the suitability of a bone graft substitute for a clinical indication.

It is sufficient to consider pore size into two general groups: micropores (<5 μm pore size) and macropores (>100 μm pore size) [52, 53]. Whereas microporosity is considered necessary for predictable bioresorption [54], it is generally accepted that macropores >100 μm are required for osteoconductivity [48, 49] the accepted "optimal" macroporosity of a bone substitute that allows for ingrowth of bone is between 150 and 500 μm [55]. An osteogenic response, including BMP-induced osteogenesis, is reportedly better when the pore size is >300 μm [52, 56, 57].

Within a scaffold, in vivo bone formation involves creating an environment that brings together the essential elements for bone formation. Where microporosity and surface roughness enhances attachment, proliferation and differentiation of anchorage dependent bone forming cells [49] and has proven benefit as a surface treatment to improve implant fixation [58, 59], higher porosity is conducive to osteogenesis [51]. Larger pores allow vascular ingrowth and support cellular activities that enhance bone ingrowth and complete integration and potential remodeling of the graft materials after surgery.

Other factors, such as the rate of degradation and the mechanical performance of the scaffold, both of which are profoundly affected by material porosity, should be taken into account when suitability is assessed. Scaffolds fabricated from ceramics with a high degradation rates should not have high porosity (>90%) as material must persist at the site long enough to conduct new bone formation otherwise the reparative process could be compromised [52].

Where the rate and extent of bone ingrowth clearly correlates with the percent porosity of a bone graft substitute, there is no consensus regarding which type of porosity provides the optimal environment for bone formation. The rate and quality of bone integration have been related to a dependence on pore size, porosity volume fraction, and interconnectivity, both as a function of structural permeability and mechanics [51].

4.6.2 Interconnectivity

Another important factor that determines the effectiveness of porosity, and therefore the effectiveness and ultimate fate of the graft material, is the way the pores connect with each other and provide a pathway for fluids, vasculature and cells to infiltrate the innermost aspect of the material. The pores may be either interconnected or "dead-ends," features that can be reproducibly introduced into the material by design and process during manufacturing. In general, bone graft materials, and specifically calcium phosphates with interconnected pores have a distinct advantage over biomaterials containing dead-end pores. Interconnectivity allows for ingrowth of new bone which in turn provides better long-term stability at the graft-hoist interface [53], provide for consistent incorporation of the bulk graft material and more uniform remodeling.

4.6.3 Mechanical Stability, Structure, and Biology

Mechanical stability in the microenvironment of bone grafting is essential, though often not afforded the consideration given the trinity of osteoconductivity, osteoinduction, and osteogenesis [60]. The unmet challenge is to bring all four elements together in a single material, achieving clinically meaningful and sustainable mechanical properties in a material that has the pore size, structure and cellularity to support bone formation. Optimization of performance must also include consideration of the morphology of a porous structure on the material and in situ mechanical properties [11]. These properties vary greatly among the various biomaterials and depend on their porosity, micro, and macro architecture. Significant parameters that differentiate the indications of the different scaffolds and biomaterials are the quality and density of the host bone bed and the local biomechanical demands of the graft site. Moreover, bone graft biomechanics evolve parallel to the progress of the graft incorporation and remodeling. All these issues have formed the bases for intense research efforts to improve initial mechanical properties of the available biomaterials as well as to guarantee the presence of a mechanically reliable construct throughout all the remodeling phase of fracture healing. However, there is an upper limit in porosity and pore size set by constraints associated with mechanical properties. An increase in the void volume results in a reduction in mechanical strength of the scaffold, which can be critical for regeneration in load-bearing bones. For example, an increase of the total porous volume from 10 to 20% results in a fourfold decrease in material mechanical strength [53]. Whether it is possible to increase pore size while at the same time maintaining requisite mechanical requirements of the material depends on many variables, including the intrinsic material properties, processing, and ultimately the practical suitability for the intended use. The operative or nonoperative techniques of fracture stabilization and fixation, and the chemical, structural, and mechanical properties of the graft material all interact and affect the bone repair process. The mechanical environment where a graft material is expected to serve as a substrate supporting bone formation has equal significance to the biologic properties of the graft itself in making decisions regarding which synthetic bone graft substitutes are appropriate for specific clinical applications.

4.6.4 Biomechanics of Synthetic Bone Graft Substitutes

The property that is most often used to characterize the mechanical behavior of bone substitutes is their compressive strength. Bone graft substitutes that are used a graft extenders or void fillers that by regulatory approval are described as a "*resorbable bone void filler for voids or gaps that ARE NOT* (emphasis added) *intrinsic to the bone structure*" usually lack in vivo mechanical strength. These materials are suitable for using as a stand-alone graft (per the Instructions for Use) for smaller, stable voids and they will eventually resorb, remodel, or become incorporated into the host bone.

The biomechanical properties of the CaP bone graft substitutes are highly variable [61]. Unless in a cementitious or highly sintered forms they are brittle, often friable with little compressive or tensile strength. They do not provide significant biomechanical support.

TCPs are less brittle than HA but resorb quickly, rapidly losing what little mechanical strength they may have when implanted. Although increased porosity and pore size facilitate bone ingrowth, porosity compromises the structural integrity of CaPs resulting in an even further reduction in mechanical properties. When used in a site that is mechanically unstable or is otherwise compromised, these nonstructural CaP void fillers generally require some form internal or external stabilization to achieve successful arthrodesis.

4.6.5 Synthetic Bone Graft Substitutes with Mechanical Strength

The structural, physiological, and biomechanical properties of calcium sulfate, calcium phosphate and methacrylate cements should guide decisions regarding their safe and effective use. These materials may provide immediate, short-term, and often long-term mechanical stability but those essential characteristics come at the expense of other properties, notably lack of porosity and ability to remodel into bone in an appropriate time frame. Importantly, when comparing bone graft substitutes, including cements, similar chemistry or material composition does not necessarily indicate that materials have identical or even similar properties.

4.6.6 Calcium Sulfate Cements

Injectable calcium sulfate (CaS) cement has been used successfully to treat a variety of clinical indications, including tibial plateau fractures [62], benign bone lesions [63], plated proximal humeral fractures [64]. CaS cements are biocompatible, biodegradable, osteoconductive and integrates well with osseous tissues. The cement cures isothermally in less than 5 min and achieve a compressive strength of approximately 40 MPa, comparable to cancellous bone. An average pore size of approximately 60 μm (range 10–250 μm) has been reported for CaS cement [65]. However, CaS cement is brittle and is absorbed more quickly (6–12 weeks) than other cements. Degradation is by dissolution and reabsorption and occurs independent of bone formation, raising concerns that absorption of the graft and subsequent loss of mechanical strength could occur before bone healing is complete, increasing the risk that healing will fail or result in a pseudoarthrosis. These concerns naturally limit most clinical use of CaS cements to filling voids where structural support is not needed.

4.6.7 Calcium Phosphate (CaP) Cements

The two principal types of calcium phosphate (CaP) cements differ based on the end product of the setting reaction: apatite [$Ca_5(PO_4)_3OH$] or brushite [$CaHPO_4.2H_2O$]. Both are regarded as osteoconductive and have the highest compressive strength (10–30 MPa) of the resorbable synthetic bone graft substitutes [65]. Apatitic CaP cements have received the most attention as they represent the chemically and mechanically stable crystalline form of CaP phase found in bone. Also, apatitic cements set at neutral pH whereas brushite cements are stable below pH 4.2 and form a metastable calcium phosphate phase at physiological pH [66]. Brushite cements resorb much more quickly than apatite cements [67].

CaP cements are generally nonporous or slightly porous upon setting. With only a few exceptions CaP cements typically have <4% total porosity with pore size ranging between 40 and 100 μm [65]. Applying the conventional wisdom that pores >100 μm are necessary for osteoconduction, CaP cement pore size and overall porosity would be considered suboptimal but still qualify as nominally osteoconductive.

CaP cements, especially apatitic cements, can remain in place for years. Lacking macroporosity they undergo surface resorption or degradation from the outside in. Incorporation of faster resorbing calcium salts (e.g., calcium sulfate or β-TCP) into apatitic cements accelerate the cement resorption by creating porosity as these more quickly resorbing materials are removed, leaving behind voids in the remaining apatitic CaP. These cements achieve their highest mechanical strength upon implantation and complete setting and lose strength as they become porous. Likewise, CaP cements that have the highest porosity on setting have a faster resorption rate and inversely have the lowest initial compressive strength, typically <1 MPa.

CaP cements have been used in several clinical indications, including fractures of the tibial plateau [68], calcaneus [69], distal radius [70], femoral neck [71], and humerus [72].

Non apatitic cements, due to primarily to their faster resorption, have clinical applications where the loss of mechanical strength over time is acceptable. Notably they have been used to stabilize traumatic fractures of the spine [73, 74].

4.6.8 Methacrylate Cements

A discussion of the mechanical properties of bone graft substitutes would not be complete without including methacrylate cements. Polymethyl methacrylate (PMMA) and bisphenol A-glycidyl methacrylate (Bis-GMA) cements are widely used in orthopedics and dentistry as they provide immediate mechanical stabilization and pain reduction. They represent a unique class of biocompatible, non-resorbable materials that do not remodel into bone, and in the case of PMMA used primarily as anchoring cement in joint arthroplasty [75]. Both PMMA and bis-GMA are used in clinical applications that overlap with CaP cements, notably vertebroplasty and balloon kyphoplasty for treating pathological fractures of the vertebral body [76, 77].

Bis-GMA cement blended with bioactive glass 45S5 [78, 79] has a compressive strength comparable to dense cancellous bone [65]. This blended material has enhanced hydrophilicity compared to PMMA or unsubstituted bis-GMA and exhibits strong bonding to bone. In addition to use in vertebroplasty [77] blended bis-GMA has been used to augment pin fixation in distal radius fractures [80].

4.7 Summary

Bone grafting is one of the most common surgical methods used to augment bone regeneration. Bone grafts and bone graft substitutes continue to evolve with the emergence of orthobiologics which continues to produce new therapies in bone healing [36]. The goal to be "as good as" autografts and provide the triad of osteoconduction, osteoinduction, and osteogenesis is still elusive in commercially available products. The various categories of options reviewed in this paper all have varying degrees of these properties (Table 4.4).

The "diamond concept" for a successful bone repair response, gives equal importance to mechanical stability and the biological environment. Overall, the diamond concept encompasses a broader appreciation of the factors such as the presence of osteoinductive mediators, osteogenic cells, an osteoconductive matrix (scaffold), optimum mechanical environment, adequate vascularity, and any existing comorbidities of the patient [60, 81]. Developments continue to describe the use of additional materials (ions), bioactive molecules, and cells which are described in numerous review articles [3, 36, 82]. and biologic approaches. All of these future developments will need to be supported by high-quality clinical trials before they become the standard of care treatments and totally replace autograft (Fig. 4.2).

Table 4.4 Key bone healing properties of bone grafting options (adapted from [36])

Bone grafting options and their specific osteoconductive, osteoinductive, and osteogenic properties					
	Osteoconductive	Osteoinductive	Osteogenic	Advantages	Disadvantages
Cortical autograft	+	+	+	• Bone healing • All components of triad • Immunocompatible • No infection	• Limited quantity • Quality donor dependent • Harvest site morbidity
Cancellous autograft	+++	+++	+++	• Bone healing • All components of triad • Immunocompatible • No infection	• Limited quantity • Quality donor dependent • Harvest site morbidity
Cortical allograft	+	±	-	• No donor site morbidity • Ability to mix with bioactive agents	• Infection transmission • Reduced biological or biomechanical properties • Sterilization may be required
Cancellous allograft	+	±	-	• No donor site morbidity • Ability to mix with bioactive agents	• Infection transmission • Reduced biological or biomechanical properties • Sterilization may be required
Demineralized bone matrix	+	++	-	• Some bone healing • Availability • Ease of use	• Variability from lot-to-lot • Bone healing variability due to processing
Calcium ceramics	+	-	-	• Availability • Ability to mix with bioactive agents • Long shelf life • No infection risk	• No biologic activity • Low mechanical strength
Bone marrow aspirate	-	+	+++	• Biologic activity • Minimal donor site morbidity	• Requires a scaffold
Bone morphogenetic protein	-	+++	-	• Bone healing • Availability	• Potential for complications
Platelet-rich plasma	-	+	+	• Biologic activity • Immunocompatible	• Lack of efficacy in bone

+, activity; −, no activity; ±, activity depends on preparation process

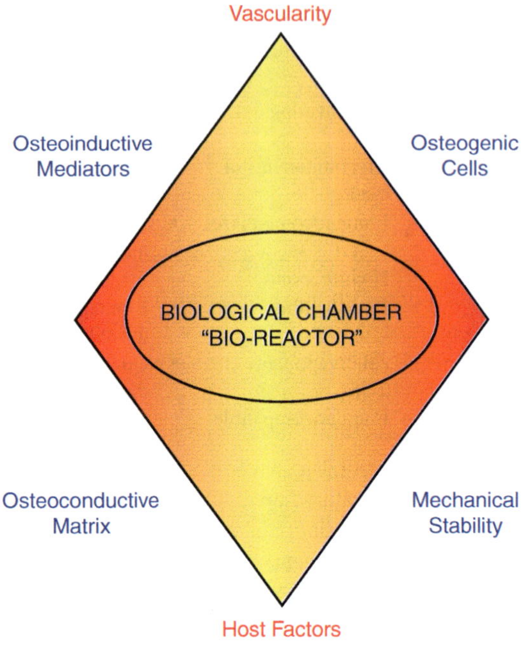

Fig. 4.2 Diamond concept of bone healing [81]

References

1. Oryan A, et al. Bone regenerative medicine: classic options, novel strategies, and future directions. J Orthop Surg Res. 2014;9:18.
2. Roberts TT, Rosenbaum AJ. Bone grafts, bone substitutes and orthobiologics: the bridge between basic science and clinical advancements in fracture healing. Organogenesis. 2012;8:115–24.
3. Wang W, Yeung KWK. Bone grafts and biomaterials substitutes for bone defect repair: a review. Bioactive Materials. 2017;2:224–47.
4. Pelker RR, Friedlaender GE. Biomechanical aspects of bone autografts and allografts. Orthop Clin North Am. 1987;18(2):235–9.
5. Egol, K.A., et al. Bone grafting: sourcing, timing, strategies, and alternatives. J Orthop Trauma. 2015;29(12).
6. Haugen HJ, et al. Bone grafts: which is the ideal biomaterial? J Clin Periodontol. 2019;46(Suppl. 21):92–102.
7. Sohn HS, Oh JK. Review of bone graft and bone substitutes with an emphasis on fracture surgeries. Biomaterials Res. 2019;23:9.
8. Campana V, et al. Bone substitutes in orthopaedic surgery: from basic science to clinical practice. J Mater Sci Mater Med. 2014;25:2445–61.
9. Brett E, et al. Biomimetics of bone implants: the regenerative road. BioResearch Open Access. 2017;6(1):1–5.
10. Oftadeh R, Perez-Viloria M, Villa-Camacho JC, Vaziri A, Nazarian A. Bioemcahnics and mechanobiology of trabecular bone: a review. J Biomech Eng. 2015;137:12–5.
11. Hannink G, Arts JJC. Bioresorbability, porosity and mechanical strength of bone substitutes: what is optimal for bone regeneration? Injury Int J Care Injured. 2011;42:S22–5.
12. An YH. Mechanical properties of bone. In: An YH, Draughn RA, editors. Mechanical testing of bone and the bone-implant interface. Boca Raton: CRC Press; 2000. p. 41–64.
13. D'Souza M, et al. Graft material and biologics for spinal interbody fusion. Biomedicine. 2019;7(75):12.
14. Fillingham Y, Jacobs J. 2016. Bone grafts and their substitutes. Bone joint J. 98B (1 Suppl a): 6-9.
15. Xu, Z.J., et al. 2011. Mechanical properties of 7-10mm bone grafts and small slurry grafts in impaction bone grafting. J. Orthop. Res., pp. 1491-1495.
16. Yamada M, Egusa H. Current bone substitutes for implant dentistry. J Prosthodont Res. 2018;62: 152–61.
17. Roffi A, et al. Does PRP enhance bone integration with grafts, graft substitutes, or implants? A systematic review. BMC Musculoskelet Disord. 2013;14:330.
18. Gupta A, et al. Bone graft substitutes for spine fusion: a brief review. World J Orthop. 2015; 6(6):449–56.
19. Gianakos AL, et al. Clinical application of concentrated bone marrow aspirate in orthopaedics: a systemic review. World J Orthop. 2017; 8(6):491–506.
20. Hernigou P, et al. Percutaneous injection of bone marrow mesenchymal stem cells for ankle non-unions decreases complications in patients with diabetes. Int Orthop. 2015;39:1639–43.
21. Kennedy JG, Murawski CD. The treatment of osteochondral lesions of the talus with autologous osteochondral Transplanation and bone marrow aspirate concentrate: surgical technique. Cartilage. 2011;2(4):327–36.
22. Chala J, et al. Bone marrow aspirate concnetrate for the treatment of osteochondral lesions of the talus: a systematic review of outcomes. J Exp Orthop. 2016;3:33.
23. DiMatteo B, et al. Adipose-derived stem cell treatments and formulations. Clin Sports Med. 2019;38(1):61–78.
24. Goldberg VM, Akhavan S. Biology of bone grafts. Beone regeneration and repair. Totowa, NJ: Springer; 2005. p. 57–65.
25. Mroz TE, et al. Musculoskeletal allograft risks and recalls in the United States. J Am Acad Orthop Surg. 2008;16(10):559–65.
26. Lomas R, Chandrasekar A, Board TN. Bone allograft in the U.K.: perceptions and realities. Hip Int. 2013;23(5):427–33.

27. Kawaguchi S, Hart RA. The need for structural allograft biomechanical guidelines. J Amer Academy Orthop Surg. 2015;23(2):119–25.
28. Mobbs RJ, Chung M, Rao PJ. Bone graft substitutes for anterior lumbar fusion. Orthop Surg. 2013;5(2):77–85.
29. Blank AT, et al. Bone grafts, substitutes, and augments in benign orthopaedic conditions - current concepts. Bull Hosp Joint Dis. 2017;75(2):119–27.
30. Pierannunzii L, Zagra L. Bone grafts, bone graft extenders, substitutes and enahncers for acetabular reconstruction in revision total hip arthoplasty. EFORT Open Reviews. 2016;1:431–9.
31. Lash NJ, et al. Bone grafts and bone substitutes for opening-wedge osteotomies of the knee: a systematic review. J of Arthroscopic Related Surgery. 2015;31(4):720–30.
32. Wee J, Thevendran G. The role of orthobiologics in foot and ankle surgery: allogenic bone grafts and bone graft substitutes. EFORT Open Reviews. 2017;2
33. Stark JR, Hsieh J, Waller D. Bone graft substitutes in single- or double-level anterior Cerivical discectomy and fusion. Spine. 2018;44(10):E618–28.
34. Werner BC, et al. Revision anterior cruciate ligament reconstruction: results of a single-stage approach using allograft dowel bone grafting for femoral defects. J Am Acad Orthop Surg. 2016; 24(8):581–7.
35. Theodorides AA, Wall OR. Two-stage revision anterior cruciate ligament reconstruction: our experience using allograft bone dowels. J Orthop Surg (Hong Kong). 2019;27(2):1–9.
36. Calcei JG, Rodeo SA. Orthobiologics for bone healing. Clin Sports Med. 2019;38:79–95.
37. Shehadi JA, Elzein SM. Review of commercially available demineralized bone matrix products for spinal fusions: a selection paradigm. Surg Neurol Int. 2017;8:203.
38. Zhang H, et al. Demineralized bone matrix carriers and their clinical applications: An overview. Orthop Surg. 2019;11(5):725–37.
39. Grabowski G, Cornett CA. Bone graft and bone graft Sustitutes in spine surgery: current concepts and controversies. J Amer Academy of Orthopaedic Surgeons. 2013;21(1):51–9.
40. Yamaguchi KT Jr, Mosich GM, Jones KJ. Arthroscopic delivery of injectable bone graft for staged revision anterior cruciate ligament reconstruction. Arthrosc Tech. 2017;6(6):e2223–7.
41. Bhamb N, et al. Comparative efficacy of commonly Avaialble human bone graft substitutes as tested for Postlateral fusion in an athymic rat model. Int J of Spine Surgery. 2019;13(5):437–558.
42. Plantz MS, Hsu WK. Recent research advances in biologic bone graft materials for spine surgery. Current Review in Musculoskeletal Medicine. 2020;13:318–25.
43. Lavender C, et al. The Lavender fertilized anterior cruciate ligament reconstruction: a quadriceps tendon all-inside reconstruction fertilized with bone marrow concentrate, demineralized bone matrix, and autograft bone. Arthrosc Tech. 2019;8(9): e1019–23.
44. Skovrlj B, et al. Cellular bone matrices: viable stem cell-containing bone graft substitutes. Spine J. 2014;14(11):2763–72.
45. Dekker, T.J, White, P. and Admas, S.B. 2017. Efficacy of a cellular bone allograft for foot and ankle arthrodesis and revision nonunion procedures. Foot Ankle Int 38(3), pp. 277–282.
46. Hak DJ. The use of osteoconductive bone graft substitutes in orthopedic trauma. J Am Acad Orthop Surg. 2007;15(9):525–36.
47. Beardmore AA, et al. Effectiveness of local antibiotic delivery with an osteoinductive and octeoconductive bone-graft substitute. J Bone Joint Surg. 2005;87(1):107–12.
48. LeGeros RZ. Properties of osteoconductive biomaterials: calcium phosphates. Clin Orthop Relat Res, pp. 2002:81–98.
49. LeGeros RZ. Calcium phosphate-based osteoinductive materials. Chem Rev. 2008;108:4742–53.
50. Buser Z, et al. Synthetic bone graft versu autograft or allograft for spinal fusion: a systematic review. J Neurosurg Spine. 2016;25:509–16.
51. Hing KA. Bioceramic bone graft substitutes: influence of porosity and chemistry. Int J Appl Ceram Technol. 2005;2(3):184–99.
52. Karageorgiou V, Kaplan D. Porosity of 3D biomaterials and osteogenesis. Biomaterials 2005. 2005;26:5474–91.
53. Blokhuis TJ, et al. Properties of calcium phosphate ceramics in relation to their in vivo behavior. J Trauma. 2000;49:179–86.
54. Driskell TD, et al. Development of ceramic and ceramic composite devices for maxillofacial applications. J Biomed Mater Res. 1972;6(1):345–61.
55. Daculsi G, Passuti N. Effect of macroporosity for osseous substitution of calcium phosphate ceramics. Biomaterials. 1990;11:86–7.
56. Tsuruga E, et al. Pore size of porous hydroxyapatite as a cell- substratum controls BMP-induced osteogenesis. J Biochem. 1997;121:317–24.
57. Kuboki Y, et al. Geometry of artificial ECM: sizes of pores controlling phenotype expression in BMP-induced osteogenesis and chondrogenesis. Connect Tissue Res. 2002;43:529–34.
58. Furlong RJ, Osborn JF. Fixation of hip prosthesis by hydroxyapatite ceramic coatings. J Bone Joint Surg. 1991;73B:741–5.
59. Cho JH, et al. Seven-year results of a tapered, titanium, hydroxyapatite-coated cementless femoral stem in primary total hip arthroplasty. Clin Orthop Surg. 2010;2:214–20.
60. Giannoudis PV, Einhorn TA, Marsh D. Fracture healing: the diamond concept. Injury. 2007;38S4:S3–6.
61. Le Huec JC, et al. Influence of porosity on the mechanical resistance of hydroxyapatite ceram-

ics under compressive stress. Biomaterials. 1995; 16(2):113–8.
62. Yu B, et al. Treatment of tibial plateau fractures with high strength injectable calcium sulphate. Int Orthop. 2009;33:1127–33.
63. Kumar Y, et al. Calcium sulfate as a bone graft substitute in the treatment of osseous defects, a prospective study. J Clin Diag Res. 2013;7(12):2926–8.
64. Somasundaram K, et al. Proximal humeral fractures: the role of calcium sulphate augmentation and extended deltoid splitting approach in internal fixation using locking plates. Injury. 2013; 44(4):481–7.
65. Van Lieshout EMM, et al. Microstructure and biomechanical characteristics of bone substitutes for trauma and orthopaedic surgery. BMC Muscoloskeletal Disorders. 2011;12(34):1–14.
66. Vereecke G, Lamaitre J. Calculations of the solubility diagrams in the system ca(OH)2-H3PO4-KOPH-HNO3-CO2-H2O. J Cryst Growth. 1990;104:820–32.
67. Ohura K, et al. Resorption of, and bone formation from, new beta-tricalcium phosphate-monocalcium phosphate cements: an in vivo study. J Biomed Mater Res. 1996;30(2):193–200.
68. Russell TA, Leighton RK. Comparison of autogenous bone graft and endothermic calcium phosphate cement for defect augmentation in tibial plateau fractures. A multicenter, prospective, randomized study. J Bone Joint Surg. 2008;90:2057–61.
69. Wee AT, Wong YS. Percutaneous reduction and injection of Norian bone cement for the treatment of displaced intra-articular calcaneal fractures. Foot Ankle Spec. 2009;2(2):98–106.
70. Liverneaux P, et al. Cement pinning of osteoporotic distal radius fractures with an injectable calcium phosphate bone substitute: report of 6 cases. Eur J Orthop Surg Traumatol. 2006;16:10–6.
71. Strauss EJ, et al. Calcium phosphate cement augmentation of the femoral neck defect created after dynamic hip screw removal. J Orthop Trauma. 2007;21:295–300.
72. Egol KA, et al. Fracture site augmentation with calcium phosphate cement reduces screw penetration after open reduction-internal fixation of proximal humeral fractures. J Shoulder Elb Surg. 2012;21:741–8.
73. C., Krop, et al. 2006. Successful posterior interlaminar fusion at the thoracic by sole use of beta-tricalcium phosphate. Arch Orthop Trauma Surg 126(3), pp. 204–210.
74. Maestretti G, et al. Prospective of stand alone balloon kyphoplasty with calcium phosphate cement augmentation in traumatic fracture. Eur Spine J. 2007;16(5):601–10.
75. Deb S. Orthopedic bone cements. Boca Raton: CRC Press; 2008.
76. Klazen C, et al. Vertebroplasty versus conservative treatment in acute osteoporotic vertebral compression fractures (VERTOS II): An open-label randomized trial. Lancet. 2010;376:1085–92.
77. Bae H, et al. A prospective randomized FDA-IDE trial comparing Cortoss with PMMS in vertebroplasty. Spine. 2012;37(7):544–50.
78. Hench LL, et al. Bonding mechanisms at the interface of ceramic prosthetic materials. J Biomed Mater Res. 1971;5(6):117–41.
79. Rahaman M. Bioactive glass in tissue engineering. Acta Biomater. 2011;7(6):2355–73.
80. Smit RS, van der Velde D, Hegeman JH. Augmented pin fixation with Cortoss for an unstable AO-A3 type distal radius fracture in a patient with manifest osteoporosis. Arch Orthop Trauma Surg. 2008;128(9):989–93.
81. Andrzejowski P, Giannoudis PV. The 'diamond concept' for long bone non-union management. J Orthop Traumatol. 2019;20(21)
82. Ho-Shui-Ling A, et al. Bone regeneration strategies: engineered scaffolds, bioactive molecules and stem cells current stage and future perspectives. Biomaterials. 2018;180:143–62.

5 Mechanical and Biologic Properties of Articular Cartilage Repair Biomaterials

George Jacob, Kazunori Shimomura, David A. Hart, Hiromichi Fujie, and Norimasa Nakamura

5.1 Introduction

Repair of articular cartilage repair has continued to be a challenge in orthopaedic practice [1, 2]. Methods to regenerate chondral tissue have involved the application of stem cells and chondrocytes, along with the development of biomaterials. Biomaterials used in chondral regeneration must eventually either possess properties similar to that of cartilage to function as an effective replacement or medium for regeneration, or be biodegradable and replaced with a hyaline cartilage-like matrix. In the past, various techniques and modalities have been utilised ranging from gel-like materials [3, 4] solid scaffolds and scaffold-free treatments [5] to achieve better cellular presence and proliferation within the defect [6]. Articular cartilage is a complex tissue in terms of structure and joint function, and can vary in different sites within a joint. From a biomechanical standpoint, the main functions of articular cartilage are load transmission, force absorption, and lubrication [7, 8]. In this chapter, we aim to highlight the characteristics of natural and synthetic biomaterials available for chondral tissue regeneration.

G. Jacob · K. Shimomura
Department of Orthopaedic Surgery, Osaka University Graduate School of Medicine, Osaka, Japan
e-mail: kazunori-shimomura@umin.net

D. A. Hart
McCaig Institute for Bone and Joint Health, University of Calgary, Calgary, AB, Canada
e-mail: hartd@ucalgary.ca

H. Fujie
Biomechanics Laboratory, Faculty of System Design, Tokyo Metropolitan University, Tokyo, Japan
e-mail: fujie@tmu.ac.jp

N. Nakamura (✉)
Institute for Medical Science in Sports, Osaka Health Science University, Osaka, Japan

Global Centre for Medical Engineering and Informatics, Osaka University, Osaka, Japan
e-mail: norimasa.nakamura@onsu.ac.jp

5.2 Components of Articular Cartilage

Cartilage is hypocellular, containing chondrocytes [7] surrounded by an extracellular matrix (ECM) made of mainly collagen, proteoglycans (PG), and water [9]. These components exhibit widely distinct and differing properties, and their presence within the tissue is variable, with the tissue nearest the articular surface having the greatest water content and a low PG content [10, 11]. Layers adjacent to the subchondral bone have greater PG content and lower water content. Collagen and chondrocyte fibre arrangement is also variable. The superficial zone has flatter and elongated chondrocytes with a surface parallel collagen fibre arrangement [12]. Into the middle zone, collagen fibres are arranged randomly with

J. Koh et al. (eds.), *Orthopaedic Biomechanics in Sports Medicine*,
https://doi.org/10.1007/978-3-030-81549-3_5

rounded chondrocytes [13]. The deeper zones have chondrocytes arranged in columns with collagen fibres perpendicular to the joint surface. The deeper zone residing chondrocytes produce a greater amount of ECM and collagen [14]. It is with these components that cartilage fulfils its biomechanical roles in joint articulation. Thus, the structural complexity of native hyaline cartilage poses challenges to its effective repair and regeneration.

5.3 Biomechanics of Cartilage

Articular cartilage is a biphasic medium with a fluid phase and a solid phase [15, 16]. The solid phase of cartilage consists of collagen, chondrocytes, PGs, and lipids. It is the fluid phase properties of cartilage that protect the solid phase from damage during loading and articulation. Cartilage maintains an aqueous environment at near neutral pH, making the PGs polyanionic with negatively charged sites [17, 18]. The molecules repel each other and disperse to occupy a larger volume which is restricted by the collagen framework. During compression, there is an increase in interstitial fluid pressure in the chondral tissue and the negatively charged PG ions are brought closer together resulting in increased repulsion forces which provide the tissue with a compressive stiffness known as the flow independent mechanism [19]. The flow-dependent mechanism of cartilage occurs when compressed fluid flows through the tissue and across the articular surface produces a frictional drag on the matrix [20, 21]. The same has been observed within the tissue where a compressed region of tissue expels fluid into the surrounding non-compressed regions of the tissue. Cartilage has low permeability and fluid does not flow through the tissue quickly resulting in a biphasic visco-elastic behaviour [22]. Within the middle zone, the random collagen fibril distribution provides resistance to shear, and intra- and intermolecular cross-links increase the strength of the fibrils, allowing them to stretch.

5.4 Biomaterials for Articular Cartilage Repair and Regeneration

A biomaterial for chondral regeneration must simulate properties similar to cartilage. An ideal biomaterial should be mechanically strong yet elastic, be biocompatible, biodegradable, porous, cyto-compatible, and offer efficient delivery. Cartilage ECM is dense, and to simulate this electrospinning has been utilised to synthesise nanofibrous scaffolds using various biomaterials. Electro spun scaffolds have a high surface area to volume ratio [23] and although they mimic the dense structure of native cartilage ([24], the increased density can hamper cell migration. Figure 5.1 enumerates the ideal attributes of a cartilage biomaterial.

5.5 Natural Biomaterials

5.5.1 Agarose

Agarose is a polysaccharide containing mainly galactose commonly derived from seaweeds [25, 26]. Agarose undergoes mild gelation with the formation of intermolecular hydrogen bonds within itself. This leads to the formation of a double helical structure forming thick bundles within the medium creating a hydrogel material [27]. Due to this mechanical stability, it is suitable for cell encapsulation [12, 28], and literature has shown chondrocytes can express their original phenotype and exhibit upregulation of glycosaminoglycan (GAG) production in an agarose medium [29]. It is also known that the use of bioreactors and functional mechanical loading and overloading of the chondrocytes in agarose resulted in tissue with Young's modulus and reparative properties similar to native cartilage [30, 31]. However, agarose does not undergo degradation due to its high chemical stability, and mammalian cells lack receptors that bind to agarose [28].

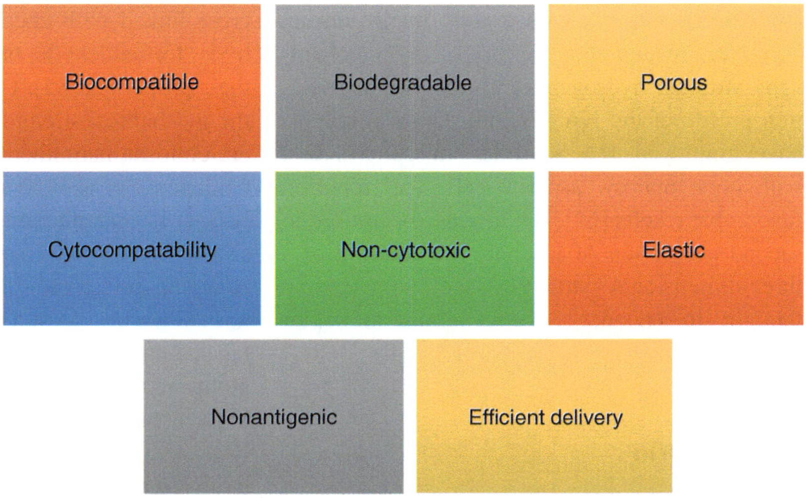

Fig. 5.1 Enumerating the ideal attributes of a biomaterial for cartilage repair and regeneration

5.5.2 Alginate

Alginates are derived from brown algae and are hydrophilic [28]. They are a linear polysaccharide copolymers made up of 1,4-linked beta-D-mannuronic acid and alpha-L-guluronic acid [32]. Alginates express effective gelling properties which encourage the development of superior chondrogenic phenotypes and mechanical properties similar to hyaline cartilage [33, 34]. The gelling properties vary depending on the species and age of algae from which the alginate is derived [35]. Ion loss through leaching is variable when implanted in vivo, leading to loss of mechanical stiffness. In addition, alginates also cannot be broken down by the body due to a lack of interaction with mammalian cell receptors [36].

5.5.3 Chitosan

Chitosan is a chitin-derived natural polysaccharide [37] made of beta- (1–4) D-glucosamine and N-acetyl-D-Glucosamine [38], similar to cartilage GAGs. It is a popular biomaterial, being biocompatible, biodegradable, non-cytotoxic, and with low immunogenicity [12, 28, 39]. Chitosan can undergo enzymatic degradation by lysosomes [40] but does not possess effective gelling properties, and therefore, it may not remain long in the intended defect site [41]. This has led to the manufacture of new drug delivery systems with the formation of chitosan gels using cross-linking methods [42–44]. In vivo, chitosan has shown superior defect fill and use of a bioreactor to induce flow shear stress yielded higher collagen II and ECM production [45, 46]. Chitosan is effective in microfracture blood clot stabilisation, preventing clot shrinkage and improving MSC density [47–51].

5.5.4 Hyaluronan

Hyaluronan (HA) is a linear GAG abundant in articular cartilage ECM, as it encapsulates native chondrocytes making it an obvious choice for a biomaterial [52, 53]. HA has an effective role in nutrient diffusion, tissue hydration, and PG organisation [54–59]. Hyaluronan also interacts with many surface proteins such as intracellular adhesion molecules, cluster differentiation molecules, and receptors for HA-mediated motility. These aid in cellular proliferation and adhesion [60–62]. HA undergoes degradation by native hyaluronidase allowing for its easy removal from the body. To improve HA for chondral repair via hydrogel applications, studies have chemically modified HA by esterification, amidation, etherification, and many more. However, results indicate producing mechanically stronger hydrogels can affect the degradation profile of the HA [52, 63]. Clinical studies have employed HA scaffolds in combination with bone marrow aspirate cells and reported favourable results [64]. HA has also been used in second generation ACI, where the addition of a three-dimensional structure such as HA is thought to encourage better chondral repair [65–67].

5.5.5 Fibrin

Fibrin is derived from fibrinogen, and is a protein in blood clot matrices. It can be manufactured into hydrogels by cross-linking fibrinogen [68]. Fibrin demonstrates good cellular compatibility [69], in vivo molecule binding, [70] and degradability in vivo [68]. It is also approved by the Food and Drug Administration (FDA) for cartilage tissue engineering [71, 72]. In chondral repair, the use of fibrin glue has been shown to promote GAG synthesis which is integral in ECM production, along with stabilising the cells in the repair site [73]. The drawbacks of fibrin glues are that they are mechanically weak with uncontrolled loss of integrity and shrinkage [74–77]. However, fibrin glues have been an effective mode of cell delivery and have been used with autologous chondrocytes [78, 79] and MSCs [80].

5.5.6 Collagen

Collagen is made up of a triple helical structure and is the main constituent of the ECM of cartilage [81]. Due to the self-aggregation properties of collagen, it is effective in cell encapsulation and provision of tensile strength [12, 82]. Additionally, collagen is bioactive exhibiting native binding motifs allowing cellular attachment, signalling, and enzymatic degradation [82, 83]. Collagens are however considered to be soft, with an elastic modulus <5 kPa [84]. Therefore, they must be manufactured at high density or cross-linked with chemicals such as glutaraldehyde, formaldehyde, or carbodiimide to enhance their properties [85]. Other materials to improve the mechanical strength of collagen preparations are chitosan nanofibers and hydroxyapatite [47]. Collagen can also be utilised as a hydrogel, as gelatin is manufactured by denaturing collagen and has been shown to promote cell proliferation, adhesion, and proliferation [86]. Preclinical data has shown benefit and good defect fill using collagen based scaffolds for chondral repair [87], along with seeding them with bone marrow derived MSCs [88] and chondrocytes [89, 90]. Kon et al. [91] reported the use of a collagen/hydroxyapatite scaffold for the repair of osteochondral defects. At 24 months, they noted 70% of lesions to have complete fill and integration at the boundaries. They concluded the implant to be safe and be a viable option for osteochondral repair. Efe et al. [92] employed a cell free strategy using only a type I collagen matrix for chondral repair. In their report, all cases had complete defect fill and integration at 24 months. Lastly, Tohyama et al. [93] utilised autologous chondrocyte implantation (ACI) combined with an atelocollagen implant in a multicentre trial. They too reported superior outcome scores but noted detachment of the implant in two cases.

5.5.7 Silk Fibroin (SF)

SF is highly biocompatible, biodegradable, permeable to water and oxygen, as well as having good mechanical properties [94–96]. SF is composed of repeating amino acid sequences (mainly glycine, alanine, and serine) and can be processed into foams, films, or electrospun into nanofibrous membranes [97]. The structure of SF is easily modifiable to improve cell adhesion and biocompatibility [98, 99]. A key drawback of silk is that it does induce a foreign body immune reaction, especially when it has not been purified to be free of sericin. In some cases, its use can result in granuloma formation [100]. Porous silk scaffolds have been used in combination with MSCs for chondral regeneration and resulted in better GAG production and collagen deposition when compared to a collagen scaffold [101, 102]. Chen et al. [97] also demonstrated in vitro silk scaffolds promoted chondrocyte proliferation with chondral ECM production and following subcutaneous implantation in mice, the presence of ectopic cartilage was confirmed.

5.5.8 Cellulose

Cellulose is a glucose polymer found in plants. It is insoluble, highly biocompatible, and can undergo enzymatic degradation [103]. It is highly modifiable, as cellulose is rich in hydroxyl groups allowing it to be manufactured into different hydrogels with varying properties. In addition to natural plant cellulose, gram negative bacteria can be used to create microbial cellulose [104]. Microbial cellulose results in a fine microfibrillar structure resulting in superior tensile strength, greater surface area, and a higher degree of polymerisation [105]. As well, it has been proven to be biocompatible [106], to have a greater surface area [107], a higher degree of polymerisation [105], and has been proven to be biocompatible [106]. Müller et al. [108] employed a cellulose scaffold seeded with chondrocytes and noted excellent adherence and homogenous distribution of the chondrocytes. Svensson et al. [109] reported on a microbial cellulose scaffold in combination with both bovine and human chondrocytes. They noted higher chondrocyte proliferation and collagen II production in the cellulose scaffold groups and concluded it could be a viable biomaterial scaffold. However, in vivo use of cellulose has not become popular and remains to be further studied in depth.

5.5.9 Tissue Engineered Construct (TEC)

A more recent option for chondral defect management has been the use of a TEC [110]. A TEC is a scaffold-free, natural 3D construct derived from synovial MSCs. These undifferentiated cells are cultured in a high density monolayer culture with the addition of ascorbate 2-phosphate for the promotion of ECM synthesis. Once detached from the base of the culture dish, TEC undergoes active tissue contraction to form the 3D structure. TEC demonstrates superior adhesiveness due to the presence of fibronectin and vitronectin noted throughout the whole tissue. This means a TEC can attach to a defect site without the need for another mode of fixation, a finding that has been demonstrated in both preclinical and clinical studies [110, 111]. The culture of TEC using a chondrogenic medium revealed increased GAG synthesis and deposition along with the presence of collagen II expression confirmed with using semiquantitative RT-PCR. Mechanical testing has revealed TEC implanted into chondral defects possess similar viscoelastic properties and coefficient of friction to that of native articular cartilage [5]. TEC has been used in multiple preclinical studies to repair isolated chondral [5] and osteochondral defects [112]. In treating osteochondral defects, TEC was attached to a hydroxyapatite block and implanted as a complete biphasic structure [112]. These studies displayed promising results, with superior defect repair characteristics, which in turn led to the use of TEC in clinical studies. A pilot clinical trial in five patients treated with autogenically manufactured TEC for the repair of chondral defects resulted in excellent results, and the defect repair tissue resembled that of hyaline cartilage both on magnetic resonance imaging and second look arthroscopy with histologic assessment [111]. Therefore, the cells can effectively differentiate in the in vivo environment after implantation. Thus, TEC has many advantages over previous implants in that they are natural, scaffold-free, allogeneic, and effective, characteristics which have led to its use in the next ongoing larger clinical trial. The only real disadvantage of TEC is the need for a large number of cells (Table 5.1).

Table 5.1 Outline of the advantages and disadvantages of natural materials

Material	Advantages	Disadvantage
Agarose	Undergoes gelation, good mechanical stability and effect cell encapsulation	Lack of cell receptors, inferior degradability
Alginate	Effective gelling properties allowing easy delivery	Lacks cellular interaction and inferior mechanical properties due to leeching in some species. Difficult to sterilise and presence of material impurities
Chitosan	Natural, contains similar constituents to cartilage (GAG and HA). Good biocompatibility, easy degradation by lysosomes	No gelling properties therefore retention at defect site may be affected. Requires cross linking and further processing
Hyaluronan	Naturally occurring, no antigenicity. Some cells possess receptors (i.e. CD44, RHAMM) for cellular interaction, readily degraded by native hyaluronidase	Soluble in water. Anionic surface at neutral pH which does not promote non-receptor-mediated cell attachment/interactions
Fibrin	Natural and blood derived, undergoes easy in vivo degradation. Easy to deliver	Inferior mechanical properties, uncontrolled degradation and clot shrinkage
Collagen	Good biocompatibility, low antigenicity, superior tensile strength. Undergoes enzymatic degradation	Mechanically soft and requires manufacture in high density, or in combination with other materials
Silk fibroin	Biocompatible, biodegradable, permeable to water and O2. Superior mechanical properties. Readily processed.	Foreign body reaction to sericin contamination
Cellulose	High purity, superior mechanical properties and increased surface area owing to nanofibrous structure, good biocompatibility.	Smaller pore diameter Not many in vivo studies
TEC	Natural, scaffold-free, can be allogeneic, easy delivery method and no need for fixation. Highly bioactive with good integration, increased GAG and collagen synthesis.	Requirement for large cell numbers to manufacture

5.6 Synthetic Biomaterials

5.6.1 Polyethylene Glycol (PEG)

PEG has been used in cartilage tissue engineering as both a hydrogel or in a rigid scaffold form. The hydrogel form has been more popular as it is non-toxic, non-immunogenic, has a high solubility in water and organic solvents, along with being FDA approved [57]. PEG has also been shown to be effective for cell encapsulation [28, 113–115]. With the presence of an end hydroxyl group, modification of PEG is made easy with functional groups [114, 116–119]. To control the degradation rate and improve the strength of PEG, polyesters such as glycolic acid, lactic acid, and caprolactone can be added to PEG. Adjusting the ratio of these hydrophobic to hydrophilic components allows for controlling the degradation rate of the biomaterial as a whole [120]. Much preclinical research has employed PEG-based hydrogels and reported good outcomes [117, 121]. Scholz et al. [121] reported that the use of a PEG albumin hydrogel prevented pathological angiogenesis in the reparative chondral tissue along with cells maintaining their phenotypes and expressing collagen I and II. Zhao et al. [117] also achieved the production of neocartilage resembling that of normal articular cartilage using a photochemically cross-linked PEG gel encapsulating porcine articular chondrocytes. More recently, Liu et al. [119] applied PEG hydrogels in combination with human MSCs, a construct that resulted in increased collagen II, better ECM production, and activation of cartilage-specific genes. These preclinical studies indicate PEG to be an effective synthetic material for chondral regeneration.

5.6.2 Poly N-Isopropylacrylamide (PNIPAM)

PNIPAM is a thermosensitive hydrogel derived from polyacrylic acid which has a phase transition above its low critical solution temperature (LCST) of 32–33 °C [122, 123]. This makes it ideal as an injectable material as at room temperature, it remains a liquid but once in in vivo where the temperature is higher, it undergoes gelation [124]. PNIMPAM can be copolymerised with other hydrophilic/hydrophobic monomers to increase or decrease the gelation temperature. For example, acrylic acid can be raised to 37 °C [12] making it suitable for in vivo use. In an in vitro study by Ibusaki et al. [125], chondrocytes were cultured for 12 weeks in a PNIPAM-gelatin hydrogel and it was found that GAG synthesis and collagen production resembled that of normal cartilage. The same author performed an in vivo study in rabbits, cartilage defects were treated with chondrocytes were embedded in PNIMPAM. At 5 weeks post-implantation, defect sites demonstrated good fill and collagen II production [126]. Recently, the use of MSCs in PNIPAM also showed good cell viability in vitro and good chondral repair in vivo [127]. The thermosensitive properties are the strength and weakness of PNIPAM, as although scaffold delivery is easier, it can change phase again if temperature changes occur in vivo [125].

5.6.3 Polylactide Acid (PLA)

PLA is a biodegradable polyester that degrades by nonenzymatic hydrolysis into a metabolic by-product of lactic acid which can be cleared via normal cell metabolism [128]. Derivatives of PLA include poly-D lactide, poly-lactic-co-glycolic, acid and poly-L-lactide [129]. PLA exhibits good cyto-compatibility and its biodegradability also makes it an attractive option for in vivo use. In terms of synthesis, it can synthesised with varying stereo-regularities allowing it to possess different mechanical properties, thermal properties, and degradation rates [130]. To make PLA more biomimetic, it has been combined with chitosan which resulted in a slower, more favourable degradation rate [131]. Also, a processing technique known as thermally induced phase separation allows for the production of nanofibrous PLA which has a higher surface area, a characteristic favourable for cell migration, as well as protein absorption and degradation [131]. However, PLA does not express any surface epitopes which react with cellular receptors for cell adhesion [132]. Tanaka et al. [133] employed PLA-based scaffolds seeded with chondrocytes and atelocollagen, and subcutaneously implanted them into nude mice. When compared to the scaffold-lacking controls, implantations that included a scaffold maintained their integrity and expressed greater levels of collagen I and II along with increased GAG synthesis. A lower macrophage presence was also noted, indicating that PLA-related scaffolds could be a good choice for cartilage regeneration. PLA has also been linked to collagen [134], gelatin [135], and fibrin [136], and shown to have positive outcomes with superior scaffold strength, chondrocyte proliferation, and collagen synthesis.

5.6.4 Polyvinylalcohol (PVA)

PVA is a synthetic polymer synthesised through partial or full hydrolysis of polyvinyl acetate [137]. The physical and mechanical properties of PVA are determined by the amount of hydroxylation, along with the processing of its molecular weight and concentration. PVA is highly soluble in water and requires cross-linking or modification to form adhesive hydrogels for application in tissue engineering [138, 139], without which it cannot be used effectively for homogenous cell encapsulation. PVA hydrogels have good cyto-compatibility, are nontoxic, hydrophilic, and do not adhere to cells and proteins making them ideal for chondrocyte encapsulation [140, 141]. When compared to PEG, PVA possesses a greater number of hydroxyl groups, making PVA more versatile for modification [142]. Compounds such as polyethylene and alginate have been combined with PVA. Polyethylene PVA combinations showed greater mechanical strength, but lower biodegradability [143], while PVA combined with alginates showed increased GAG and type II collagen presence [144]. The use of MSCs and growth factors have also been studied with a PVA-poly(caprolactone) scaffold in vitro by Mohan et al. [145]. They demonstrated that a combination of TGFβ3 and BMP2, along with their mentioned scaffold promoted better chondrogenic differentiation of MSCs, thus proving PVA to be useful as a scaffold mimicking the ECM of native cartilage.

5.6.5 Poly-Caprolactone (PCL)

PCL is a semicrystalline, hydrophobic, linear polyester. Owing to these properties, PCL has a slow degradation rate [146], making it useful for long term implants. PCL has a high permeability which also makes it useful for the delivery of growth factors and bioactive molecules [147]. On the other hand, the products of PCL degradation are acidic which may have negative effects when used in vivo, and PCL has inferior cell interaction when compared to other materials [96]. PCL is FDA approved and has found use in many medical applications [148]. PCL has been electrospun into a nanofibrous scaffold to match the fibre arrangement and dense tissue architecture of cartilage, along with the addition of polyethyleneimine. This construct has shown favourable outcomes in cartilage engineering with superior cell proliferation, collagen II expression, and GAG production [149].

5.6.6 Polyurethane

Polyurethanes can be easily processed into gel or paste form, polymerised, and have superior mechanical properties [150]. The molecular stability of polyurethanes has been the most important purpose of their use in vivo, but with the interest of increasing its biocompatibility, research has focused on developing degradable polyurethanes which have readily metabolised by-products [151]. Not many studies have utilised polyurethanes for cartilage tissue engineering. Grad et al. [143] reported the use of a developed biodegradable polyester scaffold seeded with chondrocytes. The scaffold demonstrated good cellular adhesion and ECM production. Another author reported polyurethane scaffolds seeded with chondrocytes under dynamic culture, and environment which resulted in a tissue more closely resembling native cartilage [152]. This indicates that mechanical loading of the polyurethane scaffold may result in better repair tissue, and due to the favourable mechanical properties of polyurethane, it could be a useful material for cartilage tissue engineering where mechanical strength is required [153].

5.6.7 Polypropylene Fumarate (PPF)

PPF is an unsaturated linear polyester that can be cross-linked to other monomers e.g. methyl methacrylate to create required scaffolds. PPF undergoes degradation via hydrolysis into propylene glycol and fumaric acid which can be metabolised in vivo [154, 155] but do result in an acidic environment which can stimulate an inflammatory response [156]. The mechanical properties of PPF can be altered by either cross-linking with other compounds [155, 157] or varying the molecular weight [157, 158]. PPF is a liquid, allowing for easy moulding and fabrication into customised implants or formulated for injection. PPF has good mechanical properties which are useful in cartilage and bone tissue engineering as load-bearing is a requirement of such scaffolds [96, 159]. The utilisation of PPF in bone tissue engineering has increased, even though PPF materials have been documented to have toxic reactions and thus, raise concerns about its biocompatibility [155, 156]. However, newer techniques for PPF synthesis have been developed that report less toxicity while still maintaining the mechanical properties [160] (Table 5.2).

Table 5.2 Outline of the advantages and disadvantages of synthetic materials

Material	Advantages	Disadvantage
PEG	Nontoxic, presence of end OH groups for modification	Minimal bioactivity and inferior mechanical properties
PNIPAM	Undergoes gelation at body temperature, good biocompatibility	Thermosensitive phase inversion can result in leakage of material from defects if temperature change occurs post-implantation. Deformation under physiologic loading
PLA	Good cyto-compatibility and biodegradation properties. Can be synthesised into a nanofibrous structure and can possess varying thermal and mechanical properties	Nil cell surface epitopes for enhancing cellular adhesion
PVA	Many OH groups for modification, nontoxic, good cyto-compatibility	Highly soluble in water, low biodegradability
PCL	Good for long-term implants in view of slow degradation, high permeability	Slow degradation rate with release of acidic degradation products. Poor wettability, and lack of cell adhesion and interaction
Polyurethane	High molecular stability, good mechanical strength	Slow biodegradability
PPF	Good mechanical properties, can be synthesised to be injectable. Undergoes degradation in vivo	Cross-linking agents can be toxic. Degradation products create an acidic environment in vivo that can lead to an inflammatory response

5.7 Future Directions

Despite a large number of materials currently available for articular cartilage tissue engineering, there is still no single superior material. Each material possesses its own advantages and disadvantages, and current research has focused on combining them to possess all the desired ideal properties for a scaffold. This includes combining natural and synthetic materials in order to apply the specific qualities each material may have to produce an overall better-performing graft with negligible drawbacks. Although many materials and combinations have been shown to be effective in cell encapsulation, further preclinical studies are required to make certain which material performs in the most ideal manner before translation to clinical use. Another unknown fact is the long-term effects of these materials on chondrocytes, as most data and studies in the literature only assess short term effects of the materials.

Newer methods such as 3D printing and bioreactors have been developed to improve both material characteristics and emulate in vivo conditions to ensure better cell proliferation, however, these systems are still in research phases. The addition of signalling molecules and growth factors to hydrogels has also been a consideration to improve their bioactivity. At present, it is obvious that natural materials possess better cyto-compatibility while synthetics perform better in terms of mechanical stability. More preclinical, as well as translational studies, are required to be performed to make certain the effects of these materials for cartilage regeneration are optimised, with the hope for a material that can fulfil all the necessary requirements for an effective chondral repair.

Bibliography

1. Karuppal R. Current concepts in the articular cartilage repair and regeneration. J Orthop. 2017;14(2):A1.
2. Getgood A, Brooks R, Fortier L, Rushton N. Articular cartilage tissue engineering: Today's research, tomorrow's practice? Journal of Bone and Joint Surgery - Series B. 2009;91(5):565–76.
3. Spiller KL, Maher SA, Lowman AM. Hydrogels for the repair of articular cartilage defects. Tissue Eng - Part B Rev. 2016;17(4):281–99.
4. Vega SL, Kwon MY, Burdick JA. Recent advances in hydrogels for cartilage tissue engineering. Eur Cells Mater. 2017;33:59.

5. Ando W, Tateishi K, Hart DA, Katakai D, Tanaka Y, Nakata K, et al. Cartilage repair using an in vitro generated scaffold-free tissue-engineered construct derived from porcine synovial mesenchymal stem cells. Biomaterials. 2007;28(36):5462–70.
6. Ge Z, Li C, Heng BC, Cao G, Yang Z. Functional biomaterials for cartilage regeneration. Journal of Biomedical Materials Research - Part A. 2012;100(9):2526–36.
7. Sophia Fox AJ, Bedi A, Rodeo SA. The basic science of articular cartilage: structure, composition, and function. Sports Health. 2009;1(6):461–8.
8. Williams RJ, Peterson L, Cole BJ. Cartilage repair strategies. Cartilage repair strategies. Editor. Totowa: Humana Press; 2007.
9. Buckwalter JA, Mankin HJ. Articular cartilage: tissue design and chondrocyte-matrix interactions. Instr Course Lect. 1998;47:477–86.
10. Poole CA, Flint MH, Beaumont BW. Morphological and functional interrelationships of articular cartilage matrices. J Anat. 1984;138(Pt 1):113.
11. Malda J, Benders KEM, Klein TJ, de Grauw JC, Kik MJL, Hutmacher DW, et al. Comparative study of depth-dependent characteristics of equine and human osteochondral tissue from the medial and lateral femoral condyles. Osteoarthr Cartil. 2012;20(10):1147–51.
12. Duarte Campos DF, Drescher W, Rath B, Tingart M, Fischer H. Supporting biomaterials for articular cartilage repair. Cartilage. 2012;3(3):205–21.
13. Mow VC, Guo XE. Mechano-electrochemical properties of articular cartilage: their inhomogeneities and anisotropies. Annu Rev Biomed Eng. 2002;4(1):175–209.
14. Vornehm SI, Dudhia J, Von der Mark K, Aigner T. Expression of collagen types IX and XI and other major cartilage matrix components by human fetal chondrocytes in vivo. Matrix Biol. 1996;Jul 1;15(2):91–8.
15. Mow VC, Kuei SC, Lai WM, Armstrong CG. Biphasic creep and stress relaxation of articular cartilage in compression: theory and experiments. J Biomech Eng. 1980;102(1):73–84.
16. Zhu W, Mow VC, Koob TJ, Eyre DR. Viscoelastic shear properties of articular cartilage and the effects of glycosidase treatments. J Orthop Res. 1993;11(6):771–81.
17. Gu WY, Lai WM, Mow VC. A mixture theory for charged-hydrated soft tissues containing ivjultielectrolytes: passive transport and swelling behaviors. J Biomech Eng. 1998;120(2):169–80.
18. Lai WM, Hou JS, Mow VC. A triphasic theory for the swelling and deformation behaviors of articular cartilage. J Biomech Eng. 1991;113(3):245–58.
19. Hayes WC, Bodine AJ. Flow-independent viscoelastic properties of articular cartilage matrix. J Biomech. 1978;11(8–9):407–19.
20. Frank EH, Grodzinsky AJ. Cartilage electromechanics-I. electrokinetic transduction and the effects of electrolyte pH and ionic strength. J Biomech. 1987;20(6):615–27.
21. Maroudas A, Bullough P. Permeability of articular cartilage [24]. Nature. 1968;50(1):166–77.
22. Woo SL-Y, Buckwalter JA, Fung YC. Injury and repair of the musculoskeletal soft tissues. J Biomech Eng 1989;111(1):95–95.
23. Li WJ, Laurencin CT, Caterson EJ, Tuan RS, Ko FK. Electrospun nanofibrous structure: a novel scaffold for tissue engineering. J Biomed Mater Res. 2002;60(4):613–21.
24. Li WJ, Danielson KG, Alexander PG, Tuan RS. Biological response of chondrocytes cultured in three-dimensional nanofibrous poly(ε-caprolactone) scaffolds. J Biomed Mater Res - Part A. 2003;67(4):1105–14.
25. Kelly TAN, Ng KW, Ateshian GA, Hung CT. Analysis of radial variations in material properties and matrix composition of chondrocyte-seeded agarose hydrogel constructs. Osteoarthr Cartil. 2009;17(1):73–82.
26. Ng KW, Ateshian GA, Hung CT. Zonal chondrocytes seeded in a layered agarose hydrogel create engineered cartilage with depth-dependent cellular and mechanical inhomogeneity. Tissue Eng Part A. 2009;15(9):2315–24.
27. Xiong JY, Narayanan J, Liu XY, Chong TK, Chen SB, Chung TS. Topology evolution and gelation mechanism of agarose gel. J Phys Chem B. 2005;109(12):5638–43.
28. Roberts JJ, Martens PJ. Engineering biosynthetic cell encapsulation systems. In: Biosynthetic Polymers for Medical Applications; 2016. p. 205–39
29. Mouw JK, Case ND, Guldberg RE, Plaas AHK, Levenston ME. Variations in matrix composition and GAG fine structure among scaffolds for cartilage tissue engineering. Osteoarthr Cartil. 2005;13(9):828–36.
30. Bian L, Fong JV, Lima EG, Stoker AM, Ateshian GA, Cook JL, et al. Dynamic mechanical loading enhances functional properties of tissue-engineered cartilage using mature canine chondrocytes. Tissue Eng - Part A. 2010;16(5):1781–90.
31. Tan AR, Dong EY, Ateshian GA, Hung CT. Response of engineered cartilage to mechanical insult depends on construct maturity. Osteoarthr Cartil. 2010;18(12):1577–85.
32. Lee KY, Mooney DJ. Alginate: properties and biomedical applications. Progress in Polymer Science (Oxford). 2012;37(1):106–26.
33. Shahin K, Doran PM. Improved seeding of chondrocytes into polyglycolic acid scaffolds using semi-static and alginate loading methods. Biotechnol Prog. 2011;27(1):191–200.
34. Wan LQ, Jiang J, Miller DE, Guo XE, Mow VC, Lu HH. Matrix deposition modulates the viscoelastic shear properties of hydrogel-based cartilage grafts. Tissue Eng - Part A. 2011;17(7–8):1111–22.
35. Smidsrød O, Skjåk-Bræk G. Alginate as immobilization matrix for cells. Trends Biotechnol. 1990;8:71–8.
36. Al-Shamkhani A, Duncan R. Radioiodination of alginate via covalently-bound tyrosinamide allows

monitoring of its fate in vivo. J Bioact Compat Polym. 1995;10(1):4–13.
37. Chandy T. Sharma' CP. Chitosan - as a biomaterial. Biomater Artif Cells Artif Organs. 1990;18(1):1–24.
38. Tan H, Chu CR, Payne KA, Marra KG. Injectable in situ forming biodegradable chitosan-hyaluronic acid based hydrogels for cartilage tissue engineering. Biomaterials. 2009;30(13):2499–506.
39. Jayakumar R, Menon D, Manzoor K, Nair SV, Tamura H. Biomedical applications of chitin and chitosan based nanomaterials - a short review. Carbohydr Polym. 2010;82(2):227–32.
40. Lee KY, Ha WS, Park WH. Blood compatibility and biodegradability of partially N-acylated chitosan derivatives. Biomaterials. 1995;16(16):1211–6.
41. Hao T, Wen N, Cao JK, Wang HB, Lü SH, Liu T, et al. The support of matrix accumulation and the promotion of sheep articular cartilage defects repair in vivo by chitosan hydrogels. Osteoarthr Cartil. 2010;18(2):257–65.
42. Park KM, Lee SY, Joung YK, Na JS, Lee MC, Park KD. Thermosensitive chitosan-Pluronic hydrogel as an injectable cell delivery carrier for cartilage regeneration. Acta Biomater. 2009;5(6):1956–65.
43. Hoemann CD, Sun J, Légaré A, McKee MD, Buschmann MD. Tissue engineering of cartilage using an injectable and adhesive chitosan-based cell-delivery vehicle. Osteoarthr Cartil. 2005;13(4):318–29.
44. Bhattarai N, Gunn J, Zhang M. Chitosan-based hydrogels for controlled, localized drug delivery. Adv Drug Deliv Rev. 2010;62(1):83–99.
45. Alves da Silva ML, Martins A, Costa-Pinto AR, Correlo VM, Sol P, Bhattacharya M, et al. Chondrogenic differentiation of human bone marrow mesenchymal stem cells in chitosan-based scaffolds using a flow-perfusion bioreactor. J Tissue Eng Regen Med. 2011;9:722–32.
46. Alves da Silva ML, Crawford A, Mundy JM, Correlo VM, Sol P, Bhattacharya M, et al. Chitosan/polyester-based scaffolds for cartilage tissue engineering: assessment of extracellular matrix formation. Acta Biomater. 2010;6(3):1149–57.
47. Jančář J, Slovíková A, Amler E, Krupa P, Kecová H, Plánka L, et al. Mechanical response of porous scaffolds for cartilage engineering. Physiol Res. 2007;56(1):S17–25.
48. Hoemann CD, Sun J, McKee MD, Chevrier A, Rossomacha E, Rivard GE, et al. Chitosan-glycerol phosphate/blood implants elicit hyaline cartilage repair integrated with porous subchondral bone in microdrilled rabbit defects. Osteoarthr Cartil. 2007;15(1):78–89.
49. Hoemann CD, Hurtig M, Rossomacha E, Sun J, Chevrier A, Shive MS, et al. Chitosan-glycerol phosphate/blood implants improve hyaline cartilage repair in ovine microfracture defects. J Bone Jt Surg - Ser A. 2005;87(12):2671–86.
50. Stanish WD, McCormack R, Forriol F, Mohtadi N, Pelet S, Desnoyers J, et al. Novel scaffold-based bst-cargel treatment results in superior cartilage repair compared with microfracture in a randomized controlled trial. J Bone Jt Surg - Ser A. 2013;95(18):1640–50.
51. Shive MS, Stanish WD, McCormack R, Forriol F, Mohtadi N, Pelet S, et al. BST-CarGel® treatment maintains cartilage repair superiority over microfracture at 5 years in a Multicenter randomized controlled trial. Cartilage. 2015;6(2):62–72.
52. Burdick JA, Prestwich GD. Hyaluronic acid hydrogels for biomedical applications. Adv Mater. 2011;23(12):H41–56.
53. Jin R, Teixeira LSM, Krouwels A, Dijkstra PJ, Van Blitterswijk CA, Karperien M, et al. Synthesis and characterization of hyaluronic acid-poly(ethylene glycol) hydrogels via Michael addition: an injectable biomaterial for cartilage repair. Acta Biomater. 2010;6(6):1968–77.
54. Hemshekhar M, Thushara RM, Chandranayaka S, Sherman LS, Kemparaju K, Girish KS. Emerging roles of hyaluronic acid bioscaffolds in tissue engineering and regenerative medicine. Int J Biol Macromol. 2016;86:917–28.
55. Laurent TC, Laurent UBG, Fraser JRE. The structure and function of hyaluronan: an overview. In: Immunology and Cell Biology. 1996;74(2):a1–7.
56. Reitinger S, Lepperdinger G. Hyaluronan, a ready choice to fuel regeneration: a mini-review. Gerontology. 2012;59(1):71–6.
57. Peppas NA, Hilt JZ, Khademhosseini A, Langer R. Hydrogels in biology and medicine: from molecular principles to bionanotechnology. Adv Mater. 2006;18(11):1345–60.
58. Sommarin Y, Heinegard D. Specific interaction between cartilage proteoglycans and hyaluronic acid at the chondrocyte cell surface. Biochem J. 1983;214(3):777–84.
59. Hascall VC. Interaction of cartilage proteoglycans with hyaluronic acid. Journal of Supramolecular and Cellular Biochemistry. 1977;7(1):101–20.
60. Laurent TC, Fraser JR. Hyaluronan. FASEB J. 1992;6(7):2397–404.
61. Toole BP. Hyaluronan: From extracellular glue to pericellular cue. Nat Rev Cancer. 2004;4(7):528–39.
62. Savani RC, Cao G, Pooler PM, Zaman A, Zhou Z, DeLisser HM. Differential involvement of the hyaluronan (HA) receptors CD44 and receptor for HA-mediated motility in endothelial cell function and angiogenesis. J Biol Chem. 2001;276(39):36770–8.
63. Highley CB, Prestwich GD, Burdick JA. Recent advances in hyaluronic acid hydrogels for biomedical applications. Curr Opin Biotechnol. 2016;40:35–40.
64. Gobbi A, Karnatzikos G, Scotti C, Mahajan V, Mazzucco L, Grigolo B. One-step cartilage repair with bone marrow aspirate concentrated cells and collagen matrix in full-thickness knee cartilage lesions: results at 2-year follow-up. Cartilage. 2011;2(3):286–99.
65. Kon E, Di Martino A, Filardo G, Tetta C, Busacca M, Iacono F, et al. Second-generation autologous

chondrocyte transplantation: MRI findings and clinical correlations at a minimum 5-year follow-up. Eur J Radiol. 2011;79(3):382–8.
66. Ferruzzi A, Buda R, Faldini C, Vannini F, Di Caprio F, Luciani D, et al. Autologous chondrocyte implantation in the knee joint: open compared with arthroscopic technique. Comparison at a minimum follow-up of five years. Journal of Bone and Joint Surgery - Series A. 2008;90(Suppl 4):90–101.
67. Filardo G, Kon E, Di Martino A, Iacono F, Marcacci M. Arthroscopic second-generation autologous chondrocyte implantation: a prospective 7-year follow-up study. Am J Sports Med. 2011;39(10):2153–60.
68. Wu I, Elisseeff J. Biomaterials and tissue engineering for soft tissue reconstruction. In: Natural and synthetic biomedical polymers; 2014. pp. 235–241.
69. Ahmed TAE, Dare EV, Hincke M. Fibrin: a versatile scaffold for tissue engineering applications. Tissue Engineering - Part B: Reviews. 2008;14(2):199–215.
70. Hogg PJ, Jackson CM. Fibrin monomer protects thrombin from inactivation by heparin-antithrombin III: implications for heparin efficacy. Proc Natl Acad Sci U S A. 1989;86(10):3619–23.
71. Silverman RP, Passaretti D, Huang W, Randolph MA, Yaremchuk MJ. Injectable tissue-engineered cartilage using a fibrin glue polymer. Plast Reconstr Surg. 1999;103(7):1809–18.
72. Fussenegger M, Meinhart J, Höbling W, Kullich W, Funk S, Bernatzky G. Stabilized autologous fibrin-chondrocyte constructs for cartilage repair in vivo. Ann Plast Surg. 2003;51(5):493–8.
73. Sage A, Chang AA, Schumacher BL, Sah RL, Watson D. Cartilage outgrowth in fibrin scaffolds. Am J Rhinol Allergy. 2009;23(5):486–91.
74. Rowe SL, Lee SY, Stegemann JP. Influence of thrombin concentration on the mechanical and morphological properties of cell-seeded fibrin hydrogels. Acta Biomater. 2007;3(1):59–67.
75. Devolder R, Kong HJ. Hydrogels for in vivo-like three-dimensional cellular studies. Wiley Interdiscip Rev Syst Biol Med. 2012;4(4):351–65.
76. Kneser U, Voogd A, Ohnolz J, Buettner O, Stangenberg L, Zhang YH, et al. Fibrin gel-immobilized primary osteoblasts in calcium phosphate bone cement: in vivo evaluation with regard to application as injectable biological bone substitute. Cells Tissues Organs. 2005;179(4):158–69.
77. Lee BP, Dalsin JL, Messersmith PB. Biomimetic adhesive polymers based on mussel adhesive proteins. In: Biological adhesives. 2006 (pp. 257–278). Springer, Berlin.
78. Wysocka A, Mann K, Bursig H, Dec J, Gaździk TS. Chondrocyte suspension in fibrin glue. Cell Tissue Bank. 2010;11(2):209–15.
79. Cakmak O, Babakurban ST, Akkuzu HG, Bilgi S, Ovali E, Kongur M, et al. Injectable tissue-engineered cartilage using commercially available fibrin glue. Laryngoscope. 2013;123(12):2986–92.
80. Shetty AA, Kim SJ, Shetty V, Stelzeneder D, Shetty N, Bilagi P, et al. Autologous bone-marrow mesenchymal cell induced chondrogenesis: single-stage arthroscopic cartilage repair. Tissue Eng Regen Med. 2014;11(3):247–53.
81. Lee CH, Singla A, Lee Y. Biomedical applications of collagen. Int J Pharm. 2001;221(1–2):1–22.
82. Rozario T, DeSimone DW. The extracellular matrix in development and morphogenesis: a dynamic view. Dev Biol. 2010;341(1):126–40.
83. Levental KR, Yu H, Kass L, Lakins JN, Egeblad M, Erler JT, et al. Matrix crosslinking forces tumor progression by enhancing integrin Signaling. Cell. 2009;139(5):891–906.
84. Lau YKI, Gobin AM, West JL. Overexpression of lysyl oxidase to increase matrix crosslinking and improve tissue strength in dermal wound healing. Ann Biomed Eng. 2006;34(8):1239–46.
85. Lee CR, Grodzinsky AJ, Spector M. The effects of cross-linking of collagen-glycosaminoglycan scaffolds on compressive stiffness, chondrocyte-mediated contraction, proliferation and biosynthesis. Biomaterials. 2001;22(23):3145–54.
86. Levett PA, Melchels FPW, Schrobback K, Hutmacher DW, Malda J, Klein TJ. A biomimetic extracellular matrix for cartilage tissue engineering centered on photocurable gelatin, hyaluronic acid and chondroitin sulfate. Acta Biomater. 2014;10(1):214–23.
87. Levingstone TJ, Thompson E, Matsiko A, Schepens A, Gleeson JP, O'brien FJ. Multi-layered collagen-based scaffolds for osteochondral defect repair in rabbits. Acta Biomater. 2016;32:149–60.
88. Chen WC, Yao CL, Wei YH, Chu IM. Evaluating osteochondral defect repair potential of autologous rabbit bone marrow cells on type II collagen scaffold. Cytotechnology. 2011;63(1):13–23.
89. Pulkkinen HJ, Tiitu V, Valonen P, Jurvelin JS, Rieppo L, Töyräs J, et al. Repair of osteochondral defects with recombinant human type II collagen gel and autologous chondrocytes in rabbit. Osteoarthr Cartil. 2013;21(3):481–90.
90. Lee CR, Grodzinsky AJ, Hsu HP, Spector M. Effects of a cultured autologous chondrocyte-seeded type II collagen scaffold on the healing of a chondral defect in a canine model. J Orthop Res. 2003;21(2):272–81.
91. Kon E, Delcogliano M, Filardo G, Busacca M, Di Martino A, Marcacci M. Novel nano-composite multilayered biomaterial for osteochondral regeneration: a pilot clinical trial. Am J Sports Med. 2011;39(6):1180–90.
92. Efe T, Theisen C, Fuchs-Winkelmann S, Stein T, Getgood A, Rominger MB, et al. Cell-free collagen type I matrix for repair of cartilage defects-clinical and magnetic resonance imaging results. Knee Surgery, Sport Traumatol Arthrosc. 2012;20(10):1915–22.
93. Tohyama H, Yasuda K, Minami A, Majima T, Iwasaki N, Muneta T, et al. Atelocollagen-associated autologous chondrocyte implantation for the repair of chondral defects of the knee: a prospective multicenter clinical trial in Japan. J Orthop Sci. 2009;14(5):579–88.

94. Wang Y, Kim HJ, Vunjak-Novakovic G, Kaplan DL. Stem cell-based tissue engineering with silk biomaterials. Biomaterials. 2006;27(36):6064–82.
95. Zhang Q, Yan S, Li M. Silk fibroin based porous materials. Materials. 2009;2(4):2276–95.
96. Puppi D, Chiellini F, Piras AM, Chiellini E. Polymeric materials for bone and cartilage repair. Progress in Polymer Science (Oxford). 2010;35(4):403–40.
97. Chen CH, Liu JMJ, Chua CK, Chou SM, Shyu VBH, Chen JP. Cartilage tissue engineering with silk fibroin scaffolds fabricated by indirect additive manufacturing technology. Materials (Basel). 2014;7(3):2104–19.
98. Servoli E, Maniglio D, Motta A, Predazzer R, Migliaresi C. Surface properties of silk fibroin films and their interaction with fibroblasts. Macromol Biosci. 2005;5(12):1175–83.
99. Min BM, Lee G, Kim SH, Nam YS, Lee TS, Park WH. Electrospinning of silk fibroin nanofibers and its effect on the adhesion and spreading of normal human keratinocytes and fibroblasts in vitro. Biomaterials. 2004;25(7–8):1289–97.
100. Altman GH, Diaz F, Jakuba C, Calabro T, Horan RL, Chen J, et al. Silk-based biomaterials. Biomaterials. 2003;24(3):401–16.
101. Oegema TR, Carpenter RJ, Hofmeister F, Thompson RC. The interaction of the zone of calcified cartilage and subchondral bone in osteoarthritis. Microsc Res Tech. 1997;37(4):324–32.
102. Meinel L, Hofmann S, Karageorgiou V, Zichner L, Langer R, Kaplan D, et al. Engineering cartilage-like tissue using human mesenchymal stem cells and silk protein scaffolds. Biotechnol Bioeng. 2004;88(3):379–91.
103. Märtson M, Viljanto J, Hurme T, Laippala P, Saukko P. Is cellulose sponge degradable or stable as implantation material? An in vivo subcutaneous study in the rat. Biomaterials. 1999;20(21):1989–95.
104. Mohite BV, Patil SV. A novel biomaterial: bacterial cellulose and its new era applications. Biotechnol Appl Biochem. 2014;61(2):101–10.
105. Watanabe K, Tabuchi M, Morinaga Y, Yoshinaga F. Structural features and properties of bacterial cellulose produced in agitated culture. Cellulose. 1998;5(3):187–200.
106. De Castro Pita PC, Pinto FCM, De Melo Lira MM. De Assis Dutra Melo F, Ferreira LM, De Andrade Aguiar JL. Biocompatibility of the bacterial cellulose hydrogel in subcutaneous tissue of rabbits. Acta Cir Bras. 2015;30(4):296–300.
107. Kim DY, Nishiyama Y, Kuga S. Surface acetylation of bacterial cellulose. Cellulose. 2002;9(3–4):361–7.
108. Müller FA, Müller L, Hofmann I, Greil P, Wenzel MM, Staudenmaier R. Cellulose-based scaffold materials for cartilage tissue engineering. Biomaterials. 2006;27(21):3955–6.
109. Svensson A, Nicklasson E, Harrah T, Panilaitis B, Kaplan DL, Brittberg M, et al. Bacterial cellulose as a potential scaffold for tissue engineering of cartilage. Biomaterials. 2005;26(4):419–31.
110. Ando W, Tateishi K, Katakai D, Hart DA, Higuchi C, Nakata K, et al. In vitro generation of a scaffold-free tissue-engineered construct (TEC) derived from human synovial mesenchymal stem cells: biological and mechanical properties and further chondrogenic potential. Tissue Eng - Part A. 2008;14(12):2041–9.
111. Shimomura K, Yasui Y, Koizumi K, Chijimatsu R, Hart DA, Yonetani Y, et al. First-in-human pilot study of implantation of a scaffold-free tissue-engineered construct generated from autologous synovial mesenchymal stem cells for repair of knee chondral lesions. Am J Sports Med. 2018;46(10):2384–93.
112. Shimomura K, Moriguchi Y, Ando W, Nansai R, Fujie H, Hart DA, et al. Osteochondral repair using a scaffold-free tissue-engineered construct derived from synovial mesenchymal stem cells and a hydroxyapatite-based artificial bone. Tissue Eng - Part A. 2014;20(17–18):2291–304.
113. Temenoff JS, Steinbis ES, Mikos AG. Effect of drying history on swelling properties and cell attachment to oligo(poly(ethylene glycol) fumarate) hydrogels for guided tissue regeneration applications. J Biomater Sci Polym Ed. 2003;14(9):989–1004.
114. Roberts JJ, Bryant SJ. Comparison of photopolymerizable thiol-ene PEG and acrylate-based PEG hydrogels for cartilage development. Biomaterials. 2013;34(38):9969–79.
115. Bryant SJ, Anseth KS. Controlling the spatial distribution of ECM components in degradable PEG hydrogels for tissue engineering cartilage. J Biomed Mater Res - Part A. 2003;64(1):70–9.
116. Gould ST, Anseth KS. Role of cell–matrix interactions on VIC phenotype and tissue deposition in 3D PEG hydrogels. J Tissue Eng Regen Med. 2016;10(10):E443–53.
117. Zhao X, Papadopoulos A, Ibusuki S, Bichara DA, Saris DB, Malda J, et al. Articular cartilage generation applying PEG-LA-DM/PEGDM copolymer hydrogels. BMC Musculoskelet Disord. 2016;17(1):245.
118. Rakovsky A, Marbach D, Lotan N, Lanir Y. Poly(ethylene glycol)-based hydrogels as cartilage substitutes: synthesis and mechanical characteristics. J Appl Polym Sci. 2009;112(1):390–401.
119. Liu SQ, Tian Q, Hedrick JL, Po Hui JH, Rachel Ee PL, Yang YY. Biomimetic hydrogels for chondrogenic differentiation of human mesenchymal stem cells to neocartilage. Biomaterials. 2010;31(28):7298–307.
120. Rice MA, Anseth KS. Controlling cartilaginous matrix evolution in hydrogels with degradation triggered by exogenous addition of an enzyme. Tissue Eng. 2007;13(4):683–91.
121. Scholz B, Kinzelmann C, Benz K, Mollenhauer J, Wurst H, Schlosshauer B. Suppression of adverse angiogenesis in an albumin-based hydrogel for articular cartilage and intervertebral disc regeneration. Eur Cells Mater. 2010;20(24):2010–8.
122. Gandhi A, Paul A, Sen SO, Sen KK. Studies on thermoresponsive polymers: phase behaviour, drug delivery and biomedical applications. Asian Journal of Pharmaceutical Sciences. 2015;10(2):99–107.

123. Kim S, Healy KE. Synthesis and characterization of injectable poly(N-isopropylacrylamide-co-acrylic acid) hydrogels with proteolytically degradable cross-links. Biomacromolecules. 2003;4(5):1214–23.
124. Zhang XZ, Xu XD, Cheng SX, Zhuo RX. Strategies to improve the response rate of thermosensitive PNIPAAm hydrogels. Soft Matter. 2008;4(3):385–91.
125. Ibusuki S, Fujii Y, Iwamoto Y, Matsuda T. Tissue-engineered cartilage using an injectable and in situ gelable thermoresponsive gelatin: fabrication and in vitro performance. Tissue Eng. 2003;9(2):371–84.
126. Ibusuki S, Iwamoto Y, Matsuda T. System-engineered cartilage using poly(N-isopropylacrylamide)-grafted Gelatin as in situ-formable scaffold: in vivo performance. Tissue Eng. 2003;9(6):1133–42.
127. Zhang J, Yun S, Du Y, Zannettino ACW, Zhang H. Fabrication of a cartilage patch by fusing hydrogel-derived cell aggregates onto electrospun film. Tissue Eng Part A. 2020; 1(ja).
128. Lasprilla AJR, Martinez GAR, Lunelli BH, Jardini AL, Filho RM. Poly-lactic acid synthesis for application in biomedical devices - a review. Biotechnol Adv. 2012;30(1):321–8.
129. Capito RM, Spector M. Scaffold-based articular cartilage repair. IEEE Eng Med Biol Mag. 2003;22(5):42–50.
130. Bigg DM. Polylactide copolymers: effect of copolymer ratio and end capping on their properties. Adv Polym Technol. 2005;24(2):69–82.
131. Zhao J, Han W, Tu M, Huan S, Zeng R, Wu H, et al. Preparation and properties of biomimetic porous nanofibrous poly(l-lactide) scaffold with chitosan nanofiber network by a dual thermally induced phase separation technique. Mater Sci Eng C. 2012;32(6):1496–502.
132. Narayanan G, Vernekar VN, Kuyinu EL, Laurencin CT. Poly (lactic acid)-based biomaterials for orthopaedic regenerative engineering. Adv Drug Deliv Rev. 2016;107:247–76.
133. Tanaka Y, Yamaoka H, Nishizawa S, Nagata S, Ogasawara T, Asawa Y, et al. The optimization of porous polymeric scaffolds for chondrocyte/atelocollagen based tissue-engineered cartilage. Biomaterials. 2010;31(16):4506–16.
134. Chen JP, Li SF, Chiang YP. Bioactive collagen-grafted poly-L-lactic acid nanofibrous membrane for cartilage tissue engineering. In: Journal of Nanoscience and Nanotechnology. 2010;10(8):5393–8.
135. Chen JP, Su CH. Surface modification of electrospun PLLA nanofibers by plasma treatment and cationized gelatin immobilization for cartilage tissue engineering. Acta Biomater. 2011;7(1):234–43.
136. Zhao H, Ma L, Gong Y, Gao C, Shen J. A polylactide/fibrin gel composite scaffold for cartilage tissue engineering: fabrication and an in vitro evaluation. J Mater Sci Mater Med. 2009;20(1):135–43.
137. Tubbs RK. Sequence distribution of partially hydrolyzed poly(vinyl acetate). J Polym Sci Part A-1 Polym Chem. 1966; 4(3):623–9.
138. Jones JI. Polyvinyl alcohol. Properties and applications. Edited by C. A. Finch. John Wiley, Chichester. 1973. Pp. xviii + 622. Price: £14.00. Br Polym J. 1973; 5(6):493–4.
139. Oka M, Ushio K, Kumar P, Ikeuchi K, Hyon SH, Nakamura T, et al. Development of artificial articular cartilage. Proc Inst Mech Eng Part H J Eng Med. 2000;214(1):59–68.
140. Chaouat M, Le Visage C, Baille WE, Escoubet B, Chaubet F, Mateescu MA, et al. A novel cross-linked poly(vinyl alcohol) (PVA) for vascular grafts. Adv Funct Mater. 2008;18(19):2855–61.
141. Dini L, Panzarini E, Miccoli MA, Miceli V, Protopapa C, Ramires PA. In vitro study of the interaction of polyalkilimide and polyvinyl alcohol hydrogels with cells. Tissue Cell. 2005;37(6):479–87.
142. Schmedlen RH, Masters KS, West JL. Photocrosslinkable polyvinyl alcohol hydrogels that can be modified with cell adhesion peptides for use in tissue engineering. Biomaterials. 2002;23(22):4325–32.
143. Grad S, Kupcsik L, Gorna K, Gogolewski S, Alini M. The use of biodegradable polyurethane scaffolds for cartilage tissue engineering: potential and limitations. Biomaterials. 2003;24(28):5163–71.
144. Bichara DA, Zhao X, Hwang NS, Bodugoz-Senturk H, Yaremchuk MJ, Randolph MA, et al. Porous poly(vinyl alcohol)-alginate gel hybrid construct for neocartilage formation using human nasoseptal cells. J Surg Res. 2010;163(2):331–6.
145. Mohan N, Nair PD, Tabata Y. Growth factor-mediated effects on chondrogenic differentiation of mesenchymal stem cells in 3D semi-IPN poly(vinyl alcohol)-poly(caprolactone) scaffolds. J Biomed Mater Res - Part A. 2010;94(1):146–59.
146. M PC. Poly (−caprolactone) and its copolymers. In: R L, editor. Biodegradable polymers as drug delivery systems. New York, NY; 1990. p. 71–119.
147. Coombes AGA, Rizzi SC, Williamson M, Barralet JE, Downes S, Wallace WA. Precipitation casting of polycaprolactone for applications in tissue engineering and drug delivery. Biomaterials. 2004;25(2):315–25.
148. Place ES, George JH, Williams CK, Stevens MM. Synthetic polymer scaffolds for tissue engineering. Chem Soc Rev. 2009;38(4):1139–51.
149. Wise JK, Yarin AL, Megaridis CM, Cho M. Chondrogenic differentiation of human mesenchymal stem cells on oriented nanofibrous scaffolds: engineering the superficial zone of articular cartilage. Tissue Eng - Part A. 2009;15(4):913–21.
150. Temenoff JS, Mikos AG. Injectable biodegradable materials for orthopaedic tissue engineering. In: Polymer based systems on tissue engineering, replacement and regeneration. Dordrecht: Springer; 2002. p. 299–312.

151. Gogolewski S, Pennings A. Biodegradable materials of polylactides, 4. Porous biomedical materials based on mixtures of polylactides and polyurethanes. Die Makromol Chemie, Rapid Commun. 1982;3(12):839–45.
152. Werkmeister JA, Adhikari R, White JF, Tebb TA, Le TPT, Taing HC, et al. Biodegradable and injectable cure-on-demand polyurethane scaffolds for regeneration of articular cartilage. Acta Biomater. 2010;6(9):3471–81.
153. Gorna K, Gogolewski S. Preparation, degradation, and calcification of biodegradable polyurethane foams for bone graft substitutes. J Biomed Mater Res - Part A. 2003;67(3):813–27.
154. Fisher JP, Holland TA, Dean D, Engel PS, Mikos AG. Synthesis and properties of photocross-linked poly(propylene fumarate) scaffolds. J Biomater Sci Polym Ed. 2001;12(6):673–87.
155. Timmer MD, Shin H, Horch RA, Ambrose CG, Mikos AG. In vitro cytotoxicity of injectable and biodegradable poly(propylene fumarate)-based networks: unreacted macromers, cross-linked networks, and degradation products. Biomacromolecules. 2003;4(4):1026–33.
156. Hedberg EL, Kroese-Deutman HC, Shih CK, Crowther RS, Carney DH, Mikos AG, et al. In vivo degradation of porous poly(propylene fumarate)/poly(DL-lactic-co- glycolic acid) composite scaffolds. Biomaterials. 2005;26(22):4616–23.
157. Peter SJ, Kim P, Yasko AW, Yaszemski MJ, Mikos AG. Crosslinking characteristics of an injectable poly(propylene fumarate)/β-tricalcium phosphate paste and mechanical properties of the crosslinked composite for use as a biodegradable bone cement. J Biomed Mater Res. 1999;44(3):314–21.
158. Peter SJ, Nolley JA, Widmer MS, Merwin JE, Yaszemski MJ, Yasko AW, et al. In vitro degradation of a poly(propylene fumarate)/β-tricalcium phosphate composite orthopaedic scaffold. Tissue Eng. 1997;3(2):207–15.
159. Lee JW, Ahn GS, Kim DS, Cho DW. Development of nano- and microscale composite 3D scaffolds using PPF/DEF-HA and micro-stereolithography. Microelectron Eng. 2009;86(4–6):1465–7.
160. Diciccio AM, Coates GW. Ring-opening copolymerization of maleic anhydride with epoxides: a chain-growth approach to unsaturated polyesters. J Am Chem Soc. 2011;133(28):10724–7.

Biomechanical Properties of Fibrocartilage

6

Jongkeun Seon

6.1 Introduction

Fibrocartilage has long been neglected and is considered to have a low association with hyaline cartilage. Earlier, histologists did not consider fibrocartilage as a tissue. However, recently developed technologies reveal that fibrocartilage has interesting biomechanical properties and plays an important role, especially in intervertebral discs and meniscus of the knee.

In general, the structural and functional properties of fibrocartilage lie between connective tissue and hyaline cartilage. In the triangular fibrocartilage complex of the wrist, fibrocartilage and hyaline cartilage merge together with fibrous tissue of ligaments and tendons [1–3]. On Masson's trichrome staining, the dyed collagen is red and squeezed collagen is green. When mechanically pressed, fibrocartilage is mostly dyed green [4].

The percentage of dry weight collagen in fibrocartilage is higher and the percentage of proteoglycan is lower than that in articular hyaline cartilage. Therefore, in a hydrated state, fibrocartilage contains less water. Generally, fibrocartilage is considered to be together and less resilient to some degree than articular cartilage. Fibrocartilage is mainly found in two tissues: the annulus fibrosus of the intervertebral disc, forming the flexible junctions between the vertebral bodies in the spine and the meniscus of the knee [5, 6]. The labrum in hip and shoulder joint, triangular fibrocartilage complex (TFCC) in the wrist, and some tendon insertion sites also have fibrocartilage [7].

6.2 Meniscus of the Knee

Knee menisci are crescent shape fibrocartilaginous tissues. Without deformities, a normal adult human knee usually contains lateral and medial menisci, and their average lengths are approximately 38 and 45 mm and average volumes are estimated to be 2.9 cm^3 and 3.45 cm^3, respectively [8]. The meniscus is a composite material consisting of a porous and interstitial fluid and a fiber-reinforced solid matrix; 63–75% of its total weight is composed of water [9]. The menisci has a fibrocartilaginous structure that consists of a fluid phase consisting of mainly water and a solid phase composed of highly oriented collagen fibers, cells, proteoglycans and other proteins. Collagen is the principal constituent of the solid phase, accounting for 75% of the dry weight [10]. Type I (98%) is the predominant collagen of the meniscus, with small amounts of types III and V, and a II like collagen also exists [11]. Proteoglycan constitutes only 2.5% of the solid phase. The balance of the solid phase is maintained mainly by non-collagenous proteins [12]. The mechanical

J. Seon (✉)
Chonnam National University Hospital, Gwangju, South Korea
e-mail: seonbell@jnu.ac.kr

J. Koh et al. (eds.), *Orthopaedic Biomechanics in Sports Medicine*,
https://doi.org/10.1007/978-3-030-81549-3_6

properties of fibrocartilage in the menisci and intervertebral discs generally lie between those of hyaline cartilage and tendon. The tensile strength of fibrocartilage (about 10 MPa) is greater than that of hyaline cartilage (about 4 MPa) but less than that of tendons (about 55 MPa) [13]. Considering these relationships, Smith et al. [14] suggested that fibrous epiphysial plates (e.g., at the upper end of the tibia) are subjected to tensile forces, while most epiphysial plates can only bear compression. Fibrocartilage ruptures on elongation (about 13%), which is less than that for hyaline cartilage (about 25%), but greater than that for tendons (5–9%) [13]. When compressed, the strength of fibrocartilage is similar to that of hyaline cartilage, but fibrocartilage is less stiff. Both the elastic moduli and aggregates in fibrocartilage are about half of those in articular cartilage [15, 16]. Because of their higher flexibility, tendons cannot withstand compressive forces. Fig. 6.1 is diagrammatic representation of the collagen fiber orientations within the meniscus [17].

6.3 Intervertebral Disc of the Spine

A series of vertebrae consists of the human spine and discs, separated from each other and surrounded by ligaments and muscles. Approximately one-third of the total length of the spine consists of discs, allowing deformations of the spine, and their function have roles of energy or shock absorbers. It connects to the vertebrae via cartilaginous end plates. The central volume is composed of nucleus pulposus, a jelly-like substance containing short fibers of mucopolysaccharides. This nucleus is surrounded by the annulus fibrosus, a system of fiber-reinforced layers forming kidney-shaped cross sections (Fig. 6.2). The annulus fibrosus of a lumbar spine is composed of a series of concentric encircling lamellae [18, 19]. The collagen fibers run a similar course in lamellae, and the fibers of the adjacent lamellae cross each other in opposite directions [18, 20]. In the annulus fibrosus of the intervertebral disc, seven different types of collagen have now been reported [21]. Type I collagen accounts for about 80% of the collagen of the annulus fibrosus. The distribution of Types I and II collagen has a reciprocal gradient feature in the intervertebral disc. Most peripheral part of the discs lacks Type II collagen. However, Type II collagen accounts for 80% of collagen in the nucleus pulposus and Type I collagen is relatively low. Type I collagen is abundant (tensile in function) in fibrocartilage, and the relative paucity of Type II collagen (characteristic of tissues subject to pressure) is often considered to be a key biochemical feature

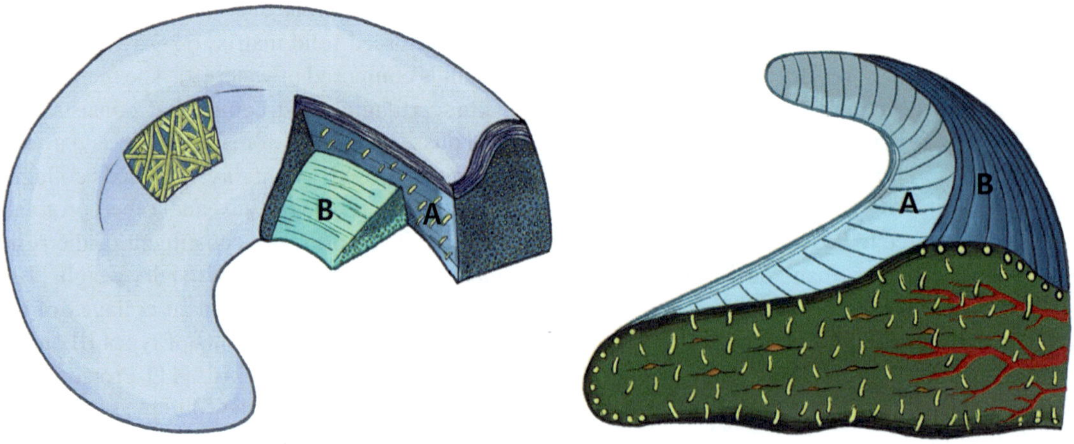

Fig. 6.1 Diagrammatic representation of the collagen fiber orientations within the meniscus. **A**: Radial fiber, **B**: Circumferential fiber

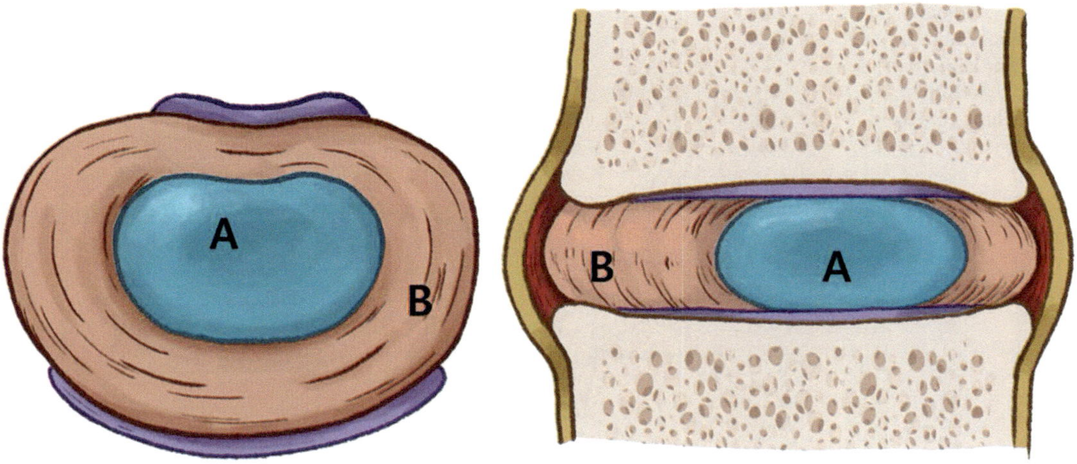

Fig. 6.2 Anatomical position of human intervertebral discs in the human body. **A**: Nucleus Pulposus, **B**: Annular Fibrosus

that makes fibrocartilage distinguishable from hyaline cartilage [15]. Other collagens are also present in fibrocartilage in small quantities. Types V and VI exist in both the menisci and intervertebral discs, but Types IX and XI have only been found in discs [15, 21]. The strength of the vertebral body has been studied extensively [22–33]. The fresh bone from the lumbar vertebrae has an average compressive strength of 0.84 Kp/mm^2 [22]. The whole vertebrae had an average compressive strength of 0.22–0.78 Kp/mm^2 [23]. Hirsch et al. [32] checked the anterior and posterior expansions of the annulus fibrosus under compressive force and reported an average expansion of 0.5 mm for 50-kg loads and 0.75 mm for 100-kg loads. Fig. 6.2 is Anatomical position of human intervertebral discs in the human body [33].

6.4 Labrum

The labrum is an annular fibrocartilage structure attached to the acetabular rim and transverse acetabular ligament [34]. The thickness of the lower part of the labrum is about 6.4 mm, which is the widest portion, and that of the upper part is about 5.5 mm [35] (Fig. 6.3). The cross section of the labrum is triangular and is divided into articular and non-articular surfaces. The articular surface joins the hyaline acetabular cartilage with an average thickness of 1.26 mm through the underlying bone, transition zone of about 1–2 mm width, and the calcified cartilage layer [36]. Fig. 6.3 is anatomy of histologic features of the adult acetabular labrum.

The labrum consists of fibrocartilage, a mixture of white fibrous and cartilaginous tissues. The orientation of collagen fibers in the transitional zone varies depending on the location [37]. While collagen fibers are attached perpendicularly at the posterior area, they are structurally vulnerable to shear forces because they are attached parallel to the rim in the anterior area. In a non-articular surface composed of dense connective tissue, it is directly attached to the bone, allowing nerves and blood vessels to pass through the tissue [35].

Histologically, the labrum includes nerve endings in the superficial articular region [34, 38]. Therefore, this site is considered to play an important role in proprioception. Blood is supplied from the peripheral anastomosis around the capsule attachment, [34] and blood supply is limited to only the outer third of the labrum [39]. As such, the deep area of the labrum is a relatively avascular tissue.

Scanning electron microscopy revealed that the labrum is divided into three distinct layers: a meshwork of thin fibrils on the articular surface,

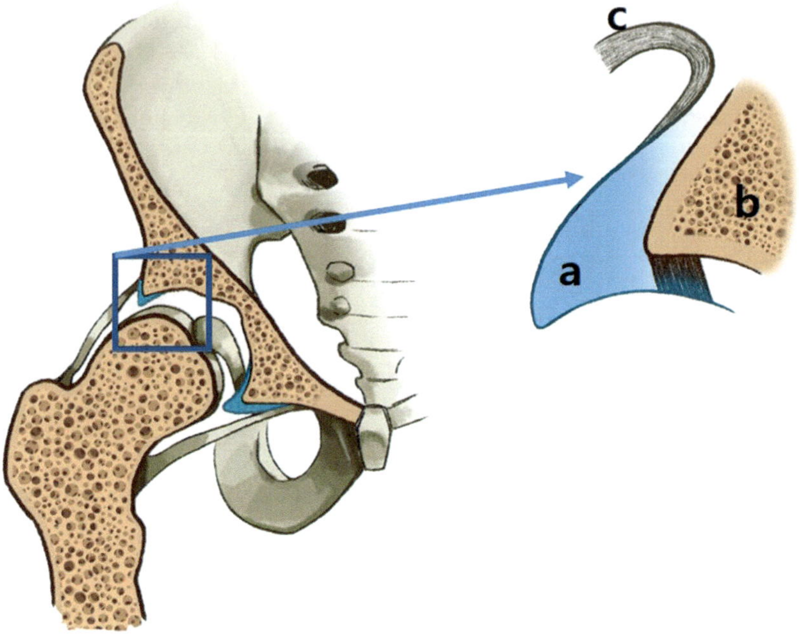

Fig. 6.3 Anatomy and histologic features of the acetabular labrum: (**a**) Labrum; (**b**) Bony Acetabulum; (**c**) Hip Capsule (Cut)

a deep region that is more layered, and the most substantial inner layer of highly circumferentially oriented fibers, which appears to carry most of the physiological load. These structures can accommodate significant physiological hoop stress [39, 40].

In the labrum circumference, the anterosuperior quadrant has a lower compressive and tensile modulus than other quadrants, suggesting that these biomechanical properties of the anterosuperior quadrant contribute to more common occurrence of tears than those of the other quadrants [41].

The most important role of the labrum is to seal the femoral head to form a pressurized layer of intra-articular fluid to distribute and support the compressive load, thereby reducing cartilage stress, strain, and consolidation [42–44]. In a human cadaveric model, Ferguson et al. [35–49] showed that labral resection resulted in a 21% increase in cartilage consolidation because of the lack of pressurized fluid. Labrum sealing plays an essential role in maintaining the stability of the hip joint through the "suction effect" between the acetabulum and femoral head [45–47]. The acetabular labrum around the femoral head is tightly attached, forming a vacuum seal to resist distraction forces. Therefore, this suction effect improves joint stability and reduces the peak cartilage stress during weight bearing by dispersing compressive loading on the joint surface [43, 44]. The labrum also plays an important role in the resistance to rotation and translation forces caused by external compressive and torsional loading. In addition, labrum sealing provides effective mechanical support by deepening the socket. Acetabular volume is known to increase by approximately 21% and surface area by 28% [35]. After labrum excision, these seals are destroyed, resulting in weakening of mechanical support, instability, and "wobbling effect" that adversely affects the cartilage. In addition, an increase in contact stress causes degenerative changes in the cartilage, leading to osteoarthritis [48, 49].

6.5 Triangular Fibrocartilage Complex (TFCC)

The TFCC is a structure in which ligamentous, fibrous, and fibrocartilaginous components are mixed as a homogenous complex; it plays an important role in wrist biomechanics. The TFCC is the main stabilizer on the wrist ulnar side and transmits about 20% of the force acting on the wrist. It also plays a role in enabling complex wrist movements [50].

The TFCC is a complex structure composed of a triangular fibrocartilage disc proper (TFC), dorsal and volar distal radioulnar ligaments (DRUL and PRUL, respectively), meniscal homolog (MH), ulnocarpal collateral ligament (UCCL), ulnotriquetral and ulnolunate ligaments (which are both volar structures), and the extensor carpi ulnaris (ECU) [51] (Fig. 6.4).

According to histological studies, DRUL and PRUL, which are longitudinal fibers extending to the ulnar insertion site and the central articular disc consisting of collagen fiber layers arranged in various layers with various oblique angles, are clearly distinguished [52]. The articular disc of the TFCC starts from the distal rim of the sigmoid notch and extends along the ulnar edge of the lunate facet and blends in the peripheral part with DRUL and PRUL.

The TFC is the largest structure among the components of the TFCC. This fibrocartilaginous disc starts from the most distal articular cartilage of the sigmoid notch of the radius and extends to the fovea and ulnar styloid processes at the base of the ulnar styloid. In the radial attachment, the TFC is a single band made of high-density fibrous tissue; however, before it attaches to the ulna, it is separated into the proximal and distal laminas. These two laminas are divided by fibrovascular tissue called ligamentum subcruentum. The proximal lamina is attached to the fovea at the base of the ulnar styloid, and the distal lamina is attached to the distal aspect of the ulnar styloid [53].

In the volar and dorsal aspects of the TFC, there is a thick fibrous band running from the volar and dorsal aspects of the sigmoid notch of the radius to the base of the ulnar styloid process, which is called volar and dorsal RULs [54].

MH, unlike fibrocartilage, is composed of fibrous tissue and contains vascular fatty tissue rich in the ulnar side [55]. MH extends from the tip and lateral aspects of the ulnar styloid process to the ulnar side of the triquetrum and has a wider distal attachment site [56, 57]. MH is a loose connective tissue with well-developed blood vessels, not containing dense collagen fibers in parallel, and can be elongated considerably; therefore, it is not an actual ligament [58].

UCCL is a thin fibrous tissue that runs superficially on MH. This ligament is relatively less important than other structures.

The ulnotriquetral ligament (UTL) and ulnolunate ligament (ULL) are extrinsic ligaments found in the volar aspect of the ulnocarpal joint [57, 59].

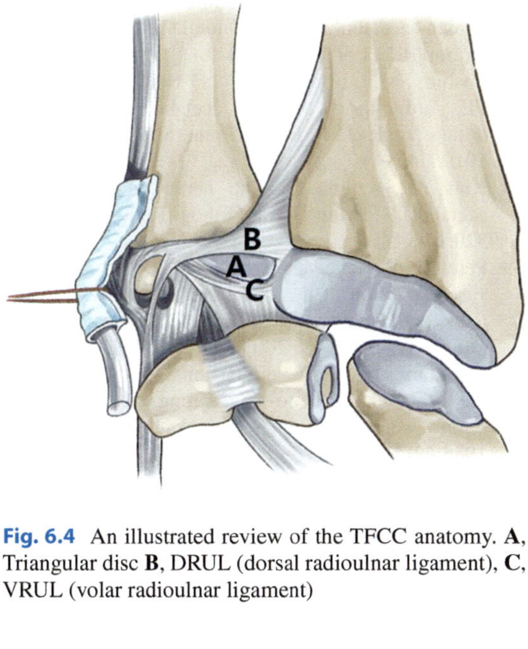

Fig. 6.4 An illustrated review of the TFCC anatomy. **A**, Triangular disc **B**, DRUL (dorsal radioulnar ligament), **C**, VRUL (volar radioulnar ligament)

References

1. Benjamin M, Evans EJ, Pemberton DJ. Histological studies on the triangular fibrocartilage complex of the wrist. J Anat. 1990;172:59–67.
2. Benjamin M, Evans EJ, Copp L. The histology of tendon attachments to bone in man. J Anat. 1986;149:89–100.
3. Woo SL, Buckwalter JA. AAOS/NIH/ORS workshop. Injury and repair of the musculoskeletal soft tissues. Savannah, Georgia, June 18-20, 1987. J Orthop Res. 1988;6(6):907–31.
4. Flint MH, Lyons MF, Meaney M, Williams D. The Masson staining of collagen—an explanation of an apparent paradox. Histochem J. 1975;7(6):529–46.
5. Panagiotacopulos ND, Knauss WG, Bloch R. On the mechanical properties of human intervertebral disc material. Biorheology. 1979;16(4–5):317–30.
6. Eyre DR, Wu JJ. Collagen of fibrocartilage: a distinctive molecular phenotype in bovine meniscus. FEBS Lett. 1983;158(2):265–70.

7. Cooper RR, Misol S. Tendon and ligament insertion. A light and electron microscopic study. J Bone Joint Surg Am. 1970;52(1):1–20.
8. Mathur PD, Mc DJ, Ghormley RK. A study of the tensile strength of the menisci of the knee. J Bone Joint Surg Am. 1949;31(3):650–4.
9. Adams ME, Muir H. The glycosaminoglycans of canine menisci. Biochem J. 1981;197(2):385–9.
10. Peters TJ, Smillie IS. Studies on the chemical composition of the menisci of the knee joint with special reference to the horizontal cleavage lesion. Clin Orthop Relat Res. 1972;86:245–52.
11. Eyre DR, Muir H. The distribution of different molecular species of collagen in fibrous, elastic and hyaline cartilages of the pig. Biochem J. 1975;151(3):595–602.
12. Ingman AM, Ghosh P, Taylor TK. Variation of collagenous and non-collagenous proteins of human knee joint menisci with age and degeneration. Gerontologia. 1974;20(4):212–23.
13. Yamada H, Evans FG. Strength of biological materials. 1970.
14. Smith JW. The structure and stress relations of fibrous epiphysial plates. J Anat. 1962;96(Pt 2):209–25.
15. Arnoczky S, Adams M, DeHaven K, Eyre D, Mow V. Meniscus in Woo S-LY, and Buckwalter J (eds): injury and repair of the musculoskeletal soft tissues. Park Rodge, IL: American Academy of Orthopaedic Surgeons; 1988.
16. Proctor CS, Schmidt MB, Whipple RR, Kelly MA, Mow VC. Material properties of the normal medial bovine meniscus. J Orthop Res. 1989;7(6):771–82.
17. Bullough PG, Munuera L, Murphy J, Weinstein AM. The strength of the menisci of the knee as it relates to their fine structure. J Bone Joint Surg Br. 1970;52(3):564–7.
18. Fick RA. Bänder, Gelenke und Muskeln: Handbuch der Anatomie und Mechanik der Gelenke unter Berücksichtigung der bewegenden Muskeln: Fischer; 1904.
19. Beadle OA. The intervertebral disc: observations on their normal and morbid anatomy in relation to certain spinal deformities. Med Res Counc Spec Rep Ser. 1931;161:7–77.
20. Straßer H. Die Bewegungsmöglichkeiten der Wirbelsäule und des Kopfes. Lehrbuch der Muskel- und Gelenkmechanik: Springer; 1913. p. 153–216.
21. Eyre D. Collagens of the disc. Ghosh P (ed) The biology of the intervertebral disc, vol I. CRC Press, Boca Raton; 1988.
22. Rauber AA. Elasticität und Festigkeit der Knochen: anatomisch-physiologische Studie: W. Engelmann; 1876.
23. Messerer O. Über Elasticität und Festigkeit der menschlichen Knochen: Cotta; 1880.
24. Lange C. Untersuchungen über Elastizitätsverhältnisse in den menschlichen Rückenwirbeln mit Bemerkungen über die Pathogenese der Deformitäten. Z Orthop Chir. 1902;10:47–110.
25. Göcke C. Traumatische Wirbelumformung im Versuch. Hefte Unfallheilk H. 1931;8:136.
26. Göcke C. Das Verhalten spongiösen Knochens im Druck-und Schlagversuch. Verh dtsch orthop Ges. 1926;47:114–29.
27. Bartelink D. The role of abdominal pressure in relieving the pressure on the lumbar intervertebral discs. J Bone Joint Surg. 1957;39(4):718–25.
28. Perey O. Fracture of the vertebral end-plate in the lumbar spine: an experimental biomechanical investigation. Acta Orthopaedica Scandinavica. 1957;28(suppl 25):1–101.
29. Decoulx P, Rieunau G, Nicoll CJ, Gerardmarchant P, Daubigne R, et al. Les fractures du rachis dorsolombaires sans troubles nerveux. Presse Medicale. 1958;66(90):2021–2.
30. Eie N. Load capacity of the low back. J Oslo City Hosp. 1966;16:73–98.
31. Weaver JK, Chalmers J. Cancellous bone: its strength and changes with aging and an evaluation of some methods for measuring its mineral content: I. age changes in cancellous bone. JBJS. 1966; 48(2):289–99.
32. Hirsch C, Nachemson A. New observations on the mechanical behavior of lumbar discs. Acta Orthop Scand. 1954;23(4):254–83.
33. Van der Veen A, Van Dieen J, Nadort A, Stam B, Smit T. Intervertebral disc recovery after dynamic or static loading in vitro: is there a role for the endplate? J Biomech. 2007;40(10):2230–5.
34. Putz R, Schrank C. Anatomy of the labrocapsular complex of the hip joint. Orthopade. 1998;27(10):675–80.
35. Seldes RM, Tan V, Hunt J, Katz M, Winiarsky R, Fitzgerald RH Jr. Anatomy, histologic features, and vascularity of the adult acetabular labrum. Clin Orthop Relat Res. 2001;382:232–40.
36. Athanasiou K, Agarwal A, Dzida F. Comparative study of the intrinsic mechanical properties of the human acetabular and femoral head cartilage. J Orthop Res. 1994;12(3):340–9.
37. Cashin M, Uhthoff H, O'Neill M, Beaule PE. Embryology of the acetabular labral-chondral complex. J Bone Joint Surg. 2008;90(8): 1019–24.
38. Kim YT, Azuma H. The nerve endings of the acetabular labrum. Clin Orthop Relat Res. 1995; 320:176–81.
39. Petersen W, Petersen F, Tillmann B. Structure and vascularization of the acetabular labrum with regard to the pathogenesis and healing of labral lesions. Arch Orthop Trauma Surg. 2003;123(6):283–8.
40. Shibutani N. Three-dimensional architecture of the acetabular labrum--a scanning electron microscopic study. Nihon Seikeigeka Gakkai Zasshi. 1988;62(4):321.
41. Smith C, Masouros S, Hill A, Amis A, Bull AM. A biomechanical basis for tears of the human acetabular labrum. Br J Sports Med. 2009;43(8): 574–8.

42. Ferguson S, Bryant JT, Ganz R, Ito K. An in vitro investigation of the acetabular labral seal in hip joint mechanics. J Biomech. 2003;36(2):171–8.
43. Ferguson S, Bryant J, Ganz R, Ito K. The acetabular labrum seal: a poroelastic finite element model. Clin Biomech. 2000;15(6):463–8.
44. Ferguson S, Bryant J, Ganz R, Ito K. The influence of the acetabular labrum on hip joint cartilage consolidation: a poroelastic finite element model. J Biomech. 2000;33(8):953–60.
45. Nepple JJ, Philippon MJ, Campbell KJ, Dornan GJ, Jansson KS, LaPrade RF, et al. The hip fluid seal—part II: the effect of an acetabular labral tear, repair, resection, and reconstruction on hip stability to distraction. Knee Surg Sports Traumatol Arthrosc. 2014;22(4):730–6.
46. Smith MV, Panchal HB, Ruberte Thiele RA, Sekiya JK. Effect of acetabular labrum tears on hip stability and labral strain in a joint compression model. Am J Sports Med. 2011;39(1 suppl):103–10.
47. Van Arkel R, Amis A, Cobb J, Jeffers J. The capsular ligaments provide more hip rotational restraint than the acetabular labrum and the ligamentum teres: an experimental study. Bone & Joint J. 2015;97(4):484–91.
48. Brand RA, Iglič A, Kralj-Iglič V. Contact stresses in the human hip: implications for disease and treatment. Hip Int. 2001;11(3):117–26.
49. Mavčič B, Iglič A, Kralj-Iglič V, Brand RA. Cumulative hip contact stress predicts osteoarthritis in DDH. Clin Orthop Relat Res. 2008;466(4):884–91.
50. Linscheid RL. Biomechanics of the distal radioulnar joint. Clin Orthop Relat Res. 1992;275:46–55.
51. Palmer AK. Triangular fibrocartilage complex lesions: a classification. J Hand Surg. 1989;14(4):594–606.
52. Chidgey LK, Dell PC, Bittar ES, Spanier SS. Histologic anatomy of the triangular fibrocartilage. J Hand Surg. 1991;16(6):1084–100.
53. Burns JE, Tanaka T, Ueno T, Nakamura T, Yoshioka H. Pitfalls that may mimic injuries of the triangular fibrocartilage and proximal intrinsic wrist ligaments at MR imaging. Radiographics. 2011;31(1):63–78.
54. Nakamura T, Takayama S, Horiuchi Y, Yabe Y. Origins and insertions of the triangular fibrocartilage complex: a histological study. J Hand Surg British Eur. 2001;26(5):446–54.
55. Buck FM, Gheno R, Nico MA, Haghighi P, Trudell DJ, Resnick D. Ulnomeniscal homologue of the wrist: correlation of anatomic and MR imaging findings. Radiology. 2009;253(3):771–9.
56. Nishikawa S, Toh S. Anatomical study of the carpal attachment of the triangular fibrocartilage complex. J Bone Joint Surg Br. 2002;84(7):1062–5.
57. Nöbauer-Huhmann I-M, Pretterklieber M, Erhart J, Bär P, Szomolanyi P, Kronnerwetter C, et al., editors. Anatomy and variants of the triangular fibrocartilage complex and its MR appearance at 3 and 7T. Seminars in musculoskeletal radiology; 2012: Thieme Medical Publishers.
58. Garcia-Elias M. Soft-tissue anatomy and relationships about the distal ulna. Hand Clin. 1998;14(2):165–76.
59. Zlatkin MB, Rosner J. MR imaging of ligaments and triangular fibrocartilage complex of the wrist. Radiologic Clin. 2006;44(4):595–623.

Tendon Biomechanics-Structure and Composition

7

Stefano Zaffagnini, Jason Koh, Umile Giuseppe Longo, Giovanna Stelitano, Farid Amirouche, and Vincenzo Denaro

7.1 Introduction

A tendon is a closely compact band of connective tissue-fibers well organized in length and strength which connects to muscle and transmits the required force developed by the muscle. The musculoskeletal is such that bones are driven by tendons and produce joint motion in coordination with muscles activities.

S. Zaffagnini
IRCCS – Rizzoli Orthopaedic Institute – 2nd Orthopaedic and Trauma Unit, University of Bologna, Bologna, Italy
e-mail: Stefano.dipaolo@ior.it

J. Koh
Department of Orthopaedic Surgery, Orthopaedic & Spine Institute NorthShore University HealthSystem, Skokie, Illinois, USA

University of Chicago Pritzker School of Medicine, Chicago, Illinois, USA

Northwestern University McCormick School of Engineering, Evanston, Illinois, USA

F. Amirouche
Department of Orthopaedic Surgery, Orthopaedic & Spine Institute NorthShore University Health System, Evanston, Illinois, USA

Department of Orthopeadic Surgery, College of Medicine, University of Illinois, Chicago, Illinois, USA
e-mail: FAmirouche@northshore.org, amirouch@uic.edu

U. G. Longo (✉) · G. Stelitano · V. Denaro
Department of Orthopaedic and Trauma Surgery, Campus Bio-Medico University, Trigoria, Rome, Italy

Tendons also play a role in joint stability, and produce the static and kinetic work-energy to perform body movements in a coordinated way to produce gait, lifting and other human joints function. Tendons are bound together in tight sheets so that when a muscle contracts, tension is created in the tendons and this will pull against the bone to cause movement. Tendons have a hierarchical arrangement composed of collagen molecules, fibrils, fibers, fascicles, and tendon unit. Tendon units are encased in epitenon, which reduces friction with neighboring tissues [1].

Tendons connective tissue are known as collagen and also found in ligaments and fascia. Tendons conform to muscle location and function and hence they take different geometrical forms and shapes. For example, they can be long or flat thin or thick and map with their corresponding muscles to fit in their perspective locations. A thin flat tendon is also called "aponeurosis." The tendon fibro-elastic composition defines its potential strength. The muscle has tendons at its ends referred to as proximal and distal. The tendon attaches to muscle and bone through the musculotendinous junction (MTJ) and the osteotendinous junction (OTJ) respectively. The proximal attachment of the tendon is also known as the origin and the distal tendon is called the insertion [1, 2].

From previous research it is seen that tendon deforms in a linear fashion within certain tensile load and if strain is less than 4%, the tendon has a tendency to return to its original length when

J. Koh et al. (eds.), *Orthopaedic Biomechanics in Sports Medicine*,
https://doi.org/10.1007/978-3-030-81549-3_7

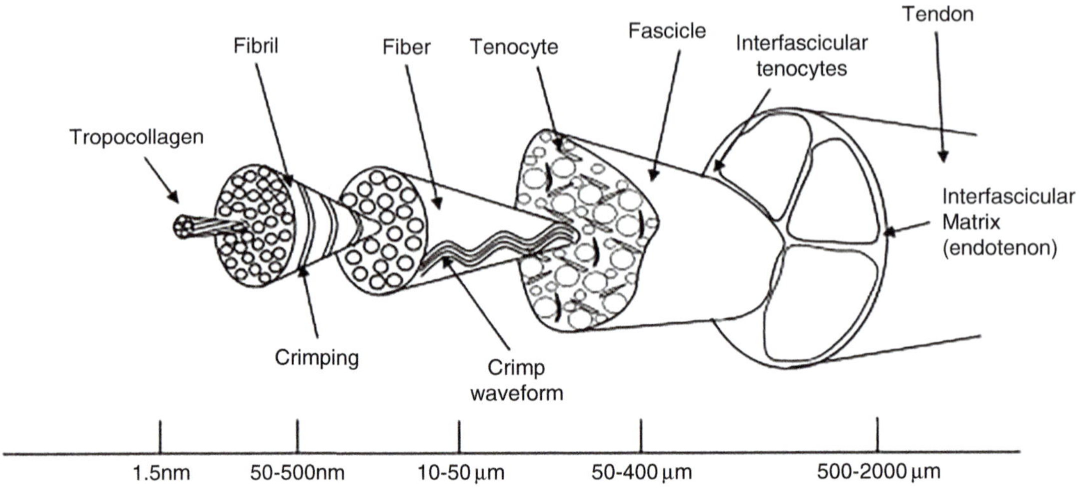

Fig. 7.1 Schematic showing the hierarchical structure of tendon, in which collagen molecules assemble to form subunits of increasing diameter. Reproduced with permission from https://doi.org/10.1016/B978-0-12-801590-2.00001-6 (Chavaunne T. Thorpe, Helen L. Birch, Peter D. Clegg, Hazel R.C. Screen, Chapter 1 - Tendon Physiology and Mechanical Behavior: Structure–Function Relationships, Editor(s): Manuela E. Gomes, Rui L. Reis, Márcia T. Rodrigues, Tendon Regeneration, Elsevier, 2015, Pages 3-39, ISBN 9780128015902)

unloaded. Hence, this portion is purely elastic (almost no loss in the hysteresis curve) and the stress-strain curve can be used to measure the slope or the Young's modulus of elasticity [3]. Above the linear range, tendon is viscoelastic and depend on the rate of at which it is loaded. Its fibers orientation, density and collagen type play a role in tendon load activation and mechanical characteristics it plays in response to the efficient transmission of large loads, and how its biomechanical structure adapts to load.

Fibers can be divided into two groups. Type 1 collagen fibers active in the elastic range and resist the capacity load and tensile strength whereas the proteoglycan fibers define the viscoelastic behavior of the tendon. These linear and nonlinear tendon material characteristics are defined by the orientation of the fibers more importantly how they bundle together as they vary from one tendon to another.

To understand further the anatomy of tendon-collagen fibers, we see that collagen molecules fibers consist of polypeptide chains, where three of these chains combined together form a densely packed helical tropocollagen molecule. Five of these combined together form a microfibril. The microfibrils then aggregate together to form fibrils (Fig. 7.1). These are then grouped into fibers and fibers into fiber bundles and then into fascicles [4].

7.2 Biochemical Composition of the Tendon

Tendon is an intricate structure made of cells and collagen and elastic fibers immersed in an extracellular matrix (ECM) [5, 6].

7.2.1 Collagen

Tendons are a modular structure made of fascicles with a diameter that can measure up to 100 μm. The fibers are constituted by collagen fibrils that represent the smallest structural unit of the tendon (see Fig. 7.1). Type I collagen is characterized by a triple-helical fibrillar structure, in which 3 left-handed polypeptide chains connected to each other creating a molecule of tropocollagen, the basic unit of the tendon matrix. Each microfibril is formed by five molecules of tropocollagen, and multiple microfibrils form the helical collagen fibrils. Collagen fibers represent the building blocks of the tendon [7]. In particular, a group of collagen fibrils make a primary band (also defined sub-band), and several primary bands circled by the endotenon define a secondary band or fascicle. Many secondary bands make up a tertiary band and several tertiary bands unitedly form the tendon, contained by the endotenon.

The amount of collagen fibers in each sub-band notably differs from tendon to tendon as well as the fiber diameter. This variability reflects tendon different functions and properties. Tendon strength and elasticity are more influenced by fibrils length rather than their diameter [8, 9].

7.2.2 Ground Substance (Extracellular Matrix)

The extracellular matrix of tendons, which circles the collagen, is made of proteoglycans, glycosaminoglycans (GAGs), glycoproteins, and other different molecules. It appears as a hydrophilic gel whose consistency changes according to the relative quantity of hyaluronic acid and chondroitin sulfate concentrations [10]. The water-binding ability of the macromolecules which compose ECM, is remarkable and it is essential to define the biomechanical characteristics of tendons. They confer tendon elasticity necessary to contrast shear and compressive strengths and, at the same time, they are helpful to stabilize of the entire collagenous structure.

7.2.3 Proteoglycans

A network of proteoglycans (PG) is distributed among the collagen fibrils. PGs are less than 1% of the dry mass of tendon. They appear as filaments regularly and orthogonally connected to the tendon-collagen fibrils made of polysaccharides, mainly in the form of glycosaminoglycans (GAGs). PGs have several different functions. On one side, the small leucine-rich PG family links collagen fibrils and has an active role in the in fibrillogenesis mechanism. In particular, decorin controls collagen apposition and distribution inside tendon fibrils. It is an essential regulator of matrix construction: it defines fibrils diameter and controls the remodeling of the tendon after tensioning. Large PGs, such as aggrecan and versican, seem to provide tendon tissues ability to resist high compressive and tensile strengths [11]. Also, aggrecans, the second macromolecules most expressed in PGs of tendons, primarily localized in those areas where compressive forces act on the tendon, function as a lubricant permitting to the fibrils gliding.

7.2.4 Glycoproteins

The glycoproteins are macromolecules made of a substantial protein component and a little glycidic part. Confronted with proteoglycans, they present quantitative variation in the protein-carbohydrate ratio, and qualitative diversity in the glycidic radicals' composition. Their specific functions are still not perfectly known. Among them, a particular subgroup is that of the so-defined adhesive glycoproteins, whose function seems to be connected macromolecules to each other or different cell surfaces [12]. Four different glycoproteins make up the tendon belly: fibronectin, thrombospondin, tenascin C, and undulin. Laminin, instead, is placed in the walls of tendons vessels and high quantity within the myotendinous junction.

7.2.5 Anorganic Components

Anorganic components represent less than 0.2% of the tendon dry mass. Among these, calcium is the most expressed. Its concentration ranges from 0.001 to 0.01%, in the tensional area of a healthy tendon, to 0.05–0.1% in the tendon insertion. Variations up to 10–20% can be observed in diseased states such as calcifying tendinopathy. The other components of the anorganic portion of tendons are magnesium, manganese, cadmium, cobalt, copper, zinc, nickel, lithium, lead, fluoride, phosphor, and silicon. Their concentration is usually of 0.02–120 ppm. The exact function of these elements is not still defined. They would seem implicated in development and metabolism of tendons. In particular, copper plays an essential function in the generation of collagen cross-linking, manganese is needed for the production of connective tissue molecules, and calcium is implicated in the formation of the OTJ [13].

7.3 Tendon Cells

About 90–95% of tendon cellular elements are represented by fibroblasts, which, in tendon tissues, are defined as tenoblasts-tenocytes. The remaining 5–10% comprises other different types of cells such as the chondrocytes, predominantly placed at the insertion sites, the synovial cells, detectable on tendon sheaths or surfaces, and the vascular cells, such as endothelial and smooth muscle cells of the arterioles, visible in the endo- and epitenon. Younger tendons are characterized by a high cell-to-matrix ratio, where predominant involved cells are the tenoblasts. Inside the tendon, tenoblasts are distributed in long and parallel chains.

Tenoblasts have also several long and thin cytoplastic filaments which throw into the matrix and they dispose of several desmosomial, tight, or gap junctions thanks to which they can create several intercellular links [14].

The tenoblasts convert to tenocytes, becoming bigger and very elongated. New morphological features appear. Among these, it is possible to observe the increase of the nucleus-to-cytoplasm ratio, the elongation of the cellular nuclei which occupy almost the entire length of the cell. At the same time, the nuclear chromatin becomes more condensate, while the actin and myosin intracytoplasmic filaments as well as the pinocytotic vesicles and lysosomes start to decrease.

Tenocytes develop longer and thinner cellular processes, required to keep close intercellular contact, useful to counterbalance for the reduced number of cells and growing quantity of tendon matrix during senescence. All of these ultrastructural changes confirm the assumption that tenocytes are metabolically active although as not as the tenoblasts.

The main role of tenocytes is to synthesize and organize the collagen matrix, supporting its process of degradation and remodeling. In particular, they regulate the synthesis of tropocollagen, the fibrillogenesis and they control fibrils orientation inside the newly synthesized matrix.

7.4 Neurovascular Supply of Tendon

7.4.1 Vascularization

Tendons are considered poorly vascularized tissues with a little and slow regeneration potential. Their vascular supply is notably less than that of muscles to which they are connected. This concept clearly explains why tendons are white and muscles red [15]. Any case, tendons vascularization remains essential for the correct functioning of tendon cells and tendons repair capacity. The blood supply of tendons derives from the perimysium, the periosteum, and the paratenon. Blood flow is slow, tendon vessels have mostly small and thin-walled and are focused on the outer surface of the tendon. Tendons present several regions of diminished blood supply where they wind nearby osseous pulleys. They press against bone predominantly at the level of the myotendinous junction and the osteotendinous junction. These areas assume a critical clinical relevance as they represent the typical regions of tendon degeneration and rupture [16]. On the other hand, peritendinous tissues such as tendon sheaths and tendon-associated adipose tissue, are characterized by more copious blood supply which increases as a consequence of improved physical performance.

7.4.2 Innervation

The innervation of tendons is scarcely conspicuous. The nerve branches run parallel to the principal axis of the tendon even if transversal and oblique anastomoses occur [17]. These branches end both in contact with corpuscular receptors such as Golgi, Pacini, Ruffini, and Golgi-Mazzoni corpuscles responsible for proprioceptive sensitivity and myotatic reflexes, both as free branches implicated in nociception, predominantly in peritendinous sheaths [18, 19]. The innervation of tendons has an essential function in the repair of damaged tendons. It is evident that nerves can develop into injured or broken tendons together with vessels and which these sites correspond to the region of tendon pain.

7.5 Mechano-Responses of Tendons

Each tendon has a capacity to withstand load. Understanding how tendons respond to load and defining their capacity limits before injury requires physical testing, evaluation of strength, and growth factors to develop a protective mechanism to allow tendons to work within their boundaries. For example, the patella tendon of a professional basketball player when compared to a normal healthy person is much stronger and would undergo high levels of loading without any detrimental effects, whereas the average person patella tendon can be injured at lower loading forces conditions. It is hypothesized that early growth of tendon relies on two types of inherent stresses: rapid muscular activity on one side and slow growth-related to bone on the other. There are two types of fibrils, large and small, and they play different role. One is activated and work as a resistance to tensile strength whereas the other works to resolve micro fiber mechanism process damage at creep. The small fibrils support the interfibrillar binding [1, 4].

Tendons are complex dynamic structures with specific functions defining their mechanical properties. For example, fresh cadavers' patella tendons tested in the laboratory and subjected to load to failure yielded a Young's modulus of elasticity in the range of 660 ± 266 MPa, whereas the tibialis anterior tendon is about 1200 MPa [3]. Tendons can transmit much higher load that can be calculated by means of inverse dynamics and modeling. For instance, the Achilles tendons can generate forces that are 12 times body weight when comparing standing versus running [20]. It is observed in the lab experiments of patella tendons that the loading rate generates different strain responses. Basically at low speed tendon are more deformable than at higher speed. Similar to shock absorbent if you apply a high force and try to distract the piston it resists effectively and generates a high stiffness. So tendons have the ability to store energy (spring type) and release energy (damper), which makes tendons unique in responding to muscle contraction, and excessive joint motions requiring dissipation of energy and maintaining the right force needed to create the joint rotation. Excessive forces lead to tendon tear or rupture. The question remains as to the adequate loading conditions required to maintain healthy tendons and for how long? What are the appropriate exercises for different tendons?

Therefore, exercise might regulate the increase of turnover of type I collagen in tendons. It is hypothesized that at later stages of exercises the anabolic processes dominate and results in net increase of collagen in tendons [6].

7.6 Mechanical Properties

The mechanical properties of the human tendons have been extensively studied both in vitro and in vivo studies [21]. The mechanical loading and, in particular, the alterations to regular loading, are recognized to play a critical role in onset and perpetuation of tendons pathologies [22].

Traditionally, the mechanical properties of tendons have been evaluated in vitro, through the use of tensile testing methodologies.

The fibrillar composition and the longitudinal structure of the tendons result in their anisotropic and nonlinear mechanical properties. Furthermore, the anisotropic and nonlinear mechanical behaviors of tendons are explained by the viscoelasticity of these structures. Viscoelasticity allows materials to display both elastic and viscous behavior. While the viscosity represents a measure of resistance to deformation, elasticity is the ability of a material to return to its original shape after forces that cause deformation are removed [9].

The anisotropic features give the tendons a directionally dependent behavior [9]. The mechanical properties of the tendons indeed, are up to 1000 times higher during tensile testing along the longitudinal axis compared with testing along the transverse axis [10].

The nonlinear properties of the tendons characterize their mechanical response to tensile testing, and are typically presented as stress-strain curves [1, 10]. The slope of the stress-strain curve is a quotient between the force per unit area (stress) and the relative measure of deformation (strain) of the tendon, and represents the Young's modulus, a material property that describes elasticity.

In the normal tendons, this curve displays three distinct regions: the toe region, the linear region, and the failure region (Figs. 7.1 and 7.2).

When the tendon is relaxed, the collagen fibers are arranged like waves and there is a nonlinear relationship between stress and strain, the so-called "toe region." The toe region, in which the sample is tensed to approximately 2% of strain, describes the mechanical behavior of tendons at low deformation, where the collagen fibril crimp is straightened.

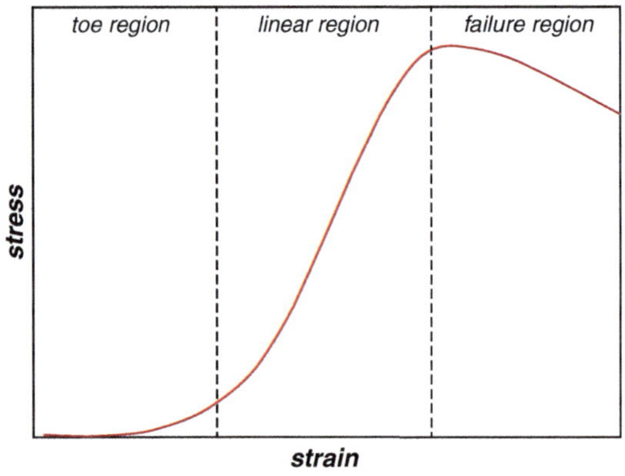

Fig. 7.2 Stress-strain curves of tendons displaying the three typical regions: the toe region, the linear region, and the failure region

As deformation increases, this pattern disappears, and the collagen fibers respond stretching itself, in linear fashion to the force. This behavior is represented by the linear region, where the loading causes elongation of fibers typically at approximately 4–8% of strain and some microscopic ruptures occur inside the tendon. As the deformation continues to increase through the linear region into the failure region the tendon accumulates irreversible damages. The macroscopic failure, with complete breakage of all fibers occurs at approximately >12% of strain [11].

The mechanical testing methods (Table 7.1) used to generate the stress-strain curves, can be adapted to analyze different types of viscoelastic properties of tendons (Table 7.2).

In a cadaveric study in which the mechanical properties of the most common types of tendinous grafts that are used to perform ligament reconstruction were compared, the elastic modulus of patellar tendon was significantly higher than all other grafts. Moreover, in the latter study, authors did not find significant differences in terms of ultimate stress and ultimate deformations between the patellar tendon graft, gracilis and semitendinosus graft, ileo-tibial band and the medial part of quadriceps tendon [12] (Fig. 7.3).

Table 7.1 Mechanical testing methods

Test	Description	Aims
Cyclic preconditioning	Low loads to stretch the tendon without causing irreversible damage (commonly performed before mechanical testing)	Standardization of measurements for comparison
Ramp to failure testing	Performed by applying a constantly increasing displacement on the tendon until failure	• To find maximum force a tendon can endure before failure • To determine the stiffness and modulus of toe and linear regions
Dynamic cyclic testing	Application of an oscillatory stress	• To measure the dynamic modulus • Evaluation of creep and stress relaxation • Analysis of phase lag
Fatigue testing	Cyclic test within a defined range of force or displacement and recording the number of cycles until failure	Useful in characterization of tendons stiffness that undergo repetitive loads, such as the Achilles tendon

However, determining the strains and mechanical behavior of tendons remain challenging. In another cadaveric study, the longitudinal and transverse strain distribution in the human Achilles tendon was examined by the use of a three-dimensional digital image correlation system (3D-DIC). Authors of the latter study found that the strain distribution of both longitudinal and transverse component was of a strongly inhomogeneous nature, both within the same specimen and among different specimens, and, moreover, showed that DIC represents a very accurate and reproducible tool for 3D strain analysis in human tendon tissue [13].

In vivo strains remain difficult to accurately quantify despite the use of various implantable devices. However, there is thought to be a large gap between the tendon strains reported in vivo and those experienced during typical daily activities, usually less than 4% [14].

Table 7.2 Tendons viscoelastic properties

Hysteresis	Amount of energy dissipated as a result of internal friction during mechanical loading and unloading
Phase lags	Delay between the applied stress and the resulting strain response
Creep	Increase in deformation under a constantly applied load
Stress relaxation	Decrease of load in tendon over time with a constant strain

In an in vivo evaluation among ten male subjects, the analysis of mechanical properties of human Achilles tendon was provided by the use of simultaneous ultrasonography-based measurements of Achilles-soleus myotendinous junction and Achilles-calcaneus osteotendinous junction displacement; Authors of the current study reported a free Achilles tendon maximal force of 1924 N, a deformation of 2,2 mm and a stiffness of 2622 N/mm during an isometric plantar flexion ramp contraction [15].

Several non-modifiable and modifiable factors are reported to alter the tendon mechanical properties and thus play a role in tendon diseases.

Differences in mechanical properties in the medial gastrocnemius muscle between a young and an elderly group were investigated using ultrasonography during a ramp and a ballistic contraction in a recent in vivo study: authors reported a significantly lower maximal elongation and strain of tendon in the elderly group compared with the young group, and a smaller

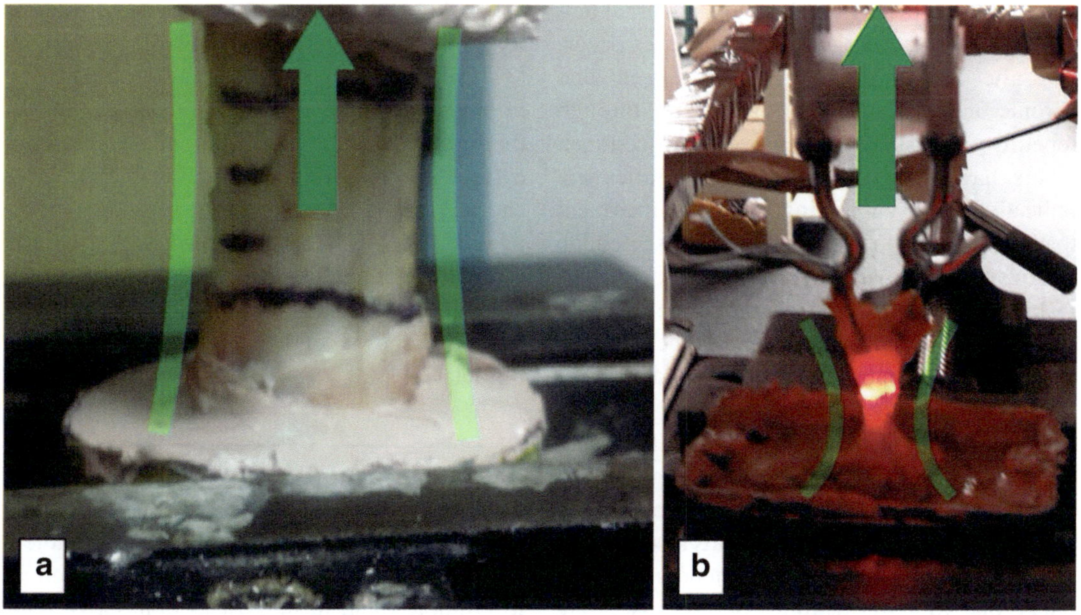

Fig. 7.3 Cadaveric experiments; experimental setup: (**a**) Patellar tendon and (**b**) Quadriceps tendon with laser used as reference to measure the cross sections

difference in hysteresis of tendon structures between ramp and ballistic contraction among the elderly compared with the young group. On the other side no difference in collagen fibers orientation of tendons was founded between the two groups [17].

Additionally, gender and anatomic location of the tendons represents non-modifiable factors that could affect tendon mechanical properties.

Disuse is also identified as altering tendon properties. Denervation and immobilization have both been found to decrease tendon stiffness and reduce final tissue strength [18, 19]. Conversely, an increase in physical activity is found to improve tendon material properties such as Young's modulus as a result of alterations to the collagen makeup [20].

7.7 Tendinopathy

A change in fiber structure can lead to load capacity decrease a tendon can exert and transmit. Tendinopathy is related to tendon injury and is characterized by a loss of normal tendon architecture, alterations in the collagen fibril distribution profile, and neovascularization [2]. tendon impairments typically result from an internal tensile overloading event, with acute injuries occurring after one isolated, overloading event and chronic injuries occurring over time through repetitive, excessive loading events. In contrast, others have suggested that acute injuries are indicative of an underlying chronic impairment that contributed to the injury. Overuse tendon injuries, commonly involving the patellar tendon, Achilles tendon, and the origin of the extensor carpi radialis brevis (tennis elbow), account for 7% (2002) of the musculoskeletal disorders in the United States [3].

While appropriate mechanical loads have anabolic effects on tendons, strengthening the tendons and improving healing quality of injured tendons, excessive mechanical loads result in tendon injuries. Such non-traumatic tendon injuries, either due to "overloading" or "overuse," are commonly referred to as tendinopathy [20].

It is also reported that excessive use and load on tendons produce high levels of inflammatory indicator PGE_2 usually found in injured tissues [23, 24]. Similarly, to bone tendons require a certain minimum of loading conditions before they become atrophied and lose some of their stiffness [25, 26]. This results in an increase in catabolic activities, stress deprivation and changes in tendons mechanical properties [27, 28].

7.8 Tendon Fibroblasts

Tendons adaptively change their structure and function in response to mechanical loads due largely to large tenocytes cells or tendon fibroblasts that are responsible for tendon homeostasis and repair. Also cells that can self-renew and differentiate into tenocytes may contribute to tendon function in this regard. These dominant tendon tenocytes can alter the ECM gene and protein expression [29] and molecular mechanism in response to the mechanoresponses [30, 31]. Other interactions between mechanical loading and cytokines, lead to stretching magnitude-dependent. This implies that repetitive mechanical stretching has two opposite effects—it is anti-inflammatory at small magnitudes, whereas it becomes pro-inflammatory at large magnitudes [31]. Tendon stem/progenitor (TSCs) also found in human recently, differ from tenocytes in that they have the ability to differentiate into tenocytes and non-tenocytes including chondrocytes and adipocytes under the proper loading conditions [24, 31, 32]. So to determine the control mechanism of loading to activate the right cells is critical to any tendon exercise. The pathophysiology of tendons is an outcome measure of tendon fibroblasts.

7.9 Clinical Applications

Where do we go from here? Can we prevent tendon failure? Can we find the appropriate exercise to promote rapid healing of tendon tears? What are the excessive limits of tendon loading capacity? Tendons transfer energy built by muscles into bone to produce the right joint motion. Normal tendon development and homeostasis is directly correlated to the mechanical load profile to which the tissue is exposed. Likewise, tendon injury and degeneration is related to changes and of this profile. Understanding the tendon, the mechanotransduction mechanism holds a promise for promoting tendon repair and restoration of tendon structure and basic function. Clinical applications of mechanobiological principles following tendon injury form the basis of rehabilitation [33].

There is a need for the development of programs which emphasize tendon loading as function of age, rehabilitation following tendon overuse injury, and in the development of postoperative physical therapy regimens that accelerate healing following surgical repair.

References

1. Physiopedia contributors, 'Tendon Biomechanics', Physiopedia, , 30 August 2018, 20:20 UTC, https://www.physio-pedia.com/index.php?title=Tendon_Biomechanics&oldid=197336. Accessed 14 Sept 2020.
2. Robi K, Jakob N, Matevz K, Matjaz V. Chapter 2: the physiology of sports injuries and repair processes. In Hamlin M (Ed.), ISBN: 978-953-51-1031-6, InTech. 2013; https://doi.org/10.5772/54234.
3. Johnson GA, Tramaglini DM, Levine RE, Ohno K, Choi NY, Woo SL. Tensile and viscoelastic properties of human patellar tendon. J Orthop Res. 1994;12(6):796–803.
4. Maganaris CN. Tensile properties of in vivo human tendinous tissue. J Biomech. 2002;35:1019–27.
5. Luyckx T, Verstraete M, De Roo K, De Waele W, Bellemans J, Victor J. Digital image correlation as a tool for three-dimensional strain analysis in human tendon tissue. J Exp Orthop. 2014;1(1):7.
6. Maganaris CN, Paul JP. In vivo human tendon mechanical properties. J Physiol. 1999;521(Pt 1):307–13.
7. Ishigaki T, Kubo K. Mechanical properties and collagen fiber orientation of tendon in young and elderly. Clin Biomech (Bristol, Avon). 2020; 71:5–10.
8. Savolainen J, Myllylä V, Myllylä R, Vihko V, Väänänen K, Takala TE. Effects of denervation and immobilization on collagen synthesis in rat skeletal muscle and tendon. Am J Phys. 1988;254(6 Pt 2):R897–902.
9. Maffulli N, Longo UG, Kadakia A, Spiezia F. Achilles tendinopathy. Foot Ankle Surg. 2020;26:240–9. https://doi.org/10.1016/j.fas.2019.03.009.
10. Trotter JA. Structure-function considerations of muscle-tendon junctions. Comp Biochem Physiol A Mol Integr Physiol. 2002;133:1127–33. https://doi.org/10.1016/s1095-6433(02)00213-1.
11. Longo UG, Loppini M, Marineo G, Khan WS, Maffulli N, Denaro V. Tendinopathy of the tendon of the long head of the biceps. Sports Med Arthrosc Rev. 2011;19:321–32. https://doi.org/10.1097/JSA.0b013e3182393e23.
12. Franchi M, Trirè A, Quaranta M, Orsini E, Ottani V. Collagen structure of tendon relates to function. Sci World J. 2007;7:404–20. https://doi.org/10.1100/tsw.2007.92.
13. Józsa L, Réffy A, Kannus P, Demel S, Elek E. Pathological alterations in human tendons. Arch Orthop Trauma Surg. 1990;110:15–21. https://doi.org/10.1007/BF00431359.
14. Longo UG, Ronga M, Maffulli N. Achilles tendinopathy. Sports Med Arthrosc Rev. 2009;17:112–26. https://doi.org/10.1097/JSA.0b013e3181a3d625.
15. Kannus P. Structure of the tendon connective tissue. Scand J Med Sci Sports. 2000;10:312–20. https://doi.org/10.1034/j.1600-0838.2000.010006312.x.
16. Maffulli N, Ajis A, Longo UG, Denaro V. Chronic rupture of tendo Achilles. Foot Ankle Clin. 2007;12:583–96., vi. https://doi.org/10.1016/j.fcl.2007.07.007.
17. Jozsa L, Kannus P, Balint JB, Reffy A. Three-dimensional ultrastructure of human tendons. Acta Anat (Basel). 1991;142:306–12. https://doi.org/10.1159/000147207.
18. Gathercole LJ, Keller A. Crimp morphology in the fibre-forming collagens. Matrix. 1991;11:214–34. https://doi.org/10.1016/s0934-8832(11)80161-7.
19. Ames PR, Longo UG, Denaro V, Maffulli N. Achilles tendon problems: not just an orthopaedic issue. Disabil Rehabil. 2008;30:1646–50. https://doi.org/10.1080/09638280701785882.
20. de Campos Vidal B. Image analysis of tendon helical superstructure using interference and polarized light microscopy. Micron. 2003;34:423–32. https://doi.org/10.1016/S0968-4328(03)00039-8.
21. Rowe RW. The structure of rat tail tendon fascicles. Connect Tissue Res. 1985;14:21–30. https://doi.org/10.3109/03008208509089840.
22. Khan KM, Cook JL, Bonar F, Harcourt P, Astrom M. Histopathology of common tendinopathies. Update and implications for clinical management. Sports Med. 1999;27(6):393–408.
23. Wang JH, Komatsu I. Tendon stem cells: Mechanobiology and development of tendinopathy. Adv Exp Med Biol. 2016;920:53–62. https://doi.org/10.1007/978-3-319-33943-6_5.
24. Langberg H, Skovgaard D, Karamouzis M, Bulow J, Kjaer M. Metabolism and inflammatory mediators in the peritendinous space measured by microdialysis

during intermittent isometric exercise in humans. J Physiol. 1999;515(Pt 3):919–27.
25. Wang JH, Guo Q, Li B. Tendon biomechanics and mechanobiology--a minireview of basic concepts and recent advancements. Journal of hand therapy: official journal of the American Society of Hand Therapists. 2012;25(2):133–41. https://doi.org/10.1016/j.jht.2011.07.004.
26. Komi PV, Fukashiro S, Jarvinen M. Biomechanical loading of Achilles tendon during normal locomotion. Clin Sports Med. 1992;11:521–31.
27. Zhang J, Li B, Wang JH. The role of engineered tendon matrix in the stemness of tendon stem cells in vitro and the promotion of tendon-like tissue formation in vivo. Biomaterials. 2011;32(29):6972–81. https://doi.org/10.1016/j.biomaterials.2011.05.088.
28. Maganaris CN, Narici MV, Almekinders LC, Maffulli N. Biomechanics and pathophysiology of overuse tendon injuries: ideas on insertional tendinopathy. Sports Med. 2004;34(14):1005–17. https://doi.org/10.2165/00007256-200434140-00005.
29. Almekinders LC, Banes AJ, Ballenger CA. Effects of repetitive motion on human fibroblasts. Med Sci Sports Exerc. 1993;25(5):603–7.
30. Wang JH, Jia F, Yang G, Yang S, Campbell BH, Stone D, Woo SL. Cyclic mechanical stretching of human tendon fibroblasts increases the production of prostaglandin E2 and levels of cyclooxygenase expression: a novel in vitro model study. Connect Tissue Res. 2003;44(3–4):128–33.
31. Jiang D, Xu B, Yang M, Zhao Z, Zhang Y, Li Z. Efficacy of tendon stem cells in fibroblast-derived matrix for tendon tissue engineering. Cytotherapy. 2014;16(5):662–73. https://doi.org/10.1016/j.jcyt.2013.07.014.
32. Connizzo BK, Grodzinsky AJ. Release of pro-inflammatory cytokines from muscle and bone causes tenocyte death in a novel rotator cuff in vitro explant culture model [published correction appears in connect tissue res. 2019 Jul;60(4):418]. Connect Tissue Res. 2018;59(5):423–36.
33. Bedi A, Maak T, Walsh C, et al. Cytokines in rotator cuff degeneration and repair. J Shoulder Elb Surg. 2012;21(2):218–27. https://doi.org/10.1016/j.jse.2011.09.020.

Biomechanical Properties of Orthopedic Materials: Muscle

8

George A. Komnos and Jacques Menetrey

8.1 Muscle Structure-Architecture

Skeletal muscle is one of the three main muscle tissues in the human body and corresponds to about 40% of body mass. Its main function is the conversion of chemical energy into mechanical energy. This energy is consumed to achieve some fundamental body tasks, such as motion production, posture maintenance, and thermoregulation [1]. The skeletal muscle contains up to 75% of all body proteins and is a huge reservoir of nutrients, especially amino acids [1].

The skeletal muscle cell is called muscle fiber, or myofiber. The number and the size of these muscle fibers are the determinants of the muscle's size. Fascicles consist of many muscle fibers bound together, with the presence of a rich network of blood vessels and nerves. The whole muscle is surrounded by a layer of connective tissue called epimysium, or fascia. Perimysium is an inner layer that surrounds each fascicle, while endomysium is the outer layer of each muscle cell. A single muscle fiber is also surrounded by a cell membrane or sarcolemma, which holds an essential role in the regulation of intracellular Ca^{++}. A band of proteins associated with the sarcolemma is coherent to the internal myofilament structure, and more specifically to the actin protein of the thin filament. Muscle weakness or atrophy derives from the absence or dysfunction of one of these proteins.

Muscle composes of force generating sarcomeres. The sarcomere is the basic contractile unit of the myofibrils. It is nowadays believed that each muscle fiber consists of thousands of myofibrils and contains billions of myofilaments. Sarcomeres are arranged in bands and lines. There are three bands, A band that contains both actin and myosin, I band that contains only actin and H band containing only myosin. The two most significant myofilaments are actin and myosin accounting for about 70–80% of the total protein content of a single fiber. Myosin is the main molecular motor, while titin is a large elastic protein that aids stabilization and alignment of the thick filament. Titin extends from the Z line to the center of the sarcomere and

G. A. Komnos
Centre de Medecine du Sport et de l'Exercice, Swiss Olympic Medical Center, Hirslanden Clinique la Colline, Geneva, Switzerland

J. Menetrey (✉)
Centre de Medecine du Sport et de l'Exercice, Swiss Olympic Medical Center, Hirslanden Clinique la Colline, Geneva, Switzerland

Orthopaedic Surgery Service, University Hospital of Geneva, Geneva, Switzerland
e-mail: Jacques.Menetrey@hirslanden.ch

J. Koh et al. (eds.), *Orthopaedic Biomechanics in Sports Medicine*,
https://doi.org/10.1007/978-3-030-81549-3_8

plays a significant role in the organization of the thick filaments in the sarcomere. Every sarcomere has three filament systems: The thick filament which is mainly myosin based, the thin filament, being actin based, and the titin filament [2]. M line interconnects the thick filaments and Z line is the site of attachment of the thin filaments.

Sarcomere length is not equal along the cell [3]. Although sarcomere lengths have a preferred operating range, skeletal muscle is extremely plastic, which makes it easier to adapt to disease or altered patterns. A matrix of connective tissue surrounds fibers and fascicles. Although these connective tissues were initially considered as a scaffold that organizes the muscle fibers into an acceptable geometry, it is recently shown that the elastic properties of this connective tissue also play an essential mechanical role [4]. Muscle's mechanical work seems to depend grossly on the extracellular matrix of connective tissue, its properties, and the flow of elastic strain energy through them (Figs. 8.1 and 8.2).

8.2 Slow-Twitch and Fast-Twitch Muscles

There are two general types of skeletal muscle fibers: slow-twitch (type I) and fast-twitch (type II). Fast-twitch muscles are further divided into two categories: type IIa, (moderate fast-twitch) and type IIb or IIx. Slow-twitch muscles are enabled in long duration and resistance exercise, such as long distance running, while fast-twitch muscles are used in forceful breakouts, such as sprinting. Moderate fast-twitch muscles contract earlier, are thicker and wear out more rapidly than slow-twitch. On the other hand, fast-twitch muscles are the most powerful and lowest in endurance.

When an athlete does aerobic exercises like running, bicycling, or swimming, slow-twitch fibers initially contract. As these fibers become tired, the fast-twitch fibers take over. In aerobic exercises, slow-twitch fibers are mainly used. This condition leads to increased stamina and a higher oxygen capacity of the muscles. A high

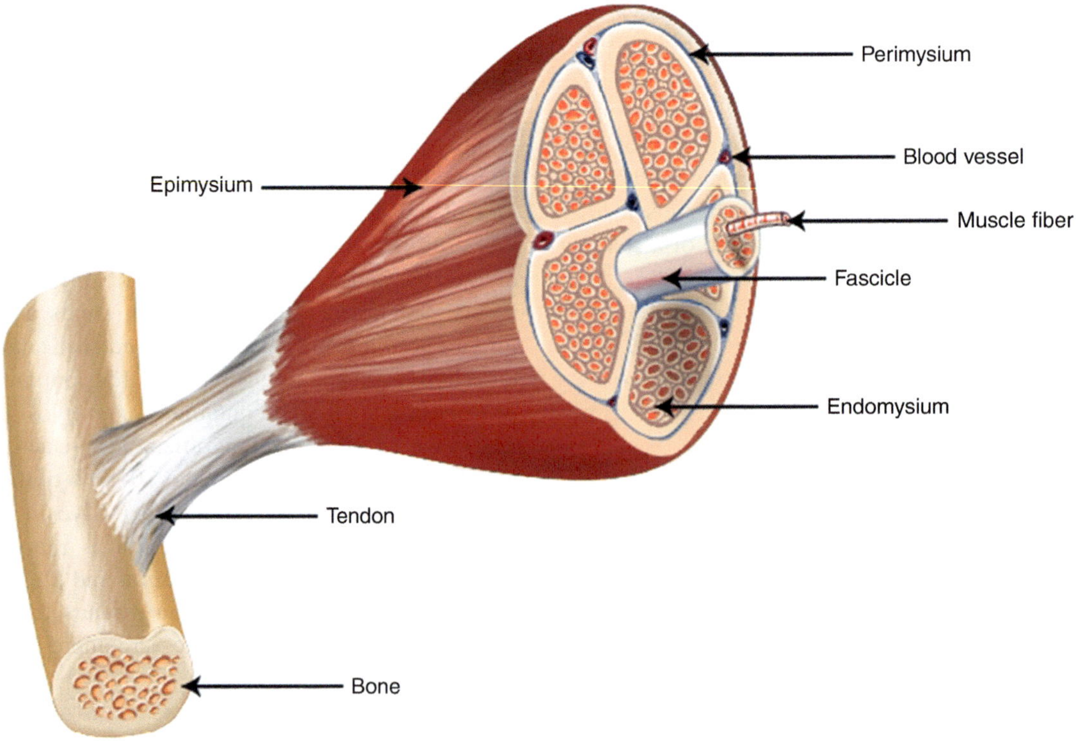

Fig. 8.1 The structure of the muscle

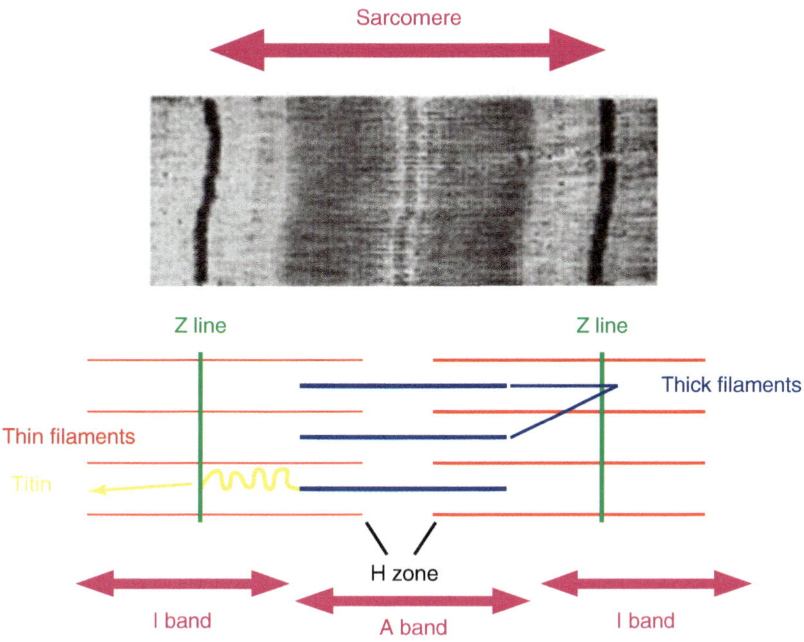

Fig. 8.2 The sarcomere

proportion of slow-twitch fibers have also been associated with low blood pressure, while it has been reported that women may have a greater distribution of the type I muscle fibers and lower distribution of type II muscle fibers than men.

Fast-twitch muscles support short, quick bursts of energy, such as sprinting or powerlifting. They have fewer blood vessels and mitochondria than slow-twitch muscles, since no need for maintaining quick, intense activities exist. Fast-twitch muscles are anaerobic and subsequently use sources of energy that are already present inside the body, such as glucose, to make adenosine triphosphate (ATP).

Type IIa fast-twitch muscles have more mitochondria than Type IIb. Due to this condition, they are more alike the slow-twitch muscles in their ability to use oxygen along with glucose and fat to consume for energy. Therefore, Type IIa muscles are not easily exhausted and can recover relatively quickly from a short, intense workout. Nevertheless, because of their ability to use oxygen and glucose for energy, they are also known as oxidative-glycolytic muscles.

Type IIb is the other type of fast-twitch muscle. These are known as non-oxidative muscles because they don't use oxygen for energy. Type IIb muscles also have a much lower number of mitochondria, since the energy needed for activity is produced by glucose. Moreover, they are much larger than other muscles and become worn out much faster.

To sum up, fast-twitch muscles are optimized for short, intense activities, such as sprinting, powerlifting, jumping, strength training, and high-intensity cycling. On the contrary, slow-twitch muscles are optimized for lower-intensity and high endurance exercises, such as running a marathon, swimming lap, or triathlon.

8.3 Muscle Power-Muscle Strength

There is controversy in sports medicine literature concerning the definition and measurement of muscular power and what it represents. It should become clear that the terms muscle strength and muscle power are related to each other, but not equivalent. Strength is the force output and power is the work produced over certain time. As a result, power measurements should be expressed in units that are equivalent to work per unit of time [5]. These power measurements can be either mean or instant values. Mean values define the average work over a definite time period, while instant values the amount of work at some definite-instant time.

8.4 Muscle Actions and Contraction

Skeletal muscles are innervated and controlled by the central nervous system. In particular, each muscle is innervated by a motor neuron. All the muscle fibers innervated by a motor nerve compose the motor unit. When the motor nerve is stimulated all the fibers of the motor unit contract, making it the functional contractile unit. Following muscle activation, muscle actions occur. These are classified into three basic types: static, concentric, and eccentric. A static (isometric) muscle action generates force with no movement of the joint or the limb. Under these circumstances, the resistance is bigger than the force generated. On the contrary, a dynamic (isotonic) action generates force in conjunction with joint movement. There are two types of dynamic actions: concentric and eccentric actions. In concentric actions, the origin and the insertion of the muscle come closer, ending up in shortening of the muscle. Opposing, in eccentric actions origin and insertion remote from each other resulting in the lengthening of the muscle. It is worth mentioning that there is also another type of dynamic action, the isokinetic one, which is generated in the laboratory using a special device with constant movement velocity. This type of muscle action cannot occur in normal human movement, where acceleration and deceleration succeed each other. Eccentric muscle actions are more prone to muscle injury if not performed carefully [6].

A term that maybe is more important and comes up before the muscle actions is muscle contraction. A muscle fiber generates tension through actin and myosin cross-bridge cycling. Under tension, the muscle may lengthen, shorten, or remain the same. Although the term contraction implies shortening, in the muscular system means the application of tension within a muscle fiber. There are different types of muscle contractions defined by the changes in the length of the muscle during contraction. The four main types are isotonic, concentric, eccentric, and isometric. In isotonic contractions, the tension is constant, while the muscle length changes. In concentric contractions, the muscle shortens while generating force and overcomes resistance. In eccentric ones, the muscle elongates while generating force. The force cannot overcome resistance in this type of contractions. Eccentric contractions can be either intentional or unintentional. In isometric contractions, the length of the muscle remains the same while force is generated. These are more common in the muscles of the upper limb and hand.

Overall, muscular contraction is a complex mechanism representing the interaction of the contractile proteins actin and myosin in the presence of calcium. Adenosine triphosphate (ATP), produced in the mitochondria, supplies the demanded energy. The sequela of events that occurs when a muscle contract depends on three other factors: calcium and two inhibitory proteins, troponin and tropomyosin. Voluntary muscle contraction begins with electrical impulses at the myoneural junction, initiating the release of calcium ions. These calcium ions bind to a protein called troponin-C that is attached to tropomyosin, and actin. The juncture of calcium with troponin-C provokes the removal of the tropomyosin molecule from its blocking position between actin and myosin. Then, the myosin head attaches to actin, and muscular contraction occurs.

As mentioned above, the main source of energy for muscular contraction is ATP, which is produced by oxidative phosphorylation. The major fuels for this process are carbohydrates (glycogen and glucose) and free fatty acids. During a rest period, the amounts of energy produced from carbohydrates and fats are almost the same. During low levels of exercise, free fatty acids are the main contributors to the energy supply, but while exercise increases and more energy is needed, carbohydrates take over. Finally, when it comes to maximal exercise, energy production relies entirely on carbohydrates.

The mitochondria can produce ATP for muscle contraction only in the presence of oxygen. At very high levels of exercise, the demand for oxygen may exceed the capacity of the cardiovascular system to deliver it. In that case, glycolysis progresses in the cytoplasm until pyruvate is formed. Electrons released during glycolysis are taken up by pyruvate to form lactic acid. Rapid diffusion of lactate from the cell inhibits the further steps in glycolysis. However oxygen-independent glycolysis is not efficient since only two ATP molecules per molecule of glucose can derive from this process.

8.5 Force of Contraction

Augmentation of the force of contraction of a muscle can be achieved by recruiting muscle fibers. As all the muscle fibers within a motor unit are activated simultaneously, a muscle can recruit more muscle fibers by recruiting more motor units. Muscle fibers are classified as fast-twitch and slow-twitch depending on their innervation. Since all fibers in a motor unit are innervated by a single a motor neuron, all of them are of the same type. Slow-twitch motor units tend to be small (100–500 muscle fibers) and are innervated by a motor neuron that is easily electrified. On the contrary, fast-twitch motor units are larger in size (1000–2000 muscle fibers) and are innervated by motor neurons that are not easily electrified. Subsequently, slow-twitch motor units are recruited first and as exercise is getting tenser, fast-twitch motor units follow. The advantage of this sequence is that the first muscle fibers recruited are those with the higher resistance to fatigue.

8.5.1 Contractility and Elasticity

Two of the most significant properties of the muscles are contractility and elasticity. Contractility is the ability of muscle cells to forcefully shorten, while elasticity is the ability to recoil to the muscle's length after being stretched. Elasticity and muscle elastic components are a critical key in the mechanism of muscle contraction and function.

The most widely acceptable model for muscle function and interaction between contractile and elastic elements is the three element Hill-type model [4, 7]. This model consists of a contractile element (CE) arranged alongside a parallel elastic element (PEE) and in series with a series elastic element (SEE). The contractile element represents the fundamental mechanical behavior of the sarcomere. Noteworthy, the length change of the contractile element is not the same with the length change of the muscle, due to the series elastic element altering the length whenever force varies.

The myotendinous junction is the part of the tendon where the ends of the muscle fibers interact with the tendon to transmit the force of the contraction and initiate movement of the skeleton. The tendon and the connective tissue layers are composed mainly of elastin and collagen fibers, and contribute to passive tension of muscle preventing them from damage in case of overstretching or contraction. The force of contraction is transmitted both longitudinally to the tendon (via myotendinous junctions) and laterally to connective tissue adjacent to the muscle fibers (via costameres).

8.5.2 Force–Velocity Curve

The velocity at which a muscle shortens depends on the amount of force that the muscle must produce. The force–velocity curve is a physical representation of the inverse relationship between force and velocity. In other words, an increase in force causes a decrease in velocity and vice versa. This balance between force and velocity is thought to occur due to a decrease in the time available for cross bridges to be formed. In the absence of any load, the shortening velocity of the muscle is maximal. A rise of the load reduces the velocity of muscle shortening until the load reaches a peak and the muscle cannot lift the load and therefore cannot shorten any more (zero velocity). Negative velocity represents the condition when further load is applied and results in

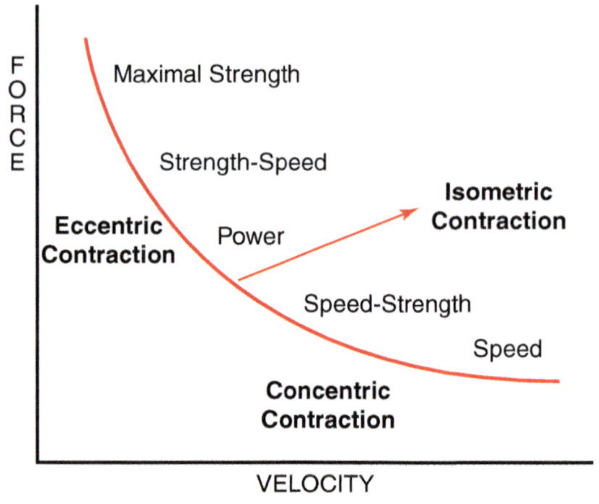

Fig. 8.3 Force–Velocity Curve

stretch of the muscle. As a result, slower velocity exercises allow the athlete to form more cross bridges and develop more force. Higher velocity exercises provide less time for cross bridges to form, and therefore results in lower force production.

In terms of contraction, the force is zero in isometric contraction, it decreases as velocity increases in concentric contraction, while the force increases as velocity decreases in eccentric contraction (Fig. 8.3).

8.6 Length–Tension Relationship

When a muscle contracts, its length is shortened and force (stress) is produced. These two parameters define if a contraction will be isotonic or isometric. So, when muscle strength is steady and the force generated is measured we have isometric contraction. On the other hand, when the force remains constant and the length of the muscle changes then it is an isotonic contraction. When stretch is applied to a muscle at rest, the muscle resists with a force that gradually increases. This gradual increase of the force that correlates to the stretch applied is due to the elasticity of the muscle. Contractile force increases as muscle length increases up to a point. Beyond that point, contractile force decreases. This represents the length-tension curve, which is consistent with the sliding filament theory.

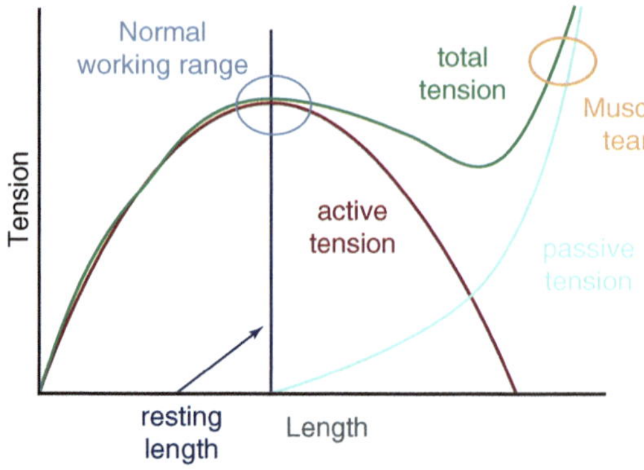

Fig. 8.4 The Length–Tension Curve

Length–tension relationship corresponds to the overlapping of microfilaments in a sarcomere. It is worth noting that the muscles besides the active properties can also produce passive tension. As length increases, passive tension follows. This is due to elastic properties of the muscle. The tissues stretch and enough force is produced to overcome the active tension deficit that occurs when a muscle lengthens. The resting length (L0) is the length that allows the higher potential active force. The active tension part of the curve is an inverted U shape with its peak at the resting length. Until the resting length active contraction generates the force. Nevertheless, beyond the resting length, when parallel elastic components are stretched, passive tension begins to contribute as active tension decreases. At the end, at more elongated lengths, passive tension contributes for the most of the force. Most of the time, muscle tear occurs at the elongated state when muscle resists further lengthening (See Fig. 8.4).

8.6.1 Exercise

It is generally accepted that there are three aspects of training. These are learning, endurance and strength training. The learning aspect includes motivational parameters and coordination skills but not adaptive changes. Nevertheless, these skills can be maintained for years, no matter the amount of exercise done. Endurance can achieve adaptive responses, mainly by performing regular exercise. This procedure occurs through an increase in the oxidative metabolic capacity of the motor units involved and results in increasing also the capacity of the heart and the respiratory muscles. Muscle strength can be increased by applying systematic efforts that involve most motor units. Such efforts recruit fast glycolytic motor units, as well as slow oxidative motor units. During these efforts, blood supply may be reduced, limiting the duration of the contraction. Strength exercise in a regular basis enhances the synthesis of more myofibrils and the growth of tendons. To summarize, muscle is a system that can adapt to most of stimuli, but that needs to be constantly stimulated to keep its conditioning.

8.7 Muscle Strain and Repair

As aforementioned, the skeletal muscle consists of muscle fibers, which derive from the fusion of myoblasts. There are three types of muscle fibers: type 1, type 2A, and type 2X. The type of myosin heavy chain protein they express differentiates these three types [8]. Satellite cells (SC) are skeletal muscle stem cells that are responsible for muscle repair after strain or injury [9, 10]. Strains commonly occur at myotendinous junction. In minor muscle strains the muscle usually heals in a short period without significant compromise and follows, as in severe muscle injuries, a cascade of three phases initiates. The first phase is the inflammation phase, during which formation of hematoma takes place and an inflammatory reaction begins. This phase happens during the first days following the injury, and is characterized by intense chemotactic changes, which result in high concentration of molecules, such as cytokines chemokine and growth factors. These are excreted by neutrophils and attract other inflammatory cells, such as monocytes and macrophages [11]. There are two types of macrophages. M1 are pro-inflammatory cells with main action the removal of cellular debris, and the M2, which are anti-inflammatory cells and promote myotube formation. Regeneration phase follows, starting at about 4 days after injury and continues until third to fourth week. Phagocytosis of damaged tissue and proliferation of satellite cells leading gradually to repair and regeneration of the myofibers characterize it. Last comes the remodeling phase. During this phase, revascularization of the muscle fibers appears and leads to maturation of regenerated myofibers, with muscle tissue formation. This healing process is a competition between muscle tissue regeneration and development of fibrosis that may end up in hypertrophic and painful scar tissue. Restoration of innervation is the final signal for muscle repair.

References

1. Frontera WR, Ochala J. Skeletal muscle: a brief review of structure and function. Behavior Genetics;2015.
2. Krüger M, Kötter S. Titin, a central mediator for hypertrophic signaling, exercise-induced mechanosignaling and skeletal muscle remodeling. Front Physiol. 2016;
3. Lieber RL, Lieber RL, Roos KP, Lubell BA, Cline JW, Baskin RJ. High-Speed Digital Data Acquisition of Sarcomere Length from Isolated Skeletal and Cardiac Muscle Cells. IEEE Trans Biomed Eng. 1983;
4. Lieber RL, Roberts TJ, Blemker SS, Lee SSM, Herzog W. Skeletal muscle mechanics, energetics and plasticity Daniel P Ferris. Journal of Neuro Engineering and Rehabilitation. 2017.
5. Sapega AA, Drillings G. The definition and assessment of muscular power. J Orthop Sports Phys Ther. 1983;
6. Fulford J, Eston RG, Rowlands AV, Davies RC. Assessment of magnetic resonance techniques to measure muscle damage 24h after eccentric exercise. Scand J Med Sci Sport. 2015;
7. Zajac FE. Muscle and tendon: properties, models, scaling, and application to biomechanics and motor control. Crit Rev Biomed Eng. 1989;
8. Ennion S, Sant' Ana Pereira J, Sargeant AJ, Young A, Goldspink G. Characterization of human skeletal muscle fibres according to the myosin heavy chains they express. J Muscle Res Cell Motil. 1995;
9. Dumont NA, Bentzinger CF, Sincennes MC, Rudnicki MA. Satellite cells and skeletal muscle regeneration. Compr Physiol. 2015;
10. Laumonier T, Menetrey J. Muscle injuries and strategies for improving their repair. Journal of Experimental Orthopaedics. 2016.
11. Toumi H, Best TM. The inflammatory response: friend or enemy for muscle injury? British Journal of sports medicine. 2003.

9 FEA Applications for Orthopedics: An Overview

Umile Giuseppe Longo, Giovanna Stelitano, Giuseppe Salvatore, Vincenzo Candela, Calogero Di Naro, Laura Risi Ambrogioni, Farid Amirouche, and Vincenzo Denaro

9.1 FEA for Orthopedics

9.1.1 Introduction

The correct use of biomechanical engineering can provide significant improvements in the reliability of orthopedic procedures leading to a better quality of life for patients. The integration of engineering principles in biomedicine has been widely promoted and continues to grow. Research in orthopedic settings needs a specific methodology which provides cost benefits, easy to use, and can be adapted to different situations. This method should work as a guide for understanding both the biomechanics as is the case of locomotor system, in physiological and pathological states, and in the study of the correct function of prosthetics and implants devices.

The introduction of simulation models from the field of bioengineering unquestionably represents the main instrument to evaluate the most suitable clinical choice, guaranteeing an accurate analysis of particular physiological conditions inside a specific picture of pathology. The most used tool in numerical simulation is the finite element method (FEM). It can be defined as a mechanical structure, discretized into elements of different types, where when the boundary conditions are applied it provides a complete analysis of stress and strain conditions at each node level. The FEM was introduced in early 1960s as an instrument for structural analysis and rapidly became the solution of choice in engineering. It is viewed as a numerical solution to problems that didn't have exact solutions and Turner who published his historic job of finite element analysis in 1956 was the first to publish the method as we know it today. Ten years later, (Zienkiewicz) wrote the book *The finite element method in structural and continuum mechanics*, which provided the basis of for using FEM in continuum systems. Finite element (FE) simulation became a standard method for the investigation of any physiological unit, notwithstanding its complexity, helping, in this way, surgeons to the understanding of the biomechanics of the human body functioning, in healthy and pathological situations. In particular, FE simulation laid the foundations for the knowledge of biomechanical changes con-

U. G. Longo · G. Stelitano · G. Salvatore · V. Candela (✉) · C. Di Naro · L. R. Ambrogioni · V. Denaro
Department of Orthopaedic and Trauma Surgery, Campus Bio-Medico University, Rome, Italy
e-mail: g.stelitano@unicampus.it; g.salvatore@unicampus.it; v.candela@unicampus.it; c.dinaro@unicampus.it

F. Amirouche
Department of Orthopaedic Surgery, Orthopaedic & Spine Institute NorthShore University Health System, Evanston, Illinois, USA

Department of Orthopeadic Surgery, College of Medicine, University of Illinois, Chicago, Illinois, USA
e-mail: FAmirouche@northshore.org, amirouch@uic.edu

J. Koh et al. (eds.), *Orthopaedic Biomechanics in Sports Medicine*,
https://doi.org/10.1007/978-3-030-81549-3_9

sequent to prosthesis or osteosynthesis implantations and biological adaptive responses of bone to biomechanical stimuli. It also has shown an added benefit in predicting the variations related to stress distribution around the implanted zones, providing prevention of expected pathological conditions, result of an inappropriate positioning of the prostheses or fixation implants. Finally, the simulation model, has been employed to predict the behavior of orthopedic splints, usually used for the correction of deformities, providing the recovering force-displacement and angle-moment curves which mark the biomechanics of the splint in the whole range of movement of joints. The fundamental concept of FE method is the division of complicated mechanical structures into finite number of elements connected by nodes which represent the geometry of the object being studied. Obviously, the number of elements will provide a more refine analysis and lead to better stress-strain distribution. Most FEA software has an automatic option for creating the so called mesh representing the discretized body. Some of the current challenges of FEA in orthopedics are the interpretation of the data, its clinical relevance, and how to make it potentially a clinical tool easy to use by surgeons. This is work in progress and successful results will need collaboration among clinicians and engineers to perhaps rewrite the output form of FEA so clinicians can interpret the data and provide feedback.

It is essential to realize the pitfalls of FEA are centered around the problem formulation, the boundary conditions and of course the assumption one makes. The FEA provides solution to the problem being formulated, so it is essential that both the clinician and engineers reach a decisive consent on the problem-solution through some validation process.

9.1.2 Methodology for the FEA in Biomechanical Systems

The generation of finite element simulation models must do the math with the geometric complexity of biomechanical systems. The old FE models lacked geometrical precision, which has been improved in current commercial software [1]. On the other hand, the recent models' challenge concerns the efficacy of the results and their adaptation to clinical practice at a reasonable time and without exorbitant costs. The combined employment of scanners and computed tomography (CT), which provides three-dimensional images design, allows the realization of great accurate models both in their external or internal interfaces definition. The method needs an adequate software tool able to process images, which, at the same time, must be compatible with the programs used next to create the FEM. Three-dimensional laser scanners, as well as 3D geometrical reconstruction programs, represent a clear example of this progress. In any case, the expected result should be a precise geometrical model for the basis of the realization of a finite element mesh. Another relevant problem in the generation of FE models is their difficulty to be experienced in living subjects [1]. Thus, it has been necessary to realize biomechanical systems with great reproducibility, versatility, in which models can be repetitively applied as many times as required without causing an aggressive alteration of the start conditions. Current research is moving toward more realistic systems with a high rate of prediction and adaptability to clinical settings. This could be achieved by formulating the geometry so that meshes fit the specific problems at hand, including, for example, the type of elements and the mesh size. The last point to consider in FE models is the interaction between surfaces and the bonding of the elements [2]. The understanding of the mechanical behavior of materials is very complicated and essentially crucial to the analysis. At the same time, it is crucial. For example, the bone shows an anisotropic behavior with selective different answers when subject to compression. Its ability to change is influenced by its composition: cortical or cancellous bone present different molecular characteristics with subsequent different biomechanical responses. Even if good results have been obtained in term of reproducibility of bone behavior, these simulation models continue to require additional changes in material properties which will the render the analysis interpretation difficult. For most cases analysts prefer to consider a linear elastic behavior when representing bone, which makes the analysis more predictable and convergence of the FEA is more attainable. This might lead to a less precise model.

Modeling of soft tissues, ligaments, cartilages, and muscles using FEM is more difficult as their intricate structural detail is not biomechanically known. These soft tissues are commonly having hyper elastic or hypo elastic properties, often associated to viscoelastic behavior, receiving the influence of the rheological effect with prolonged strains or permanent load conditions. The intrinsic complexity of biological tissues makes highly difficult and expensive to build realistic human joint models for the analysis of prostheses and implants. This unsolved problem is a future challenge to engineers and clinicians to resolve.

9.2 Example of FEA Application Models

9.2.1 FEA Application in Hip Arthroplasty

The FEA application in the field of total hip replacement is a topic worthy of attention, considering that hip replacement is the most common implant used in orthopedic surgery. Although it is widely used, this procedure still requires a number of improvements in terms of both performance and materials to ensure the design of implants can also be used for younger, active patients. Osteoarthritis is the main indication for prosthetic replacement. It is a degenerative joint disease. The hip replacement consists of two components, an acetabular and a femoral component. The acetabular component is a metal shell that includes an insert, mainly made of polyethylene or ceramic. The metal shell is pressed or threaded into the prepared acetabulum. The femoral component consists of a stem and a prosthetic head, which may be ceramic or cobalt-chrome [3]. The prosthetic stem is, in most THA, made of titanium (Ti-6Al-4 V). The bearings used are mainly ceramic-on-ceramic, ceramic-on-polyethylene and metal-on-polyethylene. Despite the current significant improvements, there are still some complications that require further studies focusing on the characteristics of the prosthetic form, veneering materials and insert [8]. When designing a joint prosthesis, it is necessary to consider the transmission of mechanical stress. The loads are extremely high both with regard to the pressure bending stresses generated in the bones (e.g., in the neck of the femur) and the pressure and the resulting deformation and friction forces transmitted between the articular surfaces [12]. Being cyclic, these loads can induce favorable responses through bone remodeling applied in bone structures [13], while they can produce fatigue phenomena in prosthetic implants. The anchoring of the uncemented prosthesis to the bone is due to the simple mechanical joint (press fit) between the femoral stem and the canal in the femur. The main purpose of these prostheses is osteointegration [14], i.e., direct communication between the prosthesis and bone, without interposed connective tissue. This direct contact must be mechanically stable: the transmission of forces to the interface must not generate relative movements between the prosthesis and the bone. Smaller contact surfaces and increased local stresses can damage bone areas [15]. It is essential to consider that, in the absence of contact, stresses are not transmitted, and bone resorption is very likely. The latter phenomenon is favored by another equally harmful mechanism: stress shielding. An uncemented prosthesis relieves forces from the distal area of the pelvis. In this way, the proximal segment of the femur is not very stressed and proceeds toward bone resorption and remineralization. This mechanism causes the loss of mechanical stability [16]. Stress shielding is also reinforced by the greater rigidity of the femoral stem compared to the bone that houses it. The prosthesis withstands loads while the bone is unloaded and tends to atrophy, causing the prosthesis to mobilize [17]. Generally, the load during walking is transferred to the femoral head, the medial cortical bone of the femoral neck, the small trochanter and the diaphyseal bone [9]. The insertion of cemented or uncemented femoral stems would significantly influence the physiological transmission of forces, because these forces are transferred centripetally from the central bone marrow cavity to the cortical bone, passing through the prosthetic stem. These variations in physiological hip biomechanics have led to the phenomenon of "adaptive remodeling" because bone tissue has to adapt to new conditions [4].

Adaptive remodeling is a multifactorial process. The physical characteristics of the implant (size, implant design and alloy) and the type of anchorage in the femur (metaphyseal, diaphyseal, hybrid) are mechanical elements involved in the load distribution resulting from the prosthetic implant in the femur [10]. Age, body weight, bone mass, and also the quality of implant fixation are biological parameters that must be considered. The initial bone mass is the most important. Several models of non-cemented stems aimed at achieving an ideal load transfer to the femur, imitating the physiological transfer from the femur to the femoral rod [5]. The main objective was to avoid stress resistance, since the lack of physiological load transfer and the lack of mechanical stimuli in this area leads to atrophy of the proximal bone. Fixation of the cemented hearth is achieved by positioning the cement in the bone, forming a cement-bone interface. A new cement-bone interface has been developed in the cement mantle. The cement mantle may appear to allow better load distribution in the femur; however, the design, material and surface of the prosthesis play an important role in bone remodeling [11]. The long-term follow-up of uncemented stems from different models shows that this has not been achieved, and that there is more or less a stress shielding phenomenon in all models, so there is proximal bone atrophy. Both the stress shielding of the bone cement handle and the resulting proximal bone atrophy and the long-term behavior of the interface between bone cement and bone cement are being studied to monitor the different stem models [6]. Finite element simulation is proposed as a tool to predict the long-term biomechanical behavior of different stems in bone cement or non-bone cement and to predict the impact of biomechanics on the femur, thus influencing femoral remodeling.

9.2.2 FEA Application in the Lumbar Spine

The spine is an intricate anatomical structure characterized by three-dimensional movements, whose purpose is to maintain the erect posture of the body, supporting the load and respecting the nervous structures contained into the spinal canal. For this reason, the spine presents a visco-elastic composition required to perform flexible movements, control stability and preserve nerves. All these characteristics make the analysis of spine biomechanics, as well as its reproduction in vivo or in vitro studies, very difficult. Usually, biomechanics of the lumbar spine is examined in cadaveric specimens although they lack flexibility making the reproduction of the whole living human range of motion complicated [7]. For in vivo studies, several strategies of analysis (such as radiography, computerized axial tomography (CAT), magnetic resonance imaging (MRI), television (TV), and computer) have been employed, leading results greatly variable if compared in subjects with different age, different disease or even in the same person during the same day. At the same time, researches conducted on animal spines, considering their differences with the human lumbar spine, have shown mixed results, not useful for the comprehension of biomechanics functioning. It's evident the necessity to generate simulation models such as FE to improve spine understanding. FEA model of the spine provides the study on the lumbar spine in physiological states to simulate varying load conditions and analyze the influence on biomechanics. It is also able to simulate the disc degeneration and evaluate its impact on the adjacent structures. Finally, FEA models may be helpful to examine several fixation systems, (pedicular screw or rigid fixations or dynamics one). The S1-L5 is the main functional district analyzed. It is studied using the methodology above described with which is possible to gain the geometrical model corresponding (Fig. 9.1). The mesh of the vertebrae is realized utilizing tetrahedrons with linear approximation inside a program with a thin size. This allows the transition from the area of the exterior cortical bone to that of the interior

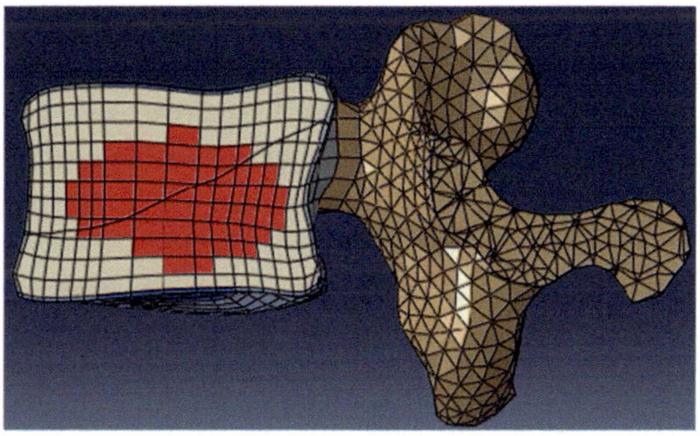

Fig. 9.1 Spine FEA model. Vertebral structure reproduction with tetrahedrons elements; dish meshes clearly divided into nucleus pulposus and annulus fibrosus

cancellous bone, realized through statistical means from CTs of vertebrae in normal people. Disc meshes represent the crucial element for the exact reproduction of the biomechanical behavior of the functional unit examined [8]. To do this, each disc is divided into nucleus pulposus and annulus fibrosus with adequate dimensions. The first functions as a non-compressible fluid, which under compression extends toward the exterior tensioning the fibers of the annulus. This last has a hyper elastic behavior, even if just in tension. Also, the reproduction of the real anatomy interaction among all different elements (vertebrae, discs, ligaments) is fundamental. To test the effectiveness of the fixation, flexion-extension movement has been examined as the most characteristic and muscle and ligament strengths are adapted to the proper values to achieve the necessary movement ranges. Calculation and post-processing are obtained applying for the Abaqus program. After the realization of the healthy model, FE systems have been adapted for different conditions based on specific levels of disc degeneration. Variations caused by surgery need to be represented in the FE model for the study of the different kind of fixations (such as screws, rigid or dynamic implants) [9]. This study has been generated by removing the elements intersected the devices and resetting the mesh along with the new interfaces. The analysis in the flexion-extension movement has been conducted to verify the changes produced by the fixations. The results permit the analysis of the variations in the local stress distribution around the screws, identifying points of potential expected pathologies, fruit of the alterations caused by the fixations.

9.3 FEA Application in the Articular Cartilage Tissue Engineering

Articular cartilage is primarily composed of chondrocytes organized into the extracellular matrix in three different layers with a respectively parallel, random and perpendicular orientation of collagen fibers. Cellular morphology and extracellular orientation depend on mechanical stimuli which, thanks to their ability to mediate conformational changes of the molecular pattern, gene expression regulation, and tissue remodeling, are vital for cartilage homeostasis and regeneration [10]. The lack of mechanical stimulus, together with senescence, inflammation, and obesity, is the main risk factors for the development of osteoarthritis (OA). The OA is one of the main diagnosed pathologies for which an effective good treatment doesn't exist. Both for the degeneration of articular cartilage derived from OA, as well as other cartilage diseases, the use of functional tissue-engineered cartilage may be successful. Scaffolds provide the proper mechanical and spatial conditions for chondrocytes proliferation, making functional tissue-engineered cartilage capable to satisfy this request. In the field of engineering tissue, scaffold design generation is a complicated process because it has to guarantee proper interactions among the cells and the scaffold. This process requires many time because it needs sequential in vitro, mechanical, and in-vivo studies to define the correct structural parameters for the expected level of mechano-transduction. To overcome these problematics, the finite element analysis (FEA) has currently introduced as a preparatory step for scaffold design. FEA makes a stress–strain analysis of the scaffold, dividing it into smaller elements with a uniform shape which can be 2D (planer triangle or quadrilateral) or 3D (tetrahedral or hexahedral). The block form result composed by placing nodes on the solid geometry. The most basic 3D element shape is a tetrahedron composed of four nodes. More tetrahedrons make an eight-node hexahedron [11]. Advanced models use higher-order 20-node hexahedral elements, providing more precise analyses. A mathematical constitutive equation is then applied and solved for the stress–strain at each node. The analysis can use both simple linear elastic formulation, which follows Hooke's Law (stress in early proportional to strain), both complex biphasic elastic once, which is a solid-fluid coupled stress–strain formulation, in which the solution is given by elastic modulus, Poisson's ratio (bulk modulus), and permeability of the matrix. FEA grants the capacity to prognosticate structural deformation, stress distribution, and cartilage tissue regeneration inside the scaffold models. Numerous researches have realized mesh models of scaffolds, cartilage, and cells, to define the mechanical parameters (stress and strain) under varying load profiles [12]. Cartilage tissue engineering and scaffold design have been strongly influenced by the use of FEA. The latter, in fact, owns the analysis capacity for the tissue engineering researchers, because they can be utilized to improve scaffolds. The ability of these models to precisely predict the stress, strain, fluid pressure, flow, and cell growth enables rigorous analyses of scaffold designs before they are actually produced.

9.3.1 FEA Application in Shoulder Complex

The shoulder is a complex articular structure which represents the perfect compromise among mobility and stability. Shoulder mobility is made possible by the glenohumeral joint that allows the greatest range of motion, while the passive stabilizers, such as ligaments, glenohumeral capsule together with the active muscular components, provide shoulder stability. The achievement of a full range of motion is guaranteed by the small glenoid articular surfaces compared to that of the humerus, but, at the same time, their low congruency challenges joint stability [13]. This shoulder specific joint configuration predisposes patients to suffer from glenohumeral articular dislocation as well as other painful soft tissue diseases. The understanding of the in vivo biomechanical functioning of the shoulder complex is still really restricted. Traditional biomechanics has resulted in very limited instruments so that the implication of FE model as advanced tools in the estimation of shoulder biomechanical behavior is not surprising. FE models of the shoulder aim to investigate as each joint element functions under different loading conditions. The fundamental step in FE shoulder model is the hard tissues, soft tissues and musculoskeletal structure reconstruction usually made in 2D or 3D geometry. To realize it, several approaches are employed. Usually, geometric reproduction of bone is realized by CT images, while geometric models of soft tissues are obtained by MR imaging, even if the representation of articular cartilage remains the hardest challenge. The reproduction of realistic materials properties of shoulder tissues is another essential key point. Several studies use the major assumption from literature to overcome the problems, assuming bones as a rigid and isotropic linear elastic material whit a big Young's modulus, and muscles and tendons as complicated structures with anisotropic behavior and viscoelastic characteristics. FE shoulder model are categorized in different sections according to the main physiological conditions and clinical problems reported as follow: glenohumeral instability, rotator cuff tears and shoulder arthroplasty. Concerning the first point, as above described, the poor congruency of shoulder articular surfaces permits a great range of motion, even if this condition commonly predisposes shoulder dislocation. In the last years, instability of the glenohumeral joint has been studied through several FE models with many strengths and limitation but with the common purpose to reach the right knowledge of instability aetiology, diagnosis, treatment and prevention. One of the most recent FE models for the study of instability focuses its attention on muscular strengths allowing shoulder a free movement of six degree. It uses tissue deformations, contact area and contact pressure to estimate joint stability [14]. These loading and boundary conditions associated with these specific parameters, permit the FE model to simulate shoulder biomechanics as close to reality as possible. Rotator cuff tears are the most diffuse disease of shoulder, whit the prevalence of injury of the supraspinatus tendon. The aetiology of the tears is multifactorial and complicated and the correct mechanism of their onset is still not clear. For this reason, several FE models have been employed to better understand the biomechanical mechanism which can provoke rupture. FE shoulder models conduct their investigation on the hypothesis that stress concentration is the main responsible for tendon damages. One of the most recent studies has been realized including rotator cuff muscles and deltoid fibers as controllers of shoulder rotator cuff biomechanics. Muscle and bone have been reconstructed by CT images, and shoulder movement has been simulated through three-dimensional programs (Fig. 9.2) [15]. Any case the results in this field need several improvements. As well as the other shoulder diseases above described, also shoulder osteoarthritis is becoming much more frequent and debilitating, requiring a higher number of shoulder arthroplasty. Total shoulder arthroplasty is made up of a humeral head and a glenoid component. The prosthesis design is the essential point of the success of the procedure and to improve its characteristics FE analysis has been widely employed. Fe models focus their attention on several crucial issues for the realization of optimal implants such as the glenoid component aseptic loosening, the conformity of components shape, and the choice of the best component's materials. In all of these cases, FEA analysis develops investigation using different parameters. Implant shape, positioning and orientation, use of bone cement, articular congruency, cement stress, central components alignment, are the main investigated.

9.3.2 FEA Application in Knee Joint

The knee joint is the largest joint which permits locomotion and supports the body weight. It is made of two different joints, the tibiofemoral and the patellofemoral, marked by complicated biomechanics functioning. The understanding of knee diffuse injuries and disease requires the employment of structured approaches as FE

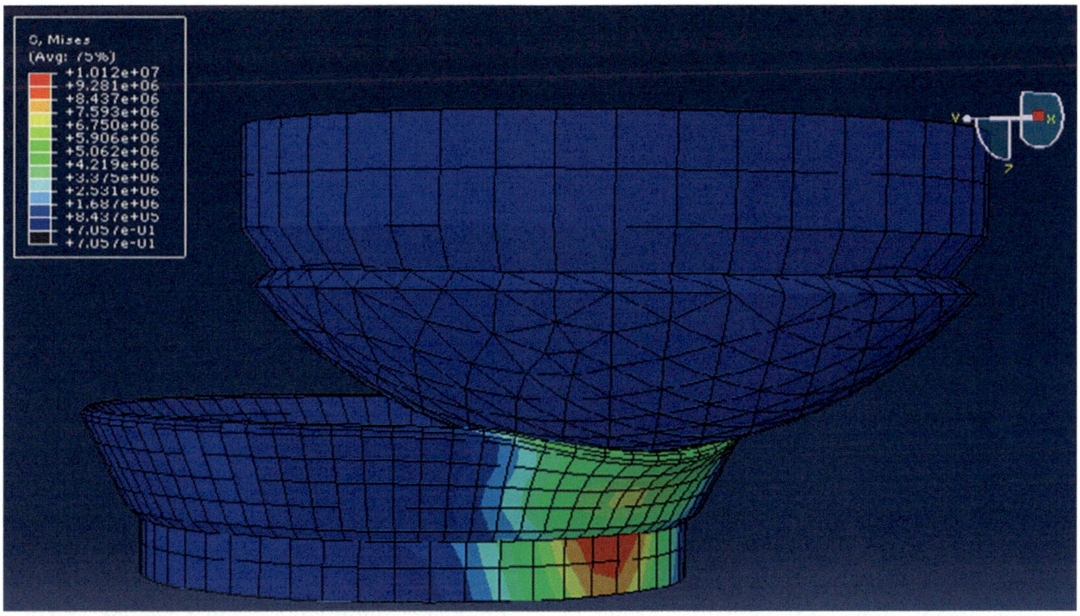

Fig. 9.2 Shoulder FEA model. Three-dimensional reproduction of glenohumeral interface

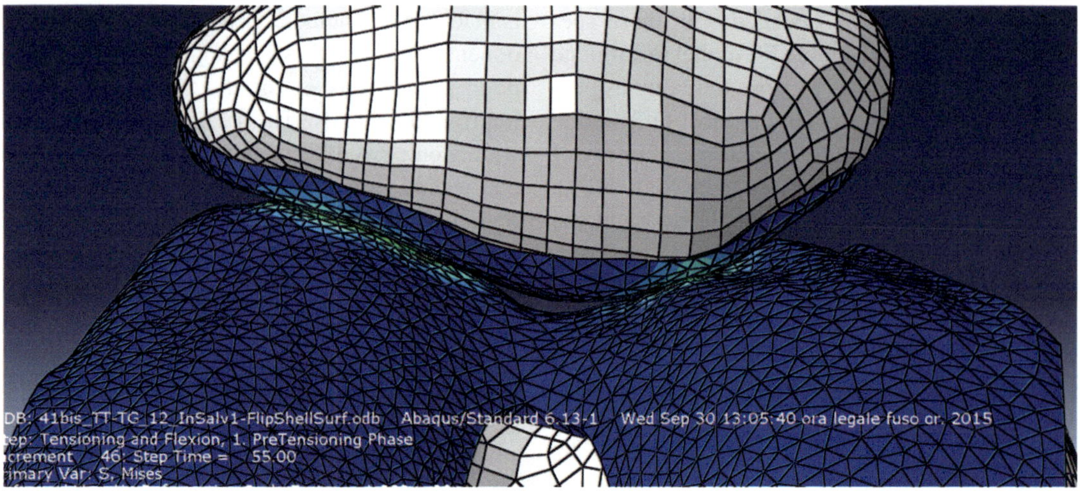

Fig. 9.3 FEA model of patellar-femoral joint. Reproduction of contact area and stress distribution on 3D model

models. Thus, also for the study of knee joint FEM is widely used, guaranteeing the investigation of knee cells, tissues, and level joints to comprehend the effect of injuries, their surgical treatment and their mechanical functions. FE analysis finds several specific applications into the field of knee physiological and pathological conditions. Ligament injuries and reconstruction, cartilage damages and knee replacement are the most investigated [16]. For the first one, the current FE models realized, employ different strategies of approaches. Ligament injuries are examined through models which simulated muscles load, or through programs which reproduce graft stiffness and initial stress on a 3D ligament-reconstructed knee. At the same way, 3D models of knee joint have been realized to better understand osteoarthritis onset and development in the contest of mechanical tissue environment (Fig. 9.3). Several studies have utilized knee models with loaded flexion of 30 degrees, evaluating the effects of the size of cartilage injuries related to contact pressure and articular strains [17]. These models include the depth-dependent properties of cartilage which is usually considered as linear elastic and isotropic material. Finally, also the mechanical properties of knee prostheses have been accurately examined through computational FE models. Current FE analysis uses a model which investigates kinematics and stress distribution of a total knee replacement during a gait cycle reproduced by a knee simulator. This model assumes the femoral component of the prosthesis as rigid and the tibial insert as an elastic and plastic material. Other types of models exploit specific scales to establish contact areas and stress distribution in prosthesis components. In these systems, the pressure is estimated from strengths resulted by kinematic models while contact areas from implants designs. Any case, the real improvement in the field of FE models is represented by the realization of patient-specific implants which should reach better compliance and patients' outcomes. Concerning this section, a patient-specific implant was designed for unicompartmental knee replacement. The imagine reproduction has been realized by a 3D model, the femoral and tibial components have been assumed to be isotropic, linear, elastic while the insert has been considered as nonlinear.

9.4 Conclusion

Undoubtedly, FE methods application in the field of orthopedics has already made an impact and will continue to play a crucial role in the design of implants, optimizing product designs, and providing an insight in clinical outcomes. While it is strictly a method for analysis it could play a significant role in orthopedics when combined with other virtual tools. It opens door for research where biomechanical material properties are not known, and provide a platform for virtual testing a number of possibilities that experiments and patient follow-up can't provide. FEA requires basic knowledge of mechanics beyond what's taught or known to clinicians. This challenge can be overcome through training and collaboration with engineering. The future holds that patient-specific FE models are readily available for patient-specific needs and prosthesis design and prototyping.

References

1. Rho JY, Hobatho MC, Ashman RB. Relations of mechanical properties to density and CT numbers in human bone. Med Eng Phys. 1995;17:347–55. https://doi.org/10.1016/1350-4533(95)97314-f.
2. Garner BA, Pandy MG. Musculoskeletal model of the upper limb based on the visible human male dataset. Comput Methods Biomech Biomed Engin. 2001;4:93–126. https://doi.org/10.1080/10255840008908000.
3. Radin EL. Biomechanics of the human hip. Clin Orthop Relat Res. 1980:28–34.
4. Markolf KL, Amstutz HC, Hirschowitz DL. The effect of calcar contact on femoral component micro-movement. A mechanical study. J Bone Joint Surg Am. 1980;62:1315–23.
5. Gibbons CE, Davies AJ, Amis AA, Olearnik H, Parker BC, Scott JE. Periprosthetic bone mineral density changes with femoral components of differing design philosophy. Int Orthop. 2001;25:89–92. https://doi.org/10.1007/s002640100246.
6. Herrera A, Panisello JJ, Ibarz E, Cegoñino J, Puértolas JA, Gracia L. Comparison between DEXA and finite element studies in the long-term bone remodeling of an anatomical femoral stem. J Biomech Eng. 2009;131:041013. https://doi.org/10.1115/1.3072888.
7. Herrera A, Ibarz E, Cegoñino J, Lobo-Escolar A, Puértolas S, López E, Mateo J, Gracia L. Applications of finite element simulation in orthopedic and trauma surgery. World J Orthop. 2012;3:25–41. https://doi.org/10.5312/wjo.v3.i4.25.
8. Guan Y, Yoganandan N, Zhang J, Pintar FA, Cusick JF, Wolfla CE, Maiman DJ. Validation of a clinical finite element model of the human lumbosacral spine. Med Biol Eng Comput. 2006;44:633–41. https://doi.org/10.1007/s11517-006-0066-9.
9. Little JP, Adam CJ, Evans JH, Pettet GJ, Pearcy MJ. Nonlinear finite element analysis of anular lesions in the L4/5 intervertebral disc. J Biomech. 2007;40:2744–51. https://doi.org/10.1016/j.jbiomech.2007.01.007.
10. Hassan CR, Qin YX, Komatsu DE, Uddin SMZ. Utilization of finite element analysis for articular cartilage tissue engineering. Materials (Basel). 2019:12. https://doi.org/10.3390/ma12203331.
11. Appelman TP, Mizrahi J, Seliktar D. A finite element model of cell-matrix interactions to study the differential effect of scaffold composition on chondrogenic response to mechanical stimulation. J Biomech Eng. 2011;133:041010. https://doi.org/10.1115/1.4003314.
12. Kang H, Lin CY, Hollister SJ. Topology optimization of three dimensional tissue engineering scaffold architectures for prescribed bulk modulus and diffusivity. Struct Multidiscipl Optim. 2010;42:633–44. https://doi.org/10.1007/s00158-010-0508-8.
13. Zheng M, Zou Z, Bartolo PJ, Peach C, Ren L. Finite element models of the human shoulder complex: a review of their clinical implications and modelling techniques. Int J Numer Method Biomed Eng. 2017;33 https://doi.org/10.1002/cnm.2777.
14. Haering D, Raison M, Begon M. Measurement and description of three-dimensional shoulder range of motion with degrees of freedom interactions. J Biomech Eng. 2014;136 https://doi.org/10.1115/1.4027665.
15. Walia P, Miniaci A, Jones MH, Fening SD. Theoretical model of the effect of combined glenohumeral bone defects on anterior shoulder instability: a finite element approach. J Orthop Res. 2013;31:601–7. https://doi.org/10.1002/jor.22267.
16. Kazemi M, Dabiri Y, Li LP. Recent advances in computational mechanics of the human knee joint. Comput Math Methods Med. 2013;2013:718423. https://doi.org/10.1155/2013/718423.
17. Walker PS, Hajek JV. The load-bearing area in the knee joint. J Biomech. 1972;5:581–9. https://doi.org/10.1016/0021-9290(72)90030-9.

Part III

Shoulder Biomechanics

10 Anatomy and Kinematics of the Shoulder Joint

Alfonso Ricardo Barnechea Rey

10.1 Introduction

The shoulder joint is the most mobile joint of the entire body. This mobility is achieved due to the lack of bony restraints and dependence on soft-tissue static and dynamic stabilizers.

The "shoulder joint" is, in itself, comprised by four "joints" and a space worth mentioning. Three of them are proper joints: glenohumeral, acromioclavicular, and sternoclavicular. On the other hand, the fourth "joint," not a joint in itself, is the scapulothoracic interface. The subacromial space, a gap between bones filled with soft tissues, completes this list.

The shoulder girdle, formed by the scapula and clavicle, connect the shoulder and upper limb to the rest of the body: the spine and thorax being the axis on which the shoulder supports itself in order to perform its main function, which is to adequately and precisely position the upper limb, and ultimately the hand, in space.

A. R. Barnechea Rey (✉)
Upper Limb Unit, Department of Orthopedics and Traumatology, Edgardo Rebagliati Martins National Hospital, Lima, Peru

10.2 Shoulder Anatomy

10.2.1 Scapula and Glenohumeral Joint

The scapula is a somewhat flat, triangular-shaped bone located posterolateral to the rib cage [1], with a "resting" orientation of 30 degrees of anteversion [2], 3–10 degrees of abduction [3] and 10–20 degrees of anterior tilt [4]. It is located vertically from the second (superior angle) to the seventh to ninth rib (inferior angle). It has three borders and three apices or angles, with a posterior ridge that goes from the spinal trigone at the medial border to the superolateral angle called the spine of the scapula; its lateral end bends an average of 78 degrees anteriorly [5] forming the acromion. The spine divides the posterior surface of the scapula into two fossae: the supraspinatus (origin of the supraspinatus muscle) and infraspinatus (origin of the infraspinatus and teres minor muscles) fossae (Fig. 10.1). Its slightly concave anterior surface, the subscapular fossa, holds the origin of the subscapularis muscle and anterior to it the scapulothoracic interface. On its lateral side, the coracoid process emerges anteriorly, superiorly and slightly laterally; medial to its origin it forms with the body of the scapula the supraglenoid notch (Fig. 10.2), covered by the superior transverse scapular ligament and traversed by the suprascapular nerve.

J. Koh et al. (eds.), *Orthopaedic Biomechanics in Sports Medicine*,
https://doi.org/10.1007/978-3-030-81549-3_10

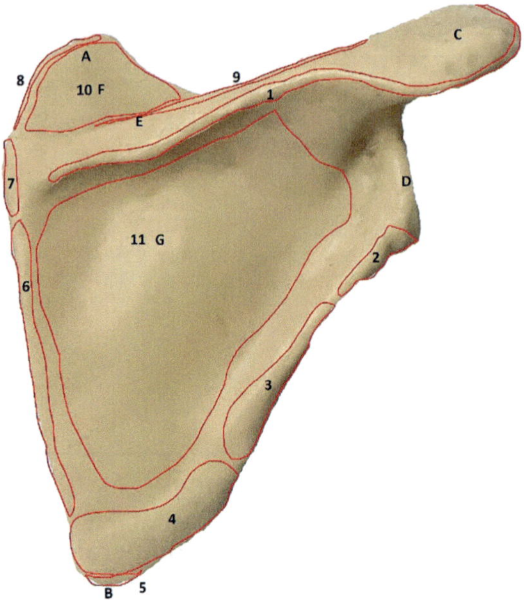

Landmarks	Muscle insertions/origins
A: Superior angle	1: Deltoid
B: Inferior angle	2: Triceps, long head
C: Acromion	3: Teres minor
D: Glenoid	4: Teres major
E: Spine	5: Latissimus dorsi
F: Supraspinatus fossa	6: Rhomboid major
G: Infraspinatus fossa	7: Rhomboid minor
	8: Levator scapulae
	9: Trapezius
	10: Supraspinatus
	11: Infraspinatus

Fig. 10.1 Osseous landmarks and muscular insertions of the posterior scapula

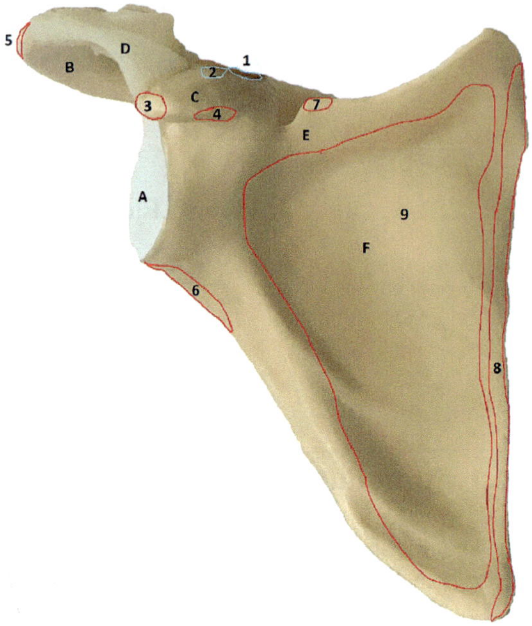

Landmarks	Muscle/ligament insertions/origins
A: Glenoid fossa	1: Conoid ligament
B: Acromion process	2: Trapezoid ligament
C: Coracoid process	3: Conjoined tendon
D: Coracoacromial ligament	4: Pectoralis minor
E: Spinoglenoid notch	5: Deltoid
F: Subscapularis fossa	6: Long head of triceps
	7: Omohyoid
	8: Serratus anterior
	9: Subscapularis

Fig. 10.2 Osseous landmarks and muscular/ligamentous insertions of the anterior scapular face

From its lateral end, the scapula projects the glenoid process (or simply "glenoid"). It arises from the body of the scapula and extends laterally, superiorly and slightly posteriorly and has a truncated "cone" shape (the articular surface constituting the base). We should remember that the scapular main transverse axis is oriented with 30/40 degrees of anteversion related to the main body axis. The glenoid surface typically has an average of 1.4–3.2 degrees of retroversion [6] related to, but it is generally considered perpendicular to, the scapular axis. The glenoid has a neck that measures approximately 2 cm from its base at the junction of the spine and the coracoid and presents at its posteromedial side a notch for the passage of the suprascapular nerve, called the spinoglenoid notch (Fig. 10.2), which in turn is covered by the inferior transverse scapular ligament (and traversed by the suprascapular artery and nerve). The glenoid widens from medial to lateral, ending on its lateral side (the "base" of the cone) as a concave, pear-shaped fossa; if we think of it as the intersection of two "circles," the superior one would have a smaller diameter than the inferior (which is considered by many to be a true circle [7]) and the center of the latter being slightly more anterior (Fig. 10.3). This fossa, covered in cartilage, has a radius of curvature 2.3 ± 0.2 mm longer than that of the humeral head (24 ± 2.1 mm) [8], but its surface is about 3–4 times smaller than the latter, resembling a golf ball over a tee (Fig. 10.4).

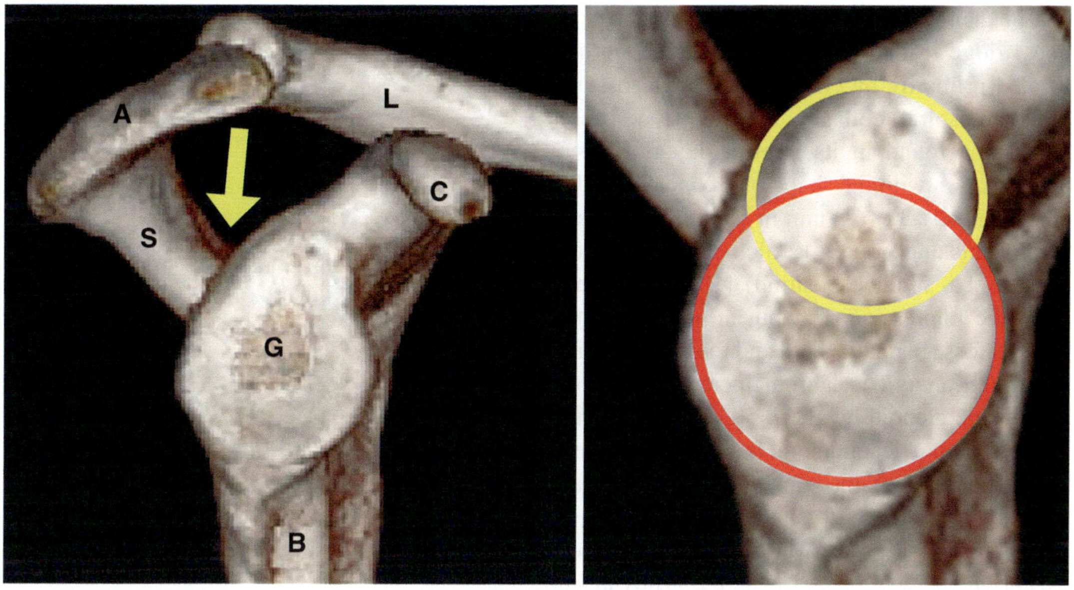

Fig. 10.3 Left: 3D-CT scan of right scapula. Note the pear-shaped glenoid surface. G: Glenoid. A: Acromion. S: Scapular spine. C: Coracoid process. L: Clavicle. B: Scapular body. Yellow arrow: spinoglenoid notch. Right: The pear-shaped glenoid can be seen as the combination of two intersecting circles, the inferior one being larger and more anterior than the superior one

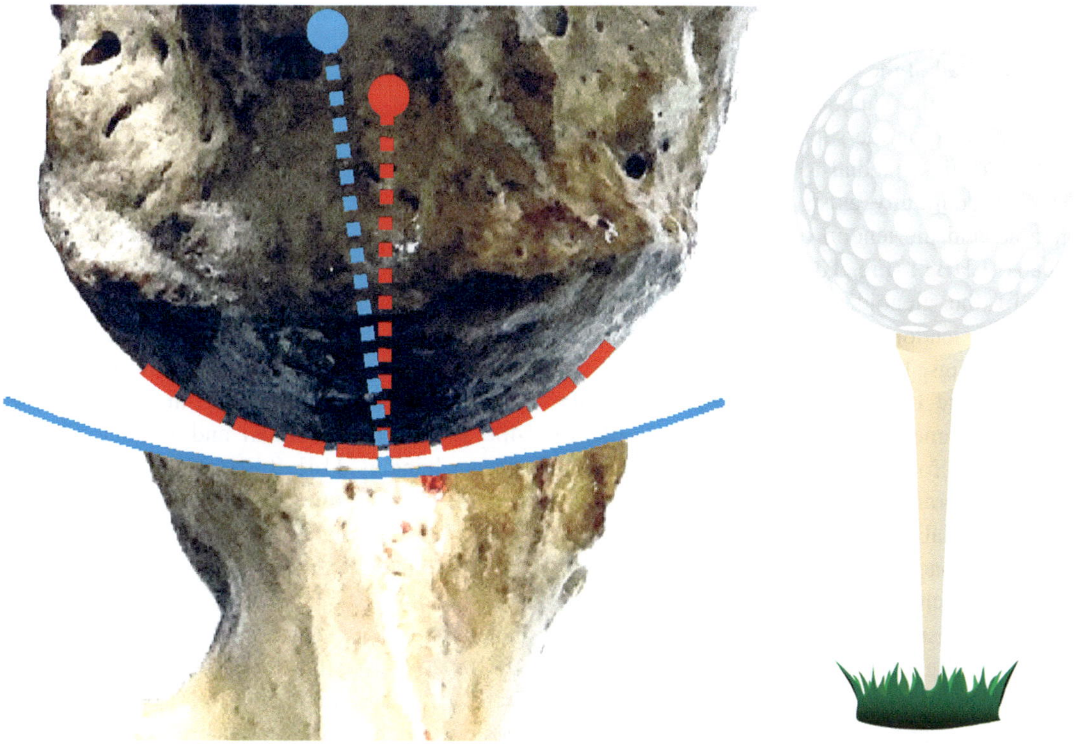

Fig. 10.4 Left: Comparison of radii of curvature of the humeral head (red) and glenoid (blue). Note how the glenoid's radius is larger than the humeral head's due to its being flatter. Right: The relationship of the sizes of the articular surfaces and shapes of the humeral head and glenoid approaches that of a golf ball over a tee (these having the same radii of curvature instead)

The glenoid fossa is circled by a fibrocartilaginous ridge called the labrum. This labrum, formed primarily by circumferential (also some radial and spiral) collagen fibers [9] (type II collagen was found in its internal layers), has a cross-section with the shape of a concave wedge approximately 3.8–6.3 mm in both thickness and height (it is thicker at 9–12 o'clock positions and thinner at 3 o'clock position) [10]. This shape gives the impression of being in a stadium; the glenoid fossa being the field and the labrum being the bench area. The labrum increases the total area of the glenoid by about 50%, adding to the stability of the glenohumeral joint. This labrum is firmly attached to the glenoid bone, including Sharpey-type fibers [9] (there is up to 4 mm of overlap of labral tissue over the glenoid surface) except on its superior part, where a synovial lining covers a recess between the posterosuperior labrum and its corresponding glenoid ridge at about 6 mm off the glenoid surface, allowing some mobility to enhance superior joint stability at elevation of the shoulder [11]. This in part is also possible by the insertion of the tendon for the long head of the biceps (LHBT), which insets partially on the superior aspect of the labrum (and partially on the supraglenoid fossa, a bony depression just medial to the superior labrum). Thus, a continuity exists between the superior glenoid and the tendon of the long head of the biceps (LHBT) (Fig. 10.5). Also, at the inferior part of the glenoid, the infraglenoid fossa is where the tendon of the long head of the triceps attaches (Figs. 10.1 and 10.2).

There are two main normal variants of the labral attachment on the anterosuperior border of the glenoid fossa [12]: the first one, called the sublabral foramen, consists of a lack of attachment of the labrum in this area, and has the aspect of an oval space. The second one, called Buford's complex, is characterized by a complete absence of the anterosuperior labrum and the middle glenohumeral ligament appears to bridge the gap in a chord-like fashion (Fig. 10.6).

The glenohumeral capsule is a serous membrane that seals the glenohumeral joint; its area is twice that of the humeral head and usually has a volume of 10–15 cm^3 [13]. It anchors itself

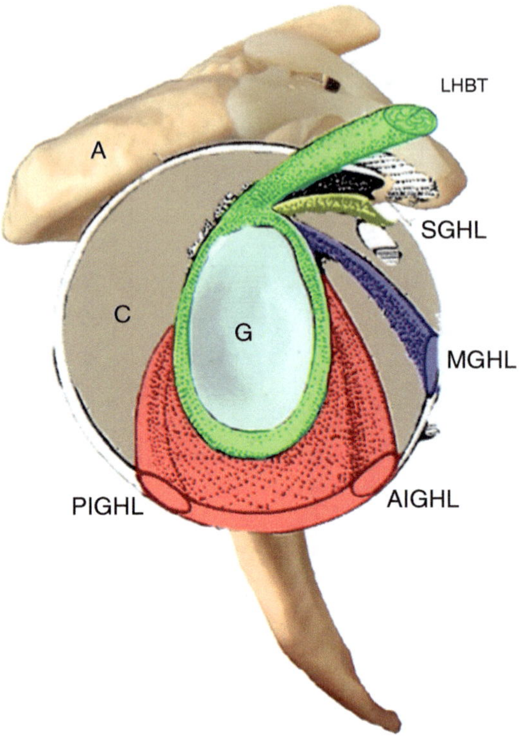

Fig. 10.5 Insertions around the glenoid. C: capsule; G: glenoid; A: acromion; LHBT: Tendon of the long head of the biceps SGHL: Superior glenohumeral ligament; MGHL: middle glenohumeral ligament; AIGHL: anterior band of the inferior glenohumeral ligament; PIGHL: posterior band of the inferior glenohumeral ligament

medially on the glenoid neck, with fibrous attachments to the labrum, and laterally on the lateral edge of the articular surface of the humeral head. This capsule, histologically comprised by type I collagen, has three recognizable fibrous layers: External (fascia and tendons), intermediate (100–400 μm fibrous bundles) and internal (20 μm fibrous bundles) [14]. It is stronger in its anterior aspect, mainly due to the presence of three "folds" or capsular ligaments that cross it from medial to lateral (Fig. 10.5). The first is the superior glenohumeral ligament (SGLH), and it originates from to the supraglenoid tubercle; it covers the superior aspect of the rotator interval (which will be discussed later) and part of its fibers go posterior to the LHBT and laterally to its humeral insertion. The second one is the middle glenohumeral ligament (MGHL), which originates from the anterior labrum and is oriented inferiorly and

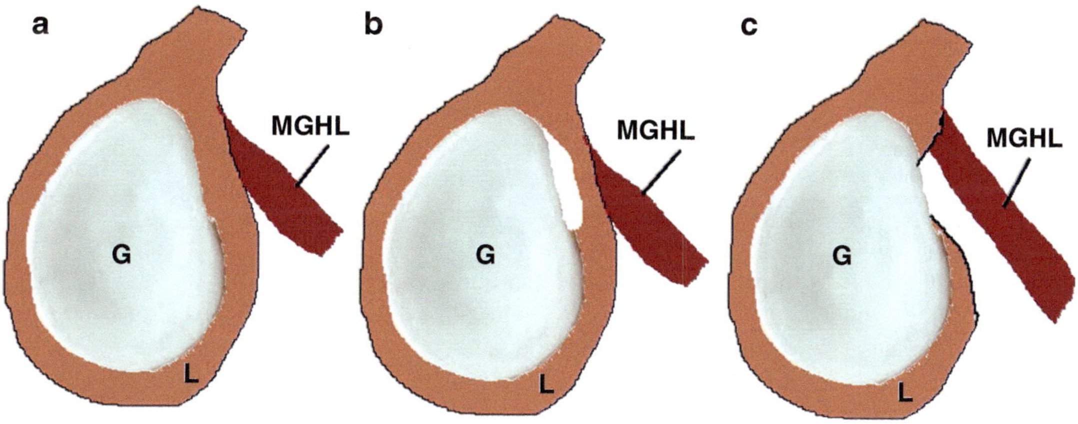

Fig. 10.6 **a:** Normal anterosuperior labrum and middle glenohumeral ligament (MGHL). **b:** Sublabral foramen. Anterosuperior glenoid rim is bare. **c:** Buford's complex. Absence of anterosuperior labrum and chord-like MGHL

laterally and can be seen next to the anterosuperior margin of the labrum. The third one, the inferior glenohumeral ligament (IGHL), thicker that the other two, starts medially on the inferior glenoid and has two bands: the anterior band (AIGHL) originates from the glenoid and labrum at the 3-to-5 o'clock position and the posterior band (PIGHL) starts from the 7-to-9 o'clock position in the glenoid and labrum. Both bands reach the humeral surgical neck and can be found at the capsular recess (a lax section of the capsule in its inferior aspect), one slightly anteriorly and the other one slightly posteriorly, the posterior band being less conspicuous (Fig. 10.7); Their disposition resembles and functions as a hammock that supports the head across the range of motion [15]. The IGHL receives a branch from the axillary nerve, which gives this ligament a proprioceptive function [16]. Also, the transverse humeral ligament (THL) originates from the capsule and makes the "roof" of the more proximal part of the bicipital groove.

The capsule is covered by synovium on its interior face and covers the interior aspect of the rotator cuff (the latter being extracapsular). The coracohumeral ligament enters through the rotator interval and merges partially with the superior face of the capsule; its superficial layer covers the articular surface of the supraspinatus and the deep layer attaches to the greater tuberosity [17]. Also, the capsule creates a fibrous band around the LHBT as it exits the intra-articular space, called the biceps pulley. Being completely sealed, the capsule, along with the labrum, have a negative pressure that creates a suction effect or the humeral head, thus adding to the stability of the glenohumeral joint.

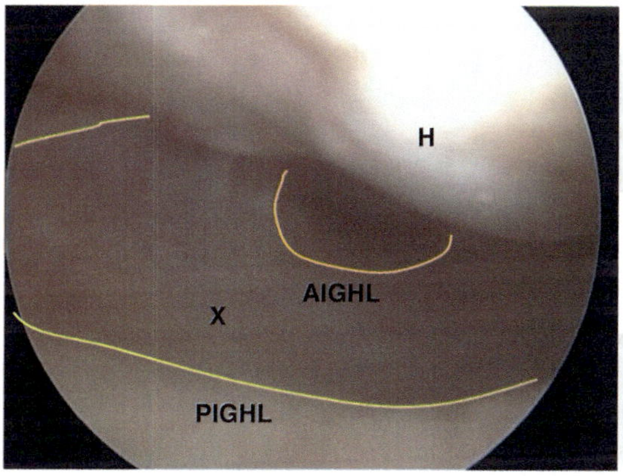

Fig. 10.7 Arthroscopic view of the anterior (AIGHL, orange) and posterior (PIGHL, yellow) bands of the inferior glenohumeral ligament. H: Humeral head. X: Axillary recess

On the lateral side of the joint, the humeral head is almost hemispherical, oriented at an average of 130–140 degrees superomedially and with 20–30 degrees of retroversion (Fig. 10.8); the sphere center has a medial offset of 6.9 mm and a posterior offset of 2.6 mm [18]. Its articular surface averages 20–30 cm^2 [19]. This hemisphere's cartilage cover is thickest in its central portion; but only 25% of its surface is in contact with the

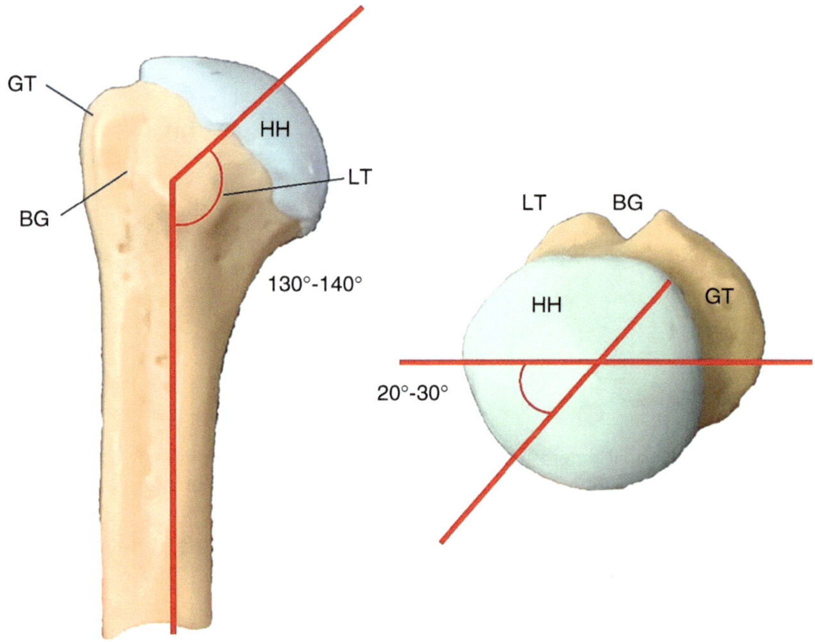

Fig. 10.8 Anatomic landmarks of the proximal humerus, inclination (related to the humeral longitudinal axis) and retroversion (related to the distal humeral transepicondylar axis) of the humeral neck. *HH* humeral head, *BG* bicipital groove, *LT* lesser tuberosity, *GT* greater tuberosity

glenoid at any point in time. Its upper border is located 5–8 mm from the upper edge of the greater tuberosity.

The greater tuberosity is a bony prominence that lies at the lateral side of the proximal humerus (Fig. 10.8); it holds the insertions of three of the rotator cuff tendons: the supraspinatus, more anterosuperiorly, the infraspinatus, postero-superiorly (which has the largest area of insertion) and the teres minor, posteroinferiorly. This area of insertion or footprint has an uneven distribution (Fig. 10.9) [20]. These three tendons' attachments are not separate; some fibers interdigitate with the adjacent ones forming a somewhat common insertion [21]. The area where these tendons insert is commonly referred to as the footprint. There is a bare area on the postero-superior aspect of the greater tuberosity that can be identified arthroscopically. The superior capsule attaches slightly medially to the cuff footprint, adjacent to it, in an area measuring 3 to 9 mm in width and spans across the greater tuberosity (Fig. 10.10) [22].

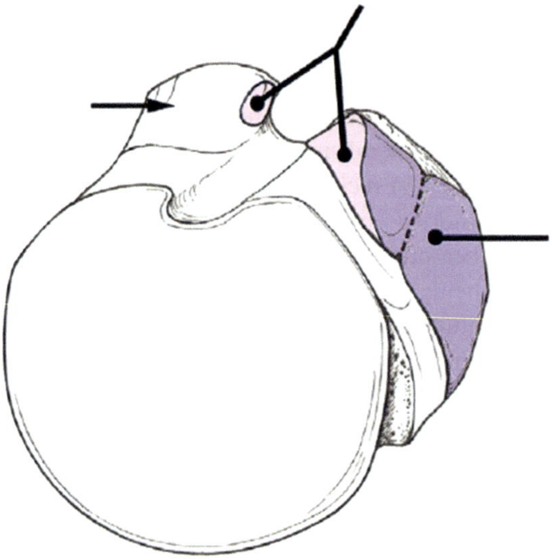

Fig. 10.9 Footprint distribution for the rotator cuff tendons in the greater tuberosity according to Mochizuki et al. [20] (reprinted with permission)

The lesser tuberosity is a bony prominence located medial to the greater tuberosity and adjacent to the anterior edge of the humeral head (Fig. 10.8). It holds the insertion of the subscapularis tendon. There is a space between both tuberosities called the bicipital groove; it is located 30 degrees medial from the sagittal plane of the humerus and 9 mm anterior to the bone's longitudinal axis (Fig. 10.8). Its dimensions are highly variable (length 3–8 cm, width 8–9 mm, depth

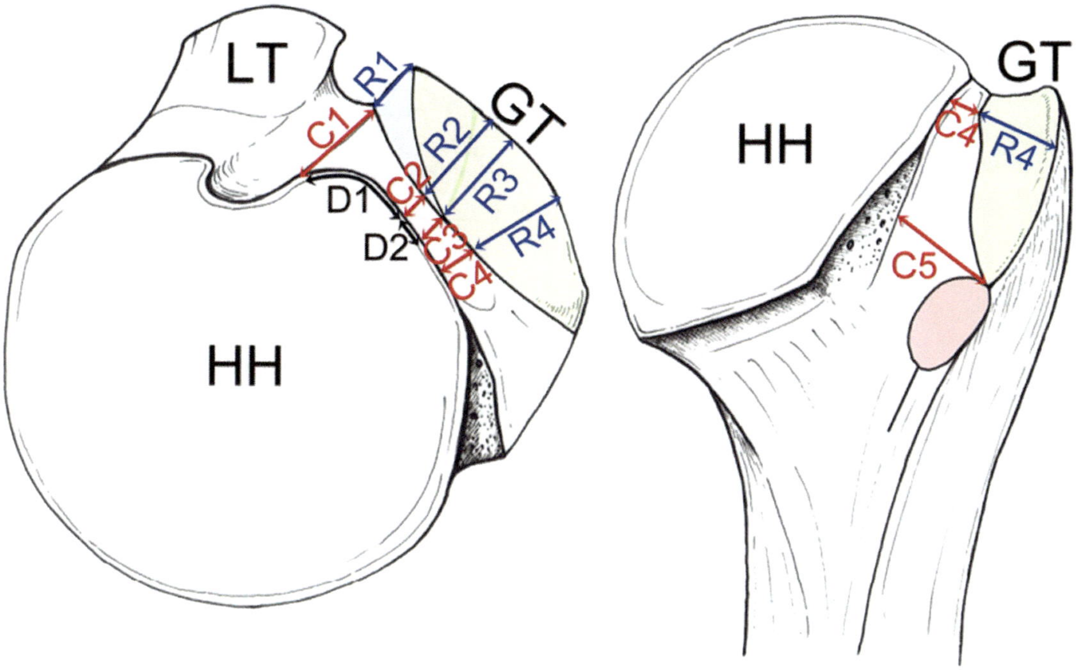

Fig. 10.10 Capsular insertion at the greater tuberosity according to Nimura et al. [22] (reprinted with permission)

4–7 mm) [23–25], and it contains the LHBT, covered by the THL. Some fibers coming from the subscapularis tendon cover its bony surface.

The humeral head receives its nutrition from both the anterior and the posterior circumflex arteries. Of note, the posterior circumflex artery enters the calcar (the stronger posteromedial area of the humeral metaphysis) within 5–8 mm inferior to the edge of the humeral head; it gives the humeral head about 64% of its blood supply [26] and includes the greater tuberosity and the head's subchondral bone. The anterior circumflex artery, mostly through its branch, the arcuate artery, supplies the lesser tuberosity and the bicipital groove [23]. Both arteries supply intraosseous branches with its terminal ones located at the subchondral head bone.

10.2.2 The Rotator Cuff

The rotator cuff is a combination of four muscles and their inserting tendons. Apart from specific movements, the cuff becomes the most important dynamic stabilizer of the glenohumeral joint.

The supraspinatus originates from the supraspinatus fossa, located on the upper third of the posterior face off the scapula. Its fibers run laterally and its tendon inserts in the anterior 1.5–2 cm of the superior aspect of the greater tuberosity (Fig. 10.11). The musculotendinous junction, located at an average of 3–4 cm from its insertion, has a mixed transition: 70% of the tendinous fibers come from intramuscular tendon and 30% from extramuscular tendon [27].

The infraspinatus muscle originates from the upper two thirds of the infraspinatus fossa on the posterior face of the scapula; its tendon originates intramuscularly and centrally and reaches its insertion on the posterosuperior aspect of the greater tuberosity (Fig. 10.11). It accounts for 60% of the external rotational force [28]. The teres minor originates on the lower third of the infraspinatus fossa and its circumpennate single tendon attaches to the posteroinferior aspect of the greater tuberosity [1]; it generates about 45% of the external rotational force [28].

Both the supraspinatus and infraspinatus have a layered composition (Fig. 10.12) [21]: Layer 1, superiorly; 1 mm thick, composed of fibers from

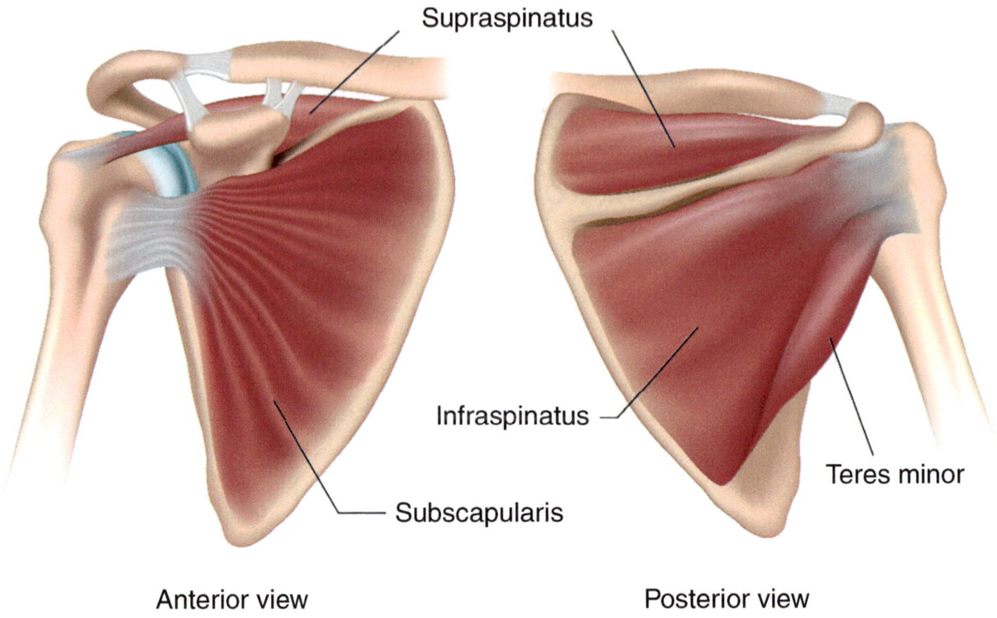

Fig. 10.11 Rotator cuff muscles and tendons (modified from Alila Medical Media/ Shutterstock.com)

the CHL. Layer 2, 3–5 mm thick, composed of parallel tendon fibers. Layer 3, 3 mm thick, composed of intercrossing fibers (at about 45 degrees); layer 4, composed of loose connective tissue and merging with the CHL at the anterior edge of the supraspinatus. Finally, layer 5, 1.5–2 mm thick, comprised by the superior capsule.

The subscapularis muscle originates from the entire anterior face of the scapula; its tendon passes under the coracoid and inserts at the lesser tuberosity (Fig. 10.11). This insertion is mixed: the upper two-thirds of it are tendinous and the lower third is entirely muscular [29]. It accounts for 50% of the dynamic stabilization force of the glenohumeral joint.

These muscles have a distinct innervation: Both the supraspinatus and the infraspinatus are innervated by the suprascapular nerve; the teres minor receives its own from the posterior branch of the axillary nerve, and the subscapularis from two branches of the subscapular nerve.

One can find a fibrous thickening lies at the lateral part of the tendons' insertions called the rotator cable [30]. This cordlike band spans from the most superolateral part of the bicipital groove all the way across the suprapinatus and to the inferior part of the infraspinatus' tendon, and originates from the deep layer of the CHL [21]. This cable seems to thicken over time and may be

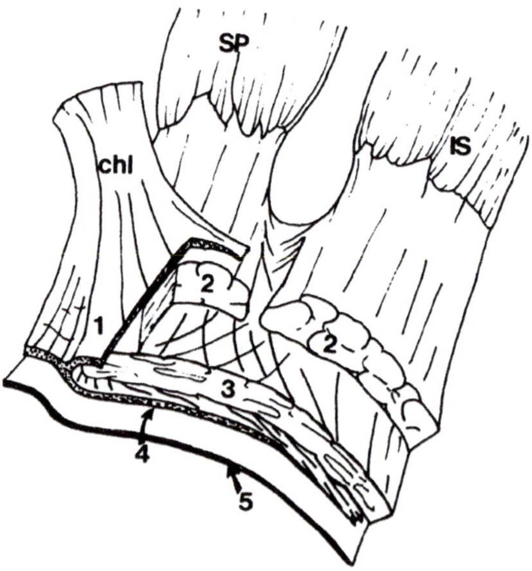

Fig. 10.12 The 5 histological layers of the supra/infraspinatus tendon [21]: 1: superficial fibers form the coracohumeral ligament (CHL); 2: parallel fibers; 3: oblique fibers of the supraspinatus and infraspinatus; 4: deep extension of the CHL; 5: articular capsule. *SP* supraspinatus, *IS* infraspinatus (reprinted with permission)

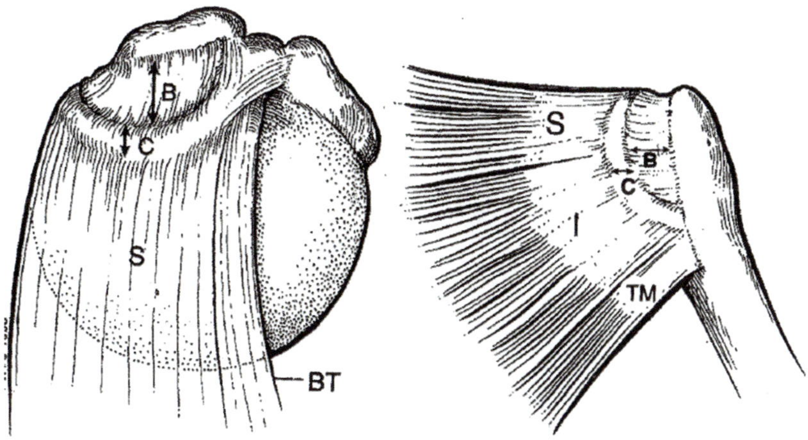

Fig. 10.13 Rotator cable (C) and rotator crescent (B) as described by Burkhart et al [30] *S* Supraspinatus tendon, *I* Infraspinatus tendon, *TM* Teres minor tendon (reprinted with permission)

related to ageing [30]. Lateral to this cable lies a crescent-shaped area called the rotator crescent, the most lateral part of the cuff tendons' insertion, which in turn seems to become thinner over the years (Fig. 10.13). The cable acts as a suspensory bridge that "holds" the lateral cuff and keeps stress away from the crescent, which is more prone to tear [21].

The gap between the supraspinatus and subscapularis tendons is known as the rotator interval. It is triangular-shaped, with its base located toward the base of the coracoid, and its tip at the anterior edge of the greater tuberosity. It is covered at the bursal side by the coracohumeral ligament, which starts to merge with the capsule, and at the articular side by fibers of the SGHL. The LHBT crosses this space from the articular side toward the bicipital groove; at this transition point, the groove portion of the CHL creates a sleeve around the LHBT and, by merging with the transverse humeral ligament (THL) at that level, creates the biceps pulley, which holds this portion of the LHBT in place as it enters the groove.

10.2.3 The Superior Shoulder Suspensory Complex (SSSC) [31]

The superior shoulder suspensory complex (SSSC) is a sort of ring-shaped rack connected to the thorax by the clavicle and scapulothoracic muscles. This ring is formed by the distal clavicle, the acromion, the glenoid process and the coracoid; these structures are connected by the acromioclavicular and coracoclavicular ligaments. This rack is connected to the trunk by two struts: the superior strut is formed by the medial clavicle and connects it to the sternum through the sternoclavicular joint, and the inferior strut is formed by the lateral scapular spine and connects to the trunk by the scapulothoracic muscles. (Fig. 10.14). In turn, this ring holds the humerus through the glenohumeral capsule and ligaments. This complex also allows for stable attachment for several soft-tissue static and dynamic glenohumeral stabilizers.

10.2.4 The Acromial Arch and Subacromial Space

The acromion process, a bony hook that originates as a lateral extension of the scapular spine, serves as an anchoring point for the deltoid and for the coracoacromial ligament. It runs from posterior to anterior and from inferior to superior, with a slight inferior concavity. It has two surfaces, superior (which is subcutaneous) and inferior; and a lateral, medial and anterior edges. The inferior surface is slightly concave and this concavity can vary: the resultant acromial shape is classified into three categories according to Bigliani [32]: Type I, which is a mainly flat inferior surface; type II which shows a slight inferior concavity; and type III, in which the concavity is more pronounced and the posteroinferior aspect shows a hooklike appearance.

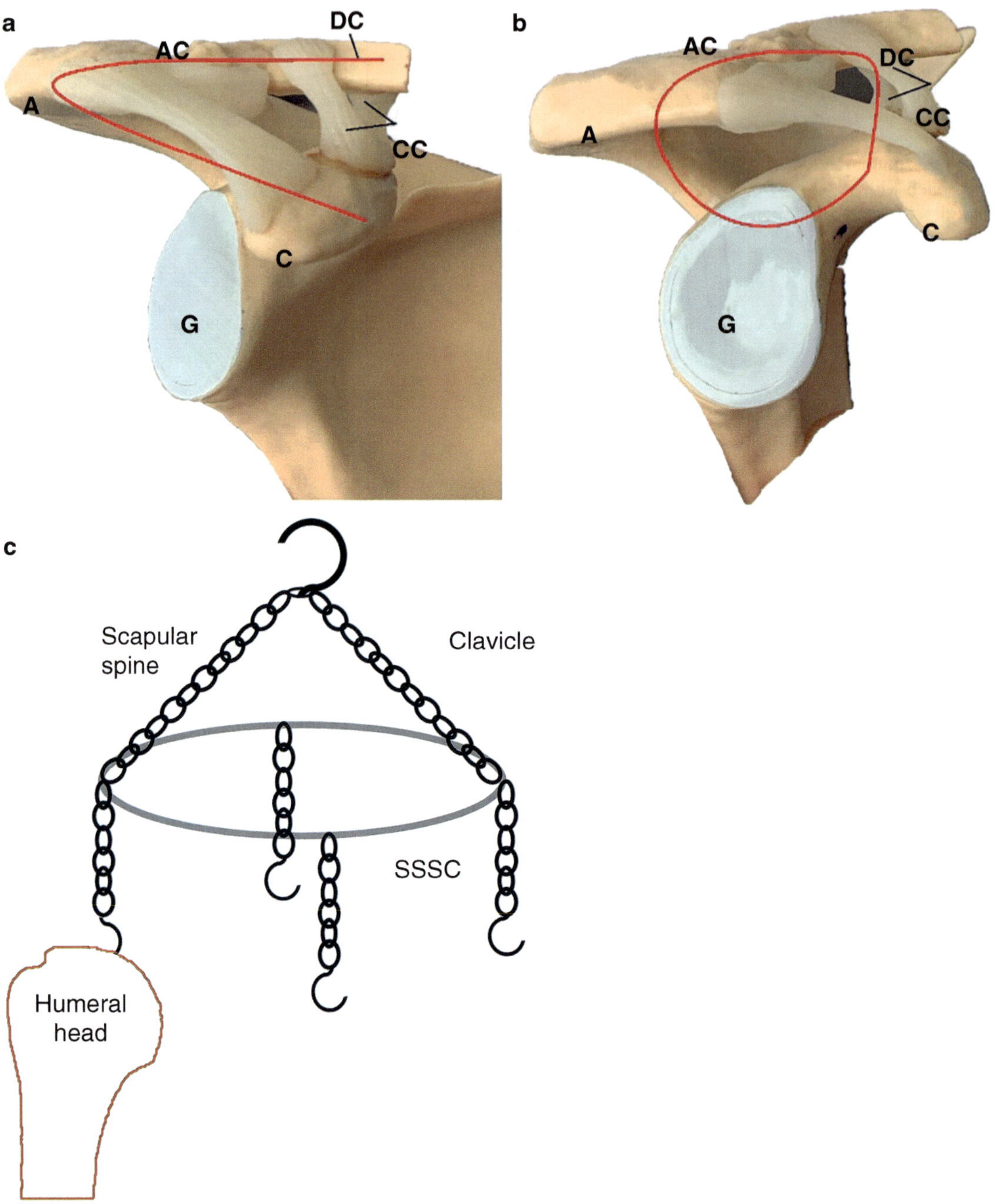

Fig. 10.14 Superior shoulder suspensory complex (SSSC), formed by the distal clavicle, acromion, glenoid, and coracoid; these structures are connected by the acromioclavicular and coracoclavicular ligaments, and the complex is connected to the trunk but the medial clavicle and scapular spine (as described by Goss et al. [31]). (**a**): Coronal view. (**b**): Sagittal view. (**c**): Representation of the SSSC as a ringed rack, connected to the trunk (ceiling) by the scapular spine and clavicle; it holds the humerus from the glenoid (hooks)

The middle portion of the deltoid anchors itself at the lateral border and the most lateral part of the anterior border if the acromion. This insertion comprises about 74% of the anterior acromial thickness and is about 5.4 mm thick [33]. On the other hand, the coracoacromial ligament

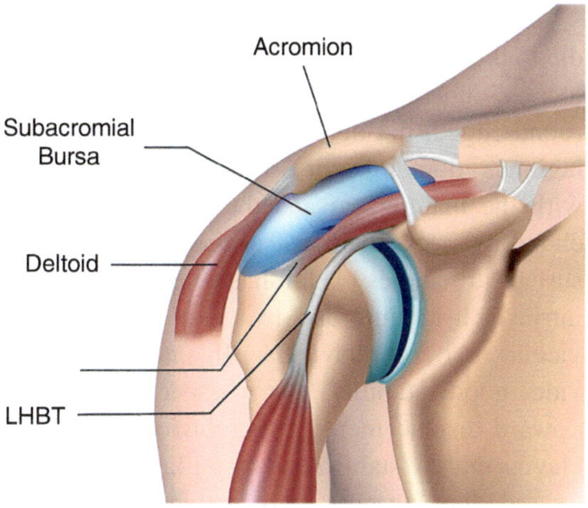

Fig. 10.15 Subacromial space and subacromiosubdetoid bursa. LHBT: long head of the biceps tendon (modified from Alila Medical Media/ Shutterstock.com)

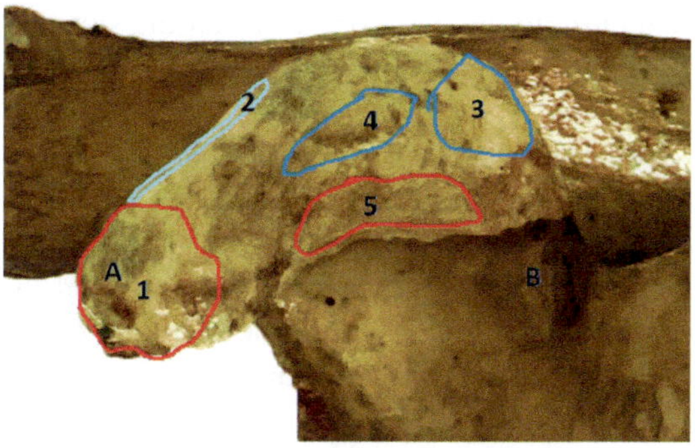

Landmarks	Muscle/Ligament insertions
A: Tip	1: Conjoined tendon
B: Base	2: Coracoacromial ligament
	3: Conoid ligament
	4: Trapezoid ligament
	5: Pectoralis minor

Fig. 10.16 Tendon and ligament insertions at the coracoid

runs from the anterolateral part of the coracoid to the medial undersurface and the most medial part of the anterior border of the acromion with an insertion area width of 7.3 mm [34]. It has two bands: the lateral (thicker) and medial bands [35]. At the acromioclavicular joint, the acromial posteromedial acromion adjacent to the joint has the greatest bone mineral density [36].

Both the acromion and the coracoacromial ligament form the coracoacromial arch, which limits upper translation of the humeral head. Under it lies a virtual space known as the subacromial space, which measures an average of 9.7 ± 1.5 mm [37], its inferior limit being the superior surface of the supraspinatus and infraspinatus tendons. This space is filled by the subacromial bursa, which covers not only the supraspinatus but most of the rotator cuff as well; sometimes it is named subacromio-subdeltoid bursa because of its extension (Fig. 10.15). Its function is to provide a sliding surface and a cushion for friction protection to the superior cuff.

10.2.5 The Coracoid and Surrounding Structures

The coracoid process is a bony hook that arises from the lateral part of the scapular body and its tip is oriented anterolaterally. A visible notch (the suprascapular notch) separates it from the scapular body; its roof is comprised by the superior transverse scapular ligament. Laterally the space between the inferolateral coracoid and the glenoid can have the following shapes: Type I, rounded (45%), type II, square-angled (34%), and type III, hooked (21%) [38]. Its average dimensions [39] are: length 4.26 ± 0.26 cm; width and height at the tip 2.11 ± 0.2 and 1.49 ± 0.12 cm. Its base lies about 1.1 to 1.3 cm distal to the inferior clavicle.

The coracoid has multiple structures attached (Fig. 10.16): on its superior face, close to the base, the two coracoclavicular ligaments: the trapezoid ligament, more laterally (its anterior edge at about 3.33 cm average from the tip of the coracoid), and the conoid ligament, more medially. On its lateral face, the two bands (lateral and medial) of the coracoacromial ligament more anteriorly, and the coracohumeral ligament, more posteriorly, can be appreciated. On its medial face one can find the insertion of the pectoralis minor tendon (distance from the tip to the anterior margin of pectoralis minor 0.1 cm and to the posterior margin 1.59 cm) [39] and, on its tip, the insertion of the conjoined tendon (formed by the tendons of the coracobrachialis and of the short head of the biceps). Adjacent to its inferior face and close to the base, one can find the subcoracoid bursa and the subscapularis tendon (Fig. 10.11).

The anteromedial margin of the coracoid is close to many major neurovascular structures [40]: the closest one to the tip is the lateral chord of the brachial plexus (28.5 ± 4.4 mm) and the closest ones to the base are the axillary (29.3 ± 5.6 mm) and the musculocutaneous nerve (36.5 ± 6.1 mm).

10.2.6 The Acromioclavicular Joint

This synovial, diarthrodial joint has two surfaces: the acromial, convex surface and the clavicular, concave surface, although this may vary. Frequently but not always, a wedge-shaped fibrocartilage disc (also called the meniscus) fills the articular space between the two [23]. The joint has an axial inclination of about 51 degrees and a coronal inclination of approximately 12 degrees. The joint has a thin, fibrous capsule that spans from an average of 2.8 mm lateral to the acromial surface to about 3.5 mm medial to the clavicular surface [34]; with four ligaments, of which the superior one is the strongest. The anterior, posterior, superior and inferior ligaments have a capsuloligamentous insertion at an average of 6.4, 6.3, 6.6 and 5.4 mm medial to the clavicular surface, and 5.6, 4.3, 5.3 and 4.0 mm lateral to the acromial surface. The trapezoid ligament is located at an average of 14 mm from the distal clavicular edge, and the conoid at an average of 32.1 mm from it [34].

10.2.7 The Clavicle and Sternoclavicular Joint

The clavicle is an S-shaped bone that connects the scapula and upper limb to the sternum. Its cross-section varies from the wide and flat distal end to the more tubular shaft and then to a thick, somewhat flatter proximal end. It holds many important muscular insertions for movement and stability of the shoulder girdle (Fig. 10.17).

The only joint that connects the shoulder girdle to the axial body is the sternoclavicular joint, located subcutaneously and saddle-shaped. The round-edged, flat distal clavicle has a fibrocartilaginous dorsoinferior surface that articulates with the manubrium and the second rib's synchondrosis; less than 50% of the distal clavicular surface articulates [41]. It has a thin capsule with intrinsic ligaments and two extrinsic (interclavicular and costoclavicular) ligaments that provide good stability (they especially restrict upward displacement of the clavicle). The subclavius muscle stabilizes it by pulling the clavicle toward the sternum and acting as a shock absorber; the sternocleidomastoid and pectoralis major's aponeuroses, along with the clavipectoral fascia, envelop the joint, adding to its stability. Also, a fibrocartilaginous disc lies within the joint.

10.2.8 The Scapulothoracic Interface

This interface between the scapular body and the rib cage does not have a capsule; six bursae [42] lie between the scapula and scapulothoracic muscles and the rib cage. The two major ones are the supraserratus (between the serratus anterior and the subscapularis) and the infraserratus (between the serratus anterior and the rib cage, Fig. 10.18) [62]; the other four minor or adventitial bursae are infrequent and respond to pathologic conditions [42].

10.2.9 External Glenohumeral and Scapulothoracic Muscles

Deltoid: The deltoid is a powerful muscle that originates at the lateral border of the clavicle anteriorly, the superolateral border of the acromion laterally and the lateral portion of the scapular spine posteriorly. There are three recognizable "portions" of the deltoid, the anterior, middle and posterior thirds (one can appreciate a soft fascial division between anterior and middle thirds). On the other hand, Sakoma et al. described up to

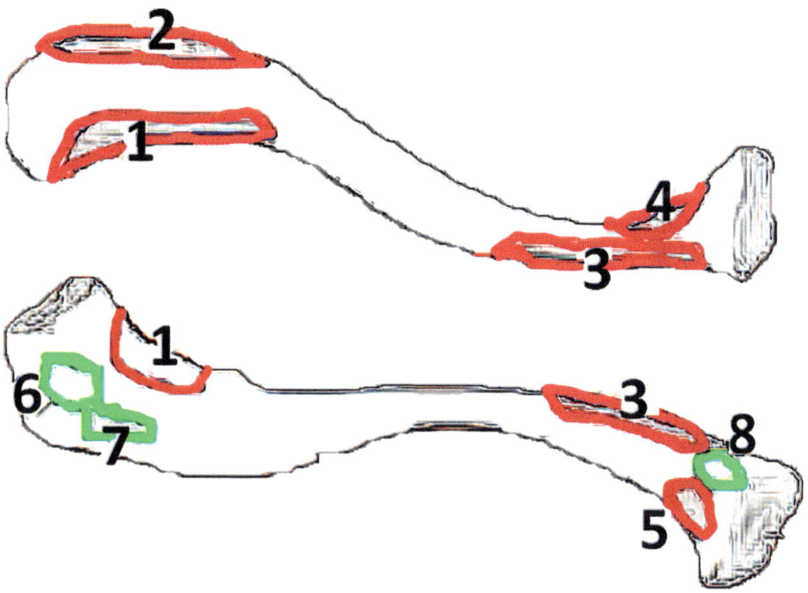

1: Deltoid	
2: Trapezius	
3: Pectoralis major	
4: Sternocleidomastoid	
5: Sternohyoid	
6: Trapezoid ligament	
7: Conoid ligament	
8: Costoclavicular ligament	

Fig. 10.17 Muscular and ligamentous insertions around the clavicle

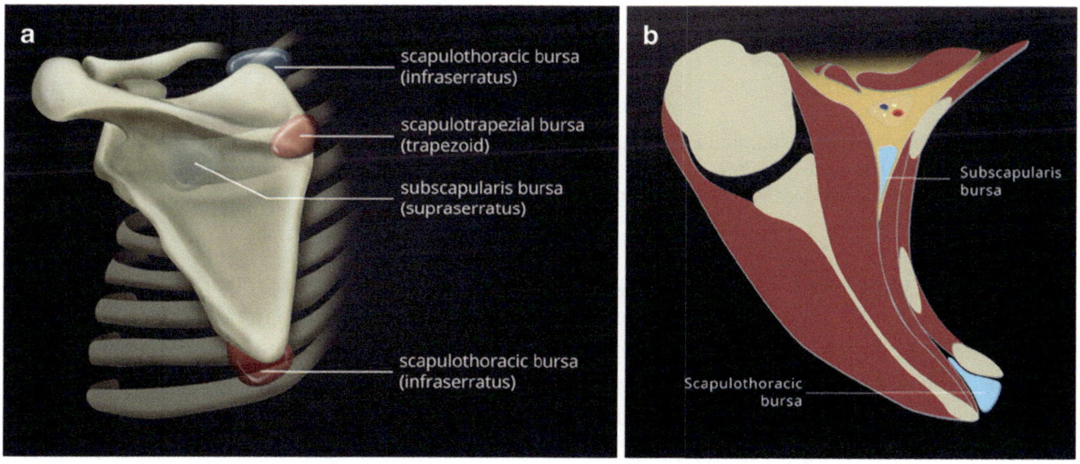

Fig. 10.18 Scapular Bursae. (**a**) Posterior and (**b**) axial illustrations of the scapula show the locations of the adjacent bursae (reproduced with permission) [62]

seven functional units (three posterior, one middle and three anterior units) (Fig. 10.19) [43]. These portions merge together into a common tendon that inserts at the deltoid tubercle, located laterally in the humeral shaft. The muscle receives its innervation from the anterior branch of the axillary nerve, which runs from anterior to posterior through the quadrilateral space (formed by

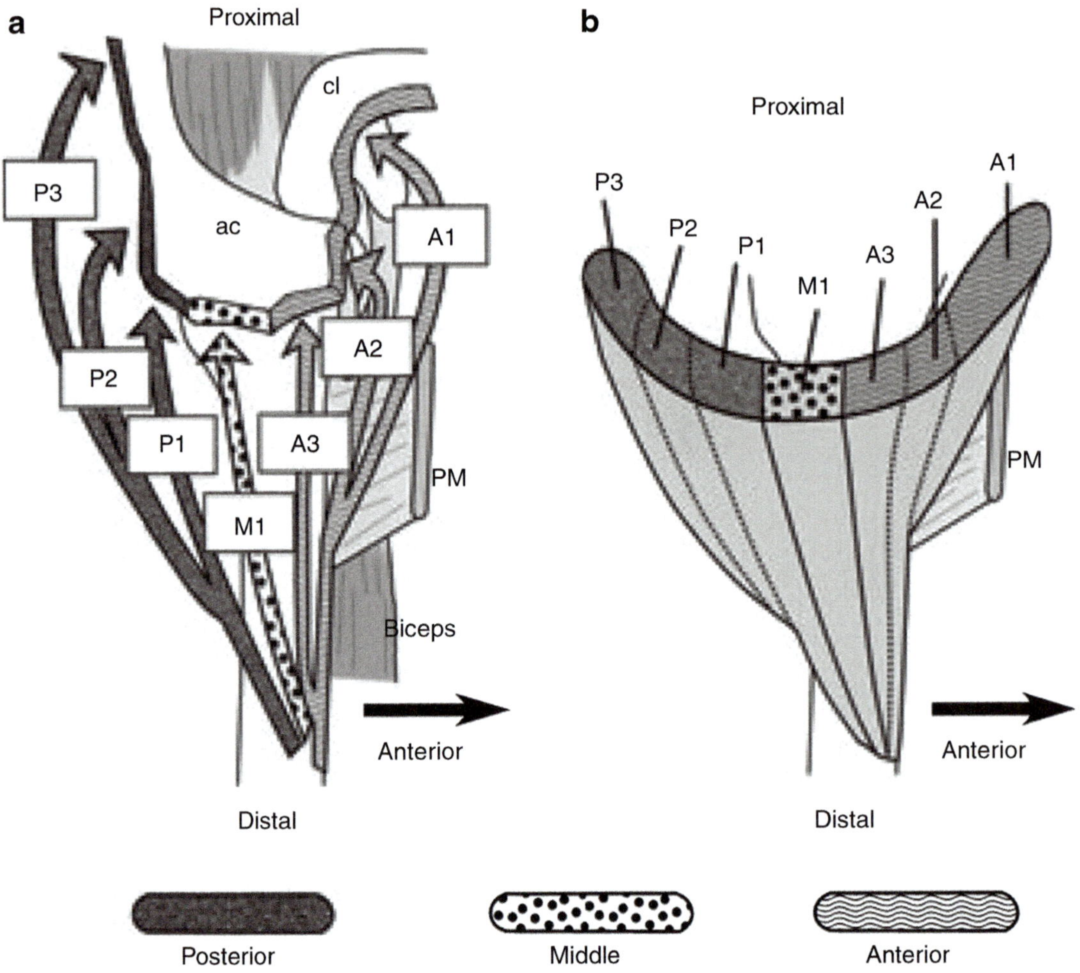

Fig. 10.19 (**a**) Directions of the intramuscular tendons of the deltoid, and (**b**) Division of deltoid segments: three anterior (A1 to A3), one middle (M1) and three posterior (P1 to P3) according to Sakoma et al. (reproduced with permission) [43]

the humeral surgical neck laterally, the long head of the triceps medially, the teres minor superiorly and the teres major inferiorly; the nerve runs along the posterior circumflex artery, which also gives irrigation to the deltoid) and divides in two branches at approximately the 6-o'clock position under the glenoid, just posterior to the long head of the triceps [44]; the anterior branch being responsible for innervating the deltoid as it travels anteromedially around the humeral surgical neck. Most times, a branch from the posterior branch reaches the posterior deltoid as well; the relative contribution from the anterior and posterior branches may take one of three patterns [45]. The anterior branch can be easily palpated under the posterior and middle thirds of the muscle at about 4.0–6.7 cm from its acromial origin [46]. Also, its position can be estimated from outside at a distance of 7.8 cm inferior to the posterolateral corner of the acromion [47] and approximately 6.08 cm inferior to the anterior margin [48], but these distances are highly variable among subjects.

The anterior portion is responsible for shoulder flexion movements and the posterior extension; the middle third is an abductor of the shoulder (the other two portions can contribute to abduction at low angles of abduction).

Pectoralis Major: This powerful muscle originates from the medial clavicle (clavicular head), the sternum and the second to fourth rib (sternocostal head). Its fibers twist 180 degrees and they form a common tendon which inserts at the lateral lip of the bicipital groove (the clavicular fibers inserting inferior to the sternocostal ones), covering it as well as the distal portion of the LHBT. It acts on the shoulder as a powerful adductor, internal rotator and (slightly) flexor. It is innervated by the lateral (clavicular fibers) and medial (sternocostal fibers) pectoral nerves (Figs. 10.20 and 10.21).

Trapezius: This enormous rhombus-shaped muscle has three distinct sections: upper or descending (goes from the occipital tubercle to the lateral clavicle), middle (from the spinous processes of C7 to D3, to the lateral scapular spine and acromion) and lower or ascending (from D3 to D12, to the medial scapular spine). It acts passively as a scapular stabilizer and actively as a scapular retractor, upward rotator and lateral elevator. Its innervation comes from the spinal or accessory nerve (cranial pair XI) (Figs. 10.20 and 10.21).

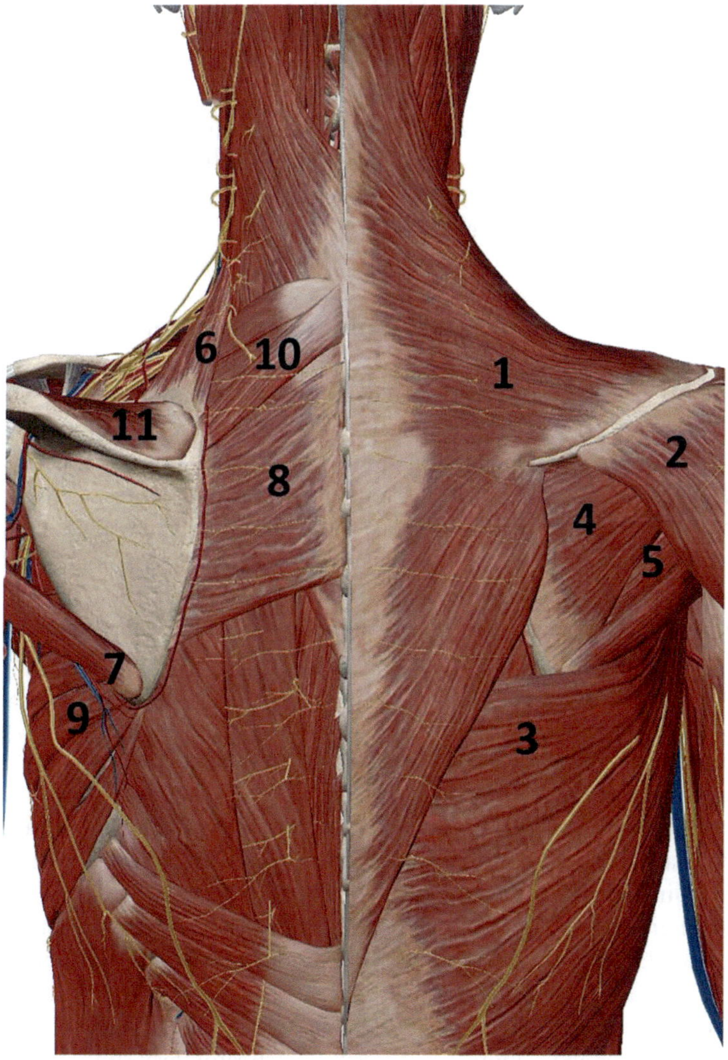

1: Trapezius	
2: Deltoid	
3: Latissimus dorsi	
4: Infraspinatus	
5: Teres minor	
6: Levator scapulae	
7: Teres major	
8: Rhomboid major	
9: Serratus anterior	
10: Rhomboid minor	
11: Supraspinatus	

Fig. 10.20 Periscapular muscles (modified from SciePro /Shutterstock.com)

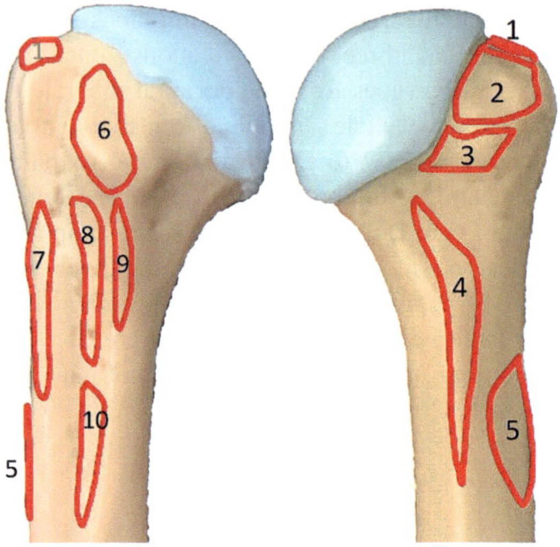

1: Supraspinatus
2: Infraspinatus
3: Teres minor
4: Triceps (lateral head)
5: Deltoid
6: Subscapularis
7: Pectoralis major
8: Latissimus dorsi
9: Teres major
10: Coracobrachialis

Fig. 10.21 Humeral insertions of periscapular and glenohumeral muscles

Latissimus Dorsi: This huge muscle originates from the thoracic spine (D7 to D12), iliac crest and the last three ribs and inserts on the medial lip of the bicipital groove. It acts as a shoulder adductor, extensor and internal rotator. Its innervation comes from the thoracodorsal nerve (Figs. Q and R).

Teres Major: It originates from the posterior surface at the inferolateral edge of the scapula and runs anterolaterally, twisting 180 degrees, toward its insertion at the medial margin of the bicipital groove, slightly posteromedial to that of the latissimus dorsi. It acts both as a scapular stabilizer and, like the latissimus, as a shoulder adductor, extensor and internal rotator. It is innervated by the subscapular nerve (Figs. 10.20 and 10.21).

Other glenohumeral muscles: The long head of the triceps acts in the shoulder as an extensor and adductor; it is innervated by the radial nerve. Also, both the coracobrachialis and the short head of the biceps (SHB) insert in a common tendon (the conjoined tendon) and act as adductors and flexors of the glenohumeral joint. They are innervated by the musculocutaneous nerve, which pierces the coracobrachialis muscle at about 2–5 cm from the tip of the coracoid (Fig. R).

Other periscapular muscles: The levator scapulae (goes from C1-C4 to the superior scapular angle) elevates and downward rotates the scapula and is innervated by the dorsal scapular nerve; the serratus anterior, which goes from the first nine ribs to the entire anterior face of the medial scapular border, acts as a scapular stabilizer and actively as a scapular upward rotator and protractor, and is innervated by the long thoracic nerve. The pectoralis minor, which goes from the second to fifth ribs to the medial coracoid, downward rotates and protracts the scapula and is innervated by the lateral and medial pectoral nerves (Fig. Q).

10.3 Shoulder Normal Kinematics

To understand normal shoulder movement, we have to understand that it is the result of the combination of simultaneous scapulothoracic motion (including sternoclavicular and acromioclavicular motion as well) and glenohumeral motion. This combination is called scapulohumeral rhythm and will be discussed later. First, we will describe glenohumeral and scapulothoracic movement individually.

10.3.1 Glenohumeral Kinematics

To understand glenohumeral kinematics we have to set a frame of reference, which in this case is the scapula.

The humeral head is almost spherical, more convex anteroposteriorly than superoinferiorly (its diameters show a difference of less than 1 mm) [18]. On the other hand, the glenoid is more convex superoinferiorly than anteroposteriorly [49]; this gives some static stability to the humeral head, especially in the superoinferior direction.

The articular surface area of the glenoid is approximately one fourth of the humeral head's surface; this resembles a golf ball laying over a tee. Also, the congruity of the humeral head and the glenoid is almost perfect resembling a ball-and-socket joint; however, the radius of curvature of the glenoid is 2.3 mm longer than that of the humeral head [8], leaving a slightly flatter glenoid relative to the head (Fig. 10.4). This congruity is reinforced by the labrum, which increases glenoid depth by 50% and surface area by 75%; it accounts for approximately 20% of the joint stability achieved by compression [50]. Due to these factors, the humeral head movement is a combination of rolling (predominant) and gliding. This gliding displaces the humeral head depending on the movement. In abduction, the head moves superiorly (3 mm in the first 30 degrees, and approximately 1 mm for every additional 30 degrees); in external rotation, it moves backward and on internal rotation, forward. In flexion, it moves 3.8 mm anteriorly in flexion and 4.9 mm posteriorly in extension [51].

The humeral head is stabilized by numerous structures during its arc of motion. The glenohumeral ligaments play a substantial role. The superior glenohumeral ligament (SGHL) resists anterior and inferior (with the arm hanging) displacement; the middle glenohumeral ligament (MGHL) also performs this function but in adduction and in the first 30 degrees of abduction [61]. The two bands of the inferior glenohumeral ligament (IGHL) act as anterior stabilizers; the anterior band opens up in abduction and external rotation, and the posterior band in flexion and internal rotation. When one band stretches, the other one narrows. In short, as the shoulder abducts, the stress on the ligaments shifts progressively from the SGLH to the MGHL and then to the AIGHL.

The coracohumeral ligament is an inferior stabilizer, especially in external rotation. It also becomes one of the main proprioceptor organs of the shoulder [23, 53].

The joint capsule seals hermetically the joint, thus producing a negative pressure that stabilizes the head, preventing inferior translation.

Shoulder muscles stabilize the head mainly by dynamic contraction, pushing the head toward the glenoid (the pressure produced can reach up to 650 N during motion); these muscles contract in coordination in such a way that the joint reaction force is redirected toward the center of the glenoid (Fig. 10.22) [54].

The rotator cuff plays a major role in this dynamic stabilization. The co-contraction of the four muscles compresses the joint, creating a fulcrum point in the glenoid so motion can occur (otherwise, there would be only humeral head translation and no motion when the powerful external muscles contract). The subscapularis is the main anterior dynamic stabilizer of the head, and accounts for 50% of the cuff's co-contraction force; it also restricts posterior displacement at 90 degrees of flexion. The co-contraction can be analyzed in two planes [55]: The vertical plane, where the deltoid and supraspinatus elevate the head and the remaining cuff muscles pull down (a coronal force couple); and the horizontal plane, where the subscapularis pulls anteriorly and the infraspinatus/teres minor posteriorly (a transverse force couple) (Fig. 10.22). These create a net resultant force that ultimately pulls the head toward the center of the glenoid.

10.3.2 Scapulothoracic Kinematics

There are two aspects that are relevant to scapulothoracic kinematics: periscapular stabilization and scapular motion. One has to understand that scapular motion involves inherently clavicular motion, both at the sternoclavicular and at the acromioclavicular joints; both bones move synchronously. Here, the clavicle moves as a strut connecting the sternum (which will be our point of reference regarding the axial body) and the scapula, and ultimately helping scapular motion.

The scapula has a "normal" resting orientation of 30 degrees of anteversion [56], 3–10 degrees of abduction [3] and 10–20 degrees of

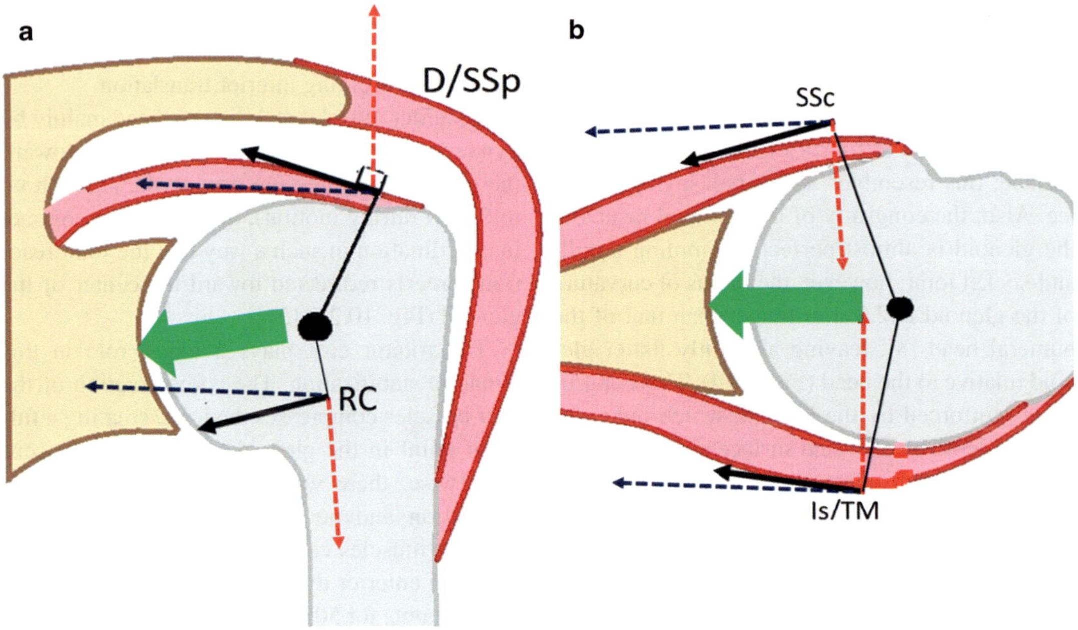

Fig. 10.22 Force couples in the shoulder. (**a**): Coronal-plane force couples: the deltoid and supraspinatus (D/SSp) pull the head upwards and the rest of the cuff (RC) downwards; (**b**): Transverse-plane force couples: the subscapularis (SSc) pulls forward and the infraspinatus and teres minor (Is/TM) pull backward. Vertical and anteroposterior forces cancel each other (red arrows) and the remaining medializing forces (blue arrows) produce a net resultant force (green arrow) that compresses the head toward the center of the glenoid (based on concepts by Lippitt et al. [54] and Saha et al. [55])

anterior tilt [4]. Usually it reaches up to 60 degrees of abduction, 20 degrees of posterior tilt and 10 degrees of external rotation (initially 6 degrees of internal rotation during the first half of elevation, to later add 16 degrees of external rotation) [57].

Scapular motion is a combination of rotational movements along three axes (anteroposterior, superoinferior, and lateromedial) with gliding movements around the posterolateral rib cage through the scapulothoracic interface. These movements can be described as follows (Fig. 10.23):

- Protraction: combination of lateral gliding (winging) away from the spine, anterotation at the acromioclavicular joint and anterior motion of the lateral clavicle.
- Retraction: combination of medial gliding toward the spine, retrorotation at the acromioclavicular joint and posterior motion of the lateral clavicle.
- Upward rotation: combination of coronal-plane rotation resulting in the glenoid facing upwards, superior displacement of the lateral clavicle and slight lateral scapular gliding.
- Downward rotation: combination of coronal-plane rotation resulting in the glenoid facing downwards, inferior displacement of the lateral clavicle and slight medial scapular gliding.
- Elevation: upward scapular gliding and superior lateral clavicular displacement.
- Depression: downward scapular gliding and inferior lateral clavicular displacement.

In elevation of the arm, scapular abduction has three effects: increase of humeral motion relative to the thorax, orients muscles for optimal function and positioning the glenoid under the humerus for load sharing.

The sternoclavicular joint works as a ball-and-socket joint. Its ranges go from 45 degrees of elevation to 15 degrees on depression, ± 30

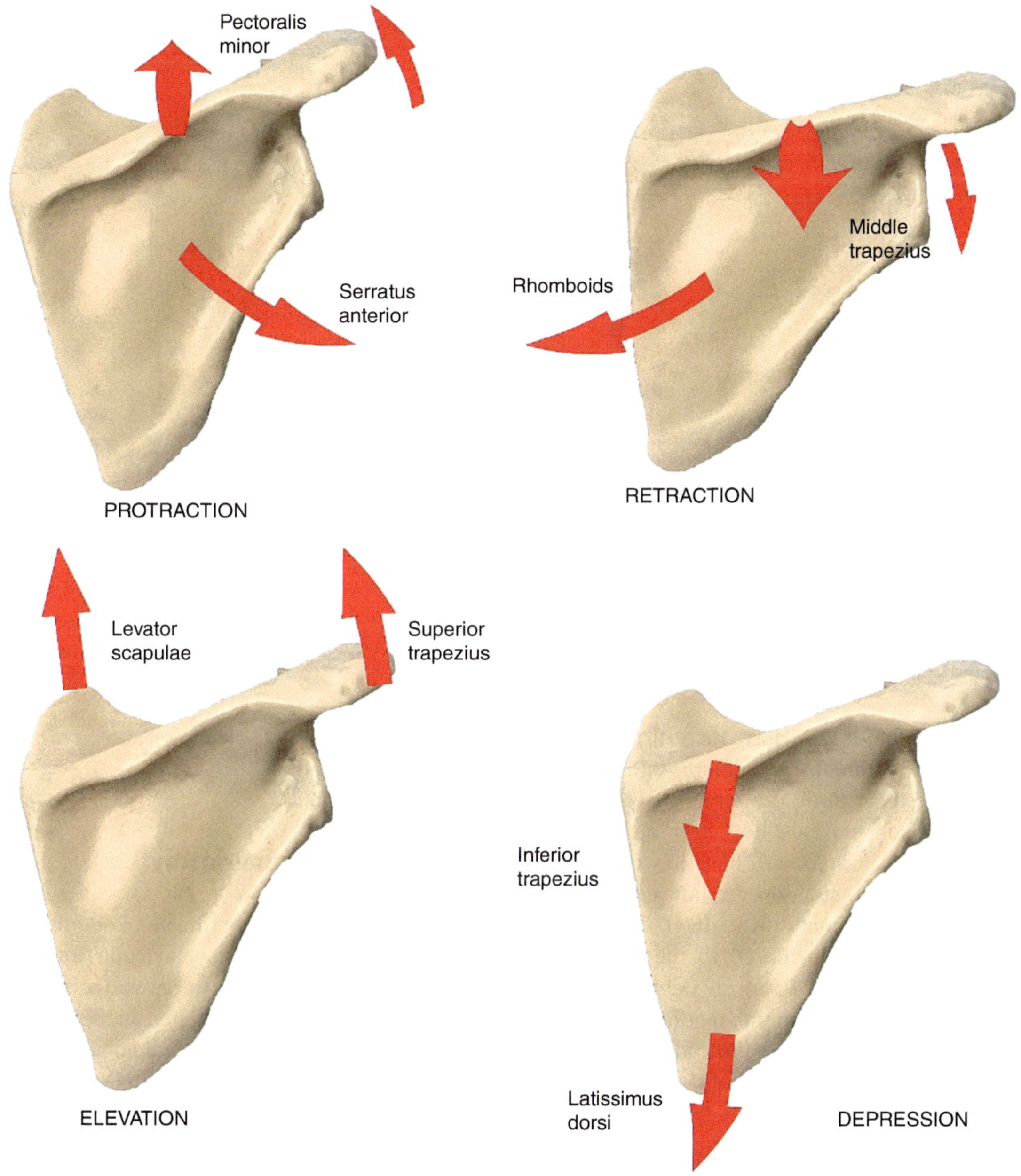

Fig. 10.23 Scapular movements and muscles

degrees on protraction/retraction, and 30–45 degrees on axial rotation. This motion allows the clavicular strut to move in coordination with the scapula. The acromioclavicular joint can rotate from 5 to 8 degrees [58] up to 40–50 degrees during elevation on a synchronous scapular and clavicular rotation.

Periscapular stability is paramount for arm motion: as mentioned before, glenohumeral movement uses the glenoid surface as a fulcrum point for the humerus to move. Thus, the glenoid must be stable when glenohumeral movement occurs. A good analogy is a fisherman (holding a fishing rod) standing on a raft in the

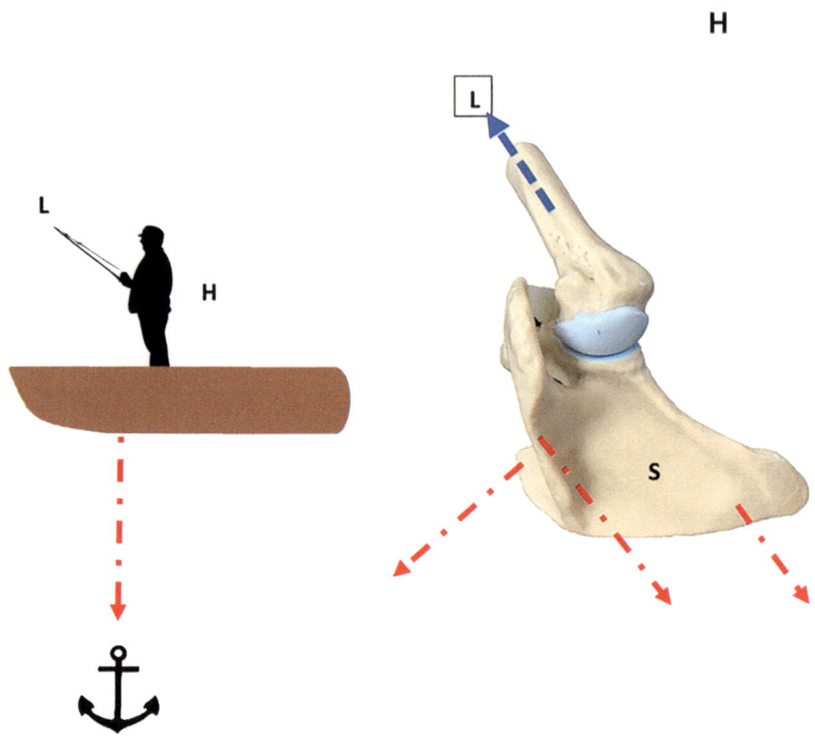

Fig. 10.24 Scapular stability contributes to efficient motion of the upper limb. The fisherman (humerus) standing on the boat (scapula) tries to move the rod (rest of upper limb) but has to expend energy to keep standing due to lack of stability of the boat on the water. If the boat is anchored (coordinated co-contraction of periscapular muscles, M), the fisherman (humerus and glenohumeral muscles) will expend less energy to remain standing (glenohumeral stability) and the rod (upper limb) will be moved more efficiently

middle of the ocean. Here, the raft is the scapula, the fisherman is the humerus and the rod is the distal upper limb; if the raft is not anchored (periscapular muscles not working appropriately), the fisherman will have to expend additional energy to stand in the raft and using the rod will be extremely difficult. If, on the other hand, the raft is anchored to the ocean floor (periscapular muscles coordinately contracted), the fisherman will remain standing (stable glenohumeral motion) and the rod can be easily used (efficient positioning of the arm and hand in space) (Fig. 10.24).

This stability is achieved by coordinated contraction of periscapular muscles: upper and lower trapezius, rhomboids, and serratus anterior, being the lower trapezius the most important scapular stabilizer because of its spinal insertions [52]. This coordinated contraction varies during arm movement to allow for coordinated combined scapulohumeral motion and still provide scapular stabilization; however, the most efficient scapular position for optimal muscle activation is of retraction and external rotation [52], thus acting as a stable "platform" for the rotator cuff muscles to produce optimal glenohumeral motion.

10.3.3 Scapulohumeral Rhythm

This term refers to the coordinated combination of scapulothoracic (including clavicular) and glenohumeral motions to achieve optimal arm positioning in space. To have a better understanding of this, one can picture the shoulder as a rack, with the shoulder girdle (scapula and clavicle as a unit) linked to the rib cage and sternum respectively, and the humerus linked to the glenoid.

Scapular motion plays a major role in maintaining scapulohumeral rhythm. First, it maintains an approximate "ball-and-socket" joint configuration by keeping glenohumeral alignment within normal limits (± 29.3 degrees) [59]. Second, it becomes a stable "platform" for optimal rotator cuff muscle activation and optimal force transmission from the core to the arm and hand, as previously discussed. Third, it keeps the acromion out of the way for the proximal humerus to elevate.

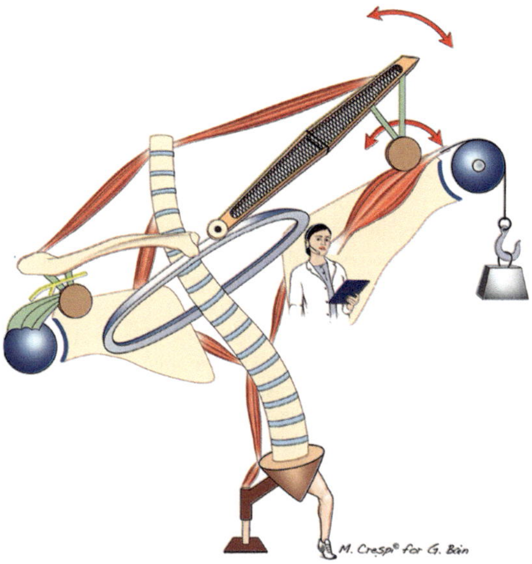

Fig. 10.25 Shoulder crane concept [53] (reprinted with permission from Dr. Gregory Bain). Images to maybe redraw (I used an app or stock pics from the web): 11, 15, 17, 20

In elevation, one can recognize the rhythm occur in two "phases": The setting phase, in which the motion is predominantly glenohumeral and occurs in the first 60 degrees of flexion or 30 degrees of abduction (the shoulder girdle barely moves); and after that the girdle starts to elevate as well. Generally speaking, after 60 degrees of elevation the ratio of glenohumeral-to-scapulothoracic motions is 1.25–1.5:1., with an average ratio for the full arc of elevation of 2:1 [60].

Acromioclavicular retrorotation around the clavicular long axis occurs after 90–100 degrees of elevation.

Scapulohumeral motion can also be understood according to the crane model proposed by Bain et al. (Fig. 10.25) [53]. In this model, the base of the crane (legs and pelvis) supports an axial tower (the spine) and a thoracic platform (the sternum and rib cage), stabilized by the core muscles. From it, the strut (clavicle) suspends from the anterior tower (sternum) and is powered by the posterior tower (trapezius). This strut in turn holds the suspensory cascade: this cascade starts from the skull to the cervical spine then to the trapezius and up to the clavicle; the force is transmitted then to the coracoclavicular ligaments to the coracoid, coracohumeral ligament and finally to the humerus. Here, the coracoid acts as a pulley, in which the scapular pivots aided by periscapular muscles, and the rotator cuff powers motion on the humeral head (supported by the scapula).

References

1. Halder AM, Itoi E, An KN. Anatomy and biomechanics of the shoulder. Orthop Clin North Am. 2000;31(2):159–76. https://doi.org/10.1016/s0030-5898(05)70138-3.
2. Dines D, Williams GR Jr, Laurencin C. Arthritis and arthroplasty: the shoulder. Philadelphia: Elsevier Health Sciences; 2009.
3. Llusá M, Merí A, Ruano D. Surgical atlas of the musculoskeletal system. Rosemont: American Academy of Orthopaedic Surgeons; 2008. p. ix, 422 p.
4. Ludewig PM, Cook TM, Nawoczenski DA. Three-dimensional scapular orientation and muscle activity at selected positions of humeral elevation. J Orthop Sports Phys Ther. 1996;24(2):57–65. https://doi.org/10.2519/jospt.1996.24.2.57.
5. Banas MP, Miller RJ, Totterman S. Relationship between the lateral acromion angle and rotator cuff disease. J Shoulder Elb Surg. 1995;4(6):454–61. https://doi.org/10.1016/s1058-2746(05)80038-2.
6. van de Bunt F, Pearl ML, Lee EK, Peng L, Didomenico P. Glenoid version by CT scan: an analysis of clinical measurement error and introduction of a protocol to reduce variability. Skelet Radiol. 2015;44(11):1627–35. https://doi.org/10.1007/s00256-015-2207-4.
7. Huysmans PE, Haen PS, Kidd M, Dhert WJ, Willems JW. The shape of the inferior part of the glenoid: a cadaveric study. J Shoulder Elb Surg. 2006;15(6):759–63. https://doi.org/10.1016/j.jse.2005.09.001.
8. Iannotti JP, Gabriel JP, Schneck SL, Evans BG, Misra S. The normal glenohumeral relationships. An anatomical study of one hundred and forty shoulders. J Bone Joint Surg Am. 1992;74(4):491–500.
9. Petersen W, Petersen F, Tillmann B. Structure and vascularization of the acetabular labrum with regard to the pathogenesis and healing of labral lesions. Arch Orthop Trauma Surg. 2003;123(6):283–8. https://doi.org/10.1007/s00402-003-0527-7.
10. Alashkham A, Alraddadi A. Soames, R er. (2019). Morphometric parameters of the glenoid labrum. Eur J Anat. 2019;23(1):41–7.
11. Bain GI, Galley IJ, Singh C, Carter C, Eng K. Anatomic study of the superior glenoid labrum. Clin Anat. 2013;26(3):367–76. https://doi.org/10.1002/ca.22145.
12. Williams MM, Snyder SJ, Buford D Jr. The Buford complex--the "cord-like" middle glenohumeral ligament and absent anterosuperior labrum complex: a normal anatomic capsulolabral variant. Arthroscopy. 1994;10(3):241–7. https://doi.org/10.1016/s0749-8063(05)80105-7.

13. Rothman RH, Marvel JP Jr, Heppenstall RB. Anatomic considerations in glenohumeral joint. Orthop Clin North Am. 1975;6(2):341–52.
14. Yamazaki S. Fibrous structure of the joint capsule in the human shoulder. Okajimas Folia Anat Jpn. 1990;67(2–3):127–39. https://doi.org/10.2535/ofaj1936.67.2-3_127.
15. Pouliart N, Gagey O. Reconciling arthroscopic and anatomic morphology of the humeral insertion of the inferior glenohumeral ligament. Arthroscopy. 2005;21(8):979–84. https://doi.org/10.1016/j.arthro.2005.04.111.
16. Gelber PE, Reina F, Monllau JC, Yema P, Rodriguez A, Caceres E. Innervation patterns of the inferior glenohumeral ligament: anatomical and biomechanical relevance. Clin Anat. 2006;19(4):304–11. https://doi.org/10.1002/ca.20172.
17. Huri G, Kaymakoglu M, Garbis N. Rotator cable and rotator interval: anatomy, biomechanics and clinical importance. EFORT Open Rev. 2019;4(2):56–62. https://doi.org/10.1302/2058-5241.4.170071.
18. Boileau P, Walch G. The three-dimensional geometry of the proximal humerus. Implications for surgical technique and prosthetic design. J Bone Joint Surg Br. 1997;79(5):857–65. https://doi.org/10.1302/0301-620x.79b5.7579.
19. Prescher A. Anatomical basics, variations, and degenerative changes of the shoulder joint and shoulder girdle. Eur J Radiol. 2000;35(2):88–102. https://doi.org/10.1016/s0720-048x(00)00225-4.
20. Mochizuki T, Sugaya H, Uomizu M, et al. Humeral insertion of the supraspinatus and infraspinatus. New anatomical findings regarding the footprint of the rotator cuff. J Bone Joint Surg Am. 2008;90(5):962–9. https://doi.org/10.2106/JBJS.G.00427.
21. Clark JM, Harryman DT 2nd. Tendons, ligaments, and capsule of the rotator cuff. Gross and microscopic anatomy. J Bone Joint Surg Am. 1992;74(5):713–25.
22. Nimura A, Kato A, Yamaguchi K, et al. The superior capsule of the shoulder joint complements the insertion of the rotator cuff. J Shoulder Elb Surg. 2012;21(7):867–72. https://doi.org/10.1016/j.jse.2011.04.034.
23. Bain GI, Itoi E, Di Giacomo G Sugaya H (Eds.). Normal and pathological anatomy of the shoulder. Springer, Berlin; 2015. doi:10.1007/978-3-662-45719-1
24. Rajani S, Man S. Review of bicipital groove morphology and its analysis in north Indian population. ISRN Anat. 2013:243780. Published 2013 Sep 11. doi:10.5402/2013/243780.
25. Khan R, Satyapal KS, Naidoo N, Lazarus L. Dimensional analysis of the bicipital groove and its associated pathology in a South African population. J Orthop. 2019;19:128–131. Published 2019 Nov 18. doi:https://doi.org/10.1016/j.jor.2019.11.019
26. Hettrich CM, Boraiah S, Dyke JP, Neviaser A, Helfet DL, Lorich DG. Quantitative assessment of the vascularity of the proximal part of the humerus. J Bone Joint Surg Am. 2010;92(4):943–8. https://doi.org/10.2106/JBJS.H.01144.
27. Minagawa H, Itoi E, Sato T, et al. Structure of the rotator cuff muscles [in Japanese]. J Jpn Orthop Assoc. 1995;69:S1642.
28. Colachis SC Jr, Strohm BR, Brechner VL. Effects of axillary nerve block on muscle force in the upper extremity. Arch Phys Med Rehabil. 1969;50(11):647–54.
29. Klapper RC, Jobe FW, Matsuura P. The subscapularis muscle and its glenohumeral ligament-like bands. A histomorphologic study. Am J Sports Med. 1992;20(3):307–10. https://doi.org/10.1177/036354659202000312.
30. Burkhart SS, Esch JC, Jolson RS. The rotator crescent and rotator cable: an anatomic description of the shoulder's "suspension bridge" [published correction appears in arthroscopy 1994 Apr;10(2):239]. Arthroscopy. 1993;9(6):611–6. https://doi.org/10.1016/s0749-8063(05)80496-7.
31. Goss TP. Double disruptions of the superior shoulder suspensory complex. J Orthop Trauma. 1993;7(2):99–106. https://doi.org/10.1097/00005131-199304000-00001.
32. Bigliani L, Morrison D, April EW. The morphology of the acromion and its relationship to rotator cuff tears. Orthop Trans. 1986;10:228.
33. Green A, Griggs S, Labrador D. Anterior acromial anatomy: relevance to arthroscopic acromioplasty. Arthroscopy. 2004;20(10):1050–4.
34. Stine IA, Vangsness CT Jr. Analysis of the capsule and ligament insertions about the acromioclavicular joint: a cadaveric study. Arthroscopy. 2009;25(9):968–74.
35. Fealy S, April EW, Khazzam M, Armengol-Barallat J, Bigliani LU. The coracoacromial ligament: morphology and study of acromial enthesopathy. J Shoulder Elb Surg. 2005;14(5):542–8.
36. Voss A, Dyrna F, Achtnich A, et al. Acromion morphology and bone mineral density distribution suggest favorable fixation points for anatomic acromioclavicular reconstruction. Knee Surg Sports Traumatol Arthrosc. 2017;25(7):2004–12. https://doi.org/10.1007/s00167-017-4539-1.
37. Petersson CJ, Redlund-Johnell I. The subacromial space in normal shoulder radiographs. Acta Orthop Scand. 1984;55(1):57–8. https://doi.org/10.3109/17453678408992312.
38. Gumina S, Postacchini F, Orsina L, Cinotti G. The morphometry of the coracoid process - its aetiologic role in subcoracoid impingement syndrome. Int Orthop. 1999;23(4):198–201. https://doi.org/10.1007/s002640050349.
39. Terra BB, Ejnisman B, de Figueiredo EA, et al. Anatomic study of the coracoid process: safety margin and practical implications. Arthroscopy. 2013;29(1):25–30.
40. Lo IK, Burkhart SS, Parten PM. Surgery about the coracoid: neurovascular structures at risk. Arthroscopy. 2004;20(6):591–5. https://doi.org/10.1016/j.arthro.2004.04.060.
41. Sewell MD, Al-Hadithy N, Le Leu A, Lambert SM. Instability of the sternoclavicular joint: current

concepts in classification, treatment and outcomes. Bone Joint J. 2013;95-B(6):721–31. https://doi.org/10.1302/0301-620X.95B6.31064.
42. Conduah AH, Baker CL 3rd, Baker CL Jr. Clinical management of scapulothoracic bursitis and the snapping scapula. Sports Health. 2010;2(2):147–55.
43. Sakoma Y, Sano H, Shinozaki N, et al. Anatomical and functional segments of the deltoid muscle. J Anat. 2011;218(2):185–90. https://doi.org/10.1111/j.1469-7580.2010.01325.x.
44. Ball CM, Steger T, Galatz LM, Yamaguchi K. The posterior branch of the axillary nerve: an anatomic study. J Bone Joint Surg Am. 2003;85(8):1497–501. https://doi.org/10.2106/00004623-200308000-00010.
45. Zhao X, Hung LK, Zhang GM, Lao J. Applied anatomy of the axillary nerve for selective neurotization of the deltoid muscle. Clin Orthop Relat Res. 2001;390:244–51. https://doi.org/10.1097/00003086-200109000-00028.
46. Liu KY, Chen TH, Shyu JF, Wang ST, Liu JY, Chou PH. Anatomic study of the axillary nerve in a Chinese cadaveric population: correlation of the course of the nerve with proximal humeral fixation with intramedullary nail or external skeletal fixation. Arch Orthop Trauma Surg. 2011;131(5):669–674. doi:https://doi.org/10.1007/s00402-010-1184-2
47. Uz A, Apaydin N, Bozkurt M, Elhan A. The anatomic branch pattern of the axillary nerve. J Shoulder Elb Surg. 2007;16(2):240–4. https://doi.org/10.1016/j.jse.2006.05.003.
48. Cetik O, Uslu M, Acar HI, Comert A, Tekdemir I, Cift H. Is there a safe area for the axillary nerve in the deltoid muscle? A cadaveric study. J Bone Joint Surg Am. 2006;88(11):2395–9. https://doi.org/10.2106/JBJS.E.01375.
49. McPherson EJ, Friedman RJ, An YH, Chokesi R, Dooley RL. Anthropometric study of normal glenohumeral relationships. J Shoulder Elb Surg. 1997;6(2):105–12. https://doi.org/10.1016/s1058-2746(97)90030-6.
50. Lippitt SB, Vanderhooft JE, Harris SL, Sidles JA, Harryman DT 2nd, Matsen FA 3rd. Glenohumeral stability from concavity-compression: a quantitative analysis. J Shoulder Elb Surg. 1993;2(1):27–35. https://doi.org/10.1016/S1058-2746(09)80134-1.
51. Harryman DT, Sidles JA, Clark JM, et al. Translation of the humeral head on the glenoid with passive glenohumeral motion. J Bone Joint Surg Am. 1990;72:1334–43.
52. Bagg SD, Forrest WJ. A biomechanical analysis of scapular rotation during arm abduction in the scapular plane. Am J Phys Med Rehabil. 1988;67:238–45.
53. Bain GI, Phadnis J, Itoi E, et al. Shoulder crane: a concept of suspension, stability, control and motion. JISAKOS. 2019;4:63–70. https://doi.org/10.1136/jisakos-2017-000187.
54. Lippitt S, Matsen F. Mechanisms of glenohumeral joint stability. Clin Orthop. 1993;291:20–8.
55. Saha AK. Mechanisms of shoulder movements and a plea for the recognition of "zero position" of the glenohumeral joint. Clin Orthop. 1983;173:3–10.
56. Bakhsh W, Nicandri G. Anatomy and physical examination of the shoulder. Sports Med Arthrosc Rev. 2018;26(3):e10–22. https://doi.org/10.1097/JSA.0000000000000202.
57. Laumann U. Kinesiology of the shoulder joint. In: Kolbel R, Helbig B, Blauth W, et al., editors. Shoulder Replacement. Berlin: Springer; 1987.
58. Rockwood CJ, Williams G, Young D. Disorders of the acromioclavicular joint. In: Rockwood CJ, Matsen F, editors. The shoulder, vol. 1. Philadelphia: Saunders; 1998. p. 483–553.
59. Nieminen H, Niemi J, Takala EP, Viikari-Juntura E. Load-sharing patterns in the shoulder during isometric flexion tasks. J Biomech. 1995;28(5):555–66.
60. Poppen NK, Walker PS. Normal and abnormal motion of the shoulder. J Bone Joint Surg Am. 1976;58:195–201.
61. O'Connell PW, Nuber GW, Mileski RA, Lautenschlager E. The contribution of the glenohumeral ligaments to anterior stability of the shoulder joint. Am J Sports Med. 1990;18(6):579–84. https://doi.org/10.1177/036354659001800604.
62. Osias W, Matcuk GR Jr, Skalski MR, et al. Scapulothoracic pathology: review of anatomy, pathophysiology, imaging findings, and an approach to management. Skelet Radiol. 2018;47(2):161–71. https://doi.org/10.1007/s00256-017-2791-6.

11 Biomechanics of Rotator Cuff Injury and Repair

Giacomo Dal Fabbro, Margherita Serra, Giuseppe Carbone, Alberto Grassi, Khalid Al-Khelaifi, and Stefano Zaffagnini

11.1 Introduction

Although the rotator cuff tears are more common in older individuals, this kind of pathology also frequently affects athletes and young workers engaged in repetitive overhead activities. The rotator cuff is a complex musculotendinous unit, which plays a major role in glenohumeral joint stability and mobilization. Injuries of the rotator cuff can lead to significant functional impairment, resulting in time loss from sport and competition, and, in the more advanced stages, in trouble in performing daily activities. Therefore, fully understand the biomechanical features of rotator cuff tears and repair is crucial to better manage and address the injuries of rotator cuff. This chapter provide an overview of the normal and injured rotator cuff biomechanics, followed by literature evidence and considerations about the biomechanical concept of the rotator cuff repair.

11.2 Biomechanical Properties of the Rotator Cuff

Rotator cuff is composed by the tendons of four muscles: the supraspinatus, the infraspinatus, the teres minor and the subscapularis. These muscles create compressive forces to stabilize the glenohumeral joint and generate torque to move the humerus [1]. In particular, the compressive and antagonistic forces of the subscapularis anteriorly and infraspinatus and teres minor posteriorly, imparts a compression on the humeral head into the glenoid, with a mechanism known as "concavity compression" [2] (Fig. 11.1). This tendinous connection, with the addition of ligaments and bone factors, allows for the articular stability and, at same time, for the humeral motion with respect to the scapula, defining the role for tendons as specific active restraints during the shoulder movement [3]. The wide range of motion of glenohumeral joint is allowed by the variety of rotational moments of the cuff muscles, whose tendons insert continuously around

G. Dal Fabbro (✉) · M. Serra · A. Grassi
IRCCS - Istituto Ortopedico Rizzoli - Clinica Ortopedica e Traumatologica II, Bologna, Italy
e-mail: giacomo.dalfabbro@studio.unibo.it; margherita.serra2@studio.unibo.it; alberto.grassi@ior.it

G. Carbone
IRCCS - Istituto Ortopedico Rizzoli – Ortopedia Bentivoglio, Bologna, Italy

K. Al-Khelaifi
Orthopaedic Surgery Department, Aspetar Orthopaedic and Sports Medicine Hospital, Doha, Qatar
e-mail: Khalid.Alkhelaifi@aspetar.com

S. Zaffagnini
IRCCS – Rizzoli Orthopaedic Institute – 2nd Orthopaedic and Trauma Unit, University of Bologna, Bologna, Italy
e-mail: Stefano.zaffagnini@unibo.it

J. Koh et al. (eds.), *Orthopaedic Biomechanics in Sports Medicine*,
https://doi.org/10.1007/978-3-030-81549-3_11

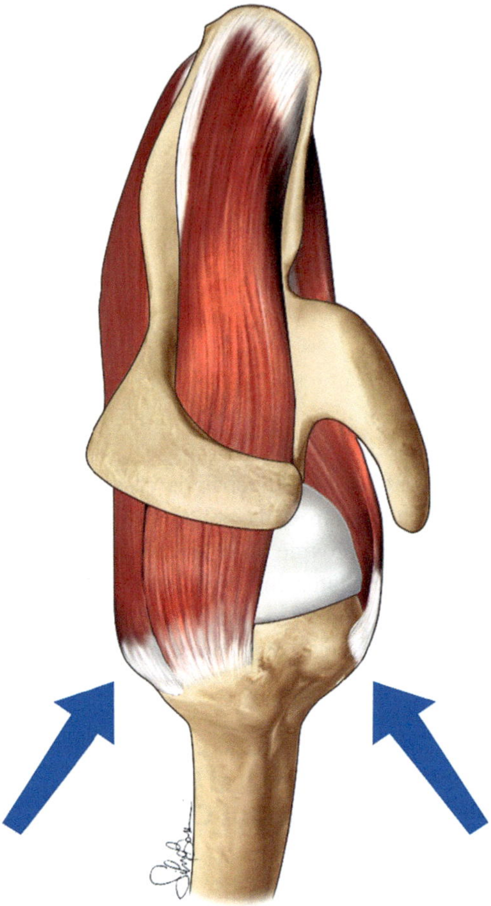

Fig. 11.1 Concavity compression mechanism. The compressive and antagonistic forces of the subscapularis anteriorly and infraspinatus and teres minor posteriorly, imparts a compression on the humeral head into the glenoid, with a mechanism known as "concavity compression"

the humeral head. The force and the torque that a muscle can generate vary with the position of the joint: at the extremes of the range of motion, the muscles of the rotator cuff are weaker, because they are fully contracted or extended, while the maximal force is in the midrange of their pretension [4]). The supraspinatus has a major role in humeral elevation at higher glenohumeral angles, while at lower angles the deltoid contribution is higher. The subscapularis and infraspinatus also play a role in abduction during internal and external humeral rotation [5]. During the arm motion, movements of the scapulothoracic joint are associated with those of glenohumeral joint, performing a complex and coordinated pattern of motions that Codman defined the "scapulohumeral rhythm" [6]: while under the 90° of humeral elevation scapular and clavicular motion are minimal, beyond the 90° there is upward rotation, posterior tilting and external rotation of the scapula to allow for the full range of motion in the upper extremity [7].

As in the shoulder there is not a fixed axis of rotation, the functions of the shoulder muscles must be balanced, working together to produce a net torque and neutralize opposed elements: the superior portion of the cuff and the deltoid neutralize the adduction of the latissimus dorsi during pure internal rotation, the posterior cuff and the posterior deltoid muscles neutralize the internal rotation moment of the anterior deltoid during forward elevation without rotation [5].

The tendons of the rotator cuff are subjected to complex tension loads. Compressive loads stress the cuff between the humeral head and the coracoacromial arch: in vivo subacromial pressure values of 8 mmHg at rest, 39 mmHg at 45° of flexion and 56 with the arm in elevated position with 1 Kg weight held in the hand were reported [8]. Concentric and eccentric tension loads are produced in abduction against resistance and in active resistance to downward respectively [8].

11.3 Biomechanics of Rotator Cuff Tears

11.3.1 Mechanical Factors in Rotator Cuff Tears

Rotator cuff tear aetiology is multifactorial: the role of extrinsic factors, which have been classically considered among the causes of rotator cuff tears, takes alongside to that of intrinsic factors, such as biologic degeneration and deficit of vascularisation [3]. The extrinsic theory dominated the pathophysiology of the rotator cuff tears and the impingement syndrome for decades. In 1931, Meyer suggested that tears of the rotator cuff could develop secondary to attrition due to the friction between the rotator cuff and the under surface of the acromion [9]. In the 1972, Neer introduced the theory of subacromial impingement, describing it as a mechanical phenomenon corresponding to impingement of the rotator cuff tendon beneath the anterior-inferior acromion [10]. Neer observed that this condition affected mainly the supraspinatus tendon insertion on the greater tuberosity, and that occurred when the shoulder was placed in forward flexion and external rotation. Moreover, he observed that this condition presupposed an anatomic irregularity of the acromion and the presence of bony spurs in the site of coracoacromial ligament insertion on the acromion [10]. The correlation between the acromial shape and the cuff disease was supported by subsequent cadaveric studies, in which a higher prevalence of full-thickness tears of the rotator cuff was noted in association with the hooked or type III acromion [11, 12]. However, later, acromial impingement was found not to be the primary cause of rotator cuff tears [3, 5, 13]: it resulted associated with changes on the bursal side of the rotator cuff tears which is less frequently affected than the intra-tendinous or articular side [14], and the acromioplasty procedures did not avoid the occurrence of rotator cuff tears, leading to damage of tendons in 20% of cases [15]. Furthermore, the morphology of the acromion seemed to be secondary to pathology of the bursal side of the cuff [16, 17]: the cuff degeneration lead to produce an acromial hook, which results in a greater load on the coracoacromial arch producing traction spurs in the coracoacromial ligament [18, 19]. A contact between rotator cuff and the coracoacromial arch was also reported in normal shoulders in both cadaveric studies and healthy human subject, suggesting that the contact phenomenon between the coracoacromial arch and the rotator cuff was not a pathological but a physiological condition [20, 21].

Further bone features has been associated with injuries of the rotator cuff; a glenohumeral joint with a higher acromial coverage index, which is the ratio between the distance from the glenoid plane to the acromion and the distance from the glenoid plane to the lateral aspect of the humeral head, is associated with an increased risk of rotator cuff tear [22]. The angle between the line connecting the inferior border to the superior border of the glenoid fossa, and a second line connecting the inferior border of the glenoid fossa to the most inferolateral point of the acromion, represents the critical shoulder angle (CSA) (Fig. 11.2). Since it combines the measurement of the glenoid inclination and the lateral extension of the acromion, CSA has been

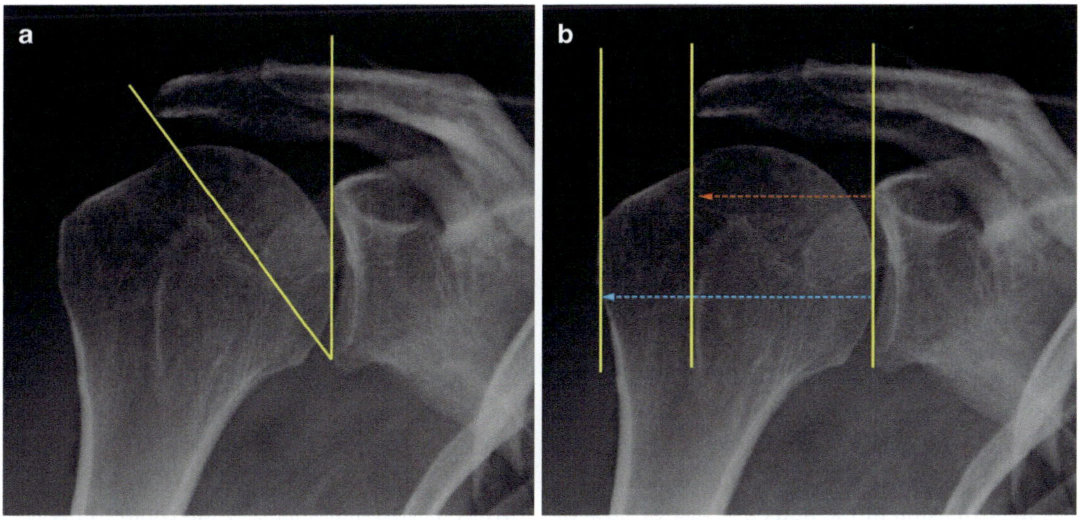

Fig. 11.2 Radiologic parameters associated with rotator cuff pathology. The critical shoulder angle (**a**) and the acromial coverage index (**b**)

used to predict the presence of degenerative cuff tears [23]. In a biomechanical simulation model, a larger CSA has been shown to increase the supraspinatus tendon load at low degrees of abduction compared to a control with a normal CSA [24].

The mechanism of rotator cuff fatigue damage starts with isolated changes in microstructures and develops into severe matrix disruption and kinked deformation; this degeneration, in addition to biological changes due to aging and overuse, predispose the tendons to failure [3]. Studies of torn rotator cuff tendons have revealed that degenerative changes also appear medially from the tear, indicating the presence of degeneration before the tear occurs [25, 26]. In particular, the critical zone of the rotator cuff, localized at 1 cm medially to the supraspinatus insertion in the tuberculus major, resulted to have a decreased vascularisation [27]. This feature, associated with increased local pressure during humeral elevation and abduction, may be related with the degeneration process, predisposing the tendons to failure [3].

In patients who require to achieve a large range of external rotation, such as throwers, defects of the deep surface of the rotator cuff are common. Among these patients, the mechanism of cuff injury is explained by the abutment of the edge of the glenoid against the under surface of the supraspinatus tendon at the extreme of the external rotation [5].

Rotator cuff tears typically start at the deep surface of the supraspinatus tendon, close to the long head of the biceps. This site is subject to greater loads, even at rest. The complex and coordinated actions of the rotator cuff muscles are coupled in a manner that produces increased strain on the surrounding tendons when a tear in one tendon occurs. For this reason, a tear of one tendon in the rotator cuff could potentially lead to an increased risk for injury in the remaining muscles, determining a pattern that induces the tear progression and extension [28]. The loads are concentred at the margin of the tear, facilitating the extension of the defect. A tear of 60% of the tendon thickness increases of 23.8% the tensile strain in the posterior tendon [29]. Tears of the articular side of the infraspinatus increase the strain in the middle and superior portions of the supraspinatus. Subsequently, the defects of supraspinatus propagate posteriorly through the reminder of the supraspinatus and the infraspinatus [5].

11.3.2 Biomechanical Effects of Rotator Cuff Tears

Tears in the rotator cuff result in altered glenohumeral joint mechanics and are frequently associated with loss of strength and stability of the shoulder. However, clear evidence about the relationship between the tear features and the functional outcomes still lack. The superior translation of the humeral head represents the most important biomechanical consequence of the rotator cuff tears, and the main radiograph signs of massive cuff deficiency. Large-to-massive tears may lead to rotator cuff tear arthropathy, which implies the development of progressive glenohumeral and acromiohumeral arthritis secondary to the tear, associated with joint pain and reduced humeral elevation [30, 31].

When the dynamic function of rotator cuff is compromised by a tear, the external rotation and the elevation are primarily decreased. Burkhart introduced the concept of the rotator cuff cable [32], a thickened area of the rotator cuff tendon which inserts anteriorly between the anterior insertion of the supraspinatus and superior half insertion of subscapularis, and posteriorly near the inferior insertion of infraspinatus tendon. Burkhart noted that rotator cuff tears that did not involve the cable, such as supraspinatus tears alone, were functional and biomechanically intact, because the cable remained intact [33].

Regarding the abduction torque, the features of rotator cuff tears which could affect the force transmission are the supraspinatus tendon detachment and the retraction of its muscle. In a cadaveric study [34] the effect of different size and shape of the rotator cuff tear on in vitro force transmission was investigated, comparing different size of supraspinatus tendon detachment with a defect of tendon substance and muscle retraction: authors reported that only after detachment of the entire width of the supraspinatus tendon, the transmitted force significantly decreased; this finding support the concept of rotator cable, pointing out that the muscle forces are effectively transmitted by the rotator cuff as long as the rotator cable is intact. Furthermore, in this cadaveric study, the simulated muscle retraction resulted in significantly greater decrease in transmitted force compared with the isolated detachment of the tendon, showing that rotator cuff muscle retraction contributes substantially to loss of shoulder strength following large rotator cuff ruptures [34].

In order to gain more insight in the effect of rotator cuff tear in shoulder biomechanics, the kinematics of the joint with a defect of the cuff was analysed in literature. In particular, the assessment of humeral head position, external humeral rotation, humeral abduction and associated scapulothoracic rotations were investigated, and the findings showed that the shoulder kinematics is associated with rotator cuff tear size [35–37]: while isolated tears of supraspinatus had no significant biomechanics consequences, massive cuff tears involving supraspinatus or subscapularis altered significantly the kinematics of the shoulder. In a cadaveric study [36] in which five shoulder specimens were subjected to different loading conditions with sequentially larger anterosuperior cuff tears, a significant increase in anterior-superior and superior translation resulted only after that the anterosuperior tears were extended to the superior half of the subscapularis. No significant biomechanical consequences, in contrast, were reported after isolated tears of supraspinatus. These results support the rotator cuff cable concept, suggesting that the preservation or the loss of the anterior cable attachment represents an important determinant of the biomechanics of anterosuperior rotator cuff tears [36]. In line with these findings, in another cadaveric study, in which the effect of 1 and 3 cm isolated supraspinatus tears on joint kinematics was examined, no significant changes were founded in glenohumeral translation [38].

Regarding the position of humeral head along the superior-inferior and anterior-posterior axis during scapular plane abduction in shoulder with rotator cuff tears, superior migration of humeral head, when the superior shear forces created by the deltoid are no longer effectively opposed, was described in studies employed in vitro static evaluation techniques [39, 40] (Fig. 11.3).

In a dynamic in vivo study, on the other hand, an inferior dynamic translation was founded during scapular plane humeral abduction [41]. During an in vitro simulation of active shoulder kinematics, the effect of three sequential simulated tears affecting the supraspinatus and the subscapularis during the unconstrained glenohumeral abduction was provided [35]: authors founded that, as the size of the simulated cuff tear increased, the plane of elevation became significantly more posterior at all angles of abduction, and the position of humeral head became more anterior for angle of abduction greater than 50°. No differences, on the other hand, were founded regarding the position of humeral head along the superior-inferior axis [35]. In an in vivo study in which a 3D to 2D model-to-image registration techniques was provided [42], a more anterior position of the humeral head centre was founded in shoulder with large-to-massive full-thickness rotator cuff tear than in controls, at internal rotation position during dynamic axial rotation in adducted position. Moreover, authors reported a more medial position of humeral head in patients with rotator cuff tear than in controls at the late phase of dynamic scapular plane abduction [42]. A recent in vivo three-dimensional bone model of ten shoulders with massive rotator cuff tears showed a significantly higher humeral head position under the 40° of abduction angle than the normal contralateral shoulder [43].

Regarding the associated scapular kinematics, Authors of the latter study reported a significant increase of scapular upward rotation in the initial position at 20° of humeral abduction and at the

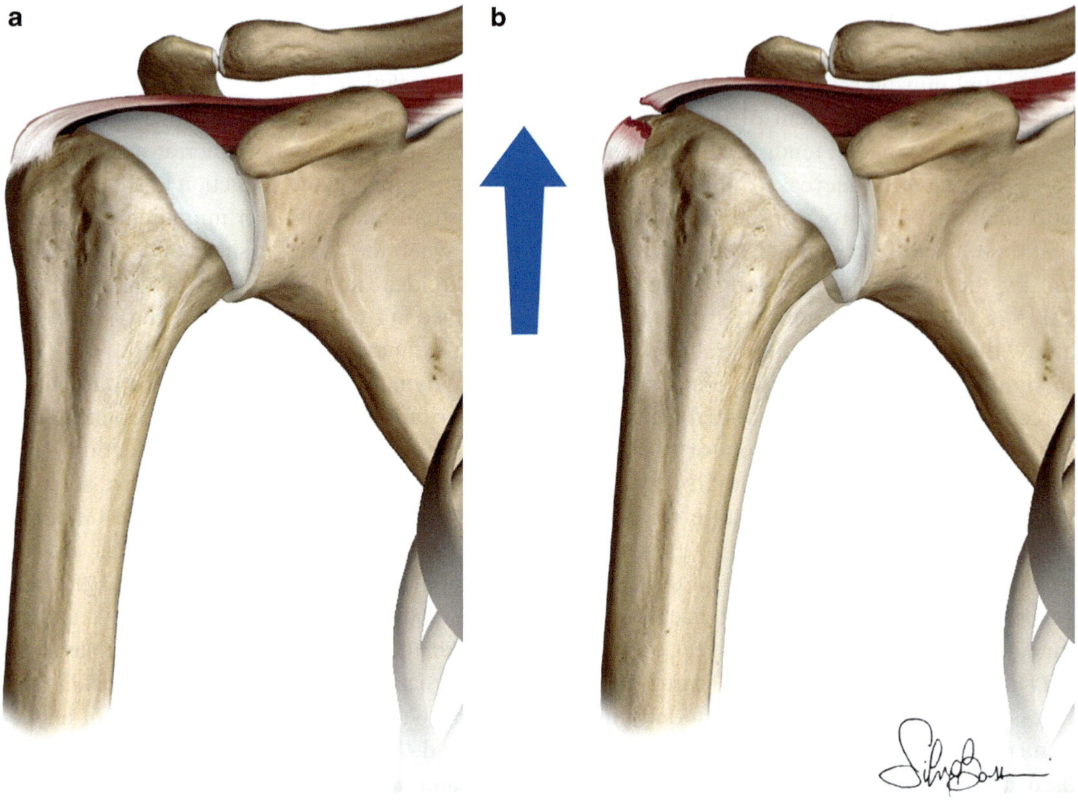

Fig. 11.3 Humeral head position in shoulder with rotator cuff tear. Rotator cuff tear resulted in superior migration of humeral head among cadaveric static evaluations. (**a**): normal cuff; (**b**): injured cuff

final elevation, over 130° of humeral abduction in shoulders with rotator cuff tears [43]. In another kinematics study, the comparison between patients with massive posterior-superior rotator cuff tear and patients with subacromial pain syndrome associated with intact rotator cuff showed a lower humero-thoracic abduction and forward flexion, associated with a marked increase in scapular upward rotation in the rotator cuff injured group [37]. Furthermore, decline in humeral abduction and increase in scapular upward rotation were founded in simulated massive posterior-superior rotator cuff tear created after a suprascapular nerve block in healthy volunteers [44]. These results about the postero-superior massive tears, since the infraspinatus muscle has a direct impact on the glenohumeral joint and does not directly control scapula-thorax motion, suggested that the increased scapular upward rotation (i.e. scapulo-thorax lateral rotation) should be compensatory in nature [37, 44] (Fig. 11.4).

11.4 Biomechanics of Rotator Cuff Repair

The rotator cuff repair must aim to fully restore the anatomy and function of the rotator cuff tendons.

Regarding isolated supraspinatus lesion, the biomechanical effect of repair was seen by Yu et al. [45] only at 10° abduction with 60 N loading with an increase in percent inferior force after repair that may represent greater concavity compression and spacer effect, which are both important functions of the supraspinatus. This same increase was not observed with the deltoid loaded to 90 N, which could be a function of the effect of loading producing an offset between the inferior

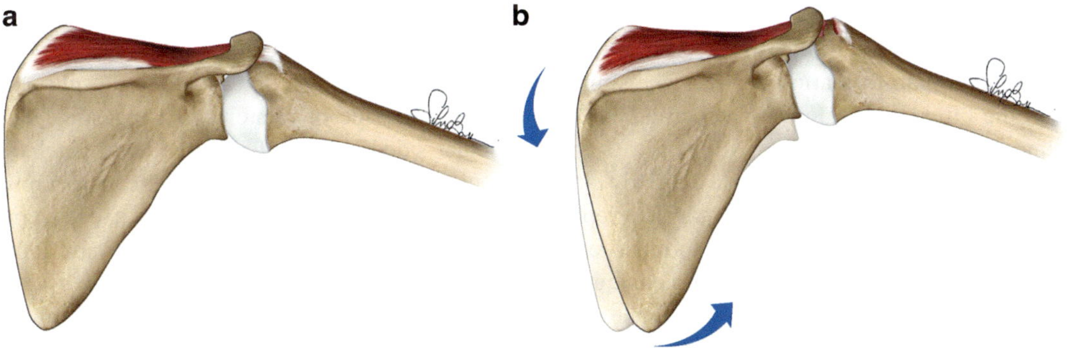

Fig. 11.4 Scapular associated kinematics in rotator cuff tears. In vivo kinematics studies reported increase in scapular upward rotation (scapula-thorax lateral rotation) associated with a decreased humeral abduction in shoulders with postero-superior massive rotator cuff tears. **a**: normal cuff; **b**: injured cuff

and superior forces. No changes were seen in the contact area, pressure, or position after repair, suggesting that repair of full-thickness rotator cuff tears does not completely recreate normal glenohumeral biomechanics or that no change existed in the pathologic tear specimens from normal.

By the way, when a tear is large, it may sometimes be difficult to bring the torn edges back to the original insertion site of the greater tuberosity. In those cases, a medial shift of the insertion site of the cuff tendon is one of the surgical options. According to biomechanical studies by Yamamoto and those by Liu et al., a medial shift of 17 mm or more should be avoided from the functional point of view because it reduces both the moment arm and the range of shoulder motion [46, 47].

Several biomechanical studies investigated the effect of different angles anchor insertion and inclination in rotator cuff repair, with the aim to achieve greater pullout strength possible. In 1995 Burkhart introduced the deadman theory [48], suggesting that an anchor inserted at 45° may show the greatest pullout strength theoretically. However, biomechanical pullout studies, have shown that anchors inserted at 90°, 135° or between 105° and 135° present the greatest pull-out strength [49–51], advancing that the deadman theory depends on the friction of the anchors. In a review associated with additional biomechanical studies, Itoi et al. [52] reported that the insertion angle of 90° is the strongest for a threated anchor. Furthermore, authors of this latter study showed that the pullout strength depends on the inclination of the anchor, friction of the anchor–bone interface and quality of bone. In particular, the insertion angle of the suture threated anchor should replicate the angle of applied load to ensure the optimum pullout strength [52]. (Fig. 11.5).

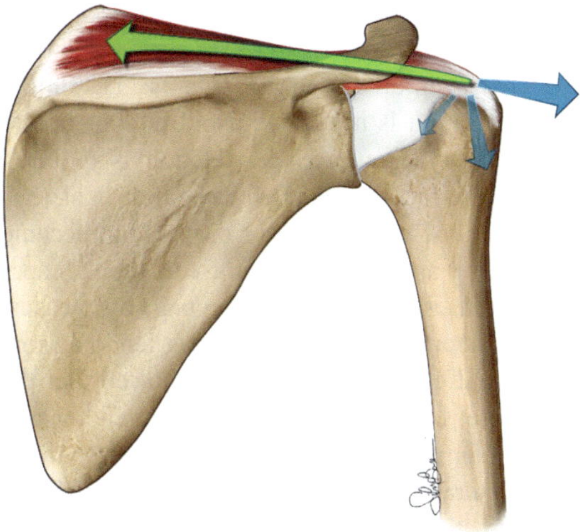

Fig. 11.5 Effect of different angle insertion of suture threated anchors in rotator cuff repair. In order to achieve the best performance of suture threated anchor with strong friction, surgeons must insert the anchor close to the line of pull

Regarding the rows of the anchor, there are single, double and triple row repair described, with associated different types of stiches configuration. In a cadaveric study, in which a simple single-row repair construct and a linked single-row construct were compared, no significant differences in ultimate load to failure and resistance to gap formation between the simple and the linked single-row construct were founded [53].

Many biomechanical studies have already clarified that the use of transosseous equivalent repair for rotator cuff repair was superior to other techniques such as the single- or dual-row technique in terms of contact area or pressure between the rotator cuff tendon and the footprint, and the initial pullout strength [54–57]. Using the transosseous equivalent technique, a greater initial fixation strength may be achieved. However, the tendon itself is inherently elastic. Fixing the torn tendon with the transosseous equivalent repair technique, the tendon may lose its inherent elasticity due to the crossover of the strong sutures. The initial strong fixation is needed to avoid pullout failure after surgery. However, the fixation is required only for the first few months. Yamamoto et al. [57] demonstrated that the rotator cuff tendon became stiffer after arthroscopic rotator cuff repair, especially transosseous equivalent repair, by measuring the strain of the supraspinatus tendon using ultrasound elastography. Results showed that the superficial layer of the supraspinatus tendon was stiffer after transosseous equivalent repair than the contralateral side at the final follow up [57]. In addition, their study revealed that a high-stress concentration was observed around the insertion sites of the medial row threads.

The tying of the medial knots during transosseous equivalent repair is controversial. There are several biomechanical reports demonstrating biomechanical improvement (ultimate load, contact pressure, and stiffness) by tying the medial row of a transosseous equivalent repair [58, 59]. Tamboli et al. [60] investigated the effects of bite-size horizontal mattress stitch: whereas a 4-mm bite fixed the tendon more tightly but at the cost of decreased ultimate strength, a 10-mm bite conveyed greater ultimate strength but with increased gap and strain. Authors concluded that for transosseous equivalent repair, large stitches are beneficial because the repaired tendon has a higher strength, and the slightly mobile medial knot can be tightened by lateral fixation.

Anatomic and biomechanical cadaveric studies highlighted the structural and functional relationship between superior shoulder capsule and the articular side of rotator cuff suggesting that articular-sided partial-thickness tears include detachment of the superior shoulder capsule from the greater tuberosity [61, 62]. In a cadaveric study a significant increase of anterior and inferior glenohumeral translation was founded after that a tear of superior capsule was performed, compared with the intact capsule state [62]. Satisfactory results were reported about the repair procedure to manage articular-sided partial-thickness rotator cuff tear [63, 64]. In a cadaveric analysis of the effect of an articular-sided partial supraspinatus and infraspinatus tear during the simulated late-cocking and acceleration phases of throwing, no significant differences in humeral head apex position, internal impingement area or glenohumeral contact pressure were reported compared with the intact condition [65]. In the same study, authors performed a transtendon repair with two anchors providing a significant decrease of external humeral rotation, posterior shift of humeral apex position, decrease of glenohumeral contact pressure at maximum external rotation and decrease of internal impingement area compared with the intact and the torn status during both the two phases of throwing. These results suggested that transtendon repair of an articular-sided partial supraspinatus tear may cause overtightening of the torn tendon and superior capsule, minimizing secondary subacromial and interior impingement, and that careful evaluation of patient individual shoulder laxity should be assessed when the transtendon repair is chosen for address an articular-sided partial tear [65].

11.4.1 Irreparable Tears

Rotator cuff tears are defined as irreparable when the lesions cannot be repaired primarily to their insertion on the tuberosities with conventional techniques of surgical release/mobilization, because of their size, retraction, and muscle impairment caused by atrophy and fatty infiltration [66].

With the aim to better address the massive irreparable rotator cuff tears, Burkhart et al. [33] introduced the concept of the "suspension bridge" which lead to the development of the functional partial repair: the repair of the whole subscapularis and the inferior half of infraspinatus, without the complete coverage of the defect, can restore the normal shoulder mechanics, recovering the transverse force couples and a stable fulcrum for glenohumeral kinematics [66].

Several biomechanical studies investigated the influence of the superior capsule reconstruction on the superior stability of the shoulder joint. Superior capsular reconstruction (SCR) potentially improves function by re-centering the humeral head and improving glenohumeral kinematics [67].

Several techniques for performing superior capsule reconstruction have been described, characterized by different type of the graft (fascia lata, extracellular matrix dermal grafts, long head of biceps and tendon allografts), whether performed in an open manner or arthroscopically, the mode of the glenoid and greater tuberosity fixation, and whether the anchors are all inserted before or after passage of the graft [67–73].

In a cadaveric study involving eight shoulders, Mihata et al. [74] compared the superior translations of the proximal humerus in five conditions: intact rotator cuff, excised supraspinatus, reconstructed supraspinatus using a bridging graft connecting the remnant of the supraspinatus to the greater tuberosity, reconstructed superior capsule with graft attached to the glenoid and the greater tuberosity, reconstructed superior capsule and supraspinatus with the patch [74]. They demonstrated that excising the supraspinatus tendon resulted in a significant increase in the superior translation of the proximal humerus, which was fully restored only when the superior capsule was reconstructed with the graft [74]. Supraspinatus reconstruction with the graft only resulted in a partial restoration of superior translation [74]. In another biomechanical study, Mihata et al. demonstrated an 8 mm thick fascia lata graft resulted in a greater superior stability than a 4 mm thick graft [75]. A significant decrease in superior translation was only witnessed with 8 mm thick graft as compared to the 4 mm thick graft [75].

When performing SCR, it was also reported that addition of posterior side-to-side suturing between the graft and residual infraspinatus tendon increases the superior stability of the proximal humerus [76]. In a biomechanical study, SCR without posterior side-to-side suturing, did not inhibit glenohumeral superior translation, whereas addition of posterior side-to-side suturing resulted in significantly reduced superior translation [76].

The superior rotator cuff and superior capsule reconstruction technique using long head of biceps (LHB) was also proposed and described in literature. In a biomechanical study by Han et al. the superior capsule reconstruction using LHB permitted to reduced humeral head translation and subacromial contact pressure [77]. Moreover, Authors did not find significant differences between the isolated SCR using LHB procedure and the side-to-side repair augmentation associated procedure [77].

Recently, performing acromioplasty with a rotator cuff repair has become a subject of controversy. There have been some clinical reports describing the excellent outcome after cuff repair without acromioplasty [78, 79]. According to the 2011 guidelines published by the American Academy of Orthopaedic Surgeons, [80] the routine acromioplasty is not required at the time of rotator cuff repair. A biomechanical study by Denard et al. [81] demonstrated that a limited acromioplasty, of 3 mm in thickness, might be enough for decompression while preserving the coracoacromial ligament. Because of variations in the distance between the undersurface of the acromion and the cuff surface, the level of acromioplasty should be decided during surgery on a patient specific level.

Although the results of routinely acromioplastic are controversial, Mihata et al., investigating the effects of acromioplasty on shoulder biomechanics associated with superior capsule reconstruction [82], illustrated that adding acromioplasty to SCR with fascia lata significantly decreased the subacromial peak contact area compared to SCR without acromioplasty, without altering the humeral head position, superior translation or subacromial peak contact pressure [82]. Based on their findings the authors suggested that when performing SCR, acromioplasty may help to decrease the postoperative risk of abrasion and tearing of the graft beneath the acromion, without increasing the risk of superior translation [82].

To treat the irreparable rotator cuff tears the subacromial balloon spacers have been proposed. There is limited biomechanical literature on the subacromial balloon spacer and its ability to manage irreparable rotator cuff tears. Literature is also lacking on the appropriate inflation volumes.

The biomechanical study by Singh et al. [83] found out that the 25-mL balloon displaced the humeral head more inferior than the intact condition with a mean difference statistically significant, but it is not likely clinically significant. Their study has some limitations and they only tested static abduction.

11.5 Conclusions

The rotator cuff tears modify shoulder biomechanics radically. Rotator cuff repair allows to restore biomechanics in order to address the clinical impairment. If the tear is not completely reparable, adding superior capsule reconstruction to partial cuff repair could be a good option for young and active patient without rotator cuff tear arthropathy, with the aim to restore the native shoulder biomechanics.

Acknowledgments Authors would like to express special thanks to Ms. Silvia Bassini for the pictures and graphic support.

References

1. Karduna AR, Williams GR, Williams JL, Iannotti JP. Kinematics of the glenohumeral joint: influences of muscle forces, ligamentous constraints, and articular geometry. J Orthop Res. 1996;14(6):986–93.
2. Soslowsky LJ, Malicky DM, Blasier RB. Active and passive factors in inferior glenohumeral stabilization: a biomechanical model. J Shoulder Elb Surg. 1997;6(4):371–9.
3. Huegel J, Williams AA, Soslowsky LJ. Rotator cuff biology and biomechanics: a review of normal and pathological conditions. Curr Rheumatol Rep. 2015;17(1):476.
4. Hersche O, Gerber C. Passive tension in the supraspinatus musculotendinous unit after long-standing rupture of its tendon: a preliminary report. J Shoulder Elb Surg. 1998;7(4):393–6.
5. Longo UG, Berton A, Papapietro N, Maffulli N, Denaro V. Biomechanics of the rotator cuff: European perspective. Med Sport Sci. 2012;57:10–7.
6. G. BJ. The Shoulder. By E. A. Codman. Reprint of the 1934 edition. 7 x 6 in. Pp. 513, with 85 figures, 7 plates (2 in colour), some tables and an epilogue of 29 pages. 1965. Brooklyn: G. Miller & Company. Price $17.50. The Journal of Bone and Joint Surgery British volume. 1967;49-B(1):206-.
7. Provencher CM, Makani A, McNeil JW, Pomerantz ML, Golijanin P, Gross D. The role of the scapula in throwing disorders. Sports Med Arthrosc Rev. 2014;22(2):80–7.
8. Sigholm G, Styf J, Herberts P. Subacromial pressure during diagnostic shoulder tests. Clin Biomech (Bristol, Avon). 1988;3(3):187–9.
9. L M. Rupture of the supraspinatus tendon. J Bone Jt Surg. 1937;19:640–2.
10. Neer CS 2nd. Anterior acromioplasty for the chronic impingement syndrome in the shoulder: a preliminary report. J Bone Joint Surg Am. 1972;54(1):41–50.
11. Bigliani LU. The morphology of the acromion and its relationship to rotator cuff tears. OrthopTrans. 1986;10:228.
12. Zuckerman JD, Kummer FJ, Cuomo F, Simon J, Rosenblum S, Katz N. The influence of coracoacromial arch anatomy on rotator cuff tears. J Shoulder Elb Surg. 1992;1(1):4–14.
13. Shi LL, Edwards TB. The role of acromioplasty for management of rotator cuff problems: where is the evidence? Adv Orthop. 2012;2012:467571.
14. Schneeberger AG, Nyffeler RW, Gerber C. Structural changes of the rotator cuff caused by experimental subacromial impingement in the rat. J Shoulder Elb Surg. 1998;7(4):375–80.
15. Hyvönen P, Lohi S, Jalovaara P. Open acromioplasty does not prevent the progression of an impingement syndrome to a tear. Nine-year follow-up of 96 cases. J Bone Joint Surg Br. 1998;80(5):813–6.
16. Ozaki J, Fujimoto S, Nakagawa Y, Masuhara K, Tamai S. Tears of the rotator cuff of the shoulder associated with pathological changes in the acromion. A study in cadavera. J Bone Joint Surg Am. 1988;70(8):1224–30.
17. Fukuda H, Hamada K, Nakajima T, Tomonaga A. Pathology and pathogenesis of the intratendinous tearing of the rotator cuff viewed from en bloc histologic sections. Clin Orthop Relat Res. 1994;304:60–7.
18. Yazici M, Kopuz C, Gülman B. Morphologic variants of acromion in neonatal cadavers. J Pediatr Orthop. 1995;15(5):644–7.

19. Flatow EL, Soslowsky LJ, Ticker JB, Pawluk RJ. Hepler M, ark J, et al. excursion of the rotator cuff under the acromion. Patterns of subacromial contact. Am J Sports Med. 1994;22(6):779–88.
20. Yamamoto N, Muraki T, Sperling JW, Steinmann SP, Itoi E, Cofield RH, et al. Contact between the coracoacromial arch and the rotator cuff tendons in non-pathologic situations: a cadaveric study. J Shoulder Elb Surg. 2010;19(5):681–7.
21. Papadonikolakis A, McKenna M, Warme W, Martin BI, Matsen FA 3rd. Published evidence relevant to the diagnosis of impingement syndrome of the shoulder. J Bone Joint Surg Am. 2011;93(19):1827–32.
22. Torrens C, López JM, Puente I, Cáceres E. The influence of the acromial coverage index in rotator cuff tears. J Shoulder Elb Surg. 2007;16(3):347–51.
23. Moor BK, Röthlisberger M, Müller DA, Zumstein MA, Bouaicha S, Ehlinger M, et al. Age, trauma and the critical shoulder angle accurately predict supraspinatus tendon tears. Orthop Traumatol Surg Res. 2014;100(5):489–94.
24. Gerber C, Snedeker JG, Baumgartner D, Viehöfer AF. Supraspinatus tendon load during abduction is dependent on the size of the critical shoulder angle: a biomechanical analysis. J Orthop Res. 2014;32(7):952–7.
25. Castagna A, Cesari E, Garofalo R, Gigante A, Conti M, Markopoulos N, et al. Matrix metalloproteases and their inhibitors are altered in torn rotator cuff tendons, but also in the macroscopically and histologically intact portion of those tendons. Muscles Ligaments Tendons J. 2013;3(3):132–8.
26. Garofalo R, Cesari E, Vinci E, Castagna A. Role of metalloproteinases in rotator cuff tear. Sports Med Arthrosc Rev. 2011;19(3):207–12.
27. Rothman RH, Parke WW. The vascular anatomy of the rotator cuff. Clin Orthop Relat Res. 1965;41:176–86.
28. Andarawis-Puri N, Kuntz AF, Jawad AF, Soslowsky LJ. Infraspinatus and supraspinatus tendon strain explained using multiple regression models. Ann Biomed Eng. 2010;38(9):2979–87.
29. Yang S, Park HS, Flores S, Levin SD, Makhsous M, Lin F, et al. Biomechanical analysis of bursal-sided partial thickness rotator cuff tears. J Shoulder Elb Surg. 2009;18(3):379–85.
30. Collins DN, Harryman DT 2nd. Arthroplasty for arthritis and rotator cuff deficiency. Orthop Clin North Am. 1997;28(2):225–39.
31. Jensen KL, Williams GR Jr, Russell IJ, Rockwood CA Jr. Rotator cuff tear arthropathy. J Bone Joint Surg Am. 1999;81(9):1312–24.
32. Burkhart SS, Esch JC, Jolson RS. The rotator crescent and rotator cable: an anatomic description of the shoulder's "suspension bridge". Arthroscopy. 1993;9(6):611–6.
33. Burkhart SS, Nottage WM, Ogilvie-Harris DJ, Kohn HS, Pachelli A. Partial repair of irreparable rotator cuff tears. Arthroscopy. 1994;10(4):363–70.
34. Halder AM, O'Driscoll SW, Heers G, Mura N, Zobitz ME, An KN, et al. Biomechanical comparison of effects of supraspinatus tendon detachments, tendon defects, and muscle retractions. JBJS. 2002;84(5):780–5.
35. Kedgley AE, Mackenzie GA, Ferreira LM, Johnson JA, Faber KJ. In vitro kinematics of the shoulder following rotator cuff injury. Clin Biomech (Bristol, Avon). 2007;22(10):1068–73.
36. Su W-R, Budoff JE, Luo Z-P. The effect of Anterosuperior rotator cuff tears on Glenohumeral translation. Arthroscopy. 2009;25(3):282–9.
37. Kolk A, Henseler JF, de Witte PB, van Zwet EW, van der Zwaal P, Visser CPJ, et al. The effect of a rotator cuff tear and its size on three-dimensional shoulder motion. Clin Biomech (Bristol, Avon). 2017;45:43–51.
38. Mueller AM, Rosso C, Entezari V, McKenzie B, Hasebroock A, Cereatti A, et al. The effect of supraspinatus tears on glenohumeral translations in passive pitching motion. Am J Sports Med. 2014;42(10):2455–62.
39. Konrad GG, Markmiller M, Jolly JT, Ruter AE, Sudkamp NP, McMahon PJ, et al. Decreasing glenoid inclination improves function in shoulders with simulated massive rotator cuff tears. Clin Biomech (Bristol, Avon). 2006;21(9):942–9.
40. Mura N, O'Driscoll SW, Zobitz ME, Heers G, Jenkyn TR, Chou SM, et al. The effect of infraspinatus disruption on glenohumeral torque and superior migration of the humeral head: a biomechanical study. J Shoulder Elb Surg. 2003;12(2):179–84.
41. Millett PJ, Giphart JE, Wilson KJ, Kagnes K, Greenspoon JA. Alterations in Glenohumeral kinematics in patients with rotator cuff tears measured with biplane fluoroscopy. Arthroscopy. 2016;32(3):446–51.
42. Kozono N, Okada T, Takeuchi N, Hamai S, Higaki H, Shimoto T, et al. Dynamic kinematics of the glenohumeral joint in shoulders with rotator cuff tears. J Orthop Surg Res. 2018;13(1):9.
43. Kim D, Lee B, Yeom J, Cha J, Han J. Three-dimensional in vivo comparative analysis of the kinematics of normal shoulders and shoulders with massive rotator cuff tears with successful conservative treatment. Clin Biomech (Bristol, Avon). 2020;75:104990.
44. McCully SP, Suprak DN, Kosek P, Karduna AR. Suprascapular nerve block disrupts the normal pattern of scapular kinematics. Clin Biomech (Bristol, Avon). 2006;21(6):545–53.
45. Yu J, McGarry MH, Lee YS, Duong LV, Lee TQ. Biomechanical effects of supraspinatus repair on the glenohumeral joint. J Shoulder Elbow Surg. 2005;14(1 Suppl S):65s–71s.
46. Yamamoto N, Itoi E, Tuoheti Y, Seki N, Abe H, Minagawa H, et al. Glenohumeral joint motion after medial shift of the attachment site of the supraspinatus tendon: a cadaveric study. J Shoulder Elb Surg. 2007;16(3):373–8.
47. Liu J, Hughes RE, O'Driscoll SW, An KN. Biomechanical effect of medial advancement of the

supraspinatus tendon. A study in cadavera. J Bone Joint Surg Am. 1998;80(6):853–9.
48. Burkhart SS. The deadman theory of suture anchors: observations along a South Texas fence line. Arthroscopy. 1995;11(1):119–23.
49. Strauss E, Frank D, Kubiak E, Kummer F, Rokito A. The effect of the angle of suture anchor insertion on fixation failure at the tendon-suture interface after rotator cuff repair: deadman's angle revisited. Arthroscopy. 2009;25(6):597–602.
50. Clevenger TA, Beebe MJ, Strauss EJ, Kubiak EN. The effect of insertion angle on the pullout strength of threaded suture anchors: a validation of the deadman theory. Arthroscopy. 2014;30(8):900–5.
51. Green RN, Donaldson OW, Dafydd M, Evans SL, Kulkarni R. Biomechanical study: determining the optimum insertion angle for screw-in suture anchors-is deadman's angle correct? Arthroscopy. 2014;30(12):1535–9.
52. Itoi E, Nagamoto H, Sano H, Yamamoto N, Kawakami J. Deadman theory revisited. Biomed Mater Eng. 2016;27(2–3):171–81.
53. Meisel AF, Henninger HB, Barber FA, Getelman MH. Biomechanical comparison of standard and linked single-row rotator cuff repairs in a human cadaver model. Arthroscopy. 2017;33(5):938–44.
54. Mazzocca AD, Bollier MJ, Ciminiello AM, Obopilwe E, DeAngelis JP, Burkhart SS, et al. Biomechanical evaluation of arthroscopic rotator cuff repairs over time. Arthroscopy. 2010;26(5):592–9.
55. Park MC, ElAttrache NS, Tibone JE, Ahmad CS, Jun BJ, Lee TQ. Part I: Footprint contact characteristics for a transosseous-equivalent rotator cuff repair technique compared with a double-row repair technique. J Shoulder Elb Surg. 2007;16(4):461–8.
56. Park MC, Tibone JE, ElAttrache NS, Ahmad CS, Jun BJ, Lee TQ. Part II: Biomechanical assessment for a footprint-restoring transosseous-equivalent rotator cuff repair technique compared with a double-row repair technique. J Shoulder Elb Surg. 2007;16(4):469–76.
57. Yamamoto N, Itoi E. A review of biomechanics of the shoulder and biomechanical concepts of rotator cuff repair. Asia Pac J Sports Med Arthrosc Rehabil Technol. 2015;2(1):27–30.
58. Busfield BT, Glousman RE, McGarry MH, Tibone JE, Lee TQ. A biomechanical comparison of 2 technical variations of double-row rotator cuff fixation: the importance of medial row knots. Am J Sports Med. 2008;36(5):901–6.
59. Kaplan K, ElAttrache NS, Vazquez O, Chen YJ, Lee T. Knotless rotator cuff repair in an external rotation model: the importance of medial-row horizontal mattress sutures. Arthroscopy. 2011;27(4):471–8.
60. Tamboli M, Hwang J, McGarry MH, Kang Y, Lee TQ, Mihata T. Biomechanical characteristics of the horizontal mattress stitch: implication for double-row and suture-bridge rotator cuff repair. J Orthop Sci. 2014;19(2):235–41.
61. Nimura A, Kato A, Yamaguchi K, Mochizuki T, Okawa A, Sugaya H, et al. The superior capsule of the shoulder joint complements the insertion of the rotator cuff. J Shoulder Elb Surg. 2012;21(7):867–72.
62. Ishihara Y, Mihata T, Tamboli M, Nguyen L, Park KJ, McGarry MH, et al. Role of the superior shoulder capsule in passive stability of the glenohumeral joint. J Shoulder Elb Surg. 2014;23(5):642–8.
63. Castagna A, Delle Rose G, Conti M, Snyder SJ, Borroni M, Garofalo R. Predictive factors of subtle residual shoulder symptoms after transtendinous arthroscopic cuff repair: a clinical study. Am J Sports Med. 2009;37(1):103–8.
64. Shin SJ. A comparison of 2 repair techniques for partial-thickness articular-sided rotator cuff tears. Arthroscopy. 2012;28(1):25–33.
65. Mihata T, McGarry MH, Ishihara Y, Bui CN, Alavekios D, Neo M, et al. Biomechanical analysis of articular-sided partial-thickness rotator cuff tear and repair. Am J Sports Med. 2015;43(2):439–46.
66. Novi M, Kumar A, Paladini P, Porcellini G, Merolla G. Irreparable rotator cuff tears: challenges and solutions. Orthop Res Rev. 2018;10:93–103.
67. Mihata T, Lee TQ, Watanabe C, Fukunishi K, Ohue M, Tsujimura T, et al. Clinical results of arthroscopic superior capsule reconstruction for irreparable rotator cuff tears. Arthroscopy. 2013;29(3):459–70.
68. Hirahara AM, Adams CR. Arthroscopic superior capsular reconstruction for treatment of massive irreparable rotator cuff tears. Arthrosc Tech. 2015;4(6):e637–41.
69. Hartzler RU, Burkhart SS. Superior Capsular Reconstruction. Orthopedics. 2017;40(5):271–80.
70. Narvani AA, Consigliere P, Polyzois I, Sarkhel T, Gupta R, Levy O. The "pull-over" technique for arthroscopic superior capsular reconstruction. Arthrosc Tech. 2016;5(6):e1441–e7.
71. Petri M, Greenspoon JA, Millett PJ. Arthroscopic superior capsule reconstruction for irreparable rotator cuff tears. Arthrosc Tech. 2015;4(6):e751–5.
72. Sutter EG, Godin JA, Garrigues GE. All-arthroscopic superior shoulder capsule reconstruction with partial rotator cuff repair. Orthopedics. 2017;40(4):e735–e8.
73. Dimock RAC, Malik S, Consigliere P, Imam MA, Narvani AA. Superior capsule reconstruction: what do we know? Arch Bone Jt Surg. 2019;7(1):3–11.
74. Mihata T, McGarry MH, Pirolo JM, Kinoshita M, Lee TQ. Superior capsule reconstruction to restore superior stability in irreparable rotator cuff tears: a biomechanical cadaveric study. Am J Sports Med. 2012;40(10):2248–55.
75. Mihata T, McGarry MH, Kahn T, Goldberg I, Neo M, Lee TQ. Biomechanical effect of thickness and tension of fascia Lata graft on Glenohumeral stability for superior capsule reconstruction in irreparable supraspinatus tears. Arthroscopy. 2016;32(3):418–26.
76. Mihata T, McGarry MH, Kahn T, Goldberg I, Neo M, Lee TQ. Biomechanical role of capsular continuity in superior capsule reconstruction for irreparable

tears of the supraspinatus tendon. Am J Sports Med. 2016;44(6):1423–30.
77. Han F, Kong CH, Hasan MY, Ramruttun AK, Kumar VP. Superior capsular reconstruction for irreparable supraspinatus tendon tears using the long head of biceps: a biomechanical study on cadavers. Orthop Traumatol Surg Res. 2019;105(2):257–63.
78. Gartsman GM, O'Connor DP. Arthroscopic rotator cuff repair with and without arthroscopic subacromial decompression: a prospective, randomized study of one-year outcomes. J Shoulder Elb Surg. 2004;13(4):424–6.
79. MacDonald P, McRae S, Leiter J, Mascarenhas R, Lapner P. Arthroscopic rotator cuff repair with and without acromioplasty in the treatment of full-thickness rotator cuff tears: a multicenter, randomized controlled trial. J Bone Joint Surg Am. 2011;93(21):1953–60.
80. Tashjian RZ. AAOS clinical practice guideline: optimizing the management of rotator cuff problems. J Am Acad Orthop Surg. 2011;19(6):380–3.
81. Denard PJ, Bahney TJ, Kirby SB, Orfaly RM. Contact pressure and glenohumeral translation following subacromial decompression: how much is enough? Orthopedics. 2010;33(11):805.
82. Mihata T, McGarry MH, Kahn T, Goldberg I, Neo M, Lee TQ. Biomechanical effects of Acromioplasty on superior capsule reconstruction for irreparable supraspinatus tendon tears. Am J Sports Med. 2016;44(1):191–7.
83. Singh S, Reeves J, Langohr GDG, Johnson JA, Athwal GS. The effect of the subacromial balloon spacer on humeral head translation in the treatment of massive, irreparable rotator cuff tears: a biomechanical assessment. J Shoulder Elb Surg. 2019;28(10):1841–7.

Biomechanics of Shoulder Instability and Repair

12

John Fritch, Andre Labbe, Jacques Courseault, and Felix Savoie

12.1 Introduction

The shoulder complex comprises over 30 muscles, two anatomic spaces, and four articulations moving through three main degrees of freedom allowing for the greatest range of motion of any joint in the human body. The shoulder demands synchronized, harmonious action through balanced mobility and stabilization to provide movement utility. Understanding the fundamental concepts of the shoulder complex's clinical kinesiology serves as a foundation for the development of recognition, treatment, and repair strategies of shoulder instability.

In understanding the closed-chain mechanism of the shoulder, the glenohumeral joint balances stability against excessive translations with mobility necessary to achieve specific tasks [1]. Pathologic glenohumeral joint instability is generally classified as "traumatic" as a result from an acute event such as a fall or "atraumatic" as a result of repetitive microtrauma or in patients with preexisting generalized ligamentous laxity. Instability is further characterized by direction: anterior, posterior, or multidirectional. The severity of instability can be variable ranging from frank joint dislocation to subtle subluxation, all of which can result in debilitating symptoms. Primary complaints may relate to dysfunction or pain, with underlying instability often contributing chondral injury and the development of glenohumeral arthritis over time [2].

J. Fritch (✉)
Department of Orthopaedic Surgery, University of Chicago, Chicago, IL, USA

A. Labbe · J. Courseault · F. Savoie
Department of Orthopaedic Surgery, Tulane University, New Orleans, LA, USA

12.2 Anatomy and Biomechanics

The complex bony and soft tissue anatomy of the shoulder is integral to its function and even small derangements can lead to pathologic dysfunction. Understanding normal biomechanics helps the clinician predict the limitation and failure of these systems. The movement of this system is directed through an intricate relationship between static and dynamic stability. Together, static and dynamic stability provide functionality throughout a wide range of motion. The inherent stability of the shoulder is complex and involves a system of structures that adapt to physiologic demands.

12.2.1 Static Shoulder Stability

Along with negative intra-articular pressure, which draws the humeral head into the glenoid surface, the bony anatomy of the shoulder, glenoid labrum, glenohumeral ligaments, and joint capsule all contribute to the inherent stability of

J. Koh et al. (eds.), *Orthopaedic Biomechanics in Sports Medicine*,
https://doi.org/10.1007/978-3-030-81549-3_12

the shoulder [3]. Even with these structural components, the glenohumeral relationship contains a certain degree of mismatch between the articulating surfaces of the proximal humerus and the glenoid fossa. In supporting the base of the glenohumeral joint, the scapula plays a significant role in the stability of the joint through its structural geometry and position. The height and width of the glenoid surface measures approximately 39 mm and 29 mm with its surface area in approximately 7 degrees of retroversion [4, 5]. The superior and inferior poles of the glenoid serve as insertion points for the long head of the biceps and the long head of the triceps, respectively. Due to the inherent offset of the humeral head and the glenoid fossa, approximately 30% of the articular surface is in contact with the articular cartilage of the glenoid throughout the maximum functional range of motion [6]. Supplementation by the fibrous glenoid labrum provides an extra level of stability to the articular surface. Howell and Galiant report anatomic studies in which the combination of the bony glenoid surface and fibrous labral tissue create a socket (~ 9 mm deep at the superior and inferior positions and ~ 5 mm deep at the anterior and posterior positions*)* for the articulation of the humeral head (~ 30 deg. *retroversion)* [7]. The glenoid labrum plays a significant role in its ability to restrict the translational motion of the humeral head over the glenoid surface, providing additional structural stability. The labrum differs in the morphology and attachment to the glenoid surface depending on location. The labrum is loosely attached superiorly and anterosuperiorly to the bony surface where it serves as an attachment site for the tendon of the long head of the biceps. Inferiorly the labrum is firmly attached to provide structural stability against inferior translational forces [8]. The acromion and the coracoid provide additional structural support for the shoulder. These structures serve as attachment points for ligamentous tissue, such as the coracoacromial ligament (CAL), which contributes to the coracoacromial arch. The coracoid, acromion, and coracoacromial ligament together provide resistance to the superior and anterosuperior translation of the humeral head, as Fagelman et al. describe in cadaveric studies [9]. Additional shoulder stability is provided by the glenohumeral ligaments which appear as thickenings of the shoulder capsule. The superior glenohumeral ligament (SGHL) originates from the labrum at its anterior superior border, then courses inferiorly in the region of the rotator interval where it inserts into the lesser tuberosity of the humerus. The SGHL is composed of oblique and direct fibers. These fibers provide additional stability to the long head of the biceps tendon and contribute to the structure of the transverse humeral ligament [10]. Capsuloligamentous restraint studies have determined that the SGHL plays a specific role in resisting the inferior translation of the humeral head in the adducted shoulder. As the angle of abduction increases from 0 to 90°, the anterior portions of the ligament engage to provide the greatest resistance to inferior translation at 45°, whereas the posterior aspect of the ligament engages to provide the greatest resistance at 90° [11]. According to a study that investigated normal variations in the glenohumeral ligament complex (GHLC), the middle glenohumeral ligament (MGHL) was found to be present in only 63% of individuals [12]. The MGHL arises from the anterior aspect of the superior portion of the labrum and courses deep to the subscapularis muscle attaching to the anatomic neck of the humerus. The MGHL provides mild resistance to anterior translation of the humeral head at approximately 45 degrees of abduction [13]. The IGHL consists of two distinct bands—the anterior and posterior band, which represent cord-like thickenings of the glenohumeral joint capsule. The anterior-inferior glenohumeral ligament (aIGHL) most commonly arises from the 2 to 4 o'clock position of the glenoid, and the posterior-inferior glenohumeral ligament (pIGHL) most commonly arises from between 6 and 9 o'clock. Interposed between the anterior and posterior bands of the IGHL is the axillary pouch. Together, these two bands form the IGHL complex at the inferior margin of the humeral neck [14]. The IGHL is thought of as one of the most critical stabilizing structures when the shoulder is externally rotated and abducted to 90° [15]. The joint capsule itself comprises a multitude of fibrous

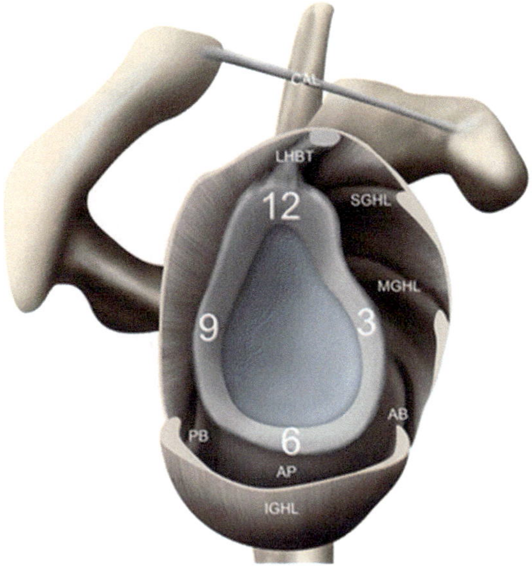

Fig. 12.1 Lateral view of the glenoid with the humerus subtracted. Demonstrates the static stabilizers including the inferior glenohumeral ligament (IGHL) complex comprised of the anterior (AB) and posterior (PB) bands and the axillary pouch (AP). The clock face positions of the glenoid labrum (12 = superior, 3 = anterior, 6 = inferior, 9 = posterior), long head of the biceps tendon (LHBT), superior (SGHL) and middle (MGHL) glenohumeral ligaments, and coracoacroimal ligament are also shown. From Passanante, et al. [16]

bands which course obliquely in alternating directions and maintains a negative pressure environment which, as previously stated, contributes to an intimate glenohumeral relationship. The contracture of this fibrous capsule affects glenohumeral motion and plays a role in glenohumeral motion (Fig. 12.1).

12.2.2 Dynamic Shoulder Stability

Muscles surrounding the glenohumeral joint and scapula provide dynamic stability. Combined with negative intra-articular pressure, the rotator cuff muscles create a joint compressive force directing the humeral head into its glenoid articulation. This directed pull is key in generating force to direct purposeful movement. The scapulothoracic musculature must maintain proper scapular orientation for full glenohumeral motion. This involves internal rotation of 30°, abduction of 3°, and anterior tilt of 20° [17]. This position is important for creating a proper interface for the humeral head and glenoid to interact as well as maintaining smooth motion of the scapulothoracic articulation in its anatomic space between the posterior thoracic wall and the anterior surface of the scapula [18]. The scapula serves as the insertion point for many of these dynamic components. Contributions come directly from attachments of the trapezius, levator scapulae, rhomboids, serratus anterior, pectoralis minor, and the subclavius. Together, these muscles contribute to proper scapulothoracic motion and facilitate harmonious scapular and glenohumeral motion. Anatomic abnormalities in any one of these attachments can alter the biomechanics of shoulder motion leading to various shoulder pathology. Scapulothoracic articulation allows for abduction beyond the 120° the glenohumeral joint allows [18]. The serratus anterior is essential in maintaining close apposition of the scapula and the thorax, ensuring smooth scapular rotation.

The rotator cuff consists of four muscles: the supraspinatus, infraspinatus, teres minor, and the subscapularis. The muscles of the rotator cuff provide an additional source of stability for the glenohumeral joint. These muscles effectively align to resist glenohumeral shear stress and facilitate compressive force at the glenohumeral interface, which corresponds to increased stability and decreased translation [19]. Wuelker et al. demonstrate that a 50% decrease in rotator cuff force results in an approximate 50% increase in anterior subluxation of the humeral head when subjected to loading at all glenohumeral joint positions [19]. Each rotator cuff muscle demonstrates assistance in anterior shoulder stability; however, there is no evidence to suggest one particular muscle's contribution over the other [20]. The supraspinatus works in conjunction with the infraspinatus as the primary cuff muscles involved in shoulder abduction. The supraspinatus contributes to approximately 50% of shoulder abduction in the scapular plane. In studies where the supraspinatus and infraspinatus were paralyzed, a loss of approximately 75% of strength in abduction was observed, highlighting their utility in force generation necessary for abduction [21]. The supraspinatus and infraspinatus are highly involved in the force-couple mechanism which provides compression into the glenoid fossa to stabilize the glenohumeral joint and facilitates abduction [22]. Weakness of the supraspinatus and infraspinatus may lead to alterations in joint reaction force to the superior posterior rim of the glenoid surface. These alterations demonstrate the significance of these muscles in the stability of the glenoid articulation and provide information on shoulder biomechanics [23].

The teres minor contributes to external rotation and balance of the glenohumeral joint. Its contribution to the stabilization of the glenohumeral join is best demonstrated in the presence of injury. The teres minor also may substitute for lost joint torque during several simulated muscle tears to maintain joint stability in a study conducted by Williams et al., joint instability demonstrated a positive correlation when compared to the size of the rotator cuff tear [24].

The origin of the subscapularis arises from the subscapular fossa on the anterior surface of the scapula and inserts on the lesser tuberosity. The subscapularis contributes to internal rotation and exhibits the largest impression at its insertion medial to the bicipital groove. At 0 degrees of abduction, the subscapularis serves as a primary stabilizing structure and is in a direct force-coupling relationship to anterior-posterior glenohumeral motion preventing posterior subluxation [25]. The complex relationship between static and dynamic stability in relation to the biomechanics of the shoulder serves as a foundation for the natural shoulder motion (Figs. 12.2 and 12.3).

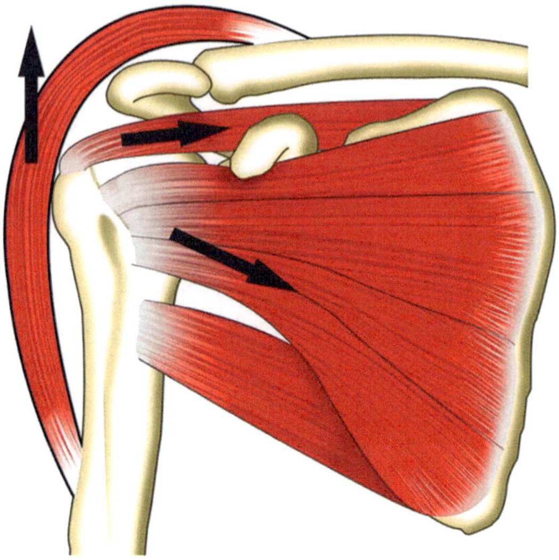

Fig. 12.2 Force couple acting on the glenohumeral joint in the coronal plane. The deltoid (upward pointing arrow) forms a force couple with the supraspinatus (left to right pointing arrow) and the infraspinatus, teres minor, subscapularis, latissimus dorsi, and teres major (left to right and slightly downwards pointing arrow). As a result of this force couple, when the deltoid contracts the arm abducts rather than pulling the humeral head upward. From Panayiotou Charalambous [26]

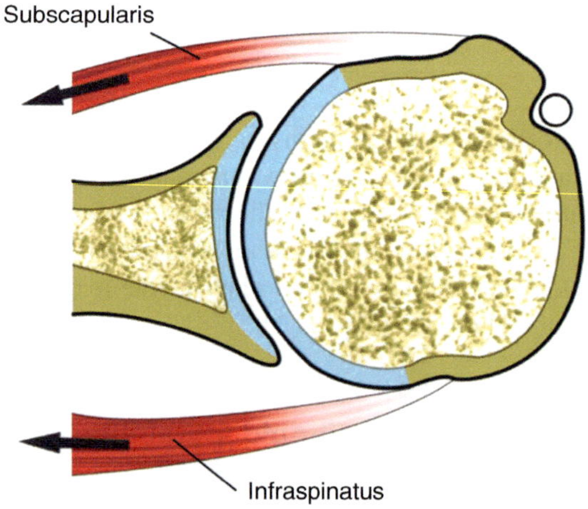

Fig. 12.3 Force couple acting on the glenohumeral joint in the horizontal plane. The infraspinatus contracts to pull the humeral head medially and posteriorly while the subscapularis contracts to pull the humeral head medially and anteriorly. The force couple acts to dynamically stabilize the humeral head in the glenoid throughout shoulder range of motion. From Panayiotou Charalambous [26]

12.3 Pathophysiology

12.3.1 Anterior Instability

Anterior shoulder dislocation is thought to occur with the arm in 90 degrees of abduction and external rotation, although this is often debated [27]. The anterior band of the inferior glenohumeral ligament is the primary restraint to dislocation in this position and once it fails, other soft tissue and bony stabilizers are disrupted depending on the degree of injury force.

Anterior instability can also present without frank dislocation. Chronic, repetitive microtrauma, such as that seen in overhead athletes, can cause subtle subluxations with remaining glenohumeral contact. Such patients will often demonstrate anterior shoulder pain and feelings of transient instability [28].

A detailed history, physical exam, and plain radiograph imaging is essential for diagnosis. MRI is useful for evaluating the soft tissue stabilizers such as the labrum, glenohumeral ligaments, rotator cuff, and capsule. MR arthrogram is particularly useful for labral injuries. CT imaging is important for evaluating for bony lesions, glenoid bone loss, and the size and orientation of humeral head lesions.

The most common injury after an anterior dislocation or subluxation is a Bankart lesion, a tear of the anterior-inferior labrum and capsule from the glenoid [28]. The intact labrum provides stability to the joint by acting as a bumper as well as creating a suction seal effect between the humeral head and glenoid [29]. Any disruption to this results in decreased stability. In cadaveric biomechanical studies, there was an increase in anterior translation of the humeral head from 6.8 mm to 9.6 mm between shoulders with an intact labrum and those with a simulated Bankart lesion [30].

Glenoid labral articular disruption (GLAD) lesions are similar to Bankart lesions but include injury to the adjacent glenoid cartilage. GLAD lesions can result in recurrent instability and bone loss [31]. A humeral avulsion of the glenohumeral ligament (HAGL) is an injury to the IGHL which can be seen in some anterior dislocations. Patients usually report pain and less commonly recurrence of instability [32]. Anterior labroligmentous periosteal sleeve avulsions (ALPSA) occur when the labrum, anterior IGHL, and periosteum are pealed back medially along the glenoid neck [33]. Importantly, they can heal displaced medially, which requires restoration of normal anatomy to restore stability. Superior labrum anterior to posterior (SLAP) tears may involve the long head of the biceps anchor and may be seen after 22% of traumatic dislocations [34]. Injuries to the rotator cuff are often seen after traumatic dislocation in patients over 40 years old.

Bony Bankart lesions, or Bankart fractures, are seen in one third of first-time anterior dislocations [35]. They can be easily missed on initial imaging and if not addressed at the time of repair can result in attrition of the fragment, glenoid bone loss, and further instability [35]. A Hill–Sachs lesion is a compression fracture of the posterior humeral head that is the result of the impact of the humeral head on the anterior glenoid rim during anterior dislocation. The size and location of the lesion as well as its interaction with a concurrent glenoid lesion (Bankart) plays an important role in the risk for recurrence of instability. The glenoid track is the contact area between the humeral head and glenoid [36]. A Hill–Sachs lesion is off-track when it engages the anterior glenoid rim, otherwise it is considered on-track. As the Hill–Sachs lesion gets larger and more medial and glenoid bone loss increases, so does the risk of recurrent instability. Yamamoto et al. developed a reliable method for calculating whether a Hill–Sachs lesion is on- or off-track based on 3D CT [36]. In this method, both glenoids are compared in en face view. The glenoid track is 83% of the uninjured glenoid's width. The glenoid defect is then subtracted from the uninjured glenoid and the calculated width of the glenoid track is applied to the posterior humeral view. The glenoid track is superimposed on the posterior humerus starting medially at the medial aspect of the cuff insertion on the greater tuberosity. If the Hill–Sachs lesion extends medially to the glenoid track, it is considered "off-track" and at higher risk of engaging and causing recurrent instability. As such, the presence of bone loss with off-track Hill–Sachs lesions drives surgical decision-making toward restoring bony anatomy to restore stability [37] (Fig. 12.4).

12.3.2 Posterior Instability

Posterior glenohumeral instability is a common cause of shoulder dysfunction. It usually presents without a traumatic event, making its diagnosis more challenging. Unlike anterior instability, posterior instability is usually caused by repetitive microtrauma from activities that load the shoulder in the flexed, adducted, and internally rotated position [39].

Several key anatomic and biomechanical features differentiate posterior from anterior glenohumeral instability. With regards to its soft tissue stabilizers, the posterior capsule and the posterior band of the IGHL are significantly thinner and less robust then their anterior counterparts [40]. In patients with posterior instability the cross-sectional area of the capsule is increased, unlike in anterior instability, suggesting that posterior capsular disruption may occur with lower energy injuries [41]. This may explain repetitive micro-trauma as a common cause of posterior instability. In addition to enlargement of the capsule, posterior Bankart lesions (posterior-inferior labrocapsular tears) and Kim lesions (incomplete avulsions of the posterior labrum) are commonly seen in these patients [42]. In a cadaveric study by Wellmann et al., posterior Bankart lesions resulted in an 86% increase in posterior humeral translation in the jerk position and 31% increase in inferior translation in the sulcus position [43]. An injury to the pIGHL resulted in 53% increase in inferior translation but no changes in posterior translation [43]. This biomechanical data suggests that posterior labral and capsular injuries can result in significant clinical translation of the humerus.

Bony congruency of the humeral head on the glenoid throughout the arc of motion is necessary for stability. Glenoid retroversion and hypoplasia have been shown to compromise this constraint allowing for recurrent posterior subluxation, but it is unclear if this pathology causes or is caused by repetitive instability [44]. In patients with traumatic or recurrent instability, posterior bone loss can be observed including reverse bony Bankart lesions (posterior-inferior glenoid) and reverse Hill–Sachs lesions (anterior humeral head impaction fracture). Similar to their anterior counterparts, bone loss in the posterior shoulder plays an important role in treatment decisions.

12.3.3 Multidirectional Instability

Multidirectional shoulder instability (MDI) is defined as symptomatic instability in two or more directions. It was initially described by Neer and Foster in 1980 as the combination of anterior and posterior instability [45]. Diagnosis of MDI challenging due to a lack of pathognomonic clinical and radiographic findings. Although patients

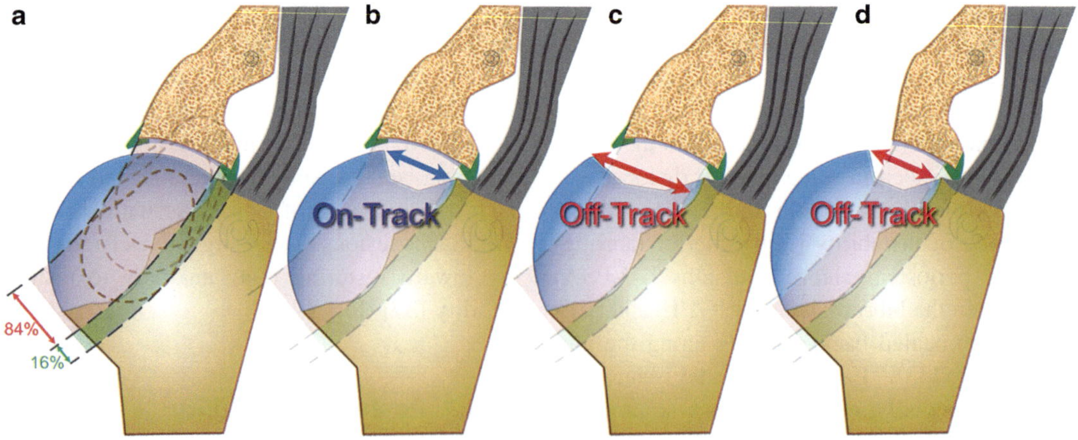

Fig. 12.4 Glenoid track theory. (**a**) The glenoid track is the contact area of the glenoid surface onto the humeral head in the ABER position (abduction, external rotation), which is approximately 83–84% of the glenoid width. (**b**) On-track: the Hill-Sachs lesion remains within the glenoid track and is not engaging. (**c**) Off-track: the Hill–Sachs lesion extends past the glenoid track medially and is engaging. (**d**) Off-track: the bony Bankart defect narrows the glenoid track on the humeral head so that the Hill–Sachs lesion is engaging. From Tuite and Pfirrmann 2021, an open access article [38]

with abnormal joint laxity are predisposed to developing instability, it is important to distinguish that for a diagnosis of MDI, patients must have the presence of symptoms of instability, with or without joint hyperlaxity.

MDI results from a failure in static and dynamic stabilizers of the glenohumeral joint. Patients with MDI have been shown to exhibit shallower glenoid cavities [46], insufficiency in the glenohumeral ligaments (particularly the bands of the IGHL) as evidenced by a patulous inferior capsule [45], and labral abnormalities [41]. Dysfunction in the dynamic shoulder stabilizers including abnormal scapular kinematics [47] and aberrant neuromuscular control of the rotator cuff [48] results in increased glenohumeral translation that is seen in patients with MDI.

12.4 Surgical Treatment and Biomechanical Considerations

12.4.1 Bankart Repair

The most common surgical procedure for anterior shoulder stability is a repair to the anterior-inferior capsulolabral complex. This can be performed arthroscopically or open with similar clinical outcomes and comparable recurrence rates of instability [49]. Advantages of arthroscopic Bankart repair include reduced surgical time, faster recovery, better visualization of intra-articular pathology, and improved cosmesis [50]. Advantages of open Bankart repair include better ability to repair large bony Bankart fractures and the ability to incorporate an open capsular shift in cases of hyperlaxity or in failed arthroscopic repairs. The most important risk factors for failure of arthroscopic repair include the number of previous dislocations, total duration of instability, glenoid bone loss, and the presence of an off-track Hill–Sachs lesion [51].

Regardless of the approach, studies have shown that repair of the anterior-inferior glenoid capsulolabral complex improves stability. In cadaveric studies, repair of a simulated Bankart lesion decreased anterior humeral translation from 9.6 mm to 6.95 mm on average [30]. Another biomechanical study showed stability of the repair increases with the number of anchors used as well as making sure the repair extends completely from 3 to 6 o'clock on the glenoid face [52]. Single versus double row techniques, single versus double loaded anchors, knotted versus knotless anchors, and simple versus horizontal mattress suture configurations have not shown any difference in biomechanical strength [53, 54].

Bony Bankart lesions are important to recognize early and incorporate into the repair to prevent fragment resorption. They can be repaired with screws or arthroscopic suture in linear or bridge techniques with similarly good results [55]. However, failure to address bony Bankart lesions can result in glenoid bone loss and recurrent instability.

Many studies have worked to develop a treatment algorithm based on the risk of recurrent instability. Most recently DiGiacomo et al. published a study validating the Glenoid Track Instability Management Score (GTIMS) (Table 12.1) [56]. This scoring system uses 3D CT to classify bone loss as on- or off-track. By using this system, investigators found more accurate prediction of failure after arthroscopic Bankart repair and reduced the number of Latarjet procedures without significant difference in outcome [56].

Table 12.1 Treatment options based on glenoid bone loss (GBL) and on-track versus off-track Hill–Sachs lesion (HSL). From Di Giacomo et al. [56].

GBL %	On-track HSL	Off-track HSL
0–13.5	Arthroscopic Bankart repair	Arthroscopic Bankart repair + remplissage
		Open inferior capsular shift
		Latarjet
13.5–25	Arthroscopic Bankart repair + remplissage	Arthroscopic Bankart repair + remplissage
	Open inferior capsular shift	Open inferior capsular shift
	Latarjet	Latarjet
>25	Latarjet	Latarjet

12.4.2 Capsular Shift

Capsular imbrication is considered in patients with excessive capsular volume, multidirectional instability, and in conjunction with Bankart repair when there is glenoid bone loss greater than 13.5% or an off-track Hill–Sachs lesion. Initially described as an open humeral-based inferior capsular shift by Neer and Foster [45], this involves a T-shaped incision in the anterior capsule between the middle and inferior glenohumeral ligaments and advancement of the flaps over each other to reduce posterior capsular redundancy and eliminate the inferior capsular pouch. Recent studies have shown good results after open Bankart repair with capsular shift with a low rate of recurrent instability, however, there is frequently loss of external rotation which patients should be counseled on prior to surgery [57]. Care must be taken to avoid overtightening as it can result in posterior subluxation, pain, and arthrosis [58].

Arthroscopic capsulorrhaphy techniques can be used in the setting of anterior, posterior, and multidirectional instability. Advantages over open capsular plication include decreased morbidity, direct visualization of the decrease in capsular volume, and avoidance of subscapularis detachment. In these techniques, suture is passed sometimes through multiple pleats of capsular tissue and often incorporated into labral repair. This can be performed in the anterior, inferior, or posterior capsule depending on the pathology. Biomechanical studies have shown use of suture anchors is superior to suture alone in this setting [59].

12.4.3 Remplissage

Remplissage means "to fill in" in French. Originally described as an open procedure in 1972, this involves transferring a portion of the infraspinatus tendon into the Hill–Sachs defect, thereby making the defect extraarticular and preventing engagement of the glenoid and recurrent instability [60]. More recently, arthroscopic techniques have been used, reducing surgical time, morbidity, and allowing for concomitant arthroscopic procedures to be performed [61]. This technique has been improved by using a double pulley suture anchor technique allowing for a larger footprint of fixation covering the entire defect while allowing for a more anatomic and tissue sparing approach [62]. This construct has been shown to have superior biomechanical strength and preserve motion. Similar to capsular shift, remplissage can result in loss of external rotation and patients should be counseled accordingly. In one study by Boileau et al., in 459 patients undergoing arthroscopic remplissage, there was a mean loss of 8 degrees of external rotation with the arm at the side and 9° with the arm abducted [63].

A similar procedure can be performed in the setting of a reverse Hill–Sachs lesion and posterior instability. The McLaughlin procedure involves open transfer of the subscapularis tendon and lesser tuberosity into the reverse Hill–Sachs defect thereby preventing engagement and instability by the same principles as the remplissage [64]. The modified McLaughlin procedure is an arthroscopic technique with good outcomes [65].

12.4.4 Coracoid Transfer

The Latarjet procedure is the most commonly used bone block procedure for anterior instability. It confers stability through restoring the joint surface in the setting of glenoid bone loss, creating a sling with the conjoint tendon and subscapularis to resist anteroinferior translation of the humeral head, and utilizing a repair of the coracoacromial ligament stump to the anterior capsule to form a bumper resisting anterior translation [66]. Two techniques for graft placement on the glenoid exist. The traditional placement uses the lateral side of the coracoid as the joint surface, whereas another method uses the inferior coracoid surface. The inferior surface offers a larger and more anatomic congruent arc, however there is a higher risk of graft fragmentation. Another coracoid transfer is the Bristow procedure, which involves a smaller graft fixed with one instead of two screws and confers less stability in the setting of significant bone loss than the Latarjet [67]. The Latarjet procedure has a relatively low recurrent instability rate even in the setting of large bone defects, engaging Hill–Sachs lesions, and preoperative recurrent instability [68]. The procedure does include a different complication profile than arthroscopic Bankart repair, most notably risk of axillary or musculocutaneous nerve injury. It is important to discuss these factors with the patient in the shared decision-making process.

12.4.5 Bone Block Procedures

Recurrence of instability after Latarjet has been reported to be as high as 12% [69]. Most often this is due to graft resorption, malunion, or hardware complications. In the setting of a failed Latarjet, there are several graft options available. Iliac crest autograft procedures include the Eden-Hybinette and J-graft procedures which have comparable and satisfactory recurrent instability rates with the most common disadvantage of donor site morbidity [70]. Distal clavicle autograft is another viable option and provides a broad cartilage-covered surface to augment the glenoid [71]. Another option is distal tibia allograft which has the advantages of avoiding donor site morbidity and restoring the native curvature of the glenoid thus restoring the congruent arc [72].

Humeral sided bone loss is most commonly managed with soft tissue procedures but in the case of large defects greater than 25% of the humeral head surface, bone augmentation may be considered. The principle behind such a procedure is to fill the defect and restore native anatomy by increasing the articular arc of the humerus and preventing engagement and instability. Case series of such procedures have shown reasonable outcomes but with complications including graft collapse and hardware prominence [73].

12.5 Conclusion

The glenohumeral joint is the most mobile joint in the body and as a result is susceptible to instability. This chapter highlights the delicate balance between static and dynamic stabilizers of the shoulder. Understanding the anatomic and biomechanical principles behind various types of shoulder instability are crucial to developing a treatment plan. Various repair techniques take into consideration such factors as direction of instability, failure of soft tissue or bony constraints, glenoid bone loss, and on versus off-track humeral lesions. Good outcomes can be expected when these variables are respected and patients are appropriately counseled on the current data regarding shoulder instability.

References

1. Veeger HE, van der Helm FC. Shoulder function: the perfect compromise between mobility and stability. J Biomech. 2007;40(10):2119–29.
2. Ludewig PM, Reynolds JF. The association of scapular kinematics and glenohumeral joint pathologies. J Orthop Sports Phys Ther. 2009;39(2):90–104.
3. Lugo R, Kung P, Ma CB. Shoulder biomechanics. Eur J Radiol. 2008;68(1):16–24.
4. Iannotti JP, Gabriel JP, Schneck SL, Evans BG, Misra S. The normal glenohumeral relationships. An anatomical study of one hundred and forty shoulders. J Bone Joint Surg Am. 1992;74(4):491–500.
5. Joseph J. Theory of shoulder mechanism. descriptive and applied. J Anat. 1962;96(Pt 3):412-.
6. Warner JJ, Bowen MK, Deng XH, Hannafin JA, Arnoczky SP, Warren RF. Articular contact patterns of the normal glenohumeral joint. J Shoulder Elb Surg. 1998;7(4):381–8.
7. Howell SM, Galinat BJ. The glenoid-labral socket. A constrained articular surface. Clin Orthop Relat Res. 1989;243:122–5.
8. Cooper DE, Arnoczky SP, O'Brien SJ, Warren RF, DiCarlo E, Allen AA. Anatomy, histology, and vascularity of the glenoid labrum. An anatomical study. J Bone Joint Surg Am. 1992;74(1):46–52.
9. Fagelman M, Sartori M, Freedman KB, Patwardhan AG, Carandang G, Marra G. Biomechanics of coracoacromial arch modification. J Shoulder Elb Surg. 2007;16(1):101–6.
10. Kask K, Põldoja E, Lont T, Norit R, Merila M, Busch LC, et al. Anatomy of the superior glenohumeral ligament. J Shoulder Elb Surg. 2010;19(6):908–16.

11. Warner JJ, Deng XH, Warren RF, Torzilli PA. Static capsuloligamentous restraints to superior-inferior translation of the glenohumeral joint. Am J Sports Med. 1992;20(6):675–85.
12. Ide J, Maeda S, Takagi K. Normal variations of the glenohumeral ligament complex: an anatomic study for arthroscopic Bankart repair. Arthroscopy. 2004;20(2):164–8.
13. Turkel SJ, Panio MW, Marshall JL, Girgis FG. Stabilizing mechanisms preventing anterior dislocation of the glenohumeral joint. J Bone Joint Surg Am. 1981;63(8):1208–17.
14. Chahla J, Aman ZS, Godin JA, Cinque ME, Provencher MT, LaPrade RF. Systematic review of the anatomic descriptions of the glenohumeral ligaments: a call for further quantitative studies. Arthroscopy. 2019;35(6):1917–26.
15. O'Brien SJ, Schwartz RS, Warren RF, Torzilli PA. Capsular restraints to anterior-posterior motion of the abducted shoulder: a biomechanical study. J Shoulder Elb Surg. 1995;4(4):298–308.
16. Passanante GJ, Skalski MR, Patel DB, White EA, Schein AJ, Gottsegen CJ, et al. Inferior glenohumeral ligament (IGHL) complex: anatomy, injuries, imaging features, and treatment options. Emerg Radiol. 2017;24(1):65–71.
17. Halder AM, Itoi E, An KN. Anatomy and biomechanics of the shoulder. Orthop Clin North Am. 2000;31(2):159–76.
18. Terry GC, Chopp TM. Functional anatomy of the shoulder. J Athl Train. 2000;35(3):248–55.
19. Wuelker N, Korell M, Thren K. Dynamic glenohumeral joint stability. J Shoulder Elb Surg. 1998;7(1):43–52.
20. Blasier RB, Guldberg RE, Rothman ED. Anterior shoulder stability: contributions of rotator cuff forces and the capsular ligaments in a cadaver model. J Shoulder Elb Surg. 1992;1(3):140–50.
21. Gerber C, Blumenthal S, Curt A, Werner CM. Effect of selective experimental suprascapular nerve block on abduction and external rotation strength of the shoulder. J Shoulder Elb Surg. 2007;16(6):815–20.
22. Howell SM, Imobersteg AM, Seger DH, Marone PJ. Clarification of the role of the supraspinatus muscle in shoulder function. J Bone Joint Surg Am. 1986;68(3):398–404.
23. Hölscher T, Weber T, Lazarev I, Englert C, Dendorfer S. Influence of rotator cuff tears on glenohumeral stability during abduction tasks. J Orthop Res. 2016;34(9):1628–35.
24. Williams MD, Edwards TB, Walch G. Understanding the importance of the Teres minor for shoulder function: functional anatomy and pathology. J Am Acad Orthop Surg. 2018;26(5):150–61.
25. Huegel J, Williams AA, Soslowsky LJ. Rotator cuff biology and biomechanics: a review of normal and pathological conditions. Curr Rheumatol Rep. 2015;17(1):476.
26. Panayiotou Charalambous C. Shoulder biomechanics. the shoulder made easy. Cham: Springer; 2019. p. 45–67.
27. Provencher MT, Frank RM, LeClere LE, Metzger PD, Ryu JJ, Bernhardson A, et al. The hill-Sachs lesion: diagnosis, classification, and management. J Am Acad Orthop Surg. 2012;20(4):242–52.
28. Owens BD, Nelson BJ, Duffey ML, Mountcastle SB, Taylor DC, Cameron KL, et al. Pathoanatomy of first-time, traumatic, anterior glenohumeral subluxation events. J Bone Joint Surg Am. 2010;92(7):1605–11.
29. Habermeyer P, Schuller U, Wiedemann E. The intra-articular pressure of the shoulder: an experimental study on the role of the glenoid labrum in stabilizing the joint. Arthroscopy. 1992;8(2):166–72.
30. Marquardt B, Hurschler C, Schneppendahl J, Witt KA, Pötzl W, Steinbeck J. Quantitative assessment of glenohumeral translation after anterior shoulder dislocation and subsequent arthroscopic bankart repair. Am J Sports Med. 2006;34(11):1756–62.
31. Pogorzelski J, Fritz EM, Horan MP, Katthagen JC, Provencher MT, Millett PJ. Failure following arthroscopic Bankart repair for traumatic antero-inferior instability of the shoulder: is a glenoid labral articular disruption (GLAD) lesion a risk factor for recurrent instability? J Shoulder Elb Surg. 2018;27(8):e235–e42.
32. Provencher MT, McCormick F, LeClere L, Sanchez G, Golijanin P, Anthony S, et al. Prospective evaluation of surgical treatment of humeral avulsions of the Glenohumeral ligament. Am J Sports Med. 2017;45(5):1134–40.
33. Neviaser TJ. The anterior labroligamentous periosteal sleeve avulsion lesion: a cause of anterior instability of the shoulder. Arthroscopy. 1993;9(1):17–21.
34. Taylor DC, Arciero RA. Pathologic changes associated with shoulder dislocations. Arthroscopic and physical examination findings in first-time, traumatic anterior dislocations. Am J Sports Med. 1997;25(3):306–11.
35. Nakagawa S, Iuchi R, Hanai H, Hirose T, Mae T. The development process of bipolar bone defects from primary to recurrent instability in shoulders with traumatic anterior instability. Am J Sports Med. 2019;47(3):695–703.
36. Yamamoto N, Itoi E, Abe H, Minagawa H, Seki N, Shimada Y, et al. Contact between the glenoid and the humeral head in abduction, external rotation, and horizontal extension: a new concept of glenoid track. J Shoulder Elb Surg. 2007;16(5):649–56.
37. Yamamoto N, Shinagawa K, Hatta T, Itoi E. Peripheral-track and central-Track Hill-Sachs lesions: a new concept of assessing an on-track lesion. Am J Sports Med. 2020;48(1):33–8.
38. Tuite MJ, Pfirrmann CWA. Shoulder: instability. In: Hodler J, Kubik-Huch RA, von Schulthess GK, editors. Musculoskeletal Diseases 2021–2024: Diagnostic Imaging. Cham: Springer International Publishing; 2021. p. 1–9.

39. Frank RM, Romeo AA, Provencher MT. Posterior Glenohumeral instability: evidence-based treatment. J Am Acad Orthop Surg. 2017;25(9):610–23.
40. Bey MJ, Hunter SA, Kilambi N, Butler DL, Lindenfeld TN. Structural and mechanical properties of the glenohumeral joint posterior capsule. J Shoulder Elb Surg. 2005;14(2):201–6.
41. Dewing CB, McCormick F, Bell SJ, Solomon DJ, Stanley M, Rooney TB, et al. An analysis of capsular area in patients with anterior, posterior, and multidirectional shoulder instability. Am J Sports Med. 2008;36(3):515–22.
42. Kim SH, Ha KI, Park JH, Kim YM, Lee YS, Lee JY, et al. Arthroscopic posterior labral repair and capsular shift for traumatic unidirectional recurrent posterior subluxation of the shoulder. Journal of Bone and Joint Surgery-American Volume. 2003;85A(8):1479–87.
43. Wellmann M, Blasig H, Bobrowitsch E, Kobbe P, Windhagen H, Petersen W, et al. The biomechanical effect of specific labral and capsular lesions on posterior shoulder instability. Arch Orthop Trauma Surg. 2011;131(3):421–7.
44. Inui H, Sugamoto K, Miyamoto T, Yoshikawa H, Machida A, Hashimoto J, et al. Glenoid shape in atraumatic posterior instability of the shoulder. Clin Orthop Relat Res. 2002;403:87–92.
45. Neer CS 2nd. Involuntary inferior and multidirectional instability of the shoulder: etiology, recognition, and treatment. Instr Course Lect. 1985;34:232–8.
46. Kim SH, Noh KC, Park JS, Ryu BD, Oh I. Loss of chondrolabral containment of the glenohumeral joint in atraumatic posteroinferior multidirectional instability. J Bone Jt Surg Am Vol. 2005;87A(1):92–8.
47. Ogston JB, Ludewig PM. Differences in 3-dimensional shoulder kinematics between persons with multidirectional instability and asymptomatic controls. Am J Sports Med. 2007;35(8):1361–70.
48. Barden JM, Balyk R, Raso VJ, Moreau M, Bagnall K. Atypical shoulder muscle activation in multidirectional instability. Clin Neurophysiol. 2005;116(8):1846–57.
49. Bottoni CR, Smith EL, Berkowitz MJ, Towle RB, Moore JH. Arthroscopic versus open shoulder stabilization for recurrent anterior instability: a prospective randomized clinical trial. Am J Sports Med. 2006;34(11):1730–7.
50. Provencher MT, Midtgaard KS, Owens BD, Tokish JM. Diagnosis and management of traumatic anterior shoulder instability. J Am Acad Orthop Surg. 2021;29(2):e51–61.
51. Lee SH, Lim KH, Kim JW. Risk factors for recurrence of anterior-inferior instability of the shoulder after arthroscopic Bankart repair in patients younger than 30 years. Arthroscopy. 2018;34(9):2530–6.
52. Bokshan SL, DeFroda SF, Gil JA, Badida R, Crisco JJ, Owens BD. The 6-O'clock anchor increases labral repair strength in a biomechanical shoulder instability model. Arthroscopy. 2019;35(10):2795–800.
53. Judson CH, Voss A, Obopilwe E, Dyrna F, Arciero RA, Shea KP. An anatomic and biomechanical comparison of Bankart repair configurations. Am J Sports Med. 2017;45(13):3004–9.
54. Nho SJ, Frank RM, Van Thiel GS, Wang FC, Wang VM, Provencher MT, et al. A biomechanical analysis of anterior Bankart repair using suture anchors. Am J Sports Med. 2010;38(7):1405–12.
55. Godin JA, Altintas B, Horan MP, Hussain ZB, Pogorzelski J, Fritz EM, et al. Midterm results of the bony Bankart bridge technique for the treatment of bony Bankart lesions. Am J Sports Med. 2019;47(1):158–64.
56. Di Giacomo G, Peebles LA, Pugliese M, Dekker TJ, Golijanin P, Sanchez A, et al. Glenoid track instability management score: radiographic modification of the instability severity index score. Arthroscopy. 2020;36(1):56–67.
57. Pagnani MJ. Open capsular repair without bone block for recurrent anterior shoulder instability in patients with and without bony defects of the glenoid and/or humeral head. Am J Sports Med. 2008; 36(9):1805–12.
58. Ahmad CS, Wang VM, Sugalski MT, Levine WN, Bigliani LU. Biomechanics of shoulder capsulorrhaphy procedures. J Shoulder Elbow Surg. 2005;14(1 Suppl S):12s–8s.
59. Gillis RC, Donaldson CT, Kim H, Love JM, Dreese JC. Arthroscopic suture anchor capsulorrhaphy versus labral-based suture capsulorrhaphy in a cadaveric model. Arthroscopy. 2012;28(11):1615–21.
60. Connolly J. Humeral head defects associated with shoulder dislocation: their diagnostic and surgical significance. Instr Course Lect. 1972;21:42–54.
61. Purchase RJ, Wolf EM, Hobgood ER, Pollock ME, Smalley CC. Hill-Sachs "Remplissage": An arthroscopic solution for the engaging hill-Sachs lesion. Arthrosc J Arthrosc Relat Surg. 2008;24(6):723–6.
62. Koo SS, Burkhart SS, Ochoa E. Arthroscopic double-pulley Remplissage technique for Engaging Hill-Sachs lesions in anterior shoulder instability repairs. Arthrosc J Arthrosc Relat Surg. 2009;25(11):1343–8.
63. Boileau P, O'Shea K, Vargas P, Pinedo M, Old J, Zumstein M. Anatomical and functional results after arthroscopic hill-Sachs remplissage. J Bone Joint Surg Am. 2012;94(7):618–26.
64. HL ML. Posterior dislocation of the shoulder. J Bone Jt Surg Am Vol. 1952;34-A(3):584–90.
65. Krackhardt T, Schewe B, Albrecht D, Weise K. Arthroscopic fixation of the subscapularis tendon in the reverse Hill-Sachs lesion for traumatic unidirectional posterior dislocation of the shoulder. Arthrosc J Arthrosc Relat Surg. 2006;22(2):227-.
66. Boileau P, Bicknell RT, El Fegoun AB, Chuinard C. Arthroscopic Bristow procedure for anterior instability in shoulders with a stretched or deficient capsule: the "belt-and-suspenders" operative technique

and preliminary results. Arthrosc J Arthrosc Relat Surg. 2007;23(6):593–601.
67. Giles JW, Degen RM, Johnson JA, Athwal GS. The Bristow and Latarjet procedures: why these techniques should not be considered synonymous. J Bone Joint Surg Am. 2014;96(16):1340–8.
68. Hardy A, Sabatier V, Laboudie P, Schoch B, Nourissat G, Valenti P, et al. Outcomes after Latarjet procedure: patients with first-time versus recurrent dislocations. Am J Sports Med. 2020;48(1):21–6.
69. An VV, Sivakumar BS, Phan K, Trantalis J. A systematic review and meta-analysis of clinical and patient-reported outcomes following two procedures for recurrent traumatic anterior instability of the shoulder: Latarjet procedure vs. Bankart repair. J Shoulder Elbow Surg. 2016;25(5):853–63.
70. Moroder P, Schulz E, Wierer G, Auffarth A, Habermeyer P, Resch H, et al. Neer award 2019: Latarjet procedure vs. iliac crest bone graft transfer for treatment of anterior shoulder instability with glenoid bone loss: a prospective randomized trial. J Shoulder Elb Surg. 2019;28(7):1298–307.
71. Kwapisz A, Fitzpatrick K, Cook JB, Athwal GS, Tokish JM. Distal clavicular osteochondral autograft augmentation for glenoid bone loss: a comparison of radius of restoration versus Latarjet graft. Am J Sports Med. 2018;46(5):1046–52.
72. Provencher MT, Peebles LA, Aman ZS, Bernhardson AS, Murphy CP, Sanchez A, et al. Management of the Failed Latarjet Procedure: outcomes of revision surgery with fresh distal Tibial allograft. Am J Sports Med. 2019;47(12):2795–802.
73. Diklic ID, Ganic ZD, Blagojevic ZD, Nho SJ, Romeo AA. Treatment of locked chronic posterior dislocation of the shoulder by reconstruction of the defect in the humeral head with an allograft. J Bone Jt Surg Brit Vol. 2010;92B(1):71–6.

Biomechanics of the Throwing Shoulder

13

John Fritch, Amit Parekh, Andre Labbe, Jacques Courseault, Felix Savoie, Umile Giuseppe Longo, Sergio De Salvatore, Vincenzo Candela, Calogero Di Naro, Carlo Casciaro, and Vincenzo Denaro

13.1 Introduction

The throwing shoulder is subject to extremes of motion and complex dynamic activity involving the entire body. It requires movement in multiple planes, repetitive torque, and extreme loads on the shoulder. For the overhead athlete, this generally results in adaptive changes to the soft tissues and bony structures of the shoulder. These adaptions give biomechanical advantages which improve performance. Despite this, the often times violent activity of throwing predisposes the shoulder to dysfunction and injury. It is important to have knowledge about the mechanics of throwing in this population to understand optimum function in performance and differentiate positive adaptions from pathology.

J. Fritch
Department of Orthopaedic Surgery, University of Chicago, Chicago, IL, USA

A. Parekh
Department of Orthopaedic Surgery, University of Illinois at Chicago, Chicago, IL, USA

A. Labbe · J. Courseault · F. Savoie (✉)
Department of Orthopaedic Surgery, Tulane University, New Orleans, LA, USA
e-mail: fsavoie@tulane.edu

U. G. Longo · S. De Salvatore · V. Candela
C. Di Naro · C. Casciaro · V. Denaro
Department of Orthopaedic and Trauma Surgery, Campus Bio-Medico University, Rome, Italy

13.2 Biomechanics of Throwing

Overhead motion is developed and regulated through a sequentially coordinated and task-specific kinetic chain of force development [1]. The kinematics of both the baseball pitch and tennis serve have been well described and may be broken down into phases [2–4]. The kinetics are not as well described but are important as they help explain the complex transfer of energy through the body resulting in the ultimate task of throwing or hitting a ball. The term, "kinetic chain," is used collectively to describe these mechanical linkages [5].

The stages of throwing in the overhead athlete are the windup, early cocking, late-cocking, acceleration, deceleration, and follow-through (Fig. 13.1). Complex coordination of various muscle groups throughout the kinetic chain are required to complete the phases resulting in ball release. The transition between the late-cocking phase and the acceleration phase is considered the critical point at which many injuries occur [5]. In late-cocking, the shoulder is abducted and externally rotated, placing the anterior capsule under significant strain as it prevents anterior translation of the humerus. In some throwers, this repetitive and excessive humeral external rotation may increase shoulder laxity or attenuate the anterior capsule [6]. Repetitive forceful throwing can also cause stretching of the coracohumeral ligament [7]. With the shoulder in the position of

J. Koh et al. (eds.), *Orthopaedic Biomechanics in Sports Medicine*,
https://doi.org/10.1007/978-3-030-81549-3_13

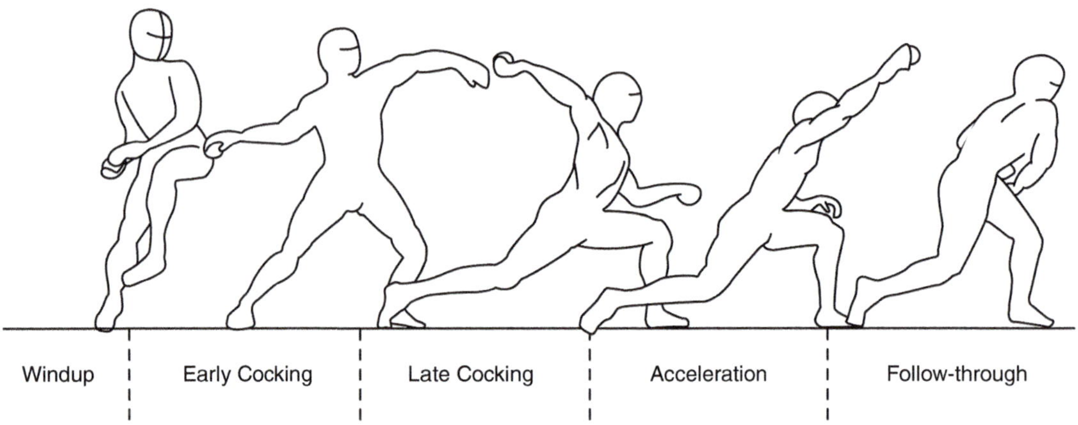

Fig. 13.1 The phases of throwing [14]

90 degrees of abduction and increasing external rotation as seen in late-cocking, the posterior superior cuff and posterior superior labrum gets pinched between the greater tuberosity and then glenoid, leading to internal impingement [8, 9].

Transitioning from late-cocking to the acceleration phase involves a complex coordination of eccentric and concentric muscle firing to rapidly accelerate in the forward vector while stabilizing the glenohumeral joint. During late-cocking, the infraspinatus and teres minor contract while the subscapularis, pectoralis major, and latissimus provide stability to the anterior aspect of the shoulder [10]. As the arm reaches it maximal posterior position and begins accelerating forward, the posterior deltoid keeps the arm abducted and the anterior cuff fires to internally rotate the shoulder and generate sufficient torque to accelerate ball. The pectoralis major and latissimus dorsi also concentrically fire to internally rotate the shoulder while the teres minor and infraspinatus eccentrically contract to provide a posterior stabilizing force [11]. Although the rotator cuff helps stabilize the shoulder in the acceleration phase, pitchers primary rely on the pectoralis major, latissimus, triceps, and serratus anterior to internally rotate the shoulder.

Conversely, deceleration primarily relies on the rotator cuff. As the arm decelerates into the follow-through phase, the rotator cuff fires eccentrically to slow forward motion. The posterior capsule and cuff are subjected to the highest stress during follow-through, which can be up to 108% of body weight [12]. Repetitive activity may lead to muscular fatigue of the cuff, transferring stress to the posterior capsule. Chronic stress resulting in microtrauma and tearing of the posterior capsule may result in stiffening of posterior cuff and capsule [13].

13.3 Kinematics and Kinetics of the Throwing Shoulder

Kinematics is the study of motion throughout a system that excludes the external forces acting to create motion. The scapula provides a base structure linking the appendicular skeleton to the axial skeleton to facilitate natural anatomic motion. Changes in the normal kinematics of any one part of the shoulder complex can alter the kinematics of the entire system. Some adaptations impart a biomechanical advantage in the thrower's shoulder, while others result in shoulder pathology [15]. As such, knowledge of "normal" shoulder kinematics is a crucial factor in understanding pathology of the shoulder complex [16].

The shoulder complex consists of three bones (humerus, clavicle, and scapula) and four functional anatomic joints (acromioclavicular, sternoclavicular, glenohumeral, and scapulothoracic) [17]. The glenohumeral and scapulothoracic joints play important roles in determining range of motion [16, 18]. Shoulder kinematics largely depends on the combined, coordinated movements of these joints to create what is known as the "kinematic chain" of the shoulder [17]. Most shoulder pathologies involve changes to one or multiple components of the kinematic chain, resulting in altered motion and eventual instability or pain [15, 19].

Scapular kinematics play a huge role in the throwing shoulder and overhead movement. The scapulothoracic articulation is unique in that it relies on the coordinated firing of many muscular attachments for movement in several planes. As a result, any derangements in its complex movement may lead to an improperly positioned or unstable base for glenohumeral motion. For example, reduction in scapular upward rotation is detrimental to maintaining inferior stability at the glenohumeral joint and increased scapular internal rotation is associated with a reduction in anterior bony stability [20, 21]. As such, scapular position and motion are integral to the functional stability of the glenohumeral joint [18].

Kinetics is the study of motion when broadly examining the forces created by active movement. The overhead throwing motion is ultimately performed through a coordinated sequence of motions collectively known as the kinetic chain [1, 5]. In order to better understand failures in the kinetic chain resulting in pathology, one must examine the force couples and force generators at the shoulder.

The most important force-couple of the shoulder involves the symbiotic relationship of the rotator cuff and the scapular stabilizers [22]. Unstable scapular movements can lead to significant shoulder instability [23, 24]. Through coordinated firing of the periscapular musculature, the stabilized scapula provides a base for the rotator cuff to direct and assist in correct positioning of the humeral head in the glenoid fossa [25, 26]. The kinetics of the humerus and scapula rely on other coupled relationships outside of these muscle groups. The serratus anterior and the upper and lower trapezius are another force-couple pair critical to stabilizing the shoulder through dynamic motion. During elevation the serratus anterior and the trapezius act to maximize the size of the subacromial space by upwardly rotating the shoulder in addition to maintaining an ideal length-tension relationship with the deltoid [27–31]. As the shoulder rotates upwards, contributions from the middle and lower trapezius act as a counterforce to scapular protraction created by actions of the serratus anterior in order to balance the forces between the axial and appendicular skeleton to facilitate fluid motion [32, 33].

The rotator cuff assists in internal and external rotation of the shoulder in addition to generating medial, inferior, anterior, and posterior stabilizing forces across the glenohumeral joint [34]. They are less involved in gross abduction and adduction strength in the shoulder. Biomechanical studies have shown the pectoralis major, teres major, rhomboid major and latissimus dorsi play a large role generating adduction force [34]. Additionally, the pectoralis major and anterior deltoid are responsible for creating torque during flexion of the arm in the sagittal plane as occurs in overhead motion [35]. The pectoralis major muscle is a strong anterior translator of the humeral head and as a result, the posterior rotator cuff muscles must fire to stabilize the glenohumeral joint [36]. This coordinated firing is an example of another force-couple.

Shear forces at the glenohumeral joint are of particular interest to overhead athletes, as they often play a role in injury. While shear forces are a normal part of shoulder kinematics they can lead to pathology as well. Shoulder peak proximal force during throwing involves high glenohumeral shear forces, particularly at the labrum, which can lead to tearing and injury. Also contributing to stress on the labrum, is pull from the long head of the biceps. The biceps contracts during overhead motion to control extension of the elbow while stabilizing the glenohumeral joint [37]. Repetitive pull of the biceps on the labrum can create microtrauma and eventual SLAP tears [38]. Understanding contributors of shear forces and identifying risk factors for pathology can allow physicians to develop early intervention protocols involving scapular stabilization, rotator cuff strengthening, and therapies aimed at preventing failures in the kinetic chain [37].

13.4 Adaptive Changes in the Overhead Athlete

High performance of the overhead athlete occurs as a result of the summation of speed principle, which states that in order to maximize the speed at the distal end of a linked system (in this scenario, the shoulder, elbow, and wrist), optimal movement and transfer of energy should start with optimal use of the proximal segments (the hips and core) and progress effectively down the kinetic chain [5, 39]. Adaptive changes occur throughout the entire kinetic chain as mechanics are improved by further decreasing the possible degrees of freedom throughout the entire motion, adding increased force while limiting instability [40–42]. These large repetitive forces challenge the involved bone and soft tissues, and place the shoulder at risk for injury. Shoulder soft tissue and bone adaptations in response to these large forces have been observed in several studies, as the shoulder must be hypermobile enough to perform overhead activity, yet stable enough to prevent pathologic instability [43–49]. This section will outline the adaptive changes that occur in the thrower's shoulder to maximize efficiency.

13.4.1 Scapular Kinematics

Due to the repetitive and often extreme nature of throwing, serving, and spiking motions, overhead athletes are required to produce precise scapular function in order to create large amounts of shoulder rotation, force and repeated overhead elevation of the arm while remaining at a competitive and injury-free level. This pre-disposes the athlete to scapular dyskinesis. Yet, changes in scapular kinematics may often occur to improve efficiency or compensate for other potentially harmful changes. For example, several clinicians have described the occurrence of increased scapular internal rotation, anterior translation, anterior tilt, and retraction among both symptomatic and asymptomatic throwing athletes [26, 49, 50].

Athletes of overhead sports seem to demonstrate an increase in scapular upward rotation on their dominant limb during elevation when compared to their non-dominant limb or with non-athlete individuals [51–53]. This increase in scapular upward rotation during arm elevation has been suggested as a chronic adaptation to preserve the subacromial space and to achieve the clearance needed during the throwing motion to prevent rotator cuff compression and injury [26, 51, 54]. Furthermore, the follow-through phase of the throwing motion requires a significant amount of glenohumeral horizontal adduction which is diminished in contractures of the posterior soft tissue structures. Athletes may adapt for this loss of cross body motion with an acquired increase in scapular internal rotation, anterior translation and anterior tilt in an effort to reach the follow-through position [49]. However, the increase in scapular internal rotation can result in a decrease of the subacromial space and an inability of the greater tuberosity to pass freely under the acromion during humeral elevation leading to impingement [55, 56]. Notably, glenohumeral contact pressure has been shown to be significantly increased with more scapular internal rotation in a cadaveric study [57]. This study suggests that athletes of overhead sports may be more vulnerable to rotator cuff tendinopathy as these changes are believed to irritate the rotator cuff tendons and the subacromial tissues during arm elevation [19]. Scapular positioning changes such as the increased internal rotation will give the athlete the appearance of a dropped shoulder, often seen in the "SICK" scapula defined by Burkhart et al. [50].

Alongside changes in upward and internal rotation, adaptations in scapular retraction also exist in throwing athletes. Studies have demonstrated increased scapular retraction with humeral elevation beyond 90° [51, 58]. The scapula moves medially about the trunk, toward the spine at these positions. During the cocking phase of the throwing motion, the arm typically achieves about 90° of humeral elevation, similar to the position at which retraction is increased [51, 59]. Kibler described how scapular retraction is necessary to achieve the cocking position during the throw and tennis serve. Thus, retraction may facilitate an increased cocking position for subsequent vigorous acceleration during the throwing motion [51, 60].

Overall, it appears that throwing athletes develop chronic adaptations of the scapula that either contribute to or result from the throwing motion. Currently, it is not well known whether these adaptations are more likely to generally result in improved throwing skill, injury prevention, or injury. Further study is needed to guide understanding and clinical management.

13.4.2 Humeral Retroversion

Increased humeral retroversion has also been studied as an adaptive osseous change in the thrower's shoulder. From birth, the proximal humerus is retroverted and de-rotates over time, with the most rapid changes occurring before adolescence and continuing until maturity. In the average adult, mean retroversion differs between 33° in the dominant arm and 29° in the non-dominant arm. In contrast, further increased humeral retroversion is found in the thrower's dominant shoulder. Crocket et al. described the bone adaptation of increased humeral head retroversion that leads to changes in measured rotation in the throwing shoulder [45]. Humeral head retroversion decreases with maturity, yet alterations in this change may be possible during growth in skeletally immature athletes [61–63]. Adaptation of the humerus to external stresses has been described in literature and is consistent with Wolff's Law [61, 64, 65]. One example of this, Little League Shoulder, is a proximal humeral epiphyseal overuse syndrome which is considered a stress reaction injury to the epiphyseal plate [66, 67]. These patients present during puberty when the proximal humeral physis undergoes rapid growth, with widening of the physeal plate. Although this is a symptomatic overuse condition, it suggests that the proximal humerus adapts and responds to the repeated stresses it experiences. This is supported by studies examining the humeral retroversion in athletes over time, which show significant changes of increased external rotation and decreased internal rotation in athletes after the age of 12 [46].

Several studies on athletes following maturity demonstrate increased retroversion. A study of collegiate throwers demonstrated an increased average retroversion of 36°(±9.8°) and a total arc of motion of 159.5° in the dominant arm versus a decreased average retroversion of 26°(±9.4°) and total arc of motion of 157.8° in the non-dominant arm [68]. Another study on collegiate pitchers demonstrated increased dominant arm external rotation when measured at zero and 90° of abduction, which was associated with greater humeral retroversion [69]. Similarly, professional baseball pitchers' dominant shoulders showed increased external rotation and humeral/glenoid retroversion [45].

The bony adaptation of increased glenohumeral retroversion observed by Crockett allows for pre-positioning of the arm in more external rotation, so that a larger effective arc of motion can be used to generate greater rotational torque and angular velocity prior to ball release without significantly increasing anterior shoulder strain on the capsule and glenohumeral ligaments [14, 45, 46]. In addition to reaching an elite level of performance, Crocket et al. postulated that baseball players adapt an increase in humeral head retroversion to reduce risk of injury [45]. Supporting this idea, a recent study found a correlation between injured professional pitchers missing more playing days and lesser retroversion of the throwing shoulder [70].

13.4.3 Peri-Articular Soft Tissue Adaptations

In addition to the above changes, soft tissue alterations occur about the shoulder. Whether these changes occur primarily in response to the demands of the throwing shoulder or as a result of the osseous changes that occur in the adaptive shoulder remains under debate. The most important noted soft tissue changes include laxity of the anteroinferior glenohumeral joint capsule as well as contracture of posterior shoulder soft tissue structures.

Many overhead athletes have been found to have increased anterior glenohumeral joint laxity, which permits excessive external range of motion, also known as external rotation gain (ERG). The hypermobility of the thrower's shoulder in particular has been referred to as an "acquired laxity" as a result of the repetitive and excessive external rotation required in the throwing shoulder [71, 72]. In late-cocking, the shoulder is abducted and the anterior capsule is under large strain to resist anterior translation of the humerus. While this allows increased transmission of force, this is also theorized to cause attenuation of the anterior capsule [6]. The identification of this laxity and possible subtle anterior shoulder instability was first reported by Jobe et al. in the late 1980s [73, 74]. While this is a generally accepted theory, it should be noted that several studies examining the shoulder in a functional test position of abduction and external rotation of the humerus all found minimal anterior translation in several populations of overhead athletes, suggesting that anterior-inferior constraints are intact and stable in the non-pathologic throwing shoulder [75–78]. Still, there is some recent evidence supporting the link between shoulder laxity and ERG [6, 47]. Ultimately, the available evidence regarding acquired hyperlaxity in the non-pathologic throwing shoulder is equivocal.

Additionally, there are multiple studies that have demonstrated posterior shoulder tightness in overhead athletes [79–82]. Contracture of the posterior joint structures has been proposed as a contributor to glenohumeral internal rotation deficit (GIRD) and decreased horizontal adduction in the overhead athlete [50, 83, 84]. In most throwing shoulders, this immobility does not appear problematic, although excessive immobility has been linked to shoulder pathology [50, 82]. Furthermore, is not currently clear which structures (posterior capsule or cuff musculature) cause the motion restriction. Two theories exist as to why this phenomenon occurs. Lombardo et al. identified the late-cocking phase of the thrower's motion as a possible cause of posterior shoulder pain. This posterosuperior impingement leads to an adaptive posterior capsular thickening [85–87]. The more commonly accepted theory relates to the previously discussed rotational instability, where the throwing shoulder over rotates into a position of increased external rotation during the late-cocking and acceleration phases. The anterior capsule undergoes increased strain during the late-cocking phase whereas the posterior capsule is widened and traumatized during the deceleration phase. The consequence is micro-instability leads to corresponding posterior capsular hypertrophy. The first step in this cascade described by Burkhart is the development of contracture of the posterior band of the inferior glenohumeral ligament (IGHL) and posteroinferior capsule [50, 79]. Posterior shoulder tightness leads to changes in the rotational axis of the humeral head. This involves moving the humeral head into a more posterior and superior direction, thus promoting a reduction in internal rotation and an increase in external rotation, which generates an alteration in the athlete's arc of motion [50].

Finally, it should be noted that the osseous and soft tissue changes that occur in overhead athletes leads to an observed posterior shift in the arc of rotational ROM. Some studies have noted a shift around 10°, which can be considered a normal adaptation in asymptomatic throwers [68, 88, 89]. Across most studies on rotational changes, contralateral extremity comparisons have demonstrated no significant changes in total ROM for many types of overhead athletes [88, 90–93].

13.5 Scapular Pathophysiology in the Thrower's Shoulder

As previously discussed, scapular dysfunction alters the position and mechanics of the scapula resulting in downstream alterations and potential injury [94]. The scapulothoracic articulation is unique in that it lacks a bony articulation point, allowing motion in many directions, and requiring many stabilizing muscles to maintain its function (rhomboid major and minor, trapezii, levator scapulae, and serratus anterior muscles) [94]. Dysfunction in these scapular stabilizing muscles is common cause of altered biomechanics and overuse injuries in the overhead athlete [95, 96]. The lower trapezius and serratus anterior muscles contribute to upward rotation and acromial elevations and are most commonly responsible for abnormal scapular movement [96]. Weakness in the serratus anterior causes the medial border of the scapula to sit elevated from the rib cage, decreasing acromial elevation and resulting in subsequent impingement [97]. In a systematic review by Keshevars et al., patients with instability and impingement were found to have increased scapular internal rotation and protraction as well as decreased upward rotation during elevation [98].

Abnormalities in scapular motion, known as scapular dyskinesia, is common among overhead athletes [18]. The exact causes of scapular dyskinesia are often difficult to discern. Multiple factors may contribute, including macrotrauma to muscles, strain induced by microtrauma, and/or inhibition of motion due to pain from injuries like labral lesions, instability, or arthrosis [26]. Ludewig and Reynolds describe a variety of biologic mechanism behind changes in scapular mechanics seen in scapular dyskinesia (Table 13.1) [19]. Some of these changes, such as posterior glenohumeral tightness, are specific to the throwing shoulder.

Table 13.1 Proposed biomechanical mechanisms of changes to scapular mechanics [19]

Biological Mechanism	Changes to Scapular Mechanics
Soft tissue tightness of the posterior glenohumeral joint	Increased anterior tilt of the scapula
Tightness of the pectoralis minor	Increased internal rotation and anterior tilt of the scapula
Increased activation of the upper trapezius	Increased elevation of the clavicle
Decreased activation of the serratus	Decreased upward rotation and posterior tilt of the scapula
Flexed thoracic posture or kyphosis	Increased internal rotation and anterior tilt of the scapula, decreased upward rotation of the scapula

13.6 Glenohumeral Pathophysiology in the Thrower's Shoulder

As a result of the unique biomechanical adaptations and stresses seen in the throwing shoulder, a certain subset of injuries is seen in the glenohumeral joint. These injuries fall into three general groups: internal impingement, instability secondary to internal impingement, and primary anterior or multidirectional instability [14].

As previously described in the biomechanics of throwing section, the shoulder is most susceptible to injury in the transition from late-cocking to acceleration phase. During this phase, the shoulder is in a position of maximal abduction and external rotation, which can cause a mechanical pinch of the posterior superior cuff and labrum between the glenoid and greater tuberosity, otherwise known as internal impingement [8, 9]. Even though this biomechanical process occurs frequently in throwing athletes, repeated impingement can lead to eventual pathologic internal impingement. In such cases, patients experience posterior shoulder pain in late-cocking and reduced pitching velocity. Repeated injury results in PASTA lesions (partial articular supraspinatus tendon avulsion) near the supraspinatus/infraspinatus junction and peel-back lesions of the posterior superior labrum which can lead to SLAP tears (superior labrum from anterior to posterior) [99].

In the follow-through phase, the posterior capsule and posterior band of the inferior glenohumeral ligament (PIGHL) experiences maximal stress and can be injured. Repeated microtrauma as a result of throwing can cause thickening and contracture of the posterior capsule and PIGHL [100]. This results in a net shift of glenohumeral rotation superiorly and posteriorly on the glenoid and is considered an adaptive response as it allows more clearance for the greater tuberosity on the glenoid resulting in increased external rotation [101]. This adaptive response is considered pathologic when the posterior contracture results in a loss of greater than 25 degrees of total arc of motion, also known as GIRD or glenohumeral internal rotation deficit [100]. Pitchers with GIRD experience poor pitching mechanics and a breakdown in the kinetic chain resulting in hyperexternal rotation, hyperhorizontal abduction, a dropped elbow, and premature trunk rotation [100]. As a result, pitchers with GIRD are twice as likely to sustain further injury [102].

Although increased external rotation is an adaptive change seen in pitchers, excessive hyperexternal rotation and abduction may result in stretching and eventual injury to the anterior capsule and anterior band of the IGHL causing anterior instability [9, 103]. These patients typically experience anterior shoulder pain in late-cocking and acceleration and can present with Bankart and PASTA lesions.

Another subset of throwing injury patients presents primarily with nonspecific pain at the rotator cuff with sustained activity. These patients suffer from primary anterior or multidirectional instability and rely heavily on their rotator cuff to provide glenohumeral stability through the pitching motion. Once the cuff fatigues, they become symptomatic and often cannot continue throwing [14].

13.7 Treatment

13.7.1 Nonoperative

Any break in the kinetic chain of movement in the overhead athlete may lead to pathology. Key to treatment of injuries of the thrower's shoulder is ensuring proper coordination of energy delivery from the legs, through the core, and into the upper extremities. The most common pathology seen is GIRD and fortunately 90% of throwers respond well to nonoperative stretching programs including the sleeper stretch [104]. One subset of patients develop SICK scapula syndrome (Scapular malposition, Inferior medial border prominence, Coracoid pain, and dysKinesia of scapular movement) and will present with asymmetric dropped scapula [100]. In these patients it is important to implement a therapy program that will encourage balanced rotator cuff and periscapular muscle strength.

13.7.2 Surgical Management

Although a variety of shoulder pathologies can be seen in the thrower's shoulder including labral tears, SLAP tears, PASTA lesions, and internal impingement, one must understand that these patients have undergone adaptive changes in their shoulder that allow them to perform their overhead activities, in some cases at an elite level. This is important because surgical intervention in such patients may result in worse outcomes than nonoperative management and may prevent a return to the previous level of performance. For example, studies have shown that repairs of type II SLAP lesions in overhead athletes have resulted in only 73% of patients returning to their previous level of activity. Rotator cuff tears seen in this population are typically partial articular sided and respond well to debridement as opposed to repair [105]. In overhead athletes with anterior or multidirectional instability, care must be taken with capsular plication and labral repair as overtightening may compromise the adaptive changes that allowed them to participate in their level of sports. One study on this population showed 77% are able to return to preinjury level of sport after surgical repair [106].

13.8 Summary

The throwing shoulder requires a complex coordination of movement across the entire body. Adaptive changes to the shoulder occur with repetitive overhead sports which can provide a biomechanical advantage and allow a high level of competition. Unique injuries are seen in this population and it is important to differentiate between beneficial adaptive changes and pathology in their treatment. Surgical treatment can sometimes compromise their biomechanical advantages and return to play.

References

1. Kibler WB, Kuhn JE, Wilk K, Sciascia A, Moore S, Laudner K, et al. The disabled throwing shoulder: spectrum of pathology-10-year update. Arthroscopy. 2013;29(1):141–61. e26
2. Lintner D, Noonan TJ, Kibler WB. Injury patterns and biomechanics of the athlete's shoulder. Clin Sports Med. 2008;27(4):527–51.
3. Kovacs M, Ellenbecker T. An 8-stage model for evaluating the tennis serve: implications for performance enhancement and injury prevention. Sports Health. 2011;3(6):504–13.
4. Sciascia A, Cromwell R. Kinetic chain rehabilitation: a theoretical framework. Rehabil Res Pract. 2012;2012:853037.
5. Kibler WB, Wilkes T, Sciascia A. Mechanics and pathomechanics in the overhead athlete. Clin Sports Med. 2013;32(4):637–51.
6. Mihata T, McGarry MH, Neo M, Ohue M, Lee TQ. Effect of anterior capsular laxity on horizontal abduction and forceful internal impingement in a cadaveric model of the throwing shoulder. Am J Sports Med. 2015;43(7):1758–63.
7. Kuhn JE, Bey MJ, Huston LJ, Blasier RB, Soslowsky LJ. Ligamentous restraints to external rotation of the humerus in the late-cocking phase of throwing - a cadaveric biomechanical investigation. Am J Sports Med. 2000;28(2):200–5.
8. Walch G, Boileau P, Noel E, Donell ST. Impingement of the deep surface of the supraspinatus tendon on the posterosuperior glenoid rim: An arthroscopic study. J Shoulder Elb Surg. 1992;1(5):238–45.
9. Jobe CM. Posterior superior glenoid impingement - expanded spectrum. Arthroscopy. 1995;11(5):530–6.
10. Rodosky MW, Harner CD, Fu FH. The role of the long head of the biceps muscle and superior glenoid labrum in anterior stability of the shoulder. Am J Sports Med. 1994;22(1):121–30.
11. Digiovine NM, Jobe FW, Pink M, Perry J. An electromyographic analysis of the upper extremity in pitching. J Shoulder Elb Surg. 1992;1(1):15–25.
12. Werner SL, Gill TJ, Murray TA, Cook TD, Hawkins RJ. Relationships between throwing mechanics and shoulder distraction in professional baseball pitchers. Am J Sports Med. 2001;29(3):354–8.
13. Thomas SJ, Swanik CB, Higginson JS, Kaminski TW, Swanik KA, Kelly JD, et al. Neuromuscular and stiffness adaptations in division I collegiate baseball players. J Electromyogr Kinesiol. 2013;23(1):102–9.
14. Thigpen C, Evans DT. Pathomechanics and injury in the overhead motion. In: Kibler WB, Sciascia AD, editors. Mechanics, Pathomechanics and injury in the overhead athlete: a case-based approach to evaluation, diagnosis and management. Cham: Springer International Publishing; 2019. p. 25–38.
15. Lefevre-Colau MM, Nguyen C, Palazzo C, Srour F, Paris G, Vuillemin V, et al. Kinematic patterns in normal and degenerative shoulders. Part II: review of 3-D scapular kinematic patterns in patients with shoulder pain, and clinical implications. Ann Phys Rehabil Med. 2018;61(1):46–53.
16. Ludewig PM, Phadke V, Braman JP, Hassett DR, Cieminski CJ, LaPrade RF. Motion of the shoulder complex during multiplanar humeral elevation. J Bone Joint Surg Am. 2009;91(2):378–89.
17. Lefevre-Colau MM, Nguyen C, Palazzo C, Srour F, Paris G, Vuillemin V, et al. Recent advances in kinematics of the shoulder complex in healthy people. Ann Phys Rehabil Med. 2018;61(1):56–9.
18. Kibler WB, Ludewig PM, McClure PW, Michener LA, Bak K, Sciascia AD. Clinical implications of scapular dyskinesis in shoulder injury: the 2013 consensus statement from the 'Scapular Summit'. Br J Sports Med. 2013;47(14):877–85.
19. Ludewig PM, Reynolds JF. The association of scapular kinematics and glenohumeral joint pathologies. J Orthop Sports Phys Ther. 2009;39(2):90–104.
20. Itoi E, Motzkin NE, Morrey BF, An KN. Scapular inclination and inferior stability of the shoulder. J Shoulder Elb Surg. 1992;1(3):131–9.
21. Weiser WM, Lee TQ, McMaster WC, McMahon PJ. Effects of simulated scapular protraction on anterior glenohumeral stability. Am J Sports Med. 1999;27(6):801–5.
22. Joshi M, Thigpen CA, Bunn K, Karas SG, Padua DA. Shoulder external rotation fatigue and scapular muscle activation and kinematics in overhead athletes. J Athl Train. 2011;46(4):349–57.
23. Cools AM, Declercq GA, Cambier DC, Mahieu NN, Witvrouw EE. Trapezius activity and intramuscular balance during isokinetic exercise in overhead athletes with impingement symptoms. Scand J Med Sci Sports. 2007;17(1):25–33.
24. Huang HY, Lin JJ, Guo YL, Wang WT, Chen YJ. EMG biofeedback effectiveness to alter muscle activity pattern and scapular kinematics in subjects with and without shoulder impingement. J Electromyogr Kinesiol. 2013;23(1):267–74.
25. Belling Sorensen AK, Jorgensen U. Secondary impingement in the shoulder. An improved terminology in impingement. Scand J Med Sci Sports. 2000;10(5):266–78.
26. Kibler WB. The role of the scapula in athletic shoulder function. Am J Sports Med. 1998;26(2):325–37.
27. Doody SG, Freedman L, Waterland JC. Shoulder movements during abduction in the scapular plane. Arch Phys Med Rehabil. 1970;51(10):595–604.
28. Lucas DB. Biomechanics of the shoulder joint. Arch Surg. 1973;107(3):425–32.
29. Ludewig PM, Hoff MS, Osowski EE, Meschke SA, Rundquist PJ. Relative balance of serratus anterior and upper trapezius muscle activity during push-up exercises. Am J Sports Med. 2004;32(2):484–93.
30. van der Helm FC. Analysis of the kinematic and dynamic behavior of the shoulder mechanism. J Biomech. 1994;27(5):527–50.
31. Mottram SL. Dynamic stability of the scapula. Man Ther. 1997;2(3):123–31.

32. Bagg SD, Forrest WJ. A biomechanical analysis of scapular rotation during arm abduction in the scapular plane. Am J Phys Med Rehabil. 1988; 67(6):238–45.
33. Johnson G, Bogduk N, Nowitzke A, House D. Anatomy and actions of the trapezius muscle. Clin Biomech (Bristol, Avon). 1994; 9(1):44–50.
34. Reed D, Halaki M, Ginn K. The rotator cuff muscles are activated at low levels during shoulder adduction: an experimental study. J Physiother. 2010;56(4):259–64.
35. Myers JB, Pasquale MR, Laudner KG, Sell TC, Bradley JP, Lephart SM. On-the-field resistance-tubing exercises for throwers: An Electromyographic analysis. J Athl Train. 2005;40(1):15–22.
36. Wattanaprakornkul D, Halaki M, Boettcher C, Cathers I, Ginn KA. A comprehensive analysis of muscle recruitment patterns during shoulder flexion: an electromyographic study. Clin Anat. 2011;24(5):619–26.
37. Keeley DW, Oliver GD, Dougherty CP. A biomechanical model correlating shoulder kinetics to pain in young baseball pitchers. J Hum Kinet. 2012;34:15–20.
38. Snyder SJ, Karzel RP, Del Pizzo W, Ferkel RD, Friedman MJ. SLAP lesions of the shoulder. Arthroscopy. 1990;6(4):274–9.
39. Putnam CA. Sequential motions of body segments in striking and throwing skills: descriptions and explanations. J Biomech. 1993;26(Suppl 1):125–35.
40. Sporns O, Edelman GM. Solving Bernstein's problem: a proposal for the development of coordinated movement by selection. Child Dev. 1993;64(4):960–81.
41. Davids K, Glazier P, Araujo D, Bartlett R. Movement systems as dynamical systems: the functional role of variability and its implications for sports medicine. Sports Med. 2003;33(4):245–60.
42. Glazier PS, Davids K. Constraints on the complete optimization of human motion. Sports Med. 2009;39(1):15–28.
43. Bigliani LU, Codd TP, Connor PM, Levine WN, Littlefield MA, Hershon SJ. Shoulder motion and laxity in the professional baseball player. Am J Sports Med. 1997;25(5):609–13.
44. Jobe CM. Superior glenoid impingement. Orthop Clin North Am. 1997;28(2):137–43.
45. Crockett HC, Gross LB, Wilk KE, Schwartz ML, Reed J, O'Mara J, et al. Osseous adaptation and range of motion at the glenohumeral joint in professional baseball pitchers. Am J Sports Med. 2002;30(1):20–6.
46. Levine WN, Brandon ML, Shubin Stein BE, Gardner TR, Bigliani LU, Ahmad CS. Shoulder adaptive changes in youth baseball players. J Shoulder Elb Surg. 2006;15(5):562–6.
47. Sethi PM, Tibone JE, Lee TQ. Quantitative assessment of glenohumeral translation in baseball players: a comparison of pitchers versus non-pitching athletes. Am J Sports Med. 2004;32(7): 1711–5.
48. Ticker JB, Fealy S, Fu FH. Instability and impingement in the athlete's shoulder. Sports Med. 1995;19(6):418–26.
49. Borsa PA, Laudner KG, Sauers EL. Mobility and stability adaptations in the shoulder of the overhead athlete: a theoretical and evidence-based perspective. Sports Med. 2008;38(1):17–36.
50. Burkhart SS, Morgan CD, Kibler WB. The disabled throwing shoulder: spectrum of pathology part III: the SICK scapula, scapular dyskinesis, the kinetic chain, and rehabilitation. Arthroscopy. 2003;19(6):641–61.
51. Myers JB, Laudner KG, Pasquale MR, Bradley JP, Lephart SM. Scapular position and orientation in throwing athletes. Am J Sports Med. 2005;33(2):263–71.
52. Hosseinimehr SH, Anbarian M, Norasteh AA, Fardmal J, Khosravi MT. The comparison of scapular upward rotation and scapulohumeral rhythm between dominant and non-dominant shoulder in male overhead athletes and non-athletes. Man Ther. 2015;20(6):758–62.
53. Downar JM, Sauers EL. Clinical measures of shoulder mobility in the professional baseball player. J Athl Train. 2005;40(1):23–9.
54. Michener LA, McClure PW, Karduna AR. Anatomical and biomechanical mechanisms of subacromial impingement syndrome. Clin Biomech (Bristol, Avon). 2003;18(5):369–79.
55. Browne AO, Hoffmeyer P, Tanaka S, An KN, Morrey BF. Glenohumeral elevation studied in three dimensions. J Bone Joint Surg Br. 1990;72(5):843–5.
56. Solem-Bertoft E, Thuomas KA, Westerberg CE. The influence of scapular retraction and protraction on the width of the subacromial space. An MRI study Clin Orthop Relat Res. 1993;296:99–103.
57. Mihata T, Jun BJ, Bui CN, Hwang J, McGarry MH, Kinoshita M, et al. Effect of scapular orientation on shoulder internal impingement in a cadaveric model of the cocking phase of throwing. J Bone Joint Surg Am. 2012;94(17):1576–83.
58. Laudner KG, Stanek JM, Meister K. Differences in scapular upward rotation between baseball pitchers and position players. Am J Sports Med. 2007;35(12):2091–5.
59. Fleisig GS, Barrentine SW, Zheng N, Escamilla RF, Andrews JR. Kinematic and kinetic comparison of baseball pitching among various levels of development. J Biomech. 1999;32(12):1371–5.
60. Kibler WB, Uhl TL, Maddux JW, Brooks PV, Zeller B, McMullen J. Qualitative clinical evaluation of scapular dysfunction: a reliability study. J Shoulder Elb Surg. 2002;11(6):550–6.
61. Krahl VE. The torsion of the humerus; its localization, cause and duration in man. Am J Anat. 1947;80(3):275–319.
62. Mair SD, Uhl TL, Robbe RG, Brindle KA. Physeal changes and range-of-motion differences in

the dominant shoulders of skeletally immature baseball players. J Shoulder Elb Surg. 2004; 13(5):487–91.
63. Yamamoto N, Itoi E, Minagawa H, Urayama M, Saito H, Seki N, et al. Why is the humeral retroversion of throwing athletes greater in dominant shoulders than in nondominant shoulders? J Shoulder Elb Surg. 2006;15(5):571–5.
64. Krahl H, Michaelis U, Pieper HG, Quack G, Montag M. Stimulation of bone growth through sports. A radiologic investigation of the upper extremities in professional tennis players. Am J Sports Med. 1994;22(6):751–7.
65. Jones HH, Priest JD, Hayes WC, Tichenor CC, Nagel DA. Humeral hypertrophy in response to exercise. J Bone Joint Surg Am. 1977;59(2):204–8.
66. Cahill BR, Tullos HS, Fain RH. Little league shoulder: lesions of the proximal humeral epiphyseal plate. J Sports Med. 1974;2(3):150–2.
67. Adams JE. Little league shoulder: osteochondrosis of the proximal humeral epiphysis in boy baseball pitchers. Calif Med. 1966;105(1):22–5.
68. Reagan KM, Meister K, Horodyski MB, Werner DW, Carruthers C, Wilk K. Humeral retroversion and its relationship to glenohumeral rotation in the shoulder of college baseball players. Am J Sports Med. 2002;30(3):354–60.
69. Osbahr DC, Cannon DL, Speer KP. Retroversion of the humerus in the throwing shoulder of college baseball pitchers. Am J Sports Med. 2002; 30(3):347–53.
70. Polster JM, Bullen J, Obuchowski NA, Bryan JA, Soloff L, Schickendantz MS. Relationship between humeral torsion and injury in professional baseball pitchers. Am J Sports Med. 2013;41(9): 2015–21.
71. Wilk KE, Arrigo CA, Hooks TR, Andrews JR. Rehabilitation of the overhead throwing athlete: there is more to it than just external rotation/internal rotation strengthening. PM R. 2016;8(3 Suppl):S78–90.
72. Wilk KE, Obma P, Simpson CD, Cain EL, Dugas JR, Andrews JR. Shoulder injuries in the overhead athlete. J Orthop Sports Phys Ther. 2009; 39(2):38–54.
73. Jobe FW, Kvitne RS, Giangarra CE. Shoulder pain in the overhand or throwing athlete. The relationship of anterior instability and rotator cuff impingement. Orthop Rev. 1989;18(9):963–75.
74. Jobe FW, Giangarra CE, Kvitne RS, Glousman RE. Anterior capsulolabral reconstruction of the shoulder in athletes in overhand sports. Am J Sports Med. 1991;19(5):428–34.
75. Ellenbecker TS, Mattalino AJ, Elam E, Caplinger R. Quantification of anterior translation of the humeral head in the throwing shoulder. Manual assessment versus stress radiography. Am J Sports Med. 2000;28(2):161–7.
76. Borsa PA, Scibek JS, Jacobson JA, Meister K. Sonographic stress measurement of glenohumeral joint laxity in collegiate swimmers and age-matched controls. Am J Sports Med. 2005; 33(7):1077–84.
77. Borsa PA, Wilk KE, Jacobson JA, Scibek JS, Dover GC, Reinold MM, et al. Correlation of range of motion and glenohumeral translation in professional baseball pitchers. Am J Sports Med. 2005;33(9):1392–9.
78. Crawford SD, Sauers EL. Glenohumeral joint laxity and stiffness in the functional throwing position of high school baseball pitchers. J Athl Train. 2006;41(1):52–9.
79. Muraki T, Yamamoto N, Zhao KD, Sperling JW, Steinmann SP, Cofield RH, et al. Effect of posteroinferior capsule tightness on contact pressure and area beneath the coracoacromial arch during pitching motion. Am J Sports Med. 2010; 38(3):600–7.
80. Burkhart SS, Morgan CD, Kibler WB. Shoulder injuries in overhead athletes. The "dead arm" revisited. Clin Sports Med. 2000;19(1):125–58.
81. Myers JB, Oyama S, Wassinger CA, Ricci RD, Abt JP, Conley KM, et al. Reliability, precision, accuracy, and validity of posterior shoulder tightness assessment in overhead athletes. Am J Sports Med. 2007;35(11):1922–30.
82. Myers JB, Laudner KG, Pasquale MR, Bradley JP, Lephart SM. Glenohumeral range of motion deficits and posterior shoulder tightness in throwers with pathologic internal impingement. Am J Sports Med. 2006;34(3):385–91.
83. Tyler TF, Nicholas SJ, Roy T, Gleim GW. Quantification of posterior capsule tightness and motion loss in patients with shoulder impingement. Am J Sports Med. 2000;28(5):668–73.
84. Tyler TF, Nicholas SJ, Lee SJ, Mullaney M, McHugh MP. Correction of posterior shoulder tightness is associated with symptom resolution in patients with internal impingement. Am J Sports Med. 2010;38(1):114–9.
85. Davidson PA, Elattrache NS, Jobe CM, Jobe FW. Rotator cuff and posterior-superior glenoid labrum injury associated with increased glenohumeral motion: a new site of impingement. J Shoulder Elb Surg. 1995;4(5):384–90.
86. Lombardo SJ, Jobe FW, Kerlan RK, Carter VS, Shields CL Jr. Posterior shoulder lesions in throwing athletes. Am J Sports Med. 1977;5(3):106–10.
87. Jobe CM. Superior glenoid impingement. Current concepts Clin Orthop Relat Res. 1996; 330:98–107.
88. Beitzel K, Zandt JF, Buchmann S, Beitzel KI, Schwirtz A, Imhoff AB, et al. Structural and biomechanical changes in shoulders of junior javelin throwers: a comprehensive evaluation as a proof of concept for a preventive exercise protocol. Knee Surg Sports Traumatol Arthrosc. 2016;24(6):1931–42.
89. Meister K, Day T, Horodyski M, Kaminski TW, Wasik MP, Tillman S. Rotational motion changes in the glenohumeral joint of the adolescent/little

league baseball player. Am J Sports Med. 2005; 33(5):693–8.
90. Almeida GP, Silveira PF, Rosseto NP, Barbosa G, Ejnisman B, Cohen M. Glenohumeral range of motion in handball players with and without throwing-related shoulder pain. J Shoulder Elb Surg. 2013;22(5):602–7.
91. Herrington L. Glenohumeral joint: internal and external rotation range of motion in javelin throwers. Br J Sports Med. 1998;32(3):226–8.
92. Nakamizo H, Nakamura Y, Nobuhara K, Yamamoto T. Loss of glenohumeral internal rotation in little league pitchers: a biomechanical study. J Shoulder Elb Surg. 2008;17(5):795–801.
93. Torres RR, Gomes JL. Measurement of glenohumeral internal rotation in asymptomatic tennis players and swimmers. Am J Sports Med. 2009;37(5): 1017–23.
94. Paine R, Voight ML. The role of the scapula. Int J Sports Phys Ther. 2013;8(5):617–29.
95. Moseley JB Jr, Jobe FW, Pink M, Perry J, Tibone J. EMG analysis of the scapular muscles during a shoulder rehabilitation program. Am J Sports Med. 1992;20(2):128–34.
96. Kuhn JE, Plancher KD, Hawkins RJ. Scapular Winging. J Am Acad Orthop Surg. 1995; 3(6):319–25.
97. Kamkar A, Irrgang JJ, Whitney SL. Nonoperative management of secondary shoulder impingement syndrome. J Orthop Sports Phys Ther. 1993;17(5):212–24.
98. Keshavarz R, Bashardoust Tajali S, Mir SM, Ashrafi H. The role of scapular kinematics in patients with different shoulder musculoskeletal disorders: a systematic review approach. J Bodyw Mov Ther. 2017;21(2):386–400.
99. Kuhn JE, Lindholm SR, Huston LJ, Soslowsky LJ, Blasier RB. Failure of the biceps superior labral complex: a cadaveric biomechanical investigation comparing the late cocking and early deceleration positions of throwing. Arthrosc J Arthrosc Relat Surg. 2003;19(4):373–9.
100. Burkhart SS, Morgan CD, Ben KW. The disabled throwing shoulder: Spectrum of pathology part III: the SICK scapula, scapular dyskinesis, the kinetic chain, and rehabilitation. Arthrosc J Arthrosc Relat Surg. 2003;19(6):641–61.
101. Grossman MG, Tibone JE, McGarry MH, Schneider DJ, Veneziani S, Lee TQ. A cadaveric model of the throwing shoulder: a possible etiology of superior labrum anterior-to-posterior lesions. J Bone Jt Surg Am Vol. 2005;87A(4):824–31.
102. Wilk KE, Macrina LC, Fleisig GS, Aune KT, Porterfield RA, Harker P, et al. Deficits in Glenohumeral passive range of motion increase risk of shoulder injury in professional baseball pitchers: a prospective study. Am J Sports Med. 2015;43(10):2379–85.
103. Burkhart SS, Lo IKY. The cam effect of the proximal humerus: its role in the production of relative capsular redundancy of the shoulder. Arthrosc J Arthrosc Relat Surg. 2007;23(3):241–6.
104. Burkhart SS, Morgan CD, Ben KW. The disabled throwing shoulder: Spectrum of pathology part 1: Pathoanatomy and biomechanics. Arthrosc J Arthrosc Relat Surg. 2003;19(4):404–20.
105. Reynolds SB, Dugas JR, Cain EL, McMichael CS, Andrews JR. Debridement of small partial-thickness rotator cuff tears in elite overhead throwers. Clin Orthop Relat Res. 2008;466(3):614–21.
106. Raynor MB, Horan MP, Greenspoon JA, Katthagen JC, Millett PJ. Outcomes after arthroscopic Pancapsular Capsulorrhaphy with suture anchors for the treatment of multidirectional Glenohumeral instability in athletes. Am J Sports Med. 2016;44(12):3188–97.

Biomechanics of Acromioclavicular Joint Injury and Repair

14

Matthew R. LeVasseur, Michael B. DiCosmo, Rafael Kakazu, Augustus D. Mazzocca, and Daniel P. Berthold

14.1 Introduction

Acromioclavicular (AC) joint injuries have historically been treated with a myriad of surgical techniques, devices, implants, and approaches. Cadaveric and biomechanical studies have highlighted the integral role of the investing capsuloligamentous structures of the AC joint to promote stability and to allow appropriate function of the upper limb. As such, the restoration of the native joint anatomy and biomechanics is key to optimal AC joint stability. Operative techniques include a variety of open and arthroscopic-assisted approaches such as ligament and tendon transfers, AC wires, coracoclavicular (CC) screws, hook plates, CC reconstruction with synthetic and biological grafts, anatomic coracoclavicular ligament reconstruction (ACCR), and AC capsuloligamentous repair and augmentation. Although significant progress has been made over the past decades in order to reduce clinical failures, controlling rotation at the AC joint remains highly challenging, which may be contributory to operative failures.

M. R. LeVasseur (✉) · M. B. DiCosmo · R. Kakazu · A. D. Mazzocca
Department of Orthopaedic Surgery, University of Connecticut, Farmington, CT, USA
e-mail: mlevasseur@uchc.edu; mdicosmo@uchc.edu; kakazu@uchc.edu; mazzocca@uchc.edu

D. P. Berthold
Department of Orthopaedic Surgery, University of Connecticut, Farmington, CT, USA

Department of Orthopaedic Sports Medicine, Technical University of Munich, Munich, Germany

14.2 Anatomy

14.2.1 Acromioclavicular Joint and the Acromioclavicular Ligament Complex

The AC joint is a diarthrodial joint containing an intra-articular disc, which functions similar to the meniscus of the knee, to improve congruence, function and reduce intra-articular pressure. Together with the glenohumeral, scapulothoracic, and sternoclavicular joints, the AC joint, including the AC and coracoclavicular (CC) ligaments are part of the chain that permits the upper limb to suspend from the torso, promoting upper extremity function and motion.

Surrounding the AC joint is a thin capsule that requires ligament reinforcement to maintain the integrity of the joint. The circumferential AC ligament complex (ACLC) has commonly been categorized into 4 parts based on anatomic direction: anterior, posterior, superior, and inferior [1, 2]. Anatomically, Stine et al. noted that the superior ligament inserts on the acromion on average 2.8 mm from the medial acromion, and the inferior ligament attaches onto the clavicle on average 3.5 mm

J. Koh et al. (eds.), *Orthopaedic Biomechanics in Sports Medicine*,
https://doi.org/10.1007/978-3-030-81549-3_14

from the distal clavicle with an average width of insertion being 1.0 to 5.1 mm [3]. Of importance, the fibers of the superior ligament are continuous with the aponeuroses of the deltoid and trapezius with a thickness from 2.0 mm to 5.5 mm. The relation of these muscles to the AC joint is shown in Figs. 14.1 and 14.2. Biomechanically, the superior ligament was demonstrated as stronger than the inferior AC ligament, thus having an important role in controlling rotational and horizontal stability [2, 4, 5]. However, as the inferior ligament is often much thinner, anatomically, it should be considered more difficult to consistently detect and measure [3–6].

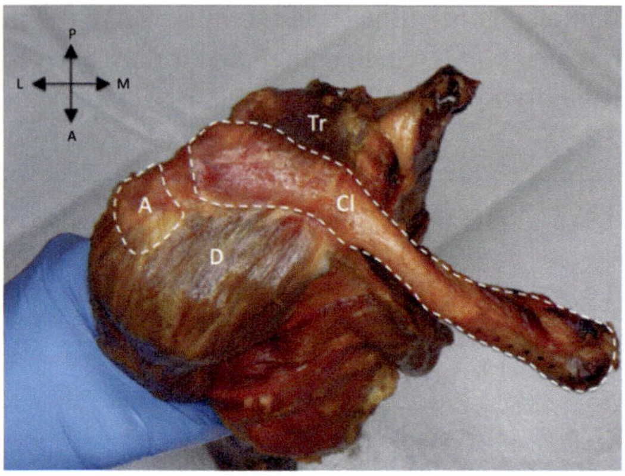

Fig. 14.1 Superior view of the convergence of the trapezius (Tr) and deltoid (D) to the distal clavicle and acromioclavicular joint. The convergence of these structures is provided by the deltotrapezial fascia. Dotted lines represent the contours of the clavicle (Cl) and acromion (A). Each muscle often has attachments to the AC capsule. Pectoralis major (not shown) has been resected

In contrast to the findings above, a more recent anatomic study of the ACLC by Nakazawa et al. demonstrated multiple variations in the structure of the inferior part of the complex. As such, the long described narrative that the ACLC runs directly across and completely surrounds the joint was questioned by the authors of this study [6]. Subsequently, the investigators proposed that the structure of the ACLC is best defined as two separate bundles, the superoposterior (SP) bundle and anteroinferior (AI) bundle [6]. The SP bundle runs at an average angle of 30° to the joint surface from the superior part of the anterior acromion to the posterior part of the distal clavicle (Fig. 14.3). Additionally, using high quality structural composition, the investigators were able to make two more significant conclusions: first, there are three different structural configurations of the AI bundle and second, there is a region of the anterior AC joint where neither the AI or SP cover the joint [6]. Nakazawa et al. found that in 42.3% of the specimens, the AI bundle extended from the anterior face of the acromion to the anterior margin of the distal clavicle. In a similar percentage, the AI bundle extended from the anterior to inferior part of the joint capsule leaving a gap on the inferior surface of the joint. However, only 15.4% of the AI bundles extended from the anterior to superior surface of the joint capsule [6].

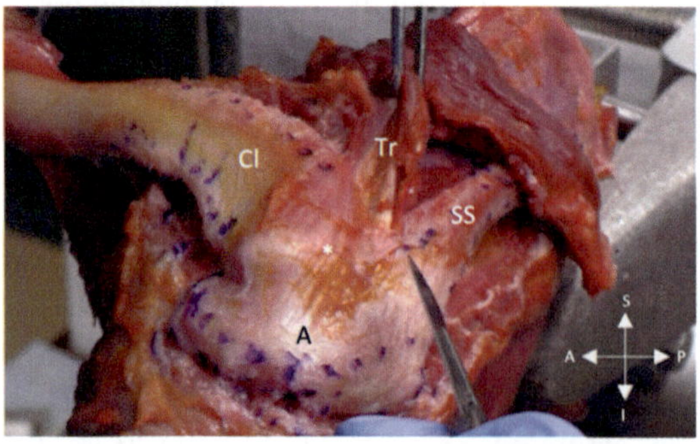

Fig. 14.2 Lateral view of the AC joint with tension applied to the trapezius (Tr) revealing muscular insertion to the posterior and superior margin of the acromioclavicular ligamentous complex (ACLC,*). The fibers of the trapezius become continuous with the ACLC. *A* acromion, *Cl* clavicle, *Co* coracoid, *SS* scapular spine

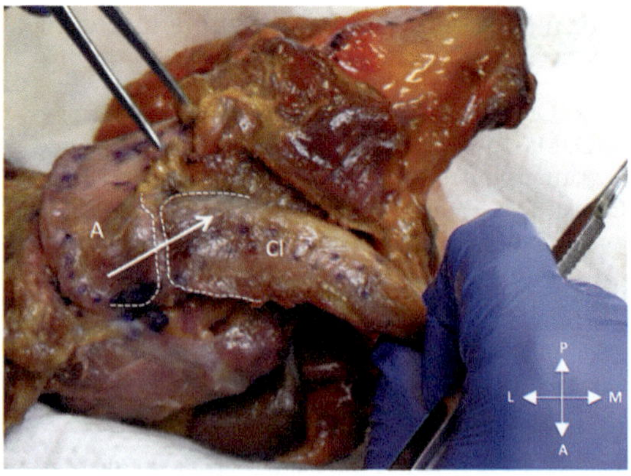

Fig. 14.3 Superior view of the acromioclavicular ligamentous complex with the superior fibers running obliquely across the AC joint. *A* acromion, *Cl* clavicle

Other recent anatomical studies, have investigated the bony morphology and bone mineral density (BMD) of the acromion in the region of the ACLC complex [7]. Voss et al. identified the greatest BMD at the posteromedial acromion nearest to the AC joint, with increasing density from lateral to medial. This finding has important clinical implications for ACLC reconstructions, which will be discussed later.

14.2.2 The Coracoclavicular Ligaments

In addition to the ACLC, the AC joint is stabilized by the CC ligaments (Fig. 14.4). The CC ligaments consist of the trapezoid and conoid ligaments spanning from the coracoid to the lateral clavicle. As such, the lateral trapezoid ligament extends from the anterolateral angle of the coracoid to the inferior clavicle in a superior, anterior, and slightly lateral direction [8]. The medial conoid ligament is attached further posteromedial with respect to the angle of the coracoid and extends superiorly and slightly medial toward its clavicular attachment at the posterior-inferior clavicle at the conoid tubercle, in the region of the posterior curvature of the clavicle [8]. The location of the CC ligaments can be seen in Fig. 14.5. The footprints of the conoid and

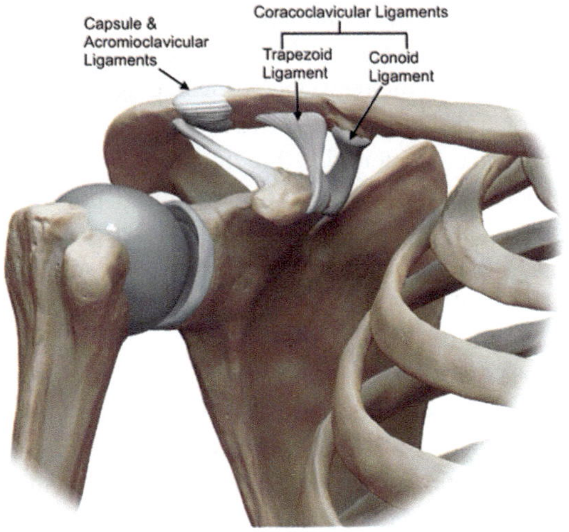

Fig. 14.4 Anatomical drawing of the ligaments (L) supporting the acromioclavicular and glenohumeral joints. From Funk and Imam [12]

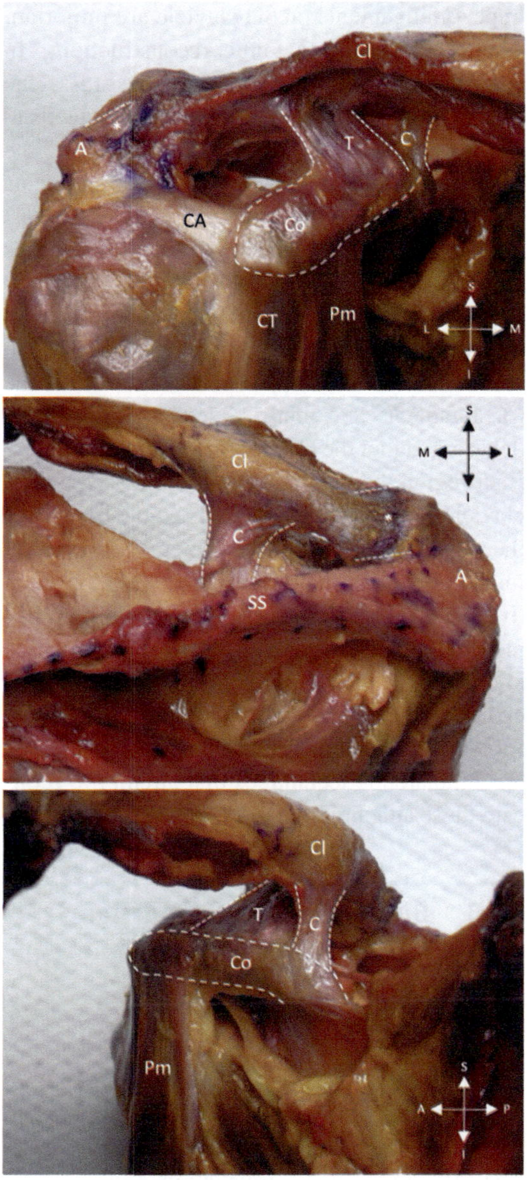

Fig. 14.5 Cadaveric dissection demonstrating the attachments of the conoid and trapezoid ligaments from 3 views: anterior (top), posterior (middle), and medial (bottom). Dotted lines outline the conoid (C) and trapezoid (T) ligaments as well as the coracoid process. The directionality of the acromioclavicular ligamentous complex (ACLC,*) is also shown with dotted lines. The conoid is more medial and posterior on the clavicle compared to the trapezoid. Top: Lateral conoid attachments are deep to the trapezoid. Middle: The conoid has broad attachments to the clavicle. The more anterior trapezoid is unable to be visualized from the posterior view. Bottom: The conoid is more proximally based on the coracoid process and its fibers are more vertically directed. *A* acromion, *Cl* clavicle, *Co* coracoid, *CT* conjoint tendon, *Pm* pectoralis minor, *SS* scapular spine

trapezoid ligaments on the clavicle are important when considering anatomic reconstructions. In particular, to appropriately reconstitute the conoid ligament, the medial bone tunnel needs to be placed posteriorly [9, 10]. Of interest, the region of the clavicle where the CC ligaments attach have also been shown to have the greatest BMD, which has important clinical implications when considering reconstruction [11].

Anatomic studies have identified reproducibility in CC ligamentous attachments. Rios et al. identified a relationship between the origin of the CC ligaments and total clavicular length. They determined that the CC ligaments attach at a constant region that is proportional to the size of the clavicle regardless of gender. The attachments of the trapezoid and conoid ligaments were determined to be consistently at 17% and 31%, respectively, of the total clavicular length when measured from the lateral edge [10]. Detailed knowledge of the location of the conoid and trapezoid ligaments and their footprints is important for surgical technique, particularly with more recent anatomic techniques that attempt to replicate these attachments [9, 11, 13, 14].

14.3 Biomechanical Properties of the Native AC Joint

Complex interactions between the AC and CC ligaments, the AC capsule including the deltotrapezial and pectoral fascia, and the scapulothoracic and sternoclavicular joints are needed to allow optimal function of the shoulder girdle. The AC joint allows translation and rotation. Subsequently, the complex three-dimensional motion of the AC joint has been described as having 6 degrees of freedom when discussing the motion of the scapula relative to the clavicle [15, 16]. Translational kinematics include anterior-posterior, superior-inferior, and distractive/compressive motion. In regards to rotation, motions include internal and external rotation on the vertical axis, upward and downward rotation relative to an axis perpendicular to the scapular plane, and anterior to posterior tilting relative to a horizontal axis in the scapular plane [15, 16]. To this, the clavicle rotates nearly 40–50° during scapular motion [17, 18], providing more complicated motion at the AC joint [19]. Additionally, rotation through the clavicle and AC joint are important for active forward elevation of the arm, particularly at higher degrees, thus allowing correct scapulothoracic function [20].

In accordance to anatomic studies consisting of experimental sectioning, the knowledge of how the aforementioned ligamentous structures relate to stability of the AC joint has significantly increased in the past years. As complex loading-and-unloading occurs, the AC and CC ligaments were shown to be to be important in resisting translational, distractive, and rotational forces [13, 21, 22].

14.3.1 Acromioclavicular Ligament Complex

The importance of the ACLC in horizontal or anterior-posterior stability has already been established in current literature, with cadaveric studies showing that AC capsulotomy results in increased horizontal translation [21–23]. Back in 1941, horizontal stability was first demonstrated in a cadaveric study by Urist et al. in which the authors manually displaced and loaded the clavicle with forceps and simply observed the resulting motion of the humerus [24]. Furthermore, the ACLC has also been demonstrated to resist superior translation of the clavicle during load displacement testing, although less so when compared to the resistance provided against posterior translation [4, 15, 21, 23, 25, 26]. In fact, Dawson et al. identified threefold greater restraint provided by the ACLC in the anterior-posterior plane compared to the superior-inferior plane [21]. In summary, the posterior and superior aspects of the ACLC are the main contributors to posterior translation of the distal clavicle [27].

Additionally, Debski et al. were among the first authors to mimic in vivo joint motion with three degrees of freedom to confirm the observations mentioned above [23, 28, 29]. Subsequently, Debski and colleagues demonstrated that the superior and inferior portions of the ACLC work in synergy only during superior joint loading, which would suggest that they do not function as two completely separate bands as previously suggested [23, 28, 29].

Recently, Morikawa et al. demonstrated that the different segments of the superior AC ligament have different contributions to posterior translation and rotational motions of the clavicle [30]. In this study, the investigators divided the superior ACLC into three regions based on a 180-degree plane that split the ACLC in half. The authors found that together the anterior and superior aspects of the ACLC contributed significantly more to posterior translational stability than did the superior and posterior aspects or the anterior and posterior aspects [30]. Additionally, the authors concluded that the anterior and posterior aspects of the superior ACLC provided the greatest stability to the AC joint during posterior rotation of the clavicle. This evidence suggests that anatomically the superior portion of the ACLC should be considered three distinct parts. Consequently, the entire superior part of the ACLC must be repaired or reconstructed in order to restore both the posterior translational and rotational constraint provided by the ACLC [30].

Similarly, Dyrna et al. identified the AC capsule as paramount in the centering of the AC joint during rotation and limiting abnormally elevated amplitudes of clavicular motion with respect to the acromion [26]. As such, not only is the ACLC important in rotation restraint, but also synchronizes the motion of the clavicle and acromion during rotation, limiting unnecessary motion. These authors identified significant reductions in resistance to translation and rotational strength to 25% and 10% of the intact state, respectively, with AC capsulotomy while keeping the CC ligaments intact [26]. Additional studies have shown even further reductions in posterior rotational restraint after AC transection, with values as low as 5.4% of the intact state [14]. The importance of the ACLC in rotational stability cannot be overstated [14, 26, 30].

14.3.2 Coracoclavicular Ligaments

The CC ligaments have been shown to control horizontal and vertical translation of the AC joint [3, 5, 21, 24, 26, 31]. As shown by historical dissection studies, transection to both AC and CC ligaments are required to cause AC joint dislocation [24, 32]. Of interest, Debski et al. have shown that the role of the CC ligaments becomes more important with concomitant ACLC injury [23]. As such, following AC capsular transection, the conoid displayed a 227% increase in force with anterior load and the trapezoid having a 66% increase in force with posterior load. These authors highlighted that the CC ligaments cannot be treated as one combined structure because of their varying mechanical functions. Yoo et al. concluded that the conoid ligament becomes a primary restraint to anterior and superior loading whereas the trapezoid ligament becomes a primary restraint to posterior loading in the setting of AC ligamentous complex transection [33].

Moreover, the relative contribution of load to the CC ligaments is dependent on the magnitude of motion. Fukuda et al. identified with larger displacements of the clavicle, the CC ligaments, primarily the conoid ligament, became the primary restraint to superior elevation of the clavicle [4]. These authors identified an important role of the conoid ligament in prevention of anterior and superior displacement of the clavicle. Seo et al. described the dynamic nature of the CC ligaments with shoulder abduction using finite element analysis, revealing progressive lengthening of the conoid and relatively unchanged length of the trapezoid ligament during gradual abduction, until the trapezoid becomes lax at full abduction [34].

Recent biomechanical investigations demonstrated that the AC and CC ligaments are both important in rotational stability [13, 22, 26, 30, 35]. Branch et al. identified significant changes in rotation after distal clavicle excision and AC ligament transection [31]. In the first cadaveric sectioning experiments attempting to understand the ligamentous restraint of the AC joint, Cadenet hypothesized that the CC ligaments act as a pivot for anterior-posterior rotation of the clavicle [32]. Fukuda et al. described the conoid ligament having a role in anterior and superior rotation [4]. As will be discussed later, more recent repair techniques have been developed to help prevent non-physiologic rotation.

14.4 Mechanism of Injury

The most common mechanism of AC joint injury is a direct force to a falling patient on an adducted arm [5, 36–38]. This causes a direct force on top of the acromion, driving it inferiorly. Since the sternoclavicular joint will resist inferior translation of the clavicle, this force will cause either the ACLC and CC ligaments to fail or the clavicle to fracture [5, 38]. In the case that the clavicle does not fracture, there is a typical sequence of events that occurs after the direct force occurs. The force will initially spread throughout the ACLC and can cause either a mild sprain or if strong enough, will cause it to tear [5, 36, 37]. After rupturing the ACLC, the force will be transmitted to the CC ligaments causing them to tear along with the clavicular attachments of the deltotrapezial fascia and thus a complete dislocation of the AC joint [5, 36, 37].

14.5 Biomechanical Considerations in AC Joint Repair

Cadaveric sectioning and biomechanical investigations have been paramount in the formulation of treatment options for AC joint injuries. As such, biomechanical testing of intact and defective ligamentous structures have been used as models for testing repair techniques and devices and comparing them to the native, uninjured state. Particularly with respect to high-grade AC joint injuries, the appropriate treatment option remains controversial and there have been an increasing number of publications in recent years attempting to identify the best treatment option. Today, there are a significant variety of treatment options, some of which are only historical today, including both arthroscopic-assisted and open approaches with techniques such as ligament and tendon transfers, AC wires, coracoclavicular (CC) screws, hook plates, CC stabilization with synthetic and biological grafts, anatomic coracoclavicular ligament reconstruction (ACCR), and AC capsuloligamentous repair and augmentation. The current belief and understanding is the restoration of native joint mechanics will improve healing, patient satisfaction, and AC joint stability.

14.5.1 Coracoacromial Ligament and Tendon Transfer

Coracoacromial ligament (CAL) transfer for AC joint injuries was first described by Cadenet in 1917, whereas Cadenet described this technique as an option for dislocations and fractures of "the outer end of the clavicle." [32] This technique was then advocated by Neviaser and later by Weaver and Dunn [39, 40]. The Weaver-Dunn (W-D) technique published in 1972 consisted of distal clavicle excision and transfer of the CAL to the distal clavicle to restore the CC ligament in Type III injuries [40]. This technique has since been modified by an abundance of surgeons, consisting of variations in transfer with and without an accompanying fragment of bone as well as supplementary fixation methods to augment the strength of the transfer (Fig. 14.6). As such, Cerciello et al. augmented ligament transfer with CC cerclage, CC ligament remnant suturing, and transarticular dual K-wire fixation of the reduced AC joint [41].

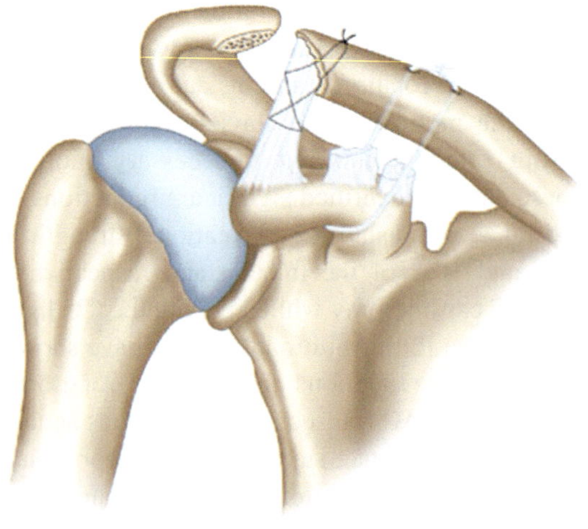

Fig. 14.6 The Weaver-Dunn (W-D) procedure, which consists of distal clavicle excision and release of the coracoacromial ligament from its attachments on the acromion and re-attachment to the distal clavicle. The construct shown has been augmented (modified W-D procedure) with a suture loop. From Yoo [42]

Tendon transfers for the treatment of AC joint injuries were first described by Dewar et al. in 1965 [43]. These techniques would cede injury to additional native ligamentous restraints that would be important in postoperative stability. Dewar et al. created an osteotomy of the distal coracoid process and transferred the bone block with the attached coracobrachialis and short head of the biceps tendon to the clavicle [43]. Since then, others have utilized the long head of the biceps tendon or a slip of the conjoined tendon to replace the injured CC ligaments [44–46].

14.5.1.1 Biomechanics

Coracoacromial ligament transfers have consistently demonstrated increased laxity and inferior strength compared to the native ligamentous restraints, thus failing to restore the native AC joint kinematics [47–49]. Similarly, by using the W-D procedure, Deshmukh et al. identified increased laxity in the anterior, posterior, and superior directions [49]. Harris et al. identified the ultimate tensile strength of the transferred CA ligament to be 145 N, which is significantly less than the native CA ligament failure load of 312 N identified by Soslowsky et al. [47, 50] Furthermore, this tensile strength is significantly lower than the intact CC ligamentous complex of 725 N [51]. Even with modifications to the original W-D procedure, there fails to be a reproduction of the load-to-failure capabilities of the native joint [9, 52]. Grutter and Peterson identified the strength of the modified W-D procedure to be nearly half the strength of the native AC joint, 483 N compared to 815 N [52]. Mazzocca et al. reported statistically significant posterior displacement with the modified W-D procedure [9]. Additionally, Kippe et al. identified poor rotational stability after modified W-D reconstruction compared to the native shoulder [19]. However, improvements in AC joint stability may be afforded with fixation supplementation of CAL transfer [49]. These improvements include AC capsule repair, CC screw fixation, CC ligament remnant repair, CC cerclage, tendon grafting, and others. Similarly, AC capsuloligamentous repair has been shown to increase anterior-posterior stability [53]. In contrast, near normal joint kinematics and improvements in strength can be expected by using the modified W-D reconstruction with addition of a coracoid transclavicular cerclage [54, 55]. Kippe et al. analyzed different coracoclavicular suture loop combinations to augment the modified W-D reconstruction and identified increased abrasive wear with sutures looped around the clavicle compared to suture placed through bone tunnels in the clavicle [19]. Harris et al. identified insufficient stiffness in the modified W-D reconstruction and recommended augmentation with another form of CC fixation such as bicortical screw or suture anchor [47]. Despite improvements with CAL transfers with these modifications/augmentations, they may be considered inferior to newer, more "anatomic" methods of repair [9, 52, 56–58].

To date, limited studies are available regarding the biomechanical testing of tendon transfers. Tendon transfers for traumatic AC joint separations were created to spare the coracoacromial arch, unlike in the W-D reconstruction [46]. To note, the coracoacromial arch is important as a secondary, static stabilizer to superior migration of the humeral head in the presence of rotator cuff tears [59]. Sloan et al. identified the strength of the lateral half of the conjoined tendon to be similar to that the coracoacromial ligament in a simulated reconstruction, 265 N compared to 246 N [59]. However, both are significantly weaker compared to the intact CC ligaments, which the authors identified as 621 N in their study. Unfortunately, there is a paucity of information on translational and rotational restraint of tendon transfers.

14.5.2 AC Joint Repair Using Transarticular K-Wire Fixation

AC joint reduction can be afforded by temporary threaded or smooth wires or pins across the AC joint using percutaneous or open techniques. AC joint repair using transarticular K-wire fixation was first introduced by Phemister in 1942 [60]. However, to date, this technique has largely been abandoned when performed in isolation

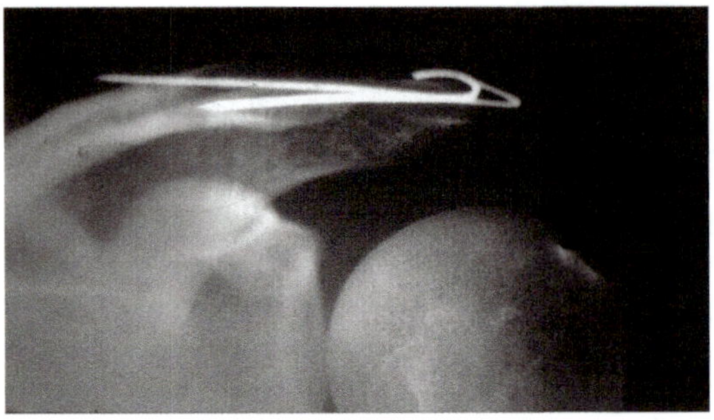

Fig. 14.7 Transarticular K-wire fixation. From Cerciello et al. [41]

secondary to complications such as subsequent procedures for removal, breakage, loss of reduction, and wires backing out [61]. Some have even described these pins penetrating the lung [62]. However, this technique may still be considered for augmentation of AC or CC ligamentous repair or reconstructions, such as in coracoacromial ligament transfers [63, 64]. This technique is shown in Fig. 14.7.

14.5.2.1 Biomechanics

Biomechanical testing of isolated transarticular K-wire fixation has been infrequently described in the literature. In a biomechanical comparison study of transarticular K-wire fixation, ligament reconstruction with a synthetic augmentation device, and coracoclavicular screw, Fialka et al. identified K-wire fixation as having the highest rate of loosening [65]. In a biomechanical comparison study, Bargren et al. compared AC wires to loop fixation [61]. These authors identified AC wire failure in 18 of 45 cases (40%) by 6 weeks. Furthermore, threaded wires had no improvements compared to smooth wires. To this, the AC wires afforded less translational restraint compared to loop fixation. Of interest, the authors believed the AC wire failure was secondary to bending moments and subsequent loosening or bending of the wires [61].

14.5.3 Coracoclavicular Screws

Coracoclavicular screws for AC joint instability were first described by Bosworth in 1941 (e.g. Bosworth screw) [66]. Since their inception as a means to provide rigid internal fixation, there have been a variety of published techniques including blind or fluoroscopic-assisted placement, closed or open techniques of AC joint reduction, non-cannulated or cannulated screws, and temporary or permanent placement [66–69]. Additionally, CC screws (Fig. 14.8) have also been advocated as augmentation to other forms of CC stabilization, such as in CAL transfer [69].

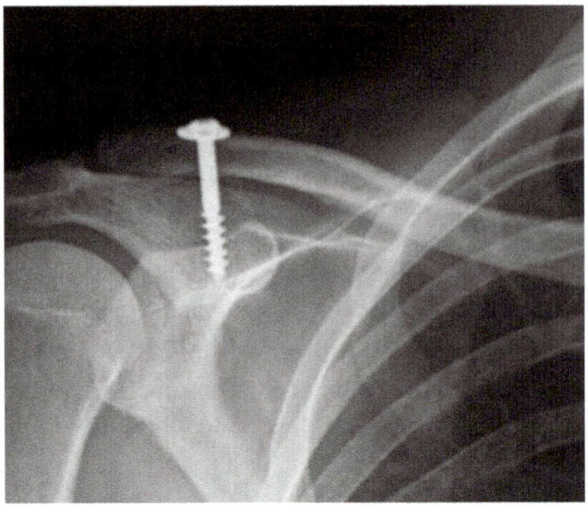

Fig. 14.8 Coracoclavicular screw. From Yoo [42]

However, these devices are often associated with complications including injury to nearby neurovascular structures, which are located immediately inferomedial to the coracoid process, screw breakage, and secondary procedures for screw removal [67]. Furthermore, to prevent screw breakage, patients often have increased postoperative restrictions and longer periods of immobilization in efforts to prevent loss of reduction or recurrent instability [67]. As will be discussed, these devices provide significant strength, however, secondary to their complications and inferiority to more anatomic reconstructions, these devices have largely fallen out of favor.

14.5.3.1 Biomechanics

The first biomechanical studies with the CC screw showed its effectiveness in providing rigid stabilization [70, 71]. Compared to CAL transfer, CC screws have shown improvement in translational restraint and overall strength. To this, excellent vertical stability with CC screws and horizontal stability better than those of CAL transfers and suture loop techniques has been shown [72]. Similarly, Harris et al. demonstrated superior strength of the Bosworth screw to the intact CC ligaments, if bicortical fixation was achieved [47]. Thus, unicortical fixation was noted to be inferior to the native strength with strengths of 229 N to 390 N compared to 927 N (bicortical) [47, 51].

Of interest, when compared to CC loop using Mersilene tape and hook plate fixation, the CC screw showed increased stiffness and highest load-to-failure [73]. In fact, the stiffness was nearly double that of the intact state. However, hook plate fixation achieved stiffness closer to the native joint state. Fialka et al. identified failure of CC screw stability with increased upper extremity abduction, particularly past 120°. These authors recommended limiting shoulder abduction to less than 90° after placement of CC screws [65].

Optimal screw size and material are still being investigated in the current literature. Various CC screw sizes have been proposed, most commonly 4.5 mm and 6.5 mm [51, 53], however, head-to-head comparisons are limited. Furthermore, the size of the screw is limited in the intra-operative setting to prevent fracture. CC screws, even if unicortical, would remain helpful for augmentation of the W-D reconstruction or CC ligament repair [51, 53].

Of interest, several authors have compared metal and bioabsorbable screws, with the latter avoiding secondary procedures for screw removal. Talbert et al. reported no difference in pullout strength comparing 4.5 mm stainless steel (721 N) and poly L-lactic acid polymer (PLLA) bioabsorbable screws (580 N) in 7 matched cadaveric pairs [74]. However, significance may have been achieved with a larger sample size. Based on these biomechanical properties, the authors suggested that biodegradable screws are a potential surgical option. Despite both devices having pullout strengths greater than the previously reported native CC ligamentous strengths of 500 N [47], other studies have reported much larger native strengths [51]. As such, it is unclear what the reliability of bioabsorbable screws would be.

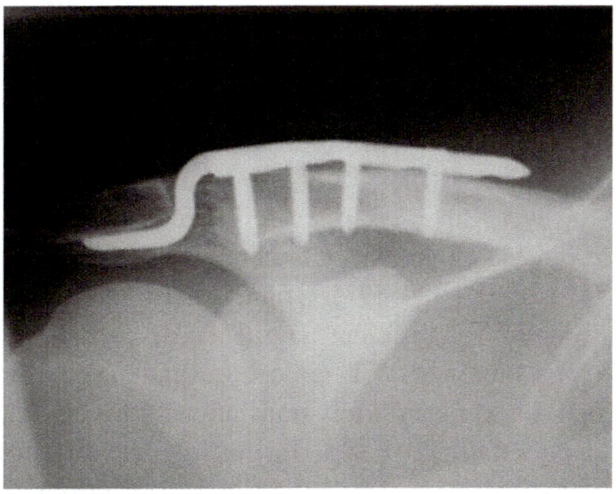

Fig. 14.9 Hook plate fixation. From Jensen et al. [78]

14.5.4 AC Joint Reconstruction with Hook Plates

Hook plate fixation (Fig. 14.9) for AC joint separation was first introduced by Balser back in 1976 [75]. Over the past decades, different hook plate designs have been presented, all based on the same principle: rigid internal fixation with a pre-contoured plate and healing, predominantly by scar tissue [76–78]. Several modifications have been described to further enhance the construct, including direct CC ligament repair and ACLC suturing as well as additional screw fixation, biological or artificial augmentation, or ligament/tendon transfers [76, 79, 80]. Unfortunately, hook plates can bend or dislocate and some have stated that they are at a higher risk of infection compared to other techniques [81], which have contributed to their limited role today. Moreover, implant removal is controversial, with some advocating implant removal in the early postoperative period [82].

14.5.4.1 Biomechanics

Hook plates have demonstrated excellent stiffness but fail to replicate the native AC joint kinematics. In a biomechanical study, McConnell et al. compared CC screw, CC sling with Mersilene tape, and hook plate under superior cyclic loading [73]. The hook plate had a stiffness of 26 ± 17 N/mm, most comparable to the intact joint stiffness of 25 ± 8 N/mm ($p = 0.785$). However, the hook plate failed at significantly lower load than the CC screw, 459 N versus 744 N ($p = 0.034$) [73]. Of interest, in a comparative study of different augmentation techniques to the modified W-D procedure under superior and anterior-posterior translation, Luis et al. identified effective superior loading stability but limitations on providing anterior-posterior restraint [53]. The hook plate augmentation provided 50% and 30% of the tensile strength and stiffness of the native ligaments, respectively. The authors also identified 3 sites of fracture associated with the hook plate, one of which was resultant from the hook bending and slipping superiorly, causing gradual deformation of the acromion [53]. Moreover, Kiefer et al. reported no to minimal rotational restraint of Balser (hook) plates, likely secondary to hook slippage underneath the acromion [70]. As a consequence to these changes in AC joint kinematics, patients undergoing hook plate fixation may demonstrate reduced internal rotation along with increased anterior translation (2 mm) of the clavicle with respect to the medial acromion, when compared to the native AC joint [83].

14.5.5 Coracoclavicular Stabilization Using Synthetic or Biological Devices

Reconstruction of the CC ligamentous complex can be performed using a variety of synthetic or biological devices around or through both the clavicle and coracoid using bone tunnels. In current literature, there is significant variety in reported materials and techniques for this procedure, which can be used either in open or arthroscopic-assisted procedures. Materials can either be absorbable or nonabsorbable, braided or unbraided, and synthetic or natural, i.e., autografts or allografts [19]. As such, single or dual tunnels may be applied to the clavicle or coracoid. Subsequently, fixation can be achieved with high-tensile suture anchors or cortical buttons (Fig. 14.10). In contrast to open, more sophisticated approaches, arthroscopic-assisted CC stabilization often requires shorter operative times, does not routinely require hardware removal, and there is a lower risk of neurovascular injury [84, 85]. Additionally, intra-articular lesions, which occur in more than 20% of the cases, can be assessed [86].

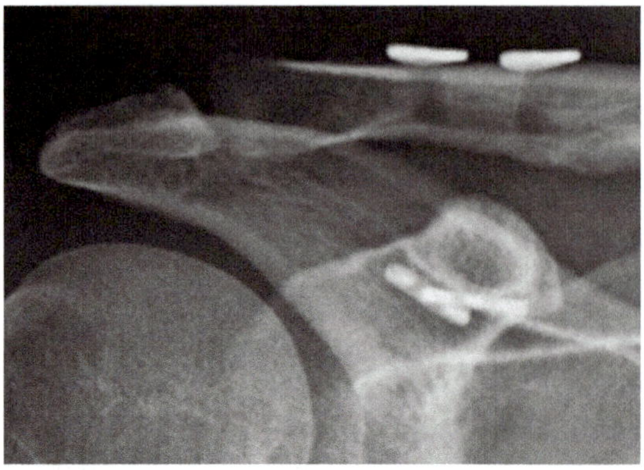

Fig. 14.10 Coracoclavicular stabilization using double cortical buttons with high-tensile sutures. From Jensen et al. [78]

Overall, the various forms of CC stabilization may be used for primary fixation or to augment CC ligamentous repair, for example during the modified W-D procedure [47, 49]. This technique allows stabilization of the CC interval until biologic healing and scarring occurs. Over the past years, CC stabilization using high-tensile suture tapes with endobutton fixation and simultaneous AC capsular stabilization have emerged as one of the most used methods for acute high-grade AC joint injuries.

14.5.5.1 Biomechanics

Biomechanically, CC sling or loop procedures are more effective in providing vertical stability compared to other loading patterns [71, 72]. Loading in the superior-inferior direction parallels the orientation of the reconstruction, maximizing the tensile capacity of sutures and tapes [61]. However, multiple studies have shown that these techniques are unable to restore the native stiffness of the CC ligaments, despite providing sufficient tensile strength [47, 71, 73]. In a biomechanical comparison study under unidirectional tensile loading, Harris et al. reported that CC slings constructed from a straight-woven polyester vascular prosthesis provide similar tensile strengths to the native ligaments but lead to greater deformation before failure compared to the native ligaments, resulting in reduced stiffness and loss of reduction [47]. In fact, the stiffness for CC sling and CC suture anchor procedures were only 22% and 27%, respectively, to that of the native ligaments [47]. Similarly, Motamedi et al. showed that braided polydioxanone sutures (PDS) were as strong as the native CC ligaments, but not as stiff [51]. As a consequence, the authors were concerned about progressive loss of reduction of the repair, as the PDS suture stretches out with time. Similar to Harris et al., others have shown comparable results with suture-sling and suture anchors for CC fixation, however suture anchor repair may be advantageous as it does not require subcoracoid dissection, helping to limit neurovascular injury [87, 88].

However, there are concerns regarding the CC sling and its ability to control anterior-posterior translation and rotation [19, 33, 51, 61, 71, 72]. This has been theorized secondary to looped constructs being prone to graft slippage along the coracoid and failure secondary to abrasion [33, 61, 89]. As a result, the inability to adequately control anterior-posterior motion, leads to anterior subluxation of the clavicle and a malreduction of the AC joint [72, 90, 91]. Jari et al. identified significantly increased anterior and posterior translation of 110% and 330%, respectively, compared to the native ligaments with the CC suture-sling procedure. In a comparative biomechanical study, Kippe et al. compared different suture-sling combinations, namely two forms of clavicular fixation (looped or bone tunnels) and three different suture materials (FiberWire, braided PDS, and Mersilene tape) [19]. These authors reported decreased mean rotation torque strengths with all repair combinations compared to the intact state, highlighting the inabilities of these procedures to control rotation.

Despite limitations on horizontal and rotational restraint, grafts or suture placed through bone tunnels in the clavicle and coracoid have been found to be more anatomic, afford better kinematics, and improve biomechanical strength [19, 33, 51]. In particular, Yoo et al. reported the CC sling was incapable of anterior restraint secondary to slippage along the coracoid [33]. Moreover, grafts or sutures placed through bone tunnels exhibit less abrasive patterns, compared to those looped around either the coracoid or clavicle, limiting bony and soft tissue injury during clavicular rotation [19, 72]. In particular, FiberWire may show the most advanced abrasive pattern [19]. With the various CC cerclage techniques, clavicular or coracoid fracture is a possible mechanism of treatment failure, which can occur from loading or during tunnel drilling [47, 51, 92].

Newer techniques involve flip or cortical buttons through drill holes in the either the coracoid, clavicle, or both. These flip buttons can be used to pass biologic grafts or synthetic materials including suture or tapes [93]. Wellmann et al. demonstrated that subcoracoid flip button fixations can provide similar strengths to the native CC ligamentous complex, while avoiding extensive subcoracoid dissection and neurovascular injury [94]. Furthermore, Grantham et al. identified superior horizontal and vertical translational restraint with double endobutton repair compared to CC soft tissue sling with tibialis anterior allograft. Moreover, the double endobutton repair had stronger load-to-failure characteristics with greater stiffness, yield load, and ultimate load [95].

Overall, CC stabilization procedures are a heterogeneous group of varying techniques and materials. Good outcomes are associated with using these procedures as means of augmentation for other repairs until biologic healing can occur.

14.5.6 Anatomic Coracoclavicular Ligament Reconstruction (ACCR)

Historically, Bunnell first described an anatomic solution for CC ligament reconstruction using a fascial graft weave [96]. Since then, various authors have proposed modified techniques for anatomic AC joint reconstruction, as

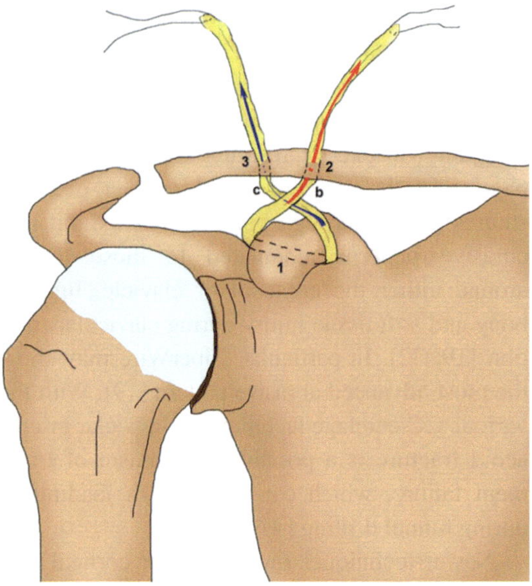

Fig. 14.11 The anatomic coracoclavicular ligament reconstruction (ACCR) utilizing allograft tendon that is looped around the coracoid and fixed to the clavicle through two drill holes and interference screws. The remaining allograft tendon can be used to reconstruct the acromioclavicular ligamentous complex (ACLC; not shown). From Saccomanno et al. [100]

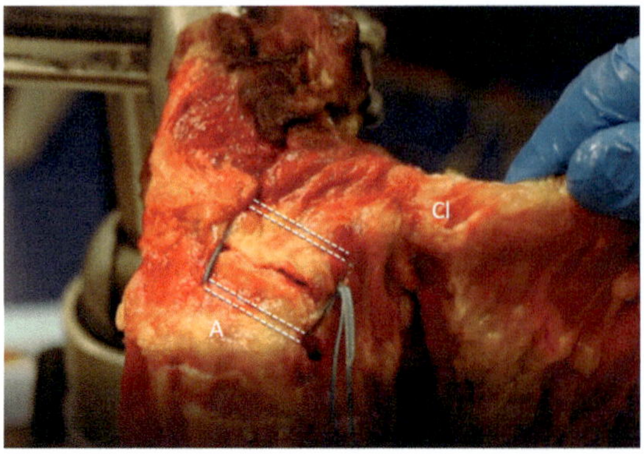

Fig. 14.12 The O-frame acromioclavicular suture configuration. The horizontal bone tunnels through the acromion (A) and clavicle (Cl) are represented by the dotted lines

the restoration of native AC joint kinematics is important to optimize patient function and comfort. The first biomechanical investigations date back to Jones and colleagues in the early 2000s [97]. Since then, several studies demonstrated the efficacy of anatomic coracoclavicular ligament reconstruction (ACCR) using autograft or allograft tendon in replicating the CC ligaments at their anatomic locations and in mimicking the properties of the native CC ligaments compared to ligament transfers [9, 48, 98, 99]. The ACCR technique is demonstrated in Fig. 14.11.

14.5.6.1 Biomechanics

The ACCR was shown to demonstrating superior load-to-failure characteristics when compared to the (modified) W-D procedure, nonanatomic allograft, anatomic suture, and graft-rope techniques [9, 56, 99]. Peak loads equivalent to that of the native CC ligaments with lower stiffness may be expected [95, 101], with anatomic reconstruction of the CC ligaments best restoring the biomechanical properties of the native CC ligaments [9, 99]. Of interest, Costic and colleagues demonstrated that the ACCR technique using a semitendinosus tendon failed to demonstrate significant graft elongation during cyclic loading, which is of clinical importance [98]. Additionally, Geaney et al. demonstrated that optimal tunnel placement in the clavicle is of high importance to achieve optimal strength and reduce potential risk of failures. As such, tunnel placement in the clavicle corresponding to the attachment of the CC ligaments has the highest BMD, and correlates to higher loads to failure [11]. More recent studies have investigated tunnel widening after ACCR, which may lead to reconstruction failure, loss of reduction, and fracture, particularly in the more lateral trapezoid tunnel where lower BMD exists [102].

As shown by Voss et al. and Morikawa et al., persistent horizontal and rotational instability of the AC joint following reconstruction [14, 85, 103], may be reduced by reconstructing the AC capsule containing the AC ligaments and the deltotrapezial fascia [7, 13, 85, 95]. As discussed by Beitzel et al., the remnant of the ACCR graft can be used for ACLC reconstruction [13]. Dyrna et al. reported that transection of only the ACLC reduced translational and rotational resistance to <25% and <10%, respectively, compared to the native state [26]. These authors later compared different AC capsule suture configurations, namely anterior, superior, posterior, O-frame (Fig. 14.12), and X-frame

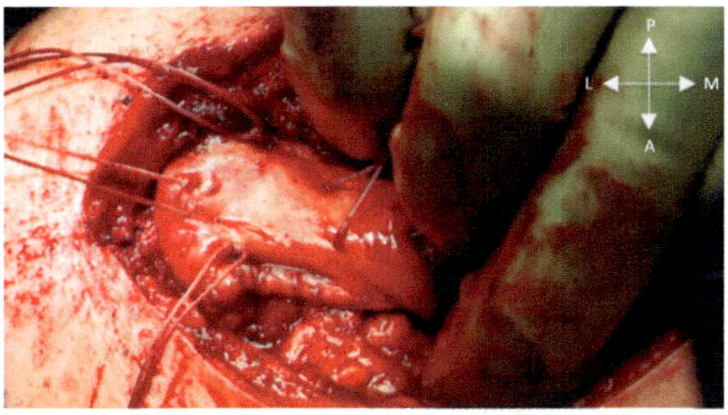

Fig. 14.13 Dermal allograft patch for reconstruction of the AC joint capsule

and identified restoration of translational stability; however rotational stability and resistance torques were significantly weaker than the native specimens, with maximum values of 42% of the native state [22]. Furthermore, recent studies have identified the importance of the superior ACLC with respect to posterior translational and posterior rotational stability [30]. As such, these authors have shown biomechanical success with dermal allografts (Fig. 14.13) for superior ACLC reconstruction [35].

Additionally, the posterior-medial acromion closest to the AC joint shows the highest BMD with increasing density from lateral to medial; thus, fixation at this location should be considered [7]. Bone tunnels should be placed at the acromion within the "safe zone" (i.e., within the anterior half of the acromion) to not affect the load-to-failure at the acromion [104].

Unfortunately, persistent rotational and translational instability remains a concern and plays an important role in AC joint reconstruction failure [22, 26, 30]. Thus, shoulder surgeons should try to replicate native joint kinematics by considering the biomechanical properties of the AC joint.

14.6 Conclusion

The AC joint is an important component of the chain that links the upper extremity to the thorax, to allow upper extremity function and motion. As such, the native AC joint allows translational and rotational motion, which has been identified as important in shoulder abduction and forward flexion. Further, the ACLC and CC ligamentous complexes provide mechanical stability during upper extremity motion. Biomechanical cadaveric studies have had pivotal roles in understanding the function of the components of the ACLC complex as well as the conoid and trapezoid ligaments, all of which have been identified as important in providing horizontal, vertical, and rotation restraint. There are a vast number of techniques and devices that can and have been used for AC joint injuries, including ligament and tendon transfers, CC screws, hook plates, CC slings, double cortical buttons, etc., with biomechanical studies suggesting limitations on controlling rotation still exist. However, more recent techniques, such as ACCR, have suggested that anatomic replication of the native ligamentous restraints may lead to better outcomes and better biomechanical strength.

Conflict of Interest The authors LeVasseur M.R., DiCosmo M.B., Kakazu R., and Berthold D.P. declare that they have no conflicts of interest. Mazzocca A.D. reports research grants and is a consultant for Arthrex, Inc.

Funding No funding has been provided from any agency in the public, commercial, or not-for-profit sectors to complete this manuscript.

References

1. Rockwood CJ, Young D, Bucholz R, Heckmen J. Rockwood and Green's fractures in adults. In: 2nd, editor. Rockwood and Green's fractures in adults. Philadelphia: JB Lippincott-Raven; 1998. p. 1341–413.
2. Salter E, Nasca R, Shelley B. Anatomical observations on the acromioclavicular joint and supporting ligaments. Am J Sports Med. 1987;15(3):199–206. https://doi.org/10.1177/036354658701500301.
3. Stine I, Vangsness C. Analysis of the capsule and ligament insertions about the acromioclavicular joint: a cadaveric study. Arthrosc - J Arthrosc Relat Surg. 2009;25(9):968–74. https://doi.org/10.1016/j.arthro.2009.04.072.
4. Fukuda K, Craig E, An K, Cofield R, Chao E. Biomechanical study of the ligamentous system of the acromioclavicular joint. J Bone Jt Surg. 1986;68(3):434–40. https://doi.org/10.2106/00004623-198668030-00019.

5. Rockwood C., Williams G., Young C. Disorders of the acromioclavicular joint. In: Rockwood C., Matsen FA, Wirth MA, Harryman DT, eds. The shoulder. 2nd ed. Philadelphia; 1998:483–499.
6. Nakazawa M, Nimura A, Mochizuki T, Koizumi M, Sato T, Akita K. The orientation and variation of the acromioclavicular ligament: An anatomic study. Am J Sports Med. 2016;44(10):2690–5. https://doi.org/10.1177/0363546516651440.
7. Voss A, Dyrna F, Achtnich A, et al. Acromion morphology and bone mineral density distribution suggest favorable fixation points for anatomic acromioclavicular reconstruction. Knee Surgery, Sport Traumatol Arthrosc. 2017;25(7):2004–12. https://doi.org/10.1007/s00167-017-4539-1.
8. Collins, DN. Disorders of the acromioclavicular joint. In: Rockwood CA Jr, Matsen FA III, Wirth MA, et al Eds. Rockwood and Matsen's the shoulder. 5th. Philadelphia, PA: Elsevier; 2017.
9. Mazzocca AD, Santangelo SA, Johnson ST, Rios CG, Dumonski ML, Arciero RA. A biomechanical evaluation of an anatomical coracoclavicular ligament reconstruction. Am J Sports Med. 2006;34(2):236–46. https://doi.org/10.1177/0363546505281795.
10. Rios CG, Arciero RA, Mazzocca AD. Anatomy of the clavicle and coracoid process for reconstruction of the coracoclavicular ligaments. Am J Sports Med. 2007;35(5):811–7. https://doi.org/10.1177/0363546506297536.
11. Geaney LE, Beitzel K, Chowaniec DM, et al. Graft fixation is highest with anatomic tunnel positioning in acromioclavicular reconstruction. Arthrosc - J Arthrosc Relat Surg. 2013;29(3):434–9. https://doi.org/10.1016/j.arthro.2012.10.010.
12. Funk L, Imam MA. Acromioclavicular Joint Injuries. In: Funk L, Walton M, Watts A, Hayton M, Ng CY, editors. Sports injuries of the shoulder. Cham, Switzerland: Springer Nature Switzerland AG; 2020. p. 153–78.
13. Beitzel K, Obopilwe E, Apostolakos J, et al. Rotational and translational stability of different methods for direct acromioclavicular ligament repair in anatomic acromioclavicular joint reconstruction. Am J Sports Med. 2014;42(9):2141–8. https://doi.org/10.1177/0363546514538947.
14. Morikawa D, Huleatt JB, Muench LN, et al. Posterior rotational and translational stability in acromioclavicular ligament complex reconstruction: a comparative biomechanical analysis in cadaveric specimens. Am J Sports Med. 2020;48(10):15. https://doi.org/10.1177/0363546520939882.
15. Kent BE. Functional anatomy of the shoulder complex: a review. Phys Ther. 1971;51(8):867–88. https://doi.org/10.1093/ptj/51.8.867.
16. Morris J. The joints of the shoulder girdle. Aust J Physiother. 1978;24(2):63–6. https://doi.org/10.1016/S0004-9514(14)60869-2.
17. Inman VT, Saunders JB, Abbott LC. Observations of the function of the shoulder joint. 1944. Clin Orthop Relat Res. 1996;330:3–12. https://doi.org/10.1097/00003086-199609000-00002.
18. Codman EA, editor. The shoulder. Boston, MA: Thomas Todd; 1934.
19. Kippe MA, Demetropoulos CK, Baker KC, Jurist KA, Guettler JH. Failure of Coracoclavicular artificial graft reconstructions from repetitive rotation. Arthrosc - J Arthrosc Relat Surg. 2009;25(9):975–82. https://doi.org/10.1016/j.arthro.2009.03.016.
20. Ludewig PM, Phadke V, Braman JP, Hassett DR, Cieminski CJ, Laprade RF. Motion of the shoulder complex during multiplanar humeral elevation. J Bone Jt Surg - Ser A. 2009;91(2):378–89. https://doi.org/10.2106/JBJS.G.01483.
21. Dawson PA, Adamson GGJ, Pink MMM, et al. Relative contribution of acromioclavicular joint capsule and coracoclavicular ligaments to acromioclavicular stability. J Shoulder Elb Surg. 2009;18(2):237–44. https://doi.org/10.1016/j.jse.2008.08.003.
22. Dyrna F, Imhoff FBB, Haller B, et al. Primary stability of an acromioclavicular joint repair is affected by the type of additional reconstruction of the acromioclavicular capsule. Am J Sports Med. 2018;46(14):3471–9. https://doi.org/10.1177/0363546518807908.
23. Debski R, Parsons IV I, Woo S.-YS, Fu FHF. Effect of capsular injury on acromioclavicular joint mechanics. J Bone Jt Surg - Ser A. 2001;83(9):1344–1351. doi:https://doi.org/10.2106/00004623-200109000-00009.
24. Urist M. Complete dislocations of the acromiclavicular joint; the nature of the traumatic lesion and effective methods of treatment with an analysis of forty-one cases. J bone Jt Surg. 1946;28(4):813–37.
25. Renfree K, Wright T. Anatomy and biomechanics of the acromioclavicular and sternoclavicular joints. Clin Sports Med. 2003;22(2):219–37. https://doi.org/10.1016/S0278-5919(02)00104-7.
26. Dyrna FGE, Imhoff FB, Voss A, et al. The integrity of the acromioclavicular capsule ensures physiological centering of the acromioclavicular joint under rotational loading. Am J Sports Med. 2018;46(6):1432–40. https://doi.org/10.1177/0363546518758287.
27. Klimkiewicz JJ, Williams GR, Sher JS, Karduna A, Des Jardins JD, Iannotti JP. The acromioclavicular capsule as a restraint to posterior translation of the clavicle: a biomechanical analysis. J Shoulder Elb Surg. 1999;8(2):119–24. https://doi.org/10.1016/S1058-2746(99)90003-4.
28. Debski R, Parsons I, Fenwick J, Vangura A. Ligament mechanics during three degree-of-freedom motion at the acromioclavicular joint. Ann Biomed Eng. 2000;28(6):612–8. https://doi.org/10.1114/1.1304848.
29. Lee K, Debski R, Chen C, Woo S, Fu F. Functional evaluation of the ligaments at the acromioclavicular joint during anteroposterior and superoinferior translation. Am J Sports Med. 1997;25(6):858–62. https://doi.org/10.1177/036354659702500622.

30. Morikawa D, Dyrna F, Cote MMP, et al. Repair of the entire superior acromioclavicular ligament complex best restores posterior translation and rotational stability. Knee Surgery, Sport Traumatol Arthrosc. 2019;27(12):3764–70. https://doi.org/10.1007/s00167-018-5205-y.
31. Branch TPT, Burdette HLH, Shahriari AS, Carter FM, Hutton WC, Huttlon WC. The role of the acromioclavicular ligaments and the effect of distal clavicle resection. Am J Sports Med. 1996;24(3):293–7. https://doi.org/10.1177/036354659602400308.
32. Cadenet F. The treatment of dislocations and fractures of the outer end of the clavicle. Int Clin. 1917;1:145–69.
33. Yoo YS, Tsai AG, Ranawat AS, et al. A biomechanical analysis of the native coracoclavicular ligaments and their influence on a new reconstruction using a coracoid tunnel and free tendon graft. Arthroscopy. 2010;26(9):1153–61. https://doi.org/10.1016/j.arthro.2009.12.031.
34. Seo Y-J, Yoo Y-S, Noh K-C, et al. Dynamic function of coracoclavicular ligament at different shoulder abduction angles: a study using a 3-dimensional finite element model. Arthroscopy. 2012;28(6):778–87.
35. Morikawa D, Mazzocca AD, Obopilwe E, et al. Reconstruction of the acromioclavicular ligament complex using dermal allograft: a biomechanical analysis. Arthrosc - J Arthrosc Relat Surg. 2020;36(1):108–15. https://doi.org/10.1016/j.arthro.2019.07.021.
36. Johansen JA, Grutter PW, Mcfarland EG, Petersen SA, Henze EP, Petersen SA. Acromioclavicular joint injuries: indications for treatment and treatment options. J Shoulder Elb Surg. 2011;20:70–82. https://doi.org/10.1016/j.jse.2010.10.030.
37. Warth RJ, Martetschläger F, Gaskill TR, Millett PJ. Acromioclavicular joint separations. doi:https://doi.org/10.1007/s12178-012-9144-9.
38. Bearn JG. Direct observations on the function of the capsule of the sternoclavicular joint in clavicular support. J Anat. 1967;101(Pt 1):159–70. http://www.ncbi.nlm.nih.gov/pubmed/6047697. Accessed 16 Aug 2020
39. Neviaser JS. Acromioclavicular dislocation treated by transference of the coracoacromial ligament. Arch Surg. 1952;64:292–7.
40. Weaver JK, Dunn HK. Treatment of acromioclavicular injuries, especially complete acromioclavicular separation. J Bone Joint Surg Am. 1972;54:1187–97.
41. Cerciello S, Edwards TB, Morris BJ, Cerciello G, Walch G. The treatment of type III acromioclavicular dislocations with a modified Cadenat procedure: surgical technique and mid-term results. Arch Orthop Trauma Surg. 2014;134(11):1501–6. https://doi.org/10.1007/s00402-014-2085-6.
42. Yoo Y-S. Acromioclavicular joint problems in athletes. In: Park J-Y, editor. Sports injuries to the shoulder and elbow. Springer: Berlin; 2015. p. 265–80.
43. Dewar FP, Barrington TW. The treatment of chronic acromioclavicular dislocation. J Bone Joint Surg Br. 1965;47:32–5.
44. Budoff J, Rodin D, Ochiai D. Conjoined tendon transfer for chronic acromioclavicular dislocation in a patient with paraplegia: a case report with 38-year follow- up. Am J Orthop. 2005;34:189–91.
45. Laing P. Transplantation of the long head of the biceps in complete acromioclavicular separations. J Bone Jt Surg Am. 1969;51:1677–8.
46. Jiang C, Wang M, Rong G. Proximally based conjoined tendon transfer for coracoclavicular reconstruction in the treatment of acromioclavicular dislocation. J Bone Jt Surg - Ser A. 2007;89(11):2408–12. https://doi.org/10.2106/JBJS.F.01586.
47. Harris RI, Wallace AL, Harper GD, Goldberg JA, Sonnabend DH, Walsh WR. Structural properties of the intact and the reconstructed coracoclavicular ligament complex. Am J Sports Med. 2000;28(1):103–8. https://doi.org/10.1177/03635465000280010201.
48. Lee SJ, Nicholas SJ, Akizuki KH, McHugh MP, Kremenic IJ, Ben-Avi S. Reconstruction of the coracoclavicular ligaments with tendon grafts a comparative biomechanical study. Am J Sports Med. 2003;31(5):648–54. https://doi.org/10.1177/03635465030310050301.
49. Deshmukh AV, Wilson DR, Zilberfarb JL, Perlmutter GS. Stability of acromioclavicular joint reconstruction: biomechanical testing of various surgical techniques in a cadaveric model. Am J Sports Med. 2004;32(6):1492–8. https://doi.org/10.1177/0363546504263699.
50. Soslowsky LJ, An CH, Johnston SP, Carpenter JE. Geometric and mechanical properties of the coracoacromial ligament and their relationship to rotator cuff disease. Clin Orthop Relat Res. 1994;304:10–7. https://doi.org/10.1097/00003086-199407000-00003.
51. Motamedi AR, Blevins FT, Willis MC, McNally TP, Shahinpoor M. Biomechanics of the coracoclavicular ligament complex and augmentations used in its repair and reconstruction. Am J Sports Med. 2000;28(3):380–4. https://doi.org/10.1177/03635465000280031701.
52. Grutter PW, Petersen SA. Anatomical acromioclavicular ligament reconstruction: a biomechanical comparison of reconstructive techniques of the acromioclavicular joint. Am J Sports Med. 2005;33(11):1723–8. https://doi.org/10.1177/0363546505275646.
53. Luis GE, Yong CK, Singh DA, Sengupta S, Choon DSK. Acromioclavicular joint dislocation: a comparative biomechanical study of the palmaris-longus tendon graft reconstruction with other augmentative methods in cadaveric models. J Orthop Surg Res. 2007;2(1):1–10. https://doi.org/10.1186/1749-799X-2-22.
54. LaPrade R, Wickum D, Griffith C, Ludewig P. Kinematic evaluation of the modified Weaver-

Dunn acromioclavicular joint reconstruction. Am J Sport Med. 2008;36(11):2216–21. https://doi.org/10.1038/jid.2014.371.
55. Lee SJ, Keefer EP, McHugh MP, et al. Cyclical loading of coracoclavicular ligament reconstructions: a comparative biomechanical study. Am J Sports Med. 2008;36(10):1990–7. https://doi.org/10.1177/0363546508324284.
56. Thomas K, Litsky A, Jones G, Bishop JY. Biomechanical comparison of coracoclavicular reconstructive techniques. Am J Sports Med. 2011;39(4):804–10. https://doi.org/10.1177/0363546510390482.
57. Beitzel K, Obopilwe E, Chowaniec DM, et al. Biomechanical comparison of arthroscopic repairs for acromioclavicular joint instability: suture button systems without biological augmentation. Am J Sports Med. 2011;39(10):2218–25. https://doi.org/10.1177/0363546511416784.
58. Beitzel K, Obopilwe E, Chowaniec DM, et al. Biomechanical properties of repairs for dislocated AC joints using suture button systems with integrated tendon augmentation. Knee Surgery, Sport Traumatol Arthrosc. 2012;20(10):1931–8. https://doi.org/10.1007/s00167-011-1828-y.
59. Sloan SM, Budoff JE, Hipp JA, Nguyen L. Coracoclavicular ligament reconstruction using the lateral half of the conjoined tendon. J Shoulder Elb Surg. 2004;13(2):186–90. https://doi.org/10.1016/j.jse.2003.12.002.
60. Phemister D. The treatment of dislocation of the acromioclavicular joint by open reduction and threaded wire fixation. J Bone Jt Surg. 1942;24: 166–8.
61. Bargren JH, Erlanger S, Dick HM. Biomechanics and comparison of two operative methods of treatment of complete acromioclavicular separation. Clin Orthop Relat Res. 1978;130:267–72. https://doi.org/10.1097/00003086-197801000-00031.
62. Mazet R. Migration of a kirschner wire from the shoulder region into the lung. J Bone Joint Surg. 1943;25:477–83.
63. Karlsson J, Árnarson H, Sigurjónsson K. Acromioclavicular dislocations treated by coracoacromial ligament transfer. Arch Orthop Trauma Surg. 1986;106(1):8–11. https://doi.org/10.1007/BF00435643.
64. Dumontier C, Sautet A, Man M, Apoil A. Acromioclavicular dislocations: treatment by coracoacromial ligamentoplasty. J Shoulder Elb Surg. 1995;4(2):130–4. https://doi.org/10.1016/S1058-2746(05)80067-9.
65. Fialka C, Michlits W, Stampfl P, et al. Biomechanical analysis of different operative techniques for complete acromioclavicular joint disruptions. Osteosynthesis Trauma Care. 2005;13(3):154–9. https://doi.org/10.1055/s-2005-836559.
66. Bosworth B. Acromioclavicular separation: new method of repair. Surg Gynecol Obs. 1941;73:866–71.
67. Tsou PM. Percutaneous cannulated screw coracoclavicular fixation for acute acromioclavicular dislocations. Clin Orthop Relat Res. 1989;243:112–21. https://doi.org/10.1097/00003086-198906000-00017.
68. Kennedy J, Cameron H. Complete dislocation of the acromioclavicular joint. J Bone Jt Surg Br. 1954;36:202–8.
69. Guy D, Wirth M, Griffin J. Rockwood, CA J. reconstruction of chronic and complete dislocations of the acromioclavicular joint. Clin Orthop Relat Res. 1998;347:138–49.
70. Kiefer H, Claes L, Burri C, Holzwarth J. The stabilizing effect of various implants on the torn acromioclavicular joint - a biomechanical study. Arch Orthop Trauma Surg. 1986;106(1):42–6. https://doi.org/10.1007/BF00435651.
71. Jari R, Costic RS, Rodosky MW, Debski RE. Biomechanical function of surgical procedures for acromioclavicular joint dislocations. Arthrosc - J Arthrosc Relat Surg. 2004;20(3):237–45. https://doi.org/10.1016/j.arthro.2004.01.011.
72. Jerosch J, Filler T, Peuker E, Greig M, Siewering U. Which stabilization technique corrects anatomy best in patients with AC-separation? An experimental study. Knee Surgery, Sport Traumatol Arthrosc. 1999;7(6):365–72. https://doi.org/10.1007/s001670050182.
73. McConnell AJ, Yoo DJ, Zdero R, Schemitsch EH, McKee MD. Methods of operative fixation of the acromio-clavicular joint: a biomechanical comparison. J Orthop Trauma. 2007;21(4):248–53. https://doi.org/10.1097/BOT.0b013e31803eb14e.
74. Talbert T, Green J 3rd, Mukherjee D, Ogden A, Mayeux R. Bioabsorbable screw fixation in coracoclavicular ligament reconstruction. J Long Term Eff Med Implant. 2003;13:319–23.
75. Balser D. Eine neue Methode zur operativen Behandlung der akromioklavikularen Luxation. Chir Prax. 1976;24:275.
76. Gstettner C, Tauber M, Hitzl W, Resch H. Rockwood type III acromioclavicular dislocation: surgical versus conservative treatment. J Shoulder Elb Surg. 2008;17(2):220–5.
77. Göhring H, Matusewicz A, Friedl W, et al. Results of treatment after different surgical procedures for management of acromioclavicular joint dislocation. Chirurg. 1993;64:565–71.
78. Jensen G, Katthagen JC, Alvarado LE, Lill H, Voigt C. Has the arthroscopically assisted reduction of acute AC joint separations with the double tight-rope technique advantages over the clavicular hook plate fixation? *Knee surgery*. Sport Traumatol Arthrosc. 2014;22(2):422–30. https://doi.org/10.1007/s00167-012-2270-5.
79. Lin HY, Wong PK, Ho WP, Chuang TY, Liao YS, Wong CC. Clavicular hook plate may induce subacromial shoulder impingement and rotator cuff lesion - dynamic sonographic evaluation. J Orthop Surg Res. 2014;9(1):1–9. https://doi.org/10.1186/1749-799X-9-6.

80. Yoon JP, Lee BJ, Nam SJ, et al. Comparison of results between hook plate fixation and ligament reconstruction for acute unstable acromioclavicular joint dislocation. CiOS Clin Orthop Surg. 2015;7(1):97–103. https://doi.org/10.4055/cios.2015.7.1.97.
81. Sim E, Schwarz N, Hocker K, Berzlanovich A. Repair of complete acromioclavicular separations using the acromioclavicular-hook plate. Clin Orthop Relat Res. 1995;314:134–42. https://doi.org/10.1097/00003086-199505000-00017.
82. Ejam S, Lind T, Falkenberg B. Surgical treatment of acute and chronic acromioclavicular dislocation Tossy type III and V using the hook plate. Acta Orthop Belg. 2008;74(4):441–5.
83. Kim YS, Yoo YS, Jang SW, Nair AV, Jin H, Song HS. Invivo analysis of acromioclavicular joint motion after hook plate fixation using three-dimensional computed tomography. J Shoulder Elb Surg. 2015;24(7):1106–11. https://doi.org/10.1016/j.jse.2014.12.012.
84. Choi SW, Lee TJ, Moon KH, Cho KJ, Lee SY. Minimally invasive coracoclavicular stabilization with suture anchors for acute acromioclavicular dislocation. Am J Sports Med. 2008;36(5):961–5. https://doi.org/10.1177/0363546507312643.
85. Dyrna F, Berthold DP, Feucht MJ, et al. The importance of biomechanical properties in revision acromioclavicular joint stabilization: a scoping review. Knee Surgery, Sport Traumatol Arthrosc. 2019;27(12):3844–55. https://doi.org/10.1007/s00167-019-05742-6.
86. Tischer T, Salzmann GM, El-Azab H, Vogt S, Imhoff AB. Incidence of associated injuries with acute acromioclavicular joint dislocations types III through V. Am J Sports Med. 2009;37(1):136–9. https://doi.org/10.1177/0363546508322891.
87. Breslow MJ, Jazrawi LM, Bernstein AD, Kummer FJ, Rokito AS. Treatment of acromioclavicular joint separation: suture or suture anchors? J Shoulder Elb Surg. 2002;11(3):225–9. https://doi.org/10.1067/mse.2002.123904.
88. Rosslenbroich SB, Schliemann B, Schneider KN, et al. Minimally invasive coracoclavicular ligament reconstruction with a flip-button technique (MINAR): clinical and radiological midterm results. Am J Sports Med. 2015;43(7):1751–7. https://doi.org/10.1177/0363546515579179.
89. Kany J, Amaravathi RS, Guinand R, Valenti P. Arthroscopic acromioclavicular joint reconstruction using a synthetic ligament device. Eur J Orthop Surg Traumatol. 2012;22(5):357–64. https://doi.org/10.1007/s00590-011-0856-0.
90. Schliemann B, Lenschow S, Schürmann P, et al. Biomechanics of a new technique for minimal-invasive coracoclavicular ligament reconstruction. Knee Surgery, Sport Traumatol Arthrosc. 2013;21(5):1176–82. https://doi.org/10.1007/s00167-012-2041-3.
91. Metzlaff S, Rosslenbroich S, Forkel PH, et al. Surgical treatment of acute acromioclavicular joint dislocations: hook plate versus minimally invasive reconstruction. Knee Surgery, Sport Traumatol Arthrosc. 2016;24(6):1972–8. https://doi.org/10.1007/s00167-014-3294-9.
92. Walz L, Salzmann GM, Fabbro T, Eichhorn S, Imhoff AB. The anatomic reconstruction of acromioclavicular joint dislocations using 2 TightRope devices: a biomechanical study. Am J Sports Med. 2008;36(12):2398–406. https://doi.org/10.1177/0363546508322524.
93. Wellmann M, Kempka JP, Schanz S, et al. Coracoclavicular ligament reconstruction: biomechanical comparison of tendon graft repairs to a synthetic double bundle augmentation. Knee Surgery, Sport Traumatol Arthrosc. 2009;17(5):521–8. https://doi.org/10.1007/s00167-009-0737-9.
94. Wellmann M, Zantop T, Weimann A, Raschke MJ, Petersen W. Biomechanical evaluation of minimally invasive repairs for complete acromioclavicular joint dislocation. Am J Sports Med. 2007;35(6):955–61. https://doi.org/10.1177/0363546506298585.
95. Grantham C, Heckmann N, Wang L, Tibone JE, Struhl S, Lee TQ. A biomechanical assessment of a novel double endobutton technique versus a coracoid cerclage sling for acromioclavicular and coracoclavicular injuries. Knee Surgery, Sport Traumatol Arthrosc. 2016;24(6):1918–24. https://doi.org/10.1007/s00167-014-3198-8.
96. Bunnell S. Fascial graft for dislocation of acromioclavicular joint. Surg Gynecol Obstet. 1928;46:563–4.
97. Jones HP, Lemos MJ, Schepsis AA. Salvage of failed acromioclavicular joint reconstruction using autogenous semitendinosus tendon from the knee. Surgical technique and case report. Am J Sports Med. 2001;29(2):234–7. https://doi.org/10.1177/03635465010290022001.
98. Costic RS, Labriola JE, Rodosky MW, Debski RE. Biomechanical rationale for development of anatomical reconstructions of coracoclavicular ligaments after complete acromioclavicular joint dislocations. Am J Sports Med. 2004;32(8):1929–36. https://doi.org/10.1177/0363546504264637.
99. Mazzocca AD, Conway JE, Johnson S, et al. The anatomic coracoclavicular ligament reconstruction. Oper Tech Sports Med. 2004;12(1):56–61. https://doi.org/10.1053/j.otsm.2004.04.001.
100. Saccomanno MF, Marchi G, Mocini F, et al. Anatomic reconstruction of the coracoclavicular and acromioclavicular ligaments with semitendinosus tendon graft for the treatment of chronic acromioclavicular joint dislocation provides good clinical and radiological results. Knee Surgery, Sport Traumatol Arthrosc. 2020;(0123456789). doi:https://doi.org/10.1007/s00167-020-06285-x
101. Staron JS, Esquivel AO, Pandhi NG, et al. Biomechanical evaluation of anatomical double-bundle coracoclavicular ligament reconstruction secured with knot fixation versus screw fixation. Orthopedics. 2013;36(8):e1047–52.

102. Berthold DP, Muench LN, Dyrna F, et al. Radiographic alterations in clavicular bone tunnel width following anatomic coracoclavicular ligament reconstruction (ACCR) for chronic acromioclavicular joint injuries. Knee Surgery, Sport Traumatol Arthrosc. April 2020:1–9. https://doi.org/10.1007/s00167-020-05980-z.
103. Saier T, Venjakob AJ, Minzlaff P, et al. Value of additional acromioclavicular cerclage for horizontal stability in complete acromioclavicular separation: a biomechanical study. Knee Surgery, Sport Traumatol Arthrosc. 2015;23(5):1498–505. https://doi.org/10.1007/s00167-014-2895-7.
104. Dyrna F, de Oliveira CCT, Nowak M, et al. Risk of fracture of the acromion depends on size and orientation of acromial bone tunnels when performing acromioclavicular reconstruction. Knee Surgery, Sport Traumatol Arthrosc. 2018;26(1):275–84. https://doi.org/10.1007/s00167-017-4728-y.

Part IV

Elbow Biomechanics

Elbow Anatomy and Biomechanics 15

Andrea Celli, Bartoli Matteo, Peruzzi Marco, and Luigi Adriano Pederzini

15.1 Introduction

The elbow is a complex structure providing an important function as the mechanical link in the upper extremity between the hand, the wrist, and the shoulder. Its primary function is positioning the hand in space—the loss of which can cause significant impairment in activities of daily living.

Elbow and wrist joints associated with the ulna and radius bones linked with the interosseous membrane constitute the anatomical and functional unit of the forearm. This unit provides the rotational movements of the forearm and it allows the transmission of forces from the hand to the elbow when the elbow joint is the stable fulcrum needed for powerful grasping and fine motions. The soft tissues are divided in passive and active stabilizers. The former include the lateral collateral ligament, the medial collateral ligament, and the capsule. The active stabilizers involve the muscles that provide joint compressive forces and functions.

A. Celli (✉)
Shoulder and Elbow Unit, Department of Orthopaedic and Traumatology Surgery, Hesperia Hospital Modena, Modena, Italy

B. Matteo · P. Marco · L. A. Pederzini
Department of Orthopaedic, Traumatology and Arthroscopic Surgeries, Nuovo Ospedale di Sassuolo, Modena, Italy

15.2 Anatomy of the Elbow

15.2.1 Bone Anatomy

The distal humerus comprises two condyles forming the articular surfaces of the capitellum laterally and the trochlea medially. The most prominent medial epicondyle is the region where the ulnar collateral ligament and the flexor-pronator muscles are attached. The least prominent lateral epicondyle is the attachment point for the lateral collateral ligament and the extensor supinator muscles. The articular surface is angled approximately 30° anterior to the humerus shaft axis. The medial ridge of the trochlea is larger than the lateral ridge. This gives the articular surface a slight valgus position—approximately 6° from the epicondylar axis [1–7]. Proximally to the articular surface, the coronoid fossa and the olecranon fossa accommodate the olecranon process during extension and the coronoid tip during flexion movements [1, 4, 8]. Laterally, a small radial fossa accepts the contour of the radial head with the elbow in full flexion (Fig. 15.1). The proximal ulna provides elbow articulation, the greater sigmoid notch is the surface from the coronoid process anteriorly to the tip of the olecranon posteriorly. The contour of the surface of the articular greater sigmoid notch is not a semicircle but rather an ellipsoid—the articular cartilage is usually discontinuous centrally (nonarticular portion) (Fig. 15.2) [3, 9, 10]. The greater sigmoid

J. Koh et al. (eds.), *Orthopaedic Biomechanics in Sports Medicine*,
https://doi.org/10.1007/978-3-030-81549-3_15

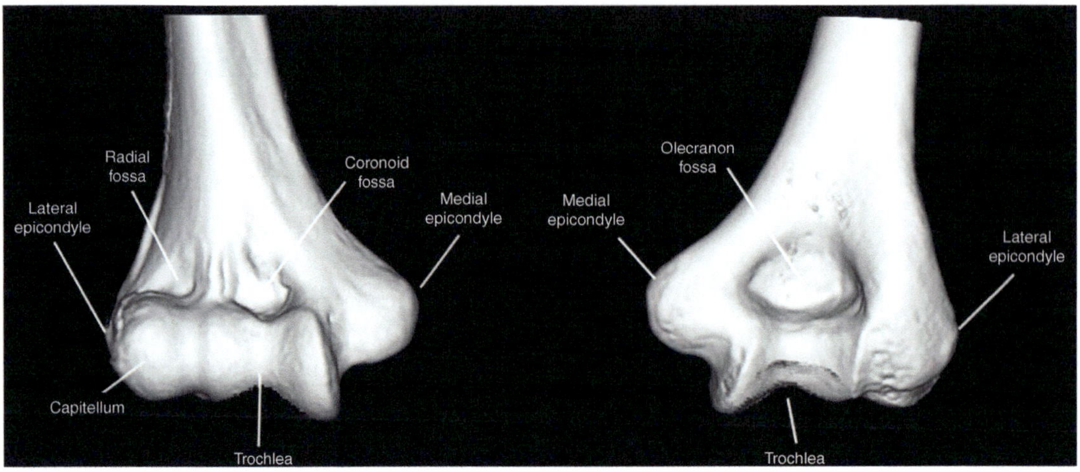

Fig. 15.1 Distal humeral anatomy

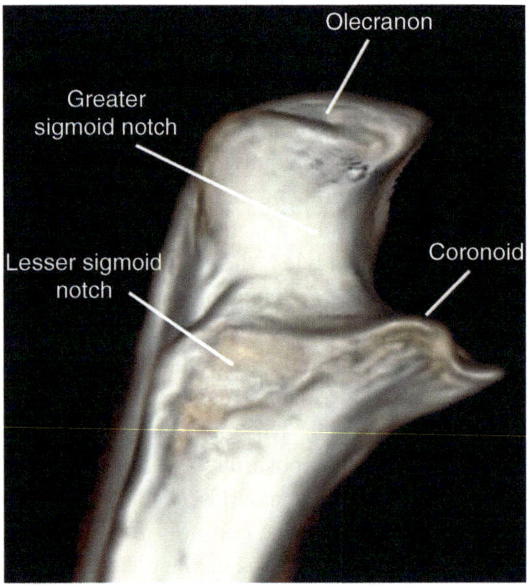

Fig. 15.2 Olecranon anatomy

notch has an approximate 30° inferior angulation to match with the anterior angulation of the distal humerus. The distal aspect of the notch is the coronoid process. In the medial aspect of the coronoid process, the sublime tubercle serves as the insertion site for the medial ulnar collateral ligament. Lateral to the coronoid process is the lesser sigmoid notch (radial notch). With a 70-degree arc, it allows the articulation with the radial head (Fig. 15.3) [3]. Distal to the radial notch is the crista supinatoris, where the lateral ulnar collateral ligament attaches. The proximal radius includes the cylindrically shape radial head with concave disc which articulates with the capitellum and with the radial sigmoid notch. The radial margin which it articulates is at approximately 240°, the anterolateral third (120°) of the radial head is void of cartilage. The transverse cross section of the radial head is not circular but elliptical in shape, the head and the neck form an approximate 15-degree angle with the long axis of the proximal radius opposite to the radial tuberosity. This serves as an insertion for the biceps tendon [3] (Fig. 15.4).

15.2.2 Soft Tissue Anatomy

Elbow stability results from the combination of articulate congruence of the ulnohumeral and of the radio-capitellum joints and its capsulo-ligamentous structures [11].

The medial collateral complex (Fig. 15.5) originates from the distal portion of the medial epicondyle. More specifically, it originates from the anteroinferior surface of the epicondyle and not from the condylar part of the trochlea [8, 12–17]. The medial collateral ligament complex (MUCL) consists of the anterior bundle, the posterior bundle, and the transverse ligament (ligament of Cooper/oblique bundle). The anterior bundle (AMCL) has been established as the main

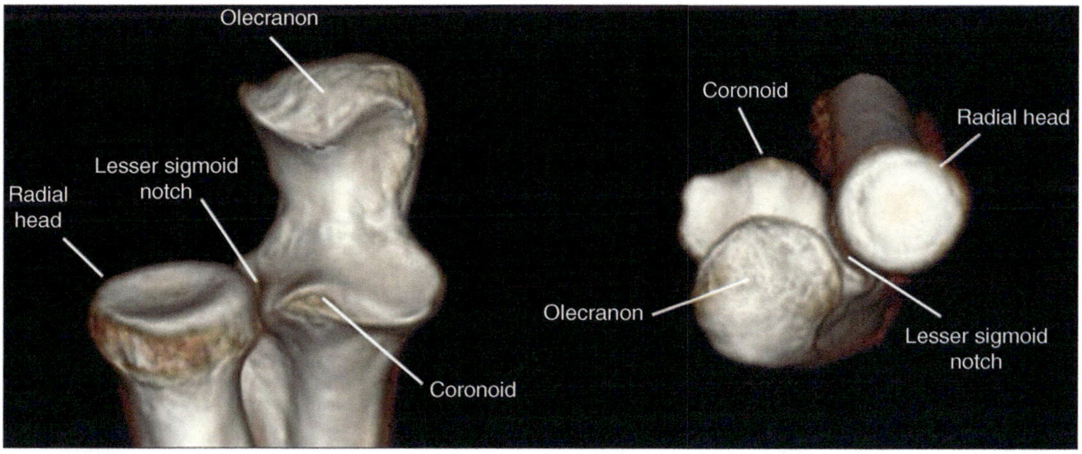

Fig. 15.3 Radial head anatomy

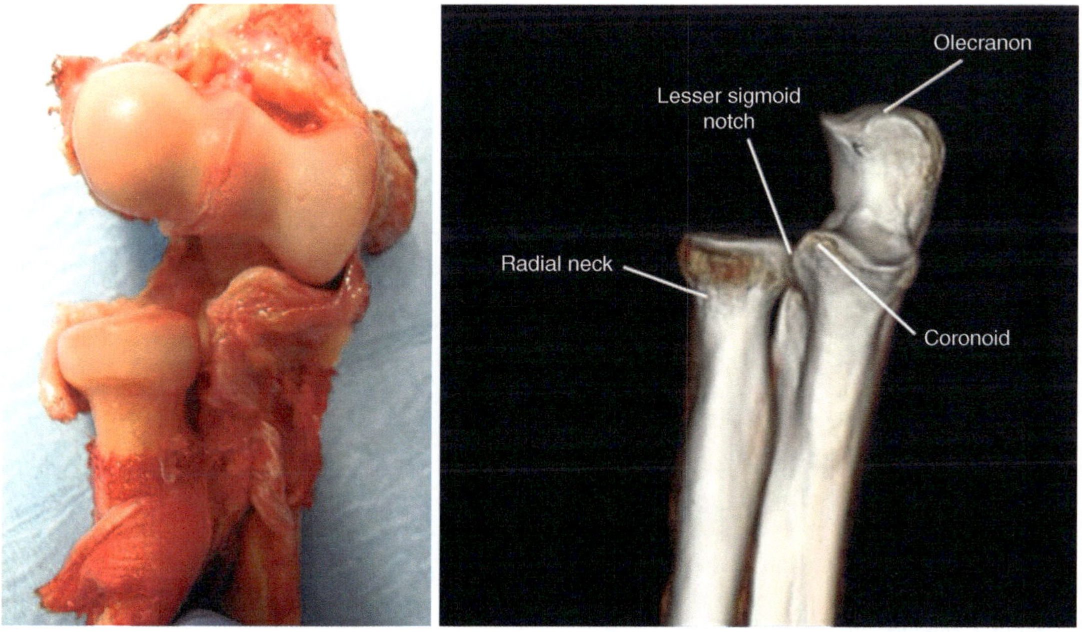

Fig. 15.4 The proximal radius ulnar joint

ligamentous stabilizer to valgus torques of the elbow [6–13]. *The AMCL* originates from the anterior inferior surface of the medial epicondyle, just posterior to the elbow axis. The posterior bundle is thin and it is inserted into the posteromedial margin of the greater sigmoid notch [8, 12, 15]—it is more posterior to the rotation axis. The posterior bundle also provides stability to valgus stress as well as constraint to hyperflexion. The transverse ligament has been differently defined and it appears to have no significant role in stabilizing the elbow.

The lateral collateral complex (Fig. 15.6) consists of three parts: the annular ligament, the radial collateral, and the ulnar collateral ligaments [18–22]. The lateral collateral complex originates on the lateral epicondyle near the axis of rotation of the elbow. The radial part terminates along the course of the annular ligaments, while the ulnar part is inserted on the crista

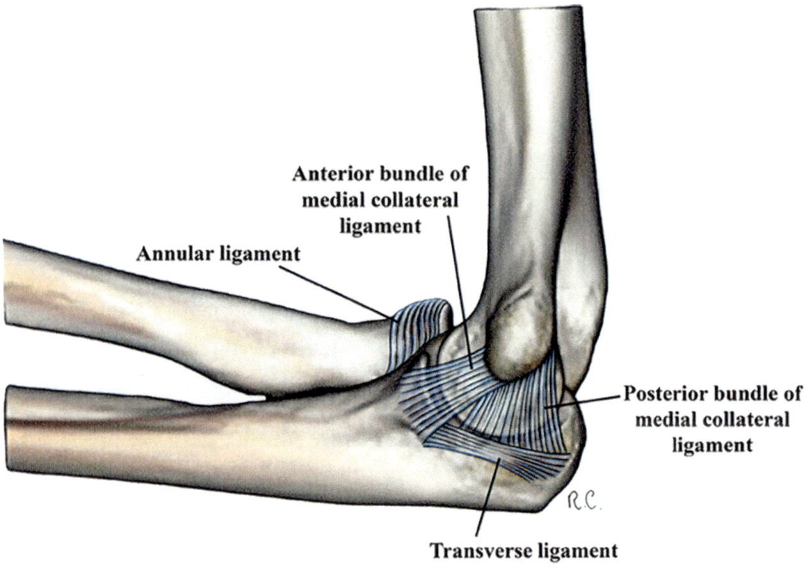

Fig. 15.5 The medial collateral ligament complex

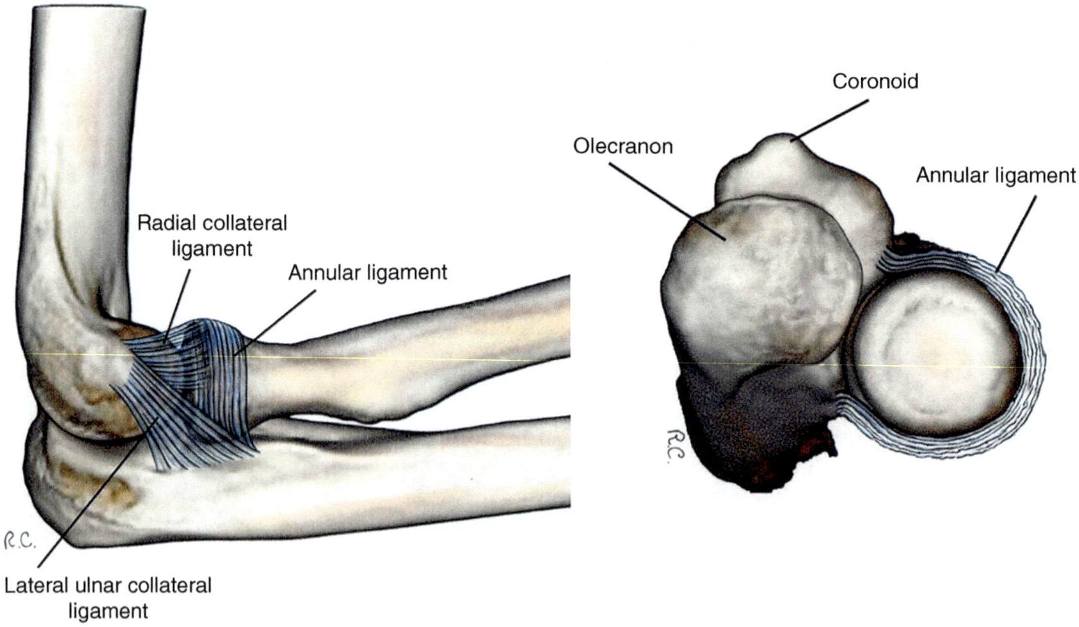

Fig. 15.6 The lateral collateral ligament complex

supinatoris of the ulna and it has been considered as an important elbow stabilizer. The lateral collateral ligament is almost uniformly taut during flexion-extension range of motion with little change in distance between origin and insertion during motion (isometric position) [23]. The annular ligament is inserted on the anterior and posterior margins of the less sigmoid notch. *Its purpose is to* maintain the radial head in contact with the ulna [15]. The anterior insertion becomes taut during supination while the posterior insertion becomes taut during pronation [16, 22] this is because the radial head is not a pure circular dish.

The oblique ligament is a small structure comprising the fascia overlying the deep head of the supinator between the radius and the ulna. It is believed to have limited functional importance [16, 22–24]. The annular ligament along with the radial part of the lateral collateral ligament is an important stabilizer of the radial head, thus avoiding posterolateral subluxation. The quadrate ligament is a thin fibrous layer between the annular ligament and the ulna. It is a stabilizer of the proximal radioulnar joint during prono-supination.

The elbow capsule is attached to the articular margins of the joint and its fibers are connected to the annular ligament [24]. Anteriorly it includes the coronoid and the radial fossae. Posteriorly it includes the olecranon fossa. Its maximum distension occurs at 70–80° elbow flexion. The contribution of the capsule as a passive stabilizer has been matter of debate. Some studies suggested no change in joint laxity after complete capsulotomy. On the other hand, Morrey et al. reported that this structure has an important function as a stabilizer to varus-valgus and distraction loading in extension but not in flexion [17]. The anterior capsule presents transverse and oblique bands that seem to provide significant stability in extension [19]. On the other hand, the posterior capsule becomes taut in flexion and it may have a role as static stabilizer.

The muscles across the elbow can be divided into four main groups. Posteriorly, the triceps tendon crosses the elbow joint, laterally; the extensor and supinator muscles (brachio-radialis, extensor carpi radialis longus, extensor radialis brevis, finger extensors, extensor carpi ulnaris and anconeus) anteriorly and medially; the flexor-pronator group (pronator teres, flexor carpi radialis, palmaris long, flexor carpi ulnaris, and flexor digitorum). The majority of the muscles crossing the elbow work to rotate the forearm and to flex and extend the wrist and the fingers. Only a few muscles play a role in moving the elbow joint. The primary elbow flexor muscles are the brachialis, the biceps brachialis, and the brachioradialis. Secondary muscles include the pronator teres, the extensor carpi radialis longus, and the flexor carpi radialis. The triceps and the anconeus muscles are the main extensor muscles of the elbow. Pronation is provided by the pronator teres and by the pronator quadratus. Supination is mainly performed by the biceps assisted by the supinator muscle. Muscle loading results in a compressive force generated across the articulation of the elbow. Muscle activities may also produce a dynamic stabilization (compressing the articular surfaces together) and they protect static ligaments constraints [1, 2, 23, 25, 26].

15.3 Biomechanics of the Elbow

The elbow is one of the most congruous joints in the body with three articulations: the ulnohumeral joint, the radio-capitellum joint, and the proximal radioulnar joint. The ulnohumeral joint is a hinge (ginglymus) joint, allowing flexion and extension (Fig. 15.7). The radio-capitellum and radioulnar joints are trochoid joints, allowing axial rotation or pivoting [3]. The elbow has two degrees of freedom: flexion-extension (0–140°) and prono-supination (75–85°). Following Morrey's study, the necessary arc of motion to perform most activities of daily living is 30–130 degrees of flexion-extension and 50 to 50 degrees of pronation-supination [3, 23]. During flexion-extension arc of motion, the elbow rotation axis passes through the capitellum in line with the bottom of the trochlea sulcus and it rotates internally by approximately

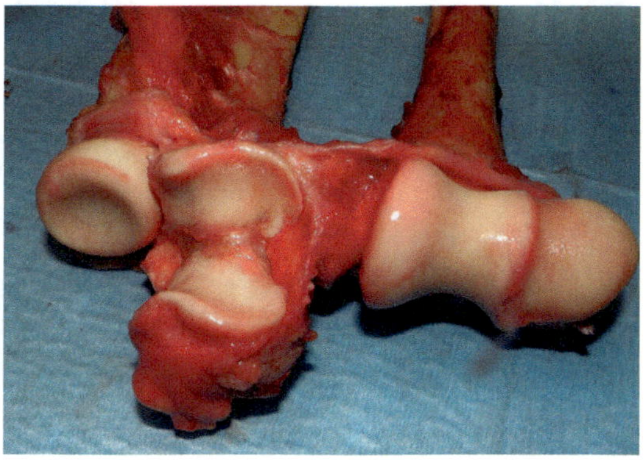

Fig. 15.7 The ulnohumeral joint is a hinge (ginglymus)

5° to 7° with respect to the plane of the medial and lateral epicondyles and 4° to 8° of valgus with respect to the long axis of the humerus [3, 15, 23, 27, 28] (Fig. 15.8). This is possible because the center of rotation of the elbow in flexion-extension is not a fixed single point but it moves with an area. The olecranon moves on the trochlea (like a screw) (Fig. 15.9) during the flexion-extension and this *explains* the movement of the elbow from valgus in extension to varus in flexion. This movement produces the carrying angle, defined as the angle between the long axis of the humerus and the long axis of the ulna measured in the frontal plane [10, 29] (Fig. 15.10). This angle is generally higher in women than in men. In females, the average valgus angle ranges from 13° to 16°, whereas in males, it ranges from 11° to 14° [19]. Forearm motion is clearly influenced by the morphology of the radial head, the axis of rotation of the forearm passes proximally through the center of the radial head and distally through the fovea of the ulna at the base of the ulnar styloid [8]. During forearm rotation, the edge of the radial head maintains contact with the lesser sigmoid notch, the radial shaft moves away and toward the ulna, in the transverse plane.

15.3.1 Articular and Capsulo-Ligamentous Stabilizers

The structures that stabilize the elbow during motion can also be divided into two main groups: passive and active stabilizers. Their actions differ over the range of motion of the elbow [30–35].

The radial head is a secondary stabilizer to valgus stress. Morrey et al. showed that selective radial head resection does not influence valgus instability as long as the medial ulnar collateral ligament is intact [9, 36]. When the ulnar collateral ligament is released, the radial head becomes the primary constraint to valgus instability [3, 9, 36, 37]. The radial head is the main longitudinal stabilizer when the interosseous membrane is injured [38]. Biomechanical tests suggest that the radial head contributes for about 30% of valgus stability in both flexion and extension [19, 23, 38]. Complete release of the ulnar collateral ligament produces an average 18° increase of valgus–varus laxity reaching 36° after radial head excision [19]. Axis loading produces 40% stress distribution across the ulna and 60% distribution across the radius. Transmission across the radial head increases when elbow flexion is 0–35° and when it is placed in supination [19, 23].

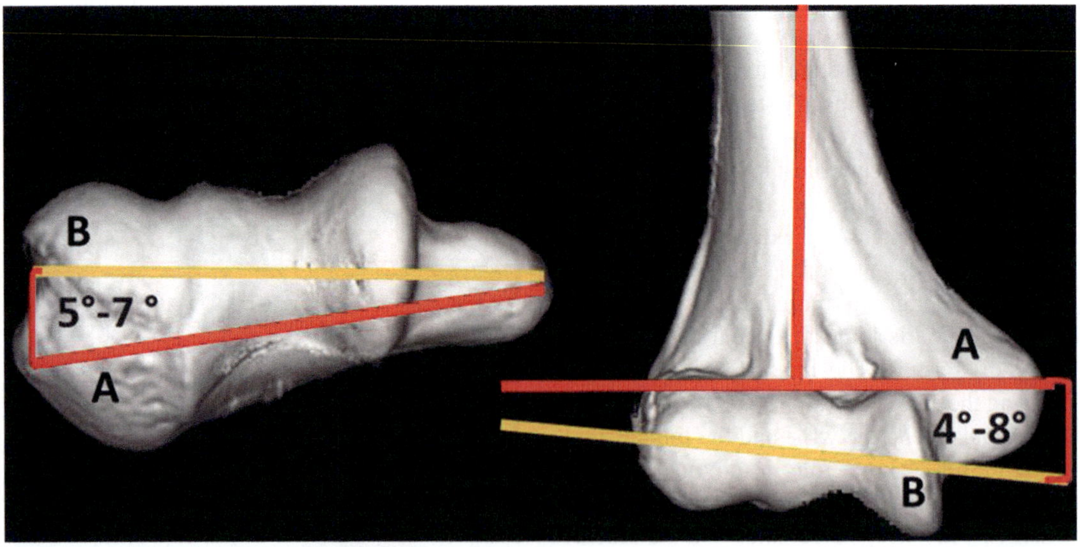

Fig. 15.8 The axis of rotation of the elbow (A) passes through the capitellum in line with the bottom of the trochlea sulcus and it moves approximately 5° to 7° of internal rotation to the plane of the medial and lateral epicondyles (B) and in 4° to 8° of valgus with respect to the long axis of the humerus

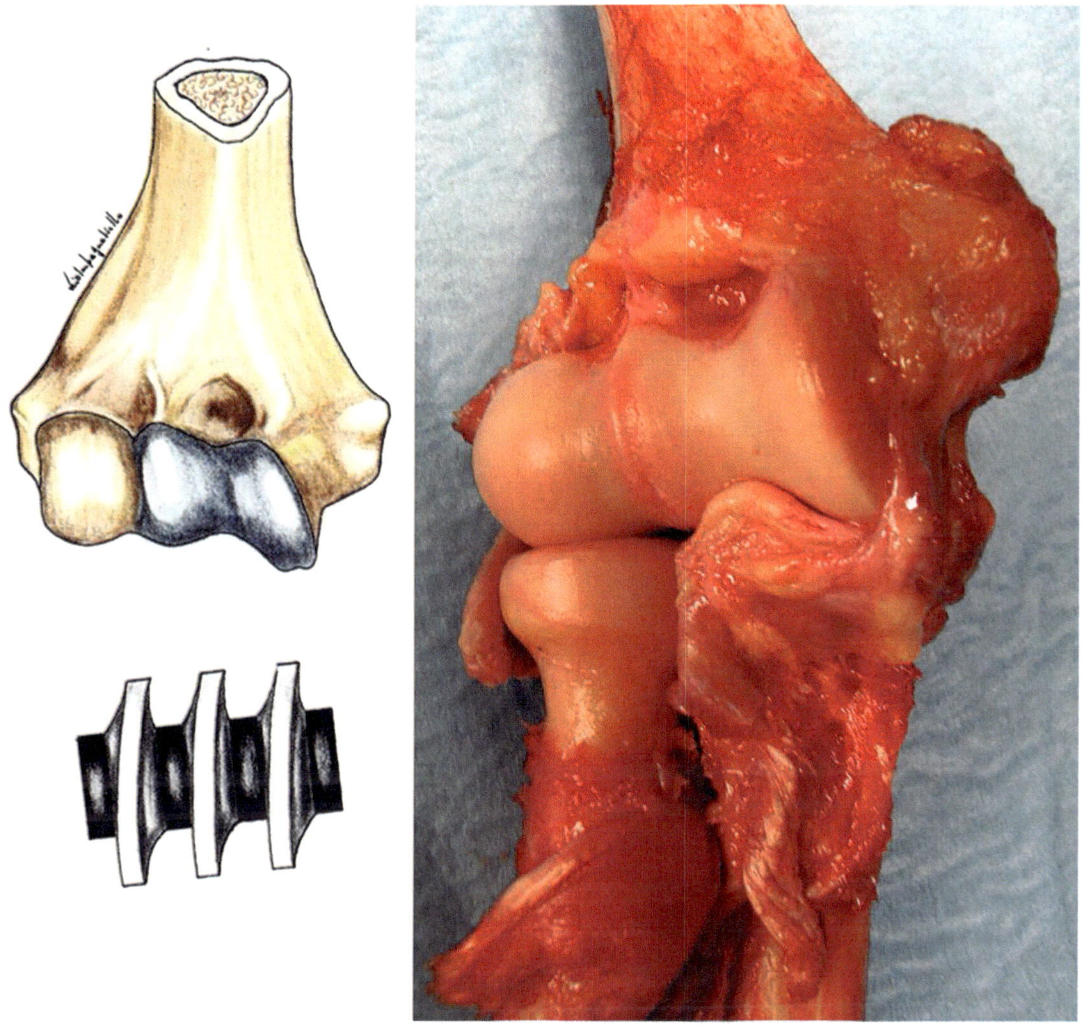

Fig. 15.9 The olecranon moves on the trochlea as a screw

The role of the olecranon has been analyzed in different biomechanical studies. An et al. showed that 75% to 85% of valgus stress was resisted by the proximal half of the olecranon and that the distal half of the sigmoid notch (coronoid) resisted 60% of varus stress in flexion and 67% in extension [23]. The olecranon engages the olecranon fossa of the humerus at 20°, at this point the elbow becomes more stable against varus-valgus laxity.

The coronoid process plays an important role in preventing posterior dislocation of the elbow. O'Driscoll et al. reported that when the collateral ligaments are injured and the radial head is compromised, a 30% resection of the coronoid produces instability [39, 40]. Fracture of the coronoid—even small fragments—can decrease the stability of the elbow, particularly when it is associated with ligament injuries. The coronoid is also essential to varus stability when the collateral ligament is intact [27]. The elbow becomes more unstable as successive portions of the coronoid are damaged, the radial head resection produces an increase in instability correlated to the type of coronoid fracture [11].

The passive soft tissue stabilizers include the medial and lateral collateral ligaments complexes [30]. Their activities during elbow motion were

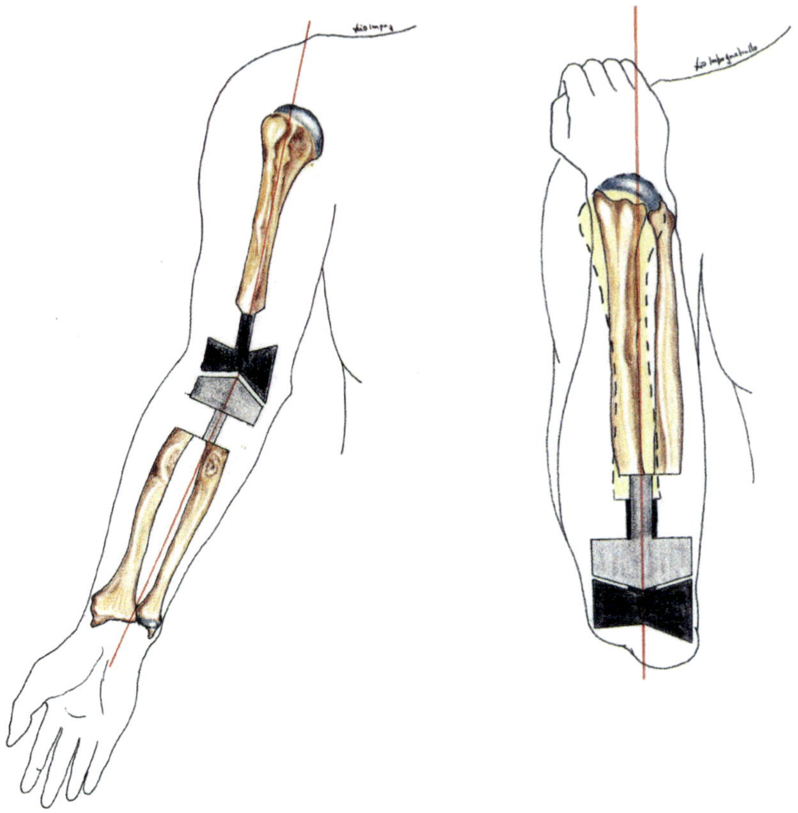

Fig. 15.10 The carrying angle (defined as the angle between the long axis of the humerus and the long axis of the ulna measured in the frontal plane)

analyzed by Morrey and An [9, 23]. At 90-degree flexion, the anterior band of the medial collateral ligament is the primary stabilizer to the valgus stress. In extension, instead, the medial collateral ligament, the anterior capsule and bony fit are equally resistant to valgus stress. Its posterior component is more important in higher degrees of flexion. During elbow motion, the midportion and the anterior band maintains tension, the posterior bundle is taut at 65° to full flexion [8, 27]. Morrey et al. showed that the primary restraint to valgus stress is the medial collateral ligament and that the radial head is the secondary stabilizer. The radial head becomes the primary restraint to the valgus in case of medial ligaments complex insufficiency [3, 23]. The lateral collateral ligament has been analyzed in different studies. The lateral collateral ligament remains taut throughout the complete range of motion, this is because its origin lies close to the axis of rotation. The anterior portion of the annular ligament is taut during supination, while the posterior portion is taut in pronation. The function of the lateral collateral ligament has been analyzed by O'Driscoll et al. They described it as the primary ligamentous stabilizer to varus and posterolateral rotatory instability [39].

The active stabilizers are the muscles crossing the elbow joint. They can be divided into four compartments as previously described [31, 33]. The pull and contraction lines of these muscle create compressive forces around the humerus, the ulna, and the radius. These forces function as dynamic stabilizers of the joint, the largest force has been seen axially at the distal humerus near full extension and it decreases when the elbow moves in flexion [23, 25, 26, 40–43].

15.3.2 Force Transmission through the Elbow

An et al. [26] found that among the muscles crossing the elbow joint, the brachialis and the triceps muscles have the largest work capacity and contractile strength. With extension and axial loading, stress is distributed 40% across the ulnohumeral joint and 60% across the radiohumeral joint. Morrey et al. measured force transmission through the radial head. Forces were the greatest from 0° to 30° flexion and they were always higher in pronation [17]. An and Morrey [9] calculated the force in the ulnohumeral joint and found that the joint force can range from one to three times body weight with strenuous lifting. Direction of the resultant joint force changes with flexion angle, pointing anteriorly with elbow extension and posteriorly with elbow flexion. Stress on the articular cartilage in the trochlear notch was evaluated by An et al. [23]. The study showed that contact pressure depends on the direction and magnitude of the compressive force. When force was oriented at the center of the articular surface, contact stress was equally distributed throughout the articular surface. When force was directed in an anterior or posterior direction towards the margin of the articulation, the weight-bearing surface was smaller, contact stresses were higher, and stress distribution was uneven [23].

References

1. Bernstein AD, Jazrawi LM, Rokito AS, Zuckerman JD. Elbow joint biomechanics: basic science and clinical applications. Orthopedics. 2000;23(12):1293–301.
2. Werner FW, An KN. Biomechanics of the elbow and forearm. Hand Clin. 1994;10(3):357–73.
3. Morrey B.F Anatomy of the elbow joint. The elbow and its disorders Ed Morrey Third edition 2000 W.B Saunders Company: 13–42.
4. Zimmerman NB. Clinical application of advances in elbow and forearm anatomy and biomechanics. Hand Clin. 2002;18(3):547.
5. Werner SI, Fleisig GS, Dillman CJ. Biomechanics of the elbow during baseball pitching. J Orthop Sports Phys Ther. 1993;17:274–8.
6. Duck TR, Dunnig CE, King GJ. Variability and repeatability of the flexion axis at the ulnohumeral joint. J Orthop Res. 2003;21:399–404.
7. Morrey BF, Tanaka S, An KN. Valgus stability of the elbow. A definition of primary and secondary constraints. Clin Orthop Relat Res. 1991;265:187–95.
8. Pomianowski S, O'Driscoll SW, Neale PG, Park MJ, Morrey BF, An KN. The effect of forearm rotation on laxity and stability of the elbow. Clin Biomech (Bristol, Avon). 2001;16(5):401–7.
9. Morrey BF, An KN. Articular and ligamentous contributions to the stability of the elbow joint. Am J Sports Med. 1983;11(5):315–9.
10. Paraskevas G, Papadopoulos A, Papaziogas B, Spanidou S, Argiriadou H, Gigis J. Study of the carrying angle of the human elbow joint in full extension: a morphometric analysis. Surg Radiol Anat. 2004;26(1):19–23.
11. Fornalski S, Gupta R, Lee T. Anatomy and biomechanics of the elbow joint. Tech Hand Upper Extremity. 2003;7(4):168–78.
12. Floris S, Olsen BS, Dalstra M, Sojbjerg JO, Sneppen O. The medial collateral ligament of the elbow joint: anatomy and kinematics. J Shoulder Elb Surg. 1998;7(4):345–51.
13. Hannouche D, Begue T. Functional anatomy of the lateral collateral ligament complex of the elbow. Surg Radiol Anat. 1999;21(3):187–91.
14. Rongieres M, Akhavan H, Mansat P, Mansat M, Vaysse P, Becue J. Functional anatomy of the medial ligamentous complex of the elbow. Its role in anterior posterior instability. Surg Radiol Anat. 2001;23(5):301–5.
15. Regan WD, Korinek SL, Morrey BF, An KN. Biomechanical study of ligaments around the elbow joint. Clin Orthop Relat Res. 1991;271:170–9.
16. Eygendaal D, Olsen BS, Jensen SL, Seki A, Sojbjerg JO. Kinematics of partial and total ruptures of the medial collateral ligament of the elbow. J Shoulder Elb Surg. 1999;8(6):612–6.
17. Morrey BF, An KN. Functional anatomy of the ligaments of the elbow. Clin Orthop Relat Res. 1985;201:84–90.
18. Dunning CE, Zarzour ZD, Patterson SD, Johnson JA, King GJ. Ligamentous stabilizers against posterolateral rotatory instability of the elbow. J Bone Joint Surg Am. 2001;83-A(12):1823–8.
19. King GJ, An KN. Biomechanics and functional anatomy of the elbow Shoulder and Elbow (orthopaedic Knowledge Update) Ed Norris T.R ASES 1997 published by AAOS.
20. Olsen BS, Sojbjerg JO, Dalstra M, Sneppen O. Kinematics of the lateral ligamentous constraints of the elbow joint. J Shoulder Elb Surg. 1996; 5(5):333–41.
21. Nielsen KK, Olsen BS. No stabilizing effect of the elbow joint capsule. A kinematic study. Acta Orthop Scand. 1999;70(1):6–8.
22. Olsen BS, Vaesel MT, Sojbjerg JO, Helmig P, Sneppen O. Lateral collateral ligament of the elbow joint: anatomy and kinematics. J Shoulder Elb Surg. 1996;5(2 Pt 1):103–12.
23. An K, Morrey BF. Biomechanics of the elbow The elbow and its disorders. Ed Morrey Third edition 2000 W.B Saunders Company: 43–60.
24. O'Driscoll SW, Morrey BF, An KN. Intraarticular pressure and capacity of the elbow. Arthroscopy. 1990;6(2):100–3.

25. An KN, Hui FC, Morrey BF, Linscheid RL, Chao EY. Muscles across the elbow joint: a biomechanical analysis. J Biomech. 1981;14(10):659–69.
26. Funk DA, An KN, Morrey BF, Daube JR. Electromyographic analysis of muscles across the elbow joint. J Orthop Res. 1987;5(4):529–38.
27. Alcid JG, Ahmad CS, Lee TQ. Elbow anatomy and structural biomechanics. Clin Sports Med. 2004;23(4):503–17.
28. Stroyan M, Wilk KE. The functional anatomy of the elbow complex. J Orthop Sports Phys Ther. 1993;17(6):279–88.
29. An KN, Morrey BF, Chao EY. Carrying angle of the human elbow joint. J Orthop Res. 1984;1(4):369–78.
30. Safran MR, Baillargeon D. Soft-tissue stabilizers of the elbow. J Shoulder Elbow Surg. 2005;14:179–85.
31. Davidson PA, Pink M, Perry J, Jobe FW. Functional anatomy of the flexor pronator muscle group in relation to the medial collateral ligament of the elbow. Am J Sports Med. 1995;23(2):245–50.
32. Dunning CE, Zarzour ZD, Patterson SD, Johnson JA, King GJ. Muscle forces and pronation stabilize the lateral ligament deficient elbow. Clin Orthop Relat Res. 2001;388:118–24.
33. Park MC, Ahmad CS. Dynamic contributions of the flexor-pronator mass to elbow valgus stability. J Bone Joint Surg Am. 2004;86-A(10):2268–74.
34. Hotchkiss RN, Weiland AJ. Valgus stability of the elbow. J Orthop Res. 1987;5(3):372–7.
35. Stein J, Murthi AM. Current concepts in elbow kinematics and biomechanics. Curr Opin Orthop. 2005;16:276–9.
36. Morrey BF, An KN. Stability of the elbow: osseous constraints. J Shoulder Elbow Surg. 2005;14(1 Suppl S):174–178.
37. Cohen MS, Bruno RJ. The collateral ligaments of the elbow: anatomy and clinical correlation. Clin Orthop Relat Res. 2001;383:123–30.
38. Morrey BF, An KN, Stormont TJ. Force transmission through the radial head. J Bone Joint Surg Am. 1988;70(2):250–6.
39. O'Driscoll SW, Bell DF, Morrey BF. Posterolateral rotatory instability of the elbow. J Bone Joint Surg Am. 1991;77:440–6.
40. O'Driscoll SW, Jaloszynski R, Morrey BF, An KN. Origin of the medial ulnar collateral ligament. J Hand Surg [Am]. 1992;17(1):164–8.
41. Callaway GH, Field LD, Deng XH, Torzilli PA, O'Brien SJ, Altchek DW, Warren RF. Biomechanical evaluation of the medial collateral ligament of the elbow. J Bone Joint Surg Am. 1997;79(8):1223–31.
42. Sojbjerg JO, Ovesen J, Nielsen S. Experimental elbow instability after transection of the medial collateral ligament. Clin Orthop Relat Res. 1987;218:186–90.
43. Askew LJ, An KN, Morrey BF, Chao EY. Isometric elbow strength in normal individuals. Clin Orthop Relat Res. 1987;222

16 Orthopaedics Biomechanics in Sports Medicine Thrower's Elbow

Luigi Adriano Pederzini, A. F. Cheli, F. Nicoletta, Giovanna Stelitano, and Andrea Celli

16.1 Principles of Elbow Anatomy: Focus on

16.1.1 Osteoarticular Anatomy

Elbow anatomy is hugely complex according to its twofold function of ensuring mobility and stability of the whole upper limb. Flexion-extension and prono-supination movements require the presence of an intricate joint construction able, at the same time, to withstand the forces which the upper limb is subjected [1]. Elbow joint is made of the ending part of three long bones: the distal humerus, proximal radius, and ulna interconnected each other creating three different articulations: the radio-humeral, the ulno-humeral, and radioulnar one. The elbow acts as a trocho-ginglymoid joint for its ability to realize the movements of flexion-extension and prono-supination, respectively, involving the ulno-humeral and radio-capitellar joints, and the proximal radioulnar joint [2].

The proximal part of the elbow involves the distal humerus surface, made by the trochlea and the capitellum. The spool-shaped trochlea is in line with the long axis of the humeral shaft, and its medial ridge is more prominent than the lateral one [3]. This conformation creates from 6° to 8° of valgus tilt with the greater sigmoid notch of the proximal ulna. The capitellum has a hemispheric shape and is localized laterally to the trochlea. It joins with the radial head, in correspondence of the lesser sigmoid notch. The radial head represents one of the leading secondary stabilizers of the elbow. As said back, it is not perfectly circular and is marked by a changeable offset from the axis of the neck. Its joint surface covers the humeral concave surface with an arc of about 280° [4]. The high congruency of the proximal ulna and trochlea surfaces is on the base of the primary constraint mechanisms of the elbow. The sagittal ridge of the greater sigmoid notch moves longitudinally, joining with the apex of the trochlea. The medial and lateral concavities to the sagittal ridge articulate with the convex medial and lateral faces of the trochlea. The lesser sigmoid notch is joined to the surface of the radial head [5].

The relevance of elbow osseous stabilization can be demonstrated through its role in the simple elbow dislocations [6]. Once reduced, these appear almost stable, despite the medial ligament complex (MCL) and the lateral ligament complex (LCL) result fully broken respectively in approximately all cases and most cases. Loads on the elbow are divided 43% over the ulno-humeral joint and 57% across the radio-capitellar joint.

L. A. Pederzini · A. F. Cheli (✉) · F. Nicoletta
G. Stelitano
Orthopeadic Unit, New Civil Sassuolo Hospital, Modena, Italy

A. Celli
Orthopaedic Unit, Hesperia Hospital, Modena, Italy

J. Koh et al. (eds.), *Orthopaedic Biomechanics in Sports Medicine*,
https://doi.org/10.1007/978-3-030-81549-3_16

Articular reaction strengths change with elbow position. At the level of radio-capitellar joint, they are highest between 0° and 30° of flexion and in pronation than in supination. The congruency of the ulno-humeral joint contributes as much as 50% of the elbow stabilization. Osseous stability is maximum in flexion, when the coronoid process is locked into the coronoid fossa, and the radial head is in the radial fossa of the distal humerus. The coronoid process has a crucial role in elbow stability [7]. Fractures involving more than 50% of the coronoid surface showed a significant increase of the varus-valgus laxity, even after elbow collateral ligaments repair. In the context of uninjured ligaments, coronoid fractures that involve more than 50% of the coronoid cause elbow posterior dislocation more quickly than those with less than 50%, particularly when the elbow is flexed more than 60°. The coronoid further plays a relevant part in posterolateral stability together with the radial head. The ulno-humeral joint, in fact, presents a higher degree of dislocation with 30% of the coronoid height removed and concomitant excision of the radial head. In this case, the replacement of the radial head alone is enough to restore the stability. Instead, if 50% of the coronoid height is removed, coronoid reconstruction is necessary to restore previous stability [8]. The radial head is an essential secondary valgus stabilizer of the elbow. It guarantees around 30% of the valgus stability of the elbow joint, and its role increases in the setting of MCL deficiency. When the MCL is ruptured, replacement of the radial head can restore valgus stability to a level comparable to that of an elbow with an intact radial head. With the elbow extended, osseous stability is reached thanks to the rotation of the olecranon tip into the olecranon fossa. It was asserted that 80% of the olecranon surface could be eliminated without compromising elbow stability.

16.1.2 Capsuloligamentous Anatomy

Precise knowledge of elbow surrounding structures is an essential factor for understanding the function of this joint, especially to comprise the different phases of sports activities. The elbow is the most stable joint of the human body, not only for its bony components but mostly for the presence of several soft tissue stabilizers. The static soft tissue stabilizers are the anterior and posterior joint capsule and the MCL and LCL complexes [9, 10].

16.1.2.1 The Articular Capsule

The capsule surrounds the articular border of the elbow. The anterior capsule proximally originates on the coronoid and radial fossae, extending distally to the coronoid process and laterally to the annular ligament. On the other hand, the posterior capsule proximally inserts on the olecranon fossa, while distally along the articular edges of the greater sigmoid notch, and laterally on the annular ligament. The capsule is taut anteriorly during elbow extension, posteriorly in flexion. Joint intra-capsular pressure is lowest at 70–80° of flexion. When wholly distended at 80° of flexion, elbow capacity is 25–30 mL. The capsule gives most of its stabilizing effects with the elbow extended [11, 12].

16.1.2.2 MCL complex

The MCL complex is made of three different components: the anterior bundle (AMCL), the posterior bundle, and the transverse ligament (Cooper ligament). The MCL originates from the anteroinferior surface of the medial epicondyle. The anterior bundle, the most distinct structure of the MCL complex, inserts on the sublime tubercle of the coronoid process. It is furthermore subdivided into an anterior and a posterior band. A third central band has been often described. The posterior bundle, which inserts on the medial surface of the olecranon, should be considered as a thickening of the joint capsule. The transverse ligament, made of horizontal fibers unseparated from the capsule, moves among the coronoid to the tip of the olecranon. It doesn't seem to contribute to joint stability [13]. The AMCL is considered the primary constraint for valgus and posteromedial elbow stability. It is composed by an anterior band taut in extension and a posterior one taut in flexion. Subsequently, the anterior band results more vulnerable to valgus stress during elbow extension, while the posterior band during elbow flexion. The anterior band is the main restraint to valgus stress at 30°, 60°, and 90°, and a coprimary restraint at 120°. The posterior band is a coprimary restraint at 120° and a secondary restraint at 30°, 60°, and 90°. A traumatic valgus force with the elbow flexed at 90° or less is more likely to damage only the anterior band, whereas if the elbow is flexed more than 90° is more likely to damage the complete AMCL. Total rupture of the AMCL provokes valgus and internal rotatory instability throughout the full arch of flexion with maximal valgus instability at 70° and maximal rotational instability at 60° [14]. The posterior bundle minimally contributes to valgus stability. Its main role is the control of the posteromedial rotatory stability. The complete rupture of the posterior band causes isolated internal rotatory laxity, maximal at 130° of flexion. The MCL origin is located posteriorly to the axis of elbow flexion, creating a cam-like effect: ligaments tension changes throughout flexion and extension of the elbow. The AMCL rises by 18% from 0° to 120° of flexion. In comparison, the middle band origin has been identified proximally to the axis of rotation of the ulno-humeral joint. This band has been termed the "guiding" band for its isometric and close position to the axis of rotation which makes it taut equally throughout the full arc of flexion. If it is sectioned, it produces important elbow instability. Single-strand reconstruction of the middle band in the MCL deficiency has been shown to return valgus stability, similar to the uninjured elbow [15].

16.1.2.3 LCL Complex

The LCL complex is made of four components. They include the radial collateral ligament, the lateral ulnar collateral ligament, the annular ligament, and the accessory collateral ligament. The LCL complex originates from the inferior surface of the lateral epicondyle. The lateral ulnar collateral ligament attaches on the crista supinatoris; the radial collateral ligament into the annular ligament. It is responsible for radial head stabilization. The annular ligament ends on the lesser sigmoid notch. The accessory collateral ligament has attachments at the annular ligament and the crista supinatoris [16]. The LCL is mainly responsible for elbow stability in external rotation and varus stress. The flexion axis of the elbow crosses the LCL origins. In this way, it results in equal tension throughout the arc of flexion. The current literature widely has confirmed that LCL complex injury is the initial damage resulting from the elbow dislocation. Instability derived by rupture of the LCL must be considered for each complex fracture-dislocations of the elbow. The total failure of the LCL is associated with varus and posterolateral rotatory instability as well as posterior radial head subluxation. Any case, the specific role of each portion of the LCL complex doesn't still understand. Recent studies have suggested that the LCL complex acts as a single operative unit rather than each part owning its specific role. The singular rupture of the annular ligament and lateral ulnar collateral ligament causes only minimal laxity [17]. When the annular ligament is uninjured, significant posterolateral rotatory and varus-valgus instability could be consequences of lateral ulnar collateral ligament and radial collateral ligament transection. The last concept is relevant for the planning of the lateral surgical approaches to the elbow for radial head fixation or replacement. If the annular ligament is intact, the radial collateral ligament or the lateral ulnar collateral ligament can be transected and restored without producing instability. When the radial head is removed in a condition of LCL deficiency, an increasing of varus and external rotatory instability occurs. Radial head replacement, in this case, raises posterolateral instability. Complete stability, however, is not replaced until the LCL complex is repaired [18].

16.1.2.4 Muscles

Muscles of the elbow give dynamic stabilization and protect the elbow against the static forces. They can be categorized into four different groups: elbow flexors, elbow extensors, forearm flexor-pronators, and forearm extensors. Each of these muscles employs a compressive load to the joint during its contraction [19]. Only a few muscles are mainly implicated into joint motion. The biceps, brachialis, and brachioradialis flex the elbow. The biceps is the principal supinator of the forearm. The triceps is the primary elbow extensor. Anconeus seems to have a minimal role in elbow extension, but it acts as a dynamic constraint to varus and posterolateral rotatory instability [20].

16.1.3 Biomechanics of the Elbow

The upper limb can be considered as a kinematic chain, where all the components are strictly interconnected to ensure its function best. Inside this system, the shoulder acts as a girdle, the elbow forms the central part, while the hand represents the effector. The shoulder and the elbow move synchronically to determine hand position in space, although elbow range of motion is comparatively constrained with compared to that of the shoulder. In the context of sportive activities, the whole upper limb, and in particular the elbow, require "particularly fit" to control specific movements. Sportspeople, in fact, need atypical strength, speed, and extreme precision of the joint motion.

Elbow movements follow different axes of action.

- *Flexion-extension.*
- The flexion-extension axis of the elbow has been described as a loose hinge that changes with forearm pronation and supination as well as with passive and active action. It is oriented from 3° to 5° of internal rotation with respect to the plane of medial and lateral epicondyles and from 4° to 8° of valgus relative to the long axis of the humerus. Notably, the flexion-extension axis is identified by the long axis of humerus and the long axis of the ulna and defines a slightly valgus angle, the so-called carrying angle, which varies from 11° to 14° respectively in men and woman [21]. This angle changes about 6° during elbow extension while cancelling out during flexion. When the carrying angle is 14°, the axis of movement is approximately 7° from the base of the coronoid process and the articular surface of the radial head against the anterior surfaces of humeral fossae. The physiological range of elbow motion in flexion and extension is 0–140°, even if for the most of daily living activities a range of 30–130° is enough [22].
- *Pronation-supination.*
- The radio-capitellar and proximal radioulnar joints are responsible for pronation and supination of the forearm. The normal range of forearm varies from 180° with pronation of 80–90° and supination of about 90°. Wrist and fingers movements allow adding other 30° of rotation. Almost all daily activities can be performed with 100° of forearm rotation (50° of pronation and 50° of supination). The physiological axis of forearm prono-supination moves from the center of the radial head to that of the distal ulna. For a long time, the axis of rotation of the elbow has been considered constant and independent with respect to flexion-extension movements [23]. Currently, it has been shown a slightly ulnar and volar shifting of the axis during supination and radial and dorsal one during pronation. The distal ulna, in fact, follows a circumduction movement, characterized by ulnar abduction and flexion during pronation and adduction and flexion during supination. This happens thanks to the congruency of the distal humerus and proximal ulna surfaces, which blocks ulna rotation on the elbow. At the same time, during forearm rotation, the radius runs proximally with the pronation and distally with the supination [24]. The elliptical shape of the radial head, together with the annular ligament, permits the sliding of the proximal radioulnar articulation. This highly organized structure avoids the radial head subluxation during everyday activities which requires elbow flexion.

The factors that stabilize the elbow joint change with the position of the arm. The forearm rotation influences varus and valgus laxity of the elbow. For this reason, athletes elbow stability should be verified in different ranges of prono-supination. During elbow extension, the anterior capsule contributes about 70 of soft tissue constraint, while the medial collateral ligament is the primary agent during elbow flexion. In full extension, the ulno-humeral articulation, the anterior joint capsule, and medial collateral ligament give the same contributions to valgus stability [25]. In flexion, the medial collateral ligament provides for about 70% to elbow resistance. Principally, the resistance to varus stress is provided by the ulno-humeral joint, together with the anterior joint capsule. In full extension, varus forces are equally controlled by articular congruency: olecranon in olecranon fossa and lateral collateral ligament. These structures provide approximately 50% of strains stabilization. In contrast, at 90° of flexion, their contribution rises to about 70%. The radial collateral ligament gives smallest varus restraint, both in flexion (9%) and extension (14%). In extension, the anterior capsule contributes 85% to contrast distraction. In flexion, the medial collateral ligament produces almost 80% of resistance to distraction [26].

16.2 Throwing Biomechanics

Injuries to the upper limb represent approximately 25% of all sports-related injuries. Several elbow damage patterns are so familiar to a precise sport that related names have been employed to define them (tennis elbow, golfer's elbow, or little leaguer's elbow) [27]. Nowadays, it is widely recognized that acute elbow dislocations primary occur in relative extension, independent from the forearm position [26]. The primary mechanism of injury seems to involve a valgus moment to an extended elbow, which causes a needed rupture of the medial collateral ligament, the principal constraint to valgus force. The biomechanics of elbow injury has been broadly analyzed in the context of the so-called overhead activities, such as the baseball pitch, the football pass, the tennis serve, and the javelin throw. Expert pitchers could throw the ball up to 1500 times a day, and every technical detail of this action needs to be meticulously studied, to avoid elbow injury. All technical mistakes during the sport could have disastrous consequences on joint function, and precise knowledge of the elbow biomechanics in sport is essential for the understanding of elbow diseases [28].

16.2.1 Baseball

The baseball pitch is usually divided into five main phases, widely studied [29]. Phase I (windup) includes initial preparation to the throwing with the elbow flexes, and the forearm slightly pronated. Phase II (early cocking) starts when the ball leaves the hand and terminates when the forward foot contacts the ground. During this phase, shoulder abduction and external rotation begin. Phase III (late cocking) is marked by complete shoulder abduction and maximal external rotation, together with the elbow flexion among 90° and 120° and the forearm pronation to 90°. Phase IV (rapid acceleration) creates an extensive forward-directed force on the extremity, followed by a quick elbow extension. It ends with ball release. During the throwing cycle, the elbow extends over 2300°/s. This kind of movement creates a medial shear force of about 300 N and a lateral compressive force of approximately 900 N. Highest valgus force on the elbow is produced during the phases of late cocking and acceleration when the elbow, flexed to 95°, is subjected to valgus strengths up to 64 N. When the ball is released, the lateral side of the elbow is exposed to forces greater than 500 N [30]. The presence of these extreme forces on the medial and lateral elbow compartments is responsible for producing dramatical injuries that could wholly jeopardize the career of the athletes. Enormous valgus stress is created over the medial side of the elbow during the acceleration phase, and the ulnar collateral ligament primarily supports it [31]. The secondary stabilizing structures of the medial elbow, in particular the flexor-pronator musculature, contrast the remaining part of this force. Phase V (follow-through) is characterized by the dissipation of all excess kinetic energies, while the elbow resumes full extension and ends the arch of motion. These excessive forces produced on the elbow by the overhead athlete make the elbow extremely susceptible to damage. The characteristic pattern of injury supported is due to both repetitive microtrauma or chronic stress overload [26].

16.2.2 Football

The throwing movement of football is very similar to throwing a baseball. During arm cocking, the elbow flexion of a quarterback is higher than that of pitchers, with an average of 113°. Then, during arm cocking, a maximum medial force of 280 N and a maximum varus strength of 54 N are generated around the elbow joint. In the arm acceleration phase, the elbow shows a maximum extension velocity of 1760°/s. The following deceleration involves the generation of a flexion torque of 41 nm, with an accompanying compressive force of 620 N. The slower elbow mechanism of extension can justify for fewer elbow damages in quarterbacks than pitchers [26].

16.2.3 Tennis

Elbow joint provides 15% of the strength necessary during a tennis serve, in which tennis ball can reach a speed of 4650 km/h. Considering the end of the racquet, in a backhand position, with a wrist diameter of about 9 cm, and a distance from the center of the grip to the ball of 45 cm, the energy produced is comparable to an effort of lifting 25 kg for the elbow. These concepts represent the base of understanding why the extensor mechanism is responsible for "tennis elbow" condition. As with the overhand throw, the tennis serve produces significant angular velocity. During elbow extension, it can reach 982°/s, while during pronation 347°/s [32, 33].

16.3 Summary

Elbow pain in the overhead-throwing athlete has become more usual in the last years. Appropriate knowledge of the skeletal, ligamentous, nervous, and musculotendinous structures of the elbow joint, together with an accurate understanding of elbow biomechanics in sport, are required to comprehend the pathogenic mechanism of elbow injuries, providing to prevent them.

References

1. Karbach LE, Elfar J. Elbow instability: anatomy, biomechanics, diagnostic maneuvers, and testing. J Hand Surg Am. 2017;42:118–26. https://doi.org/10.1016/j.jhsa.2016.11.025.
2. Prasad A, Robertson DD, Sharma GB, Stone DA. Elbow: the trochleogingylomoid joint. Semin Musculoskelet Radiol. 2003;7:19–25. https://doi.org/10.1055/s-2003-41082.
3. Bryce CD, Armstrong AD. Anatomy and biomechanics of the elbow. Orthop Clin N Am. 2008;39:141–54., v. https://doi.org/10.1016/j.ocl.2007.12.001.
4. King GJ, Zarzour ZD, Patterson SD, Johnson JA. An anthropometric study of the radial head: implications in the design of a prosthesis. J Arthroplast. 2001;16:112–6. https://doi.org/10.1054/arth.2001.16499.
5. Morrey BF, An KN. Stability of the elbow: osseous constraints. J Shoulder Elb Surg. 2005;14:174S–8S. https://doi.org/10.1016/j.jse.2004.09.031.
6. Josefsson PO, Gentz CF, Johnell O, Wendeberg B. Dislocations of the elbow and intraarticular fractures. Clin Orthop Relat Res. 1989:126–30.
7. Hull JR, Owen JR, Fern SE, Wayne JS, Boardman ND. Role of the coronoid process in varus osteoarticular stability of the elbow. J Shoulder Elb Surg. 2005;14:441–6. https://doi.org/10.1016/j.jse.2004.11.005.
8. Beingessner, D.M.; Dunning, C.E.; Stacpoole, R.A.; Johnson, J.A.; King, G.J. The effect of coronoid fractures on elbow kinematics and stability. Clin Biomech (Bristol, Avon) 2007, 22, 183–190, doi:https://doi.org/10.1016/j.clinbiomech.2006.09.007.
9. Safran MR, Baillargeon D. Soft-tissue stabilizers of the elbow. J Shoulder Elb Surg. 2005;14:179S–85S. https://doi.org/10.1016/j.jse.2004.09.032.
10. Morrey, B.F.; Tanaka, S.; An, K.N. Valgus stability of the elbow. A definition of primary and secondary constraints. Clin Orthop Relat Res 1991, 187–195.
11. Gallay SH, Richards RR, O'Driscoll SW. Intraarticular capacity and compliance of stiff and normal elbows. Arthroscopy. 1993;9:9–13. https://doi.org/10.1016/s0749-8063(05)80336-6.
12. O'Driscoll SW, Morrey BF, An KN. Intraarticular pressure and capacity of the elbow. Arthroscopy. 1990;6:100–3. https://doi.org/10.1016/0749-8063(90)90007-z.
13. Callaway GH, Field LD, Deng XH, Torzilli PA, O'Brien SJ, Altchek DW, Warren RF. Biomechanical evaluation of the medial collateral ligament of the elbow. J Bone Joint Surg Am. 1997;79:1223–31. https://doi.org/10.2106/00004623-199708000-00015.

14. Morrey BF, An KN. Functional anatomy of the ligaments of the elbow. Clin Orthop Relat Res. 1985:84–90.
15. Floris S, Olsen BS, Dalstra M, Søjbjerg JO, Sneppen O. The medial collateral ligament of the elbow joint: anatomy and kinematics. J Shoulder Elb Surg. 1998;7:345–51. https://doi.org/10.1016/s1058-2746(98)90021-0.
16. Seki A, Olsen BS, Jensen SL, Eygendaal D, Søjbjerg JO. Functional anatomy of the lateral collateral ligament complex of the elbow: configuration of Y and its role. J Shoulder Elb Surg. 2002;11:53–9. https://doi.org/10.1067/mse.2002.119389.
17. Jensen SL, Olsen BS, Tyrdal S, Søjbjerg JO, Sneppen O. Elbow joint laxity after experimental radial head excision and lateral collateral ligament rupture: efficacy of prosthetic replacement and ligament repair. J Shoulder Elb Surg. 2005;14:78–84. https://doi.org/10.1016/j.jse.2004.05.009.
18. Erickson BJ, Harris JD, Chalmers PN, Bach BR, Verma NN, Bush-Joseph CA, Romeo AA. Ulnar collateral ligament reconstruction: anatomy, indications, techniques, and outcomes. Sports Health. 2015;7:511–7. https://doi.org/10.1177/1941738115607208.
19. Buchanan TS, Delp SL, Solbeck JA. Muscular resistance to varus and valgus loads at the elbow. J Biomech Eng. 1998;120:634–9. https://doi.org/10.1115/1.2834755.
20. Johnson JA, Rath DA, Dunning CE, Roth SE, King GJ. Simulation of elbow and forearm motion in vitro using a load controlled testing apparatus. J Biomech. 2000;33:635–9. https://doi.org/10.1016/s0021-9290(99)00204-3.
21. Duck TR, Dunning CE, King GJ, Johnson JA. Variability and repeatability of the flexion axis at the ulnohumeral joint. J Orthop Res. 2003;21:399–404. https://doi.org/10.1016/S0736-0266(02)00198-5.
22. Boone DC, Azen SP. Normal range of motion of joints in male subjects. J Bone Joint Surg Am. 1979;61:756–9.
23. Kapandji A. Biomechanics of pronation and supination of the forearm. Hand Clin. 2001;17:111–22. vii
24. Moore DC, Hogan KA, Crisco JJ, Akelman E, Dasilva MF, Weiss AP. Three-dimensional in vivo kinematics of the distal radioulnar joint in malunited distal radius fractures. J Hand Surg Am. 2002;27:233–42. https://doi.org/10.1053/jhsu.2002.31156.
25. Kancherla VK, Caggiano NM, Matullo KS. Elbow injuries in the throwing athlete. Orthop Clin North Am. 2014;45:571–85. https://doi.org/10.1016/j.ocl.2014.06.012.
26. Loftice J, Fleisig GS, Zheng N, Andrews JR. Biomechanics of the elbow in sports. Clin Sports Med. 2004;23:519–30., vii-viii. https://doi.org/10.1016/j.csm.2004.06.003.
27. Plancher KD, Minnich JM. Sports-specific injuries. Clin Sports Med. 1996;15:207–18.
28. Jobe FW, Moynes DR, Tibone JE, Perry J. An EMG analysis of the shoulder in pitching. A second report. Am J Sports Med. 1984;12:218–20. https://doi.org/10.1177/036354658401200310.
29. Erickson BJ, Gupta AK, Harris JD, Bush-Joseph C, Bach BR, Abrams GD, San Juan AM, Cole BJ, Romeo AA. Rate of return to pitching and performance after Tommy John surgery in Major League Baseball pitchers. Am J Sports Med. 2014;42:536–43. https://doi.org/10.1177/0363546513510890.
30. Petty DH, Andrews JR, Fleisig GS, Cain EL. Ulnar collateral ligament reconstruction in high school baseball players: clinical results and injury risk factors. Am J Sports Med. 2004;32:1158–64. https://doi.org/10.1177/0363546503262166.
31. Chen FS, Rokito AS, Jobe FW. Medial elbow problems in the overhead-throwing athlete. J Am Acad Orthop Surg. 2001;9:99–113. https://doi.org/10.5435/00124635-200103000-00004.
32. Kibler WB. Clinical biomechanics of the elbow in tennis: implications for evaluation and diagnosis. Med Sci Sports Exerc. 1994;26:1203–6.
33. Sprigings E, Marshall R, Elliott B, Jennings L. A three-dimensional kinematic method for determining the effectiveness of arm segment rotations in producing racquet-head speed. J Biomech. 1994;27:245–54. https://doi.org/10.1016/0021-9290(94)90001-9.

Elbow Tendon Injury and Repair: Triceps and Biceps Tendons

17

Andrea Celli, Cheli Andrea, Bartoli Matteo, and Luigi Adriano Pederzini

17.1 Triceps Tendon

17.1.1 Introduction

Triceps tendon injuries are probably among the rarest tendon injuries in the human body. Evidence shows male predominance (3:1, male to female) with wide age variance at occurrence [1]. The most common rupture site is at tendon insertion into the olecranon and more rarely at the myotendinous junction or intramuscularly [2, 3, 4]. Anzel et al. reviewed more than 1000 tendon injuries. They reported approximately 0.8% triceps injuries, half of which were associated with open lacerations in the posterior aspect of the upper extremity [5]. Mair et al. described 21 cases of triceps tendon rupture observed over a 6-year period with contact athletes (e.g., professional football players). [6]. Kibuule et al. reported that triceps injuries can also occur in adolescents with incompletely fused physis and avulsions of the olecranon apophysis [7]. Several triceps tendon injuries risk factors have been reported. The main injury mechanism is falling on an outstretched hand [8, 9]. Other conditions that can increase risk of triceps ruptures are: sport activities (such as contact activities or body building) [10], use of anabolic steroid drugs or local steroid injection, olecranon bursitis [11]. Alternatively, risk of triceps ruptures can arise secondary to surgical exposure (such as following total elbow replacement) or secondary to release for stiff elbow [12, 13]. The lesions can also occur following direct penetrating trauma in the posterior aspect of the elbow. It is not rare to observe associated injuries, such as fractured radial heads or complex fractures of the distal humerus.

17.1.2 Anatomy

Triceps brachii muscle has three heads:

- The long head arises from the infraglenoid tubercle of the scapula.
- The lateral head has a linear attachment from the upper margin of the radial grove of the humerus.
- The medial head originates below the lateral margin of the radial groove that contains the radial nerve. Its insertion covers the entire rear surface of the lower part of the humerus.

In the distal third of the posterior aspect of the arm, the lateral head joins with the long head from the superficial tendinous part of the insertion

A. Celli (✉)
Shoulder and Elbow Unit, Department of Orthopaedic and Traumatology Surgery, Hesperia Hospital, Modena, Italy

C. Andrea · B. Matteo · L. A. Pederzini
Department of Orthopaedic, Traumatology and Arthroscopic Surgeries, Nuovo Ospedale di Sassuolo, Modena, Italy

J. Koh et al. (eds.), *Orthopaedic Biomechanics in Sports Medicine*,
https://doi.org/10.1007/978-3-030-81549-3_17

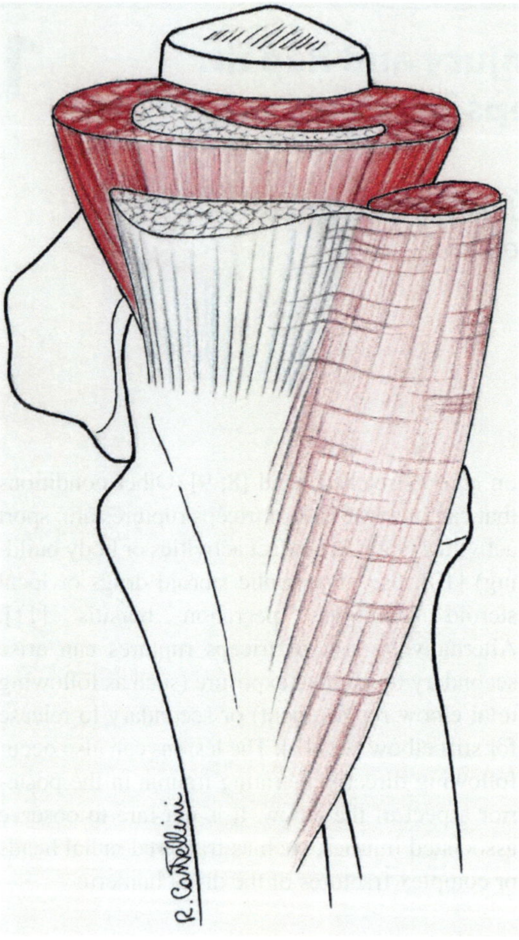

Fig. 17.1 Distal triceps tendon insertion (superficial and deep components): The lateral part is more expansive and relatively thinner in continuity with anconeus muscle and fascia. The medial part is thicker than the lateral aspect like a proper triceps tendon and inserts directly onto olecranon

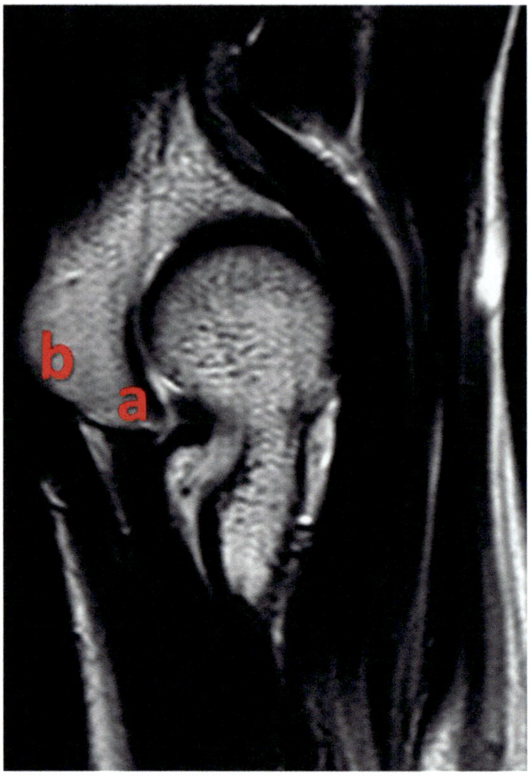

Fig. 17.2 Magnetic resonance imaging describes a bipartite insertion between the deep (a) and superficial (b) components of the triceps tendon into the olecranon

on the posterior surface of the olecranon. The medial head (deep part of the triceps) inserts through muscular and tendinous fibers directly onto the olecranon. There are anatomical differences between the lateral and medial side of the superficial tendon and between the superficial and deep tendon anatomy (Fig. 17.1). As it approaches its insertion area, the superficial tendon forms two components:

- A lateral part which is more expansive and relatively thinner in continuity with the anconeus muscle and fascia.
- A medial part which is thicker than the lateral aspect *like* a proper triceps tendon. *This* inserts directly onto the olecranon.

All three heads of the triceps contribute to forming the triceps olecranon dome-shaped footprint [5] (Figs. 17.2 and 17.3).

17.1.3 Olecranon Footprint (Figs. 17.2 and 17.3).

The mean medial to lateral width of the insertion area is about 20 mm and the mean proximal to distal length is about 13 mm [14]. The mean distance from the olecranon tip to the most proximal aspect of the medial head insertion ranges between 14.8 [14] and 16 mm [15]. The mean width (medial to lateral distance) of the medial head footprint is 16 mm and the mean thickness

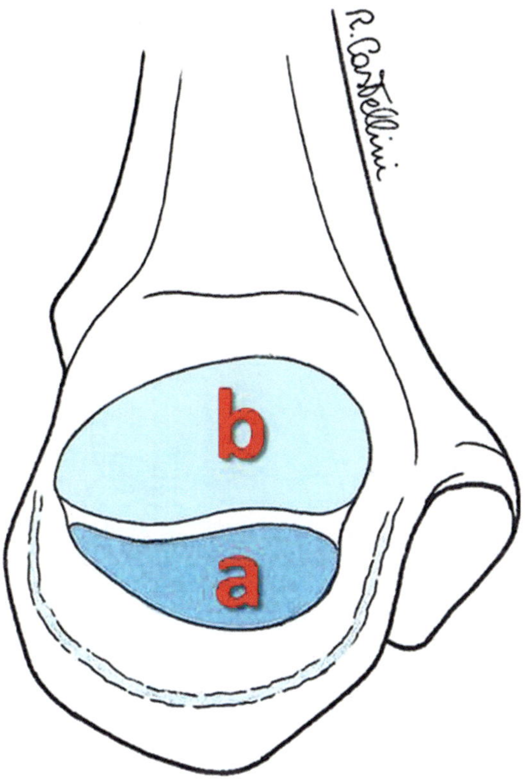

Fig. 17.3 Mean width (medial to lateral distance) (a) of the medial head footprint is about 16 mm and the mean thickness 4 mm. Mean width of the common superficial tendon footprint is 19 mm and the mean thickness is 8 mm (b)

is 4 mm. The mean width of the common superficial tendon footprint is 19 mm and the mean thickness is 8 mm. [15].

17.1.4 Pathogenesis

In healthy tendons, traumatic lesions are the most common mechanisms of triceps distal tendon ruptures. In most cases, tendon ruptures more frequently occur with intrinsic pathological disorders such as overuse injuries or previous surgical treatment decreasing tendon tensile strength.

A simplified classification categorizes triceps tendon ruptures in four groups:

- Traumatic lesion.
- Spontaneous rupture.
- Overuse injuries.
- Following total elbow arthroplasty.

17.1.4.1 Traumatic Lesions

Acute triceps tendon ruptures most commonly occur when a patient falls and a forced load is applied to the contracted triceps on an outstretched hand with elbow in flexion. Alternatively, the injury mechanism may be a direct trauma to the posterior aspect of the arm. With flexed elbow, if at the same time the triceps is contracted, the injury is more severe.

Traumatic tears can occur at several different anatomical regions but, at the musculotendinous junction, they are extremely rare just like those that occur within the muscle belly [16, 17, 18]. In general, tears are most commonly observed at tendon insertion like an olecranon avulsion.

Traumatic distal tendon lesions can be partial or total [17, 19] and they are often isolated. Associated lesions were described at the radial head [20, 21], at the medial collateral ligament (MCL) [7], and at the capitellum [22].

Partial traumatic tendo-osseous avulsions with one small fragment proximal to the olecranon (flake sign) often occur (Fig. 17.4). They are associated with radial head fractures [21]. Radial head fractures or coronal fractures of the capitellum result as a consequence of direct axial forces being transmitted on the lateral elbow compartment when the lateral elbow is in a valgus and semi-flexed position while the forearm is pronated. Disruption of the MCL increases triceps medial head eccentric contraction which may induce avulsion “flake” fractures from the olecranon leading to radial head or capitellum fractures.

Triceps avulsion, radial head fractures, and MCL rupture were reported in combination by Yoon as triad injuries [22].

17.1.4.2 Spontaneous Ruptures

Triceps injury can occur spontaneously due to attrition if tendon integrity is compromised. Several risk factors have been studied for triceps muscle and tendon pathological disorders. These include rheumatoid arthritis, chronic renal failure, endocrine disorders, metabolic bone diseases, as well as steroid use.

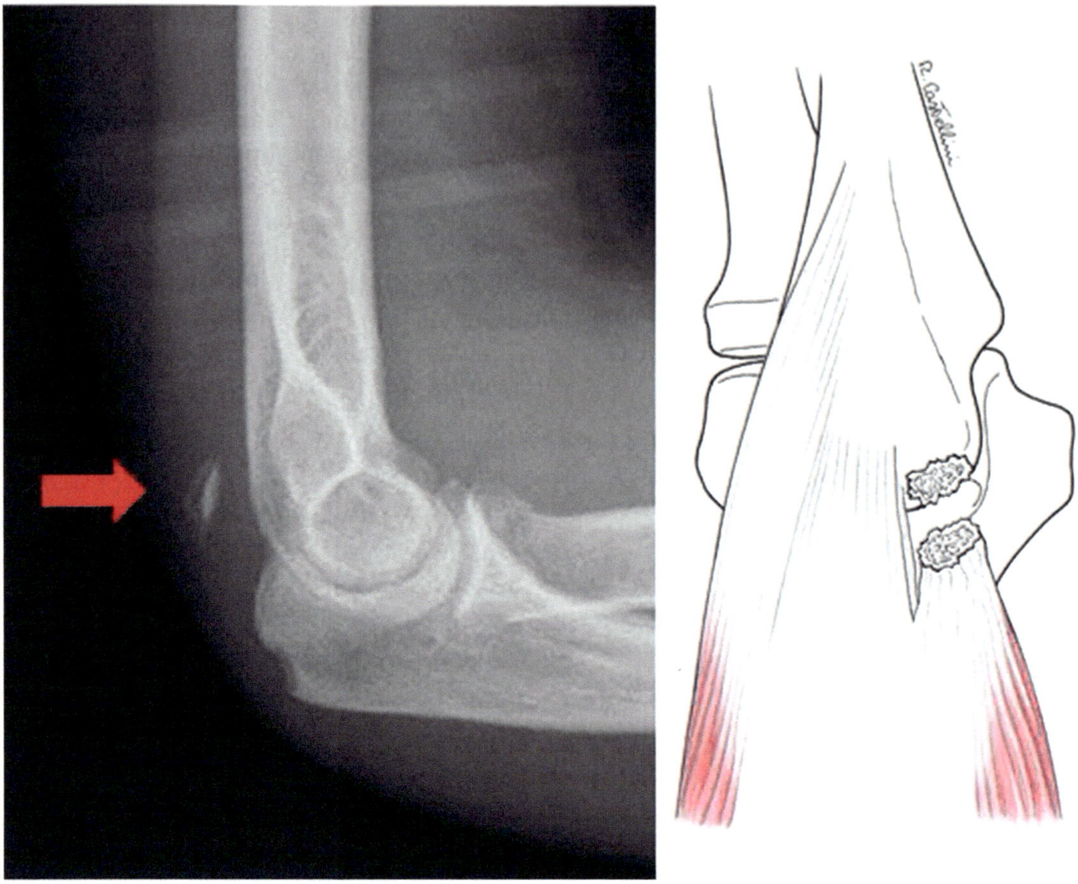

Fig. 17.4 Lateral elbow X-rays are useful in confirming the diagnosis if a small extra-articular avulsion fracture of the olecranon (flake sign) is present

Use of local or systemic corticosteroid or anabolic steroids may also predispose to tendon rupture by decreasing tendon tensile strength. It is widely accepted that the biosynthesis of collagen is inhibited by glucocorticoids [23].

17.1.4.3 Overuse Injuries

Differently from traumatic acute lesion injuries, sport-related injuries are frequently ascribed to repetitive motions that result in pain and inability to take part in sport activities.

Cumulative submaximal loading of the tissue is referred to as "overuse injuries." Complete rupture often occurs through abnormal tendon with intrinsic pathological disorders.

Triceps tendinosis during sport activities is not infrequent. In cases of chronic tendon pain, the pathological lesion is typical of a degenerative process with areas of marked degeneration and lack of local vascular supply [24].

17.1.4.4 Following Total Elbow Arthroplasty

Failure of triceps reattachment can be seen following surgical treatment in which triceps takedown is performed.

At Mayo Clinic, with over 887 total elbow arthroplasties, this complication was observed in 16 elbows—about 2% of all procedures [12].

Predisposing factors include inflammatory arthropathy with poor tissue quality. Surgical elevation of fragile triceps tendon risks becoming damaged when performing reinsertion into the olecranon. Also, triceps weakness is to be expected.

17.1.5 Physical Findings

Acute triceps tendon ruptures most commonly occur, but chronic lesions have been reported [25].

Post-traumatic presenting signs and symptoms are correlated to:

- Type of lesions (tendinous, tendon avulsion or within muscle belly),
- Degree (partial or total),
- Time (acute or chronic).

In general, triceps lesions are characterized by:

- Spontaneous and evocate pain on palpation at indicated site of lesion. Tenderness and palpable defect in the tendon proximally to the olecranon are usually observed.
- Swelling, ecchymosis, and body habitus often impede tendon defect to be palpable in its acute stage. Once swelling subsides, most patients show a palpable gap in the tendon.
- Active extension is typically diminished or absent depending on whether a complete or partial tear is there [26, 27].

Partial ruptures of the medial head are frequently undiagnosed. Diagnosis is difficult, but care should be taken when radiographs assess presence of flake signs. Fragment avulsions from the olecranon are also associated with radial head fractures which is frequently reported along with medial collateral ligament (MCL) injury. However, when radial head fractures are present and patients present pain and swelling in the medial compartment of the elbow, the MCL and triceps tendon integrity should always be assessed.

A universal physical finding with triceps ruptures is inability to extend the arm against gravity. (Fig. 17.5).

Discerning between partial or complete ruptures can be a diagnostic challenge. Partial tendon lesions manifest themselves as weakness but ability to actively extend the elbow against gravity—not against resistance. These findings are likely secondary to intact lateral expansion or compensating anconeus muscle.

Total tendon tear manifests itself as loss of extension strength also against gravity and inability to extend the elbow. Triceps extension tests can also be performed by observing the patients' ability to extend the elbow over their head against gravity.

17.1.6 Diagnostic Tests

Viegas [28] described a provocative test similar to Thompson's test (used to help in diagnosing Achilles tendon ruptures) that can be employed in triceps tendon rupture diagnosis. In prone

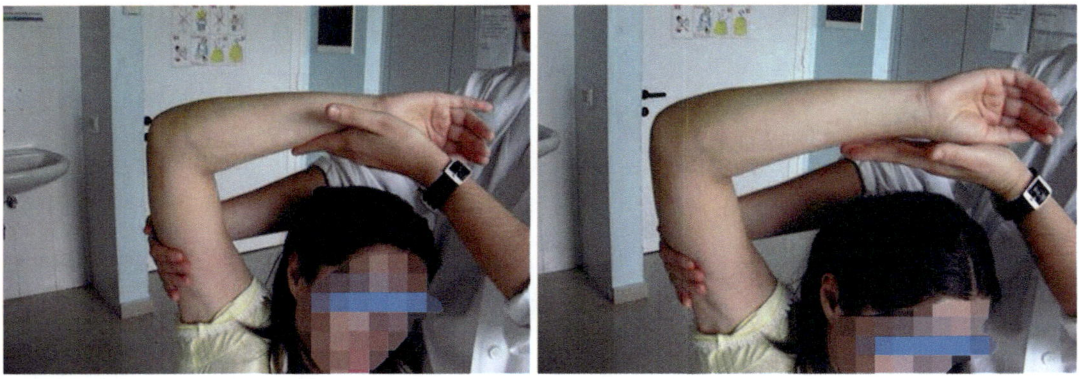

Fig. 17.5 Tendon tear is manifested by loss of extension strength also against gravity and impossibility of elbow extension. Triceps extension tests can also be performed observing the ability of the patient to extend the elbow over his/her head against gravity

position, the patient lets their relaxed forearm hang over the table while the physician squeezes the triceps muscle belly. When the tendon lesion is partial, this should produce a slight elbow extension. Conversely, no motion will occur in case of complete rupture.

Celli et al. [29] defined a test called "fall down triceps test" for triceps insertion tendon rupture. This test assesses inability of patients to keep their forearm in maximum extension against gravity. While standing with their shoulder at 90° abduction and internal rotation, the patient's forearm is kept in full passive extension by the examiner placed behind the subject. Upon dropping the forearm, if the triceps tendon presents a complete rupture, the patient will not be able to maintain the initial position and their elbow will drop down to 90° flexion. Conversely, in case of partial rupture the patient's forearm will only slightly drop down given the patient's effort to maintain limited elbow extension. (Fig. 17.6).

17.1.7 Imaging Studies

Imaging studies help to identify the level of lesion (olecranon insertion, myotendinous junction, or intramuscular), to discriminate between partial and complete tear, to estimate the amount of tendon retraction, and to exclude associated osseous injuries.

Lateral elbow X-rays are useful in confirming diagnosis when small extra-articular avulsion fracture of the olecranon (flake sign) is present. The bony fragment is usually small and easy to ignore, but its presence is pathognomonic of distal triceps avulsion (Fig. 17.4).

X-rays (AP and lateral) and Computer tomography (CT) are also helpful in aiding diagnosis of injuries associated with triceps rupture, such as ipsilateral radial head and capitellum fractures.

Ultrasonography may be used, although it provides limited anatomical detail. Nonetheless,

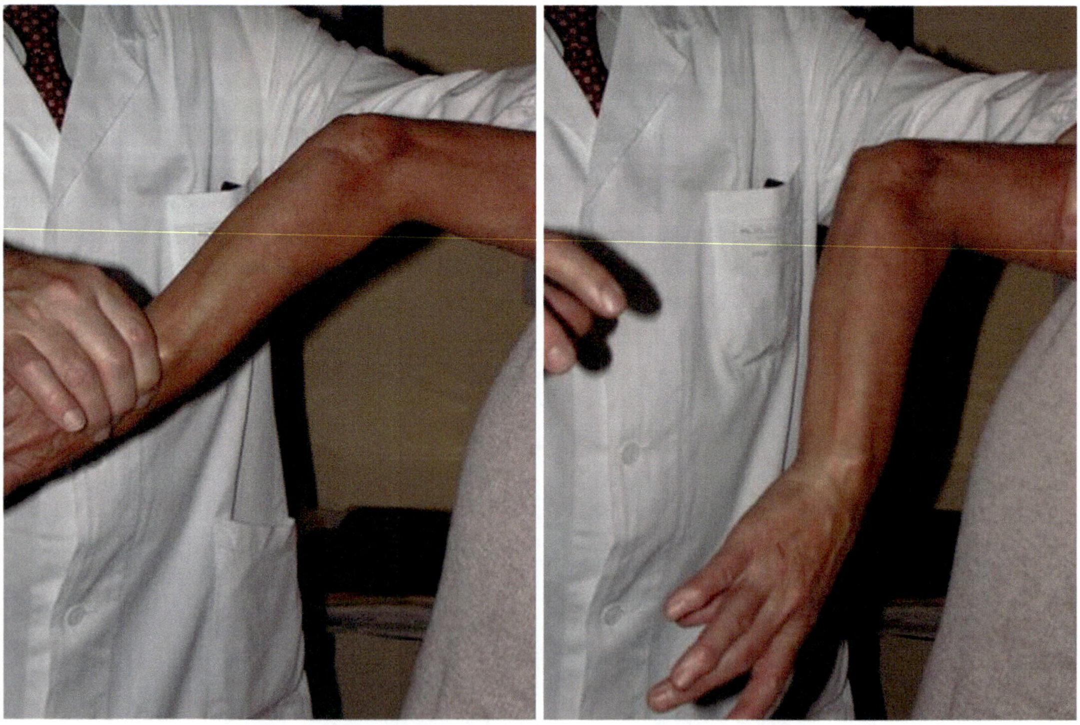

Fig. 17.6 Fall down triceps test. The forearm falls down when the examiner leaves the maximum extension and the patient is not able to maintain this position, if the triceps tendon presents a complete rupture the elbow drops down to 90 ° of flexion. In case of partial ruptures, the forearm drops down only partially given effort of the patient to maintain limited extension of the elbow

it may be useful immediately after the injury [30] when the diagnosis is doubtful.

MRI is the best technique to assess tendon lesions [31, 32] because it provides more details to distinguish between (Fig. 17.2):

- Sides of the injury,
- Partial and complete lesions,
- Degrees of tendon retraction,
- Muscular atrophies.

MRIs on the sagittal planes demonstrated the integrity of the triceps tendon: partial tears most commonly occur distal at the olecranon insertion and become visible in the form of a small fluid-filled gap within the ruptured distal triceps tendon [33].

Complete rupture is characterized by a large fluid-filled gap between the retracted triceps tendon stump and the olecranon process.

In overuse degenerative tendinosis MR imaging is characterized by thickening and signal alteration of the distal tendon fibers.

17.1.8 Nonoperative Management

Conservative management plays a role in partial triceps injuries involving the muscle belly or musculotendinous junction, or in partial ruptures of the distal tendon insertion when there is no significant extension power loss against gravity and *no* resistance according to the patient's age and lifestyle.

This nonoperative group can include non-dominant arm injuries, sedentary lifestyles, and elderly patients with high complications rates for surgical managements.

17.1.9 Surgical Management

Several surgical techniques and different approaches offer variable options for the surgical management of acute complete or chronic tears. Different surgical methods can be used to restore the extensor mechanism.

The choice depends on tissue quality and on the entity of muscle retraction. Chronic lesions older than 6 weeks can influence the surgical choice.

Quality of the olecranon has to be considered, particularly following total elbow arthroplasty, *because it is* a condition that can influence the surgical choice.

Tendon repair has to be performed between 90° and 70° elbow extension without tension at the tendon-bone reinsertion. Direct repair under tension is related to risk of secondary rupture and stiffness in flexion.

Surgical options include:

- Direct olecranon repair.
- Augmentation with auto- or allograft.
- Anconeus rotation flap.
- Achilles tendon allograft with or without calcaneus bone.

17.1.9.1 Direct Reattachment to the Olecranon

- The partial lesion reattachment.

The triceps tendon usually retracts no more than 3 to 5 cm. Preserved lateral continuity to the anconeus triceps aponeurotic fascia avoids significant proximal migration of the tendon. Proximal stump of partial or complete lesion is identified and debrided recovering the normal tendon which allows *its* direct reinsertion *into* the olecranon. Without bony fragments, in acute lesions the partial lesion is reinserted directly into the olecranon bone. (Fig. 17.7).

In chronic cases, with bony fragments attached to the tendon, *it* is debrided and reinserted with tendon-bone sutures.

- complete lesion reattachment.

Tendon and muscle belly are extracted and heavy nonabsorbable sutures no.5 performing (locking fashion technique) as Bunnell or Krakow-type suture repair are passed through the tendon pulling it down to the olecranon without losing passive flexion (Fig. 17.8). This is

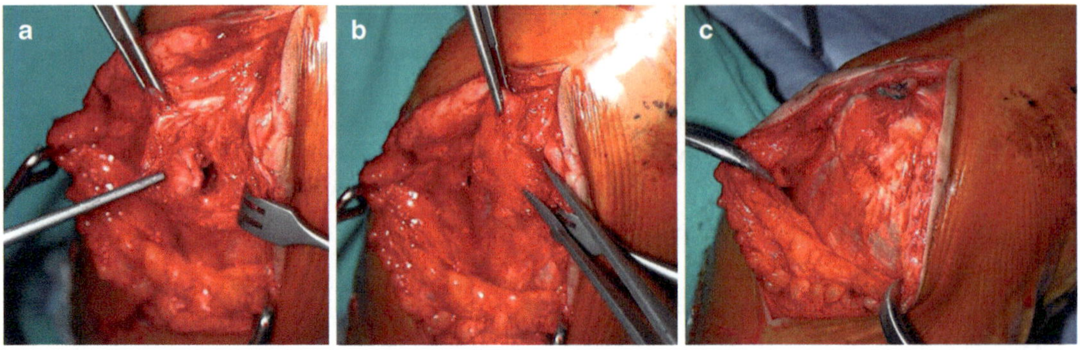

Fig. 17.7 The proximal stump of the partial lesion is identified (**a**) and debrided recovering the normal tendon (**b**) that allows its direct reinsertion into the olecranon (**c**)

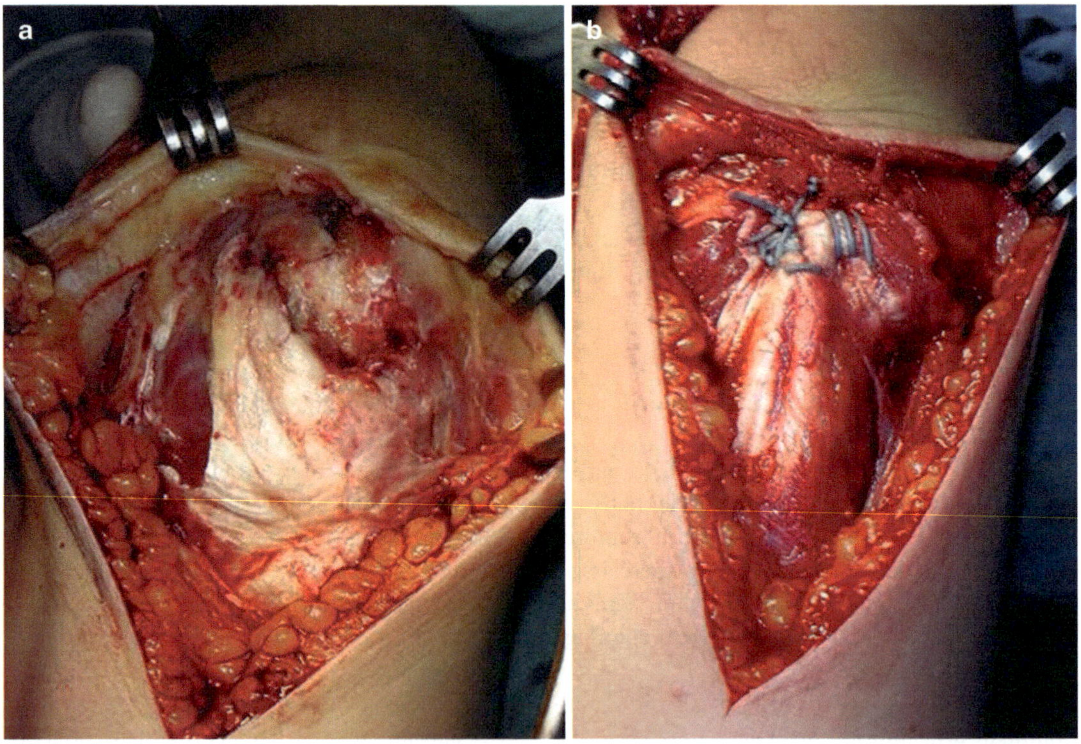

Fig. 17.8 The proximal stump of the complete lesion is identified (**a**). Heavy nonabsorbable sutures no.5 performing locking fashion technique are passed through the tendon pulling it down to the olecranon and reinsertion is performed using suture anchors and bony tunnels (**b**). Direct reattachment to the olecranon is possible when the tendon is reinserted with 90° of elbow flexion

possible when the tendon is reinserted with 90° elbow flexion. An additional fixation is obtained using a proximal suture anchor (double row repair) to improve fixation of the anatomical footprint of the tendon. Before reattaching the tendon to the olecranon footprint, the bone is decorticated with a burr. Two transverse holes are drilled (2.5 mm drill hole) starting at the triceps insertion and exiting at the dorsal side of the olecranon.

17.1.9.2 Tendon Augmentation

In chronic lesions or following total elbow arthroplasty, when direct triceps tendon to bone reattachment is possible only at 50–60 degrees of elbow flexion, tendon augmentation is advised. After excising the scarring tissue, the proximal portion of the tendon and muscle belly is mobilized and advanced distally to be reinserted on the posterior olecranon. In case of a degenerative tendon and retraction of the triceps, to avoid elbow stiffness and to reduce risk of tendon re-rupture, tendon to olecranon continuity can be recovered using different auto- or allograft augmentations.

Small size of the defect can be recovered using the palmaris longus, plantaris tendon autograft, or flexor carpi radialis allograft. Rotation forearm fascia can be used to reinforce tendon augmentation. Alternatively, for larger defects, an Achilles tendon allograft is useful.

17.1.9.3 Anconeus Rotational Flap

Anconeus rotational flap described by Morrey [34, 35, 36] is a useful procedure when direct reattachment is impossible for degenerative and fragile triceps tendons. This procedure is indicated when the triceps defect is small and the lateral triceps fascia and anconeus are preserved, or for triceps lateral dislocation in continuity with the anconeus muscle.

17.1.9.4 Achilles Tendon Allograft

When chronic ruptures have significant muscle retraction and there is tendon tissue deficiency, Achilles allograft reconstruction is indicated (Figs. 17.9). Achilles tendon allograft with a small piece of calcaneus can be used to reconstruct the continuity of the triceps tendon and also as an osseous graft to the olecranon if this is deficient [34, 35, 36].

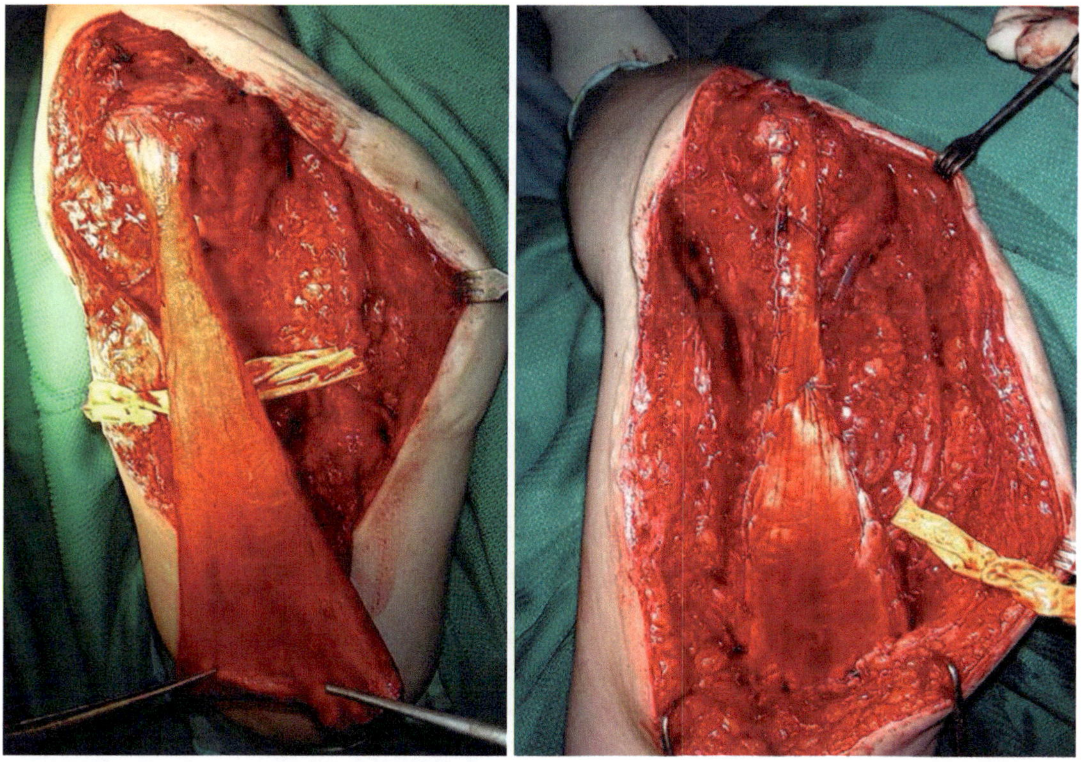

Figs. 17.9 Achilles allograft reconstruction is indicated in case of chronic rupture with significant muscle retraction and tendon tissue deficiency. Allograft with calcaneal bone provides an ideal reconstructive unit especially if the olecranon is deficient. The proximal expansion of the allograft is used to wrap the remaining triceps muscle as well as the remaining tendon using nonabsorbable sutures

- Two reconstructive techniques may be used:
  - Attaching the allograft tendon directly onto the olecranon through drill holes.
  - Fixing the calcaneal portion of the allograft to the remaining olecranon with a screw or a tension-band wire. The allograft with calcaneal bone provides an ideal reconstructive unit especially if the olecranon is deficient [12, 13].

17.2 Distal Biceps Tendon

17.2.1 Introduction

Distal biceps tendon injuries are relatively uncommon, accounting for about 3% of all tendon ruptures [37, 38]. However, a demographic study suggests that incidence of these injuries may be closer to 10% [39]. Rupture can be acute or chronic, complete or partial. The average patient is 50 years old. The dominant arm is affected in 86% of patients [38, 40]. Most of them are men (93%) [38, 41].

The pathogenesis of distal biceps rupture is unclear. A variety of local causes including mechanical compression, attrition, impingement, hypovascularity, and avascularity have been implicated. Hypertrophic radial tuberosity (RT), similar to a subacromial spur, may erode the tendon during forearm rotation due to the narrowing of the space between radius and ulna. Anabolic steroids have been implicated in degenerative changes affecting the distal biceps tendon [42]. Tendon rupture may result from a combination of some or all of these factors.

17.2.2 Anatomy

Anatomical studies [43, 44] demonstrated that the short and long head of the biceps muscle possess two distinct tendons with distinct radial tuberosity insertions [45] (Fig. 17.10). The two ribbon-like bicipital tendons and the short tendon that remains on the ulnar side of the long head insert distally in the footprint, whereas the insertion of the long head occupies most of the proximal RT.

In a cadaveric study, Mazzocca et al. [46] demonstrated that the RT has a rough posterior ridge that varies in size (large, medium, small) and that it can even be absent in some individuals. Morphologically, the footprint is oval in the proximal portion and ribbon-shaped in the distal portion. Its average length and width range between 22–24 mm and 15–19 mm, respectively [43].

Proximally, at the musculotendinous junction, the two heads lie side by side on the brachialis muscle, the short head being medial to the long head. When the tendons cross the bicipital tunnel, the short tendon makes a 90° bend; as a result, its course begins medially and ends distal to the long head on the radial footprint [43, 47, 48].

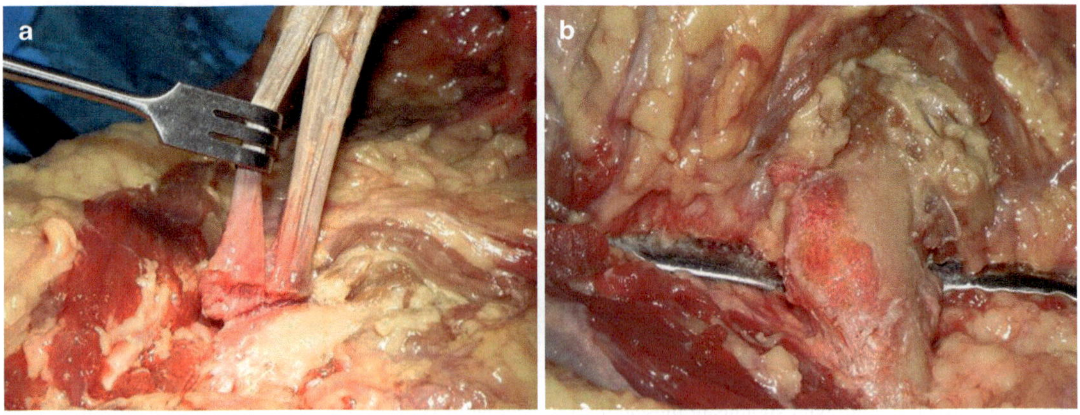

Fig. 17.10 The short and long head of the biceps muscle (**a**) possess two distinct tendons with distinct radial tuberosity insertions (**b**)

The two tendons are surrounded by the bicipital aponeurosis, which begins muscle-tendon junction level. The distal biceps tendon is stabilized by the aponeurosis, which merges distally with the flexor muscle mass through strong fascial adhesions.

17.2.3 Pathogenesis

Distal biceps tendon ruptures are partial or complete. Complete ruptures are divided into acute and chronic, depending on whether patients come to attention more or less than 6 weeks after the rupture. In turn, chronic lesions are distinguished between those where the bicipital aponeurosis is intact and those where it is injured. Patients with complete rupture (tendon and lacertus) present a palpable defect in the antecubital fossa.

17.2.4 Physical Findings

Typical patients are men in their fifth or sixth decade. Injury mechanisms involve sudden, forceful, eccentric contractions of the biceps.

Patients with acute ruptures often report severe pain with tearing sensation at the front of the elbow and an audible pop or snap accompanied by immediate loss of elbow strength in flexion and supination. After the trauma, physical examination discloses an ecchymosis around the antecubital fossa that may extend proximally and distally. In patients with complete rupture, there often is a palpable absence of biceps tendon compared to the contralateral arm. When the bicipital aponeurosis is torn, the biceps muscle is usually retracted by more than 8 cm and its anatomical contour changes to a "reversed Popeye deformity." (Fig. 17.11).

Diagnosing partial tendon ruptures or subacute injuries is more difficult, due to absence of deformity and proximal migration of the biceps muscle-tendon unit; the course of the tendon is frequently palpable but painful [49]. In patients with partial tears elbow flexion and supination strength are slightly reduced compared to the contralateral limb, but forearm supination power is usually markedly reduced.

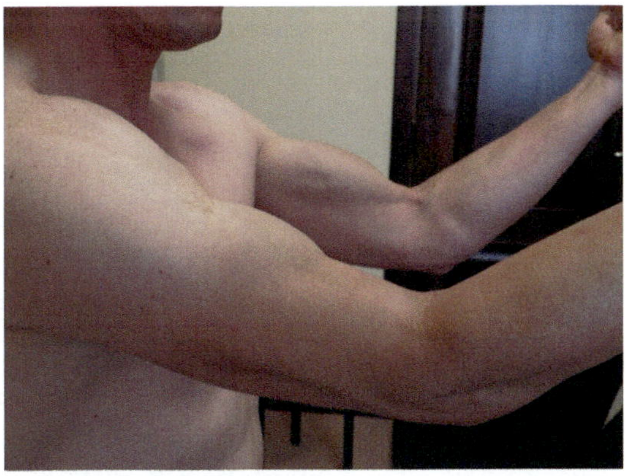

Fig. 17.11 When the bicipital aponeurosis is torn, the biceps muscle is usually retracted and its anatomical contour changes to a "reversed Popeye deformity"

Pain in the forearm and in the antecubital fossa associated with intact bicipital aponeurosis and a palpable tendon indicate a partial rupture. This is typical in men and in manual workers with a history of forced eccentric contracture. Anterior elbow pain is exacerbated by resisted forearm supination and by full elbow extension and pronation.

After 6 weeks, tear is considered chronic. When the bicipital aponeurosis is intact, and the tendon is injured at bone interface, retraction is minimal; in such cases the tendon is loosely curled in the antecubital fossa and shows good elasticity. In patients with chronic rupture, torn bicipital aponeurosis, and muscle retraction, the tendon is coiled inside the muscle belly and cannot be reduced, requiring elongation with an autograft or an allograft. Patients with chronic complete rupture typically feel little pain, but report persistent weakness in flexion and supination; the biceps tendon cannot be palpated in the antecubital fossa and the muscle belly shows proximal retraction compared to the contralateral limb.

17.2.5 Diagnostic Tests

The hook test described by O'Driscoll et al. [50] can be used to assess biceps tendon integrity and it is considered as a specific and sensitive method to detect complete rupture.

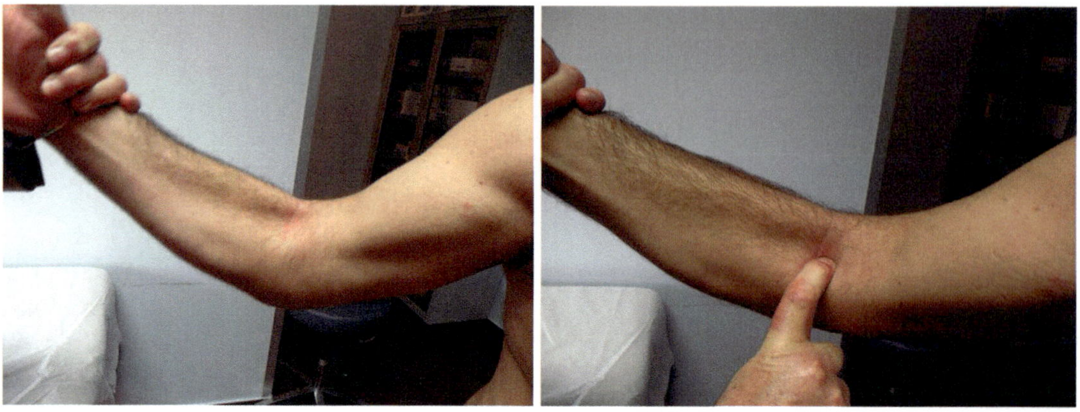

Figs. 17.12 Hook test (the examiner tries to insert his/her index finger under the biceps tendon from lateral to medial and hook it)

With the patient's elbow flexed at 90° and the forearm supinated, the examiner tries to insert his/her index finger under the biceps tendon from lateral to medial and hook it. If the tendon is intact, the fingertip will make some headway under the tendon before meeting with resistance, while inability to hook it reflects complete rupture (Fig. 17.12).

In the biceps squeeze test [51], the examiner squeezes the biceps muscle belly. An intact tendon induces forearm supination just like in Thompson's test for Achilles tendon rupture. The test is positive if squeezing does not result in supination.

Fatigue in forearm supination and inability to turn a screwdriver can be considered as a functional test of biceps rupture. In partial tears, supination against resistance is associated with pain or weakness.

17.2.6 Imaging Studies

Routine X-rays seldom depict abnormalities, except some soft tissue swelling, and they may show some calcific changes in the RT associated with bone irregularity and enlargement. However, they are not particularly helpful in diagnosing distal biceps tendon rupture [52].

Ultrasound can depict tendon ruptures, but it is less reliable than MRI.

An MRI scan shows the tendon in the antecubital fossa as far as its RT insertion [53]. In the case of biceps tendons, longitudinal views—generally employed to examine tendons—are less informative, due to the oblique course of the biceps tendon in the elbow and proximal forearm. Partial tears may manifest themselves with subtle clinical signs; in such cases, FABS view (flexion of the elbow, abduction of the shoulder, supination of the forearm: the patient lies prone with the affected arm above their head, shoulder abducted at 80°, elbow flexed at 90°, and forearm supinated with their thumb pointing upwards) described by Giuffre and Moss [54] provides full-length images of the biceps tendon from the muscle-tendon junction to its RT insertion (Fig. 17.13).

Distal biceps tendon endoscopy involves introducing the arthroscope directly into the bicipital bursa through a skin incision over the distal biceps tendon. Eames and Bain [55] used this technique to assess partial tears or biceps bursitis, which can be managed by endoscopic debridement.

17.2.7 NONOPERATIVE Management

Conservative treatment is usually associated with fatigue and weakness in elbow flexion and forearm supination. Clinical studies [38, 56] demonstrated a 21–30% loss of elbow flexion strength

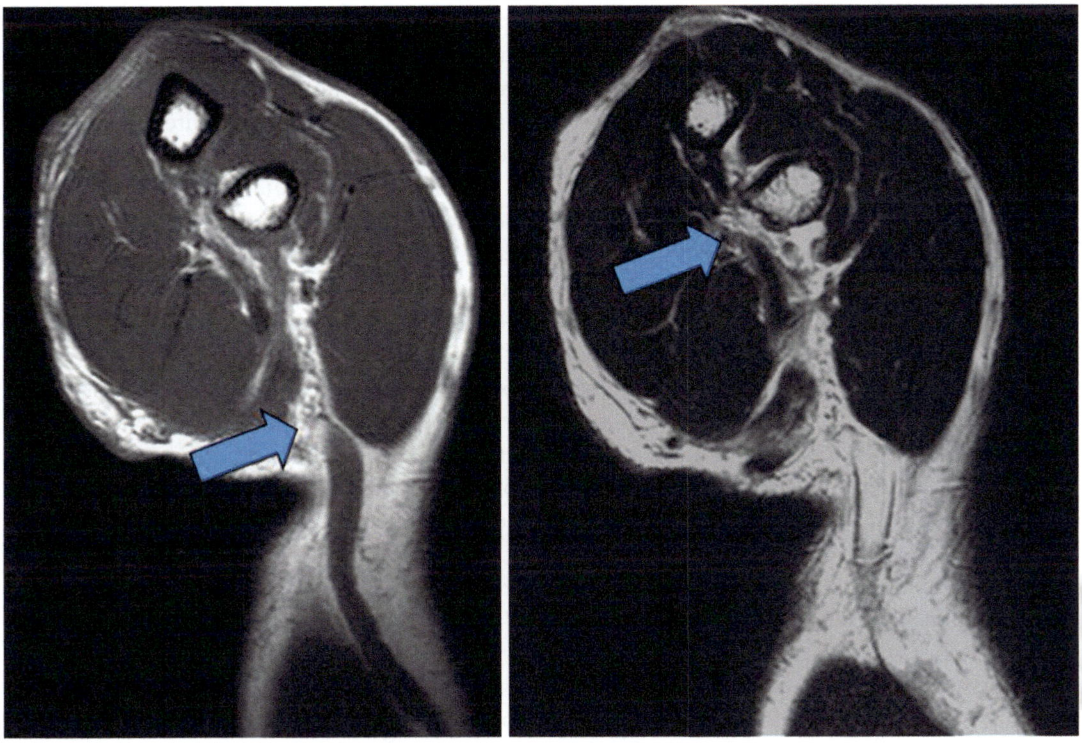

Fig. 17.13 MRI using FABS view (flexion of the elbow, abduction of the shoulder, supination of the forearm) provides full-length images of the biceps tendon from the muscle-tendon junction to its radial tuberosity insertion

and a 27–40% reduction in supination strength compared to the contralateral arm.

17.2.8 Surgical Management

The aim of distal biceps rupture surgical repair in healthy, active patients is restoring elbow flexion strength and, above all, forearm supination strength and endurance.

Current surgical indications and techniques for functional recovery are primary surgical repair or secondary tendon reconstruction with anatomical reinsertion of the ruptured tendon into the RT.

Surgical approach and reconstruction technique depend on rupture site and on time elapsed since rupture (acute or chronic). In relation to aponeurosis, distal biceps tendon injuries are divided [55] into: pre-aponeurosis (muscle-tendon junction); aponeurosis, post-aponeurosis (tendon-bone interface)—the most common lesion.

17.2.8.1 Partial Rupture

Partial rupture of the distal biceps tendon involves an incomplete tear with or without cubital bursitis, i.e., inflammation of the bursa which lies between the tendon and the RT which can be managed by operative and nonoperative approaches.

According to the literature, tendon detachment, debridement, and reattachment provide results which are comparable to acute complete ruptures repairs [57, 58]. Anatomical reattachment has been reported to reduce pain and to enable patients to return to their customary activities [57, 59].

Kelly and O'Driscoll described an approach involving a single posterior incision to repair partial tears of the distal biceps tendon [60, 61]. It is a simple technique entailing low risk of intraoperative complications but moderate risk of radioulnar synostosis or posterior interosseous nerve (PIN) injury, which, however, can be avoided by careful procedure execution.

17.2.8.2 Acute Complete Rupture

Boyd and Anderson [62] and Morrey [63] proposed a two-incision technique, where the RT is exposed through a posterior incision near the ulna. However, the approach is still associated with complications, especially proximal radioulnar synostosis [64] and PIN palsy [65]. Subsequently, a further modification—a muscle-splitting approach—has been introduced to minimize incidence of radioulnar synostosis secondary to periosteal injury [64, 66]. (Figs. 17.14A–F).

In recent years, advanced instrumentation enabling tendon reinsertion with anchors or with an endobutton has been providing significantly shorter anterior incision and consequent reduction of dissection.

Two approaches have been devised to minimize risk of complications related to RT exposure for tendon reinsertion:

- A single anterior incision [67–73],
- A two-incision technique [62–64, 66].

Respective advantages and disadvantages have extensively been discussed, particularly which of the two approaches enables anatomical reattachment to restore normal forearm supination, strength, and endurance.

The main tendon fixation techniques are:

- Transosseous suture.
- Suture anchor.
- Intraosseous screw.
- Suspensory cortical button [74–76].
- Tension-slide technique with a bicortical endobutton and a tenodesis screw.

17.2.8.3 Chronic Complete Rupture

It often results from missed diagnosis or assessment that conservative treatment is the first best option after injury.

Chronic tears pose additional problems compared to acute ruptures, since tissue retraction, atrophy of the muscle-tendon unit, and loss of elasticity hamper or prevent direct tendon reattachment to the RT.

Retraction prevents tendon to bone reattachment with acceptable flexion, which may involve reduced postoperative elbow extension. Retraction largely depends on whether the bicipital aponeurosis is torn or intact.

Surgical treatment is indicated in case of persistent weakness and dysfunction in activities involving repetitive elbow movements, particularly in patients with high functional demand in forearm supination and in those where persistent weakness interferes with daily living activities.

The traditional treatment, reattachment to the brachialis, has been described as a salvage option [77, 78], but it does not restore supination strength and it appears to provide limited improvement in flexion strength compared to anatomical repair [77, 78].

An option to manage chronic complete rupture of the distal biceps tendon with retraction and intact bicipital aponeurosis involves tendon mobilization and reinsertion into the RT.

Some chronic ruptures are amenable to primary reattachment, which can be performed through a single anterior incision or by means of the two-incision technique. However, while these measures may achieve some effect, tendon mobilization for anatomical repair is usually not feasible long after rupture (e.g., 2 months), and augmentation procedure is required.

When the biceps stump cannot be reapproximated to the RT, which requires flexing the elbow to >45°, augmentation procedure with a tendon autograft or allograft is recommended. Various types of grafts have been described:

- Fascia lata [78, 79],
- Semitendinosus tendon [80, 81],
- Flexor carpi radialis (FCR) tendon [82],
- Achilles tendon allograft [71, 72, 83–85].

When tendon augmentation is required, the authors prefer an Achilles tendon allograft with or without a fleck of calcaneal bone. Achilles ten-

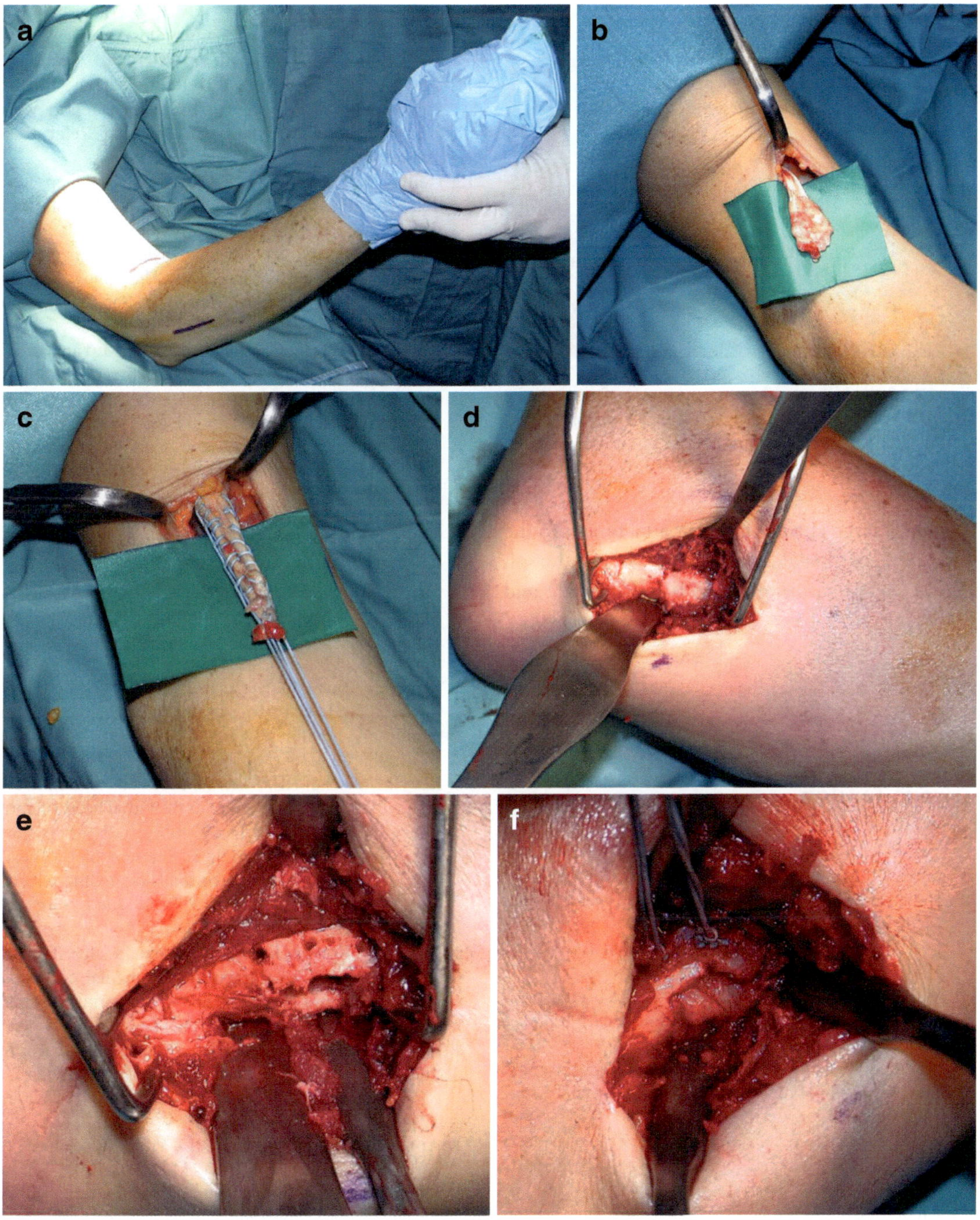

Figs. 17.14 (**a**-**f**) Two-incision surgical technique (Morrey's technique): skin incisions (**a**), distal biceps tendon prepared with double nonabsorbable sutures (**b**, **c**), exposure of the radial tuberosity thought the lateral skin incision and muscle split (**d**), radial tuberosity preparation (**e**), the distal biceps tendon is inserted into the tuberosity (**f**)

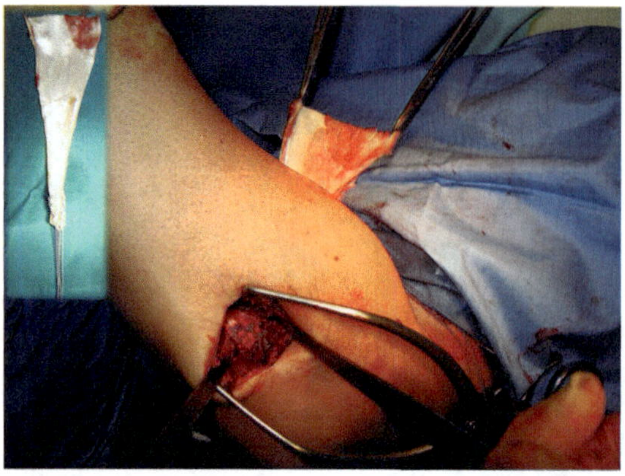

Fig. 17.15 Achilles tendon has a large surface area for attachment to the native biceps muscle and tendon, and its length allows secure suturing to the biceps muscle and the radial tuberosity

don has a large surface area for attachment to the native biceps muscle and tendon, and its length allows secure suturing to the biceps muscle and the RT. (Fig. 17.15).

References

Triceps Tendon

1. Yeh PC, Dodds SD, Smart LR, et al. Distal triceps rupture. J Am Acad Orthop Surg. 2010;18(1):31–40.
2. Halleb P.D., Bach B.R. Triceps Brachii Injuries Sports Med. 1990: 10 /273–276/.
3. Lee ML. Rupture of the triceps tendon Br. Med J. 1960;2(51–93):197.
4. Wagner JR, Cooney WP. Rupture of the triceps muscle at the musculotendinous junction: a case report. J Hand Surg Am Vol. 1997;22:341–3.
5. Anzel SH, Convey KW, Weiner AD, Lipscomb PR. Disruption of muscles and tendons: an analysis of 1014 cases. Surgery. 1959;45:406–14.
6. Mair SD, Isbell WM, Gill TJ, et al. Triceps tendon ruptures in professional football players. Am J Sports Med. 2004;32(2):431–4.
7. Kibuule LK, Fehringer EV. Distal triceps tendon rupture and repair in an otherwise healthy pediatric patient: a case report and review of the literature. J Shoulder Elb Surg. 2007;16(3):e1–3.
8. Strauch RJ. Biceps and triceps injuries of the elbow. Orthop Clin North Am. 1999;30:95–107.
9. Van Riet RP, Morrey BF, Ho E, O'Driscoll SW. Surgical treatment of distal triceps ruptures. J Bone Joint Surg. 2003;85-A(10):1961–7.
10. Sherman OH, Snyder SJ, Fox JM. Triceps tendon avulsion in a professional body builder. A case report. Am J Sports Med. 1984;12(4):328–9.
11. Clayton ML, Thirupathi RG. Rupture of the triceps tendon with olecranon bursitis. A case report with a new method of repair. Clin Orthop Relat Res. 1984;184:183–5.
12. Celli A, Arash A, Adams RA, Morrey BF. Triceps tendon insufficiency following total elbow arthroplasty. J Bone Jt Surg. Am. 2005;87:1957–4964.
13. Celli A., Morrey BF. Triceps insufficiency following total elbow arthroplasty. In: The Elbow and Its Disorders 4° Edition Morrey B.F- Sanchez Sotelo J. Ed. 2009 /873–879.
14. Keener JD, Chafik D, Kim M, Galatz LM, Yamaguchi K. Insertional anatomy of the triceps brachii tendon. J Shoulder Elb Surg. 2010;19:399–405.
15. Athwal GS, McGill RJ, Rispoli DM. Isolated avulsion of the medial head of the triceps tendon: an anatomic study and arthroscopic repair in 2 cases. J Arthrosc Relat Surg. 2009;25(9):983–8.
16. Aso K, Torisu T. Muscle belly tear of the triceps. Am J SportsMed. 1984;12:485–7.
17. Bach BR Jr, Warren RF, Wickiewicz TL. Triceps rupture. A case report and literature review. Am J Sports Med. 1987;15(3):285–9.
18. O'Driscoll SW. Intramuscular triceps rupture. Can J Surg. 1992;35:203–7.
19. Khiami F, Tavassoli S, De Ridder Bauer L, Catonne' Y, Sariali E. Distal partial ruptures of triceps Brachii tendon in an athlete. Orthop Traumatol Surg Res 2012; 98:242–246.
20. Levy M, Fischel RE, Stern GM. Triceps tendon avulsion with or without fracture of the radial head – a rare injury J. Trauma. 1978;18:677–9.
21. Levy M, Goldberg I, Meir I. Fracture of the head of the radius with a tear or avulsion of the triceps tendon. A new syndrome? J Bone Joint Surg Br. 1982;64(1):70–2.
22. Yoon MY, Koris MJ, Ortiz JA, Papandrea RF. Triceps avulsion, radial head fracture, and medial collateral ligament rupture about the elbow: a report of 4 cases. J Shoulder Elb Surg. 2012;21:12–7.
23. Huxley AF, Niedergerke R. Structural changes in muscle during contraction: interference microscopy of living muscle fibers. Nature. 1954;173: 971–3.
24. Apple DV, O'Toole J, Annis C. Professional basketball injuries. Physician Sports Med. 1982;10:81–6.
25. Inhofe PD, Manein MS. Late presentation of triceps rupture: a case report and review of literature. Am J Orthop. 1996;11:290–2.
26. Bos CF, Nelissen RG, Bloem JL. Incomplete rupture of the tendon of the triceps brachii. A case report. Int Orthop. 1994;18:273–5.
27. Pina A, Garcia I, Sabater M. Traumatic avulsion of the triceps brachii. J Orthop Trauma. 2002;16:273–6.
28. Viegas SF. Avulsion of the triceps tendon. Orthop Rev. 1990;19(6):533–6.

29. Celli A. Triceps tendon ruptures. Volume 3 Shoulder and elbow part 7 in Operative techniques in orthopaedic surgery Ed Sam W. Wiesel Wolters Kluwer. Chapter 38; 2015: 3917–3933.
30. Kaempffe FA, Lerner RM. Ultrasound diagnosis of tricep tendon rupture: A report of 2 cases. Clin Orthop Relat Res. 1996;332:138–42.
31. Gaines ST, Durbin RA, Marsalka DS. The use of magnetic resonance imaging in the diagnosis of triceps tendon ruptures. Contemp Orthop. 1990;20:607–11.
32. Wenzke DR. MR imaging of the elbow in the injured Athlete. Radial Clin N Am. 2013;51:195–213.
33. Kijowski R, Tuite M, Sanford M. Magnetic resonance imaging of the elbow. Part II: abnormalities of the ligament, tendons and nerves. Skeletal Radial. 2003;34:1–18.
34. Morrey BF. Rupture of the triceps tendon. In: Morrey BF, editor. The elbow and its disorders. 3rd ed. Philadelphia: WB Saunders; 2000. p. 479–548.
35. Morrey BF. Open treatment of acute and chronic triceps tendon ruptures. In: Yamaguchi K, et al., editors. Advanced reconstruction elbow. Rosement, IL: American Academy of Orthopaedic Surgeons; 2007. p. 1007–13.
36. Morrey BF. Rupture of the triceps tendon. In: Morrey BF, Sanchez-Sotelo J, editors. The elbow and its disorders. New York: Saunders Elsevier; 2009. p. 536–46.

Distal Biceps Tendon

37. Gilcreest EL. The common syndrome of rupture, dislocation and elongation of the long head of the biceps brachii. An analysis of one hundred cases. Surg Gynaecol Obstest 1934: 58/322–40.
38. Morrey BF, Askew LJ, An KN, Dobyns JH. Rupture of the distal tendon of the biceps brachii: A biomechanical study. J Bone Joint Surg Am. 1985:67/418–21.
39. Safran MR, Graham SM. Distal biceps tendon ruptures: incidence demographics and the effect of smoking. Clin Orthop Relat Res. 2002:404/275–83.
40. Chavon PR, Duquin TR, Bisson LJ. Repair of the ruptured distal biceps tendon a systematic review. Am J Sport Med. 2008:36/1618–24.
41. Bauman J.T., Sotereanos D.G., Weiser R.W. Complete rupture of the distal biceps tendon in a women. Case report. J Hand Surg Am. 2006: 31/798–800.
42. Davis W.M., Yassine Z. An etiological factor in tear of the distal biceps brachii. A report of two cases. J Bone Joint Surg. 1956: 38A/1365–1368.
43. Athwal GS, Steinnann SP, Rispoli DM. The distal biceps tendon: foot-print and relevant clinical anatomy. J Hand Surg Am. 2007:32/1225–9.
44. Hutchinson HL, Gloystein D, Gillespil M. Distal biceps tendon insertion: An anatomic study. J Shoulder Elb Surg. 2008:17/342–6.
45. Eames M, Bain GI, Fogg QA, Van Riet RP. Distal biceps tendon anatomy: a cadaveric study. J Bone Joint Surg. 2007: 89A/1044.
46. Mazzocca AD, Cohen M, Berkson E, et al. The anatomy of the bicipital tuberosity and distal biceps tendon. J Shoulder Elb Surg. 2007:16/122–7.
47. Schmidt CC, Jarrett CD, Brown BT. The distal biceps tendon. J Hand Surg Am. 2013:38/811–21.
48. Schmidt CC, Savoie FH III, Steinmann SP, Hausman M, Voloshin I, Morrey BF, Sotereanos DG, Bero EH, Brown BT. Distal biceps tendon history, updates and controversies: from the closed American Shoulder and Elbow Surgeons Meeting – 2015. J Shoulder Elb Surg. 2016;25:1717–30.
49. Bourne MH, Morrey BF. Partial ruptures of the distal biceps tendon. Clin Orthop. 1991:271/143–8.
50. O'Driscoll SW, Goncalves LB, Dietz P. The hook test for distal biceps avulsion. Am J Sports Med. 2007;35:1865–9.
51. Rutland RT, Dunbar RP, Bowen JD. The biceps squeeze test for diagnosis of distal biceps tendon ruptures. Clin Orthop Relat Res. 2005;437:128–31.
52. Chew ML, Giuffre BM. Disorders of the distal biceps brachii tendon. Radiographics. 2005;25(5):1227–37.
53. Le Huec J.C., Moinard M., Liquois F., Zipoli B., Chaveaux D., Lebeller A. Distal rupture of the tendon of biceps brachii: evaluation by MRI and the results of repair. J Bone Joint Surg. 1996:78b: 767–70.
54. Giuffre BM, Moss MJ. Optimal positioning for MRI of the distal biceps brachii tendon: flexed abducted supinated view. Asr A J Roèntgenal. 2004;182(4):944–6.
55. Eames MH, Bain GI. Distal biceps tendon endoscopy and anterior elbow arthroscopy portal. Techniques Shoulder Elbow Surg. 2006;7(3):139–42.
56. Baker BE, Bierwagon D. Rupture of the distal tendon of the biceps brachii: Operative versus non operative treatment. J Bone Joint Surg Am. 1985;67(3): 414–7.
57. Becks RB, Claessen F, Oh LS, Ring D, Chen NL. Factors associated with adverse events after distal biceps tendon repair or reconstruction. J Shoulder Elb Surg. 2016:25/1229–34.
58. Bain GL, Johnson LJ, Turner PC. Treatment of partial distal biceps tendon tears. Sports Med Arthroscop. 2008:16/154–61.
59. Rokito AS, McLaughlin MD, Gallagher MA, Zuckerman JD. Partial rupture of the distal biceps tendon. J Shoulder Elb Surg. 1996:5/73–5.
60. Kelly EW, O'Driscoll SW. Exploration and repair of partial distal biceps tendon ruptures through a single posterior incision. Tech Shoulder Elb Surg. 2002;3:68–72.
61. Kelly EW, Steinman S, O'Driscoll SW. Surgical treatment of partial distal biceps tendon ruptures through a single posterior incision. J Shoulder Elb Surg. 2003:12/456–61.
62. Boyd HB, Anderson LD. A method for reinsertion of the distal biceps brachii tendon. J Bone Joint Surgery AM. 1961;43:1041–3.
63. Morrey BF, Askewl J, An KN, et al. Rupture of the distal biceps tendon: biomechanical assessment of different treatment options. J Bone Jt Surg. 1985;67A:418.

64. Failla JM, Amadio PC, Morrey BF, et al. Proximal radioulnar synostosis after repair of distal biceps brachii rupture by two-incision technique. Report of four cases. Clin Orthop Relat Res. 1990;253:133–6.
65. Stearns KL, Sarris I, Sotereanos DG. Permanent posterior interosseus nerve palsy following a two-incision distal biceps tendon repair. Orthopedics. 2004;27:867–8.
66. Kelly EW, Morrey BF, O'Driscoll SW. Complications of repair of the distal biceps tendon with the modified two-incision technique. J Bone Joint Surg Am. 2000:82/1575–81.
67. Galatz CM, Jani M, Yamaguchi K. Single anterior incision exposure for distal biceps tendon repair. Techniques in shoulder and elbow surgery. 2002;3(1):63–7.
68. Lintner S, Fischer T. Repair of the distal biceps tendon using suture anchors and an anterior approach. Clin Orthop. 1996;322:116–9.
69. Jon CK, Field LD, Weiss KS, Savoie FH III. Single incision repair of acute distal biceps ruptures by use of suture anchors. J Shoulder Elb Surg. 2007: 16/78–83.
70. Khan AD, Penna S, Yin Q, Sinopidis C, Brownson P, Frostick SP. Repair of distal biceps tendon ruptures using suture anchors trough a single anterior incision. Arthroscopiy. 2008:24/39–45.
71. Hughes JS, Morrey BF. Injury of the flexors of the elbow: Biceps tendon injury. In: Morrey BF, Sanchez-Sotelo J, editors. The elbow and its disorders. 4th ed. New York: Sanders-Elsevier; 2009. p. 518–35.
72. Morrey BF. Distal biceps tendon rupture in the elbow: master techniques. In: Morrey BF (eds) Orthopaedic surgery. Second Edition. New York: Sanders-Elsevier; 2002. p. 173–191.
73. Goljan P, Patel N, Stull JD, Donnelly BP, Culp RW. Single incision distal biceps repair with Hemi-Krackow suture technique and early outcomes. Hand. 2016;11(2):238–44.
74. Bain GI, Prem H, Heptinstall RJ, Verhellen R, Paix D. Repairs of distal biceps tendon rupture: a new technique using the endo button. J Shoulder Elb Surg. 2000;9:120.126.
75. Greenberg JA, Fernandez JJ, Wang T, Turner C. Endobutton—assisted repair of distal biceps tendon ruptures. J Shoulder Elb Surg. 2003:484–90.
76. Kodda IF, Van Den Bekeron MP, Eygendaal D. Reconstruction of distal biceps tendon ruptures with a cortical button. Knee Surg Sports Traumatol Arthroscop. 2013:25/580–5.
77. Hovelins L, Josefsson G. Rupture of the distal bicep tendon: report of five cases. Acta Orthop Scand. 1977:48/280–2.
78. Klonz A, Loitz D, Wohler P, Reilmann H. Rupture of distal biceps brachii tendon isokinetic power analysis and complications after anatomic reinsertion compared with fixation to the brachialis muscle. J Shoulder Elb Surg. 2003:12/607–11.
79. Kaplan FT, Rokito AS, Birdzell MG, Zuckerman JD. Reconstruction of chronic distal biceps tendon rupture with use of fascia lata combined with a ligament augmentation device: a report of 3 cases. J Shoulder Elb Surg. 2002:11/633–6.
80. Hang DW, Bach BR, Bojchuk J. Repair of chronic ruptures of the distal biceps brachii tendon rupture using free autogenous semitendinosus tendon. Clin Orthop. 1996:323/188–91.
81. Wiley WB, Noble JS, Dulaney TD, Bell RH, Noble DD. Late reconstruction of chronic distal biceps tendon ruptures with a semitendinosus autograft technique. J Shoulder Elb Surg. 2006:15/440–4.
82. Levy HJ, Mashoof AA, Morgan D. Repair of chronic ruptures of the distal biceps tendon using flexor carpi radialis tendon graft. Am J Sports Med. 2000:28/538–40.
83. Sanchez-Sotelo J, Morrey BF, Adams RN, O'Driscoll SW. Reconstruction of chronic ruptures of the distal biceps tendon with use of an Achilles tendon allograft. J Bone Joint Surgery. 2002:84A/999–1005.
84. Sanchez-Sotelo J. Reconstruction of chronic distal biceps tendon ruptures. In: Yamaguchi K, King GJW, McKee MD, O'Driscoll SW (eds) Advanced reconstruction elbow. Rosement, IL: AAOS 2007:167–172.
85. Darlis NA, Sotereanos DG. Distal biceps tendon reconstruction in chronic ruptures. J Shoulder Elb Surg. 2006:5/614–9.

Part V

Hip Biomechanics

Biomechanics of Hip Function

18

Kyle R. Sochacki and Marc R. Safran

18.1 Anatomy

18.1.1 Bony Anatomy

The acetabulum is composed of the triradiate cartilage and the acetabular cartilage complex formed by the fusion of the ilium, ischium, and pubis [1]. The triradiate cartilage will form the non-articular medial wall, and the acetabular cartilage complex, composed mainly of hyaline cartilage, will form the articular portion of the acetabulum [1]. Physeal growth occurs through the triradiate cartilage with appropriate height and depth of the socket developing in response to the presence of the femoral head [2]. Acetabular maturation continues until the fusion of the triradiate cartilage, usually from 16 to 18 years of age [2]. The normal acetabulum is anteverted 15–20° with a mean depth and diameter of 29.49 ± 4.2 millimeters (mm) and 54.29 ± 3.8 mm, respectively [3]. A deep acetabulum (profunda or protrusio) may result in pincer type impingement, while failure of the secondary ossification centers to develop will result in a shallow socket known as hip dysplasia [2, 4, 5].

The femoral head develops simultaneously with the acetabulum with growth occurring through the longitudinal (between the femoral head and the neck), trochanteric (between the femoral neck and the greater trochanter), and femoral neck isthmus physes [1, 2]. The neck shaft angle is the angle measured between the axis of the femoral neck and the femoral shaft. The angle is highest at birth, but decreases to an average adult value of 125 ± 5° [1, 6]. Femoral neck version is the angle between the femoral neck and the axis that crosses the distal femur epicondyles with the normal amount of adult anteversion ranging from 12 to 14° for a mean of 15.4° [1, 3, 7, 8]. The mechanical advantage of the gluteus maximus muscle increases while the hip abductor mechanical advantage decreases with increasing proximal femur anteversion [9].

The femoral head-neck junction is normally shaped with the femoral neck narrower than its head. The head-neck junction morphology can be quantified by the anterior offset or the alpha angle [10–13]. The offset is measured as the ratio between the femoral head and neck radii, or as an absolute distance, which is normally about 10 mm [11, 14]. The alpha angle is a method to quantify the concavity at femoral head-neck junction with normal alpha angle values less than 50 or 55°. The two bony prominences on the superior-lateral and posterior-medial aspects of the proximal femur are named the greater and lesser trochanters, respectively, and serve as the insertion sites for a variety of muscles that contribute to hip motion.

K. R. Sochacki · M. R. Safran (✉)
Stanford University, Redwood City, CA, USA
e-mail: msafran@stanford.edu

J. Koh et al. (eds.), *Orthopaedic Biomechanics in Sports Medicine*,
https://doi.org/10.1007/978-3-030-81549-3_18

18.1.2 The Hip Joint Capsule

The hip joint is surrounded by a thick fibrous capsule that extends from the acetabular rim to the proximal femur attaching anteriorly to the intertrochanteric line, superiorly to the base of the femoral neck, superomedially to the intertrochanteric crest, and inferiorly to the femoral neck near the lesser trochanter [15, 16]. Three reinforcing ligaments are confluent with the hip capsule that help to provide hip stability. These include the iliofemoral ligament (Y ligament of Bigelow), pubofemoral ligament, and ischiofemoral ligament (Fig. 18.1). The ligaments serve as the primary hip rotational restraints depending on the position and rotation of the hip during specific motions [17, 18]. The iliofemoral ligament originates from the anterior inferior iliac spine (AIIS) and splits distally into two distinct arms with the lateral arm inserting on the anterior prominence of the greater trochanteric crest and the medial arm inserting on a subtle angulated prominence of the anterior-inferior femur at the level of the lesser trochanter. This acts to restrict external rotation in all hip positions and both internal and external rotation with the hip in extension. The medial arm was most dominant in hip neutral flexion or extension while the lateral arm dominated in all other hip positions [17, 19, 20]. The pubofemoral ligament attaches to the superior pubic ramus proximally and then blends with the ischiofemoral and iliofemoral ligaments distally as it lacks a true bony femoral attachment [15]. It was previously believed to control external rotation in extension; however, a recent study by Martin et al. demonstrated the key function of the pubofemoral ligament as limiting internal rotation with increasing hip flexion [20, 21]. The ischiofemoral ligament extends from the ischium to the femoral neck-trochanteric junction acting as a primary restrictor of internal rotation in both high and low hip flexion [17, 20]. The iliofemoral ligament is the strongest of these ligaments while the posterior ischiofemoral ligament is the thinnest and weakest [22]. Several studies have demonstrated the important role of the hip capsule in providing stability to the joint synergistically with other static soft tissue stabilizers such as the acetabular labrum and transverse acetabular ligament throughout physiologic and supraphysiologic range of motion [23–26].

18.1.3 Intra-Articular Structures

The acetabular labrum is a horseshoe shaped structure attached to the acetabular rim that lies just deep to the hip capsule. The capsular side of the labrum is composed of mainly type I and III collagen, while the articular side is composed of fibrocartilage [27]. The labrum functions to deepen the acetabulum with labral size inversely proportional to the depth of the bony acetabular socket. It also acts to increase hip stability by increasing the acetabular surface area and volume while creating a suction seal that opposes the flow of synovial fluid in and out of the central compartment [28, 29]. Weight bearing activities with a functioning labral seal leads to an increase in intra-articular pressure that reduces intra-articular friction by improving joint lubrication [30, 31]. Additionally, Safran et al. demonstrated that the labrum has strain at rest, which can increase and decrease in different locations of the labrum depending on hip position [32].

The ligamentum teres is composed of well-organized collagen bundles that attaches the femoral head to the acetabulum. The biomechanical role of the ligamentum teres has been a widely debated topic with proposed functions including hip stabilizer, fluid and force distributor of the acetabulum, and embryonic remnant without a specific function in adults [33–36]. However, recent studies have shown the ligamentum teres to be taut in flexion, adduction, and external rotation, leading some authors to believe that it does provide some degree of hip stability, resisting dislocation, and micro-instability [33, 37–39].

Hip articular cartilage has been shown to be highly inhomogeneous in thickness distribution on both the acetabular and femoral sides. A study by Von Eisenhart et al. demonstrated that maximum cartilage thickness was found ventro-superiorly in the acetabulum and in the femoral head with a maximal thickness that ranged from 2.6 to 4.3 mm in the acetabulum and from 2.4 to

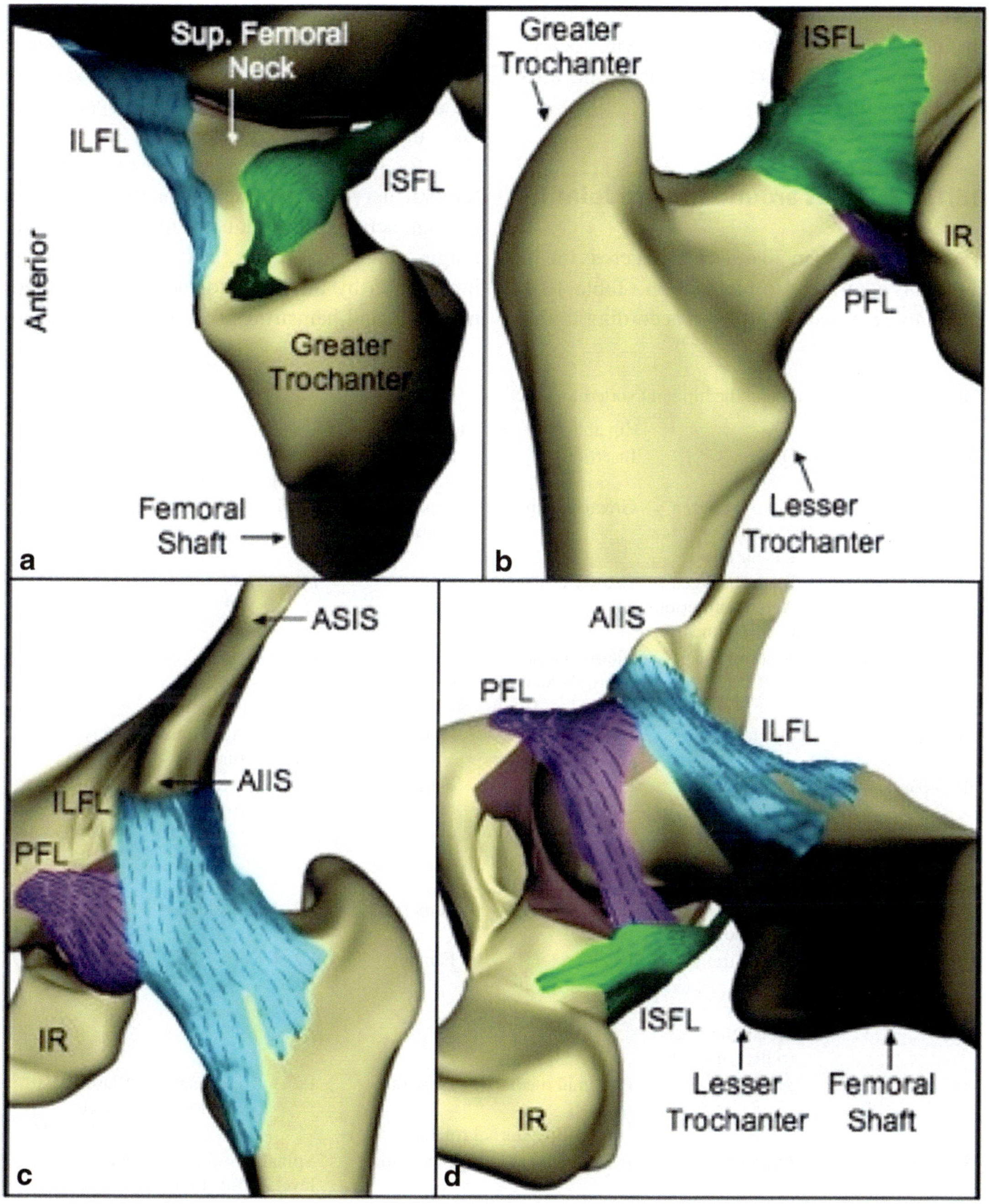

Fig. 18.1 Ligamentous relationships of the hip capsule. Computer model (**a**) demonstrating the relationship of the distal lateral iliofemoral ligament (ILFL) and the distal ischiofemoral ligament (ISFL), viewed from a superior position looking down the femoral shaft. Computer model showing (**b**) the posterior blend of the pubofemoral ligament (PFL) and ISFL, and (**c**) the anterior blend of the PFL and ILFL. Fig. (**d**) demonstrates the relationships of all three ligaments as viewed from an inferior position looking upwards at the inferior aspect of the femoral head. Figure used with permission from the senior author. (ASIS, anterior superior ischial spine; AIIS, anterior inferior ischial spine; IR, ischial ramus). Reprinted from DeLee Drez and Miller's Orthopaedic Sports Medicine, Fifth Edition, Mark D. Miller, MD and Stephen R. Thompson, MD, MEd, FRCSC, Hip Anatomy and Biomechanics, 907–924, 2020, with permission from Elsevier

5.3 mm in the femoral head [40]. This location of maximum thickness corresponded with location of maximum pressure recorded during the walking cycle.

18.1.4 Muscles around the Hip Joint

There are more than 20 muscles that cross the hip joint and contribute to hip motion (Table 18.1). These muscles can be grouped according to their main function and innervation. The hip abductors and internal rotators (gluteus medius, gluteus minimus, and tensor fascia lata) are innervated by the superior gluteal nerve [6]. The iliopsoas, rectus femoris, sartorius, and pectineus are responsible for hip flexion and are innervated by the femoral nerve [6]. Adduction of the hip occurs through action of the adductor magnus, adductor longus, adductor brevis, and gracilis through innervation by the obturator nerve. The gluteus maximus and hamstrings (biceps femoris, semi-

Table 18.1 Muscles around the hip joint sorted by function

Hip abductors and internal rotators						
Muscle	Origin	Insertion	Innervation	Spinal Level	Additional Function	Notes
Gluteus Medius	Ilium (between posterior-anterior gluteal lines)	Greater trochanter	Superior gluteal	L4-S1		
Gluteus Minimus	Ilium (between Anterior/Inferior Gluteal Lines)	Greater trochanter	Superior gluteal	L4-S1		
Tensor fasciae Latae	Anterior iliac crest	Iliotibial band (Gerdy's tubercle)	Superior gluteal	L4-S1		
Hip flexors						
Muscle	Origin	Insertion	Innervation	Spinal level	Additional function	Notes
Iliopsoas	Iliac Fossa (iliacus) Transverse process L1-L5 (psoas)	Lesser trochanter	Femoral	L2–4	External rotation	Strongest hip flexor
Pectineus	Pectineal line of pubis bone	Pectineal line of femur	Femoral	L2–4	Adduction	
Rectus Femoris	AIIS (straight head) Anterior acetabular rim (reflected head)	Patella	Femoral	L2–4		Biarthrodial muscle
Sartorius	ASIS	Proximal medial tibia	Femoral	L2–4	External rotation	Biarthrodial muscle
Hip external rotators						
Muscle	Origin	Insertion	Innervation	Spinal level	Additional function	Notes
Gluteus Maximus	Ilium along crest posterior to posterior gluteal line	Iliotibial band/ posterior femur	Inferior gluteal	L5-S2	Extension	
Piriformis	Anterior sacrum, through sciatic notch	Proximal greater trochanter (piriformis Fossa)	Piriformis	S1-S2		

Table 18.1 (continued)

Obturator externus	Ischiopubic rami and external surface of the obturator membrane	Medial greater trochanter	Obturator (posterior branch)	L2–4	Adduction	
Obturator internus	Ischiopubic rami/ obturator membrane	Medial greater trochanter	Obturator internus	L5-S2		
Superior Gemellus	Ischial spine	Medial greater trochanter	Obturator internus	L5-S2		
Inferior Gemellus	Ischial tuberosity	Medial greater trochanter	Quadratus Femoris	L4-S1		
Quadratus Femoris	Ischial tuberosity	Quadrate line of femur	Quadratus Femoris	L4-S1		
			Hip extensors			
Muscle	Origin	Insertion	Innervation	Spinal level	Additional function	Notes
Gluteus Maximus	Ilium along crest posterior to posterior gluteal line	IT band/posterior femur	Inferior gluteal	L5-S2	External rotation	
Long head of biceps Femoris	Medial ischial tuberosity	Fibular head/lateral tibia	Tibial	L5-S2		Biarthrodial muscle, also flex the knee
Semitendinosus	Distal medial ischial tuberosity	Anterior Tibial Crest	Tibial	L5-S2		Biarthrodial muscle, also flex the knee
Semimembranosus	Proximal lateral ischial tuberosity	Posterior/medial tibia, posterior capsule, medial meniscus, popliteus, popliteal ligament	Tibial	L5-S2		Biarthrodial muscle, also flex the knee
			Hip adductors			
Muscle	Origin	Insertion	Nerve supply	Spinal level	Additional function	Notes
Adductor Magnus	Inferior pubic ramus/ischial tuberosity	Linea Aspera/ adductor tubercle	Obturator (posterior branch) sciatic (Tibial)	L2–4	Flexion External rotation Extension	
Adductor brevis	Inferior pubic ramus	Linea Aspera/ pectineal line	Obturator (posterior branch)	L2–4	Flexion External rotation	
Adductor longus	Anterior pubic ramus	Linea Aspera	Obturator (anterior branch)	L2–4	Flexion Internal rotation	
Gracilis	Inferior symphysis/pubic arch	Proximal medial tibia	Obturator (anterior branch)	L2–4	Flexion Internal rotation	Biarthrodial muscle, also flex the knee

membranosus, and semitendinosus) function as the hip extensors with innervation by the inferior gluteal nerve and tibial branch of the sciatic nerve, respectively [6].

18.2 Hip Motion

Although the femoral head moves relative to the acetabulum, the hip functions as a ball-in-socket joint [41, 42]. The acetabulum acts as the socket with the femoral head serving as the ball. As such, it has six degrees of freedom with three planes of motion (flexion-extension, abduction-adduction, and internal-external rotation) and three of translation (anterior-posterior, medial-lateral, and proximal-distal).

Hip range of motion is generally greatest in the sagittal plane. However, this may be affected by the bony morphology and/or laxity of ligaments and muscles around the hip. Knee position can also have a significant impact on hip motion as several muscles are biarthrodial, crossing both the hip and knee joints. As such, hip flexion ranges from 120° to 140° with the knee flexed and 90° with the knee fully extended due to increased tension across the hamstring muscles [9, 43]. Normal hip extension ranges from 10° to 30°, but is limited by the iliofemoral ligament, anterior capsule, and hip flexors [9, 43]. Normal hip abduction and adduction is at least 50° and 30°, respectively. Internal and external rotation of the hip is dependent upon the degree of hip motion in the sagittal plane (flexion or extension) with considerably less internal and external rotation possible with the hip extended due to increased soft tissue tension. Internal rotation is limited by the short external rotator muscles (obturator internus and externus, superior and inferior gemelli, quadratus femoris, and piriformis) and the ischiofemoral ligament [44]. External rotation is limited by the lateral band of the iliofemoral ligament, the pubofemoral ligament, the internal rotator muscles, and the degree of femoral neck anteversion [45]. During hip flexion, hip internal rotation ranges from 0° to 70° and external rotation can range from 0° to 90° [46].

Abnormalities of the proximal femur or acetabular morphology in conditions such as femoroacetabular impingement can also lead to reductions in hip range of motion. It is often limited because of abnormal bony contact between the proximal femur and acetabulum at the extremes [47]. However, surgical correction of these bony abnormalities has been shown to reduce this impingement and lead to improved hip range of motion similar to normal values [48, 49].

It is important to note that hip motions are the result of combined hip joint, pelvic, and lumbar spine motion with increased contribution from the pelvis and lumbar spine when there is bony impingement at the hip [50]. A previous study by Dewberry et al. determined that 26–39% of hip flexion comes from lumbopelvic rotation depending on whether the knees were flexed or extended, respectively [43]. Pelvic rotation has also been found to contribute 18% of hip flexion during weight-bearing [51]. Additionally, lumbar spine motion significantly contributes to hip flexion with the majority of the contribution occurring early in the forward bending process [52]. This intimate connection between the hip, pelvis, and spine may lead to "hip-spine syndrome" as stress can be transferred from the hip to the spine and vice versa when there is abnormal sagittal balance, bony morphology, or irregular gait [53].

18.3 Gait Cycle

The human bipedal gait cycle consists of the stance and swing phases as measured from heel strike to heel strike. The stance phase is defined by the period of time that the foot is on the ground. During walking, the stance phase accounts for about 60% of the gait cycle with both feet on the ground (double-support) occurring for approximately 20% of the time [54]. This double-support phase defines walking. It is eliminated with running and replaced by the addition of the float phase, in which both legs are in the air at the same time [54–56]. As the velocity of running increases, the stance phase shortens to less than 22% of the cycle at maximum velocity [54].

Hip motion during gait is dependent upon the different phases of the gait cycle with the main motion of the hip occurring in the sagittal plane (flexion and extension). The extent to which the hip flexes and extends is dependent upon the rate of ambulation as it increases from walking to jogging to running. The hip is extended, adducted, and internally rotated in the stance phase while it is flexed, abducted, and externally rotated in the swing phase of the gait cycle [44, 45, 56]. During normal walking, the hip flexes to about 30° and extends to around 10° [44, 45]. Hip flexion increases with running and sprinting to approximately 50° and 65°, respectively [54]. Extension of the hip has also been found to increase with running but paradoxically decreases with sprinting [54, 57].

The amount of hip abduction and adduction also differs between walking, jogging, and running with maximum values of both occurring with running. Hip adduction is 5–10° while walking and increases to 15–20° during running just before heel strike [44, 45]. Maximum abduction, on the other hand, occurs after toe-off during the swing phase of running.

The muscles around the hip joint work in conjunction with each other during the gait cycle. Hip flexors are most active during the swing phase, while extensors increase activity during the stance phase. In terminal swing, however, the gluteus maximus and the hamstrings also function to decelerate the swinging thigh [54, 58]. Hip adductor muscles are active throughout all phases while running, but only activate from the swing phase to mid-stance when walking [56]. The gluteus medius and tensor fascia lata also help to stabilize the pelvis in normal gait. During running, they are active in the swing and early stance phases. However, while walking, they are mainly active during the stance phase only [58].

18.4 Forces around the hip

The human hip is biomechanically complicated with several forces contributing to the joint reaction forces across the hip. Direct measurement can be difficult. As such, free body diagrams have been developed to estimate these forces based on several assumptions regarding the soft tissue structures around the hip and their individual contributions (Fig. 18.2) [44]. The most commonly used free body diagram makes estimates using single limb stance [46, 59, 60]. Under static conditions, the gravitational force, force from the abductor muscles (A), and force exerted by the femoral head on the acetabulum (F, joint reaction force) act on the hip to keep the hip level [60]. The gravitational force is the weight of the body (W) minus the weight of the contralateral lower limb (1/6 W) or 5/6 W. It is possible to determine the joint reaction force (F) after the force from the abductor muscles (A) is calculated. This can be calculated with knowledge of a person's weight, moment arm of the gravitational force (*d*), and moment arm of the abductor muscles (*l*) using the following equation [46]:

$$A = \frac{5/6W \times d}{l}$$

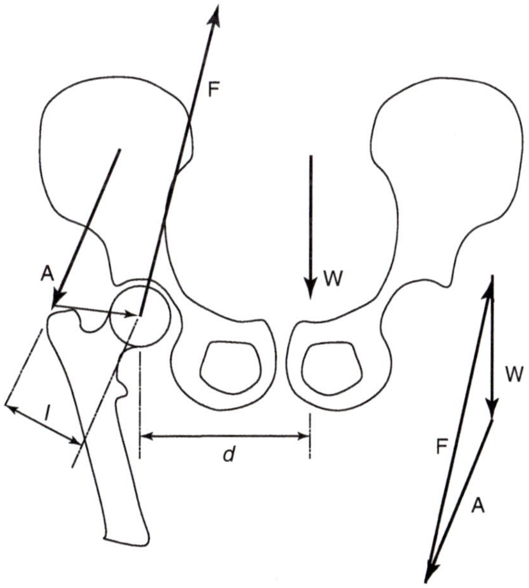

Fig. 18.2 Forces acting on the hip joint during single leg stance under conditions of equilibrium. Gravitational force W, abductor muscle force A, hip joint reaction force F, abductor muscle moment arm *l*, and force of gravity moment arm *d*. Reprinted from DeLee Drez and Miller's Orthopaedic Sports Medicine, Fifth Edition, Mark D. Miller, MD and Stephen R. Thompson, MD, MEd, FRCSC, Hip Anatomy and Biomechanics, 907–924, 2020, with permission from Elsevier

At equilibrium, according to Newton's first law of motion, the joint reaction force (F) is equal to the sum of the gravitational force and force from the abductors. As such, F is calculated to be 2.7 times the body weight during the single leg stance phase of walking with the pelvis parallel to the floor [46].

These principles can also be applied to estimate the forces exhibited on a hip joint in motion, which may be more applicable and of greater interest in the treatment of athletes. Previous studies using kinetic and kinematic data have estimated that the hip is loaded up to 4 times body weight while slow walking [61]. The forces seen by the hip increase as the speed of gait increases with forces 7–8 times body weight transmitted to the hip during running [61]. Unfortunately, these estimates are based on straight line motion. As such, sports with cutting, pivoting, and twisting motions would be expected to have even higher forces across the hip. This has been demonstrated in mogul skiers with as much as 12.4 times body weight seen by the hip joint [62].

Direct measurements of these activities can be difficult in athletes due to the requirement for surgical implantation of a force transducer. In order to account for this, Bergmann et al. was able to determine the forces acting on the hip during several activities after total hip arthroplasty by implanting pressure transducers at the time of surgery [63]. The results of the study confirmed previous estimates with increasing hip joint forces as gait speed increases. The forces transmitted to the hip were 300% body weight during slow walking, 350% to 400% with quick walking, and up to 500% during jogging [63]. This in vivo measurement has the advantage of including all the forces acting on the hip during activities, but is likely dependent upon hip implant position to a certain degree [64].

18.5 Hip Joint Surface Pressure

Focal increases in hip articular cartilage pressure has been thought to contribute to the development of hip osteoarthritis. As such, understanding of this mechanism and the pathology that contributes is paramount for the sports medicine surgeon and hip preservation specialist. Joint pressure (P) can be estimated based on the joint reaction force (F) and surface area (A) over which this force is distributed using the following equation [60]:

$$P = F / A.$$

Based on the above calculations, the average pressure seen by the hip joint is 75 N/cm^2 assuming a body mass of 60 kilograms, femoral head diameter of 5 centimeters (cm), and joint reaction force of 1500 Newtons (N) [60, 65].

However, the pressure distribution across the hip articular cartilage is not uniform due to the bearing surface lacking perfect sphericity. In a native hip, the femoral head is slightly out-of-round and the acetabulum is a horseshoe shaped, nonuniform hemisphere. This leads to low pressures at the most constrained aspects of the hip and increasing pressures on the articular cartilage near the uncovered rim of the acetabulum [60]. This principle has been further reinforced by Greenwald and Brinckmann. Decreased acetabular coverage, as in hip dysplasia, leads to significantly greater and more laterally (along the acetabular rim) based pressure on the articular cartilage compared to normal hips [60, 66].

In addition to the acetabular bony morphology, cartilage and soft tissue structural integrity have been shown to contribute to hip joint pressure. Day et al. have shown up to 5 times normal pressure in areas with thin fibrocartilage, located mainly at the top of the acetabulum [67]. Another study by Song et al. measured the resistance to rotation, which reflects articular cartilage friction, in an intact hip and after focal and complete labrectomy. Resistance to rotation, and likely resultant joint pressure, was significantly increased by up to 20% following focal and complete labrectomy indicating the importance of the labrum in maintaining joint homeostasis and normal hip joint biomechanics [30].

Hip position can also affect hip pressure distribution by decreasing the surface area between the femoral head and acetabulum [68]. Deep hip flexion, as seen in sitting with subsequent rising from a chair as well as stair climbing produces the highest joint pressure predominantly in the posterior acetabulum [68, 69]. This is in contrast to normal walking where the hip joint pressure is the lowest of all full weight bearing activities and is maximally focused on the superior aspect of the acetabulum [69].

18.6 Summary

Several studies of basic hip joint anatomy and biomechanics have been published over the years. These mostly involve straight line motion, and therefore likely underestimate the complex movements and resultant interactions between the bony anatomy, capsule, labrum, ligaments, and muscles of this joint in high level athletes. As such, future research and biomechanical studies should focus on abnormalities of these structures and how injury may affect changes to the biomechanics of the athletic hip.

References

1. Lee MC, Eberson CP. Growth and development of the child's hip. Orthop Clin North Am. 2006;37(2):119–32. v
2. Ponseti IV. Growth and development of the acetabulum in the normal child. Anatomical, histological, and roentgenographic studies. J Bone Joint Surg Am. 1978;60(5):575–85.
3. Tonnis D, Heinecke A. Acetabular and femoral anteversion: relationship with osteoarthritis of the hip. J Bone Joint Surg Am. 1999;81(12):1747–70.
4. Ganz R, Klaue K, Vinh TS, Mast JW. A new periacetabular osteotomy for the treatment of hip dysplasias: Technique and preliminary results. Clin Orthop Relat Res. 1988;2004(418):3–8.
5. Byrd JW, Jones KS. Hip arthroscopy in the presence of dysplasia. Arthroscopy. 2003;19(10):1055–60.
6. Fagerson TL. The hip handbook. Boston: Butterworth-Heinemann; 1998.
7. Hamill J, Knutzen K. Biomechanical basis of human movement. 3rd ed. Philadelphia: Wolters Kluwer Health/Lippincott Williams and Wilkins; 2009.
8. Fabry G, MacEwen GD, Shands AR Jr. Torsion of the femur. A follow-up study in normal and abnormal conditions. J Bone Joint Surg Am. 1973;55(8):1726–38.
9. Radin EL. Biomechanics of the human hip. Clin Orthop Relat Res. 1980;152:28–34.
10. Notzli HP, Wyss TF, Stoecklin CH, Schmid MR, Treiber K, Hodler J. The contour of the femoral head-neck junction as a predictor for the risk of anterior impingement. J Bone Joint Surg Br. 2002;84(4):556–60.
11. Ito K, Minka MA 2nd, Leunig M, Werlen S, Ganz R. Femoroacetabular impingement and the cam-effect. A MRI-based quantitative anatomical study of the femoral head-neck offset. J Bone Joint Surg Br. 2001;83(2):171–6.
12. Clohisy JC, Carlisle JC, Trousdale R, Kim YJ, Beaule PE, Morgan P, et al. Radiographic evaluation of the hip has limited reliability. Clin Orthop Relat Res. 2009;467(3):666–75.
13. Beaule PE, Zaragoza E, Motamedi K, Copelan N, Dorey FJ. Three-dimensional computed tomography of the hip in the assessment of femoroacetabular impingement. J Orthop Res. 2005;23(6):1286–92.
14. Tannast M, Siebenrock KA, Anderson SE. Femoroacetabular impingement: radiographic diagnosis--what the radiologist should know. AJR Am J Roentgenol. 2007;188(6):1540–52.
15. Telleria JJ, Lindsey DP, Giori NJ, Safran MR. A quantitative assessment of the insertional footprints of the hip joint capsular ligaments and their spanning fibers for reconstruction. Clin Anat. 2014;27(3):489–97.
16. Bedi A, Galano G, Walsh C, Kelly BT. Capsular management during hip arthroscopy: from femoroacetabular impingement to instability. Arthroscopy. 2011;27(12):1720–31.
17. van Arkel RJ, Amis AA, Cobb JP, Jeffers JR. The capsular ligaments provide more hip rotational restraint than the acetabular labrum and the ligamentum teres: an experimental study. Bone Joint J. 2015;97-B(4):484–91.
18. van Arkel RJ, Amis AA, Jeffers JR. The envelope of passive motion allowed by the capsular ligaments of the hip. J Biomech. 2015;48(14):3803–9.
19. Myers CA, Register BC, Lertwanich P, Ejnisman L, Pennington WW, Giphart JE, et al. Role of the acetabular labrum and the iliofemoral ligament in hip stability: an in vitro biplane fluoroscopy study. Am J Sports Med. 2011;39(Suppl):85S–91S.
20. Martin HD, Savage A, Braly BA, Palmer IJ, Beall DP, Kelly B. The function of the hip capsular ligaments: a quantitative report. Arthroscopy. 2008;24(2):188–95.
21. Martin HD, Khoury AN, Schroder R, Johnson E, Gomez-Hoyos J, Campos S, et al. Contribution of the Pubofemoral ligament to hip stability: a biomechanical study. Arthroscopy. 2017;33(2):305–13.
22. Nicholas JA, Hershman EB. The lower extremity & spine in sports medicine. 2nd ed. St. Louis: Mosby; 1995.
23. Shindle MK, Ranawat AS, Kelly BT. Diagnosis and management of traumatic and atraumatic hip instability in the athletic patient. Clin Sports Med. 2006;25(2):309–26. ix-x
24. Domb BG, Philippon MJ, Giordano BD. Arthroscopic capsulotomy, capsular repair, and capsular plication of

the hip: relation to atraumatic instability. Arthroscopy. 2013;29(1):162–73.
25. Boykin RE, Anz AW, Bushnell BD, Kocher MS, Stubbs AJ, Philippon MJ. Hip instability. J Am Acad Orthop Surg. 2011;19(6):340–9.
26. Bowman KF Jr, Fox J, Sekiya JK. A clinically relevant review of hip biomechanics. Arthroscopy. 2010;26(8):1118–29.
27. Petersen W, Petersen F, Tillmann B. Structure and vascularization of the acetabular labrum with regard to the pathogenesis and healing of labral lesions. Arch Orthop Trauma Surg. 2003;123(6):283–8.
28. Seldes RM, Tan V, Hunt J, Katz M, Winiarsky R, Fitzgerald RH Jr. Anatomy, histologic features, and vascularity of the adult acetabular labrum. Clin Orthop Relat Res. 2001;382:232–40.
29. Tan V, Seldes RM, Katz MA, Freedhand AM, Klimkiewicz JJ, Fitzgerald RH Jr. Contribution of acetabular labrum to articulating surface area and femoral head coverage in adult hip joints: an anatomic study in cadavera. Am J Orthop (Belle Mead NJ). 2001;30(11):809–12.
30. Song Y, Ito H, Kourtis L, Safran MR, Carter DR, Giori NJ. Articular cartilage friction increases in hip joints after the removal of acetabular labrum. J Biomech. 2012;45(3):524–30.
31. Ferguson SJ, Bryant JT, Ganz R, Ito K. The acetabular labrum seal: a poroelastic finite element model. Clin Biomech (Bristol, Avon). 2000;15(6):463–8.
32. Safran MR, Giordano G, Lindsey DP, Gold GE, Rosenberg J, Zaffagnini S, et al. Strains across the acetabular labrum during hip motion: a cadaveric model. Am J Sports Med. 2011;39(Suppl):92S–102S.
33. Wenger D, Miyanji F, Mahar A, Oka R. The mechanical properties of the ligamentum teres: a pilot study to assess its potential for improving stability in children's hip surgery. J Pediatr Orthop. 2007;27(4):408–10.
34. Savory WS. The use of the ligamentum Teres of the hip-joint. J Anat Physiol. 1874;8(Pt 2):291–6.
35. Kapandji IA. The physiology of the joints. 6th ed. Edinburgh. New York: Churchill Livingstone; 2007.
36. Wenger DR, Mubarak SJ, Henderson PC, Miyanji F. Ligamentum teres maintenance and transfer as a stabilizer in open reduction for pediatric hip dislocation: surgical technique and early clinical results. J Child Orthop. 2008;2(3):177–85.
37. Martin HD, Hatem MA, Kivlan BR, Martin RL. Function of the ligamentum teres in limiting hip rotation: a cadaveric study. Arthroscopy. 2014;30(9):1085–91.
38. Cerezal L, Kassarjian A, Canga A, Dobado MC, Montero JA, Llopis E, et al. Anatomy, biomechanics, imaging, and management of ligamentum teres injuries. Radiographics. 2010;30(6):1637–51.
39. Bardakos NV, Villar RN. The ligamentum teres of the adult hip. J Bone Joint Surg Br. 2009;91(1):8–15.
40. von Eisenhart R, Adam C, Steinlechner M, Muller-Gerbl M, Eckstein F. Quantitative determination of joint incongruity and pressure distribution during simulated gait and cartilage thickness in the human hip joint. J Orthop Res. 1999;17(4):532–9.
41. Safran MR, Lopomo N, Zaffagnini S, Signorelli C, Vaughn ZD, Lindsey DP, et al. In vitro analysis of peri-articular soft tissues passive constraining effect on hip kinematics and joint stability. Knee Surg Sports Traumatol Arthrosc. 2013;21(7):1655–63.
42. Charbonnier C, Kolo FC, Duthon VB, Magnenat-Thalmann N, Becker CD, Hoffmeyer P, et al. Assessment of congruence and impingement of the hip joint in professional ballet dancers: a motion capture study. Am J Sports Med. 2011;39(3):557–66.
43. Dewberry MJ, Bohannon RW, Tiberio D, Murray R, Zannotti CM. Pelvic and femoral contributions to bilateral hip flexion by subjects suspended from a bar. Clin Biomech (Bristol, Avon). 2003;18(6):494–9.
44. Polkowski GG, Clohisy JC. Hip biomechanics. Sports Med Arthrosc Rev. 2010;18(2):56–62.
45. Hughes PEHJ, Matava MJ. Hip anatomy and biomechanics in the athlete. Sports Med Arthrosc Rev. 2002;10(2):103–14.
46. Nordin M, Frankel VH. Basic biomechanics of the musculoskeletal system. 3rd ed. Philadelphia: Lippincott Williams & Wilkins; 2001.
47. Ganz R, Parvizi J, Beck M, Leunig M, Notzli H, Siebenrock KA. Femoroacetabular impingement: a cause for osteoarthritis of the hip. Clin Orthop Relat Res. 2003;417:112–20.
48. Bedi A, Dolan M, Magennis E, Lipman J, Buly R, Kelly BT. Computer-assisted modeling of osseous impingement and resection in femoroacetabular impingement. Arthroscopy. 2012;28(2):204–10.
49. Bedi A, Dolan M, Hetsroni I, Magennis E, Lipman J, Buly R, et al. Surgical treatment of femoroacetabular impingement improves hip kinematics: a computer-assisted model. Am J Sports Med. 2011;39(Suppl):43S–9S.
50. Ross JR, Nepple JJ, Philippon MJ, Kelly BT, Larson CM, Bedi A. Effect of changes in pelvic tilt on range of motion to impingement and radiographic parameters of acetabular morphologic characteristics. Am J Sports Med. 2014;42(10):2402–9.
51. Murray R, Bohannon R, Tiberio D, Dewberry M, Zannotti C. Pelvifemoral rhythm during unilateral hip flexion in standing. Clin Biomech (Bristol, Avon). 2002;17(2):147–51.
52. Esola MA, McClure PW, Fitzgerald GK, Siegler S. Analysis of lumbar spine and hip motion during forward bending in subjects with and without a history of low back pain. Spine (Phila Pa 1976). 1996;21(1):71–8.
53. Offierski CM, MacNab I. Hip-spine syndrome. Spine (Phila Pa 1976). 1983;8(3):316–21.
54. Mann RA, Hagy J. Biomechanics of walking, running, and sprinting. Am J Sports Med. 1980;8(5):345–50.
55. Novacheck TF. The biomechanics of running. Gait Posture. 1998;7(1):77–95.

56. Nicola TL, Jewison DJ. The anatomy and biomechanics of running. Clin Sports Med. 2012;31(2):187–201.
57. Franz JR, Paylo KW, Dicharry J, Riley PO, Kerrigan DC. Changes in the coordination of hip and pelvis kinematics with mode of locomotion. Gait Posture. 2009;29(3):494–8.
58. Cappellini G, Ivanenko YP, Poppele RE, Lacquaniti F. Motor patterns in human walking and running. J Neurophysiol. 2006;95(6):3426–37.
59. Pauwels F. Biomechanics of the locomotor apparatus: contributions on the functional anatomy of the locomotor apparatus. Berlin. New York: Springer; 1980.
60. Brinckmann P, Frobin W, Leivseth G. Musculoskeletal biomechanics. Stuttgart. New York: Thieme; 2002.
61. Kadaba MP, Ramakrishnan HK, Wootten ME, Gainey J, Gorton G, Cochran GV. Repeatability of kinematic, kinetic, and electromyographic data in normal adult gait. J Orthop Res. 1989;7(6):849–60.
62. van den Bogert AJ, Read L, Nigg BM. An analysis of hip joint loading during walking, running, and skiing. Med Sci Sports Exerc. 1999;31(1):131–42.
63. Bergmann G, Deuretzbacher G, Heller M, Graichen F, Rohlmann A, Strauss J, et al. Hip contact forces and gait patterns from routine activities. J Biomech. 2001;34(7):859–71.
64. Delp SL, Maloney W. Effects of hip center location on the moment-generating capacity of the muscles. J Biomech. 1993;26(4–5):485–99.
65. Watanabe RS. Embryology of the human hip. Clin Orthop Relat Res. 1974;98:8–26.
66. Brinckmann P, Frobin W, Hierholzer E. Stress on the articular surface of the hip joint in healthy adults and persons with idiopathic osteoarthrosis of the hip joint. J Biomech. 1981;14(3):149–56.
67. Day WH, Swanson SA, Freeman MA. Contact pressures in the loaded human cadaver hip. J Bone Joint Surg Br. 1975;57(3):302–13.
68. Martin DE, Greco NJ, Klatt BA, Wright VJ, Anderst WJ, Tashman S. Model-based tracking of the hip: implications for novel analyses of hip pathology. J Arthroplast. 2011;26(1):88–97.
69. Hodge WA, Carlson KL, Fijan RS, Burgess RG, Riley PO, Harris WH, et al. Contact pressures from an instrumented hip endoprosthesis. J Bone Joint Surg Am. 1989;71(9):1378–86.

Biomechanics of Femoroacetabular Impingement

19

Seper Ekhtiari, Luc Rubinger, Aaron Gazendam, and Olufemi R. Ayeni

19.1 Overview of FAI

Femoroacetabular impingement (FAI) is a condition of the hip with a complex presentation of irregular anatomic morphology, symptomatology, and clinical signs [1]. FAI has typically been defined as an "abutment" between the femoral head-neck junction and the acetabular rim secondary to these morphologic irregularities [2]. This repetitive abutment occurs mainly during activities involving range of motion including flexion and internal rotation, which can lead to secondary pathologies such as lesions of the acetabular labrum and the adjacent cartilage [2]. Two types of FAI have been classically described: a cam type (lesion of the proximal femur) and a pincer type (lesion of the acetabular rim). Most cases of FAI are now recognized as mixed cam-pincer impingement [3].

There is a wide and variable clinical presentation of patients with FAI. However, most patients will present with deep, intermittent, activity or rotationally related hip or groin pain. The onset of the pain is usually insidious, but in the context of secondary pathologies such as labral tears, some patients may report a specific injury or an acute precipitating event [4]. There are a multitude of physical exam maneuvers that have been documented for the screening and diagnosis of FAI. While an in-depth description of these maneuvers is beyond the scope of this chapter, the principles are to move the affected hip into a position of possible impingement (whether it be anterior, lateral, or posterior) to reproduce pain.

Imaging studies are an important component in the workup for suspected FAI. A standard plain radiographic series for the preliminary evaluation of a patient with potential FAI would include a well-centered anteroposterior (AP) view to assess both hips, a 90° Dunn lateral, and a false-profile view. These radiographic views permit an initial assessment of the cam and/or pincer lesions associated with impingement, while also evaluating for potential secondary pathologies. Anterior acetabular overcoverage, or anterior pincer type impingement, is evaluated by the presence of the crossover sign, posterior wall sign, and the prominent ischial spine sign. Lateral overcoverage can be assessed with the lateral center-edge angle (LCEA), with values greater than 40° indicative of lateral pincer type morphology (Fig. 19.1). Alpha angles can be calculated on the Dunn views to assess for cam type morphology (Fig. 19.2). Cam type FAI can be diagnosed with an alpha angle greater than 55° and reduced femoral head and neck offset [5]. Magnetic resonance imaging

S. Ekhtiari (✉) · L. Rubinger · A. Gazendam
O. R. Ayeni
Division of Orthopaedics, Department of Surgery, McMaster University, Hamilton, ON, Canada
e-mail: seper.ekhtiari@medportal.ca; luc.rubinger@medportal.ca; aaron.gazendam@medportal.ca; ayenif@mcmaster.ca

J. Koh et al. (eds.), *Orthopaedic Biomechanics in Sports Medicine*,
https://doi.org/10.1007/978-3-030-81549-3_19

(MRI) is useful in the assessment of pincer and cam type morphologies, as well as the aforementioned secondary pathologies, and compensatory soft tissue injuries. MRI is also useful in alpha angle measurement and quantification of the cam type deformity (Fig. 19.2).

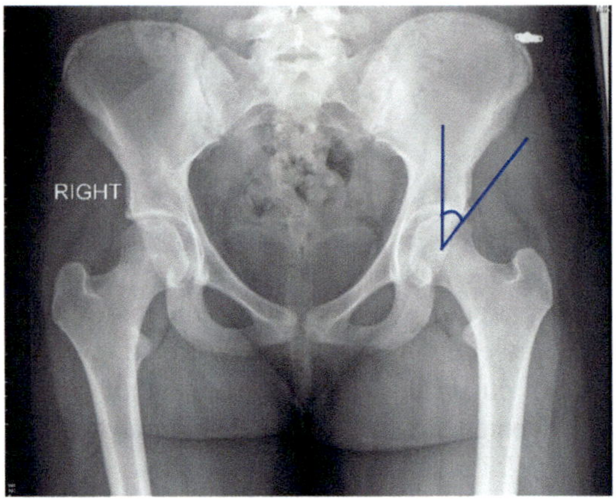

Fig. 19.1 AP radiograph is shown for a patient with mixed type FAI and increased lateral overcoverage of the acetabulum. The LCEA is measured with the vertex of the angle at the center of the femoral head, the vertical limb perpendicular to the horizontal and the second limb tangent to the lateral-most aspect of the acetabular rim

19.2 Biomechanics of Cam Type Impingement

19.2.1 Development of Cam Type Morphology

A cam is a rotating part in a machine that translates rotational movement into linear motion through an eccentrically shaped sphere, or cylinder. Thus, cam type morphology of the femoral head-neck junction is characterized by an aspherical femoral head moving inside the spherical acetabulum causing impingement and levering against the acetabular rim, thereby impacting on the acetabular labrum, and subsequently damaging the underlying articular cartilage (Fig. 19.3) [6, 7]. The cam type morphological change can be conceptualized as a bony prominence typically at the femoral head-neck junction or alternatively as a reduced offset between the femoral head and neck [7, 8]. Some also use the term "pistol grip deformity" to describe the cam type morphology, because of the appearance of the proximal femur on an AP radiograph (Fig. 19.4).

There are several hypothesized etiologies of the cam type morphology in the literature. The

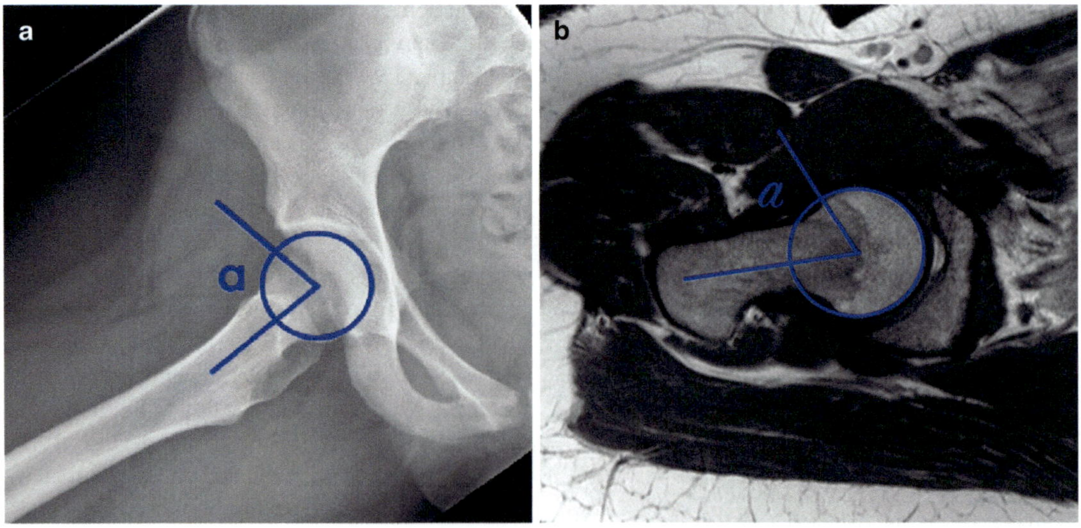

Fig. 19.2 (**a**) Dunn lateral radiograph is shown for the same patient with mixed type FAI with the alpha angle drawn. A circle is outlined over the femoral head, using the posterior aspect of the head as the reference to place the circle. The angle is drawn with the vertex at the center of the circle and one line down the center of femoral neck. The other line bisects the femoral head-neck junction as it exits the contour of the circle. (**b**) Measurement of alpha angle on T2 mid-axial MRI view of the same patient

development of the cam lesion may be due to eccentric premature closure of the capital femoral physis, ultimately resulting in an aspherical femoral head [9]. A cohort study of pediatric patients, followed with MRI both before and after capital physeal closure, showed that the cam type morphology was present exclusively in the scans completed after the physis had closed [10]. This finding implied that the cam lesion developed during physeal closure, and it was also hypothesized that increased physical activity during this period was associated with cam lesion development [10–12].

A variant of cam type morphology is also a long-term sequela of a slipped capital femoral epiphysis, Legg–Calvé–Perthes disease, and epiphyseal dysplasias, and can be secondary to malunited femoral neck fractures [13–16].

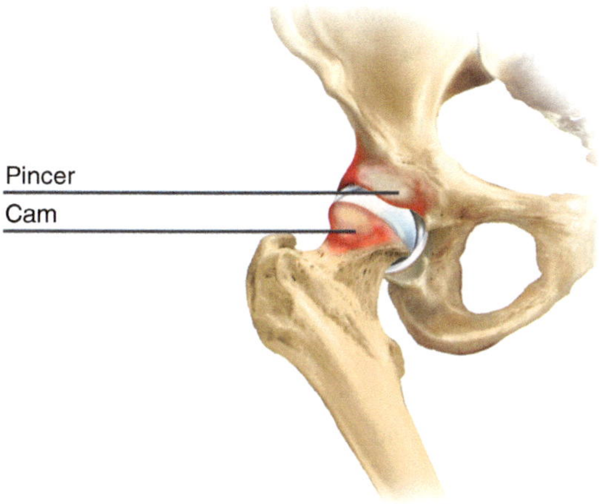

Fig. 19.3 The bony deformity at the head-neck junction present in the cam type morphology enters the acetabulum in hip flexion and can cause cartilage degeneration over time. The acetabular overcoverage present in the pincer type morphology leads to an impaction type injury and can lead to labral pathologies. Reprinted with permission from Smith and Nephew Commons:Free Art License 1.3 - Wikimedia Commons

19.2.2 Altered Biomechanics: Cam Type Morphology

There is a wide range of incidence of cam type morphology reported in the athletic population (60–90%); however, not all those affected develop, or will develop symptoms [11, 17–19]. Elevated alpha angles, low femoral neck-shaft angles, and increased pelvic range of motion have all been associated with symptom development [20]. FAI is a condition inherently linked to movement, and end range of motion, thus biomechanical impairments associated with the cam type morphology are likely to play some role in symptom development and persistence.

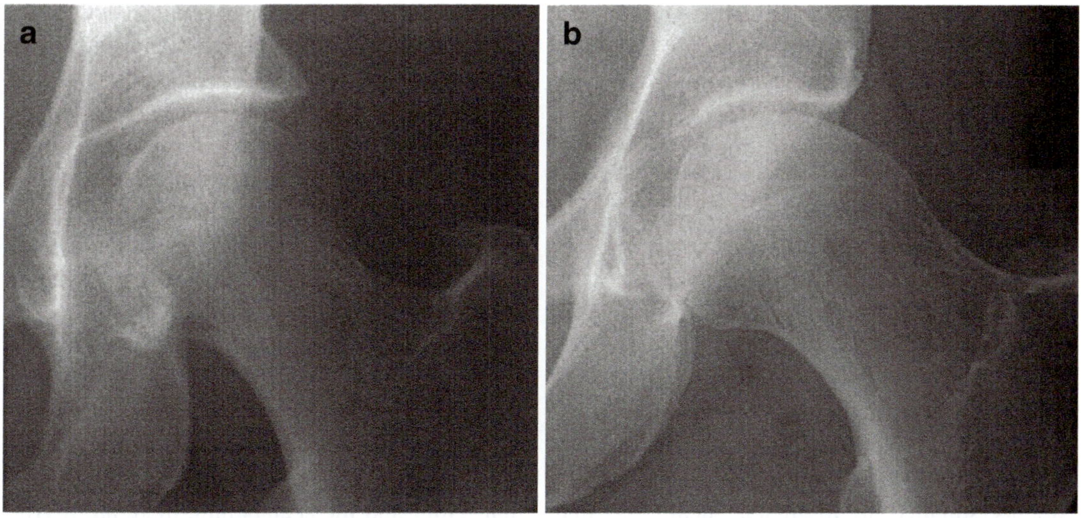

Fig. 19.4 AP radiographs of the left hip, showing normal morphology (**a**) and a cam type morphology or "pistol grip deformity" (**b**)

In the context of a squat movement, Bagwell et al. compared a control group to patients with isolated unilateral symptomatic cam type morphology. They demonstrated decreased peak hip internal rotation and decreased mean hip extensor moments [21]. In addition, the cam type FAI group demonstrated decreased posterior pelvis tilt during squat descent compared to the control group, resulting in a more anteriorly tilted pelvis at the time of peak hip flexion. Lamontagne et al. compared patients diagnosed with cam type FAI and matched control participants when performing unloaded squats while collecting 3-D motion analysis [22]. Patients with FAI had no differences in hip motion during squatting but had decreased sagittal pelvic range of motion compared to the control group. Further, the FAI group also could not squat as low, as a percentage of leg length, compared to the control group.

With regard to gait mechanics, Kennedy et al. used 3D kinematics, and 3D kinetics during gait to compare a control group to those with isolated unilateral symptomatic cam type morphology group. The cam type group had significantly lower peak hip abduction, frontal range of motion, and a lower sagittal ROM [23]. Rutherford et al. compared symptomatic isolated unilateral cam type morphology patients to a control group, but also compared affected and unaffected limbs [24]. No differences in muscle strength were found between the symptomatic and contralateral legs in the FAI group. However, there were greater amplitudes of gluteus maximus activation in the FAI symptomatic leg compared with the asymptomatic group.

19.2.3 Biomechanics After Surgical Correction: Cam Type Morphology

The current standard of care for addressing symptomatic cam type impingement is hip arthroscopy with bony resection of the cam lesion. Postoperatively, in a study by Lamontagne et al., patients squatted to a greater mean maximal depth with greater knee flexion and ankle dorsiflexion angles maximal depth compared to preoperative values [25]. The squat performance improved postoperatively, was postulated to be secondary to the combined effects of increased knee and ankle angles as well as a greater acetabular opening and thus reduced anterior femoral head coverage, allowing increased posterior pelvic pitch during the descent phase of the squat. Rylander et al. [26] compared hip and pelvic kinematics in cam type FAI patients preoperatively to postoperative values and controls. Postoperatively, gait kinematics returned to normal values. However, improved but below-normal hip internal rotation and hip sagittal plane range of motion remained. These studies point to the benefit of quality operative management in improving hip and lower extremity biomechanics associated with cam type morphology; however, the lack of complete normalization of functional movements leaves much more to be elucidated through larger and more detailed biomechanical studies.

19.3 Biomechanics of Pincer Type Impingement

19.3.1 Development of Pincer Type Morphology

The term "pincer" evokes the imagery of a lobster or crab's claw, which is reminiscent of how pincer type impingement usually appears on an AP radiograph. Three-dimensionally, however, this morphology is considerably more complex. Pincer type impingement refers to acetabular overcoverage of the femoral head (Fig. 19.3) [27]. This overcoverage leads to repetitive contact between the acetabulum and the femoral neck, which can lead to a number of pathologies, including most commonly damage to the acetabular chondrolabral junction (Fig. 19.3), usually at the anterosuperior portion of the acetabulum where the overcoverage is most pronounced [28].

Given that pincer type impingement is rooted in a pathological interaction between the proximal femur and the acetabulum, there are myriad combinations in femoral and acetabular morphology which can lead to this type of FAI. Genetic factors have been hypothesized as one potential etiology for the development of pincer type FAI. Specifically, Sekimoto et al. reported that HOX9 single nucleotide polymorphisms (SNPs) had a significant association with acetabular overcoverage in Japanese individuals [29]. Furthermore, familial associations between siblings and parents of individuals with pincer type morphology have been demonstrated, though it is unclear whether this is an exclusively genetic relationship or associated with environmental factors such as types of athletic activity and lifestyle [30]. Post-traumatic pincer type impingement can also occur from malunion or secondary changes following pelvic and/or femoral neck and shaft fractures. Overcorrection of hip dysplasia, particularly anterolateral coverage and acetabular retroversion can also result in iatrogenic pincer type FAI [31]. Ultimately, all of the above etiologies lead to some combination of acetabular malorientation and femoral neck/shaft torsion, leading to relative overcoverage [28].

19.3.2 Altered Biomechanics: Pincer Type Morphology

There are three main considerations when conceptualizing the biomechanics of pincer type impingement: depth, coverage, and orientation [32]. In various combinations, these three characteristics can combine to lead to one of three "subtypes" of pincer type impingement: focal anterior overcoverage, relative anterior overcoverage, and global overcoverage. Focal anterior overcoverage is primarily an issue of coverage, with mostly normal depth and orientation. Relative anterior overcoverage is a combination of altered coverage (excess anterior overcoverage and posterior undercoverage), orientation (acetabular retroversion), sometimes with minor changes in depth (coxa profunda). Finally, global overcoverage, which is less common, is often primarily a depth issue, with profound coxa profunda or acetabular protrusio.

The acetabular overcoverage seen in pincer type morphology often leads to focal chondral damage, particularly at the chondrolabral junction of the anterosuperior acetabulum. Arthroscopically, this manifests as the "wave sign" [28]. Given that isolated pincer type morphology is relatively rare, the majority of biomechanical FAI research has focused on cam type FAI or mixed FAI. Typical activities which cause symptoms in patients with pincer type impingement involve cutting, pivoting, and torsional activities, and pain is often intermittent and activity-related, likely related to episodic contact between the femoral neck and the acetabular rim. In contrast to cam impingement, there is typically minimal mechanical block to range of motion in pincer type FAI. Another biomechanical consequence of pincer type impingement is subtle posterior subluxation of the femoral head, leading to posterior acetabular chondromalacia [33].

Biomechanics After Surgical Correction: Pincer Type Morphology.

Given that pincer type FAI is a disease of repeated abutment of the acetabular rim on the femoral neck, any joint preserving surgical procedure must include techniques which prevent future recurrence. Given the complexity of pincer type pathology as described above, the corrective techniques must be carefully considered and aimed at the underlying pathomechanics. For example, if the primary issue is acetabular retroversion (i.e., relative anterior overcoverage), it is unlikely that arthroscopic osteochondroplasty alone will achieve a stable and impingement-free hip.

When focal overcoverage is the primary issue, arthroscopic osteochondroplasty can be undertaken. With adequate resection, the contact between the femoral neck and the acetabular rim is eliminated. Furthermore, osteochondroplasty reduces the loading surface of the acetabulum against the femoral head [28]. Thus, while under-resection is among the most common reasons for revision surgery [34], over-resection can lead to micro- or gross instability [28].

In cases requiring periacetabular osteotomy (PAO), the goal is ultimately to improve coverage of the femoral head by adjusting the center of hemipelvic rotation, while preserving the posterior column [35]. Following successful PAO, there is an increase in the size of the contact area between the femoral head and the acetabulum (~30%), a decrease in peak contact pressures (~50% decrease), and an increase in center-edge angle (~120% increase) [36]. Given that range of motion limitations are often minimal or absent in pure pincer type FAI, surgical correction does not usually result in a noticeable change in range of motion, except by removing pain as a limiting factor.

19.3.3 Associated Kinematics

The spinal column is inextricably linked with pelvic and hip anatomy and kinematics; thus, it is not surprising that there are associated findings with respect to pelvic and spinal alignment in patients with FAI. In a prospective study of 548 participants, Mascarenhas et al. compared symptomatic patients awaiting hip surgery with asymptomatic volunteers [37]. In addition to larger cam deformities and smaller acetabular coverage, they also reported that symptomatic patients had significantly larger pelvic incidence (the axis of spinopelvic sagittal balance) and sacral slope (an angle between a horizontal line and a line parallel to the superior sacral endplate on a sagittal view) [37]. Grammatopoulos et al. reported similar findings of increased pelvic incidence in patients with cam type FAI [38]. As with any hip pathology, it is always important to assess and consider potential spinal causes as part of the diagnostic workup.

19.4 Consequences of Altered Biomechanics in FAI

19.4.1 Osteoarthritis

The relationship between FAI and development of osteoarthritis was hypothesized by Ganz et al. in 2003 [6]. Since, there has been a growing body of literature demonstrating a link between FAI, specifically cam deformities, and the development of osteoarthritis [39]. Prospective cohort data has demonstrated a link between cam deformities and the development of hip osteoarthritis (OA) [40–42]. The Cohort Hip and Cohort Knee (CHECK) study was a Dutch nationwide prospective cohort study of 1002 individuals with early symptomatic osteoarthritis. Moderate and severe cam deformities, defined as an α angle >60° and >83°, resulted in an adjusted odds ratio (OR) of 3.67 and 9.66, respectively, for development of end stage osteoarthritis.

The same relationship has not been found with pincer type deformities [43]. The CHECK cohort demonstrated no associated risk for development of osteoarthritis in patients with a pincer type deformity, defined as an LCEA >40°. These findings have been corroborated by other prospective cohorts across other populations [41, 44].

The biomechanical abnormalities differ between cam and pincer type deformities. Cam impingement is thought of as an inclusion-type injury as the aspherical head found in cam deformities enters and engages in the spherical acetabulum (Fig. 19.3) [6, 13]. This occurs most commonly in hip flexion or with deep flexion and internal rotation given that cam deformities most commonly arise on the anterior and anterolateral junction of the femoral head and neck. The cam lesion pushes the acetabular cartilage centrally and creates a tensile load at the chondrolabral junction [45]. Repeated impingement can result in disruption at the chondrolabral junction, a chondral flap and progress to cartilage delamination and degeneration [45–47]. The literature has consistently demonstrated that acetabular damage occurs consistently in the anterior and superolateral aspects of the rim [45, 48, 49].

Much of our understanding of the biomechanical differences and abnormal stresses secondary to cam deformities come from finite element analyses [50–53]. A systematic review of finite element analyses demonstrated that a cam morphology elevated the shear stress on the bone and elevated contact pressure at the acetabular cartilage [54].

What is still lacking in the literature is an understanding of how, why, and when this patient population transitions from cam FAI development through to end stage osteoarthritis. A deeper understanding of this natural history would help guide timing and strategy of interventions.

19.4.2 Labral Tears

Labral tears are a common associated finding in patients with FAI, in particular in patients with a pincer type morphology [55–57]. In patients with pincer type FAI, there is repetitive abutment of the normal femoral head-neck junction onto the abnormal acetabular rim. The labrum is crushed between the acetabular rim and the femoral neck and can lead to intrasubstance tearing and can progress to labral ossification and worsening of the acetabular overcoverage [58]. Similarly, patients with cam type impingement can develop a junctional pattern chondrolabral injury, at the meeting point between the acetabular cartilage and labrum [59, 60]. Labral damage in the setting of the pincer type morphology occurs most commonly anterosuperiorly secondary to the linear impact occurring between the acetabulum and femur. This anatomical zone is predisposed to shearing stresses and injury given its collagen fiber orientation [61].

There has been a growing interest in the role of the labrum on the biomechanical function of the hip joint. The intact labrum provides a seal around the joint, creating a pressurized layer of intra-articular fluid that distributes compressive loads [62–64]. The intact labrum also increases the total contact area of the hip joint resulting in lower cartilage contact pressure [65]. Chondrolabral tears have been shown to compromise the sealing capacity of the labrum in cadaveric studies [66]. Microinstability is increasingly becoming a recognized pathology and cause of pain and dysfunction, even in patients without bony deficiency [67]. Given the stabilizing role the labrum plays on the hip joint, labral tears are suggested to be a contributor to hip instability [67, 68]. These findings have correlated to clinical outcomes as labral repair or reconstruction have shown improved outcomes when compared to debridement or resection [69–74].

19.4.3 Chondral Injury

Chondral damage in pincer type FAI is less significant and is limited to the peripheral acetabular rim. However, there is a subset of patients who develop a contrecoup pattern of cartilage loss on both the acetabulum and femoral head with continued impingement [16, 58, 75]. This is thought to result from flexion of the hip past the engagement of the rim lesion which leads to levering of the femoral head on the posterior chondral surface [16].

With respect to cam type FAI, Kumar et al. [76] undertook a cumbersome study and compared peak kinematic and kinetic variables during walking, deep-squat, and drop-landing tasks between patients with FAI, FAI, and MRI confirmed cartilage lesions, and control subjects. Subjects who had cartilage lesions in the presence of a cam lesion demonstrated greater adduction, greater internal rotation moment, and lower transverse plane range of motion during the deep-squat task; and greater adduction and lower internal rotation during the drop-landing task compared with FAI subjects who did not have cartilage lesions. Ultimately, patients with FAI associated cartilaginous lesions have amplified abnormal movement patterns during functional tasks.

19.5 Conclusions

An understanding of the biomechanical consequences secondary to FAI allows for a greater understanding of the commonly associated intra-articular pathologies. Clinicians should be aware of these associated pathologies and should investigate on clinical examination, imaging and intraoperatively. Diagnosis and appropriate management of associated pathologies provides an opportunity to improve treatment and outcomes in patients presenting with symptomatic FAI.

References

1. King MG, Lawrenson PR, Semciw AI, Middleton KJ, Crossley KM. Lower limb biomechanics in femoroacetabular impingement syndrome: a systematic review and meta-analysis. Br J Sports Med. 2018;52(9):566.
2. Lavigne M, Parvizi J, Beck M, Siebenrock KA, Ganz R, Leunig M. Anterior femoroacetabular impingement: part I. Techniques of joint preserving surgery. Clin Orthop Relat Res. 2004 Jan;418:61–6.

3. Kaya M, Suzuki T, Emori M, Yamashita T. Hip morphology influences the pattern of articular cartilage damage. Knee Surg Sports Traumatol Arthrosc. 2016 Jun;24(6):2016–23.
4. Clohisy JC, Knaus ER, Hunt DM, Lesher JM, Harris-Hayes M, Prather H. Clinical presentation of patients with symptomatic anterior hip impingement. In: Clinical Orthopaedics and Related Research. 2009. p. 638–44.
5. Nemtala F, Mardones RM, Tomic A. Anterior and posterior femoral head-neck offset ratio in the cam impingement. Cartilage. 2010;
6. Ganz R, Parvizi J, Beck M, Leunig M, Nötzli H, Siebenrock KA. Femoroacetabular impingement: a cause for osteoarthritis of the hip. Clin Orthop Relat Res 2003;(417):112–120.
7. Ito K, Minka MA 2nd, Leunig M, Werlen S, Ganz R. Femoroacetabular impingement and the cam-effect. A MRI-based quantitative anatomical study of the femoral head-neck offset. J Bone Joint Surg Br. 2001 Mar;83(2):171–6.
8. Pfirrmann CWA, Mengiardi B, Dora C, Kalberer F, Zanetti M, Hodler J. Cam and pincer femoroacetabular impingement: characteristic MR arthrographic findings in 50 patients. Radiology. 2006 Sep;240(3):778–85.
9. Siebenrock KA, Wahab KHA, Werlen S, Kalhor M, Leunig M, Ganz R. Abnormal extension of the femoral head epiphysis as a cause of cam impingement. Clin Orthop Relat Res. 2004 Jan;418:54–60.
10. Carsen S, Moroz PJ, Rakhra K, Ward LM, Dunlap H, Hay JA, et al. The Otto Aufranc Award. On the etiology of the cam deformity: a cross-sectional pediatric MRI study. Clin Orthop Relat Res. 2014 Feb;472(2):430–6.
11. Siebenrock KA, Ferner F, Noble PC, Santore RF, Werlen S, Mamisch TC. The cam-type deformity of the proximal femur arises in childhood in response to vigorous sporting activity. Clin Orthop Relat Res. 2011 Nov;469(11):3229–40.
12. Tibor LM, Leunig M. The pathoanatomy and arthroscopic management of femoroacetabular impingement. Bone Joint Res. 2012;
13. Leunig M, Casillas MM, Hamlet M, Hersche O, Nötzli H, Slongo T, et al. Slipped capital femoral epiphysis: early mechanical damage to the acetabular cartilage by a prominent femoral metaphysis. Acta Orthop Scand. 2000;71(4):370–5.
14. Eijer H, Podeszwa DA, Ganz R, Leunig M. Evaluation and treatment of young adults with femoro-acetabular impingement secondary to Perthes' disease. Hip Int J Clin Exp Res hip Pathol Ther. 2006;16(4): 273–80.
15. Eijer H, Myers SR, Ganz R. Anterior femoroacetabular impingement after femoral neck fractures. J Orthop Trauma. 2001;15(7):475–81.
16. Ganz R, Leunig M, Leunig-Ganz K, Harris WH. The etiology of osteoarthritis of the hip: an integrated mechanical concept. Clin Orthop Relat Res. 2008 Feb;466(2):264–72.
17. Johnson AC, Shaman MA, Ryan TG. Femoroacetabular impingement in former high-level youth soccer players. Am J Sports Med. 2012 Jun;40(6):1342–6.
18. Agricola R, Bessems JHJM, Ginai AZ, Heijboer MP, van der Heijden RA, Verhaar JAN, et al. The development of Cam-type deformity in adolescent and young male soccer players. Am J Sports Med. 2012 May;40(5):1099–106.
19. Diamond LE, Dobson FL, Bennell KL, Wrigley TV, Hodges PW, Hinman RS. Physical impairments and activity limitations in people with femoroacetabular impingement: a systematic review. Br J Sports Med. 2015 Feb;49(4):230–42.
20. Ng KCG, Lamontagne M, Adamczyk AP, Rakhra KS, Beaulé PE. Patient-specific anatomical and functional parameters provide new insights into the pathomechanism of cam FAI. Clin Orthop Relat Res. 2015 Apr;473(4):1289–96.
21. Bagwell JJ, Snibbe J, Gerhardt M, Powers CM. Hip kinematics and kinetics in persons with and without cam femoroacetabular impingement during a deep squat task. Clin Biomech. 2016;31:87–92.
22. Lamontagne M, Kennedy MJ, Beaulé PE. The effect of cam FAI on hip and pelvic motion during maximum squat. Clin Orthop Relat Res. 2009 Mar;467(3):645–50.
23. Kennedy MJ, Lamontagne M, Beaulé PE. Femoroacetabular impingement alters hip and pelvic biomechanics during gait Walking biomechanics of FAI. Gait Posture. 2009 Jul;30(1):41–4.
24. Rutherford DJ, Moreside J, Wong I. Differences in Hip Joint Biomechanics and Muscle Activation in Individuals With Femoroacetabular Impingement Compared With Healthy, Asymptomatic Individuals: Is Level-Ground Gait Analysis Enough? Orthop J Sport Med. 2018;6(5):2325967118769829.
25. Lamontagne M, Brisson N, Kennedy MJ, Beaulé PE. Preoperative and postoperative lower-extremity joint and pelvic kinematics during maximal squatting of patients with cam femoro-acetabular impingement. J Bone Joint Surg Am. 2011;93(Suppl 2):40–5.
26. Rylander J, Shu B, Favre J, Safran M, Andriacchi T. Functional testing provides unique insights into the pathomechanics of femoroacetabular impingement and an objective basis for evaluating treatment outcome. J Orthop Res. 2013;31(9):1461–8.
27. Chaudhry H, Ayeni OR. The Etiology of Femoroacetabular Impingement: What We Know and What We Don't. Sports Health. 2014;
28. Sabetta E, Scaravella E. Treatment of pincer-type femoroacetabular impingement. Joints. 2015;
29. Sekimoto T, Kurogi S, Funamoto T, Ota T, Watanabe S, Sakamoto T, et al. Possible association of single nucleotide polymorphisms in the 3′ untranslated region of HOXB9 with acetabular overcoverage. Bone Jt Res. 2015;
30. Pollard TCB, Villar RN, Norton MR, Fern ED, Williams MR, Murray DW, et al. Genetic influences in the aetiology of femoroacetabular impingement: A sibling study. J Bone Jt Surg - Ser B. 2010;

31. Klein C, Fontanarosa A, Khouri N, Bellity J, Padovani JP, Glorion C, et al. Anterior and lateral overcoverage after triple pelvic osteotomy in childhood for developmental dislocation of the hip with acetabular dysplasia: Frequency, features, and medium-term clinical impact. Orthop Traumatol Surg Res. 2018;
32. Hadeed MM, Cancienne JM, Gwathmey FW. Pincer Impingement. Clin Sports Med. 2016;
33. Larson CM. Arthroscopic management of pincer-type impingement. Sports Med Arthrosc Rev. 2010;
34. Ekhtiari S, Coughlin RP, Simunovic N, Ayeni OR. Strategies in revision hip arthroscopy. Ann Jt. 2018;
35. Kamath AF. Bernese periacetabular osteotomy for hip dysplasia: Surgical technique and indications. World J Orthop. 2016;
36. Lee KJ, Park SJ, Lee SJ, Naito M, Kwon SY. Biomechanical study on the efficacy of the periacetabular osteotomy using Patient-specific finite element analysis. Int J Precis Eng Manuf. 2015;
37. Mascarenhas VV, Rego P, Dantas P, Caetano AP, Jans L, Sutter R, et al. Can we discriminate symptomatic hip patients from asymptomatic volunteers based on anatomic predictors? A 3-dimensional magnetic resonance study on cam, pincer, and spinopelvic parameters. Am J Sports Med. 2018;
38. Grammatopoulos G, Speirs AD, Ng KCG, Riviere C, Rakhra KS, Lamontagne M, et al. Acetabular and spino-pelvic morphologies are different in subjects with symptomatic cam femoro-acetabular impingement. J Orthop Res. 2018;
39. Kowalczuk M, Yeung M, Simunovic N, Ayeni OR. Does femoroacetabular impingement contribute to the development of hip osteoarthritis? a systematic review. Sports Med Arthrosc Rev. 2015;
40. R. A, M. H, S. B-Z, J. V, H. W, E. W. Cam impingement causes end-stage osteoarthritis of the hip: A nationwide prospective study (CHECK). HIP International. 2012.
41. Thomas GER, Palmer AJR, Batra RN, Kiran A, Hart D, Spector T, et al. Subclinical deformities of the hip are significant predictors of radiographic osteoarthritis and joint replacement in women. A 20 year longitudinal cohort study. Osteoarthr Cartil. 2014;
42. Wyles CC, Norambuena GA, Howe BM, Larson DR, Levy BA, Yuan BJ, et al. Cam Deformities and Limited Hip Range of Motion Are Associated With Early Osteoarthritic Changes in Adolescent Athletes: A Prospective Matched Cohort Study. Am J Sports Med. 2017;
43. Agricola R, Heijboer MP, Bierma-Zeinstra SM, Verhaar JA, Weinans H, Waarsing JH. Pincer deformities and mild acetabular dysplasia: the relationship between acetabular coverage and development of hip OA in the nationwide prospective check cohort. Osteoarthr Cartil. 2013;
44. Clohisy JC, Dobson MA, Robison JF, Warth LC, Zheng J, Liu SS, et al. Radiographic structural abnormalities associated with premature, natural hip-joint failure. J Bone Jt Surg - Ser A. 2011;
45. Beck M, Kalhor M, Leunig M, Ganz R. Hip morphology influences the pattern of damage to the acetabular cartilage. Femoroacetabular impingement as a cause of early osteoarthritis of the hip. J Bone Jt Surg - Ser B. 2005;
46. Anderson LA, Peters CL, Park BB, Stoddard GJ, Erickson JA, Crim JR. Acetabular cartilage delamination in femoroacetabular impingement. Risk factors and magnetic resonance imaging diagnosis. J Bone Jt Surg - Ser A. 2009;
47. Tannast M, Goricki D, Beck M, Murphy SB, Siebenrock KA. Hip damage occurs at the zone of femoroacetabular impingement. In: Clinical Orthopaedics and Related Research; 2008.
48. Bhatia S, Nowak DD, Briggs KK, Patterson DC, Philippon MJ. Outerbridge Grade IV Cartilage Lesions in the Hip Identified at Arthroscopy. Arthrosc - J Arthrosc Relat Surg. 2016;
49. Pascual-Garrido C, Li DJ, Grammatopoulos G, Yanik EL, Clohisy JC. The Pattern of Acetabular Cartilage Wear Is Hip Morphology-dependent and Patient Demographic-dependent. In: Clinical Orthopaedics and Related Research. 2019.
50. Chegini S, Beck M, Ferguson SJ. The effects of impingement and dysplasia on stress distributions in the hip joint during sitting and walking: A finite element analysis. J Orthop Res. 2009;
51. Ng KCG, Mantovani G, Lamontagne M, Labrosse MR, Beaulé PE. Cam FAI and Smaller Neck Angles Increase Subchondral Bone Stresses during Squatting: A Finite Element Analysis. In: Clinical Orthopaedics and Related Research. 2019.
52. Cooper RJ, Williams S, Mengoni M, Jones AC. Patient-specific parameterised cam geometry in finite element models of femoroacetabular impingement of the hip. Clin Biomech. 2018;
53. Hellwig FL, Tong J, Hussell JG. Hip joint degeneration due to cam impingement: a finite element analysis. Comput Methods Biomech Biomed Engin. 2016;
54. Ng KCG, Lamontagne M, Labrosse MR, Beaulé PE. Hip joint stresses due to cam-type femoroacetabular impingement: A systematic review of finite element simulations. PLoS One. 2016;
55. Wenger DE, Kendell KR, Miner MR, Trousdale RT. Acetabular labral tears rarely occur in the absence of bony abnormalities. Clinical Orthopaedics and Related Research. 2004.
56. Groh MM, Herrera J. A comprehensive review of hip labral tears. Curr Rev Musculoskelet Med. 2009;
57. Peelle MW, Della Rocca GJ, Maloney WJ, Curry MC, Clohisy JC. Acetabular and femoral radiographic abnormalities associated with labral tears. Clin Orthop Relat Res. 2005.
58. Ito K, Leunig M, Ganz R. Histopathologic features of the acetabular labrum in femoroacetabular impingement. Clin Orthop Relat Res. 2004;
59. Seldes RM, Tan V, Hunt J, Katz M, Winiarsky R, Fitzgerald RH. Anatomy, histologic features, and vascularity of the adult acetabular labrum. Clin Orthop Relat Res. 2001;

60. Beck M, Leunig M, Parvizi J, Boutier V, Wyss D, Ganz R. Anterior Femoroacetabular Impingement: Part II. Midterm Results of Surgical Treatment. In: Clinical Orthopaedics and Related Research. 2004.
61. Cashin M, Uhthoff H, O'Neill M, Beaulé PE. Embryology of the acetabular labral-chondral complex. J Bone Jt Surg - Ser B. 2008;
62. Ferguson SJ, Bryant JT, Ganz R, Ito K. An in vitro investigation of the acetabular labral seal in hip joint mechanics. J Biomech. 2003;
63. Ferguson SJ, Bryant JT, Ganz R, Ito K. The acetabular labrum seal: A poroelastic finite element model. Clin Biomech. 2000;
64. Ferguson SJ, Bryant JT, Ganz R, Ito K. The influence of the acetabular labrum on hip joint cartilage consolidation: A poroelastic finite element model. J Biomech. 2000;
65. Lee S, Wuerz TH, Shewman E, McCormick FM, Salata MJ, Philippon MJ, et al. Labral reconstruction with iliotibial band autografts and semitendinosus allografts improves hip joint contact area and contact pressure: An in vitro analysis. Am J Sports Med. 2015;
66. Philippon MJ, Nepple JJ, Campbell KJ, Dornan GJ, Jansson KS, LaPrade RF, et al. The hip fluid seal-Part I: The effect of an acetabular labral tear, repair, resection, and reconstruction on hip fluid pressurization. Knee Surgery, Sport Traumatol Arthrosc. 2014;
67. Kalisvaart MM, Safran MR. Microinstability of the hip--it does exist: etiology, diagnosis and treatment. J Hip Preserv Surg. 2015;
68. Shu B, Safran MR. Hip Instability: Anatomic and Clinical Considerations of Traumatic and Atraumatic Instability. Clinics in Sports Medicine. 2011.
69. Schilders E, Dimitrakopoulou A, Bismil Q, Marchant P, Cooke C. Arthroscopic treatment of labral tears in femoroacetabular impingement: A comparative study of refixation and resection with a minimum two-year follow-up. J Bone Jt Surg - Ser B. 2011;
70. Philippon MJ, Briggs KK, Yen YM, Kuppersmith DA. Outcomes following hip arthroscopy for femoroacetabular impingement with associated chondrolabral dysfunction: Minimum two-year follow-up. J Bone Jt Surg - Ser B. 2009;
71. Ayeni OR, Alradwan H, de Sa D, Philippon MJ. The hip labrum reconstruction: Indications and outcomes-a systematic review. Knee Surgery, Sports Traumatology, Arthroscopy. 2014.
72. Domb BG, El Bitar YF, Stake CE, Trenga AP, Jackson TJ, Lindner D. Arthroscopic labral reconstruction is superior to segmental resection for irreparable labral tears in the hip: A matched-pair controlled study with minimum 2-year follow-up. Am J Sports Med. 2014;
73. Krych AJ, Thompson M, Knutson Z, Scoon J, Coleman SH. Arthroscopic labral repair versus selective labral debridement in female patients with femoroacetabular impingement: A prospective randomized study. Arthrosc - J Arthrosc Relat Surg. 2013;29(1):46–53.
74. Larson CM, Giveans MR, Stone RM. Arthroscopic debridement versus refixation of the acetabular labrum associated with femoroacetabular impingement: Mean 3.5-year follow-up. Am J Sports Med. 2012;
75. Bedi A, Dolan M, Leunig M, Kelly BT. Static and dynamic mechanical causes of hip pain. Arthrosc J Arthrosc Relat Surg. 2011.
76. Kumar D, Dillon A, Nardo L, Link TM, Majumdar S, Souza RB. Differences in the association of hip cartilage lesions and cam-type femoroacetabular impingement with movement patterns: a preliminary study. PM R. 2014 Aug;6(8):681–9.

Biomechanics of Soft Tissue Injuries about the Hip

20

Ran Atzmon and Marc R. Safran

20.1 Introduction

The hip is an intrinsically stable joint due its bony congruence and spheroidal morphology, through the architecture of the femoral head and acetabulum. This congruency is further enhanced by an optimized skeletal design including adequate acetabular and femoral offset, version, inclination, and coverage [1–6]. Soft tissue contributions to hip stability include static structures, such as the capsuloligamentous structures and the labrum, as well as dynamic contributions provided by the concavity-compression effect produced by the bony anatomy in concert with the peri-articular muscles, the negative intra-articular pressure that exists in all joints, as well as adhesion–cohesion provided by the joint fluid and smooth articulating surfaces.

Together, the hip joint, comprised of bony and soft tissue elements which are obligated to work together in synchronization, create smooth and stable kinesis throughout the functional range of motion and under various loads. Damage to any one of these components might interfere with this synchronization and put the hip joint at risk for developing abnormal movement which can result in further joint damage, and may manifest as pain, instability (or microinstability), and disability [7].

While there is another chapter on the normal anatomy and biomechanics of the hip, this chapter explores the biomechanical consequences of soft tissue damage about the hip. This chapter will focus on what is known in this recently growing body of literature about the biomechanical effects of damage to the primary soft tissue elements about the hip, mainly the labrum and capsuloligamentous structures with a significant focus on hip microinstability. Hip microinstability is an evolving concept which is becoming the focus of interest in recent years, and has been gaining recognition for being a prevalent source of pain and disability among people participating in sports activities.

20.2 Functional Anatomy, Kinematics, and General Biomechanics

As was previously described, the hip joint serves to balance the upper body while standing or walking, as well as strenuous activities such as running and jumping, due to its unique ball and socket like design which allows it to become a multiaxial joint. In order to achieve that balance, the joint has to be stable at any position. This unique stability is achieved through a delicate

R. Atzmon · M. R. Safran (✉)
Division of Sports Medicine, Department of Orthopaedic Surgery, Stanford University, Stanford, CA, USA
e-mail: msafran@stanford.edu

J. Koh et al. (eds.), *Orthopaedic Biomechanics in Sports Medicine*,
https://doi.org/10.1007/978-3-030-81549-3_20

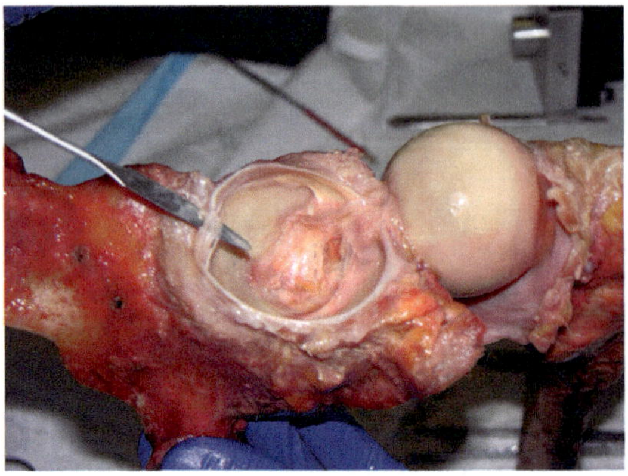

Fig. 20.1 Labrum. Photograph of cadaver hip dislocated, with an elevator underneath a degenerated labral tear, laterally, while the head is pulled inferiorly with ligamentum teres attached. Note the acetabulum is not entirely a sphere, with the inferior most aspect being the transverse acetabular ligament, which connects the anterior–inferior aspect of the labrum with the posterior inferior aspect

interplay between a high degree of bony articular congruency of the femoral head within the bony acetabulum, the acetabular labrum, the capsule (with its ligamentous contributions), ligamentum teres, and the encompassing soft tissue elements, being mainly the muscles [6, 8, 9]. However, in reality, the hip is a gimbal joint, where there is also translation and rotation of the femoral head relative to the acetabulum, resulting in a moving center of motion of the femoral head relative to the acetabulum through range of motion [6, 10, 11]. Damage to any one of these elements can lead to interruption of this delicate balance, increasing femoral head motion relative to the acetabulum, and may jeopardize normal hip function. Familiarity with the underlying factors predisposing to hip instability (and microinstability), and the general principles of hip biomechanics, should be considered with respect to the normal hip, and in regard to a pathological hip (Fig. 20.1).

20.3 The Labrum

The labrum has many functions, but two key functions are contributions to (1) joint stability and (2) pressure distribution within the femoroacetabular articulation. Thus, damage to the labrum, either the result of injury or iatrogenically through surgical debridement or excision, can disrupt labral function.

The labrum aids in hip stability through a couple of mechanisms. First, the labrum deepens the acetabular socket, serving as a barrier to excessive translation, like a car tire at the curb on a street. The femoral head has to translate further to get beyond the labrum to dislocate. Second, the labrum helps seal the joint, maintaining the negative intra-articular pressure. This negative pressure effect helps resist distraction of the femoral head from the acetabulum. Like a suction cup, the more the hip is distracted, the greater the negative pressure to resist that translation, when the labrum is intact and the seal is maintained. However, when a labral tear occurs, and the seal is breached, there is an increased femoral head motion relative to the acetabulum and the forces to distract the hip diminish [12, 13]. Crawford et al. [14] explored the effect of labral tears on hip kinematics at the extremes of joint motion. The authors showed that following joint venting (thus eliminating the negative intra-articular pressure), the force necessary for hip distraction diminished compared to the unvented state, and that a labral tear also decreased the force required to distract the hip, with an increase in hip rotation and displacement. Signorelli and colleagues [13, 15] studied the effect of hip pivoting movement and distraction on the hip joint center on eight cadaveric hips with five different labral conditions. The labrum of each hip was tested in its intact state, labral-chondral separation, vertical mattress suture repair, cerclage (simple) suture repair (around the labrum), and partial labrectomy. The authors found a significantly greater displacement of the femoral head with respect to the hip joint center after partial labrectomy when compared with labral repair and the intact labrum, and a significantly less force was required to distract the femur after cerclage repair (25%), vertical mattress repair (22%), and labrectomy (40%).

Of note, the joint resistance to distraction between the intact and vented cases is not linear, with the highest distraction occurring in a relatively small joint displacement of 1–3 mm [14, 16]. When examining the femoral head subluxation relative to the center of the acetabulum, venting of the joint produced a 1 mm shift during hip internal rotation, and 0.5 mm shift with hip abduction, whereas a labral tear did not have any additional impact with these maneuvers [14]. Additionally, a full thickness labral tear did cause excessive external rotation and abduction under constant torque conditions [14]. Other studies [17, 18] have also shown that venting the central compartment with a needle through the capsule breaks the sealing effect and significantly reduces the required force needed for hip distraction. Of note, solely venting the capsule by placing the needle in the peripheral compartment did not achieve similar results, thus showing the importance of the labrum in maintaining hip stability, and the consequences of labral tear.

Ito et al. [16] dissected seven cadaveric hip specimens and reached similar conclusions. The authors noticed that after a complete removal of the capsule, the required distraction load for the hip was further reduced after labral tear or resection was performed, and concluded that the isolated acetabular labrum contributes to hip stability when the joint is distracted in smaller distances (1 and 3 mm). However, no significant difference was found at 5-mm displacement. When comparing distraction forces after the capsule was partially resected with a completely resected capsule, the reduction of the force for distraction was greatest at 1, 3, and 5 mm. The authors attributed phenomenon to the effect of the zona orbicularis which was found to be a fundamental structure for hip stability in distraction, and will be discussed in more detail below. Another fundamental study, by Nepple et al. [19], evaluated the suction seal in eight cadaveric hips. These investigators measured the labral sealing effect in different conditions: an intact capsule, intact labrum, minor labral tear, complete labral resection, and labral reconstruction with iliotibial band graft. In each of these conditions, the hip sealing ability was measured by the negative pressure produced though the force required to distract the femoral head. The results showed that the relative contribution of the labrum to distractive stability with an intact capsule was the greatest at the first 1 and 2 mm, where it accounted for up to 77% of the total distractive stability. The capsule contribution to distractive stability increased with the progression of the femoral head displacement, mainly from the first 3 mm to 5 mm. The contribution of the labrum to distractive stability was significantly greater than the capsule in the first 1 and 2 mm ($p < 0.05$), with no significant difference at 3 mm to 5 mm. Further results showed that partial and complete labral resection resulted in a significant decrease in the distractive resistance, whereas labral repair and reconstruction clearly improved the force to distract the hip, with the latter also being significant. The labrum continued to resist the distraction force until greater than 6 mm of distraction, after which the capsule became the primary stabilizer. This study demonstrates that the labrum is the most important limiting factor against distraction, by preserving the suction effect, especially at small displacements (1–2 mm). Further, labral repair or reconstruction improves the distractive stability of the hip suction seal. Moreover, the contribution of other factors apart from the capsule or the labrum was less than 5% at all the measured conditions [19]. This excessive femoral head motion may result in further damage to the labrum, stretching of the capsuloligamentous structures resulting in pain, and greater femoral head motion which leads to more damage. A paradigm for this pathway of damage is represented in Fig. 20.2 and explained in more detail below.

Another critical function of the labrum is unified distribution of forces within the femoroacetabular articulation. The labrum contributes to this distribution through a couple of different mechanisms as well. First, the labrum helps seal the joint to resist extravasation of synovial fluid from inside the central compartment articulation. This allows a thin layer of synovial fluid to be maintained between the ball and the socket during weight bearing. This enhances the joint lubrication and transfers part of the load to the synovial fluid through pressurization effect, and hence reduces the load on the articular cartilage [12]. Soltz et al. [20] also found the fluid pressurization reduces the joint friction between the two surfaces significantly, and shields the cartilage matrix from as much as 90% of the applied load and this is reduced with labral tear. Cadet et al.

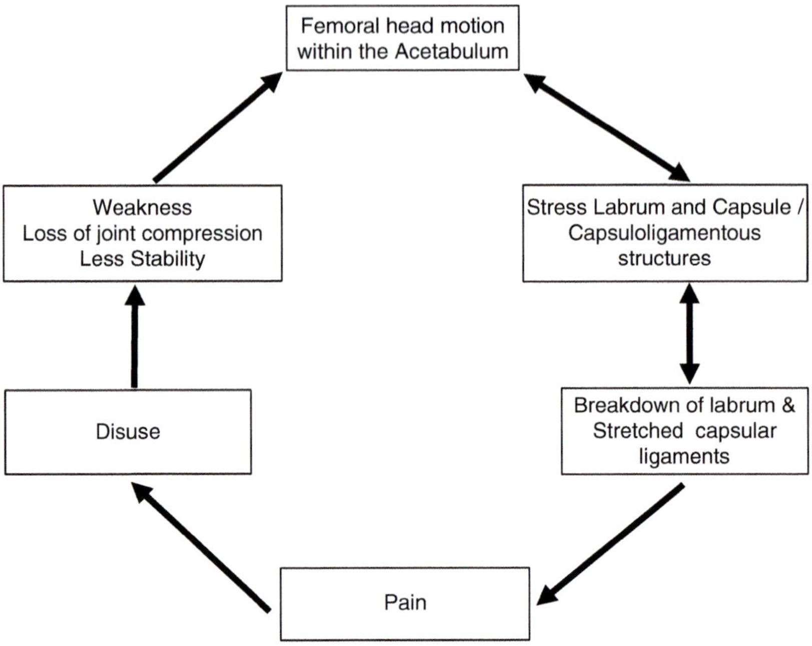

Fig. 20.2 Paradigm demonstrating how repeated femoral head translation may result in increasing stress to the labrum and capsule, resulting in injury to these structures, which, while producing pain and possible dysfunction, also increases the femoral head translations, that result in further labral and capsular damage

[21] compared an intact labrum state to four other labral pathologies, examining the fluid seal properties of each pathology. The authors found that labral repair surpassed partial labral resection and reconstruction in preserving the joint fluid seal; however, labral repair did not restore the fluid seal characteristics as effectively as in intact labral condition. This phenomenon of maintaining the fluid layer contributes to the joint gliding surface by making it smoother, more efficient and equally distributing the forces throughout the joint. Song et al. [22] looked at the acetabular labrum's sealing effect in maintaining a low friction environment at the hip joint. The authors hypothesized that the hip resistance to rotation represents the amount of friction at the articular cartilage surface. Five cadaveric hips with a normal osseous anatomy were tested under different body weight cycles, and after partial and complete labrectomy. The results demonstrated that the hip resistance to rotation in an intact hip was significantly increased following partial labrectomy and complete labrectomy. The authors felt the sealing effect, which prevents fluid exudation, also maintains a low friction environment at the joint articular surface, and that even focal labrectomy may result in increased joint friction and ultimately to osteoarthritis.

The labrum also helps reduce strain within the articular cartilage through another mechanism. Fluid flow within articular cartilage occurs with joint loading. The labrum is more impervious to fluid than articular cartilage. Thus, in the normal hip, the labrum resists or limits fluid extravasation from within the acetabular articular cartilage with joint loading, such as weight bearing. This allows for a uniform distribution of compressive forces over the articular surface [4, 5, 12, 18, 23, 24]. However, when the labrum is damaged or missing, the intra-chondral fluid can more easily be displaced and extravasate from the articular cartilage. Hence, during joint loading, the fluid within the articular cartilage may extrude outside and cause the cartilage to deform as a result [4, 12, 23, 24]. Greaves et al. [25] also proved that a repaired labral tear significantly decreased the chondral stress on the femoral head when compared with chondral stresses when a torn labrum is present. Further, they showed that labral repair significantly reduces the mean and maximum chondral stresses compared with labral resection. Thus, labral tears may lead to increased articular cartilage friction, deformation, and strain, which subsequently may lead to osteoarthritis (OA) [4, 5, 23, 24].

20.4 Labral Pathologies

Acetabular labral tears are quite common and may not necessarily cause symptoms. The labrum may be torn via many different reported mechanisms. Dy et al. [26] investigated the exact mechanism leading to labral injuries on 7 human cadaver specimens, and found that external rotation and abduction with moderate hip extension or flexion can produce considerable tensile strains in the anterior part of the acetabular labrum due to an anterior translation of the femoral head, causing labral separation and tear from its articular margin [18, 26]. Safran and colleagues [27] studied twelve cadaveric hips which were dissected free of soft tissues, except the capsuloligamentous structures and the labrum. Each hip was tested in 36 different positions and monitored for strain across the intact labrum. The authors found that the most considerable change in strain was measured in the posterior labrum, whereas the majority of labral tears occurred at the anterior or anterolateral region. The results matched the clinical impingement tests for tears in this location [28]. Thus, taking the hip through extremes of motion may result in strain injury to the labrum. Additionally, femoroacetabular impingement (FAI) may result in unique patterns of labral damage, based on the type of impingement. In cam impingement, the non-spherical femoral head is jammed into the anterosuperior acetabulum area, creating shear and compression forces at the junction between the labrum and the cartilage and at the subchondral tidemark, resulting in initially sparing the labrum as the injury is to the labral-chondral junction and tends to occur in the anterolateral area of the joint [29–32]. In pincer type impingement, the acetabular overcoverage limits the hip's range of motion and results in crushing of the labral tissue between the femoral head neck junction and acetabular rim and tends to be more diffuse, around the acetabulum [29–32]. Additionally, with continued motion, the femoral neck may lever against the acetabulum, and the femoral head may subluxate posteroinferiorly, which can lead to posteroinferior cartilage abrasion and development of contrecoup lesion in approximately one third of pincer cases [30, 31].

And lastly, in dysplasia, where the acetabular bony coverage is insufficient, there is usually associated labral hypertrophy. However due to the lack of bony support, more force generally is imparted on the labrum, as it contributes more to the acetabular volume and shares a greater amount of joint loading forces. The forces on the labrum are even greater, particularly when the muscles about the hip are weak, affecting concavity compression, resulting in increased femoral head motion relative to the acetabulum, which in turn increases the load on the labrum. This femoral head motion may occur anteriorly or laterally, resulting in shear forces and labral-chondral separation [32]. Additionally, hip microinstability with normal acetabular depth may have similar mechanisms to that of dysplasia, resulting in anterior or lateral labral-chondral separation, initially [33]. Traumatic labral tears are often more localized and confined to a particular region of the labrum, in relation to the force vector acting on the joint, with the most prevalent cause for acute labral tear occurring with external rotation and hyperextension movement and can be intra-substance (particularly when degenerative) or at the labral-chondral junction. One pathomechanism of labral tears is that extreme range of motion puts the entire joint under substantial load, and in this extreme position, the labrum takes on a weight-bearing role [34, 35]. Finally, a number of studies [36, 37] have found that the majority of patients with labral tears have an underlying anatomical deformity of the femur or acetabulum which predisposed their hips to labral tears. As discussed above, the underlying causes may include hip dysplasia and FAI, but also Legg–Calve–Perthes disease and coxa valga. Athletes may have a repetitive microtraumatic, traumatic or underlying bony dysmorphology as the cause.

20.5 Consequence of Labral Tears

Labral injuries might trigger a cascade which ultimately leads to hip osteoarthritis by affecting the joint congruency and the femoral head translation within the acetabulum, which induces shear forces to the articular cartilage and influences joint stability. Deficient labra may alter the hip biomechanics and shift the contact area from being global to a more focal area at the acetabular lateral margin, thus increasing the contact stress on the cartilage by as much as 92% [5, 6, 14]. As a result, the articular cartilage hypertrophies and becomes more condensed and consolidates by up to 40%. Furthermore, the labral damage significantly reduces the joint surface's resistance to distraction, increases the friction to initiate motion, and likewise increases the femoral head translational within the joint [5, 6, 14, 22]. Further, as previously discussed, labral tear may increase cartilage consolidation due to extravasation of intra-chondral fluid from the articular cartilage after the labral seal is broken, thus increasing chondral strain and cartilage deformation [4, 5, 23, 24].

20.6 The Capsule

Recent research has increasingly demonstrated the importance of the capsuloligamentous structures about the hip. The dynamic stability of the hip which stems from the joint congruency and the surrounding labrum is further amplified by the enclosing capsule and ligaments. The joint capsule provides stabilization by restricting excessive translation and rotation of the femoral head. Traditionally it has been thought that the stabilization effect was passive; however, some recent research has suggested that the capsuloligamentous structures may function dynamically as well [1, 38–40]. The fibrotic capsule consists of 3 main confluence ligaments that provide static constraint to the joint throughout the range of physiological motions [1, 2]. Additional capsular reinforcement derives from the zona orbicularis, which is found at the distal part of the capsule, horizontal to the femoral head, coils and tightens around the femoral neck with the hip range of motion [16]. Moreover, the capsule transmits proprioceptive and pain sensations which aid the body to protect the joint, and fortify the labral sealing effect, in order to preserve the extravasation of synovial fluid. The integrity of the hip capsule is imperative in cases of soft tissue laxity, instability, and hip dysplasia [41].

Van Arkel and colleagues [42] have shown that each ligament within the capsuloligamentous structure about the hip provides stability and resistance to certain motions. The authors particularly noted in their in vitro study of 9 cadaveric hip joints, that the capsular ligaments provide the primary restraint to internal rotation and external rotation. Cutting or stretching these ligaments results in increased femoral head motion, translation, hip range of motion, and joint stability [42]. Particularly, the iliofemoral ligament restricts external rotation in flexion and extension, as well as internal rotation in extension, and further still limits hip extension. As the strongest ligament in the body, and its position anteriorly, it resists anterior hip microinstability. Thus, damage to this ligament, such as may occur with an "interportal" capsulotomy for hip arthroscopy, increases hip external rotation, extension, and anterior translation of the femoral head. The ischiofemoral ligament limits internal rotation and adduction of the hip when in flexion. This ligament is injured in posterior hip dislocation [6, 7, 42–44]. The pubofemoral ligament resists hyperabduction, and to some degree, external rotation [39, 45]. The zona orbicularis, as noted above, resists distraction of the femoral head from the acetabulum [16].

Recent work by Tsutsumi and colleagues [46] suggests that the iliofemoral ligament may also act as a dynamic stabilizer. The authors dissected fourteen hips and examined the relationships between the anterosuperior region of the joint capsule and the iliofemoral ligament, and the deep aponeurosis of the gluteus minimus and iliopsoas tendon. Parts of the specimens were also macroscopically and histologically analyzed.

The authors found a connection between the gluteus minimus tendon and the joint capsule and an osseous connection at the intertrochanteric line. Further, the deep aponeurosis of the iliopsoas was also connected to the joint capsule, as was observed by capsular thickening at these connection sites. The gluteus minimus and iliopsoas fibers were also arranged according to these connections. The authors concluded that based on these findings the iliofemoral ligament could be viewed as a dynamic stabilizer with the ability to transmit the muscular power to the joint via the capsular complex [46]. Other cadaveric studies also demonstrate a distinct capsular thickening near the acetabular origin, especially at the posterosuperior and superior hemi-quadrants, with a maximum thickness reaching up to 15 mm. This thickened area corresponds with the location of the iliofemoral ligament and connected the capsule to other surrounding structures such as the iliocapsularis, indirect head of the rectus femoris and the conjoint tendons, the obturator externus, and the gluteus minimus tendons [47, 48]. Thus, while the capsuloligamentous structures are generally thought to be static stabilizers, they may, in fact, also provide dynamic stability. Hence, injury to the capsule or the connected muscles may result in excessive femoral head translation relative to the acetabulum. This may result in shear forces to the articular cartilage, as well as increased stress to the labrum and capsule, resulting in labral tears, capsular stretching, and further increased motion resulting in hip microinstability.

In addition, the senior author of this chapter has noted during his dissections about the hip that the capsuloligamentous structures are quite thickened near its acetabular origin. This particularly thick and unforgiving tissue may function to augment or protect the labrum from a sudden increase in the load, such as femoral head translation due to external rotation or hyperextension. Failure of this tissue may result in greater femoral head motion, labral tearing and capsular stretching… more hip microinstability.

While adequate bony support is important for stability of the hip, it has also been shown that capsular laxity can be a major cause for atraumatic hip microinstability, particularly from repeated stretching events in extension and external rotation. This theory has been supported by Jackson et al. [49] who created a cadaveric model of hip capsule laxity by stretching the capsule in extension using six different capsular conditions: intact, vented, instability, capsulotomy, side-to-side repair, and capsular shift. Their results demonstrated greater total range of motion in external rotation and extension when compared with the intact condition. Similarly, Han et al. [50] examined hip capsular laxity under tension through various hip positions by creating numerous incisions ("pie crusting") in the capsule. They also concluded that capsular laxity leads to microinstability of the hip, as was indicated by the significantly increased joint rotations and femoral head translations relative to the femoral head center. Although these models exhibit a clear relationship between static capsular stretching and capsular laxity leading to microinstability, the investigators failed to reproduce the dynamic pathologic laxity state. Hence, Johannsen et al. [51] developed a model that simulates the pathomechanism of microinstability in the young hip. To that end, the authors devised a protocol consisting of controlled dynamic cyclic stretching of the anterior hip capsule in 7 dissected hip specimens with intact, vented and stretched capsular states, in otherwise normal hips. They showed that cyclic stretching of the anterior hip capsule results in capsular insufficiency and a significant increase in hip rotation and femoral head displacement relative to the vented state in all six planes. In a subsequent study the same group, Johannsen and colleagues [52], dissected 6 cadaveric pelvii leaving only the hip capsule and labrum. The twelve dissected hips were then stretched in extension and external rotation with simultaneous rotation at the hip mechanical axis, creating a capsular laxity state. The investigators then produced labral and capsular tears laterally. When comparing the intact state to labral vented state, and to labral and capsular tear, a noticeable increase in hip instability in rotation and translation were noted, respectively. The authors concluded that the labrum has a minor, but significant role in controlling the arc of rotation when the capsule is intact, but when there is a capsular injury, the status of the labrum is critical in impacting femoral head motion and translation. Myers et al. [7] also studied the relationships and contribution of the iliofemoral ligament and the acetabular labrum for hip stability. The authors dissected fifteen fresh-frozen cadaveric hips using a biplane fluoroscopy in order to measure the hip translations and rotation with an intact and pathologic iliofemoral ligament and acetabular labrum. They found that the iliofemoral ligament had a significant role in limiting external rotation and anterior translation of the femoral head of 12.9° ± 5.2° and 1.4 ± 0.5 mm respectively, compared with the acetabular labrum which provided only a secondary stabilizing role for these motions. Further biomechanical analysis have proven the iliofemoral ligament to be the strongest ligament of the all three ligaments, capable of resisting anterior translation and substantial forces prior to failing [6, 44].

Hence, the capsule can be considered the primary soft tissue stabilizer of the hip, whereas the labrum acts as a secondary stabilizer. Of note, most significant changes in femoral head arc of rotation and translation were in the mediolateral plane in the combined capsular laxity and labral insufficiency state, which is in accordance with previous studies [11].

Many surgeons utilize a capsulotomy to obtain an atraumatic entry into the joint and facilitate instrument maneuverability within the hip joint capsule for arthroscopic surgical procedures. Several capsular incisions have been described in the literature, with the most common being either an interportal (IP) capsulotomy or a T-capsulotomy. The incision for the interportal capsulotomy is often made in the anterior aspect of the joint in order to connect the anterolateral and mid-anterior main portals, between 12-o'clock and 3-o'clock position (right hip), parallel to the acetabular rim [1, 53]. Anatomically, this capsulotomy cuts the entire proximal extent of the iliofemoral ligament [40]. The benefit of the IP capsulotomy is that it lessens the need for hip distraction and improved visualization and maneuverability of instruments within the hip joint. The T-capsulotomy utilizes the interportal capsulotomy proximally, and then from the mid-extent of the IP capsulotomy, an incision is made perpendicularly, distally, forming a "T" shape. This capsulotomy allows a wider arthroscopic view, making peripheral compartment surgery easier, but may also incise the zona orbicularis ligament, when the T-capsulotomy is taken more distally. Several studies have shown that IP- and T-capsulotomies may produce increased hip range of motion and reduce the required force hip distraction, but also increased femoral head translation and hip instability [8, 43, 54–56].

Martin et al. [39] conducted a cadaveric study, examining the stabilizing roles of the hip capsular ligaments. The authors concluded that the lateral arm of the iliofemoral ligament has dual control of external rotation in flexion and both internal and external rotation in extension, and that releasing of both the medial and lateral limbs of the iliofemoral ligament produced the greatest increase in motion for external rotation. Abrams et al. [53] tested the effect of different types of capsulotomies on hip rotational biomechanical characteristics on 7 cadaveric hip specimens. Each hip was examined under neutral flexion and 40 degrees of flexion in various capsular states. A significant increase in hip external rotation was shown with T-capsulotomy compared with the intact and interportal capsulotomy states in both flexion positions. The repaired T-capsulotomy restored the rotational profile back to the native state. Likewise, Philippon et al. [55] found that capsulotomy procedures can result in increases in external, internal, abduction, and adduction rotations throughout a full range of hip flexion. Whereas, repaired or reconstructed capsule succeeded in partially reducing the increased rotational range of motion. Wuerz and colleagues [57] studied the effect of capsulotomy size on hip stability in various position in eight cadaveric hips. Each hip was tested under torsional loads in four different conditions which included intact capsule, a 4 cm and a 6 cm capsulotomy, and a repaired capsule. Measures indicating joint kinematics such as hysteresis area, range of motion, and neutral zone were also obtained for each condition. The authors showed that capsulotomy led to a significantly increase in all three measures of joint mobility range of motion, and that a subsequent capsular repair effectively restored these measures to the intact capsular state at time zero. Clinically, Frank et al. [58] demonstrated an improved clinical outcome in patients who underwent hip arthroscopy and had a complete capsular repair at the end of the surgery as compared with those that only underwent a partial repair (closing the distal limb) after T-capsulotomy.

As a result of these biomechanics changes with capsulotomy, iatrogenic pain or instability (including microinstability) during certain hip movements may occur, especially in patients with generalized ligamentous laxity or osseous dysplasia. Due to the growing evidence indicating the capsule's major contribution to hip stability, various capsular closure procedures have been developed. Options for closure include complete repair or partial repair. In patients with microinstability and/or laxity, capsular plication and thermal shrinkage have also been advocated [43, 59]. Studies have shown that capsular plication improved the patient's outcomes scores with symptomatic microinstability [52, 60, 61]. Moreover, capsular plication reduces the joint intra-articular volume, thus increasing hip stability, especially in hypermobility state or redundant capsule [43].

Though it can be inferred from the aforementioned discussion that capsular closure procedures are significantly superior to capsular release, a recently published meta-analysis [59] which evaluated the effect of different capsular management strategies on post-surgical outcomes following hip arthroscopies, found no consensus in the literature regarding the superiority of capsular closure procedures over capsular release. However, the authors did find a consistent tendency toward superior outcomes in the capsular repair cohorts, and concluded that further randomized controlled studies are needed. Economopoulos et al. [62] recently published a randomized clinical trial comparing T-capsulotomy and interportal capsulotomy without closure, and interportal capsulotomy with closure. The authors used common clinical outcomes scores and revision surgeries in order to assess the surgical outcome. The results showed improved patient-reported and surgical outcomes in the capsular repaired group in a two-year follow-up period.

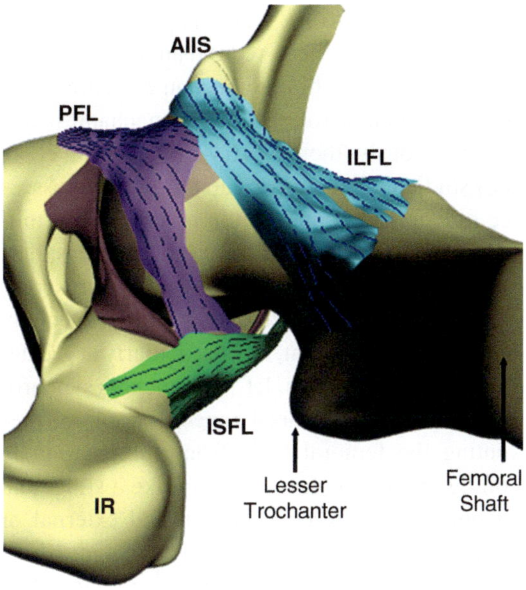

Fig. 20.3 Capsular ligaments of the hip. Computer reconstruction from an anatomical dissection demonstrating the 3 main capsular ligaments about the hip. The reconstruction is looking from the inferior aspect of the hip. The blue ligament anteriorly is the iliofemoral ligament, the purple ligament is the pubofemoral ligament, and the green ligament is the ischiofemoral ligament

In light of these findings, evidence continues to demonstrate that iatrogenic hip microinstability may occur as a result of increased femoral head translation after arthroscopic capsulotomy. As a result, many surgeons have evolved their techniques to repairing the capsule. However, the cadaveric studies only demonstrate that time zero mechanics are restored [8, 43, 53–56, 58]. Further, many questions remain, such as whether the capsular repairs heal, whether the kinematics of the hip joint are normal/restored after healing of the repaired ligaments, and the optimal method of capsular repair (Fig. 20.3).

20.7 Supplementary Key Ligaments about the Hip

Two hip ligaments that are not discussed as frequently as the capsular ligaments (i.e., the iliofemoral, ischiofemoral, and pubofemoral ligaments) are the zona orbicularis and the ligamentum teres. The zona orbicularis is discussed above.

The ligamentum teres (LT), which was viewed in the past as a redundant structure with no contribution to hip biomechanics or function, has a substantial role in hip biomechanics than was previously thought [63–65]. Martin et al. [66] studied 12 cadaveric hips in order to assess the function and contribution of the LT in limiting hip rotation in 18 distinct hip positions. The authors found that when the LT was arthroscopically cut, hip instability was produced, especially hip rotation movement, and particularly at greater than 90° of hip flexion. LT movement is synchronized with femoral head range of motion, preventing the femoral head from subluxating by coiling around the femoral head like a sling, and becoming taut with hip flexion and external or internal rotation, thus creating a "ball and string" effect [63, 65, 67]. Moreover, the coiling effect also pulls the femoral head into the acetabulum and increases the end range motion of the hip. This model is further accentuated in hip pathological conditions such as hip dysplasia or capsular laxity when the osseous and other primary soft tissue restrains are deficient [63–65, 68]. Long-term follow-up of patients with LT transection during open hip preservation surgery demonstrates mild hip instability with symptoms of giving way in about a quarter of patients. Van Arkel et al. [42] found the LT acts as a secondary restraint in external rotation during high flexion movement above 60°, with neutral or full adduction. Hence it can be concluded that the LT assumes a secondary stabilizing role in the presence of intact capsular ligaments, but its importance in hip stability (and thus femoral head translation and its effects noted above) is even more important when these capsular ligaments are injured, stretched, cut, or lax [42, 68].

20.8 Additional Soft Tissue Elements about the Hip and Associated Pathologies

The majority of the extra-capsular elements surrounding the hip serve as dynamic hip stabilizers and are neuromuscular in nature. The muscles react to the different forces acting on the hip joint, thus preserving normal joint kinematics and force-coupled compression that further augment hip stability [1, 8, 69]. Some muscles such as the iliocapsularis originate or insert on to the joint osseous surface, while others such as the rectus femoris merely cross the joint to insert more distally. In addition to a bulk effect of the thick soft tissue envelop about the hip reducing excessive translations of the femoral head, muscular co-contractions about the hip help provide stability through the concavity-compression mechanism [11]. Additionally, as noted above, some muscles such as the iliocapsularis and the gluteus medius muscle tendon units attach to the capsuloligamentous structures resulting in dynamic tensioning of the ligaments, also providing stability. And lastly, some muscles are anatomically positioned to prevent femoral head translations/subluxations, such as the iliopsoas muscle anteriorly and the gluteus medius and minimus laterally. Thus weakness or injury to these peri-articular muscles may result in abnormal femoral head translations, resulting in injury to the labrum and/or capsuloligamentous structures, resulting in hip microinstability [70, 71]. Further, due to the required synchronization of the different structure surrounding the hip, every injury may affect the joint stability and cause a "domino effect" influencing other structures.

Excessive femoral head motion, as seen with microinstability may also be associated with iliopsoas tendinitis and/or internal snapping hip syndrome (coxa saltans). The iliopsoas is positioned as a key dynamic anterior stabilizer of the hip. In hip microinstability, with anterior translation of the femoral head, the iliopsoas may become overused trying to maintain femoral head position within the acetabulum. Fatigue and overuse of the iliopsoas may result in iliopsoas bursitis, tendinitis, and potentially internal snapping hip [6, 70, 72, 73]. In the presence of hip dysplasia, the gluteus medius and minimus work harder to help stabilize the hip, either through its contributions to the iliofemoral ligament and/or attempting to stabilize the hip from lateral subluxation or translation. As a result, the gluteus muscles may become overused, fatigued and weakened, leading to greater trochanteric pain syndrome—particularly gluteus medius syndrome, gluteus medius tendinitis, and often associated trochanteric bursitis [6, 70, 72–75]. Further, in the presence of hip dysplasia, the iliopsoas may impinge on the femoral head, acting as a secondary dynamic stabilizer during range of motion, leading to pain, inflammation, and overload [76].

20.9 Hip Instability Leading to Osteoarthritis

As noted above, the hip was previously thought to be an inherently stable joint due to the innate bony anatomy; however, recent research has shown that hip joint stability exists as a result of a delicate harmony and synchronization between the static and dynamic stabilizers of the hip. Hip microinstability is a term used to describe as extraphysiologic hip motion which is manifested as pathological femoral head motion without an obvious hip dislocation, accompanied by hip pain [52, 61]. Hip laxity is defined as excessive motion between the femoral head and acetabulum without concomitant symptoms [53]. When the demanded range of motion exceeds the physiologically allowed motion of the hip joint, symptoms may arise. This extreme range of motion can result in anterior impingement followed by posterior instability, or anterior or lateral hip subluxation without impingement.

Symptomatic hip microinstability is a relatively new and evolving concept which was dismissed in the past as a cause for hip pain, and only recently was acknowledged as a significant cause of pain and disability in young patients and athletes [8, 61, 69, 76]. The two prevalent hypotheses and suggested pathomechanics of hip microinstability are: (1) that subtle anatomic abnormalities in the presence of repetitive rotational motion and stretching, along with axial loading, may lead to microinstability. This range of motion may be seen in various fields of sports such as football, soccer, gymnastics, golf, martial arts, ballet, and many more. The second hypothesis (2) advocates that soft tissue abnormalities such as capsular and innate ligamentous laxity, or muscle weakness, may lead to excessive femoral head motion relative to the acetabular center, resulting in soft tissue breakdown, culminating in microinstability. These two hypotheses are not mutually exclusive, and there may be overlap in some patients, where both mechanisms contribute to result in microinstability. Other causes, such as iatrogenic microinstability caused by arthroscopic capsulotomies, as described above, is an increasing concern. As demonstrated in Fig. 20.2, there is a self-perpetuating downward spiral of femoral head motion that can result in increasing soft tissue injury, leading to pain and increasing femoral motion and microinstability. As the femoral head motion is increased relative to the acetabulum, there is increased stress on the labrum and capsule which may result in tears of the labrum and stretching of the capsule. In turn, these tears may result in further femoral head motion that may result in greater tears of the labrum and stretching of the capsule. Additionally, this may affect the articular cartilage through edge loading, as well as loss of the labral protection of the articular cartilage strains. The microinstability that results may result in pain and affect activities of daily living, sports, and quality of life [50, 61, 77, 78]. A suggested classification [9, 78] for hip microinstability based on the predisposing known factors was recently suggested. It divides the patients into six distinct categories which include: connective tissue disorders, athletics activities or microtrauma, iatrogenic causes, significant bony abnormalities or developmental dysplasia of the hip, posttraumatic processes, and idiopathic causes. Patients suffering from hip microinstability often experience vague anterior hip pain which exacerbate with hip extension and external rotation activities [15, 78]. Risk factors predisposing to capsular laxity are congenital generalized ligamentous laxity and joint hypermobility, traumatic capsular injury, repetitive forceful stretching due to sports, iatrogenic laxity following arthroscopic capsulotomy or additional procedures. Other studies tried to appraise the iliofemoral ligament's function in restraining the femoral head motion, and concluded that capsular laxity further increases the femoral head motion [50, 52].

Sports that involve repetitive twisting and pivoting of the hip often lead to hip microinstability due to combination of anterior labral tears and elongation of the iliofemoral ligament [7, 79]. One theory claims that this combination by itself is not enough, and requires an underlying abnormal bony anatomy in order to initiate this pathological process. Alternatively, initial stretching of the iliofemoral ligament caused by the twisting motions might lead to increased tensile forces and possible tears of the anterior labrum. As suggested by Myers et al. [7] and Johannsen et al. [52], femoral head translation is increased when the iliofemoral ligament is stretched or cut, and increased further after subsequent labral tear.

20.10 Conclusion

While it has been classically taught that the hip is a ball in socket joint and its kinematics are based entirely on bony morphology, recent research has proven otherwise. Injury to the labrum results in increased femoral head translation relative to the acetabulum, as well as increased stresses and friction to the articular cartilage. Injury to the capsuloligamentous structures about the hip can also result in increased femoral head translations relative to the acetabulum. The increased femoral head translations may place increasing stress of the stabilizing labral and capsuloligamentous structures, causing more damage to these structures which, in turn, results even more increased femoral head motion, and thus more damage to these structures. The contact stresses as well as the focal chondral strains that result may result in clinical hip microinstability, and ultimately hip osteoarthritis. Additionally, patients with microinstability may develop tendinitis due to overload of the dynamic stabilizers of the hip, such as the iliopsoas and gluteus minimus and medius (Fig. 20.2).

Though hip microinstability is a well-known and established entity, clear diagnostic criteria for atraumatic instability based on physical examination or advanced imaging have yet to be widely recognized.

References

1. Domb BG, Philippon MJ, Giordano BD. Arthroscopic capsulotomy, capsular repair, and capsular plication of the hip: relation to atraumatic instability. Arthroscopy. 2013;29(1):162–73. https://doi.org/10.1016/j.arthro.2012.04.057.
2. Kohnlein W, Ganz R, Impellizzeri FM, Leunig M. Acetabular morphology: implications for joint-preserving surgery. Clin Orthop Relat Res. 2009;467(3):682–91. https://doi.org/10.1007/s11999-008-0682-9.
3. Bergmann G, Deuretzbacher G, Heller M, Graichen F, Rohlmann A, Strauss J, et al. Hip contact forces and gait patterns from routine activities. J Biomech. 2001;34(7):859–71. https://doi.org/10.1016/s0021-9290(01)00040-9.
4. Ferguson SJ, Bryant JT, Ganz R, Ito K. An in vitro investigation of the acetabular labral seal in hip joint mechanics. J Biomech. 2003;36(2):171–8. https://doi.org/10.1016/s0021-9290(02)00365-2.
5. Ferguson SJ, Bryant JT, Ganz R, Ito K. The influence of the acetabular labrum on hip joint cartilage consolidation: a poroelastic finite element model. J Biomech. 2000;33(8):953–60. https://doi.org/10.1016/s0021-9290(00)00042-7.
6. Bowman KF Jr, Fox J, Sekiya JK. A clinically relevant review of hip biomechanics. Arthroscopy. 2010;26(8):1118–29. https://doi.org/10.1016/j.arthro.2010.01.027.
7. Myers CA, Register BC, Lertwanich P, Ejnisman L, Pennington WW, Giphart JE, et al. Role of the acetabular labrum and the iliofemoral ligament in hip stability: an in vitro biplane fluoroscopy study. Am J Sports Med. 2011;39 Suppl:85s–91s. doi: https://doi.org/10.1177/0363546511412161.
8. Boykin RE, Anz AW, Bushnell BD, Kocher MS, Stubbs AJ, Philippon MJ. Hip instability. J Am Acad Orthop Surg. 2011;19(6):340–9. https://doi.org/10.5435/00124635-201106000-00004.
9. Shu B, Safran MR. Hip instability: anatomic and clinical considerations of traumatic and atraumatic instability. Clin Sports Med. 2011;30(2):349–67. https://doi.org/10.1016/j.csm.2010.12.008.
10. Delp SL, Maloney W. Effects of hip center location on the moment-generating capacity of the muscles. J Biomech. 1993;26(4–5):485–99. https://doi.org/10.1016/0021-9290(93)90011-3.
11. Safran MR, Lopomo N, Zaffagnini S, Signorelli C, Vaughn ZD, Lindsey DP, et al. In vitro analysis of peri-articular soft tissues passive constraining effect on hip kinematics and joint stability. Knee Surg Sports Traumatol Arthrosc. 2013;21(7):1655–63. https://doi.org/10.1007/s00167-012-2091-6.
12. Ferguson SJ, Bryant JT, Ganz R, Ito K. The acetabular labrum seal: a poroelastic finite element model. Clin Biomech (Bristol, Avon). 2000;15(6):463–8. doi: https://doi.org/10.1016/s0268-0033(99)00099-6.

13. Signorelli C, Lopomo N, Colle F, Bontempi M, Visani A, Zaffagnini S, et al. Restoration of the seal function of the acetabular labrum: in vitro study. J Mech Med Biol. 2015;15(02):1540036. https://doi.org/10.1142/S0219519415400369.
14. Crawford MJ, Dy CJ, Alexander JW, Thompson M, Schroder SJ, Vega CE, et al. The 2007 Frank Stinchfield award. The biomechanics of the hip labrum and the stability of the hip. Clin Orthop Relat Res. 2007;465:16–22. https://doi.org/10.1097/BLO.0b013e31815b181f.
15. Signorelli C, Lopomo N, Bonanzinga T, Marcheggiani Muccioli GM, Safran MR, Marcacci M, et al. Relationship between femoroacetabular contact areas and hip position in the normal joint: an in vitro evaluation. Knee Surg Sports Traumatol Arthrosc. 2013;21(2):408–14. https://doi.org/10.1007/s00167-012-2151-y.
16. Ito H, Song Y, Lindsey DP, Safran MR, Giori NJ. The proximal hip joint capsule and the zona orbicularis contribute to hip joint stability in distraction. J Orthop Res. 2009;27(8):989–95. https://doi.org/10.1002/jor.20852.
17. Takechi H, Nagashima H, Ito S. Intra-articular pressure of the hip joint outside and inside the limbus. Nihon Seikeigeka Gakkai Zasshi. 1982;56(6):529–36.
18. Safran MR. The acetabular labrum: anatomic and functional characteristics and rationale for surgical intervention. J Am Acad Orthop Surg. 2010;18(6):338–45. https://doi.org/10.5435/00124635-201006000-00006.
19. Nepple JJ, Philippon MJ, Campbell KJ, Dornan GJ, Jansson KS, LaPrade RF, et al. The hip fluid seal--part II: the effect of an acetabular labral tear, repair, resection, and reconstruction on hip stability to distraction. Knee Surg Sports Traumatol Arthrosc. 2014;22(4):730–6. https://doi.org/10.1007/s00167-014-2875-y.
20. Soltz MA, Ateshian GA. Experimental verification and theoretical prediction of cartilage interstitial fluid pressurization at an impermeable contact interface in confined compression. J Biomech. 1998;31(10):927–34. https://doi.org/10.1016/s0021-9290(98)00105-5.
21. Cadet ER, Chan AK, Vorys GC, Gardner T, Yin B. Investigation of the preservation of the fluid seal effect in the repaired, partially resected, and reconstructed acetabular labrum in a cadaveric hip model. Am J Sports Med. 2012;40(10):2218–23. https://doi.org/10.1177/0363546512457645.
22. Song Y, Ito H, Kourtis L, Safran MR, Carter DR, Giori NJ. Articular cartilage friction increases in hip joints after the removal of acetabular labrum. J Biomech. 2012;45(3):524–30. https://doi.org/10.1016/j.jbiomech.2011.11.044.
23. Haemer JM, Carter DR, Giori NJ. The low permeability of healthy meniscus and labrum limit articular cartilage consolidation and maintain fluid load support in the knee and hip. J Biomech. 2012;45(8):1450–6. https://doi.org/10.1016/j.jbiomech.2012.02.015.
24. Kim Y, Giori NJ, Lee D, Ahn KS, Kang CH, Shin CS, et al. Role of the acetabular labrum on articular cartilage consolidation patterns. Biomech Model Mechanobiol. 2019;18(2):479–89. https://doi.org/10.1007/s10237-018-1097-5.
25. Greaves LL, Gilbart MK, Yung AC, Kozlowski P, Wilson DR. Effect of acetabular labral tears, repair and resection on hip cartilage strain: a 7T MR study. J Biomech. 2010;43(5):858–63. https://doi.org/10.1016/j.jbiomech.2009.11.016.
26. Dy CJ, Thompson MT, Crawford MJ, Alexander JW, McCarthy JC, Noble PC. Tensile strain in the anterior part of the acetabular labrum during provocative maneuvering of the normal hip. J Bone Joint Surg Am. 2008;90(7):1464–72. https://doi.org/10.2106/JBJS.G.00467.
27. Safran MR, Giordano G, Lindsey DP, Gold GE, Rosenberg J, Zaffagnini S, et al. Strains across the acetabular labrum during hip motion: a cadaveric model. Am J Sports Med. 2011;39 Suppl:92–102. doi: https://doi.org/10.1177/0363546511414017.
28. Freehill MT, Safran MR. The labrum of the hip: diagnosis and rationale for surgical correction. Clin Sports Med. 2011;30(2):293–315. https://doi.org/10.1016/j.csm.2010.12.002.
29. Tannast M, Siebenrock KA, Anderson SE. Femoroacetabular impingement: radiographic diagnosis--what the radiologist should know. AJR Am J Roentgenol. 2007;188(6):1540–52. https://doi.org/10.2214/ajr.06.0921.
30. Ganz R, Parvizi J, Beck M, Leunig M, Nötzli H, Siebenrock KA. Femoroacetabular impingement: a cause for osteoarthritis of the hip. Clin Orthop Relat Res. 2003;417:112–20. https://doi.org/10.1097/01.blo.0000096804.78689.c2.
31. Beck M, Kalhor M, Leunig M, Ganz R. Hip morphology influences the pattern of damage to the acetabular cartilage: femoroacetabular impingement as a cause of early osteoarthritis of the hip. J Bone Joint Surg. 2005;87(7):1012–8. https://doi.org/10.1302/0301-620x.87b7.15203.
32. Kraeutler MJ, Goodrich JA, Fioravanti MJ, Garabekyan T, Mei-Dan O. The "outside-in" lesion of hip impingement and the "inside-out" lesion of hip dysplasia: two distinct patterns of acetabular chondral injury. Am J Sports Med. 2019;47(12):2978–84. https://doi.org/10.1177/0363546519871065.
33. Shibata KR, Matsuda S, Safran MR. Is there a distinct pattern to the acetabular labrum and articular cartilage damage in the non-dysplastic hip with instability? Knee surgery, sports traumatology. Arthroscopy. 2017;25(1):84–93. https://doi.org/10.1007/s00167-016-4342-4.
34. Schmerl M, Pollard H, Hoskins W. Labral injuries of the hip: a review of diagnosis and management. J Manip Physiol Ther. 2005;28(8):632. https://doi.org/10.1016/j.jmpt.2005.08.018.
35. McCarthy J, Noble P, Aluisio FV, Schuck M, Wright J, Lee JA. Anatomy, pathologic features, and treatment of acetabular labral tears. Clin Orthop Relat Res. 2003;406:38–47. https://doi.org/10.1097/01.blo.0000043042.84315.17.

36. Wenger DE, Kendell KR, Miner MR, Trousdale RT. Acetabular labral tears rarely occur in the absence of bony abnormalities. Clin Orthop Relat Res. 2004;426:145–50. https://doi.org/10.1097/01.blo.0000136903.01368.20.
37. Kelly BT, Weiland DE, Schenker ML, Philippon MJ. Arthroscopic labral repair in the hip: surgical technique and review of the literature. Arthroscopy. 2005;21(12):1496–504. https://doi.org/10.1016/j.arthro.2005.08.013.
38. Kuhns BD, Weber AE, Levy DM, Bedi A, Mather RC 3rd, Salata MJ, et al. Capsular Management in hip Arthroscopy: an anatomic, biomechanical, and technical review. Front Surg. 2016;3:13. https://doi.org/10.3389/fsurg.2016.00013.
39. Martin HD, Savage A, Braly BA, Palmer IJ, Beall DP, Kelly B. The function of the hip capsular ligaments: a quantitative report. Arthroscopy. 2008;24(2):188–95. https://doi.org/10.1016/j.arthro.2007.08.024.
40. Telleria JJ, Lindsey DP, Giori NJ, Safran MR. An anatomic arthroscopic description of the hip capsular ligaments for the hip arthroscopist. Arthroscopy. 2011;27(5):628–36. https://doi.org/10.1016/j.arthro.2011.01.007.
41. Ekhtiari S, de Sa D, Haldane CE, Simunovic N, Larson CM, Safran MR, et al. Hip arthroscopic capsulotomy techniques and capsular management strategies: a systematic review. Knee Surg Sports Traumatol Arthrosc. 2017;25(1):9–23. https://doi.org/10.1007/s00167-016-4411-8.
42. van Arkel RJ, Amis AA, Cobb JP, Jeffers JR. The capsular ligaments provide more hip rotational restraint than the acetabular labrum and the ligamentum teres: an experimental study. Bone Joint J. 2015;97(4):484–91. doi: https://doi.org/10.1302/0301-620x.97b4.34638.
43. Nho SJ, Beck EC, Kunze KN, Okoroha K, Suppauksorn S. Contemporary management of the hip capsule during arthroscopic hip preservation surgery. Curr Rev Musculoskelet Med. 2019:260–70. https://doi.org/10.1007/s12178-019-09564-4.
44. van Arkel RJ, Amis AA, Jeffers JR. The envelope of passive motion allowed by the capsular ligaments of the hip. J Biomech. 2015;48(14):3803–9. https://doi.org/10.1016/j.jbiomech.2015.09.002.
45. Fuss FK, Bacher A. New aspects of the morphology and function of the human hip joint ligaments. Am J Anat. 1991;192(1):1–13. https://doi.org/10.1002/aja.1001920102.
46. Tsutsumi M, Nimura A, Akita K. New insight into the iliofemoral ligament based on the anatomical study of the hip joint capsule. J Anat. 2020;236(5):946–53. https://doi.org/10.1111/joa.13140.
47. Walters BL, Cooper JH, Rodriguez JA. New findings in hip capsular anatomy: dimensions of capsular thickness and pericapsular contributions. Arthroscopy. 2014;30(10):1235–45. https://doi.org/10.1016/j.arthro.2014.05.012.
48. Philippon MJ, Michalski MP, Campbell KJ, Rasmussen MT, Goldsmith MT, Devitt BM, et al. A quantitative analysis of hip capsular thickness. Knee Surg Sports Traumatol Arthrosc. 2015;23(9):2548–53. https://doi.org/10.1007/s00167-014-3030-5.
49. Jackson TJ, Peterson AB, Akeda M, Estess A, McGarry MH, Adamson GJ, et al. Biomechanical effects of capsular shift in the treatment of hip microinstability: creation and testing of a novel hip instability model. Am J Sports Med. 2016;44(3):689–95. https://doi.org/10.1177/0363546515620391.
50. Han S, Alexander JW, Thomas VS, Choi J, Harris JD, Doherty DB, et al. Does capsular laxity Lead to microinstability of the native hip? Am J Sports Med. 2018;46(6):1315–23. https://doi.org/10.1177/0363546518755717.
51. Johannsen AM, Behn AW, Shibata K, Ejnisman L, Thio T, Safran MR. The role of anterior capsular laxity in hip microinstability: a novel biomechanical model. Am J Sports Med. 2019;47(5):1151–8. https://doi.org/10.1177/0363546519827955.
52. Johannsen AM, Ejnisman L, Behn AW, Shibata K, Thio T, Safran MR. Contributions of the capsule and labrum to hip mechanics in the context of hip microinstability. Orthop J Sports Med. 2019;7(12):2325967119890846. https://doi.org/10.1177/2325967119890846.
53. Abrams GD, Hart MA, Takami K, Bayne CO, Kelly BT, Espinoza Orias AA, et al. Biomechanical evaluation of capsulotomy, Capsulectomy, and capsular repair on hip rotation. Arthroscopy. 2015;31(8):1511–7. https://doi.org/10.1016/j.arthro.2015.02.031.
54. Beck EC, Suppauksorn S, Nho SJ. The role of comprehensive capsular Management in hip Arthroscopy for the treatment of Femoroacetabular impingement syndrome. Arthroscopy. 2020;36(1):9–11. https://doi.org/10.1016/j.arthro.2019.10.028.
55. Philippon MJ, Trindade CAC, Goldsmith MT, Rasmussen MT, Saroki AJ, Loken S, et al. Biomechanical assessment of hip capsular repair and reconstruction procedures using a 6 degrees of freedom robotic system. Am J Sports Med. 2017;45(8):1745–54. https://doi.org/10.1177/0363546517697956.
56. Weber AE, Neal WH, Mayer EN, Kuhns BD, Shewman E, Salata MJ, et al. Vertical extension of the T-capsulotomy incision in hip arthroscopic surgery does not affect the force required for hip distraction: effect of capsulotomy size, type, and subsequent repair. Am J Sports Med. 2018;46(13):3127–33. https://doi.org/10.1177/0363546518800710.
57. Wuerz TH, Song SH, Grzybowski JS, Martin HD, Mather RC 3rd, Salata MJ, et al. Capsulotomy size affects hip joint kinematic stability. Arthroscopy. 2016;32(8):1571–80. https://doi.org/10.1016/j.arthro.2016.01.049.
58. Frank RM, Lee S, Bush-Joseph CA, Kelly BT, Salata MJ, Nho SJ. Improved outcomes after hip arthroscopic surgery in patients undergoing T-capsulotomy with complete repair versus partial repair for femoroacetabular impingement: a comparative matched-pair analysis. Am J Sports Med. 2014;42(11):2634–42. https://doi.org/10.1177/0363546514548017.

59. Acuna AJ, Samuel LT, Roth A, Emara AK, Kamath AF. How capsular management strategies impact outcomes: a systematic review and meta-analysis of comparative studies. J Orthop. 2020;19:237–43. https://doi.org/10.1016/j.jor.2020.02.002.
60. Chandrasekaran S, Darwish N, Martin TJ, Suarez-Ahedo C, Lodhia P, Domb BG. Arthroscopic capsular plication and labral seal Restoration in borderline hip dysplasia: 2-year clinical outcomes in 55 cases. Arthroscopy. 2017;33(7):1332–40. https://doi.org/10.1016/j.arthro.2017.01.037.
61. Kalisvaart MM, Safran MR. Hip instability treated with arthroscopic capsular plication. Knee Surg Sports Traumatol Arthrosc. 2017;25(1):24–30. https://doi.org/10.1007/s00167-016-4377-6.
62. Economopoulos KJ, Chhabra A, Kweon C. Prospective randomized comparison of capsular management techniques during hip arthroscopy. Am J Sports Med. 2020;48(2):395–402. https://doi.org/10.1177/0363546519894301.
63. O'Donnell JM, Devitt BM, Arora M. The role of the ligamentum teres in the adult hip: redundant or relevant? A review. J Hip Preserv Surg. 2018;5(1):15–22. https://doi.org/10.1093/jhps/hnx046.
64. O'Donnell JM, Pritchard M, Salas AP, Singh PJ. The ligamentum teres-its increasing importance. J Hip Preserv Surg. 2014;1(1):3–11. https://doi.org/10.1093/jhps/hnu003.
65. Martin R, McDonough C, Enseki K, Kohlrieser D, Kivlan B. Clinical relevance of the Ligamentum teres: a literature review. Int J Sports Physical Ther. 2019;14:459. doi: https://doi.org/10.26603/ijspt20190459.
66. Martin HD, Hatem MA, Kivlan BR, Martin RL. Function of the ligamentum teres in limiting hip rotation: a cadaveric study. Arthroscopy. 2014;30(9):1085–91. https://doi.org/10.1016/j.arthro.2014.04.087.
67. Martin RL, Palmer I, Martin HD. Ligamentum teres: a functional description and potential clinical relevance. Knee Surg Sports Traumatol Arthrosc. 2012;20(6):1209–14. https://doi.org/10.1007/s00167-011-1663-1.
68. Amenabar T, O'Donnell J. Successful treatment of isolated, partial thickness ligamentum teres (LT) tears with debridement and capsulorrhaphy. Hip Int. 2013;23(6):576–82. https://doi.org/10.5301/hipint.5000072.
69. Shindle MK, Ranawat AS, Kelly BT. Diagnosis and management of traumatic and atraumatic hip instability in the athletic patient. Clin Sports Med. 2006;25(2):309–26., ix-x. https://doi.org/10.1016/j.csm.2005.12.003.
70. Redmond JM, Chen AW, Domb BG. Greater trochanteric pain syndrome. J Am Acad Orthop Surg. 2016;24(4):231–40. https://doi.org/10.5435/jaaos-d-14-00406.
71. Sunil Kumar KH, Rawal J, Nakano N, Sarmento A, Khanduja V. Pathogenesis and contemporary diagnoses for lateral hip pain: a scoping review. Knee Surg Sports Traumatol Arthrosc. 2020. doi: https://doi.org/10.1007/s00167-020-06354-1.
72. Winston P, Awan R, Cassidy JD, Bleakney RK. Clinical examination and ultrasound of self-reported snapping hip syndrome in elite ballet dancers. Am J Sports Med. 2007;35(1):118–26. https://doi.org/10.1177/0363546506293703.
73. Yen YM, Lewis CL, Kim YJ. Understanding and treating the snapping hip. Sports Med Arthrosc Rev. 2015;23(4):194–9. https://doi.org/10.1097/jsa.0000000000000095.
74. Mallow M, Nazarian LN. Greater trochanteric pain syndrome diagnosis and treatment. Phys Med Rehabil Clin N Am. 2014;25(2):279–89. https://doi.org/10.1016/j.pmr.2014.01.009.
75. Chi AS, Long SS, Zoga AC, Read PJ, Deely DM, Parker L, et al. Prevalence and pattern of gluteus medius and minimus tendon pathology and muscle atrophy in older individuals using MRI. Skelet Radiol. 2015;44(12):1727–33. https://doi.org/10.1007/s00256-015-2220-7.
76. Kraeutler MJ, Garabekyan T, Pascual-Garrido C, Mei-Dan O. Hip instability: a review of hip dysplasia and other contributing factors. Muscles, Ligaments Tendons J. 2016;6(3):343–53. doi: https://doi.org/10.11138/mltj/2016.6.3.343.
77. Kalisvaart MM, Safran MR. Microinstability of the hip-it does exist: etiology, diagnosis and treatment. J Hip Preserv Surg. 2015;2(2):123–35. https://doi.org/10.1093/jhps/hnv017.
78. Safran MR. Microinstability of the hip-gaining acceptance. J Am Acad Orthop Surg. 2019;27(1):12–22. https://doi.org/10.5435/jaaos-d-17-00664.
79. Epstein DM, Rose DJ, Philippon MJ. Arthroscopic management of recurrent low-energy anterior hip dislocation in a dancer: a case report and review of literature. Am J Sports Med. 2010;38(6):1250–4. https://doi.org/10.1177/0363546509358318.

Part VI

Knee Biomechanics

Biomechanics of the Knee

21

Farid Amirouche and Jason Koh

21.1 Introduction

Accurate knowledge of knee kinematics is critical to evaluate the knee joint and how it relates to changes in cartilage, ligament injuries, reconstruction or repairs and function. The knee is characterized by its geometry, forms and shapes of femoral condyles and their articulation on the tibia plateau, cartilage role in lubrication and support, and the balancing act of ligaments and muscle forces to accommodate different surface conditions during walking, and finally how the stability of the knee is maintained during flexion-extension under different loading conditions.

F. Amirouche
Department of Orthopaedic Surgery, Orthopaedic & Spine Institute NorthShore University Health System, Evanston, Illinois, USA

Department of Orthopeadic Surgery, College of Medicine, University of Illinois, Chicago, Illinois, USA
e-mail: FAmirouche@northshore.org, amirouch@uic.edu

J. Koh (✉)
Department of Orthopaedic Surgery, Orthopaedic & Spine Institute NorthShore University HealthSystem, Skokie, Illinois, USA

University of Chicago Pritzker School of Medicine, Chicago, Illinois, USA

Northwestern University McCormick School of Engineering, Evanston, Illinois, USA

Clinical interest in the knee joint remains an interesting topic of research as biologics, nanotechnology, stem cells regenerative medicine is changing the fundamental role each of the knee substructure work and the ability to potentially being able to have them regenerated [32, 38, 20, 76].

The kinematics of the knee and its relation to different pathologies has been the subject of numerous research with different perspectives in light of the experiments conducted, models proposed, and outcome measures related to gait motion tracking analysis [43, 65, 50, 72, 53, 71, 45, 89]. To extract kinematic information in a single and dual plane, several researchers used fluoroscopy imaging techniques to analyze a quasi-static situation where a knee flexion is performed through a range of motion incrementally and radiographs were taken [100, 2, 27, 102]. Moreover, mobile fluoroscopic systems emerged and analysis of the knee during full dynamic flexion movements were examined during gait [42, 34].

The different joint rotations of the knee have been described from data collected by 3D camera and markers posted along the lower and upper part of the leg during gait. Most studies describe the knee flexion with a peak value observed after toe-off and the internal tibia rotation yielding its max at heel strike [44, 56, 34, 63, 64]. The tibia is shown to rotate externally through mid-stance, before rotating internally once again prior to toe-off [56, 54, 10, 81].

J. Koh et al. (eds.), *Orthopaedic Biomechanics in Sports Medicine*,
https://doi.org/10.1007/978-3-030-81549-3_21

The tibiofemoral joint is still at the center stage of knee kinematic research and importantly, we still see different methods being used in describing the condylar anterior posterior translation following either the contact path, the condylar geometric center or other points on the condylar axis [88, 57, 73, 85, 91, 48].

The kinematics of a healthy knee or a knee with different pathologies or reconstruction depend great deal on the connective tissues around the knee in balancing and maintaining a stable and well-performing knee. The variation in knee analysis is also affected by other factors such as gender, anatomy, and articular surface behaviors which are all contributing factors to knee kinematics and kinetics.

Additional studies have investigated the tibia femoral joint contact surfaces during flexion-extension to investigate the cartilage effect on knee kinematics using both in vitro and in vivo experimental studies [35].

In silico 3D knee joint models were developed [37, 87], along with in vivo imaging measurements to provide a reliable validation of the results [13, 18, 92, 94, 61, 46]. These techniques have opened the door to advancing our understanding of modeling limitations, assumptions that clinically are relevant and acceptable versus additional knowledge needed to build better computational models on knee joint biomechanics.

The articular surface geometry as stated above at the contact area is also an important variable that affects the articular contact behaviors, such as contact stress and knee joint stability. Any changes to the cartilage contact area trigger a change in kinematics. Osteoarthritis (OA) of the knee is also a contributing factor to knee loading unbalance of forces and moments. The modulation of the knee joint results in large knee adduction moment and higher joint dynamic load during gait causally related to the severity of tibiofemoral OA [6, 52].

The complexity of the knee can also be seen in the role muscles in lower limb play in the knee joint stability and locomotion. To assess the contribution of different muscles to the knee, one needs to rely on 3D modeling techniques to evaluate contact pressure and forces generated from supporting structures at the knee using Opensim, life modeler among others [79, 84, 29]. Popular methods in gait analysis rely on knee joint flexion moment and knee joint power which are the predictors of load distribution across the tibiofemoral joint reflecting a special time point during gait stance phase.

The knee biomechanics teach us that optimization and coordination of the tibia femoral bones, patella-femoral articulating surface and muscles, ligament, and tendons interaction work together to achieve the desired joint movement. The complex interaction of these structures allows the knee to withstand tremendous forces during various normal movements.

Knee kinematics and its connective tissues was studied by Masouros, Parker, Hill, Amis, & Bull [67] and they pointed out its importance to diagnosis and treatments. Modeling and simulation methods used in the simulation of the knee during different walking conditions are studied by several researchers including to name a few [4, 106, 3, 1].

This chapter discusses the importance of the knee anatomy in joint biomechanics, provides a description of the knee joint kinematics and different methods used to assess knee performance and discusses the importance of the biomechanics of the knee, the current state of FE modeling and its clinical benefits. Understanding the knee biomechanics is an essential tool in designing prosthesis, provides guidance to rehabilitation exercises programs that will assist the patient knee stability, mobility and regain his or her knee normal function. A comprehensive understanding of the knee joint kinematics could significantly improve the future of patient's knee injuries and treatment outcomes.

21.2 Basic Knee Anatomy

The knee joint consists of two basic articulations coordinated in a fashion to provide the knee muscles and ligaments the ability to power and drive its desired movements and function. The knee articulations are between the distal femur, proximal tibia, and patella riding on femoral groove. The geometry of the femoral condyles is asymmetric, with the medial condyle projecting more distal than the lateral femoral condyle. The medial condyle is also larger and wider and conforms to the tibia plateau condyle where it articulates, on the other hand, the lateral condyle projects more anteriorly. The condyles join through a femoral groove anteriorly and the femoral notch at their distal aspect to allow for the patella to slide, glide, and rotate to maintain the direction of the quadriceps and patella tendon forces. The lateral condyle can be identified by its terminal sulcus and groove of the popliteus insertion [26, 33].

The femoral condyles articulate on a tibia plateau that is also divided in two conforming surfaces: the medial tibial plateau is concave, whereas the lateral plateau has an anteroposterior convexity. This topography accounts for the so-called screw-home mechanism, or internal rotation, of the femur on the fixed tibia as the knee approaches extension [55, 104]. Load bearing on the tibia cartilage is usually due to excess stress on the knee which might lead to arthritis [23, 16, 47]. The distribution of the load on the tibia plateau is asymmetric, and occurs depending on the loading conditions not only centrally on the medial and lateral sides but also on the sloping edges on the medial and lateral part of the tibial eminences [83].

On each side of the tibia plateau, medial and lateral menisci help maintain conformity of the condyles during the knee flexion-extension and eliminate the high impact forces at the edges of the tibia plateau when subjected to high loads. Further description of the menisci shows that the anterior horn geometrical shape of the medial meniscus, the anterior cruciate ligament (ACL), and the anterior part of the lateral meniscus attach anterior to the tibial spine. The posterior cruciate ligament (PCL) and the posterior part of the medial and lateral menisci attach posterior to the tibial spine [33].

The patella consists of medial and lateral facets, separated by a vertical ridge to allow it to move within the prescribed femoral groove. The patella is totally embedded within the tendon. The front face of the patella is convex in shape and is divided up into 3 parts. The top surface is where the patella and the quadriceps tendon intersect and this lightly covers the anterior surface of the patella and forms a deep fascia, which acts as a protector to the bone [77]. The patella tendons and femoral contact reaction forces need to be balanced during knee motion to maintain knee stability. The patella contact stress makes it vulnerable to wear and potential injury [7, 70, 103].

21.3 Knee Joint Axis of Rotation

The primary knee joint motion is easily explained in the sagittal plane during flexion-extension. A joint used for such description is referred to as a hinge joint (see Fig. 21.1). In mechanical term describing such a joint with the context of reference Cartesian frame the knee is also a revolute joint as used in a few multibody dynamics modeling of locomotion where the angle of rotation is defined as the tibia rotation relative to the femur. The knee has six degrees of freedom and can rotate along three different planes as shown in Fig. 21.2. The complex three-dimensional motion also allows for translation along each of the axis of rotation. The ligaments act as supporting structure and constraint the knee joint motion by keeping it stable and limiting its range of motion.

Full extension is usually defined when the tibia is extended to become fully aligned with the femur in the sagittal plane. Active knee flexion is possible primarily through hamstring contraction and usually reaches 130°, whereas passive flexion can reach 160°.

To better understand the kinematics of the knee researchers used gait analysis to describe both human locomotion, and its relation to joint forces used to control such a movement. The position of the body upper part has always been an indicator of body posture in relation the body center of gravity. During gait ankle, knee and hip joint are actuators driven by muscles and hence generate torque necessary to move the body for-

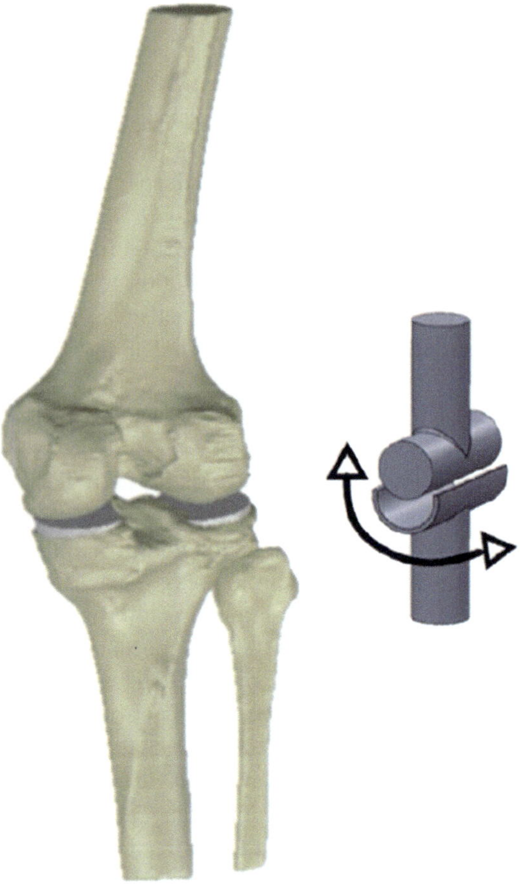

Fig. 21.1 Knee hinge joint model

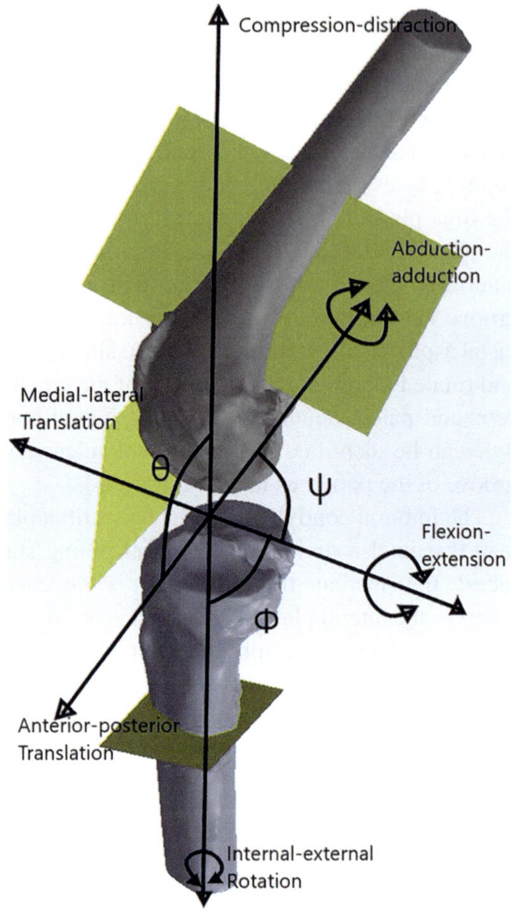

Fig. 21.2 The definition of knee joint angles and translation along 3 axes. Both tibia and femur translation are explained by the compression-distraction and each of the axis as shown. Internal-external, flexion-extension, and Abduction-adduction are the possible rotations at the knee joint

ward. Knee motion is dictated by the consideration of energy that requires the body center of gravity to move forward with ease.

During the swing phase of gait, the knee flexes to roughly 60° so the toe of the swinging leg does not be dragged on the ground. Also, during gait, as the swinging leg passes the standing leg and just before heel strike, the quadriceps muscle contracts which brings the knee to full extension and the foot forward. Flexion is not the only rotation that takes place during the gait cycle, as the knee extends 30° to 0°, the tibia externally rotates up to 30° before heel strike. The term used to describe this is "screw-home mechanism." This mechanism occurs to tighten the soft tissue structures as well as locking the knee geometry before the impact load of weight bearing [66, 67].

21.4 Functional Role of Ligaments

The ACL is the main restraint to anterior tibial movement. The viscoelastic properties of the ligaments allow it to act against hyperextension to prevent injury, and act as a secondary restraint to prevent internal and valgus rotation when the knee is at full extension. The ACL essentially controls the "screw-home motion" of the knee joint [66]. The primary function of the ACL is to prevent anterior translation of the tibia and provide further knee stability. In full extension, the ACL absorbs 75% of the anterior translation load, and 85% between 30° and 90° of flexion [74, 59].

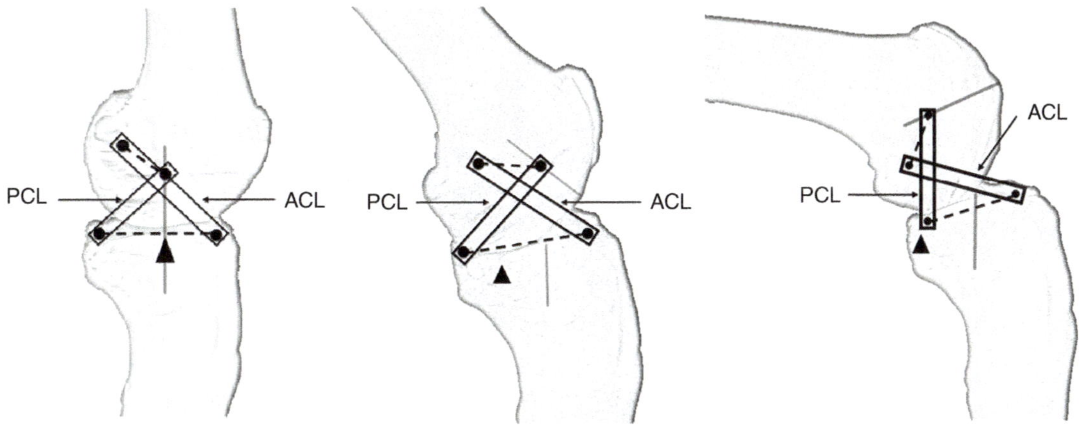

Fig. 21.3 Four-bar linkage system interplay of the ACL and PCL at 0°, 45°, and 90° degrees of flexion

Knee injury can result into an ACL tear which lead to anterior knee instability and decrease in knee performance. ACL under load to failure testing reaches the ultimate stress at approximately 15% strain, and complete failure occurs between 15% and 30% strain, with a stretch close to 1 cm [60, 33]. The ACL has been reported to have an average maximal tensile stress to failure of 2100 N to 2500 N, but this is less under cyclic loading, creep, and age [93, 82, 19, 5].

The posterior tibia translation is controlled by the posterior cruciate ligament (PCL) and is most effective in mid to deep flexion. As stated by [66] the ACL and PCL together control the anterior posterior rolling and sliding kinematics of the TF joint during flexion and extension.

The PCL has an average tensile strength of 6.1 N at 0°, and a tensile strength of 112.3 N at 90°. [82]. The orientation of the PCL which inserts onto the lateral part of the medial femoral condyle allows for adequate tensioning of the PCL during the rolling of the lateral femoral condyle posteriorly in early flexion.

The interplay between the ACL and PCL is often referred to as the "**four-bar cruciate linkage system**" [39]. The intersection of the ACL and PCL shows that the center of joint rotation moves posterior with knee flexion. This allows both sliding and rolling movements of the femur during flexion and prevents the femur from rolling off the tibial plateau at extremes of flexion [11].

The axes of rotation of the knee are fundamental to kinematic models. The hinge model is not necessarily used for the current knee motion analysis as it is contradicted by the geometrical shape of the femoral condyles and the fact that the knee center of rotation is not fixed. The concept of "**instant center of motion**" has been linked to **4-bar mechanism** where the ACL and PCL are viewed as rigid bars linking the tibia and femoral part the knee (see Fig. 21.3). The four-bar linkage theory can then be used to describe the knee motion and defines the instantaneous center of rotation associated with such mechanism. Assuming that the four bars are obliquely planar we can refer to such a system as a planar 4-bar linkage. The ligaments tension and stiffness alter the bars length and orientation in space which makes the 4-bar linkage not a reliable model for the knee. The cruciate ligaments can be assumed inextensible fibers and the IC can be tracked in relation to the tibia femoral joint as the knee flexes.

Different aspects of knee rotations and contact in both sagittal and coronal planes can be described assuming a center of rotation for the femoral component and tibia during the flexion-extension of the knee as shown in Fig. 21.4. The contact area is highlighted by the intersection of the tibial and femoral curvature radiuses. If the circles are tangent, we have a point contact otherwise there are two points of intersection defining a line contact. A surface contact is more realistic

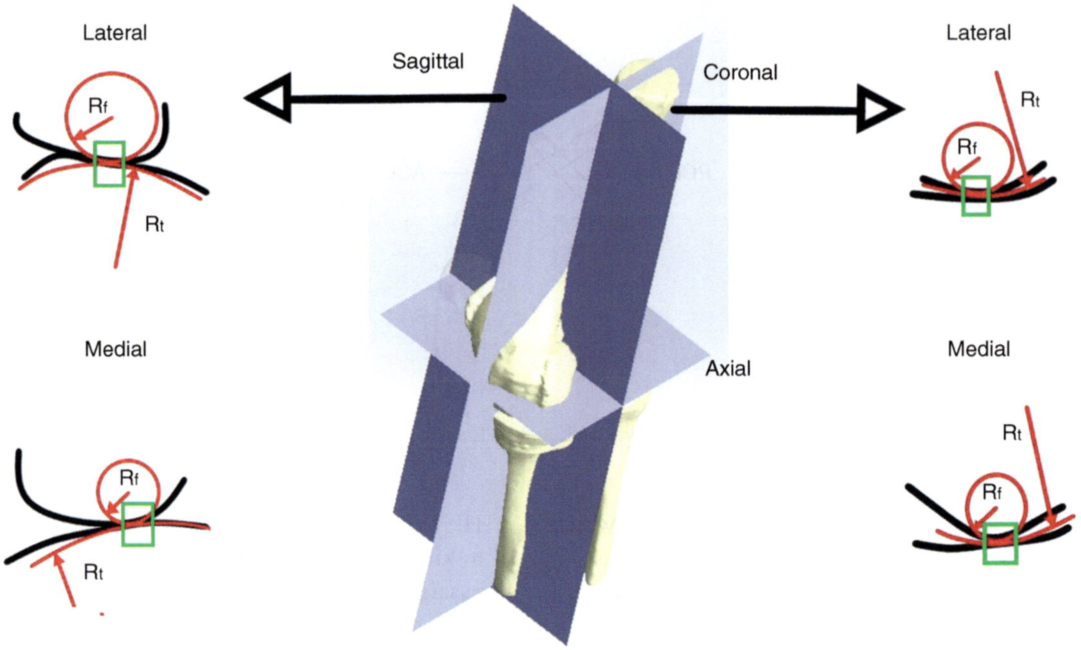

Fig. 21.4 Description of the medial and lateral contact points between the femoral and tibial in the sagittal and coronal planes

as both the line and point contacts are simply a representation of the contact area.

The role of the medial and lateral collateral ligaments can be described as follows: medial collateral ligament (MCL) is the restraint force to valgus angulation and internal tibial rotation and is a secondary restraint to external tibial rotation. It is also a secondary restraint to anterior tibial translations when the tibia is externally rotated. The lateral collateral ligament (LCL) on the hand is the restraint to varus angulation. It also assists in restraining the posterior translation, and becomes a restraining element with the PCL to external tibial rotation [66].

21.5 Function of the Menisci

The menisci wedge shape aids in providing conformity to the femoral and tibial articular surfaces. The primary functions of the menisci are to bear loads, to distribute the load of the knee joint medially and laterally passing the tibia–femur joint [105]. It also plays the role of a damper absorbing and damping energy during impact. This is done by the menisci inherent structure to deform under compressive loads by increasing the contact area [25]. The tensile stiffness and strength of the menisci are approximately 10 times greater than those of articular cartilage [101]. This allows the menisci to withstand the large hoop stresses generated by the knee joint [41, 68]. The menisci may also function as a secondary restraint to anterior translation in an ACL-deficient knee.

21.6 Gait Analysis of the Knee Joint

Gait analysis has gained popularity in the last decade due in part to better mathematical techniques and graphics display of the human musculoskeletal system with reliable solvers for inverse dynamics. Clinically gait analysis provides an insight into the motion limitations of the knee under different conditions, which when com-

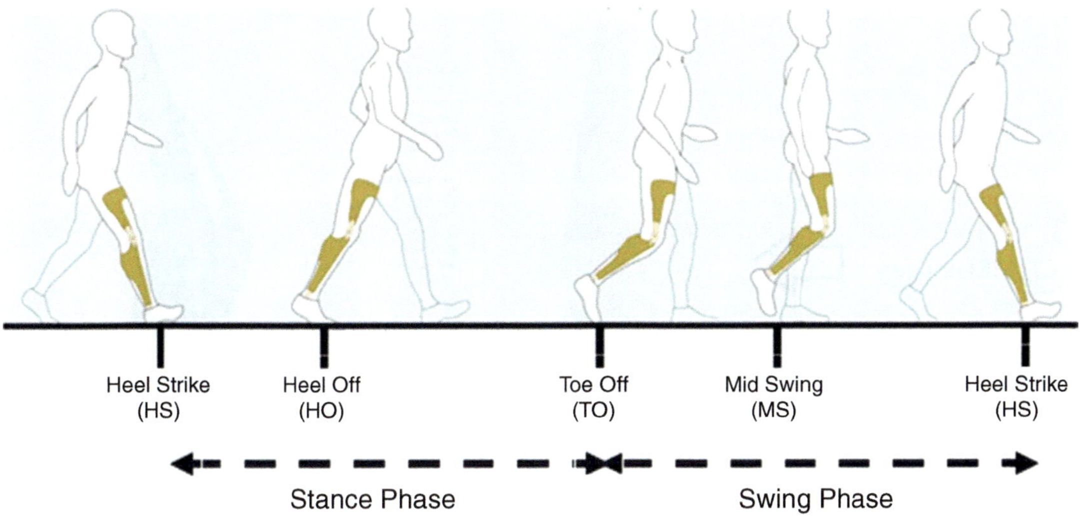

Fig. 21.5 Gait stance and swing phases defined as a percentage of 100% full gait cycle (Y. Qi, C. B. Soh, E. Gunawan, K. Low and R. Thomas, IEEE Transactions on Neural Systems and Rehabilitation Engineering, vol. 24, no. 1, pp. 88–97, Jan. 2016)

bined with imaging helps the physician develop a comprehensive idea on motion effects on patient problems. Hence, biomechanical gait analysis objective is the evaluation of the knee joint movement for a better understanding and diagnosis of knee joint injuries and pathologies. Several studies of OA have recognized the importance of biomechanical gait analysis in the pathogenesis of the knee joint and its supporting soft tissue structure.

Gait analysis combines both kinematical data of the lower and upper limbs collected through vision cameras and markers and uses mathematical models based on multibody dynamics to compute the inverse dynamics problem where it provides an estimation of the joint forces and moments. For this reason, biomechanical assessment has become important for knee joint problem diagnosis; it provides quantitative information about the structure and motion of the knee joint to complement the common orthopedic physical evaluation exam for more accurate diagnosis. Moreover, the experiments can be tailored such that gait phases can become indicators for sports and rehabilitation.

Walking, running, and stair climbing are activities that have been investigated by several researchers using gait analysis. While the main function of the knee joint is complex, gait simplified models can evaluate the joints forces and moments in relation to the body weight (BW) and walking speed. This of course is done for both the stance and swing phase of a gait cycle. The peak load at the knee joint varies from person to person and is estimated at 2–3 BW during walking, 4–6 BW during stair climbing, and 7–12 BW during running [23, 86, 40, 69].

The walking gait is characterized by two phases stance phase and swing phase as shown in Fig. 21.5. A full gait cycle is denoted by 100% and starts from the time foot heel strikes the ground and going through the swing phase to the time it strikes the ground again. The stance is the initial 65% of the gait cycle and the swing phase is the rest 35% of the cycle [49]. The stance phase consists of three sub-phases: initial (heel strike to foot flat), middle (foot flat to opposite heel strike), and terminal stance (opposite heel strike to toe-off) [22, 30, 49].

21.7 FEA Modeling of the Knee

Finite Element (FE) methods are widely used in orthopedics implant and prosthesis design. FEA is becoming a clinical tool for simulation of com-

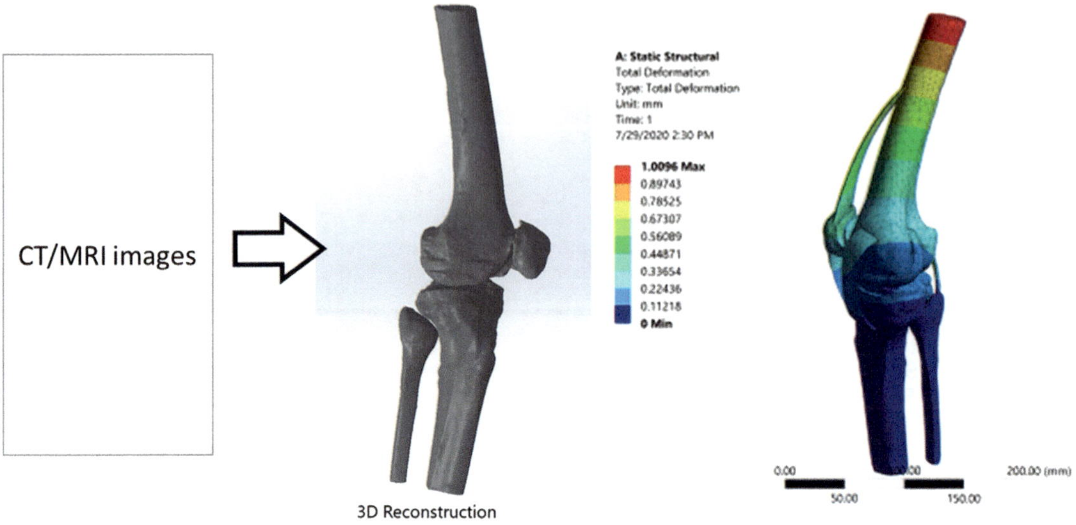

Fig. 21.6 FE modeling steps using CT/MRI scans (images), CAD model generated using MIMICS, and mesh and assembled model with connective tissues using ANSYS

plex cases where new concepts or design are tested virtually before implementation. FE combined with imaging have provided considerable insight into understanding knee joint biomechanics, including ligament function, ligament reconstruction technique, and implant design. FEA is a mathematical tool that uses the law of mechanics to provide solutions to clinical problems in terms of stress-strains, force-deformations, work-energy, and other parameters used to describe the problem at hand. Due to inherent challenges associated with experiments (in vivo and ex vivo), FEA has long been recognized and trusted by researchers and government agencies such as the FDA as a reliable alternative method to the study of human joints-implant testing and validation.

Subject-specific FE modeling based on models obtained from CT-MRI data of that specific patient is useful in the study of OA as it can investigate the true interaction between multiple tissues and how changes in one can lead to implications in an adjacent tissue, which may lead to disease initiation or progression [17, 75]. Specifically, this FEA can investigate through different iterations of the same model by changing the material properties, the boundary conditions, and loading conditions (walk versus run) to assess such change on the model function. Sensitivity of design parameters associated with prostheses is studied without the burden of experiments cost [8, 14, 95].

FEA relies great deal on building realistic models and making the right assumptions to build high fidelity into the solutions provided by the FEA simulation studies (see Fig. 21.6). In the context of joint biomechanics, modeling of soft tissues such as ligaments becomes crucial to understanding how, to design corrective therapeutics and restore joint function [90, 21, 36].

FEA is still being developed to address multitude of questions and whether current models can be used to investigate a multiscale of substructure elements simultaneously while performing flexion-extension of the knee. Since multiple measurements are difficult and most experiments are designed for a specific purpose, additional techniques must be developed for validation purposes where both rigid, deformations and fluid-structure interaction are possible.

21.8 FEA Modeling of Knee Ligaments

Ligaments are complex viscoelastic structures that generate the necessary forces to maintain knee stability and smooth articulations within the kinematical range of motion they can sustain within

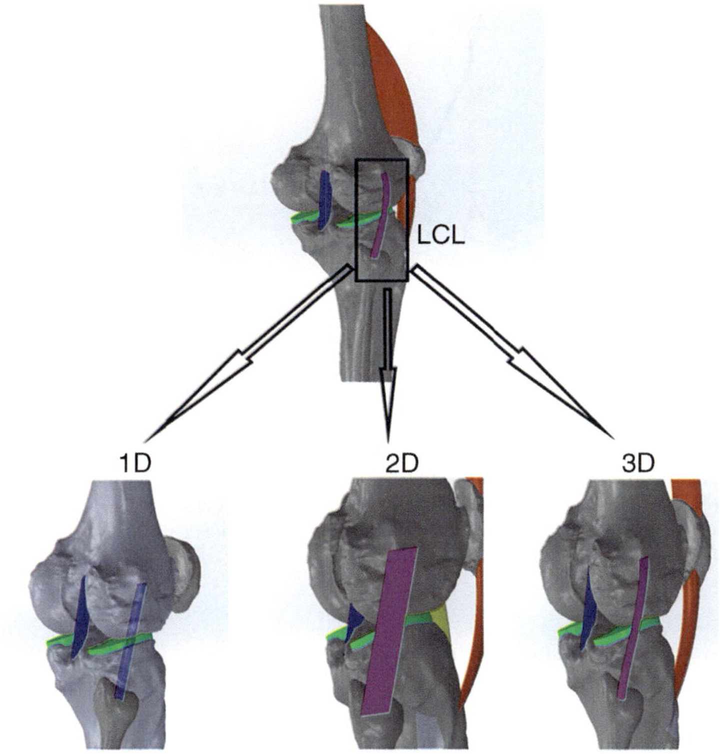

Fig. 21.7 CAD modeling and FEA of different LCL models (1D, 2D, and 3D)

injuries. How are these supporting structures-ligaments modeled in FEA. Models must be realistic, simple and capture the essential information for the task explored and analyzed. Early FE studies of the knee joint used uniaxial discrete line elements which can be viewed as springs with assigned material properties such as young modulus of elasticity found through testing and cadaveric experiments [9]. These models limitations can be overcome by high resolution imaging (MRI), and close look into the fibers alignments and mechanical testing to build a 3D ligament model that can be incorporated into the 3D FEA of the knee [9, 24]. A combination of accurate geometry, isotropic and transverse hyperelastic constitutive material models [51, 80] were developed in the study of the knee [97, 28, 12, 99]. The three ligaments models stated above are shown in Fig. 21.7.

The knee mathematical modeling seems to rely great deal on the understanding of the anatomy of ligaments, menisci, tendons, and patellofemoral and tibia articulations. While FEA and other dynamic tools have advanced the state of the art of modeling, the problem has become obviously important that additional computational tools are needed to appropriately diagnose and treat pathologies at early stage of their development.

21.9 Knee Biomechanics and Joint Acoustic Assessments

An area that needs further studies in knee biomechanics is acoustic and vibration. Knee joint sounds provide unique characteristics of how biomaterial structures respond to load. During the joint motion, the knee ligaments, tendons, cartilage and menisci inherent natural frequencies associated with their healthy geometry can be altered in the presence of injuries. Acoustic emissions are part of the knee biomechanics and their properties can be used as indicators when examining the knee (See Fig. 21.8). Vibration methods, such as acoustic and modal analysis, have been used in other industries with a lot of success to diagnose structural faults such as rattling effects, stress risers, and structural designs. Better understanding of the vibroacoustic characteristics of the knee must be developed. This work can pave the

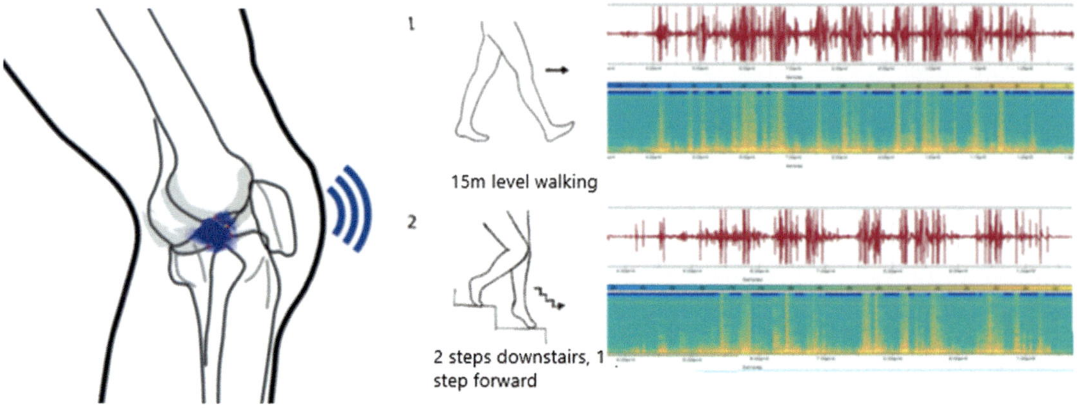

Fig. 21.8 **Knee sound recording,** Knee acoustic emissions were recorded using two high-performance, top port silicon acoustic sensors (size: 3 cm in diameter and 1 cm in depth, SPU0414HR5H-SB, Knowles Electronics, LLC. Itasca, IL, USA). One microphone was placed on the medial tibial plateau and one in the center of the patella. The sound signals were digitized with a sampling rate of 16,000 Hz. (Kalo et al. Journal of Biomechanics 109, 2020)

way for future studies aimed at employing acoustic emission and modal analysis approaches for knee health monitoring outside of clinical settings, such as for field-deployable diagnostics [15]. Other imaging techniques used to diagnose knee lesions expose the patient to potential radiation and are not suitable for outside clinical settings, such as on the sports field. Another example, the meniscus has shown to be an important factor impacting knee acoustic emissions, by influencing the width of joint space and, thus, the contact pressure of the tibiofemoral joint [98, 62].

This acoustic tool is still a new research tool in knee biomechanics but it is a well-developed method in other industries. Possible predictors to the signal at the knee are still unknown due to the complexity of anatomy and how other elements can be elicited during the knee motion. Only when most of the contributing factors to the sound signal are identifiable and become known, proper diagnosis of the knee can be performed.

21.10 Advances, Challenges, and Future Directions in Knee Biomechanics

Today surgeons are regularly performing major complex reconstructive knee surgery, with highly encouraging results. Navigation and robotics are leading the way along with new operating scopes and devices assisting the surgeon achieve the most desirable outcome. Virtual reality and augmented reality are both being tested in the education and preparedness of surgeons in a world of complexity never seen before. Interactive surgery and access to vital information in real time is key to future advances.

Some of the knee challenges tend to be related to soft tissue repairs and regenerative cartilage. A simple ligament or menisci tear tends to raise questions on what procedure is most suitable.

The menisci can cause pain, clicking, giving way, locking, and swelling in the knee. If a meniscal tear is symptomatic, then it is likely to need surgery, and unfortunately only a minority of meniscal tears are repairable. **The literature suggests that maybe only about 15 per cent of tears are repairable** [58, 78, 31, 96]. The alternative of waiting then having knee replacement needs to be revisited. The idea of replacing the torn menisci with a new one is what is done by surgeons who perform menisci transplantation. This complex procedure allograft must be sized correctly and match the side of the knee. The meniscus itself is made of collagen, and everyone's collagen is the same so there is no rejection. This area of research is improving, and each knee component plays an important role in the biomechanics of the knee.

The future of knee repair and surgery lies in advances in biological reconstruction, with allografts, with 3D printed biological scaffolds, with tissue engineering, growth factors and stem cells, augmented reality and artificial intelligence and simulation assisted surgeries. Knee biomechanics will play an important role in the future of knee repair and surgery which is set to be innovative and different from current practice.

Finite element methods together with other advance techniques in multibody dynamics, multiphysics will advance the state of simulation of the mechanical response of the complete knee under different loading and pathological conditions. FEA have benefited and will continue to benefit from increased computational power. However, the computer power never seems to be sufficient for real-time simulation of the load response of a knee joint. Improved numerical procedures or brand-new techniques are still necessary for better and faster understanding of surface contact solutions. The challenge to verify and validate a knee joint model is still work in progress.

Integrating FEA into gait analysis is required to determine the contact pressures in the knee in order to understand an abnormal gait. An FE knee model with a single-phase incompressible material law may be sufficient for the analysis of gait cycles but will not provide any information on the nutrient transport in articular cartilage that is performed by fluid flow in the tissue. A better FEA model will be needed to help understand the load share between the solid matrix and fluid pressurization, and the stress in the collagen.

Biomechanics in general is becoming a field of multi-disciplinary sciences and physics and mechanics working together to build anatomical realistic models to help us simulate conditions that are impossible in vivo or vitro. Another challenge remains on developing computationally and efficient parallel processing computers to meet future demands and challenges in knee biomechanics including surgical procedures.

Another aspects of biomechanics future studies are the classification of knee data into different diseases classes. Both clinical and modeling techniques including gait will benefit from the use of these large data samples. The use of AI (artificial intelligence) and deep learning neural networks could formulate a new knee joint biomechanical data classification that could easily be ready for access when diagnosis and treatments for complex cases are not available. Augmented reality, AI, biologics, and surgical innovations are the future for orthopedics.

References

1. Aburto-Barrera JM, Egure-Hidalgo M, Díaz-León C, Vázquez-Feijoo JA, Urriolagoitia-Sosa G. Mechanical design and numerical analysis of a femorotibial implant in patients with medial knee osteoarthritis; 2020. https://link.springer.com/chapter/10.1007/978-3-030-20801-1_23
2. Acker S, Li RM, Murray H, John PS, Banks SA, Mu S, et al. Accuracy of single-plane fluoroscopy in determining relative position and orientation of total knee replacement components. J Biomech. 2011;44(4):784–7. https://sciencedirect.com/science/article/pii/s0021929010005993
3. Adams MJ. Multi-scale modeling and analysis via surrogate modeling techniques for in vivo knee loading predictions; 2014. http://gradworks.umi.com/15/68/1568107.html
4. Amami A, Azouz ZB. Weakly supervised automatic segmentation and 3D modeling of the knee joint from MR images; 2013. https://spiedigitallibrary.org/conference-proceedings-of-spie/9067/1/weakly-supervised-automatic-segmentation-and-3d-modeling-of-the-knee/10.1117/12.2050908.full
5. Andrade A, Stock D, Costi JJ, Stanley RM, Kelly N, HearnTC, … Spriggins A. The in vivo intra-operative biomechanical properties of ACL grafts; 2002. https://online.boneandjoint.org.uk/doi/10.1302/0301-620x.84bsupp_iii.0840258
6. Astephen J, Deluzio KJ, Caldwell GE, Dunbar MJ. Biomechanical changes at the hip, knee, and ankle joints during gait are associated with knee osteoarthritis severity. J Orthop Res. 2008;26(3):332–41. https://onlinelibrary.wiley.com/doi/pdf/10.1002/jor.20496
7. Atsumi S, Arai Y, Kato K, Nishimura A, Nakazora S, Nakagawa S, et al. Transverse stress fracture of the proximal patella: a case report. Medicine. 2016;95(6) https://ncbi.nlm.nih.gov/pmc/articles/pmc4753884
8. Baldwin MA, Clary CW, Clary CW, Fitzpatrick CK, Deacy JS, Maletsky LP, Rullkoetter PJ. Dynamic finite element knee simulation for evaluation of knee replacement mechanics. J Biomech. 2012;45(3):474–83. https://sciencedirect.com/science/article/pii/s0021929011007469
9. Beidokhti HN, Janssen D, Groes SV, Boogaard TV, Verdonschot N. Subject-specific knee ligaments modeling approaches in finite element

analysis: 1D and 3D; 2016. https://narcis.nl/publication/recordid/oai:ris.utwente.nl:publications/e30742a6-9658-45b3-8a8f-34c2d5878554
10. Bergfeld JA, McAllister DR, Parker RD, Valdevit A, Kambic H. The effects of tibial rotation on posterior translation in knees in which the posterior cruciate ligament has been cut. J Bone Joint Surg (Am Vol). 2001;83(9):1339–43. https://ncbi.nlm.nih.gov/pubmed/11568196
11. Bertin KC, Komistek RD, Dennis DA, Hoff W, Anderson DT, Langer T. In vivo determination of posterior femoral rollback for subjects having a NexGen posterior cruciate–retaining total knee arthroplasty. J Arthroplasty. 2002;17(8):1040–8. https://sciencedirect.com/science/article/pii/s0883540302002619
12. Beynnon BD, Yu J, Huston DR, Fleming BC, Johnson RJ, Haugh LD, Pope MH. A sagittal plane model of the knee and cruciate ligaments with application of a sensitivity analysis. J Biomech Eng Trans Asme. 1996;118(2):227–39. https://asmedigitalcollection.asme.org/biomechanical/article/118/2/227/397992/a-sagittal-plane-model-of-the-knee-and-cruciate
13. Bingham JT, Papannagari R, Velde SV, Gross C, Gill TJ, Felson DT, et al. In vivo cartilage contact deformation in the healthy human tibiofemoral joint. Rheumatology. 2008;47(11):1622–7. https://ncbi.nlm.nih.gov/pmc/articles/pmc2569133
14. Bo G, Laurent A. Measure TKA prosthesis constraint using computer simulation; 2018. https://online.boneandjoint.org.uk/doi/abs/10.1302/1358-992x.95bsupp_34.ista2013-062
15. Bolus NB, Jeong HK, Whittingslow DC, Inan OT. A glove-based form factor for collecting joint acoustic emissions: design and validation. Sensors. 2019;19(12):2683. https://mdpi.com/1424-8220/19/12/2683
16. Calder KM, Acker SM, Arora N, Beattie KA, Callaghan JP, Adachi JD, Maly MR. Knee power is an important parameter in understanding medial knee joint load in knee osteoarthritis. Arthritis Care Res. 2014;66(5):687–94. https://ncbi.nlm.nih.gov/pmc/articles/pmc4282060
17. Carey R, Zheng L, Aiyangar AK, Harner CD, Zhang X. Subject-specific finite element Modeling of the tibiofemoral joint based on CT, magnetic resonance imaging and dynamic stereo-radiography data in vivo. J Biomech Eng-Trans ASME. 2014;136(4):041004. https://ncbi.nlm.nih.gov/pubmed/24337180
18. Carter TE, Taylor K, Spritzer CE, Utturkar GM, Taylor DC, Moorman CT, et al. In vivo cartilage strain increases following medial meniscal tear and correlates with synovial fluid matrix metalloproteinase activity. J Biomech. 2015;48(8):1461–8. https://ncbi.nlm.nih.gov/pmc/articles/pmc4558182
19. Chakraborty S, Mondal D, Motalab M. Constitutive modeling of the human Anterior Cruciate Ligament (ACL) under uniaxial loading using viscoelastic prony series and hyperelastic five parameter Mooney-Rivlin model; 2016. https://aip.scitation.org/doi/abs/10.1063/1.4958358
20. Chevalier X, Kemta-Lepka F. Are biologics a treatment option in osteoarthritis. Therapy. 2010;7(6):675–83. https://openaccessjournals.com/articles/are-biologics-a-treatment-option-in-osteoarthritis.pdf
21. Grecu D, Pucalev I, Negru M, Tarniţă DN, Ionovici N, Diţă R. Numerical simulations of the 3D virtual model of the human hip joint, using finite element method. Romanian J Morphol Embryol. 2010;51(1):151–5. https://ncbi.nlm.nih.gov/pubmed/20191136
22. Davis R. Clinical gait analysis. IEEE Eng Med Biol Mag. 1988;7(3):35–40. https://sciencedirect.com/science/article/pii/b9780123969613000251
23. D'Lima DD, Fregly BJ, Patil S, Steklov N, Colwell CW. Knee joint forces: prediction, measurement, and significance; 2012. Retrieved March 10, 2020, from https://ncbi.nlm.nih.gov/pmc/articles/pmc3324308
24. Dousteyssier B, Molimard J, Hamitouche C, Han WS, Stindel E. Patient dependent knee modeling at several flexion angles: a study on soft tissues loadings. Ann Phys Rehabil Med. 2016;59. https://sciencedirect.com/science/article/pii/s1877065716301579
25. E, P. Studies on the tensile strength of knee joint menisci. Beitr Orthop Traumatol. 1965;12(10):668. http://europepmc.org/abstract/med/5881888
26. Flandry F, Perry CC. The anatomy and biomechanics of the posteromedial aspect of the knee; 2001. https://link.springer.com/chapter/10.1007/978-0-387-21601-0_3
27. Galvin CR, Galvin CR, Galvin CR, Perriman DM, Perriman DM, Perriman DM, et al. Age has a minimal effect on knee kinematics: a cross-sectional 3D/2D image-registration study of kneeling. Knee. 2019;26(5):988–1002. https://sciencedirect.com/science/article/pii/s0968016018306744
28. Germain F, Rohan P-Y, Rochcongar G, Rouch P, Thoreux P, Pillet H, Skalli W. Role of ligaments in the knee joint kinematic behavior: development and validation of a finite element model; 2016. https://link.springer.com/chapter/10.1007/978-3-319-28329-6_2
29. Ghazwan A. Muscle strategies and mechanical loading in patients with osteoarthritis; 2017. https://ethos.bl.uk/orderdetails.do?uin=uk.bl.ethos.732272
30. Gilchrist L, Winter DA. A multisegment computer simulation of normal human gait. IEEE Trans on Rehab Eng. 1997;5(4):290–9. Retrieved 10 4, 2020, from https://ncbi.nlm.nih.gov/pubmed/9422454
31. Gill TJ, R. M. Biologics and tissue engineering about the knee: biological repair. Foreword Knee Surgery. 2009:34–5.
32. Gill TJ, Randolph MA. Biologics and tissue engineering about the knee: biological repair. Foreword. J Knee Surg. 2009;22(1):34. https://ncbi.nlm.nih.gov/pubmed/19216351

33. Goldblatt JP, Richmond JC. Anatomy and biomechanics of the knee. Oper Tech Sports Med. 2003;11(3):172–86. https://sciencedirect.com/science/article/pii/1060187203800114
34. Guan S, Gray H, Schache AG, Feller JA, Steiger R d, Pandy MG. In vivo six-degree-of-freedom knee-joint kinematics in overground and treadmill walking following total knee arthroplasty. J Orthop Res. 2017;35(8):1634–43. https://onlinelibrary.wiley.com/doi/pdf/10.1002/jor.23466
35. Guettler JH, Demetropoulos CK, Yang KH, Jurist KA. Dynamic evaluation of contact pressure and the effects of graft harvest with subsequent lateral release at osteochondral donor sites in the knee. Arthroscopy. 2005;21(6):715–20. https://ncbi.nlm.nih.gov/pubmed/15944629
36. Guo H, Guo H, Maher SA, Spilker RL. A 3D biphasic finite element model of the human knee joint for the study of tibiofemoral contact and fluid pressurization; 2013. https://asmedigitalcollection.asme.org/sbc/proceedings/sbc2013/55614/v01bt55a005/287718
37. Halonen K, Mononen ME, Jurvelin JS, Töyräs J, Salo J, Korhonen RK. Deformation of articular cartilage during static loading of a knee joint – experimental and finite element analysis. J Biomech. 2014;47(10):2467–74. https://ncbi.nlm.nih.gov/pubmed/24813824
38. Hassan F, Murrell WD, Refalo A, Maffulli N. Alternatives to biologics in Management of Knee Osteoarthritis: a systematic review. Sports Med Arthrosc Rev. 2018;26(2):79–85. https://insights.ovid.com/sports-medicine-arthroscopy-review/smart/2018/06/000/alternatives-biologics-management-knee/9/00132585
39. Hernáiz-Alzamora A, Arza JI, Zarranz JU, Renovales FD. Anthropometry of the cruciate ligaments of the knee: MRI study of the human four-bar linkage device. Eur J Anat. 2017;21(1):1–11. https://dialnet.unirioja.es/servlet/articulo?codigo=6282266
40. Hirokawa S, MotookaT, Akiyama T, Morizono R, Tanaka R, Mawatari M, ... Hotokebuchi T. Calculation of the forces acting on the knee joint during ascent from kneeling; 2012. https://online.boneandjoint.org.uk/doi/abs/10.1302/1358-992x.94bsupp_xxv.ista2010-094
41. Hopgood, P., Monk, J., & Nokes, L. D. (2004). O3074 Tensile testing of meniscal repair techniques. Retrieved 10 3, 2020, from http://bjjprocs.boneandjoint.org.uk/content/86-b/supp_iii/344.2.short
42. Hourlier H, Fennema P. Intraoperative fluoroscopy improves surgical precision in conventional TKA. Knee Surg Sports Traumatol Arthrosc. 2014;22(7):1619–25. https://ncbi.nlm.nih.gov/pmc/articles/pmc4059969
43. Jacofsky, D. J., McCamley, J. D., Bhowmik-Stoker, M., Jacofsky, M. C., & Shrader, M. (2011). A1017. Advanced osteoarthritic gait kinematics and kinetics. http://bjjprocs.boneandjoint.org.uk/content/jbjsbrproc/93-b/supp_iv/399.1.full.pdf
44. Jayabalan P, Gustafson J, Huang W, Piva SR, Sowa GA, Farrokhi S. The influence of continuous versus interval walking exercise on joint loading and serum biomarker profile in patients with knee osteoarthritis. Pm&r. 2015;7(9) https://pmrjournal.org/article/s1934-1482(15)00343-3/abstract
45. Jb S, R, K., & Da, D. A novel approach to knee kinematics. Am J Orthop. 2001;30(4):287. https://ncbi.nlm.nih.gov/pubmed/11334450
46. Jones MD, Tran CW, Li G, Maksymowych WP, Zernicke RF, Doschak MR. In vivo microfocal computed tomography and micro-magnetic resonance imaging evaluation of antiresorptive and antiinflammatory drugs as preventive treatments of osteoarthritis in the rat. Arthritis Rheum. 2010;62(9):2726–35. https://ncbi.nlm.nih.gov/pubmed/20533290
47. Jones R, Chapman GJ, Findlow AH, Forsythe L, Parkes MJ, Sultan J, Felson DT. A new approach to prevention of knee osteoarthritis: reducing medial load in the contralateral knee. J Rheumatol. 2013;40(3):309–15. http://jrheum.org/content/40/3/309
48. Kang K-T, Kwon SK, Kwon O-R, Lee J-S, Koh Y-G. Comparison of the biomechanical effect of posterior condylar offset and kinematics between posterior cruciate-retaining and posterior-stabilized total knee arthroplasty. Knee. 2019;26(1):250–7. https://ncbi.nlm.nih.gov/pubmed/30577956
49. Kasović, M., Mrđen, I., & Mejovšek, M. (2009). Biomechanics of normal gait. Retrieved 10 4, 2020, from http://bib.irb.hr/datoteka/409911.biomehanika_normalnog_hoda_zadar2009.pdf
50. Kaufman KR, Hughes CA, Morrey BF, Morrey MA, An KN. Gait characteristics of patients with knee osteoarthritis. J Biomech. 2001;34(7):907–15. https://sciencedirect.com/science/article/pii/s0021929001000367
51. Kiapour AM, Kaul V, Kiapour A, Quatman CE, Wordeman SC, Hewett TE, et al. The effect of ligament Modeling technique on knee joint kinematics: a finite element study. Appl Mathematics. 2013;4(5):91–7. https://ncbi.nlm.nih.gov/pmc/articles/pmc4160050
52. Knoop J, Dekker J, Leeden M v, Esch M v, Klein J-P, Hunter DJ, et al. Is the severity of knee osteoarthritis on magnetic resonance imaging associated with outcome of exercise therapy. Arthritis Care Res. 2014;66(1):63–8. https://ncbi.nlm.nih.gov/pubmed/23982988
53. Koo S, Andriacchi TP, Andriacchi TP. The knee joint Center of Rotation is predominantly on the lateral side during Normal walking. J Biomech. 2008;41(6):1269–73. https://ncbi.nlm.nih.gov/pmc/articles/pmc2481385
54. Kozanek M, Brodsky JW, Kane JM. The effect of ankle arthrodesis on the biomechanical function of the foot: a prospective, three-dimensional, multi-segment gait analysis with clinical correlation. Foot & Ankle Orthopaedics. 2017;2(3) https://journals.sagepub.com/doi/abs/10.1177/2473011417s000247

55. Lafortune MA, Cavanagh PR. 'Screw home' mechanism of the knee during walking. J Biomech. 1985;18(7):531. https://api.elsevier.com/content/article/pii:002192908590733x?httpaccept=text/xml
56. Lake M, Lafortune MA, Perry S. Heel plantar pressure distortion caused by discrete sensors. J Biomech. 1992;25(7):768. https://api.elsevier.com/content/article/pii:002192909290485j?httpaccept=text/xml
57. Landes CA, Sterz M. Evaluation of condylar translation by sonography versus Axiography in orthognathic surgery patients. J Oral Maxillofac Surg. 2003;61(12):1410–7. https://ncbi.nlm.nih.gov/pubmed/14663805
58. LaPrade CM, James EW, LaPrade RF, Engebretsen L. How should we evaluate outcomes for use of biologics in the knee. J Knee Surg. 2014;28(01):035–44. https://ncbi.nlm.nih.gov/pubmed/25260033
59. Laskowski ER. ACL injury and rehabilitation. Curr Phys Med Rehabil Rep. 2014;2(1):35–40. https://link.springer.com/article/10.1007/s40141-013-0036-8
60. Lu M, Johar S, Veenema K, Goldblatt JP. Patellar tendon rupture with underlying systemic lupus erythematosus: a case report. J Emerg Med. 2012;43(1) https://ncbi.nlm.nih.gov/pubmed/19959318
61. Lustig S, Scholes C, Balestro J-C, Parker D. In vivo assessment of weight-bearing knee flexion reveals compartment-specific alterations in meniscal slope. Arthroscopy. 2013;29(10):1653–60. https://sciencedirect.com/science/article/pii/s0749806313008153
62. Madeleine, P., Andersen, R. E., & Arendt-Nielsen, L. (2018). Spatial dependencies of knee vibroarthrograms during knee flexion-extension movement. https://vbn.aau.dk/da/publications/spatial-dependencies-of-knee-vibroarthrograms-during-knee-flexion
63. Mandeville D, Osternig LR, Chou L-S. The effect of total knee replacement surgery on gait stability. Gait Posture. 2008a;27(1):103–9. https://sciencedirect.com/science/article/abs/pii/s0966636207000719
64. Mandeville D, Osternig LR, Lantz BA, Mohler CG, Chou L-S. The effect of total knee replacement on the knee varus angle and moment during walking and stair ascent. Clin Biomech. 2008b;23(8):1053–8. https://sciencedirect.com/science/article/pii/s0268003308001599
65. Marcum ZA, Zhan HL, Perera S, Moore CG, Fitzgerald GK, Weiner DK. Correlates of gait speed in advanced knee osteoarthritis. Pain Med. 2014;15(8):1334–42. https://academic.oup.com/painmedicine/article-lookup/doi/10.1111/pme.12478
66. Masouros SD, Bull AM, Amis AA. A biomechanics of the knee joint. Orthop Trauma. 2010;24(2):84–91. https://sciencedirect.com/science/article/abs/pii/s1877132710000308
67. Masouros SD, Parker KH, Hill AM, Amis AA, Bull AM. Testing and modelling of soft connective tissues of joints: a review. J Strain Anal Eng Des. 2009;44(5):305–18. http://journals.sagepub.com/doi/10.1243/03093247jsa507
68. Mathur PD, McDonald JR, Ghormley RK. A study of the tensile strength of the menisci of the knee. J Bone Joint Surg (Am Vol). 1949;31(3):650–4. https://ncbi.nlm.nih.gov/pubmed/18153906
69. McLean SG, Su A, Bogert A v. Development and validation of a 3-D model to predict knee joint loading during dynamic movement. J Biomech Eng-Trans ASME. 2003;125(6):864–74. https://asmedigitalcollection.asme.org/biomechanical/article/125/6/864/459606/development-and-validation-of-a-3-d-model-to
70. Minns R, Birnie A, Abernethy P. A stress analysis of the patella, and how it relates to patellar articular cartilage lesions. J Biomech. 1979;12(9):699–711. https://sciencedirect.com/science/article/pii/0021929079900198
71. Na A, Buchanan TS. Self-reported walking difficulty and knee osteoarthritis influences limb dynamics and muscle co-contraction during gait. Hum Mov Sci. 2019;64:409–19. https://sciencedirect.com/science/article/pii/s0167945718303427
72. Naili JE, Broström EW, Clausen B, Clausen B, Holsgaard-Larsen A, Holsgaard-Larsen A. Measures of knee and gait function and radiographic severity of knee osteoarthritis – a cross-sectional study. Gait Posture. 2019;74:20–6. https://ncbi.nlm.nih.gov/pubmed/31442818
73. Nakamura S, Nakamura S, Sharma A, Ito H, Nakamura K, Zingde SM, Komistek RD. Kinematic difference between various geometric Centers and contact points for tri-condylar bi-surface knee system. J Arthroplasty. 2015;30(4):701–5. https://sciencedirect.com/science/article/pii/s0883540314008961
74. Nesbitt RJ, Herfat ST, Boguszewski DV, Engel AJ, Galloway MT, Shearn JT. Primary and secondary restraints of human and ovine knees for simulated in vivo gait kinematics. J Biomech. 2014;47(9):2022–7. https://ncbi.nlm.nih.gov/pmc/articles/pmc4032813
75. Nicolella, D. P., Bichon, B. J., Francis, W. L., & Eliason, T. D. (2011). Dynamic Modeling of Knee Mechanics. https://asmedigitalcollection.asme.org/imece/proceedings/imece2011/54884/517/354790
76. Parker D. ISAKOS knee committee: biologics in orthopaedics. Arthroscopy. 2015;31(4):714. https://ncbi.nlm.nih.gov/pubmed/25842232
77. Pastides P, Shenoy R, Nathwani D. The patella in total knee replacement. Orthopaedics and Trauma. 2013;27(6):372–8. https://sciencedirect.com/science/article/pii/s1877132713001231
78. Pot, M. W., Gonzales, V. K., Buma, P., IntHout, J., Kuppevelt, T. H., Vries, R. B., & Daamen, W. F. (2016). Improved cartilage regeneration by implantation of acellular biomaterials after bone marrow stimulation: a systematic review and meta-analysis of animal studies. Peer J, 4. https://peerj.com/articles/2243

79. Potvin, B. (2016). Predicting Muscle Activations in a Forward-Inverse Dynamics Framework Using Stability-Inspired Optimization and an In Vivo-Based 6DoF Knee Joint. http://ruor.uottawa.ca/bitstream/10393/34647/3/potvin_brigitte_2016_thesis.pdf
80. Reese, S. P., Ellis, B. J., & Weiss, J. A. (2013). Multiscale Modeling of Ligaments and Tendons. https://link.springer.com/chapter/10.1007/8415_2012_157
81. Reischl SF, Powers CM, Rao S, Perry J. Relationship between foot pronation and rotation of the tibia and femur during walking. Foot Ankle Int. 1999;20(8):513–20. https://journals.sagepub.com/doi/abs/10.1177/107110079902000809
82. Ristaniemi A, Stenroth L, Mikkonen S, Korhonen RK. Comparison of elastic, viscoelastic and failure tensile material properties of knee ligaments and patellar tendon. J Biomech. 2018;79:31–8. https://sciencedirect.com/science/article/pii/s0021929018306134
83. Rossom SV, Wesseling M, Smith CR, Smith CR, Thelen DG, Thelen DG, et al. The influence of knee joint geometry and alignment on the tibiofemoral load distribution: a computational study. Knee. 2019;26(4):813–23. https://sciencedirect.com/science/article/pii/s0968016018305258
84. Seipel, J. (2012). Towards Robustly Stable Musculo-Skeletal Simulation of Human Gait: Merging Lumped and Component-Based Modeling Approaches. https://proceedings.asmedigitalcollection.asme.org/proceeding.aspx?articleid=1736295
85. Sharma A, Dennis DA, Dennis DA, Dennis DA, Zingde SM, Mahfouz MR, Komistek RD. Femoral condylar contact points start and remain posterior in high flexing patients. J Arthroplasty. 2014;29(5):945–9. https://sciencedirect.com/science/article/pii/s0883540313007183
86. Shelburne KB, Torry MR, Pandy MG. Muscle, ligament, and joint-contact forces at the knee during walking. Med Sci Sports Exerc. 2005;37(11):1948–56. https://ncbi.nlm.nih.gov/pubmed/16286866
87. Shim VB, Shim VB, Besier TF, Lloyd DG, Lloyd DG, Mithraratne K, Fernandez J. The influence and biomechanical role of cartilage split line pattern on tibiofemoral cartilage stress distribution during the stance phase of gait. Biomech Model Mechanobiol. 2016;15(1):195–204. https://link.springer.com/article/10.1007/s10237-015-0668-y
88. Stiehl JB, Dennis DA, Komistek RD, Crane HS. In vivo determination of condylar Lift-off and screw-home in a Mobile-bearing Total knee arthroplasty. J Arthroplasty. 1999;14(3):293–9. https://ncbi.nlm.nih.gov/pubmed/10220182
89. Stiehl JB, Komistek RD, Haas BD, Dennis DA. Frontal plane kinematics after mobile-bearing total knee arthroplasty. Clin Orthop Relat Res. 2001;392(392):56–61. https://ncbi.nlm.nih.gov/pubmed/11716425
90. Sultan, S., Abdel-Malek, K., Arora, J. S., Bhatt, R., & Marler, T. (2017). An Integrated Computational Simulation System for Injury Assessment. https://link.springer.com/chapter/10.1007/978-3-319-41627-4_3
91. Takashi S, Yoshio K, Ten S, Go O, Yuji T. Three-dimensional knee kinematics analysis after tka-does the tilting of femoral component in coronal plane always indicate "condylar lift-off"?; 2018. https://online.boneandjoint.org.uk/doi/abs/10.1302/0301-620x.90bsupp_i.0880187
92. Tardy N, Marchand P, Kouyoumdjian P, Blin D, Demattei C, Asencio G. A preliminary in vivo assessment of anterior cruciate ligament–deficient knee kinematics with the KneeM device: a new method to assess rotatory laxity using open MRI. Orthop J Sports Med. 2014;2(3):–2325967114525583. https://ncbi.nlm.nih.gov/pmc/articles/pmc4555568
93. Torre M, Feo FD, Angelis GD, Ruspantini I, Frustagli G, Chistolini P. ACL reconstruction by bone-patellar tendon-bone graft: mechanical evaluation of the elastic modulus and failure modes. J Orthop Traumatol. 2003;4(2):69–75. https://link.springer.com/article/10.1007/s10195-003-0012-7
94. Wang L, Wu Y, Chang G, Oesingmann N, Schweitzer ME, Jerschow A, Regatte RR. Rapid isotropic 3D-sodium MRI of the knee joint in-vivo at 7T. J Magn Reson Imaging. 2009;30(3):606–14. https://ncbi.nlm.nih.gov/pmc/articles/pmc2759273
95. Wawro M, Fathi-Torbaghan M. A parallel framework for the FE-based simulation of knee joint motion. IEEE Trans Biomed Eng. 2004;51(8):1490–4. https://ncbi.nlm.nih.gov/pubmed/15311837
96. Wernecke GC, Constantinidis A, Harris IA, Seeto BG, Chen DB, MacDessi SJ. The diameter of a single bundle, hamstring autograft does not significantly influence revision rate or clinical outcomes after anterior cruciate ligament reconstruction. Knee. 2017;24(5):1033–1038. https://doi.org/10.1016/j.knee.2017.05.011.
97. Weiss JA, Weiss JA, Gardiner JC, Ellis BJ, Ellis BJ, Lujan TJ, et al. Three-dimensional finite element modeling of ligaments: technical aspects. Med Eng Phys. 2005;27(10):845–61. http://coen.boisestate.edu/ntm/files/2012/07/weiss_2005_mep-fe-techniques.pdf
98. Whittingslow DC, Whittingslow DC, Jeong H-K, Ganti VG, Kirkpatrick NJ, Kogler GF, Inan OT. Acoustic emissions as a non-invasive biomarker of the structural health of the knee. Ann Biomed Eng. 2020;48(1):225–35. https://link.springer.com/article/10.1007/s10439-019-02333-x
99. Woo SL, Abramowitch SD, Kilger R, Liang R. Biomechanics of knee ligaments: injury, healing, and repair. J Biomech. 2006;39(1):1–20. https://sciencedirect.com/science/article/pii/s0021929004005317
100. Yamazaki, T., Watanabe, T., Tomita, T., Sugamoto, K., Ogasawara, M., Sato, Y., … Tamura, S. (2007). 3D kinematics of normal knee using X-ray fluo-

roscopy and CT images. https://link.springer.com/chapter/10.1007/978-3-540-36841-0_706
101. Yan S-H, Ou-Yang H-K, Shan Y-L, Luo D-Z, Wang H, Zhang K. Tensile biomechanical characteristics of human meniscus. Emerg Mater Res. 2016;5(1):44–9. https://icevirtuallibrary.com/doi/abs/10.1680/jemmr.15.00031
102. Zeller, I., Grieco, T., Meccia, B., Sharma, A., & Komistek, R. D. (2018). Development and implementation of a mathematical model to successfully predict normal knee kinematics. https://online.boneandjoint.org.uk/doi/10.1302/1358-992x.2018.5.024
103. Zhang, J., Jiang, K., & Chen, R. (1991). Stress analysis in patella by three-dimensional photoelasticity. https://spiedigitallibrary.org/conference-proceedings-of-spie/1554/0000/stress-analysis-in-patella-by-three-dimensional-photoelasticity/10.1117/12.49563.full
104. Zhang L-K, Wang X-M, Niu Y, Liu H-X, Wang F. Relationship between patellar tracking and the "screw-home" mechanism of tibiofemoral joint. Orthop Surg. 2016;8(4):490–5. https://onlinelibrary.wiley.com/doi/pdf/10.1111/os.12295
105. Zhou T. Analysis of the biomechanical characteristics of the knee joint with a meniscus injury. Healthc Technol Lett. 2018;5(6):247–9. https://ieeexplore.ieee.org/document/8556606
106. Zhu Y, Chen JX, Xiao S, Mahon EM. 3D knee modeling and biomechanical simulation. Comput Sci Eng. 1999;1(4):82–7. https://aip.scitation.org/doi/full/10.1109/5992.774845

Anatomy and Biomechanics of the Anterior Cruciate Ligament

22

Daniel Guenther, Elmar Herbst, and Volker Musahl

22.1 Introduction

Anterior cruciate ligament (ACL) injuries continue to occur at a high rate, affecting over 120,000 people in the United States annually with an estimated cost of $1 billion [1–3]. Due to the high prevalence and cost, ACL reconstruction continues to be a highly studied and debated topic, with over 22,000 publications in the literature, and over 1400 publications in 2018 alone.

In order to understand the etiology of ACL tears and ACL reconstruction, surgeons and researchers must familiarize themselves with the anatomy and biomechanics of the ACL. This includes differentiating the role of the anteromedial (AM) and posterolateral (PL) bundles, as well as understanding the synergistic role of the two bundles as a unit during complex knee motion. The key function of the ACL is a primary restraint to anterior tibial translation, as well as rotatory knee stability [4–6]. Injury to the ACL leads to alterations in knee kinematics and increases the risk of injury to the menisci, cartilage, and surrounding soft tissues [7]. Surgical treatment for ACL injuries has been advocated to restore anterior-posterior and rotatory knee laxity [8, 9]. The graft choice is an important component of surgical decision making. The graft choices include autograft and allograft, each of which have their own unique advantages and disadvantages.

The goal of this chapter will be to discuss the relevant anatomy and biomechanics of the ACL in an intact knee, the effects of ACL deficiency on knee kinematics, and graft options for surgical planning.

22.2 Anatomy

The ACL is a ligamentous structure composed of predominantly Type I collagen and parallel rows of fibroblasts [10]. Dissections of human fetal knee specimens demonstrated the ACL is composed of two distinct bundles which are separated by a septum consisting of vascularized connective tissue [11]. The ACL originates on the medial and posterior aspect of the lateral femoral condyle and inserts anterior and lateral to the medial intercondylar spine of the tibia [12]. On the femoral side, the AM bundle originates more proxi-

D. Guenther (✉)
Department of Trauma and Orthopaedic Surgery, Cologne Merheim Medical Center, University Witten/Herdecke, Cologne, Germany
e-mail: GuentherD@kliniken-koeln.de

E. Herbst
Department of Trauma, Hand and Reconstructive Surgery, University Hospital Muenster, Muenster, Germany

V. Musahl
Department of Orthopaedic Surgery, UPMC Freddie Fu Sports Medicine Center, University of Pittsburgh, Pittsburgh, PA, USA
e-mail: musahlv@upmc.edu

J. Koh et al. (eds.), *Orthopaedic Biomechanics in Sports Medicine*,
https://doi.org/10.1007/978-3-030-81549-3_22

mally and the PL bundle originates more distally [13]. The lateral intercondylar ridge, aka "resident's ridge," represents the anterior edge of the AM and PL bundle attachments on the medial facet of the lateral femoral condyle with the knee flexed to 90°. The lateral bifurcate ridge, which runs perpendicular to the lateral intercondylar ridge, separates the AM and PL bundles [14]. On the tibial side, the AM bundle inserts anteromedially, while the PL bundle inserts posterolaterally [13].

Together, the AM and PL bundles form a functional unit which is indispensable for anteroposterior and rotatory knee stability [6]. Research has shown that the length of the ACL is shortest at 90° of knee flexion and increases with knee extension. On the other hand, the cross-sectional area of the ACL isthmus behaves inversely and becomes smaller in knee extension compared to flexion [15]. This illustrates the dynamic nature of the ACL with a changing stress distribution throughout the physiologic range-of-motion. Taking the distinct femoral and tibial insertion sites of each bundle into account, it becomes evident that the AM and PL bundles act differently but are synergistic [6, 16]. The AM bundle acts as the primary restraint against anterior tibial translation (ATT) in higher flexion angles, while the PL bundle acts as the primary restraint against ATT in lower flexion angles and against combined rotatory loads [6].

22.3 Biomechanics of the Native ACL

Tensile testing of the native ACL in young, healthy individuals has shown the stiffness and ultimate load of the native ACL to be 242 N/mm and 2160 N, respectively [17]. Notably, the ultimate load of the ACL taken from patients aged 60 years or older demonstrated that the ACL had a tensile strength less than one-third of that of an ACL from a young, healthy individual. Furthermore, cadaveric tests in which the ACL was mounted to a materials testing machine have demonstrated the ultimate load and elongation to ultimate load is greater when the strain rate, or the rate at which the ligament is elongated, is increased [18].

The ACL is the primary restraint to ATT in relation to the femur, providing about 85% of the total resisting force [19, 20]. The ACL also functions to limit internal rotation of the tibia, as cadaveric studies have demonstrated that when sectioning the ACL with the knee positioned near full extension, there is an increase in internal rotation of the tibia [21].

The force in the ACL has been shown to be highest at 15–30° of knee flexion and decreases with greater knee flexion [22, 23]. Using a 6-degree of freedom (DOF) robotic manipulator and universal force-moment sensor, the resultant in situ force in the ACL was found to be highest at 15° of knee flexion (111 N) compared to 90° of knee flexion (71 N) in cadaveric specimens [22]. Under applied loads to the quadriceps tendon, the force in the ACL was found to be highest (72 N) at 30° of knee flexion [24].

As the ACL consists of two distinct bundles, the biomechanics and contribution of each bundle to the kinematics of the knee have been studied. The AM bundle has been shown to be lengthened with increasing knee flexion while the PL bundle is at greatest length when the knee is extended [24].

To study the biomechanics of the AM and PL bundles of the ACL, a 6-DOF robotic manipulator was used to measure resultant forces under an applied anterior load on the tibia after successful sectioning of each bundle [22]. The in situ force of the PL bundle was found to be higher than the AM bundle at knee flexion angles between 0 and 45°, while the AM bundle was found to be relatively constant throughout all knee flexion angles. On the other hand, cadaveric studies using loads applied with cables and an optical tracking system found no difference in knee kinematics with isolated AM or PL bundle deficiency, and concluded that a clinically identifiable increase in knee laxity requires injury to both bundles of the ACL [25].

In addition to anterior instability, the role of each bundle in rotatory instability has also been studied. Using a cadaveric model, ATT was measured for intact, AM bundle-deficient, PL bundle-deficient, and ACL-deficient knees under applied anterior load and rotatory torque [6]. With applied anterior loads, after transection of the AM bundle, ATT was found to double as compared to the intact state at knee flexion angles greater than 60° of knee flexion, while sectioning of the PL bundle resulted in a significant increase at 30° of knee flexion. Under rotatory torques, only transection of the PL bundle resulted in increased ATT compared to the intact knee with the knee positioned in less than 30° of knee flexion. Additional cadaveric studies have demonstrated that the PL bundle plays a role in rotatory knee stability particularly when the knee is near full extension, leading the authors to advocate for anatomic reconstruction of the ACL to restore native knee kinematics [26]. In summary, the AM bundle is thought to be primarily responsible for anterior knee stability at higher flexion angles while the PL bundle contributes to rotatory knee stability and provides restraint to ATT at lower flexion angles.

22.4 Biomechanics of the ACL-Deficient Knee

Normal knee function in everyday activities such as walking is determined by the synergic relationship between the dynamic and static constraints, joint morphology, and the forces acting on the knee. The typical mechanism for ACL injury involves a quick deceleration and pivoting or valgus collapse of the knee [27–29]. When the ACL ruptures, the synergic interaction is disrupted, and abnormal knee kinematics and load sharing occurs. Abnormal kinematics may have deteriorative effects, as 51% of ACL injured knees have osteoarthritis (OA) compared to 7% in the contralateral, uninjured knee 12 years after injury [30]. Therefore, it is important to understand how the biomechanics of the knee are altered after ACL deficiency.

22.4.1 Altered Kinematics after ACL Rupture

After ACL injury, increases in ATT and internal rotation occur in response to external loads [9–13]. To observe abnormal kinematics, studies utilizing 6-DOF robotic testing systems have been performed on cadaveric knees. ACL-deficient knees had significantly greater internal rotation and ATT in response to an isolated internal torque as well as a simulated pivot shift when compared to the intact knee [31]. Knee kinematics after ACL injury during walking can be analyzed using a 3-dimensional motion capture system. One study found that ACL-deficient knees had significantly less ATT and more internal rotation compared to the contralateral uninjured knees; this was hypothesized to likely be due to walking strategies to enable gait on a stable, less painful knee [32, 33]. Another study utilizing biplane fluoroscopy showed increased internal rotation as well as a more posterior tibiofemoral contact point in ACL-deficient knees [34]. Though these studies are employing different methods or motions, the overall premise is that ACL deficiency causes altered knee kinematics. The increased kinematics in response to external loads may be due to the absent restraint of the ACL, while the altered walking that decreases ATT may be adaptive gait strategies to achieve knee stability despite injury in which too much ATT may cause instability and discomfort in ACL injured patients.

22.4.2 Major Effects on Other Structures of the Knee

After ACL injury, meniscal tears and other soft tissue injuries at the knee occur more frequently [35, 36]. This increase in meniscus and soft tissue injury occurs because the other structures of the knee are placed under greater forces and moments than before injury to allow ACL-deficient knees to take on the same loads as the intact knee. Using 6-DOF robotic testing systems, researchers have been able to determine differences in the in situ forces of the soft tissue structures of the knee before and after ACL transection. One study found that the in situ forces in the MCL and posterolateral structures significantly increased after ACL deficiency in response to an anterior load (Fig. 22.1) [37]. Another study analyzing the anterolateral capsule and lateral collateral liga-

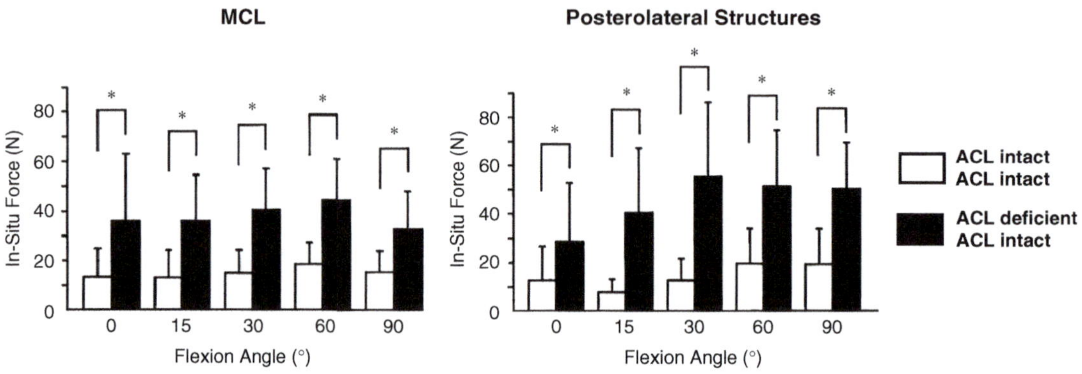

Fig. 22.1 In situ forces of the medial collateral ligament (MCL) and posterolateral structures of the knee for the anterior cruciate ligament (ACL) intact (white bars) and ACL-deficient (black bars) knee in response to a 134 N anterior tibial load. *statistically significant difference ($p < 0.05$) [37]. Reprinted from Journal of orthopaedic science, 5 (6), Kanamori A, Sakane M, Zeminski J, Rudy TW, Woo SL. In-situ force in the medial and lateral structures of intact and ACL-deficient knees, 567–571, 2000, with permission from Elsevier

ment (LCL) found that in ACL-deficient knees, the forces between the regions of the anterolateral capsule were significantly greater than the forces in the LCL and in the anterolateral capsule in response to an anterior load [38]. As meniscal injuries are often concomitant with ACL injuries [35, 39], studies have analyzed the effect of ACL deficiency on the menisci [40, 41]. After ACL transection, the strain in the meniscus significantly increased and the force in the medial meniscus doubled compared to the intact knee [40, 41]. These alterations in knee kinematics and forces in the other soft tissue structures after ACL injury explain why injuries to the menisci and other structures are often concomitant with or more common after ACL injury, and may explain the future development of OA.

22.5 Biomechanics of the ACL Reconstructed Knee

ACL reconstruction requires the use of graft tissue as a replacement for the native ACL. The biomechanical profile and performance of the reconstructed knee is influenced by the physiology of the graft itself as well as its placement in the knee. The biological and mechanical behaviors of various autografts and allografts have been studied extensively in vitro and in vivo using animal and human models, but interpretation of findings across these investigations is challenging due to the heterogeneity of human and animal subjects, graft sizes, and testing methods used. Further, the biomechanical properties of each graft choice and associated clinical outcomes may be modified by preoperative rehabilitation, timing of reconstruction, concomitant injuries, graft processing, preconditioning, fixation method, patient characteristics, and postoperative rehabilitation protocols. Currently, both the surgeon and patient have access to several acceptable graft options categorized by autograft or allograft and comprised of soft tissue alone or with a bone plug.

22.5.1 Autograft

Appropriate graft choice is driven by the tradeoff between autograft's biological advantages (faster incorporation and maturation, avoidance of immune reaction or disease transmission) and allograft's decreased morbidity and surgical time. Patellar tendon (BPTB), quadriceps tendon (QT), and hamstring tendon (HT) are the autografts commonly employed currently. In human cadaveric studies, each has been shown to have stiffness and ultimate tensile load approximating that of the native ACL (Table 22.1) [42–46]. As such, graft choice often involves consideration of additional factors.

Table 22.1 Comparison of stiffness and ultimate tensile load of the native anterior cruciate ligament (ACL) and autografts. Native ACL stiffness and ultimate tensile load vary greatly between younger and older donors [17]. Bone-patellar tendon-bone (BPTB) and hamstring tendon (HT) (double-looped semitendinosus-gracilis) grafts were obtained from 15 matched pairs in a human cadaveric study with 87% male donors between ages 17 and 53 years [47]. BPTB and quadriceps tendon (QT) specimens were obtained from 23 unmatched cadavers aged 19–55 years [48]

	Native ACL	BPTB	QT	HT
Stiffness (N/mm)	180 (age 60–97) [17] –242 (age 22–35) [17]	210 [47]–278 [48]	466 [48]	238 [47]
Ultimate tensile load (N)	658 (age 60–97) [17] –2160 (age 22–35) [17]	1581 [48]–1784 [47]	2186 [48]	2422 [47]

BPTB autograft has historically been considered the gold standard due to its strength characteristics, relative ease of harvest, compatibility with rigid fixation, bone-to-bone healing, and favorable outcomes, but donor-site morbidity has remained a concern [49]. BPTB has been consistently associated with more anterior knee pain and stiffness requiring manipulation under anesthesia or lysis of adhesions when compared to HT. [50, 51] HT has similarly been associated with donor-site morbidity as well as reduced initial fixation strength, higher rates of failure, and increased laxity, particularly in younger and in female patients [50, 52, 53]. A recent meta-analysis confirmed these findings but found the differences in laxity and graft failure to be subtle [54], and a 15-year follow-up of patients randomized to BPTB or HT demonstrated comparable extension deficit and laxity at 15 years [55]. In summary, BPTB may exhibit slightly superior biomechanical outcomes at early timepoints when compared to HT, but this advantage seems to decrease with time. Furthermore, BPTB may be associated with more undesirable sequelae.

More recently, QT has reemerged as a compelling alternative to BPTB. Several studies comparing QT to BPTB in vivo have demonstrated non-inferior biomechanics (as measured by Lachman, pivot shift, and instrumented laxity testing) and similar patient satisfaction with less donor-site morbidity, thereby avoiding the most significant shortcoming of BPTB while maintaining the benefits of healing and fixation afforded by a bone plug [43, 56, 57]. Compared to HT, QT has been associated with less residual laxity and better patient-reported outcomes [58].

All autologous graft materials and fixation techniques currently employed are thought to provide sufficient initial fixation strength [59, 60], but the bone-to-bone healing and rigid initial fixation possible with BPTB and QT with a bone plug are intuitive strengths of these grafts. Ultimately, current evidence from randomized controlled trials comparing BPTB, QT, and HT does not show a consistent difference in postoperative laxity, rate of graft failure, or clinical outcome [51, 55, 60, 61]. In the absence of definitive biomechanical or clinical evidence supporting a single graft choice for any population, graft selection should be made by the surgeon and the patient on an individual basis after discussion of the surgical options and the associated risks and benefits.

22.5.2 Allograft

Various types of allografts are readily available for ACL reconstruction including Achilles tendon, hamstring tendons, patellar tendon, quadriceps tendon, tibialis anterior or posterior tendon, peroneus longus or brevis tendon, and fascia lata allografts [62–64]. Achilles, patellar, and quadriceps tendon allografts are also available with an attached bone block, which is believed to improve graft incorporation due to the presence of a bone-to-bone interface compared to a tendon-to-bone interface in soft tissue-only allografts. However, there is no evidence to support this hypothesis for allografts [65]. The allograft type, procurement, processing and cleansing method, the duration and type of storage, preparation, and donor and graft characteristics have been shown to influence the biomechanical and viscoelastic properties of the allogenic tissue [64]. A recent systematic review showed a high variability in load-to-failure and stiffness between different allograft types, with non-looped tibialis anterior or posterior allografts demonstrating the most unfavorable results [64]. However, in general, allografts provide equivalent or even superior biomechanical properties compared to the native ACL [64].

Apart from the type of allograft, several more parameters have to be considered. It is important to be aware of the details of graft procurement, since it is known that increased donor age may weaken the tensile strength of allografts [64, 66]. Additionally, a negative correlation between the dose of gamma irradiation and the biomechanical properties of allografts has been observed [67–69]. Taken together, it is recommended to use fresh-frozen and non-irradiated or minimally irradiated grafts (<2.5 Mrd) from donors younger than 65 years of age when considering allograft for ACL reconstruction [62, 64, 70]. Knowledge about the mechanical properties of allogenic ACL grafts and its associated functional and clinical outcomes after ACL reconstruction is essential, since up to 42% of primary ACL reconstructions and almost 79% of revision ACL reconstructions are performed using allografts in the United States [71]. Given the high variability in the biomechanical properties of allografts provided by the available literature, no final conclusions on the use of allografts for ACL reconstruction can be drawn at this time. More high-quality studies are needed to determine the ideal allograft type, procurement, cleansing and storage method best suited for ACL reconstruction.

22.5.3 Double-Bundle Vs. Single-Bundle ACL Reconstruction

Given the anisometric behavior of the AM and PL bundle, different length change patterns of the bundles can be observed during knee movement. The AM bundle becomes taut in knee flexion and more lax in extension, while the PL bundle conversely is taut in extension and becomes more lax in flexion [72]. Therefore, an anatomical, rather than an isometric ACL reconstruction, is preferred to restore native knee kinematics [73].

Double bundle (DB) ACL reconstruction has been shown to increase graft forces compared to single bundle (SB) reconstruction, which could negatively affect clinical outcomes [74]. However, it is believed that DB-ACLR can better restore the complex anatomy of the native ACL. Cadaveric biomechanical studies have shown that DB-ACLR more closely restores native knee kinematics compared to SB-ACLR, especially when considering rotatory knee laxity as measured by the pivot-shift test [75]. These biomechanical and kinematic findings have only been partially supported by clinical outcomes. A prospective randomized controlled trial showed no differences between SB- and DB-ACLR in terms of patient-reported outcomes (IKDC function score, Lysholm score) and laxity testing (manual pivot-shift, KT-1000) at 10 years follow-up, but a significantly lower failure rate with DB-ACLR [76]. Additionally, a retrospective review of ACL reconstructions performed in the Swedish ACL registry also demonstrated a decreased failure rate with DB compared to SB-ACL reconstructions [77].

22.6 Conclusion

Understanding the anatomy and biomechanics of the ACL is essential, especially when planning operative ACL reconstruction strategies. The AM and PL bundles of the ACL each have their own role in knee kinematics, while also acting synergistically to control anterior tibial translation and rotatory knee stability. While injury to one bundle may not demonstrate noticeable differences in kinematics and stability, complete injury alters knee mechanics and increases risk to the other soft tissue structures in the knee. The synergetic relationship between bundles is especially important during ACL reconstruction, as both bundles must be accounted for and reconstructed anatomically. There are various graft options, both autograft and allograft, that should be individualized to the patient based on activity level, personal desires, and risk of morbidity associated with each graft type. Single- and double-bundle ACL reconstruction demonstrate similar clinical outcomes, while the most important aspect of ACL reconstruction is respecting patient individual anatomy.

References

1. Kaeding CC, Léger-St-Jean B, Magnussen RA. Epidemiology and diagnosis of anterior cruciate ligament injuries. Clin Sports Med. 2017;36(1):1–8.
2. Joseph AM, Collins CL, Henke NM, Yard EE, Fields SK, Comstock RD. A multisport epidemiologic comparison of anterior cruciate ligament injuries in high school athletics. J Athl Train. 2013;48(6):810–7.
3. Mall NA, Chalmers PN, Moric M, et al. Incidence and trends of anterior cruciate ligament reconstruction in the United States. Am J Sports Med. 2014;42(10):2363–70.
4. Lord BR, El-Daou H, Zdanowicz U, Śmigielski R, Amis AA. The role of fibers within the tibial attachment of the anterior cruciate ligament in restraining tibial displacement. Arthrosc J Arthrosc Relat Surg. 2019;35(7):2101–11.
5. Markatos K, Kaseta M, Lallos S, Korres D, Efstathopoulos N. The anatomy of the ACL and its importance in ACL reconstruction. Eur J Orthop Surg Traumatol. 2013;23(7):747–52.
6. Zantop T, Herbort M, Raschke MJ, Fu FH, Petersen W. The role of the anteromedial and posterolateral bundles of the anterior cruciate ligament in anterior tibial translation and internal rotation. Am J Sports Med. 2007;35(2):223–7.
7. Borchers JR, Kaeding CC, Pedroza AD, et al. Intra-articular findings in primary and revision anterior cruciate ligament reconstruction surgery: a comparison of the MOON and MARS study groups. Am J Sports Med. 2011;39(9):1889–93.
8. Moksnes H, Engebretsen L, Eitzen I, Risberg MA. Functional outcomes following a non-operative treatment algorithm for anterior cruciate ligament injuries in skeletally immature children 12 years and younger. A prospective cohort with 2 years follow-up. Br J Sports Med. 2013;47(8):488–94.
9. Musahl V, Karlsson J. Anterior Cruciate Ligament Tear. N Engl J Med. 2019;380(24):2341–8.
10. Amiel D, Frank C, Harwood F, Fronek J, Akeson W. Tendons and ligaments: a morphological and biochemical comparison. J Orthop Res. 1983;1(3):257–65.
11. Ferretti M, Levicoff EA, Macpherson TA, Moreland MS, Cohen M, Fu FH. The fetal anterior cruciate ligament: an anatomic and histologic study. Arthroscopy. 2007;23(3):278–83.
12. Kopf S, Musahl V, Tashman S, Szczodry M, Shen W, Fu F. A systematic review of the femoral origin and tibial insertion morphology of the ACL. Knee Surg Sports Traumatol Arthrosc. 2009;17(3):213–9.
13. Chhabra A. Anatomic, radiographic, biomechanical, and kinematic evaluation of the anterior cruciate ligament and its two functional bundles. J Bone Jt Surg (Am). 2006;88(suppl 4):2–2.
14. Ferretti M, Ekdahl M, Shen W, Fu FH. Osseous landmarks of the femoral attachment of the anterior cruciate ligament: an anatomic study. Arthrosc J Arthrosc Relat Surg. 2007;23(11):1218–25.
15. Fujimaki Y, Thorhauer E, Sasaki Y, Smolinski P, Tashman S, Fu FH. Quantitative in situ analysis of the anterior cruciate ligament: length, Midsubstance cross-sectional area, and insertion site areas. Am J Sports Med. 2016;44(1):118–25.
16. Petersen W, Zantop T. Anatomy of the anterior cruciate ligament with regard to its two bundles. Clin Orthop Relat Res. 2007;454:35–47.
17. Woo SL, Hollis JM, Adams DJ, Lyon RM, Takai S. Tensile properties of the human femur-anterior cruciate ligament-tibia complex. The effects of specimen age and orientation. Am J Sports Med. 1991;19(3):217–25.
18. Kennedy J, Hawkins R, Willis R, Danylchuck K. Tension studies of human knee ligaments. Yield point, ultimate failure, and disruption of the cruciate and tibial collateral ligaments. JBJS. 1976;58(3):350–5.
19. Markolf KL, Kochan A, Amstutz HC. Measurement of knee stiffness and laxity in patients with documented absence of the anterior cruciate ligament. J Bone Joint Surg Am. 1984;66(2):242–52.
20. Butler DL, Noyes FR, Grood ES. Ligamentous restraints to anterior-posterior drawer in the human knee. A biomechanical study. J Bone Joint Surg Am. 1980;62(2):259–70.
21. Lipke JM, Janecki CJ, Nelson CL, et al. The role of incompetence of the anterior cruciate and lateral ligaments in anterolateral and anteromedial instability. A biomechanical study of cadaver knees. JBJS. 1981;63(6):954–60.
22. Sakane M, Fox RJ, Woo SL, Livesay GA, Li G, Fu FH. In situ forces in the anterior cruciate ligament and its bundles in response to anterior tibial loads. J Orthop Res. 1997;15(2):285–93.
23. Li G, Zayontz S, Most E, DeFrate LE, Suggs JF, Rubash HE. In situ forces of the anterior and posterior cruciate ligaments in high knee flexion: an in vitro investigation. J Orthop Res. 2004;22(2):293–7.
24. Hollis JM, Takai S, Adams DJ, Horibe S, Woo SL. The effects of knee motion and external loading on the length of the anterior cruciate ligament (ACL): a kinematic study. J Biomech Eng. 1991;113(2):208–14.
25. Kondo E, Merican AM, Yasuda K, Amis AA. Biomechanical analysis of knee laxity with isolated anteromedial or posterolateral bundle-deficient anterior cruciate ligament. Arthroscopy. 2014;30(3):335–43.
26. Gabriel MT, Wong EK, Woo SL, Yagi M, Debski RE. Distribution of in situ forces in the anterior cruciate ligament in response to rotatory loads. J Orthop Res. 2004;22(1):85–9.
27. Goldstein J, Bosco JA III. The ACL-deficient knee: natural history and treatment options. Bull NYU Hosp Jt Dis. 2002;60(3–4):173.
28. Kiapour AM, Kiapour A, Goel VK, et al. Uni-directional coupling between tibiofemoral frontal and axial plane rotation supports valgus collapse mechanism of ACL injury. J Biomech. 2015;48(10):1745–51.
29. Shin CS, Chaudhari AM, Andriacchi TP. Valgus plus internal rotation moments increase anterior cruciate

ligament strain more than either alone. Med Sci Sports Exerc. 2011;43(8):1484–91.
30. Lohmander L, Östenberg A, Englund M, Roos H. High prevalence of knee osteoarthritis, pain, and functional limitations in female soccer players twelve years after anterior cruciate ligament injury. Arthrit Rheumat. 2004;50(10):3145–52.
31. Kanamori A, Woo SL, Ma CB, et al. The forces in the anterior cruciate ligament and knee kinematics during a simulated pivot shift test: a human cadaveric study using robotic technology. Arthrosc J Arthrosc Relat Surg. 2000;16(6):633–9.
32. Andriacchi TP, Dyrby CO. Interactions between kinematics and loading during walking for the normal and ACL deficient knee. J Biomech. 2005;38(2):293–8.
33. DeFrate LE, Papannagari R, Gill TJ, Moses JM, Pathare NP, Li G. The 6 degrees of freedom kinematics of the knee after anterior cruciate ligament deficiency: an in vivo imaging analysis. Am J Sports Med. 2006;34(8):1240–6.
34. Li G, Moses JM, Papannagari R, Pathare NP, DeFrate LE, Gill TJ. Anterior cruciate ligament deficiency alters the in vivo motion of the tibiofemoral cartilage contact points in both the anteroposterior and mediolateral directions. JBJS. 2006;88(8):1826–34.
35. Daniel DM, Stone ML, Dobson BE, Fithian DC, Rossman DJ, Kaufman KR. Fate of the ACL-injured patient: a prospective outcome study. Am J Sports Med. 1994;22(5):632–44.
36. Noyes FR, Mooar P, Matthews D, Butler D. The symptomatic anterior cruciate-deficient knee. Part I: the long-term functional disability in athletically active individuals. JBJS. 1983;65(2):154–62.
37. Kanamori A, Sakane M, Zeminski J, Rudy TW, Woo SL. In-situ force in the medial and lateral structures of intact and ACL-deficient knees. J Orthop Sci. 2000;5(6):567–71.
38. Guenther D, Rahnemai-Azar AA, Bell KM, et al. The anterolateral capsule of the knee behaves like a sheet of fibrous tissue. Am J Sports Med. 2017;45(4):849–55.
39. Kilcoyne KG, Dickens JF, Haniuk E, Cameron KL, Owens BD. Epidemiology of meniscal injury associated with ACL tears in young athletes. Orthopedics. 2012;35(3):208–12.
40. Hollis JM, Pearsall AW, Niciforos PG. Change in meniscal strain with anterior cruciate ligament injury and after reconstruction. Am J Sports Med. 2000;28(5):700–4.
41. Papageorgiou CD, Gil JE, Kanamori A, Fenwick JA, Woo SL, Fu FH. The biomechanical interdependence between the anterior cruciate ligament replacement graft and the medial meniscus. Am J Sports Med. 2001;29(2):226–31.
42. Hamner DL, Brown CH Jr, Steiner ME, Hecker AT, Hayes WC. Hamstring tendon grafts for reconstruction of the anterior cruciate ligament: biomechanical evaluation of the use of multiple strands and tensioning techniques. J Bone Joint Surg Am. 1999;81(4):549–57.
43. Harris NL, Smith DA, Lamoreaux L, Purnell M. Central quadriceps tendon for anterior cruciate ligament reconstruction. Part I: morphometric and biomechanical evaluation. Am J Sports Med. 1997;25(1):23–8.
44. Markolf KL, Burchfield DM, Shapiro MM, Cha CW, Finerman GA, Slauterbeck JL. Biomechanical consequences of replacement of the anterior cruciate ligament with a patellar ligament allograft. Part II: forces in the graft compared with forces in the intact ligament. J Bone Joint Surg Am. 1996;78(11):1728–34.
45. Noyes FR, Butler DL, Grood ES, Zernicke RF, Hefzy MS. Biomechanical analysis of human ligament grafts used in knee-ligament repairs and reconstructions. J Bone Joint Surg Am. 1984;66(3):344–52.
46. Staubli HU, Schatzmann L, Brunner P, Rincon L, Nolte LP. Mechanical tensile properties of the quadriceps tendon and patellar ligament in young adults. Am J Sports Med. 1999;27(1):27–34.
47. Wilson TW, Zafuta MP, Zobitz M. A biomechanical analysis of matched bone-patellar tendon-bone and double-looped semitendinosus and gracilis tendon grafts. Am J Sports Med. 1999;27(2):202–7.
48. Shani RH, Umpierez E, Nasert M, Hiza EA, Xerogeanes J. Biomechanical comparison of quadriceps and patellar tendon grafts in anterior cruciate ligament reconstruction. Arthrosc J Arthrosc Relat Surg. 2016;32(1):71–5.
49. West RV, Harner CD. Graft selection in anterior cruciate ligament reconstruction. J Am Acad Orthop Surg. 2005;13(3):197–207.
50. Freedman KB, D'Amato MJ, Nedeff DD, Kaz A, Bach BR Jr. Arthroscopic anterior cruciate ligament reconstruction: a metaanalysis comparing patellar tendon and hamstring tendon autografts. Am J Sports Med. 2003;31(1):2–11.
51. Samuelsson K, Andersson D, Karlsson J. Treatment of anterior cruciate ligament injuries with special reference to graft type and surgical technique: an assessment of randomized controlled trials. Arthroscopy. 2009;25(10):1139–74.
52. Barrett AM, Craft JA, Replogle WH, Hydrick JM, Barrett GR. Anterior cruciate ligament graft failure: a comparison of graft type based on age and Tegner activity level. Am J Sports Med. 2011;39(10):2194–8.
53. Barrett GR, Noojin FK, Hartzog CW, Nash CR. Reconstruction of the anterior cruciate ligament in females: a comparison of hamstring versus patellar tendon autograft. Arthroscopy. 2002;18(1):46–54.
54. Samuelsen BT, Webster KE, Johnson NR, Hewett TE, Krych AJ. Hamstring autograft versus patellar tendon autograft for ACL reconstruction: is there a difference in graft failure rate? A meta-analysis of 47,613 patients. Clin Orthop Relat Res. 2017;475(10):2459–68.
55. Webster KE, Feller JA, Hartnett N, Leigh WB, Richmond AK. Comparison of patellar tendon and hamstring tendon anterior cruciate ligament reconstruction: a 15-year follow-up of a randomized controlled trial. Am J Sports Med. 2016;44(1):83–90.

56. Sheean AJ, Musahl V, Slone HS, et al. Quadriceps tendon autograft for arthroscopic knee ligament reconstruction: use it now, use it often. Br J Sports Med. 2018;52(11):698–701.
57. Slone HS, Romine SE, Premkumar A, Xerogeanes JW. Quadriceps tendon autograft for anterior cruciate ligament reconstruction: a comprehensive review of current literature and systematic review of clinical results. Arthroscopy. 2015;31(3):541–54.
58. Cavaignac E, Coulin B, Tscholl P. Nik Mohd Fatmy N, Duthon V, Menetrey J. is quadriceps tendon autograft a better choice than hamstring autograft for anterior cruciate ligament reconstruction? A comparative study with a mean follow-up of 3.6 years. Am J Sports Med. 2017;45(6):1326–32.
59. Dargel J, Gotter M, Mader K, Pennig D, Koebke J, Schmidt-Wiethoff R. Biomechanics of the anterior cruciate ligament and implications for surgical reconstruction. Strategies Trauma Limb Reconstr. 2007;2(1):1–12.
60. Shea KG, Carey JL, Richmond J, et al. The American Academy of Orthopaedic surgeons evidence-based guideline on management of anterior cruciate ligament injuries. J Bone Joint Surg Am. 2015;97(8):672–4.
61. Hurley ET, Calvo-Gurry M, Withers D, Farrington SK, Moran R, Moran CJ. Quadriceps tendon autograft in anterior cruciate ligament reconstruction: a systematic review. Arthroscopy. 2018;34(5):1690–8.
62. Hulet C, Sonnery-Cottet B, Stevenson C, et al. The use of allograft tendons in primary ACL reconstruction. Knee Surg Sports Traumatol Arthrosc. 2019;27(6):1754–70.
63. Tisherman R, Wilson K, Horvath A, Byrne K, De Groot J, Musahl V. Allograft for knee ligament surgery: an American perspective. Knee Surg Sports Traumatol Arthrosc. 2019;27(6):1882–90.
64. Lansdown DA, Riff AJ, Meadows M, Yanke AB, Bach BR Jr. What factors influence the biomechanical properties of allograft tissue for ACL reconstruction? A systematic review. Clin Orthop Relat Res. 2017;475(10):2412–26.
65. Noh JH, Yang BG, Yi SR, Roh YH, Lee JS. Single-bundle anterior cruciate ligament reconstruction in active young men using bone-tendon achilles allograft versus free tendon achilles allograft. Arthroscopy. 2013;29(3):507–13.
66. Swank KR, Behn AW, Dragoo JL. The effect of donor age on structural and mechanical properties of allograft tendons. Am J Sports Med. 2015;43(2):453–9.
67. Hoburg A, Keshlaf S, Schmidt T, et al. Fractionation of high-dose electron beam irradiation of BPTB grafts provides significantly improved viscoelastic and structural properties compared to standard gamma irradiation. Knee Surg Sports Traumatol Arthrosc. 2011;19(11):1955–61.
68. Curran AR, Adams DJ, Gill JL, Steiner ME, Scheller AD. The biomechanical effects of low-dose irradiation on bone-patellar tendon-bone allografts. Am J Sports Med. 2004;32(5):1131–5.
69. DiBartola AC, Everhart JS, Kaeding CC, Magnussen RA, Flanigan DC. Maximum load to failure of high dose versus low dose gamma irradiation of anterior cruciate ligament allografts: a meta-analysis. Knee. 2016;23(5):755–62.
70. Greaves LL, Hecker AT, Brown CH Jr. The effect of donor age and low-dose gamma irradiation on the initial biomechanical properties of human tibialis tendon allografts. Am J Sports Med. 2008;36(7):1358–66.
71. Maletis GB, Inacio MC, Funahashi TT. Analysis of 16,192 anterior cruciate ligament reconstructions from a community-based registry. Am J Sports Med. 2013;41(9):2090–8.
72. Amis AA, Dawkins GP. Functional anatomy of the anterior cruciate ligament. Fibre bundle actions related to ligament replacements and injuries. J Bone Joint Surg Br. 1991;73(2):260–7.
73. Musahl V, Plakseychuk A, VanScyoc A, et al. Varying femoral tunnels between the anatomical footprint and isometric positions: effect on kinematics of the anterior cruciate ligament-reconstructed knee. Am J Sports Med. 2005;33(5):712–8.
74. Markolf KL, Park S, Jackson SR, McAllister DR. Anterior-posterior and rotatory stability of single and double-bundle anterior cruciate ligament reconstructions. J Bone Joint Surg Am. 2009;91(1):107–18.
75. Musahl V, Bedi A, Citak M, O'Loughlin P, Choi D, Pearle AD. Effect of single-bundle and double-bundle anterior cruciate ligament reconstructions on pivot-shift kinematics in anterior cruciate ligament- and meniscus-deficient knees. Am J Sports Med. 2011;39(2):289–95.
76. Jarvela S, Kiekara T, Suomalainen P, Jarvela T. Double-bundle versus single-bundle anterior cruciate ligament reconstruction: a prospective randomized study with 10-year results. Am J Sports Med. 2017;45(11):2578–85.
77. Svantesson E, Sundemo D, Hamrin Senorski E, et al. Double-bundle anterior cruciate ligament reconstruction is superior to single-bundle reconstruction in terms of revision frequency: a study of 22,460 patients from the Swedish National Knee Ligament Register. Knee Surg Sports Traumatol Arthrosc. 2017;25(12):3884–91.

Biomechanics of Extra-Articular Ligaments of the Knee and Extra-Articular Tenodesis

23

Pablo Besa, Timothy Lording, and Sebastián Irarrázaval

23.1 Introduction

Residual rotatory knee instability following anatomical anterior cruciate ligament reconstruction has highlighted the role of the anterolateral structures of the knee [1]. Extra-articular procedures were once commonly performed for anterolateral rotatory instability, however, intra-articular ACL reconstructions (ACL-R) displaced lateral procedures as the focus of ACL surgery. The main reason was initial and sustained success of intra-articular ACL reconstructions, with clinical series showing excellent results and acceptable re-rupture rates [2, 3].

Slowly, concerns around persistent instability started to increase [4]. While some authors described groups of patients with less success and high re-rupture rates, others described residual pivot shift and even sustained risk of long-term articular degeneration [5]. Therefore, there was a renewed interest in the anatomy and biomechanical role of anterolateral structures of the knee. In 2013, Claes [6] reopened the field with the description of the anterolateral ligament (ALL) and this description was quickly followed by multiple techniques to reconstruct it. Meanwhile, lateral extra articular tenodesis techniques, once used in isolation for the treatment of ACL ruptures, regained ground as an addition to the anatomic ACL [7].

A biomechanical understanding of the role of anterolateral structures in the native, ACL deficient, and ACL reconstructed knee is paramount to determine the role of added procedures in the ACL reconstructed knee. The objective of this chapter is to highlight the main biomechanical components of the anterolateral structures and their role in the aforementioned scenarios, trying to determine the role of lateral extra-articular tenodesis techniques and anterolateral ligament reconstructions.

23.2 Anatomy

The anatomy of lateral aspect of the knee was described by Seebacher et al. [8] as consisting of three well-defined layers: (1) the superficial layer consisting of the superficial fascia, iliotibial band (ITB), and biceps femoris; (2) the second layer with the quadriceps retinaculum and the patellofemoral ligaments; and the (3) deepest layer formed by the lateral collateral ligament and the lateral capsule. Key biomechanical contributors are the ITB, the anterolateral capsule and the ALL (Fig. 23.1).

P. Besa · S. Irarrázaval (✉)
Orthopaedics Department, School of Medicine, Pontifical Catholic University of Chile, Santiago, Región Metropolitana, Chile
e-mail: sirarraz@uc.cl

T. Lording
Melbourne Orthopaedic Group, Windsor, VIC, Australia

J. Koh et al. (eds.), *Orthopaedic Biomechanics in Sports Medicine*,
https://doi.org/10.1007/978-3-030-81549-3_23

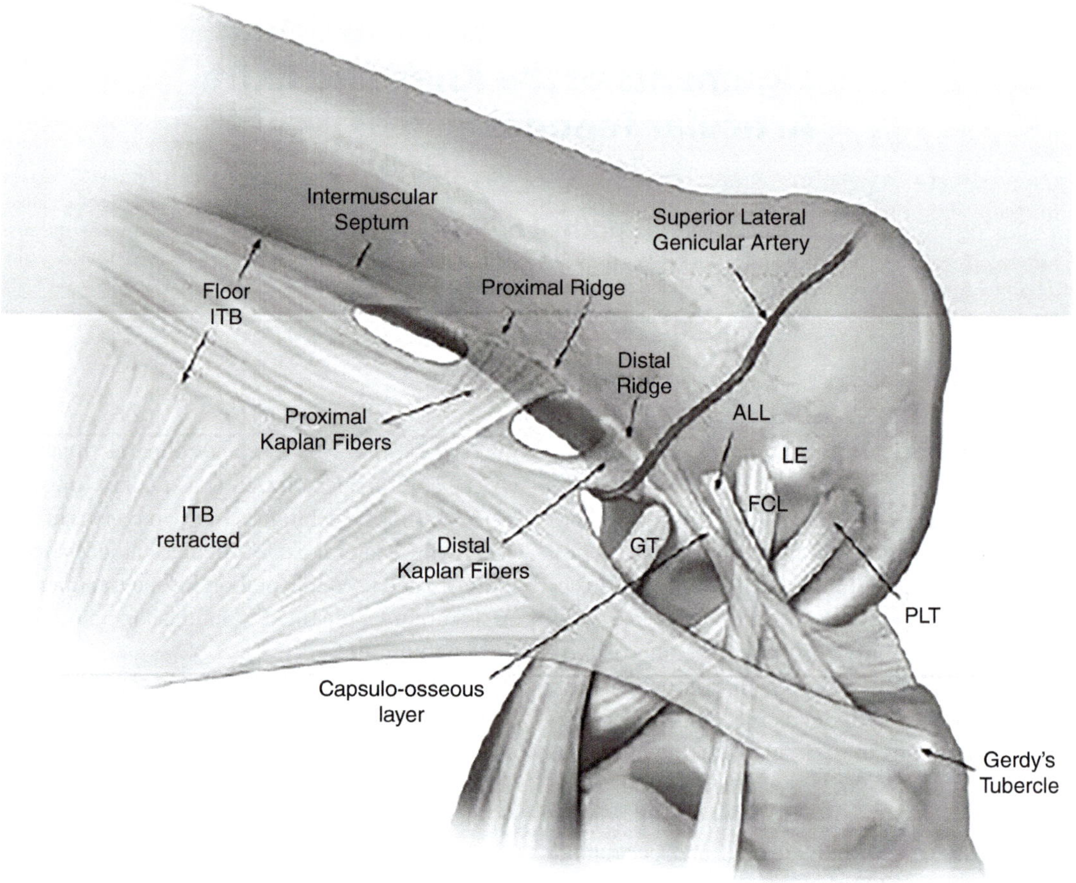

Fig. 23.1 Anatomy of the anterolateral complex of the knee, showing the insertions of the proximal and distal Kaplan fibers on the proximal and distal ridges, respectively, in a right knee. ALL: anterolateral ligament; FCL: fibular collateral ligament; GT: lateral gastrocnemius tendon; ITB: iliotibial band; LE: lateral epicondyle; PLT: popliteus tendon. From Godin et al. [15], license number 5087391062609

The ITB has two distinct layers. The most superficial, fascia-like layer inserts at Gerdy's tubercle on the anterolateral aspect of the tibia. It has an anterior reflection to the patella (iliopatellar band) and a posterior reflection that reinforces the biceps femoris tendon. Kaplan [9] described femoral attachments initiated from the superficial ITB (Kaplan fibers). These Kaplan fibers were later divided into three segments by Lobenhoffer [10] (from proximal to distal): transverse fibers near the septum inserting in the dorsolateral femur; proximal-lateral to distal-medial fibers inserting in the supracondylar area of the lateral femoral condyle; and retrograde fibers running from the Gerdy's tubercle insertion toward the dorsolateral femur bridging the knee joint. The deep ITB, also named the capsulo-osseous layer, [11–13] runs deep and medial to the superficial ITB until it merges distally with the superficial layer before inserting in Gerdy's tubercle [11]. The distal segment of the deep ITB is reinforced by the biceps femoris tendon [14].

The anterolateral capsule is also divided into a superficial and deep component [8] based on the direction of the fibers in relation with the fibular collateral ligament (FCL). Hughston, who introduced the concept of anterolateral rotatory instability (ALRI), described the anterolateral capsule ligaments [1]. He determined that the mid-third portion had distinct femoral and tibial insertions, as well as meniscal attachments (also known as coronary ligaments).

Although the biomechanical contribution of the anterolateral structures had been studied previously, an anatomical structure distinct from the ITB and the anterolateral capsule was described by Claes in 2013, denominated the anterolateral ligament (ALL) [6]. Most authors agree that the tibial attachment of the ALL lies mid-way between Gerdy's tubercle and the fibular head, 5–10 mm below the joint line [16, 17]. The femoral attachment, on the other hand, has been debated, with some authors showing a common insertion with the popliteus tendon; [18] others an epicondylar insertion next to the FCL; [6, 19] and others a supracondylar femoral insertion site [20, 21]. Musahl [11] proposed that the differences found were due to cadaver preparation techniques (frozen vs embalmed) and proposed that the ALL was either confused with the ITB deep capsulo-osseous layer described initially by Segond, and that Terry said acted "as an anterolateral ligament" (Terry); the mid-third portion of the capsule described by Hughston; a combination of both; [11] or the anterior arm of the short head of the biceps femoris muscle [22]. Regardless of the existence of a distinct ligament or the combination of previously defined anatomical structures, the anterolateral complex has a known biomechanical behavior both natively and when surgically reconstructed.

The lack of consensus on the shape and form of the ALL, and the overlap of descriptions with layers of the ITB and capsule, warrant careful interpretation of its individual role in knee biomechanics. As suggested by Amis, [23] it may be more appropriate to analyze the anterolateral complex (ALC) as a whole and avoid distinction. When discussing biomechanics of anterolateral rotatory instability of the knee, one must state the degree of knee flexion and if they are ACL competent or deficient knees. Further studies have shown that ligament engagement (i.e., involvement in knee kinematics) is also highly susceptible to tibial displacement [24–26]. When available, we will state these variables, assuming neutral position when not mentioned.

23.3 Native Biomechanics of the Anterolateral Complex

Few studies have evaluated the function of the ALC with an intact ACL and an in situ ITB, as many reflect the ITB [24, 26] to isolate the ALC. This complicates interpretation as we have previously noted the robust anatomy of the ITB and its biomechanical relevance. Kittl [27] determined that the superficial ITB was the primary restraint (contributes more that 50%) to tibial internal rotation (IR) above 30 degrees of knee flexion (5-Nm of IR; 0°, 30°, 60° and 90°); with at least 40% of restraint under 30 degrees coming from the superficial and deep ITB together. No other structures were found to contribute more than 20%. Although the ACL seems to have a poor biomechanical situation to control tibial rotation, under 30° the ACL tightens and forces tibial external rotation, locking the knee at full extension, [13] known as the screw home mechanism [28]. Huser [29] similarly found that the contribution of ALL and the deep capsulo-osseous layer of the ITB to robotic pivot shift testing (100-N of anterior load +7-Nm of valgus +1 or 5-Nm of IR) was minimal and statistically equivalent between intact knees and knees with sectioned ALL and the deep layer of the ITB. Therefore, in the native knee, the ALL seems to have a minor role in controlling IR, as primary stabilization is dependent on the ITB (mainly its superficial layer).

Femoral point	Position (from lateral femoral epicondyle)	Tibial point G (Gerdy tubercle)	Tibial point A (area of the Segond avulsion)
1	2 mm anterior, 2 mm distal	Anterior part of the Losee [30] reconstruction	Mid-third lateral capsular ligament ALL defined by Claes et al. [6]
2	10 mm posterior, 4 mm distal	Isometric point of Draganich et al. [31]	
3	4 mm posterior, 8 mm proximal	Lemaire reconstruction [32]*	ALL defined by Dodds et al. [20]
4	6 mm posterior, 10 mm proximal	Isometric point of Sidles et al. [33]	
5	Over-the-top position	Zarins-Rowe [34] reconstruction* Isometric point of Krackow and Brooks [35] Posterior part of the Losee [30] reconstruction*	
6	Posterior femoral cortex at the distal termination of the intermuscular septum	Anterior fibers of the ITT MacIntosh reconstruction [36]*	Posterior fibers of the ITT

There were four native tissue structures, four reconstructions, and three femoral isometric points. ALL: anterolateral ligament; ITT: iliotibial tract. *Indicates course deep to the lateral collateral ligament.

23.4 Biomechanics in the ACL-Deficient Knee

Contrary to ACL intact studies, most studies in ACL sectioned knees maintain the ALC and ITB intact in an attempt to individualize their roles. We have already mentioned that at lower degrees of flexion, the ACL initiates the screw home mechanism (tibial external rotation at knee full extension). Therefore, the lack of ACL reduces the screw home locking mechanism and leaves the tibia to internally rotate [13]. Spencer [37] showed that ALL sectioning determined an increase in IR when submitted to a simulated pivot shift (10-Nm of valgus +5-Nm of IR; 0° of flexion). That same year, Rasmussen [38] proved that this increase in IR was across the whole range of motion in simulated pivot shift (10-Nm of valgus +5-Nm of IR; 0 to 60° in 15° increments) and plain IR testing (5-Nm of IR; 0 to 120° in 15° increments). Albeit they found statistically significant increments in IR after ACL and ALL sectioning, the increases were up to 3.5 degrees. This was pointed out by Kittl [27] who found that ALL contribution to IR restraint (5-Nm of IR; 0°, 30°, 60°, and 90°) in the ACL-deficient knee was statistically significant but perhaps not clinically relevant. Furthermore, when studying a lower force simulated pivot shift (8-Nm of valgus +4-Nm of IR; 15°, 30° and 45°) Kittl found no significant contribution of the ALL (Fig. 23.2).

Interestingly, Noyes [39] found that neither the ALL nor the ITB contributed significantly to resisting IR in the ACL-deficient knee. They tested ALL and/or ITB sectioning under three different forces: IR (5-Nm of IR; 25°, 60° and 90°); and two pivot shift Schemes (100-N of anterior load +7-Nm of valgus +1 or 5-Nm of IR). All scenarios showed no increase with isolated ALL sectioning and minimal but significant increase in IR with isolated ITB section as well with the combined section of ITB and ALL. Similarly, Saiegh [40] tested simulated pivot shifts (8.4-Nm of anterior load +2.5-Nm of valgus; 30° and 90°) and found no statistical difference between ACL-deficient and ACL + ALL deficient knees. More recently Lagae [41] confirmed these findings, with minor differences observed between the ACL-deficient and ACL + ALC deficient knees. They also tested

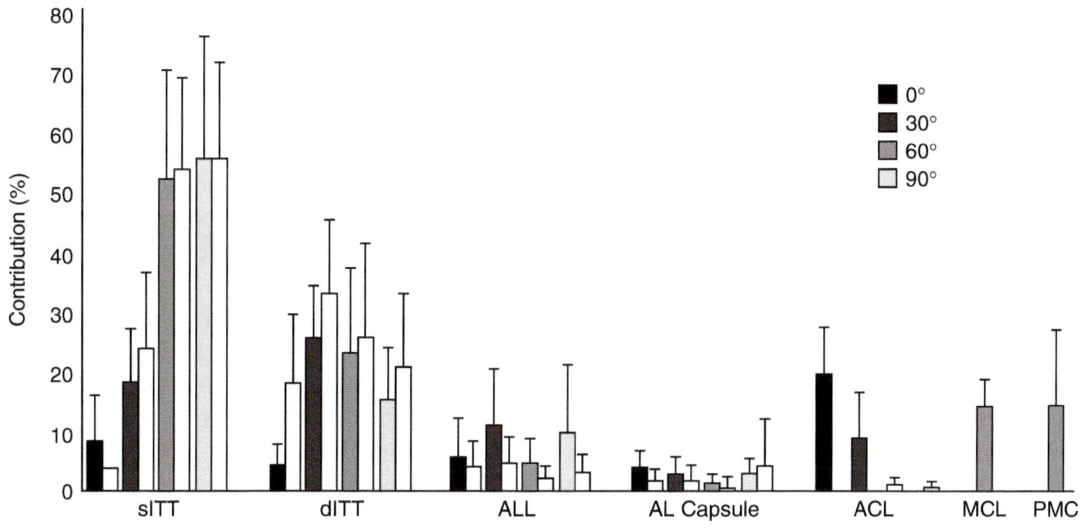

Fig. 23.2 Contribution of different anatomic structures in controlling a 5 Nm internal rotation torque at 0°, 30°, 60°, and 90° of knee flexion, according to Kittl et al. [13]. Adjacent white areas indicate results from the ACL-deficient knee in every condition. The superficial fibers (sITT) and the deep fibers (dITT) of the ITT present the primary restraint to internal rotation at 30° to 90° of flexion. The ALL and the anterolateral capsule (Cap) show a nonsignificant contribution in restraining internal rotation. Adapted from Kittl et al. [13], license number 5087390086875

anterior translation when subjected to anterior forces (90-N anterior drawer force; 0 to 110° in 10° increments) and internal torque (5-Nm of IR, 0 to 110° in 10° increments). Sectioning of the ALL in the ACL-deficient knee only increased anterior translation at 20° and 30° (by a mean of 0.8 and 0.7 mm respectively); with no difference in the IR tests.

Further studies confirmed these findings, [24, 26] but complemented the data by including varying degrees of tibial translation. Thein [42] showed that the ALL engages during pathological tibial anterior translation of 10 to 12 mm. These degrees of anterior tibial translation are only seen in ACL-deficient knees. Lording [43] showed similar results in ACL and ALL sectioned knees, with minimal but significant increases in IR (5-Nm of IR; 0 to 90° in 15° increments). Increases in IR were higher at higher degrees of knee flexion. Lording also showed that lateral meniscus posterior root played a similar role in restraining IR but during lower degrees of knee flexion (0 to 30°).

In summary, when submitted to IR or simulated pivot shift testing of ACL-deficient knees, the ALL contributes minimally to restrain IR in most cadaveric studies. This contribution is higher above 30° of knee flexion. More recent studies suggest that mere IR or pivot shift simulation are not capable of engaging the ALL within the envelope of normal knee kinematics. Additionally, engagement of the ALL has proven to have a high interspecimen variability [26].

23.5 Rotatory Instability Following ACL Reconstruction

Anterolateral extra-articular tenodesis (LET) procedures, as popularized by Lemaire (1967), [44] MacIntosh (1976), [36] Losee (1978), [30] Arnold and Cocker (1979), [45] and Ellison (1979), [46] were once used as isolated treatment options for ACL ruptures. They were superceded by the novel free patellar tendon bone (BTB) graft allowing anatomic ACL reconstruction (ACL-R), as developed by Franke [47]. The BTB quickly became the gold-standard of ACL-R [48]. Even though many proposed the addition of LET procedures to ACL-R [49], Dejour 1999, [50] the potential added morbidity lacked evident benefit and was mainly discontinued. The rebirth of anterolateral tenodesis techniques in recent years has been associated with two major concerns following anatomic ACL-R: groups with high re-rupture rates, and patients with residual rotatory instability. Recently, the Stability Group [51] has published the results of a multicenter study randomizing high-risk patients to isolated ACL (autologous hamstring tendon graft) vs ACL + anterolateral tenodesis (modified Lemaire). They defined high-risk patients as younger than 25 years old and having two of the following three risk factors: (1) grade 2 or greater pivot shift; (2) generalized ligamentous laxity; and (3) returning to cutting or pivoting sports. After 2 years, they found that while the isolated ACL group has significantly more graft ruptures (11% vs 4% of the ACL + LET group) and residual instability (29 and 21% respectively), there was no difference in patient reported outcomes nor level of activity.

The rationale behind the augmentation in these high-risk groups is that an additional constraint of anterolateral rotatory instability through LET addition could translate into decreased forces through the ACL graft tissue as suggested by Noyes in 1991 [52]. Although (as shown further below), LET does increase rotational stability, there is a lack of high-level evidence regarding re-rupture rates and a 2015 meta-analysis showed no difference regarding this outcome in ACL vs ACL + LET [53]. Therefore, the outcomes of the aforementioned study [51] could change the clinical landscape of LET procedures.

More controversial even than the former, residual rotatory instability has been one of the main arguments for anterolateral procedures in association with ACL-R [7, 54]. Biomechanical logic gifts anterolateral procedures with an increased lever to restrain IR when compared with ACL insertion sites, whom are close to the rotational axis, thus warranting poor conditions to resist rotation [23]. Nevertheless, previously we stated that in native knees, the ACL does play a key role in restraining IR, secondary only to the ITB [27]. Many have studied if ACL-R are able to replicate this degree of restraint and therefore return the knee to normal kinematics. Samuelson et al. [55] first showed that patients with a combined injury of the ACL and anterolateral structures (he incised the capsular structures from the patella medially up to the fibular collateral ligament laterally, leaving the latter and the ITB intact), failed to regain normal knee kinematics with an isolated ACL-R (autologous BTB). He found that anterior translation was normalized at all angles tested except 0° (100-N of anterior load; 30° 60° and 90°). At 0° the anterior translation was increased 2 mm in the combined injury with isolated ACL-R. This same setup failed to replicate native IR restraint (3-Nm of IR; 0 to 90° in 30° increments). Inderhaug [54] confirmed the earlier publication of Samuelson, showing that in a knee with combined ACL and anterolateral injury, isolated ACL-R (autologous BTB) did not restore normal knee kinematics. Additionally, they found a significant increase in IR when submitted to torsional torque (5-Nm IR; 0 to 90° in 10° increments) and simulated pivot shift (90-N of anterior load +5-Nm IR; 0 to 90° in 10° increments) (Fig. 23.3). Finally, Geeslin et al. [56] confirmed these findings with a similar cadaveric setup, also showing that an isolated ACL-R (BTB allograft) could not replicate native kinematics when subjected to a simulated pivot shift (88-Nm of anterior load +10-Nm of valgus +5-Nm of IR; 60 and 90° of flexion). When applying only IR (5-Nm IR; 45°, 60°, 75° and 90°), they found no difference between the ACL intact and the ACL reconstructed knees. Interestingly, more recent tests conducted by Lagae [41] found that an isolated ACL (4 strand hamstring graft) had an anterior translation similar to the intact knee across the whole range of motion (90-N anterior drawer force; 0 to 110° in 10° increments); but IR remained minimally increased when compared with the native knee between 30 and 100° (5-Nm of IR, 0 to 110° in 10° increments).

These findings should be expected, as we previously stated that when sequentially injuring the ACL followed by the anterolateral complex structures, Spencer, [37] Rasmussen [38] and others had shown a small (up to 3.5 degrees) but statistically significant contribution of the ALC

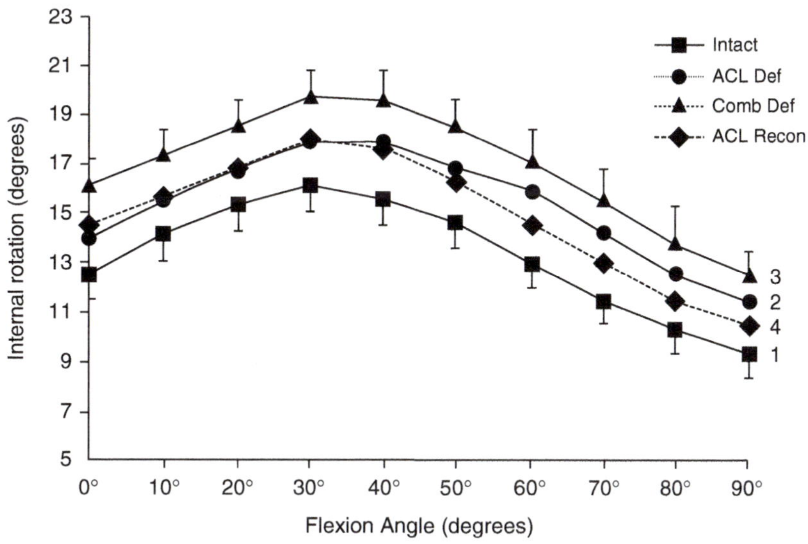

Fig. 23.3 Tibial internal rotation response to a 5-N-m internal torque for intact, ACL reconstruction injury (ACL-Def), combined lesion (Comb-Def) and ACL reconstruction (ACL-R). Reproduced from Inderhaug et al. [54], license number 5087411043004

to rotational laxity. This marginal contribution is confirmed by the lack of normal knee kinematics with the isolated reconstruction of the ACL in combined injuries (ACL and ALC) shown by Samuelson, [55] Inderhaug, [54] and Geeslin [56] (for pivot shift only). Furthermore, the differences found by the former are a maximum 2.1° of IR [54, 56], very similar to the studies of sectioning the ALC in ACL-deficient knees [27, 37, 38]. In other words, the ALC plays a small but significant role in restraining IR. This role is more notorious in ACL-deficient knees, but also alters knee kinematics in isolated ACL-R following combined ACL and ALC injuries. Additionally, Guenther [57] proposed that the inability of regaining normal knee kinematics was not a fault of the isolated ACL-R, and that the lack of normal knee kinematics was caused by the combined injury most biomechanical setups used. He tested an intact knee against isolated then reconstructed ACL against anterior loading (134-N of anterior force; 30°, 60° and 90°) and interior rotation torque (7-Nm of IR; 30°, 60° and 90°); and found no differences regarding anterior translation nor IR respectively. He furthered the point by then transecting the ALC and found that the isolated ACL-R with and ALC transection did not replicate native knee kinematics.

23.6 Biomechanics of Anterolateral Tenodesis

As mentioned previously, lateral extra-articular tenodesis (LET) techniques have over 40 years since their first modern description by Lemaire [44]. Over time, more than 12 techniques have been described for LET [58, 59]. Therefore, we will try to focus on the main biomechanical properties and results they share, detailing their main differences regarding their role in the ACL-deficient and ACL-reconstructed knee.

Engebretsen in 1990 [60] studied the force ratios of an ACL graft (10 mm BTB) in the presence of a modified Andrewes LET and the force ratios of the same tenodesis but in the ACL-deficient knee. He showed that the addition of the LET decreased ACL graft forces in over 40%, and up to 30% below the forces tested in the native ACL when subjected to anteroposterior forces (90-N of anterior load; 0 to 90° in 30° increments). Interestingly, when inverting the reconstruction sequence (LET first, then ACL), the reduction in ACL graft forces was not significant when compared to the intact ACL. Additionally, LET significantly increased external rotation of the tibia (in both sequences) of the unloaded knee, possibly interfering with the screw home mechanism.

Sharing anteroposterior loads with the ACL was confirmed later by Spencer [37]. He found that a modified Lemaire not only controlled IR against a simulated pivot shift (10-Nm of valgus +5-Nm of IR; 0° of flexion), but also recreated the anterior translation constraint of the intact ACL when subjected to a standardized anterior drawer test (90-N anterior force; 90°). Interestingly, this control over anterior translation was lost when testing a simulated Lachman test (90-N anterior force; 30°). He also noted that anteroposterior control was not shared by the ALL reconstruction he performed as described by Claes [6]. Spencer's findings were later replicated by Geeslin [56] confirming the role of LET in controlling tibial anterior translation and IR. He also tested a modified Lemaire, but he compared fixing the LET at 30 and 70° of knee flexion. When they applied IR (5-Nm of IR; 0 to 90° in 15° increments), simulated pivot shift (10-Nm valgus +5-Nm of IR: 15 and 30°), and anterior translation simulating an anterior drawer test (88-Nm of anterior force; at 90°), the modified Lemaire fixed at 30° significantly decreased tibial anterior translation and IR compared to the isolated ACL-R, obtaining values similar to the intact knees. These findings were not obtained for the LET fixed at 70° nor for the simulated Lachman test on either fixation angle (88-Nm of anterior force; at 30°), similar to Spencer. Finally, Inderhaug [54] compared three LET techniques under similar setups: MacIntosh, modifies Lemaire [37] and a new modification to Lemaire (superficial to the FCL). He also tested a simulated anterior drawer (90-N anterior drawer force; 0 to 90° of flexion at 10° increments); forced IR (5-Nm IR; 0 to 90° of flexion at 10° increments); and simulated pivot shift (90-Nm anterior drawer force +5-Nm IR; 0 to 90° of flexion at 10° increments). The results obtained confirmed previous findings, with both MacIntosh and the standard modified Lemaire (deep to the FCL) behaving similarly, basically replicating native knee kinematics. The novel modification to Lemaire (superficial to the FCL) tended to overconstraint IR over a wide range of knee flexion.

Recently, Devitt [61] has complemented these results with a novel biomechanical setup. To avoid confusion over ACL-R, he only transected the anterolateral capsule deep to the ITB and anterior to the FCL, leaving the native ACL intact. Then, he tested the ALC deficient knee against a modified Ellison LET with and without posterior closure of the ITB band, submitted to anterior force translation (90-N of anterior force; 0°, 30°, 60° and 90° of flexion), IR (5-Nm of IR torque; 0°, 30°, 60° and 90° of flexion) and simulated pivot shift (8-Nm of valgus +4-Nm IR; 0°, 15°, 30° and 45° of knee flexion). As already demonstrated, transection of the anterolateral capsule increases anterior translation and IR in all tested states (except for anterior translation in simulated pivot shift at 30 and 45° of knee flexion). No difference was found between leaving the ITB open or closing it after the modified Ellison procedure. Both procedures did not modify anterior translation against an isolated anterior force but did significantly decrease anterior translation and IR compared to the sectioned ALC state in all degrees of flexion on the IR and simulated pivot shift testing. Also noteworthy, they found that the modified Ellison technique produced a significant (albeit slight) constraint of normal IR and anterior translation at lower degrees of flexion (30° and below).

There isn't sufficient evidence to compare the outcomes of each individual technique described for performing LET, [58, 59] but the main differences observed in the cited studies favor techniques deep to the FCL and fixed at 30°. This configuration has most closely replicated native knee biomechanics, by restraining tibial anterior translation and IR, as well as load sharing with ACL grafts, in multiple time zero cadaveric studies.

Four systematic reviews and meta-analysis [7, 53, 62, 63] searched for studies that clinically measured the addition of LET to primary ACL-R. The main results Hewison and Song found were that across multiple LET techniques, all deep to the FCL, there was a significant reduction in residual pivot shit testing in favor of the combined group (ACL + LET) (14 and 7 studies respectively, level of evidence I–III), but no difference in IKDC scores. Devitt found similar results, but divided studies where surgery was done in the 12 months following ACL injury and studies were surgery was after 12 months of injury. They found that the reduction in pivot shift testing was only observed in delayed surgery (3 studies, levels of evidence I–III) and not early surgery (5 studies, levels of evidence I–III), suggesting attenuation of the anterolateral structures with prolonged instability. Similar to Hewison, they found no benefit in other clinical outcomes. In summary, clinical evidence is scarce regarding the addition of LET in primary ACL-R, showing minor improvements in residual pivot shift testing [7, 62, 63] but no demonstrated effect over patient reported outcomes or in re-rupture rates [53]. Future studies, including the Stability Group protocol are relevant in determining if the addition of LET is justified, probably in high-risk groups only [64]. Finally a recent systematic review of LET in revision ACL [65] found 12 case series (level of evidence IV), five of them (170 patients) showing a mean 2.3 mm side to side difference KT-1000 evaluation, with 45% of patience showing more than a 3 mm difference; and seven studies (628 patients) showing 80% or more patients with a negative pivot shift.

23.7 Biomechanics of Anterolateral Ligament Reconstructions

Anterolateral ligament reconstruction must follow the conviction that there is an anterolateral ligament (thus the word reconstruction). Although the description of the ALL made by Claes [6] is relatively recent, a systematic review in 2017 by DePhillipo [66] found six different techniques (Fig. 23.4) [18, 67–71]. Interestingly, none of the them have the same femoral insertion, but all of them have roughly the same tibial insertion (midway between the Gerdy tubercle and the fibular head).

Spencer's 2015 publication [37] was the first ALL reconstruction biomechanical description. He tested an isolated ALL reconstruction in an ACL-deficient knee, using an unpublished technique described by Arthrex [72] with a femoral fixation on the lateral epicondyle, fixed at 70° of knee flexion using a FiberTape© suture as graft. They chose not to follow Claes [6] description of femoral origin due to concerns with over constraining the knee. The result was that when tested against a simulated pivot shift (10-Nm of valgus +5-Nm of IR, 0° of flexion) and an anterior drawer test (90-N on anterior force; 90°), the ALL reconstruction did not change results when compared to the previous state (ACL and ALL sectioned). This could be explained by the lack of engagement as proposed previously or the limitation of a non-anatomic reconstruction of the ALL.

Later studies tested the ALL in a more relevant setting: the ACL reconstructed knee [54, 73–75]. Nitri [73] tested the ALL reconstruction described by Chahla [71] against simulated pivot shift (10-Nm of valgus +5-Nm of IR; 0 to 60° in 15° increments), IR (5-Nm of IR; 0 to 120° in 15° increments) and anterior drawer testing (88-N of anterior force; 0 to 120° in 15° increments). He found that, in an ACL reconstructed knee (10 mm allograft BTB), the ALL reconstruction had no impact over anteroposterior translation when compared to the ALL sectioned knee, neither on the anterior drawer testing nor the simulated pivot shift. On the other hand, when evaluating IR, ALL reconstruction significantly restricted IR compared to the ALL deficient knee, over 30° of flexion in both simulated pivot shift and isolated IR torque. Schon [74] tested the same ALL reconstruction described by Chahla [71] but tested the role of knee flexion at the time of fixation (0°, 15°, 30°, 45°, 60°, 75° and 90°). Although the main conclusion of his article is that the anatomic ALL described by Chahla produces IR overconstraint independent of the fixation angle; he also found that the addition of an ALL reconstruction significantly limited IR and anterior translation of the tibia compared to an isolated ACL reconstruction(10 mm allograft BTB) when submitted to IR (5-Nm of IR torque; 0 to 120° in 15° increments), anterior drawer test (88-N of anterior load; 0 to 120° in 15° increments) and simulated pivot shift (10-Nm of valgus force +5-Nm of IR torque +10-N axial compression force; 0 to 60° in 15° increments). It is noteworthy that they are the first group to add axial compressive forces in

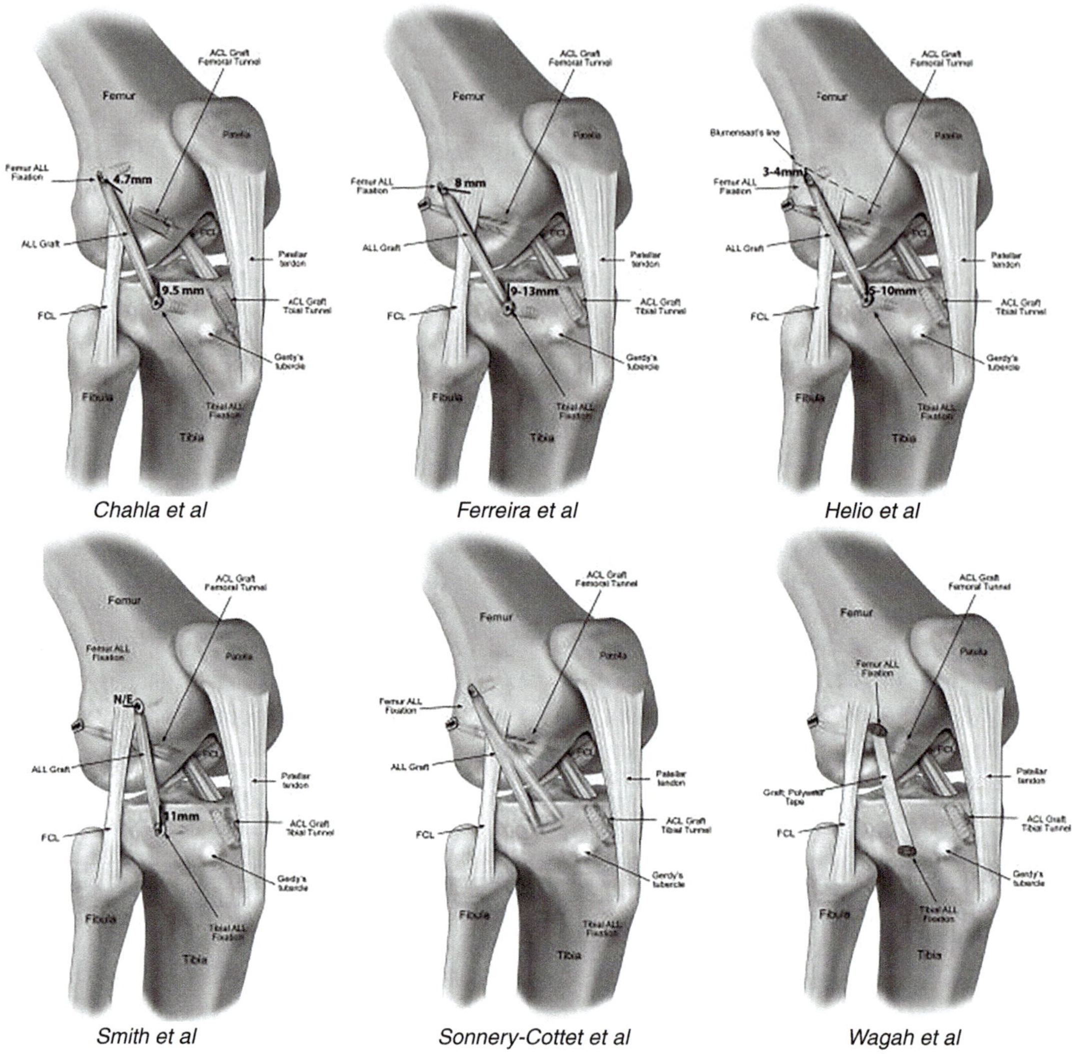

Fig. 23.4 Different anterolateral ligament reconstruction techniques. Reproduced from DePhillipo et al. [66], license number 5087401150047

the simulated pivot shift to ensure tibiofemoral contact and increase fidelity with clinical testing of the pivot shift. Additionally, Nielsen [75] tested a different ALL reconstruction [69] fixed between 10 and 20° as an addition to an isolated ACL (9 mm autograft BTB). Interestingly, they manually tested for Lachman and pivot shift, showing that the addition of the ALL reconstruction significantly eliminated residual pivot shift and Lachman when compared with the isolated ACL-R.

Inderhaug [54] also tested the addition of an ALL reconstruction to an isolated ACL (10 mm autograft BTB). He followed the description by Kennedy [76] of the anatomy of the ALL but fixed the graft at 40°. He had already reported that the isolated ACL was insufficient, but that a Lemaire LET obtained near normal knee kinematics. The addition of the ALL reconstruction to the ACL led to a persistent IR laxity. These results were replicated when testing tensioning the ALL graft at 20 and 40 N.

Anterolateral ligament reconstruction techniques have less biomechanical evidence to support their use compared to lateral extra articular tenodesis. The lack of consistency found in the few studies shown could be associated to the diverse surgical techniques and testing scenarios; and should lead to careful interpretation.

23.8 Concerns for the Future

Lateral extra-articular tenodesis and anterolateral ligament reconstruction raise some concerns for the future. Concerns include that adding a tensed structure to the lateral aspect of the knee may constrain the knee range of motion and increase lateral compartment contact pressures. Another concern is that although time zero studies may show acceptable knee kinematics, many early studies suggested graft or tenodesis stretching with time [77].

Lateral compartment constraint concerns arise from insertions sites that increase tension at higher degrees of flexion, [27] and many authors have stated their concerns of clinical implications [54, 56]. Most LET techniques proposed pass deep to the FCL [58, 59]. Kittl [27] suggested that this configuration had "desirable length changes" associated with their passage deep to the FCL. His experimental design correlated insertion sites for diverse LET and ALL reconstructions in an unloaded knee from 0 to 90° in 10° increments, a testing setup described previously by Krackow [35]. He found that deep to the FCL configurations had minor length changes and they tended to tighten near full extension. The increase in tension associated with higher degrees of knee flexion was later confirmed by Inderhaug [54]. He compared the standard modified Lemaire (deep to the FCL) [37] with a novel modification passing superficial to the FCL. He found that this change constrained the knee reducing its IR beyond the intact knee. Devitt [61] also found that a modified Ellison LET (a deep to the FCL technique) could minimally (but significantly) constrain the IR (up to 1.6 less) and anterior translation (up to 0.7 mm less) compared to the native knee. These studies show that the addition of LET not necessarily replicates native knee kinematics, and as well as the minimal residual knee laxity observed in the isolated ACL-R, the clinical impact of minimal knee constraint has not been studied.

Interestingly, most of the aforementioned studies of residual instability following ACL-R (or intact ACL as tested by Devitt) artificially generate an ALL or ALC injury to test the isolated ACL-R kinematics. It seems important to determine what percentage of ACL injuries are associated with an injury of the ALC, specially following the results of Guenther, [57] who proved that an isolated ACL-R did regain intact knee kinematics when the test knee had an isolated ACL injury (and not a combined ACL + ALC injury). In-vivo or ex-vivo studies to determine the degree of association have not been published and seem an unlikely way to answer this question although some early studies did attempt [78]. Imaging studies have pushed to determine the degree of association between both injuries. Regarding imaging studies, ALL studies have been controversial, with low reproducibility and inter- and intra-observer agreement [11]. Therefore, we will focus mainly on studies that consider the whole ALC.

In a study conducted by Musahl [79] he studied the magnetic resonance images of 41 patients with complete ACL rupture. He found that only 51% of patients had an ALC injury (with a high interobserver agreement among two blinded radiologists; kappa 0.75). He also found that in the ACL-deficient knee, patients with the combined injury had a significantly higher pivot shift. Musahl's findings are in agreement with the studies shown previously but raise a concern regarding how many patients have a combined injury as tested in biomechanical studies, and how many have an isolated ACL injury and do not behave as the subjects tested assuming combined ACL and ALC injury.

Regarding lateral compartment contact pressures, Inderhaug [80] measured lateral contact pressures in the intact knee, ALC sectioned knee and MacIntosh LET, fixed at 20-N and 80-N and fixed at neutral rotation or free rotation. He showed that transecting the ALC significantly increased lateral compartment pressures, and that the LET normalized these pressures for all fixation protocols except 80-N free rotation. Similarly, Shimakawa [81] tested contact pressures in ACL and ALC deficient knees and the changes caused by ACL-R and modified Lemaire LET. He found that ACL-R and LET did not significantly change mean nor peak contact pressures. Furthermore, Zaffagnini [82] published a long-term follow-up (20 years) of patients with ACL-R and LET (Marcacci procedure [83]) and reported no association with lateral compartment nor patellofemoral osteoarthritis.

Finally, even though biomechanical testing has raised relevant but unanswered questions, short and mid-term clinical results have not associated the addition of LET with increased complications or secondary events, suggesting the addition to be a safe procedure.

23.9 Conclusions

Anterolateral knee anatomy is complex and its main biomechanical role in the native knee is to control IR and secondarily tibial anterior translation. The main structure to control rotation is the iliotibial band, mainly its superficial capsulo-osseous layer. ACL-R can lead to persistent residual rotatory knee instability in some patients. This minor residual laxity could be secondary to injuries in the deepest layers of the anterolateral structures of the knee. Patients with injuries to the anterolateral complex may benefit from additional procedures to ACL-R to regain native knee kinematics. The addition of lateral extra articular tenodesis techniques produces more native-like kinematics with less risk of IR constraints than anterolateral ligaments reconstructions. There is not enough evidence to differentiate between lateral extra articular tenodesis techniques, but evidence suggests that deep to the FCL techniques fixed at 30° and below 40-N have less risk of over-constraint. Lateral extra articular tenodesis have proven to unload ACL grafts and clinical data strongly supports a reduction in residual pivot shift. While historical data shows no difference in re-rupture rates between isolated ACL reconstruction and the addition of lateral extra articular tenodesis techniques, the recent Stability Group findings show a decrease of re-rupture rates associated with the addition of LET in high-risk patients. Finally, concerns for rotational constraints, increase in lateral compartment contact pressures and long-term kinematics of lateral extra articular tenodesis and anterolateral ligament reconstruction techniques have not been sufficiently answered.

References

1. Hughston JC, et al. Classification of knee ligament instabilities. Part II. The lateral compartment. J Bone Joint Surg. 1976;58(2):173–9.
2. Fu FH, et al. Anatomic anterior cruciate ligament reconstruction: a changing paradigm. Knee Surg Sports Traumatol Arthrosc. 2015;23(3):640–8.
3. Anderson MJ, et al. A systematic summary of systematic reviews on the topic of the anterior cruciate Ligament. Orthop J Sports Med. 2016;4(3):2325967116634074.
4. Ayeni OR, et al. Pivot shift as an outcome measure for ACL reconstruction: a systematic review. Knee Surg Sports Traumatol Arthrosc. 2012;20(4):767–77.
5. Ajuied A, et al. Anterior cruciate ligament injury and radiologic progression of knee osteoarthritis: a systematic review and meta-analysis. Am J Sports Med. 2014;42(9):2242–52.
6. Claes S, et al. Anatomy of the anterolateral ligament of the knee. J Anat. 2013;223(4):321–8.
7. Hewison CE, et al. Lateral extra-articular tenodesis reduces rotational laxity when combined with anterior cruciate ligament reconstruction: a systematic review of the literature. Arthroscopy. 2015;31(10):2022–34.
8. Seebacher JR, et al. The structure of the posterolateral aspect of the knee. J Bone Joint Surg. 1982;64(4):536–41.
9. KAPLAN EB. The iliotibial tract; clinical and morphological significance. J Bone Joint Surg. 1958;40-A(4):817–32.
10. Lobenhoffer P, et al. Distal femoral fixation of the iliotibial tract. Arch Orthop Trauma Surg. 1987;106(5):285–90.
11. Musahl V, et al. The anterolateral complex and anterolateral Ligament of the knee. J Am Acad Orthop Surg. 2018;26(8):261–7.
12. Landreau P, et al. Anatomic study and reanalysis of the nomenclature of the anterolateral complex of the knee focusing on the distal iliotibial band: identification and description of the condylar strap. Orthop J Sports Med. 2019;7(1):232596711881806–9.
13. Kittl C, et al. Biomechanics of the anterolateral structures of the knee. Clin Sports Med. 2018;37(1):21–31.
14. Terry GC, Hughston JC, Norwood LA. The anatomy of the iliopatellar band and iliotibial tract. Am J Sports Med. 1986;14(1):39–45.
15. Godin JA, et al. A comprehensive reanalysis of the distal iliotibial band: quantitative anatomy, radio-

graphic markers, and biomechanical properties. Am J Sports Med. 2017;45(11):2595–603.
16. !!! INVALID CITATION !!! [8,9].
17. Kennedy MI, et al. An overview of clinically relevant biomechanics of the anterolateral structures of the knee. Tech Orthop. 2018;33(4):213–8.
18. Wagih AM, Elguindy AMF. Percutaneous reconstruction of the anterolateral Ligament of the knee with a polyester tape. Arthrosc Tech. 2016;5(4):e691–7.
19. Daggett M, et al. Femoral origin of the anterolateral ligament: an anatomic analysis. Arthroscopy. 2016;32(5):835–41.
20. Dodds AL, et al. The anterolateral ligament: anatomy, length changes and association with the Segond fracture. Bone Joint J. 2014;96-B(3):325–31.
21. Vieira ELC, et al. An anatomic study of the iliotibial tract. Arthroscopy. 2007;23(3):269–74.
22. Urban S, Pretterklieber B, Pretterklieber ML. The anterolateral ligament of the knee and the lateral meniscotibial ligament - Anatomical phantom versus constant structure within the anterolateral complex. Ann Anat. 2019;226:64–72.
23. Amis AA. Anterolateral knee biomechanics. Knee Surg Sports Traumatol Arthrosc. 2017;25(4):1015–23.
24. Thein R, et al. Biomechanical assessment of the anterolateral Ligament of the knee. J Bone Joint Surg. 2016;98(11):937–43.
25. Imhauser CW, et al. New parameters describing how knee ligaments carry force in situ predict interspecimen variations in laxity during simulated clinical exams. J Biomech. 2017;64:212–8.
26. Kent RN, et al. High interspecimen variability in engagement of the anterolateral ligament: an in vitro cadaveric study. Clin Orthop Relat Res. 2017;475(10):2438–44.
27. Kittl C, et al. Length change patterns in the lateral extra-articular structures of the knee and related reconstructions. Am J Sports Med. 2015;43(2):354–62.
28. Hallén LG, Lindahl O. The screw-home movement in the knee-joint. Acta Orthop Scand. 1966;37(1):97–106.
29. Eng, L.E.H.M., et al. Anterolateral ligament and iliotibial band control of rotational stability in the anterior cruciate ligament-intact knee: defined by tibiofemoral compartment translations and rotations. YJARS, 2016: 21 1–10.
30. Losee RE, Johnson TR, Southwick WO. Anterior subluxation of the lateral tibial plateau. A diagnostic test and operative repair. J Bone Joint Surg. 1978;60(8):1015–30.
31. Draganich LF, Hsieh YF, Reider B. Iliotibial band tenodesis: a new strategy for attachment. Am J Sports Med. 1995;23(2):186–95.
32. Lemaire, M. and F. Combelles, Technique actuelle de plastie ligamentaire pour rupture ancienne du ligament croisé antérieur. 1980.
33. Sidles JA, et al. Ligament length relationships in the moving knee. J Orthop Res. 1988;6(4):593–610.
34. Zarins B, Rowe CR. Combined anterior cruciate-ligament reconstruction using semitendinosus tendon and iliotibial tract. J Bone Joint Surg. 1986;68(2):160–77.
35. Krackow KA, Brooks RL. Optimization of knee ligament position for lateral extraarticular reconstruction. Am J Sports Med. 1983;11(5):293–302.
36. MacIntosh, D.L., Lateral substitution reconstruction. In proceedings of the Canadian Orthopaedic Association. J Bone Joint Surg. 1976; 58: 142.
37. Spencer L, et al. Biomechanical analysis of simulated clinical testing and reconstruction of the anterolateral Ligament of the knee. Am J Sports Med. 2015;43(9):2189–97.
38. Rasmussen MT, et al. An in vitro robotic assessment of the anterolateral Ligament, part 1. Am J Sports Med. 2016;44(3):585–92.
39. Noyes FR, Huser LE, Levy MS. Rotational knee instability in ACL-deficient knees. J Bone Joint Surg. 2017;99(4):305–14.
40. Al Saiegh, Y., et al. Sectioning the anterolateral ligament did not increase tibiofemoral translation or rotation in an ACL-deficient cadaveric model. Knee Surg Sports Traumatol Arthrosc. 2015; 12: 1–7.
41. Lagae KC, et al. ACL reconstruction combined with lateral monoloop tenodesis can restore intact knee laxity. Knee Surg Sports Traumatol Arthrosc. 2020;28(4):1159–68.
42. Tanaka MJ, et al. Passive anterior tibial subluxation in anterior cruciate ligament-deficient knees. Am J Sports Med. 2013;41(10):2347–52.
43. Lording, T., et al. Rotational laxity control by the anterolateral ligament and the lateral meniscus is dependent on knee flexion angle: a cadaveric biomechanical study. Clin Orthop Relat Res 2017: 1–8.
44. Lemaire M. Ruptures anciennes du ligament croisé anterieur du genou. J Chir (Paris). 1967;93:311–20.
45. Arnold JA, et al. Natural history of anterior cruciate tears. Am J Sports Med. 1979;7(6):305–13.
46. Ellison AE. Distal iliotibial-band transfer for anterolateral rotatory instability of the knee. J Bone Joint Surg. 1979;61(3):330–7.
47. Franke K. Clinical experience in 130 cruciate ligament reconstructions. Orthopedic Clinics of NA. 1976;7(1):191–3.
48. Chambat P, et al. The evolution of ACL reconstruction over the last fifty years. Int Orthop. 2013;37(2):181–6.
49. Clancy WG, et al. Anterior cruciate ligament reconstruction using one-third of the patellar ligament, augmented by extra-articular tendon transfers. J Bone Joint Surg. 1982;64(3):352–9.
50. H, D., D. D, and A.S.S. T, Laxité chronique du genou traité par une greffe de tendon rotulien libre et une plastie extra articulaire antérolatérale. 10 ans de recul. 148 cas. Rev Chir Orthop Réparatrice Appar Locomot., 1999. 85: p. 777–789.
51. Getgood AMJ, et al. Lateral extra-articular Tenodesis reduces failure of hamstring tendon autograft anterior cruciate Ligament reconstruction: 2-year outcomes from the STABILITY study randomized clinical trial. Am J Sports Med. 2020;48(2):285–97.
52. Noyes FR, Barber SD. The effect of an extra-articular procedure on allograft reconstructions for chronic ruptures of the anterior cruciate ligament. J Bone Joint Surg. 1991;73(6):882–92.

53. Rezende FC, et al. Does combined intra- and extraarticular ACL reconstruction improve function and stability? A meta-analysis. Clin Orthop Relat Res. 2015;473(8):2609–18.
54. Inderhaug E, et al. Biomechanical comparison of anterolateral procedures combined with anterior cruciate Ligament reconstruction. Am J Sports Med. 2017;45(2):347–54.
55. Samuelson M, et al. The effects of knee reconstruction on combined anterior cruciate ligament and anterolateral capsular deficiencies. Am J Sports Med. 1996;24(4):492–7.
56. Geeslin AG, et al. Anterolateral knee extra-articular stabilizers: a robotic study comparing anterolateral Ligament reconstruction and modified Lemaire lateral extra-articular Tenodesis. Am J Sports Med. 2018;46(3):607–16.
57. Guenther D, et al. The role of extra-articular Tenodesis in combined ACL and anterolateral capsular injury. J Bone Joint Surg Am Vol. 2017;99(19):1654–60.
58. Slette EL, et al. Biomechanical results of lateral extra-articular tenodesis procedures of the knee: a systematic review. Arthroscopy. 2016;32(12):2592–611.
59. Wascher DC, Lording TD, Neyret P. Extra-articular procedures for the ACL-deficient knee: a state of the art review. J ISAKOS. 2016;1(3):174–82.
60. Engebretsen L, et al. The effect of an iliotibial tenodesis on intraarticular graft forces and knee joint motion. Am J Sports Med. 1990;18(2):169–76.
61. Devitt BM, et al. Biomechanical assessment of a distally fixed lateral extra-articular augmentation procedure in the treatment of anterolateral rotational laxity of the knee. Am J Sports Med. 2019;47(9):2102–9.
62. Devitt BM, et al. The role of lateral extra-articular Tenodesis in primary anterior cruciate Ligament reconstruction: a systematic review with meta-analysis and best-evidence synthesis. Orthop J Sports Med. 2017;5(10):2325967117731767.
63. Song G-Y, et al. Clinical outcomes of combined lateral extra-articular tenodesis and intra-articular anterior cruciate ligament reconstruction in addressing high-grade pivot-shift phenomenon. Arthroscopy. 2016;32(5):898–905.
64. Batty L, Lording Clinical T. results of lateral extra-articular tenodesis. Tech Orthop. 2018;33(4):232–8.
65. Grassi A, et al. Good mid-term outcomes and low rates of residual rotatory laxity, complications and failures after revision anterior cruciate ligament reconstruction (ACL) and lateral extra-articular tenodesis (LET). Knee Surg Sports Traumatol Arthrosc. 2020;28(2):418–31.
66. DePhillipo NN, et al. Anterolateral ligament reconstruction techniques, biomechanics, and clinical outcomes: a systematic review. Arthroscopy. 2017;33(8):1575–83.
67. Helito CP, et al. Combined intra- and extra-articular reconstruction of the anterior cruciate Ligament: the reconstruction of the knee anterolateral Ligament. Arthrosc Tech. 2015;4(3):e239–44.
68. Ferreira C, et al. Reconstruction of anterior cruciate ligament and anterolateral ligament using interlinked hamstrings - technical note. Revista brasileira de ortopedia. 2016;51(4):466–70.
69. Sonnery-Cottet B, et al. Minimally invasive anterolateral Ligament reconstruction in the setting of anterior cruciate Ligament injury. Arthrosc Tech. 2016;5(1):e211–5.
70. Smith JO, et al. Combined anterolateral ligament and anatomic anterior cruciate ligament reconstruction of the knee. Knee Surg Sports Traumatol Arthrosc. 2015;23(11):3151–6.
71. Chahla J, et al. Anterolateral Ligament reconstruction technique: an anatomic-based approach. Arthrosc Tech. 2016;5(3):e453–7.
72. Arthrex, Anterolateral ligament reconstruction. 2014.
73. Nitri M, et al. An in vitro robotic assessment of the anterolateral Ligament, part 2: anterolateral Ligament reconstruction combined with anterior cruciate Ligament reconstruction. Am J Sports Med. 2016;44(3):593–601.
74. Schon JM, et al. Anatomic anterolateral Ligament reconstruction of the knee leads to Overconstraint at any fixation angle. Am J Sports Med. 2016;44(10):2546–56.
75. Nielsen ET, et al. Influence of the anterolateral Ligament on knee laxity: a biomechanical cadaveric study measuring knee kinematics in 6 degrees of freedom using dynamic Radiostereometric analysis. Orthop J Sports Med. 2018;6(8):232596711878969–13.
76. Kennedy MI, et al. The anterolateral Ligament: an anatomic, radiographic, and biomechanical analysis. Am J Sports Med. 2015;43(7):1606–15.
77. Khan R, et al. RSA can measure ACL graft stretching and migration: development of a new method. Clin Orthop Relat Res. 2006;448:139–45.
78. Terry GC, et al. How iliotibial tract injuries of the knee combine with acute anterior cruciate ligament tears to influence abnormal anterior tibial displacement. Am J Sports Med. 1993;21(1):55–60.
79. Musahl V, et al. The influence of meniscal and anterolateral capsular injury on knee laxity in patients with anterior cruciate Ligament injuries. Am J Sports Med. 2016;44(12):3126–31.
80. Inderhaug E, et al. The effects of anterolateral Tenodesis on tibiofemoral contact pressures and kinematics. Am J Sports Med. 2017;45(13):3081–8.
81. Shimakawa T, et al. Lateral compartment contact pressures do not increase after lateral extra-articular Tenodesis and subsequent subtotal meniscectomy. Orthop J Sports Med. 2019;7(6):2325967119854657.
82. Zaffagnini S, et al. Over-the-top ACL reconstruction plus extra-articular lateral Tenodesis with hamstring tendon grafts: prospective evaluation with 20-year minimum follow-up. Am J Sports Med. 2017;45(14):3233–42.
83. Marcacci M, et al. Arthroscopic intra- and extra-articular anterior cruciate ligament reconstruction with gracilis and semitendinosus tendons. Knee Surg Sports Traumatol Arthrosc. 1998;6(2):68–75.

Anatomy and Biomechanics of the Collateral Ligaments of the Knee

24

Kanto Nagai, Yuta Nakanishi, Kohei Kamada, Yuichi Hoshino, and Ryosuke Kuroda

24.1 Medial Collateral Ligament (MCL)

Comprehensive knowledge of medial knee joint biomechanics is valuable for the assessment of which injured structures should be repaired or reconstructed. An understanding of abnormal joint motion that occurs when a structure is injured greatly assists with the interpretation of the results from clinical examination and helps to determine the presence of concurrent ligament injury.

24.1.1 Anatomy

The medial collateral ligament (MCL) complex consists of superficial and deep MCL, and posterior oblique ligament (POL). Despite the distinct names for the separate components of the MCL, they exist in continuity [1, 2].

24.1.1.1 Superficial MCL

The superficial MCL is the largest structure of the medial aspect of the knee, which consists of one femoral and two tibial attachments with an average length of 94.8 mm [3, 4]. The femoral attachment is oval and located 3.2 mm proximal and 4.8 mm posterior to the medial epicondyle on average [3]. The distal tibial attachment of the superficial medial collateral ligament is broad and is directly to bone at an average of 61.2 mm distal to the tibial joint line; it is located just anterior to the posteromedial crest of the tibia [3].

24.1.1.2 Deep MCL

The deep MCL is a thickening of the medial joint capsule that is deep to the superficial MCL [3, 4]. It is most distinct along its anterior border, where it is roughly parallel to the anterior aspect of the superficial MCL [3]. Some reports described bursal tissue separating the superficial and deep components [5], a useful landmark for isolating the components during surgical exposure [6]. The deep MCL is divided into meniscofemoral and meniscotibial components [1, 7, 8].

24.1.1.3 POL

The POL is a fibrous extension off the distal aspect of the semimembranosus that blends with and reinforces the posteromedial aspect of the joint capsule [3, 5, 9–11]. The central arm is considered to be the main component of the POL, arising from the main semimembranosus tendon, reinforcing the deep MCL, directly attaching to the posterior joint capsule and posterior meniscus, and blending with the semimembranosus attachment on the tibia [4, 11]. Although the POL may appear to be an extension of the superficial

K. Nagai · Y. Nakanishi · K. Kamada · Y. Hoshino
R. Kuroda (✉)
Department of Orthopaedic Surgery, Kobe University Graduate School of Medicine, Kobe, Japan
e-mail: kurodar@med.kobe-u.ac.jp

J. Koh et al. (eds.), *Orthopaedic Biomechanics in Sports Medicine*,
https://doi.org/10.1007/978-3-030-81549-3_24

MCL, it is biomechanically distinct in its function and relationship with other structures in the posteromedial corner.

24.1.2 Biomechanics

24.1.2.1 Biomechanics of Normal MCL

The MCL complex is the primary stabilizer that combines static and dynamic resistance to direct valgus stress while contributing significant restraints to rotatory motion and anterior-posterior translation [1, 11, 12]. A previous study investigating the structural property of the MCL has shown that the MCL had maximum loads of 534 N (superficial MCL), 194 N (deep MCL), 425 N (posteromedial capsule) and failed when extended by a mean of 10.2 mm, 7.1 mm, and 12.0 mm, respectively [13]. The level of strain in the MCL varies with location (femoral insertion, mid-substance, tibial insertion) and the amount of flexion being tested [14, 15]. The largest strain on the MCL occurs during valgus loading, with forces concentrated near the femoral insertion site [14, 15].

24.1.2.2 Valgus Restraint

It is well established that the MCL complex acts as the primary restraint to valgus motion of the knee [1, 4, 16–18]. Among three linked components of the MCL complex, the superficial MCL serves as the primary restraint to valgus laxity from full extension through 90° of knee flexion, and appears to be dominant from 30° to 90° of knee flexion [12, 19, 20]. More specifically, the proximal division of the superficial MCL has been shown to function as a primary valgus stabilizer at all flexion angles [12, 21]. Isolated sectioning of the superficial MCL results in 3–7 mm of medial joint space opening to valgus stress, and is dependent on knee flexion angles [19, 20, 22, 23].

The meniscofemoral portion of the deep MCL is the secondary stabilizer at all knee flexion angles, while the meniscotibial portion of the deep MCL is the secondary stabilizer at 60° of knee flexion [12]. Although the superficial MCL remains tight when the knee flexes and extends [16], the POL also becomes tight as the knee approaches full extension [17, 24]; at knee extension, the contribution of the POL to valgus motion is greater compared to the condition at knee flexed [11, 20]. A recent study investigated the contribution of each component against valgus load, and showed that the superficial MCL was the primary restraint (40–54%) across 0–90° of knee flexion, the deep MCL 12%, and POL + posteromedial capsule 16% in extension for valgus load [25].

Therefore, while the superficial MCL is certainly the primary restraint to valgus load, the deep MCL and the POL including posteromedial capsule contribute as secondary stabilizer against valgus stress.

24.1.2.3 Internal and External Restraint

The MCL complex is also an important static stabilizer for internal and external rotation [11, 26]. Studies have shown that the POL is a primary stabilizer for internal rotation at all knee flexion angles [12, 27] although the most load response occurs at full extension [11, 21]. This finding correlates well with a recent study, which showed that the POL fibers become taut by internal rotation and slackened with knee flexion [28]. It has also been reported that, with applied internal rotation torque at 0° of knee flexion, the loads on the POL are significantly higher than those on the superficial MCL [21]. There is a reciprocal load response to internal rotation torque between the POL and superficial MCL as knee flexes [4], demonstrating complementary relationship between the POL and the superficial MCL with regard to the resistance of internal rotation torques that depends on the knee flexion angles.

Earlier studies showed that sectioning the superficial MCL significantly increases tibial external rotation, with the greatest increase occurring at 90° of flexion [17, 23]. The deep MCL also has been shown to provide restraint against external rotation torque in knees flexed between 30° and 90° [12, 20, 29]. One recent study demonstrated that the deep MCL resisted 23–13% across 0–90° and the superficial MCL 13–22% for external rotation [25] while another study has shown that the deep MCL was a minor restraint to external rotation [27].

24.1.2.4 Anteromedial Rotatory (AMR) Restraint

It is well accepted that the MCL complex and posteromedial corner of the knee is a stabilizer preventing AMR instability (AMRI) [9, 11, 25, 27, 30, 31], which is defined as anterior subluxation and external rotation of the medial tibial plateau with respect to the femur [11]. The posteromedial corner is functionally composed of several anatomical structures: the posterior horn of medial meniscus, the POL, the semimembranosus expansions, the meniscotibial portion of deep MCL, and the oblique popliteal ligament [5, 9]. Slocum and Larson [7] first described the term AMRI. In Hughston's series describing patients with AMRI, patients had an injury to the midportion of the superficial and deep MCLs or to the POL, often (but not always) with an associated ACL injury [30, 31]. Recent biomechanical studies have shown that superficial MCL and anterior cruciate ligament (ACL) contributed in restraining AMR motion, and that the superficial MCL appears to be the most important restraint to AMRI, and deep MCL and POL are minor restraints to AMRI [27].

24.1.2.5 Injury Mechanism

Isolated injury to the MCL complex primarily occurs by two main mechanisms [1, 18]. The most common injury is a direct blow to the outside of the thigh or leg while the foot is planted, producing direct valgus movement of the knee. This pattern is common in contact sports such as football and rugby. Biomechanical studies show higher strains near the femoral insertion site of the superficial MCL, which may suggest the reason why most lesions in clinical practice are seen in this proximal lesion [14, 15]. The second injury pattern involves valgus stress coupled with tibial external rotation, which often occurs in skiing and other cutting and pivoting sports, such as basketball and soccer. Again, the order of injury is unknown, but it has been postulated that the POL and posteromedial corner are injured first, followed by the deep and superficial MCL [1].

24.1.2.6 Biomechanics of MCL Reconstructed Knee

Treatment for an isolated, incomplete injury to the MCL (grade I or II) is generally nonoperative [1, 18]. The superficial MCL has been reported to have an abundant vascular supply and can heal. However, there are certain circumstances in which surgical treatment is necessary in acute and chronic settings [1, 18]. In an acute setting, the superficial MCL can be torn from its tibial insertion and displaced outside the pes anserinus tendons. As a result, the ligament is unable to reattach to its insertion on the tibia; this injury warrants surgical repair. Chronic pain and residual instability after nonoperative treatment of isolated grade II or III injury can be indicated for reconstructive surgery. Studies showed that clinical outcomes following MCL reconstruction appear to be satisfactory [32–37] although numerous reconstructive techniques have been used [38]. From a biomechanical perspective, several studies have shown that MCL reconstruction restored nearly normal stability to the knee [39–41].

In the study by Coobs B et al. [39], anatomical reconstruction of superficial MCL and POL was performed by using two separate hamstring grafts with four tunnels. The results showed that this reconstruction restored near-normal stability to a knee with a complete superficial MCL and POL tears. Another biomechanical study showed anatomical superficial MCL reconstruction restored knee kinematics and stability in the superficial MCL deficient knee [41]. That finding appears to be in line with the study by Wijdicks et al., which reported that both anatomic superficial MCL augmented repairs and anatomic superficial MCL reconstructions were able to improve knee stability and provide less than 2 mm of medial joint gapping at 0° and 20° of flexion with a superficial MCL injury [40].

24.1.3 Take Home Message

The MCL complex including superficial MCL, deep MCL, and POL is an essential stabilizer of valgus rotation, and contributes to restraining internal and external rotation. The MCL structure also has significant role in anteromedial rotatory laxity. Various techniques have been described for surgical intervention and biomechanical studies suggest anatomical MCL reconstruction appears to restore nearly normal knee stability, which correlated with reasonable clinical outcomes reported in the literature. Thorough understanding of the MCL biomechanics is essential for the assessment of which injured structures should be repaired or reconstructed.

24.2 Lateral Collateral Ligament (LCL) and Posterolateral Corner (PLC)

This section will introduce the biomechanical features of the lateral collateral ligament (LCL) and other structures in the posterolateral corner (PLC). For the purpose of this chapter, PLC, besides the LCL, will refer to the popliteofibular ligament (PFL) and the popliteus tendon (PLT) as those structures are described to be the most clinically relevant in providing static stability to the PLC of the knee [42–44].

24.2.1 Anatomy

24.2.1.1 LCL

The LCL is a cord-like ligament structure that is approximately 4.7 mm wide, 2.6 mm thick, and 69.9 mm long [45]. The LCL originates 1.4 mm proximal and 3.1 mm posterior to the lateral epicondyle [46, 47]. It courses deep through the superficial layer of the iliotibial band and attaches anterior to the lateral aspect of the fibular head, covering approximately 38% of the fibular head width [47]. The innervation of the LCL could be from a single branch or several branches with the common fibular nerve mainly innervating the LCL while some fibers are from the tibialis nerve [48]. The two main vessels which supplies blood to the LCL are the inferior lateral genicular and the anterior tibial recurrent arteries [49].

24.2.1.2 PLT

The PLT is approximately 54.4 mm long from the femoral attachment to the musculotendinous junction. The cross-sectional area is in average 0.59 cm^2 at the attachment site [46]. The PLT is attached to the anterior fifth and proximal half of the popliteal sulcus [46]. It is important to recognize that the PLT is attached anterior in relation to the LCL femoral attachment as it is helpful to ascertain the anatomic reconstruction site [46, 47]. As the PLT courses distally, it becomes extra-articular after it passes through the popliteal hiatus anterolaterally around the posterolateral aspect of the femur. Finally, the PLT transitions to the popliteal muscle on the posteromedial aspect of the tibia [46].

24.2.1.3 PFL

The PFL is a trapezoidal shaped ligament that is 14.8 mm in anterior length and 11.4 mm in posterior length. The width is 10.8 mm and cross section area is 6.9 mm [50]. Proximally, the PLT originates from the musculotendinous junction of the PLT [46]. It has anterior and posterior divisions with the anterior division attaching at the anterior downslope of the medial aspect of the fibular styloid process, and the posterior division attaching to the tip and posteromedial aspect of the fibular styloid process [46].

24.2.2 Biomechanics

24.2.2.1 Biomechanics of Normal LCL and PLC

Varus Restraint

It is well established from biomechanical studies that the LCL acts as the primary restraint to varus motion [47, 51–54]. A cadaveric sectioning study of the LCL showed that significant increase in varus rotation was seen compared to the intact knee at several angles between 0° and 90° [53]. Grood et al. [52] also found that cutting of the LCL constantly caused significant increase in varus angulation, which was largest at 30°of knee flexion with an angle of 5.7 ± 0.2°. In combination with other studies, it is important to note that varus laxity is most significant at 30° of knee flexion [52, 55, 56].

The PLT secondarily resists varus laxity in conjunction with the LCL [56, 57]. In one study, restraint to varus stress of the PLT was found between 0°and 90° [58]. In a different cadaveric study, small but significant increases in varus angulation was found after sectioning the PLT between 20°, 30°, and 60° [59]. Although a recent study showed conflicting results, that the PLC (PCL, PLT, PFL, surrounding fibers) had no varus stabilizing function in the LCL-intact knee, [60] the PLT may have minor contribution to resisting varus motion particularly at knee flexion angles 90°or less.

The PFL also has contribution to restraining varus motion [50, 56]. A cadaveric study that applied varus force to the knee in order to determine the sequence of failure of posterolateral structures showed that the PFL failed subsequently to the LCL and before the PLT, pointing out the contribution that the PFL has secondary to the LCL upon restriction of varus motion [50]. A different cadaveric study sectioned the LCL, PFL, and PLT and reconstructed only the LCL and PLT without the PFL. As a result, residual varus gapping was seen at 0°, 20°, and 60° of knee flexion, but with subsequent reconstruction of the PFL, stability to statistically similar measurement as to the intact knee was restored [61].

Therefore, while the LCL is certainly the primary restraint to varus motion, cadaveric studies show that the PLT and PFL may have small but significant contribution to resisting varus stress. In respect to restriction of valgus motion, there is no evidence for any significant involvement of LCL and PLC structures [47, 62–64].

Anterior and Posterior Restraint

The LCL, PLT, and PFL secondarily contribute to tibial posterior translation [45, 56, 57]. Harner et al. [65] reported LCL and PLT/PFL to serve as secondary restraints to posterior translation in a study with PCL-deficient knees. This was shown by sectioning the LCL, PLT, and PFL after PCL reconstruction. The result was an increase of posterior translation by 6.0 ± 2.7 mm at 30° and 4.6 ± 1.5 mm at 90° of knee flexion [65]. A similar study [52] showed an increased posterior translation when PLC was cut and PCL intact at 0 to 20°. PLT is also reported to provide restraint to posterior translation [56]. LCL sectioning resulted in 0.6 mm increase in posterior translation at 30° flexion, and 1.3 mm at 90° flexion, but additional sectioning of PLT further increased posterior translation by 0.2 mm at 90° [56].

Studies show contribution of LCL and PLC on anterior restraint as well. In a cadaveric study with sectioning of the PLT, an applied anterior load significantly increased anterior translation 1.5 mm at 0°, 2.6 mm at 20°, and 2.6 mm at 30° of knee flexion [59]. Wrobe et al. [64] showed larger anterior translation in extension than in flexion in anterior cruciate ligament (ACL) deficient knees. Similarly, Zantop et al. [66] showed sectioning of LCL after sectioning of ACL increased anterior translation against combined rotatory loading at 0, 30, and 60°of flexion [66]. Conversely, a study by Veltri et al. [56] did not show significant changes in anterior translation with posterolateral (LCL, PFL, and PLT) sectioning.

In this respect, while there is culminating evidence for the LCL and PLC to secondarily contribute to restraining posterior motion, its contribution to strict anterior translation may be small.

Internal and External Rotation Restraint

LCL, PLT, and PFL are reported to have contribution to restraining external rotation of the tibia. Sequential sectioning of the LCL, PLT, and PFL showed increase in external rotation of the tibia after each step of sectioning [56]. A different study found cutting of the LCL and PLT in PCL-deficient knees caused significant increase in external rotation between 0 and 100° [57]. LCL with contributions from the PLT, the PFL and LCL play equally important roles in limiting external rotation of the tibia when the knee is extended or mostly extended at 0° and 30°. When the knee is in increasingly flexed positions, the contribution of the LCL to controlling and preventing external tibial rotation decreases relative to that of the PFL [53, 54].

The PLT and PFL are the main contributors to resist external rotation [67]. Nielson et al. [58] reported that the PLT provided primary restraint to tibial external rotation when the knee is in 20–130° of flexion. Similar studies have shown the PLC to provide initial restraint to external rotation. LaPrade et al. [59] identified significant increases in external rotation after sectioning the PLT at knee flexion angles of 30°, 60°, and 90°. Similarly, Grood et al. [52] showed that while no significant change in limit of tibial external rotation occurred after cutting of the PCL, additional cutting of the PLC causes increase in external rotation. Only when the knee was flexed to 90° did the PCL engage in restraint of external rotation. Therefore, it was concluded that the PLC provides the initial lateral restraint to external rotation of the tibia [52]. Other sectioning studies have also shown that the PLC has contribution to external rotation at smaller flexion angles, especially around 30°of knee flexion [52, 56, 68].

The role the LCL and PLC in restraining internal rotation is small. Nonetheless, recent studies demonstrate that after sectioning of the LCL, increase in internal rotation can be observed between 0 and 90° of knee flexion [53]. A different study showed that cutting of the LCL caused increase in internal rotation at higher flexion angles [63]. Therefore, the LCL seems to have some contribution to internal rotation stability [69]. PLT and PFL also are reported to have some contribution to restraining internal rotation. Small but significant, increases in internal rotation after sectioning the popliteus tendon were found at knee flexion angles between 0° and 90° [59]. The contribution of the PFL to restraining internal rotation was demonstrated by a cadaveric study that sectioned the LCL, PFL, and PLT, and reconstructed the LCL and PLT and left the PFL alone [61]. As a result, increase internal rotation at 60° and 90° of knee flexion in comparison to the knee with intact LCL, PLT, and PFL. Subsequent reconstruction of the PFL restored stability to statistically similar measurement as to the intact knee, indicating the ability for the PFL to resist internal rotation [61].

24.2.2.2 Injury Mechanism of LCL and PLC

Injury Mechanism

LCL and PLC injuries can occur in practically any sports or accidents, either by contact or noncontact, that involve forced varus motion, hyperextension, or rotational stress to the knee [69, 70]. However, injuries to the lateral structures of the knee are rare and often occur with other knee injuries. A frequency of less than 1 player per year was reported for isolated grade III LCL injury [71], and only 1% national football league players were reported to have history of PLC injury [72]. Usually, LCL/PLC injuries involve more than one structure in the PLC or have concomitant ligamentous injuries. More than half of the PLC injuries involved more than one structure in the PLC [73]. Examination of 162 patients with MRI following ACL injury found 19.7% to have concomitant PLC injury [74]. Therefore, when LCL or PLC injury is detected, thorough examination or other injuries is warranted.

24.2.2.3 Biomechanics of LCL and PLC Reconstructed Knee

Surgical treatment, either ligament repair or reconstruction, is required for a complete midsubstance tear or intrasubstance stretch injury [75–77]. LCL reconstruction should be performed when anatomic reduction is not possible [78]. Chronic cases are also indication for reconstruction rather than repair. Improved outcomes have been reported in multi-ligamentous knee injuries managed with reconstruction of the LCL and PLC compared with repair, with substantially higher failure rates reported when the LCL is repaired [76]. PLC reconstruction techniques can be broadly categorized into a fibula-based approach or a fibula- and tibia-based approach [69, 79].

LCL Reconstruction

Satisfactory results have been attained by reconstruction of the LCL to restore a near intact state to the LCL injured knee. Reconstruction using patellar tendon allograft [80] showed 9 out of 10 patients obtained decrease in varus laxity and 8 out of 10 patients restored external rotation equal to the contralateral uninjured knee at 30° of knee flexion. A cadaveric study of an anatomic reconstruction of the LCL using semitendinosus tendon autograft restored near-normal stability to the knee with only small increase in varus motion detected at 15° and 30° flexion. External rotation showed minor over-constraint with decrease in external rotation compared to intact state at 0° of knee flexion but no significant difference at 15°, 30°, 60°, and 90° [53].

PLC Reconstruction

Fibula-based approach and combined tibia-and fibula-based approach reconstructions significantly reduce external rotation and varus to levels consistent with or only slightly less than the intact condition [81]. The modified Larson technique, a fibula-based approach, reconstructs the LCL and PFL restoring normal varus laxity, tibial posterior translation, and external rotation throughout the arc of knee flexion [57]. As a fibula- and tibia-based approach, LaPrade introduced a two-graft technique which restored external rotation and varus static stability for grade III PLC injury [82]. Apsingi et al. [83] compared the modified Larson's technique to the anatomic reconstruction. The results showed that both reconstruction methods did not show any significant difference in restoration of external rotation and varus laxity in combined PCL and PLC injured knees at time zero. Therefore, certain conditions may only require a fibula-based approach while others require a more extensive anatomical approach. Further studies are necessary to establish surgical indications and elucidate clinical relevancy.

24.2.3 Take Home Message

The LCL and other PLC structure are essential stabilizers especially for varus and external rotation. The structures may also have small but significant role in restraining posterior/anterior stability and internal rotation. A variety of techniques has been described for surgical intervention, while anatomic reconstruction has been increasingly advocated in order to achieve physiologic knee function. It may also cause stress on ACL and PCL grafts upon reconstruction which can contribute to failure of ACL and PCL reconstruction. Therefore, it is clinically important that PLC injury is thoroughly examined whenever suspicious.

References

1. Marchant MH, Tibor LM, Sekiya JK, et al. Management of medial-sided knee injuries, part 1: medial collateral ligament. Am J Sports Med. 2011;39(5):1102–13.
2. Wymenga AB, Kats JJ, Kooloos J, et al. Surgical anatomy of the medial collateral ligament and the posteromedial capsule of the knee. Knee Surg Sports Traumatol Arthrosc. 2006;14(3):229–34.
3. LaPrade RF, Engebretsen AH, Ly TV, et al. The anatomy of the medial part of the knee. J Bone Joint Surg Am. 2007;89(9):2000–10.
4. Wijdicks CA, Griffith CJ, Johansen S, et al. Injuries to the medial collateral ligament and associated medial structures of the knee. J Bone Joint Surg Am. 2010;92(5):1266–80.
5. Sims WF, Jacobson KE. The posteromedial corner of the knee: medial-sided injury patterns revisited. Am J Sports Med. 2004;32(2):337–45.
6. Indelicato PA. Isolated medial collateral ligament injuries in the knee. J Am Acad Orthop Surg. 1995;3(1):9–14.
7. Slocum DB, Larson RL. Rotatory instability of the knee. Its pathogenesis and a clinical test to demonstrate its presence. J Bone Joint Surg Am. 1968;50(2):211–25.
8. De Maeseneer M, Van Roy F, Lenchik L, et al. Three layers of the medial capsular and supporting structures of the knee: MR imaging-anatomic correlation. Radiographics. 2000;20(Spec):S83–9.
9. Engebretsen L, Lind M. Anteromedial rotatory laxity. Knee Surg Sports Traumatol Arthrosc. 2015;23(10):2797–804.
10. Dold AP, Swensen S, Strauss E, et al. The posteromedial corner of the knee: anatomy, pathology, and management strategies. J Am Acad Orthop Surg. 2017;25(11):752–61.
11. Tibor LM, Marchant MH, Taylor DC, et al. Management of medial-sided knee injuries, part 2: posteromedial corner. Am J Sports Med. 2011;39(6):1332–40.
12. Griffith CJ, LaPrade RF, Johansen S, et al. Medial knee injury: part 1, static function of the individual components of the main medial knee structures. Am J Sports Med. 2009;37(9):1762–70.
13. Robinson JR, Bull AMJ, Amis AA. Structural properties of the medial collateral ligament complex of the human knee. J Biomech. 2005;38(5):1067–74.
14. Gardiner JC, Weiss JA, Rosenberg TD. Strain in the human medial collateral ligament during valgus loading of the knee. Clin Orthop Relat Res. 2001;391:266–74.
15. Luyckx T, Verstraete M, De Roo K, et al. High strains near femoral insertion site of the superficial medial collateral ligament of the knee can explain the clinical failure pattern. J Orthop Res. 2016;34(11):2016–24.
16. Warren LA, Marshall JL, Girgis F. The prime static stabilizer of the medical side of the knee. J Bone Joint Surg Am. 1974;56(4):665–74.
17. Haimes JL, Wroble RR, Grood ES, et al. Role of the medial structures in the intact and anterior cruciate ligament-deficient knee: limits of motion in the human knee. Am J Sports Med. 1994;22(3):402–9.
18. Miyamoto RG, Bosco JA, Sherman OH. Treatment of medial collateral ligament injuries. J Am Acad Orthop Surg. 2009;17(3):152–61.

19. Grood ES, Noyes FR, Butler DL, et al. Ligamentous and capsular restraints preventing straight medial and lateral laxity in intact human cadaver knees. J Bone Joint Surg Am. 1981;63(8):1257–69.
20. Robinson JR, Bull AMJ, Thomas RRDW, et al. The role of the medial collateral ligament and posteromedial capsule in controlling knee laxity. Am J Sports Med. 2006;34(11):1815–23.
21. Griffith CJ, Wijdicks CA, LaPrade RF, et al. Force measurements on the posterior oblique ligament and superficial medial collateral ligament proximal and distal divisions to applied loads. Am J Sports Med. 2009;37(1):140–8.
22. LaPrade RF, Bernhardson AS, Griffith CJ, et al. Correlation of valgus stress radiographs with medial knee ligament injuries: an in vitro biomechanical study. Am J Sports Med. 2010;38(2):330–8.
23. Warren LF, Marshall JL. The supporting structures and layers on the medial side of the knee: an anatomical analysis. J Bone Joint Surg Am. 1979;61(1):56–62.
24. Robinson JR, Sanchez-Ballester J, Bull AMJ, et al. The posteromedial corner revisited. An anatomical description of the passive restraining structures of the medial aspect of the human knee. J Bone Joint Surg Br. 2004;86(5):674–81.
25. Ball S, Stephen JM, El-Daou H, et al. The medial ligaments and the ACL restrain anteromedial laxity of the knee. Knee Surg Sports Traumatol Arthrosc. 2020;28(12):3700–8.
26. Marchant MH, Willimon SC, Vinson E, et al. Comparison of plain radiography, computed tomography, and magnetic resonance imaging in the evaluation of bone tunnel widening after anterior cruciate ligament reconstruction. Knee Surg Sports Traumatol Arthrosc. 2010;18(8):1059–64.
27. Wierer G, Milinkovic D, Robinson JR, et al. The superficial medial collateral ligament is the major restraint to anteromedial instability of the knee. Knee Surg Sports Traumatol Arthrosc 2021;29(2):405–16.
28. Willinger L, Shinohara S, Athwal KK, et al. Length-change patterns of the medial collateral ligament and posterior oblique ligament in relation to their function and surgery. Knee Surg Sports Traumatol Arthrosc. 2020;28(12):3720–32.
29. Cavaignac E, Carpentier K, Pailhé R, et al. The role of the deep medial collateral ligament in controlling rotational stability of the knee. Knee Surg Sports Traumatol Arthrosc. 2015;23(10):3101–7.
30. Hughston JC. The importance of the posterior oblique ligament in repairs of acute tears of the medial ligaments in knees with and without an associated rupture of the anterior cruciate ligament. Results of long-term follow-up. J Bone Joint Surg. 1994;76(9):1328–44.
31. Hughston JC, Barrett GR. Acute anteromedial rotatory instability. Long-term results of surgical repair. J Bone Joint Surg. 1983;65(2):145–53.
32. Funchal LFZ, Astur DC, Ortiz R, et al. The presence of the arthroscopic "floating meniscus" sign as an Indicator for surgical intervention in patients with combined anterior cruciate ligament and grade II medial collateral ligament injury. Arthroscopy. 2019;35(3):930–7.
33. Koga H, Muneta T, Yagishita K, et al. Surgical management of grade 3 medial knee injuries combined with cruciate ligament injuries. Knee Surg Sports Traumatol Arthrosc. 2012;20(1):88–94.
34. Yoshiya S, Kuroda R, Mizuno K, et al. Medial collateral ligament reconstruction using autogenous hamstring tendons: technique and results in initial cases. Am J Sports Med. 2005;33(9):1380–5.
35. Lind M, Jakobsen BW, Lund B, et al. Anatomical reconstruction of the medial collateral ligament and posteromedial corner of the knee in patients with chronic medial collateral ligament instability. Am J Sports Med. 2009;37(6):1116–22.
36. Kitamura N, Ogawa M, Kondo E, et al. A novel medial collateral ligament reconstruction procedure using semitendinosus tendon autograft in patients with multiligamentous knee injuries: clinical outcomes. Am J Sports Med. 2013;41(6):1274–81.
37. LaPrade RF, Wijdicks CA. Surgical technique: development of an anatomic medial knee reconstruction. Clin Orthop Relat Res. 2012;470(3):806–14.
38. DeLong JM, Waterman BR. Surgical techniques for the reconstruction of medial collateral ligament and posteromedial corner injuries of the knee: a systematic review. Arthroscopy. 2015;31(11):2258–72.
39. Coobs BR, Wijdicks CA, Armitage BM, et al. An in vitro analysis of an anatomical medial knee reconstruction. Am J Sports Med. 2010;38(2):339–47.
40. Wijdicks CA, Michalski MP, Rasmussen MT, et al. Superficial medial collateral ligament anatomic augmented repair versus anatomic reconstruction: an in vitro biomechanical analysis. Am J Sports Med. 2013;41(12):2858–66.
41. Van den Bogaerde JM, Shin E, Neu CP, et al. The superficial medial collateral ligament reconstruction of the knee: effect of altering graft length on knee kinematics and stability. Knee Surg Sports Traumatol Arthrosc. 2011;19(S1):60–8.
42. Swinford S, LaPrade R, Engebretsen L, et al. Biomechanics and physical examination of the posteromedial and posterolateral knee: state of the art. J ISAKOS 2020;5:378–88.
43. Cooper JM, McAndrews D, LaPrade RF. Posterolateral corner injuries of the knee: anatomy, diagnosis, and treatment. Sport Med Arthrosc Rev. 2006;14:213–20.
44. Shon O-J, Park J-W, Kim B-J. Current concepts of posterolateral corner injuries of the knee. Knee Surg Relat Res. 2017;29(4):256–68.
45. Wilson WT, Deakin AH, Payne AP, et al. Comparative analysis of the structural properties of the collateral ligaments of the human knee. J Orthop Sports Phys Ther. 2012;42(4):345–51.
46. LaPrade RF, Ly TV, Wentorf FA, et al. The posterolateral attachments of the knee. A qualitative and quantitative morphologic analysis of the fibular collateral ligament, popliteus tendon, Popliteofibular ligament,

and lateral gastrocnemius tendon. Am J Sports Med. 2003;31(6):854–60.
47. Moorman CT, LaPrade RF. Anatomy and biomechanics of the posterolateral corner of the knee. J Knee Surg. 2005;18(2):137–45.
48. Yan J, Sasaki W, Hitomi J. Anatomical study of the lateral collateral ligament and its circumference structures in the human knee joint. Surg Radiol Anat. 2010;32(2):99–106.
49. Hannouche D, Duparc F, Beaufils P. The arterial vascularization of the lateral tibial condyle: anatomy and surgical applications. Surg Radiol Anat. 2006;28(1):38–45.
50. Maynard MJ, Deng X, Wickiewicz TL, et al. The popliteofibular ligament: rediscovery of a key element in posterolateral stability. Am J Sports Med. 1996;24(3):311–6.
51. Sugita T, Amis AA. Anatomic and biomechanical study of the lateral collateral and popliteofibular ligaments. Am J Sports Med. 2001;29(4):466–72.
52. Grood ES, Stowers SF, Noyes FR. Limits of movement in the human knee. Effect of sectioning the posterior cruciate ligament and posterolateral structures. J Bone Joint Surg Am. 1988;70(1):88–97.
53. Coobs BR, LaPrade RF, Griffith CJ, et al. Biomechanical analysis of an isolated fibular (lateral) collateral ligament reconstruction using an autogenous semitendinosus graft. Am J Sports Med. 2007;35(9):1521–7.
54. Lim HC, Bae JH, Bae TS, et al. Relative role changing of lateral collateral ligament on the posterolateral rotatory instability according to the knee flexion angles: a biomechanical comparative study of role of lateral collateral ligament and popliteofibular ligament. Arch Orthop Trauma Surg. 2012;132(11):1631–6.
55. LaPrade RF, Tso A, Wentorf FA. Force measurements on the fibular collateral ligament, popliteofibular ligament, and popliteus tendon to applied loads. Am J Sports Med. 2004;32(7):1695–701.
56. Veltri DM, Deng XH, Torzilli PA, et al. The role of the popliteofibular ligament in stability of the human knee: a biomechanical study. Am J Sports Med. 1996;24(1):19–27.
57. Apsingi S, Nguyen T, Bull AMJ, et al. Control of laxity in knees with combined posterior cruciate ligament and posterolateral corner deficiency: comparison of single-bundle versus double-bundle posterior cruciate ligament reconstruction combined with modified Larson posterolateral corner reco. Am J Sports Med. 2008;36(3):487–94.
58. Nielsen S, Helmig P. The static stabilizing function of the popliteal tendon in the knee - an experimental study. Arch Orthop Trauma Surg. 1986;104(6): 357–62.
59. LaPrade RF, Wozniczka JK, Stellmaker MP, et al. Analysis of the static function of the popliteus tendon and evaluation of an anatomic reconstruction. Am J Sports Med. 2010;38(3):543–9.
60. Domnick C, Frosch KH, Raschke MJ, et al. Kinematics of different components of the posterolateral corner of the knee in the lateral collateral ligament-intact state: a human cadaveric study. Arthroscopy. 2017;33(10):1821–30.
61. McCarthy M, Camarda L, Wijdicks CA, et al. Anatomic posterolateral knee reconstructions require a popliteofibular ligament reconstruction through a tibial tunnel. Am J Sports Med. 2010;38(8):1674–81.
62. Gollehon DL, Torzilli PA, Warren RF. The role of the posterolateral and cruciate ligaments in the stability of the human knee. A biomechanical study. J Bone Joint Surg Am. 1987;69(2):233–42.
63. Nielsen S, Rasmussen O, Andersen K. Rotatory instability of cadaver knees after transection of capsule and ligaments. Arch Orthop Trauma Surg. 1984;103(3):165–9.
64. Wroble RR, Grood ES, Cummings JS, et al. The role of the lateral extraarticular restraints in the anterior cruciate ligament-deficient knee. Am J Sports Med. 1993;21(2):257–63.
65. Harner CD, Vogrin TM, Höher J, et al. Biomechanical analysis of a posterior cruciate ligament reconstruction: deficiency of the posterolateral structures as a cause of graft failure. Am J Sports Med. 2000;28(1):32–9.
66. Zantop T, Schumacher T, Diermann N, et al. Anterolateral rotational knee instability: role of posterolateral structures. Arch Orthop Trauma Surg. 2007;127(9):743–52.
67. Covey DC. Injuries of the posterolateral corner of the knee. J Bone Joint Surg Am. 2001;83(1):106–18.
68. Ranawat A, Baker CL, Henry S, et al. Posterolateral corner injury of the knee: evaluation and management. J Am Acad Orthop Surg. 2008;16(9):506–18.
69. Grawe B, Schroeder AJ, Kakazu R, et al. Lateral collateral ligament injury about the knee: anatomy, evaluation, and management. J Am Acad Orthop Surg. 2018;26(6):e120–7.
70. Majewski M, Susanne H, Klaus S. Epidemiology of athletic knee injuries: a 10-year study. Knee. 2006;13(3):184–8.
71. Bushnell BD, Bitting SS, Crain JM, et al. Treatment of magnetic resonance imaging-documented isolated grade III lateral collateral ligament injuries in National Football League Athletes. Am J Sports Med. 2010;38(1):86–91.
72. Chahla J, Kennedy NI, Cinque ME, et al. Posterolateral corner injuries of the knee at the National Football League Combine: an imaging and outcomes analysis. Arthroscopy. 2018;34(3):687–92.
73. LaPrade RF, Wentorf FA, Fritts H, et al. A prospective magnetic resonance imaging study of the incidence of posterolateral and multiple ligament injuries in acute knee injuries presenting with a Hemarthrosis. Arthroscopy. 2007;23(12):1341–7.
74. Temponi EF, de Carvalho Júnior LH, Saithna A, et al. Incidence and MRI characterization of the spectrum of posterolateral corner injuries occurring in association with ACL rupture. Skelet Radiol. 2017;46(8):1063–70.

75. Stannard JP, Brown SL, Farris RC, et al. The posterolateral corner of the knee: repair versus reconstruction. Am J Sports Med. 2005;33(6):881–8.
76. Veltri DM, Deng XH, Torzilli PA, et al. The role of the popliteofibular ligament in stability of the human knee. A biomechanical study. Am J Sports Med. 2010;24(1):19–27.
77. Moulton SG, Matheny LM, James EW, et al. Outcomes following anatomic fibular (lateral) collateral ligament reconstruction. Knee Surg Sports Traumatol Arthrosc. 2015;23(10):2960–6.
78. Geeslin AG, LaPrade RF, Urgery S, et al. 16 7 21. J Bone Joint Surg Am. 2011;93(18):1672–83.
79. Blackman AJ, Engasser WM, Krych AJ, et al. Fibular head and tibial-based (2-tailed) posterolateral corner reconstruction. Sports Med Arthrosc. 2015;23(1):44–50.
80. Latimer HA, Tibone JE, Elattrache NS, et al. Reconstruction of the lateral collateral ligament of the knee with patellar tendon allograft: report of a new technique in combined ligament injuries. Am J Sports Med. 1998;26(5):656–62.
81. Rauh PB, Clancy WG, Jasper LE, et al. Biomechanical evaluation of two reconstruction techniques for posterolateral instability of the knee. J Bone Joint Surg Br. 2010;92(10):1460–5.
82. LaPrade RF, Johansen S, Wentorf FA, et al. An analysis of an anatomical posterolateral knee reconstruction: an in vitro, biomechanical study and development of a surgical technique. Am J Sports Med. 2004;32(6):1405–14.
83. Apsingi S, Nguyen T, Bull AMJ, et al. A comparison of modified Larson and "anatomic" posterolateral corner reconstructions in knees with combined PCL and posterolateral corner deficiency. Knee Surg Sports Traumatol Arthrosc. 2009;17(3):305–12.

PCL Biomechanics

25

Leonard Tiger Onsen and Jason Koh

25.1 Introduction

The posterior cruciate ligament (PCL) is one of two cruciate ligaments of the knee and a key contributor to overall knee stability. Made up of two bundles, the PCL provides resistance to posterior tibial translation and also provides rotational stability. PCL injury incidence has been studied in various populations and yielded variable results. Incidence has been reported as 3% of the general population [1]. As one might expect, higher incidences have been reported amongst patients reporting to the emergency department with acute knee hemarthrosis [2]. Majority of these cases were related to trauma or sports injury, with a common mechanism involving a posterior force to the proximal tibia with the knee flexed such as a fall on a flexed knee [3]. Additionally, PCL injuries are commonly associated with other ligamentous injuries of the knee such as posterolateral corner injuries [4]. The anatomy and biomechanics of the PCL are inherently complex. Despite this, it is essential that providers understand these concepts in order to properly evaluate, diagnose, and treat these injuries. It is the goal of this chapter to review PCL anatomy, native biomechanics, injury mechanisms, biomechanics altered by injury, and reconstruction biomechanics in order to aid in management of PCL injuries.

L. T. Onsen (✉)
Department of Orthopaedic Surgery, College of Medicine, University of Illinois at Chicago, Chicago, IL, USA

J. Koh
Department of Orthopaedic Surgery, Orthopaedic & Spine Institute NorthShore University HealthSystem, Skokie, Illinois, USA

University of Chicago Pritzker School of Medicine, Chicago, Illinois, USA

Northwestern University McCormick School of Engineering, Evanston, Illinois, USA

25.2 Anatomy

Anatomy of the PCL has been extensively researched and described. The PCL is an intra-articular structure of the knee and considered an extrasynovial structure as it is enclosed in a synovial sheath. The ligaments can be further subdivided into anterolateral (ALB) and posteromedial (PMB) bundles based on location and function (Fig. 25.1) [5]. These bundles will be further described later in this section. Average PCL length ranges from 32 to 38 mm with average midsubstance diameter noted from 11 to 13mm^2 [6, 7]. The PCL originates from the anterolateral aspect of the medial femoral condyle within the intercondylar notch and inserts on the posterior aspect of the tibia 1–1.5 cm distal to the joint line [8, 9]. Attachments are broad in nature with surface area averages found as 209 mm^2 for the femoral origin and 243 mm^2 at the tibial insertion. This demonstrates how the ligamentous attachments are sig-

J. Koh et al. (eds.), *Orthopaedic Biomechanics in Sports Medicine*,
https://doi.org/10.1007/978-3-030-81549-3_25

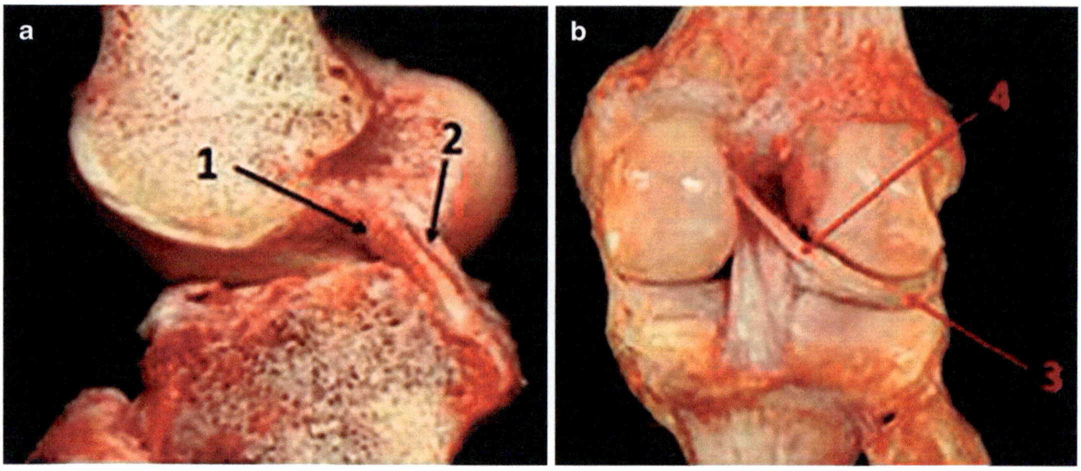

Fig. 25.1 (**a**) Lateral view of PCL anatomy showing the anterolateral [1] and posterolateral [2] bundles as well as their tibial and femoral attachments. (**b**) Posterior view of the knee showing the PCL in relation to the lateral meniscus [3] and meniscofemoral ligament [4]. (image courtesy of Logterman et al.) [8]

nificantly larger than the ligament in a whole. The PCL is innervated by the tibial nerve and receives its blood supply from the middle geniculate artery.

Recent studies have put a greater emphasis on further describing the bundles of the PCL. Much like the PCL in general, average lengths, diameters, and attachment size have been described for the individual bundles. Additionally, the relationships between the bundles have been elucidated. The two bundles are separated by a horizontal bony prominence known as the bundle ridge on the tibia. The anterolateral bundle tightens in knee flexion. Average length of the ALB has been reported as 31.79 mm with an average diameter of 6.50 mm [10]. Cross-sectional area for the ALB femoral footprint is noted to range from 112 to 118 mm^2 [11, 12]. Tibial footprint for the ALB has been reported as 88 mm^2 [11]. In accordance with the PCL as whole, the femoral footprint is larger on average and both are significantly larger than the ligament diameter. Bundle location in relationship to other landmarks has also been extensively described to aid in identification and reconstruction. Anderson et al. provided an arthroscopic description of the ALB found its femoral origin to be 7.9 mm from the distal articular cartilage of the intercondylar notch [11]. This same study found its tibial insertion to be 6.1 mm from the posterior medial meniscus root and 4.9 mm from bundle ridge. Morgan et al. described the ALB femoral origin as 13 mm posterior to the medial articular cartilage intercondylar wall interface and 13 mm inferior to the articular cartilage intercondylar roof interface [13].

Much like the ALB, the posteromedial bundle has been described in detail. The PMB tightens in knee extension. Average length of the PMB is reported as 32.42 mm with an average diameter of 5.62 mm [10]. In comparison, it is slightly longer and thinner on average than the ALB. Its femoral footprint area is noted to range from 60 to 90mm^2 and tibial footprint has been reported as 105mm^2 [11]. Again footprints are larger than the ligament diameter. Furthermore, the PMB has a smaller femoral and larger tibial footprint areas compared to the ALB. Arthroscopic description of the PMB showed its femoral origin to be 8.6 mm from the distal articular cartilage of the intercondylar notch and the tibial insertion to be 11.1 mm from the posterior medial meniscus root [11]. Morgan et al. described the ALB femoral origin as 8 mm posterior to the medial articular cartilage intercondylar wall interface and 20 mm inferior to the articular cartilage intercondylar roof interface [13]. These studies have greatly added to the understanding of PCL anatomy with the goal of improving reconstruction methods.

25.3 Biomechanics

The PCL is a key static stabilizer of the knee. Its primary function is to resist posterior translation of the tibia relative to the femur. PCL removal has shown to increase posterior tibial translation throughout knee range of motion from full extension to 120 degrees of flexion [14]. Additionally, it has also been found to serve as a secondary stabilizer to resist rotation at the knee between 90 and 120 degrees of flexion [15]. The role of the PCL in stabilizing the knee against various forces will be further discussed later in this section. Previously it was thought that both PCL bundles functioned independent of each other with the ALB being dominant in knee flexion and PMB taking over in extension [6, 16]. On the contrary, recent research has found that the two bundles instead work together changing the previous paradigm. Studies have demonstrated this in a number of ways. Cadaver studies have shown that isolated single bundle tears do not result in clinically significant increases in posterior tibial translation [17, 18]. Another cadaveric study by Ahmad et al. demonstrated that changes in bundle orientation and length throughout knee range of motion accommodate each other so that no bundle is dominant at any specific position [19]. This concept was also supported in vivo by Papannagari et al. [20]

Baseline mechanical properties of the PCL and its bundles have also been described. Tensile strength of the PCL has a reported range of 739 to 1627 newtons [21–23]. However, given the PCL resists forces in multiple directions its true tensile strength is likely greater. The ALB is reported as the primary contributor of PCL tensile strength with it reported at 1620 N as opposed to 258 N for the PMB [24]. Loading and strain patterns of the PCL in various movements have been extensively studied. In vivo studies have found 24–33% elongation of the PCL with passive knee flexion to 90 degrees [25, 26]. Similar strain patterns were noted in active knee flexion through squats or lunges [27, 28]. Furthermore, these studies found no significant elongation with flexion past 90. In vitro studies have reported differing results on PCL strain throughout knee flexion [29, 30]. Forces on the PCL have largely been studied on cadaver models. In passive flexion, low tension forces have been reported in the PCL ranging from 4 to 19 newtons at various angles of flexion [31, 32]. Active knee movement was associated with far greater forces on the PCL. A peak force of 3330 newtons was recorded with isometric knee flexion [33]. Overall, significantly greater forces were measured in active movement involving knee flexion compared to isolated extension movements. Increased PCL forces have been found when a posterior tibial force is applied at various angles of flexion. Peak forces were noted when this force were applied at 90 degrees of flexion where PCL force near 140 newtons were recorded [27, 31, 34, 35]. Understanding of forces and strain applied to the native PCL in various movements is essential in the rehab treatment protocols of PCL injuries or following PCL reconstruction.

As previously stated, the ALB is tight in knee flexion, whereas the PMB tightens in extension. It is well documented that the PCL resists posterior tibial translation through studies where the PCL is removed in cadaver knees. However, the knee position which serves this role has been debated in cadaveric and in vivo studies. One study found increased posterior tibial translation with 25 degrees of flexion and greater [36]. Other studies have found that increased posterior tibial translation occurs at 60 degrees of flexion and greater [37, 38]. Finally, multiple studies have reported increased posterior tibial translation throughout range of motion in PCL deficient knees [14, 38, 39]. Despite their differences, these studies agree on the PCL resisting posterior tibial translation in general and that it may provide increased resistance with knee flexion. These findings further support clinical exams that test the PCL at 90 degrees of knee flexion such as the posterior drawer, posterior sag sign, and quadriceps active tests.

Additionally, the PCL is an important contributor to rotational stability of the knee. Similar to posterior translation, this concept has been studied in cadaver knees with the PCL removed. Increased external rotation is noted when a physiologic posterior tibial force equivalent to the quadriceps and hamstrings is applied to PCL deficient knees [37, 40]. The ligament has also been shown to resist internal rotation beyond 90 degrees of knee flexion [18]. Furthermore, increased external and internal rotation is demonstrated in PCL deficient knees when force is applied in either direction [14]. Li et al. studied this in vivo with weight bearing MRIs finding altered internal rotation throughout knee range of motion; however, the difference was not significant [41]. These studies suggest that the PCL plays an important role in rotational stability of the knee and justify rotational exams such as the dial test in diagnosing PCL injuries.

Another function of the PCL is in balancing joint contact forces about the knee. PCL deficiency primarily impacts the medial compartment of the knee. More specifically, increased contact forces have been demonstrated in PCL deficient knees. Macdonald et al. found significant increases in medial compartment contact forces at 60 and 90 degrees of flexion but no change at 0 or 30 degrees of flexion [42], whereas Skyhar et al. found a 52% increase in medial compartment contact force on average regardless of flexion angle in this setting [43]. Natural history studies of untreated PCL injuries have found increased rates of medial compartment cartilage degeneration which can be explained by the altered joint contact forces [44, 45].

25.4 Mechanisms of Injury

As previously stated PCL injuries can occur both in isolation and are often associated with other ligamentous injuries of the knee. Multiligamentous injury patterns of the knee involving the PCL include combined PCL with anterior cruciate ligament (ACL) and PCL with posterior lateral corner (PLC) injuries. Given its anatomy and biomechanics, the most common mechanism of isolated PCL injury involves posterior translation of the tibia relative to the femur, which places increasing tension and strain on the ligament until it fails (Fig. 25.2). This most commonly occurs with a direct posterior blow to the proximal tibia such as between athletes in contact athletics or a fall on a flexed knee with the foot in plantarflexion. A higher energy mechanism includes motor vehicle collision dashboard injuries where the driver or passenger's flexed knee is driven into dashboard. Less frequent mechanisms include hyperflexion and hyperextension.

Multiligamentous injuries of the knee involving the PCL include a variety of different pat-

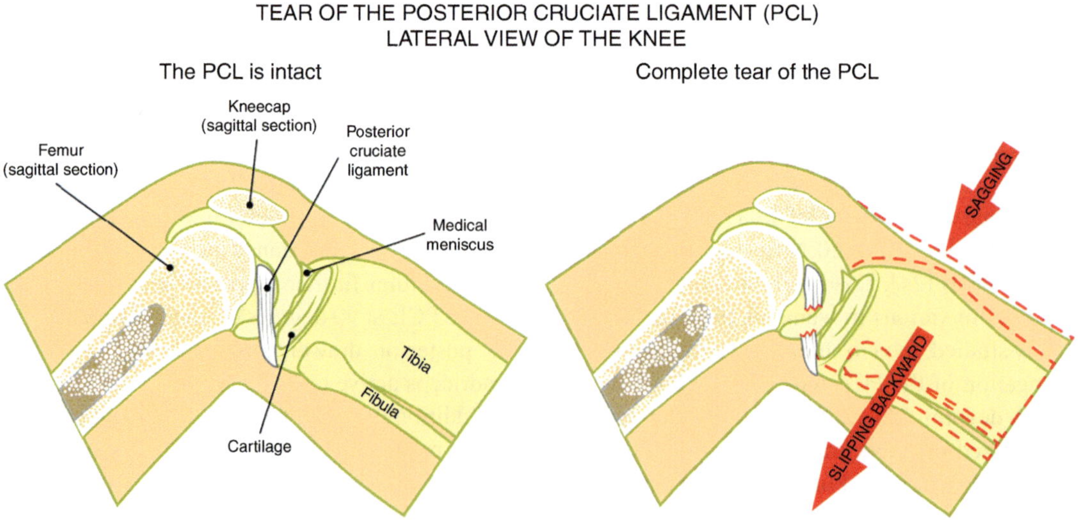

Fig. 25.2 Demonstrating normal PCL anatomy and then a PCL tear resulting from a proximal tibia force producing posterior tibial translation. (public domain image from shutterstock)

terns. Common patterns include ACL-PCL-PLC, ACL-PLC, and ACL-PCL-MCL [46, 47]. These injuries are frequently associated with high energy mechanisms such as motor vehicle accidents, pedestrian struck by vehicle, or fall from a height. Low energy mechanisms such as fall from standing height have been reported less frequently but have been associated increased body mass index (BMI) and obese patients [48]. A PCL tear is commonly associated with a posterolateral corner injury [49]. Again these injuries can result from direct blow to the proximal tibia. However, rather than a direct posterior force, the anteromedial aspect of the tibia is struck with the knee in extension producing hyperextension and varus. This pattern can also occur with a posterior force to the proximal tibial with the knee flexed or tibia in external rotation [50]. Overall, whether in isolation or multiligamentous settings the mechanisms required to injure the PCL remain similar. Understanding of these mechanisms can better heighten provider awareness to these injuries so that it can be promptly identified and treated appropriately.

25.5 Biomechanics after Injury

The primary role of the PCL is to resist posterior tibial translation relative to the femur. Cadaveric studies have shown increased posterior tibial translation in PCL deficient knees throughout range of motion from 0 to 120 degrees of flexion [14, 51, 52]. This concept has also been studied in the in vivo setting. In a self-controlled study of patients with unilateral PCL deficiency, increased posterior tibial translation was noted throughout flexion on lateral radiographs but greater translation was noted from 70 to 90 degrees of flexion as compared to lower angles (Fig. 25.3) [53]. MRI studies on a similar patient population further demonstrated posterior tibial translation in PCL deficient knees in weight bearing flexion and with posterior drawer testing [54]. The importance of the posterolateral corner, more specifically the popliteus, in resisting posterior tibial translation has been described. A cadaveric study found significantly

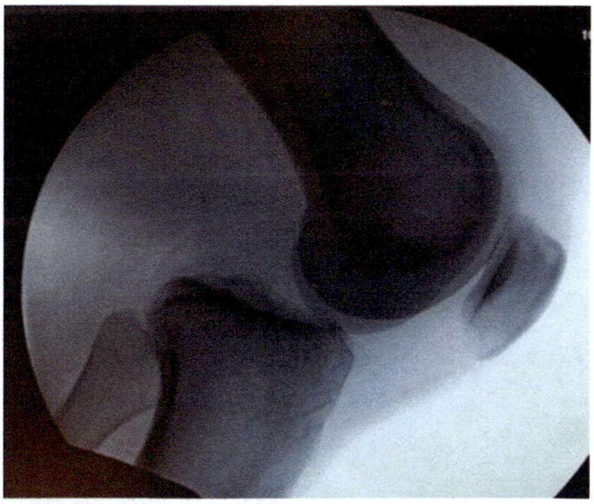

Fig. 25.3 Lateral knee X-ray demonstrating increased posterior tibial translation with posterior drawer test performed. (Gill and Gwathmey [56])

less posterior tibial translation in PCL deficient knees when popliteus contraction was simulated throughout knee flexion [55]. Thus in knees with both PCL and PLC injuries there is even less resistance against posterior tibial translation. A consequence of posterior translation is increased stress on the knee extensor mechanism. Theoretically the increased stress could result in increased patellofemoral joint contact forces and increase the risk for subsequent arthrosis.

The PCL also contributes to rotational stability of the knee. Cadaveric studies have shown mixed results on the extent of PCL restraint to external and internal tibial rotation [14, 37, 39, 51]. However, one study did show increased rotary instability when a combined PCL and PLC injury was present [57]. Specifically, there will be reduced resistance to tibial external rotation in a combined PCL and PLC injury. An in vivo study did find altered internal rotation in PCL deficient knees that was not significant [41]. Given these findings, the exact impact of PCL deficiency on rotational stability is still uncertain and further study is warranted. Additional coronal plane instability can be present in combined PCL and PLC injuries. This combination reduces stability to varus stresses on the knee when bearing. As a result, patients with this injury will ambulate with varus thrust gait where the knee will go into

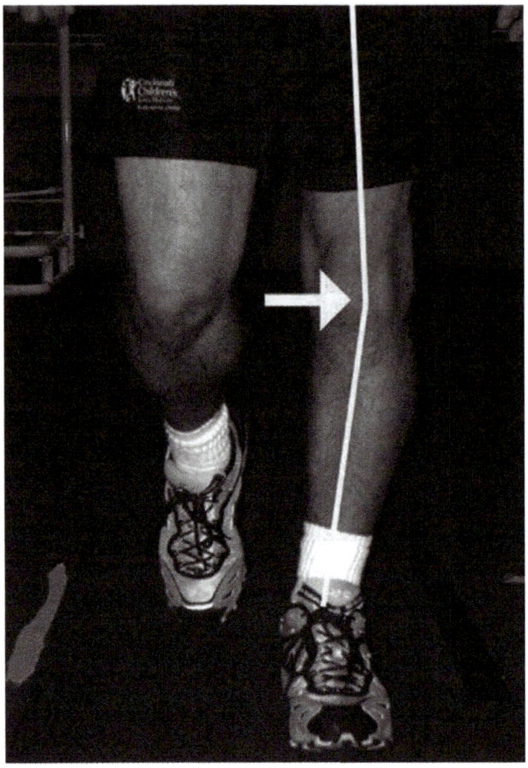

Fig. 25.4 Demonstrates varus thrust gait when the affected limb bears weight in the gait cycle. Used with permission of The International Journal of Sports Physical Therapy (formerly The North American Journal of Sports Physical Therapy) [58]

varus upon weight bearing during the gait cycle (Fig. 25.4). This can further contribute to increased knee medial compartment contact pressures.

Joint contact pressures are also impacted by PCL deficiency. The knee and patellofemoral joints are primarily affected by these changes. More specifically the medial compartment of the knee sees increased contact pressures in PCL deficiency. This has been demonstrated in multiple cadaveric studies [42, 43]. These studies also found increased contact pressures in the patellofemoral joint, likely secondary to posterior tibial translation placing increased stress on the knee extensor mechanism. Increased contact pressure helps to explain the increased frequency of patellofemoral and medial knee compartment arthritis found in PCL deficient knees.

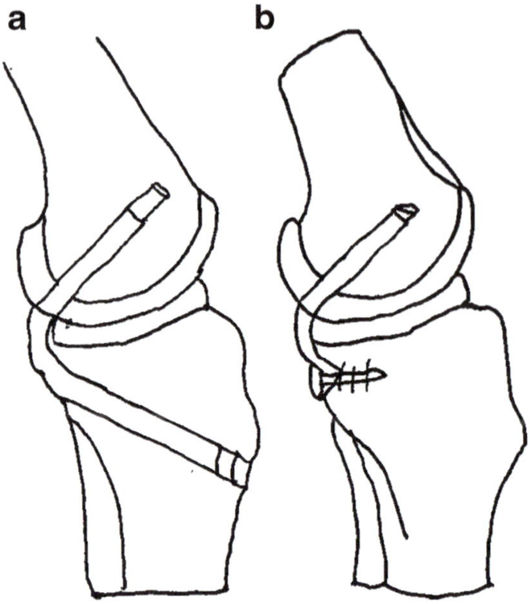

Fig. 25.5 Representation of the two common surgical techniques for PCL reconstruction. (**a**) Arthroscopic transtibial technique. (**b**) Tibial inlay technique

25.6 Reconstruction Techniques and their Biomechanical Properties

Multiple surgical techniques exist for PCL reconstruction. These include both arthroscopic and open tibial inlay techniques. Arthroscopic techniques involve transtibial tunnels to reconstruct the ligament (Figure 25.5). In arthroscopic techniques, there are concerns over the acute angle the graft takes when exiting the tibial tunnel [59]. This has been shown to cause abrasion and degradation to the graft at this site during range of motion which can result in laxity or failure after cyclic loading [59, 60]. Open tibial inlay techniques avoid this acute angle and utilize the posteromedial approach to the PCL via the interval between the medial head of the gastrocnemius and semimembranosus. Rather than a tunnel, in tibial inlay techniques bone trough is made where the graft is secured avoiding a sharp angle. Within these approaches there exist single and double bundle reconstruction techniques with various graft options.

Single bundle techniques focus on anatomic reconstruction of the anterolateral bundle by placing tunnel at the bundles femoral origin and tibial insertion [61]. Previously an isometric reconstruction was the preferred focus. However, isometric reconstruction has been found to lead to joint over constraint and eventually increased laxity [62]. Many studies have compared single bundle transtibial and tibial inlay techniques. Most studies demonstrate that at the time of initial graft fixation no difference exists in the anterior-posterior laxity of the knee between the two techniques [63–65]. However, tibial inlay techniques have been found to have superior strength after cyclical loading [59, 60, 64, 66]. This can be explained by the graft abrasion and degradation due to the acute graft angle at the tibia in transtibial techniques. Other variables in the strength of these constructs are related to graft choice. Autograft remains the most common graft choice either bone patellar tendon bone (BTB) or hamstring [67]. Allografts are used less often with achilles being the most common allograft of choice [67]. The biomechanical properties of these grafts have been tested and compared to each other. Chen et al. found quadrupled hamstrings to have the highest load to failure of the three, but was not able to stretching as well as BTB grafts [68]. Thus the technique and graft choice can impact the overall strength and longevity of a given PCL reconstruction.

Double bundle techniques aim for an anatomic reconstruction of the anterolateral and posteromedial bundles. It is the belief that this technique most closely reproduces native PCL anatomy. In the transtibial technique, a single tibial tunnel is drilled with two femoral tunnels to match the ALB and PMB origins. Similar to native anatomy, the ALB tunnel is made larger than the PMB tunnel [17]. The concerns with acute angle at the graft at the tibial tunnel again exist in this technique as described above. Also, the same shortcomings of isometric graft reconstruction apply in double bundle techniques leading to an anatomic reconstruction preference. Comparison of double versus single bundle techniques for both transtibial and inlay approaches have been extensively studied. In transtibial techniques, double bundle reconstructions were found to increase overall stability and more reproduce native PCL function [17, 61]. However, it was noted that this could be associated with increased graft forces than normal [61]. Differences are less apparent between single and double bundle tibial inlay techniques. Studies have shown mixed results in posterior tibial translation between the groups. Wiley et al. found no difference between either construct and the native knee, but did not less posterior translation in the double bundle when compared to single [69]. Another study reported no difference between the two groups in terms of posterior translation [59]. Double bundle reconstruction has shown benefit in setting of an associated posterolateral corner (PLC) injury [70]. A greater variety of auto- and allografts exist for this technique with the additional PMB reconstruction. Cited grafts include BTB, hamstrings, achilles tendon, tibialis posterior, and quadriceps tendon. Again, technique and graft selection influence overall construct strength with varying differences compared to single bundle constructions; however, an associated PLC injury may provide a scenario where double bundle provides particular benefit.

Finally, the fixation method can impact the biomechanical properties of a PCL reconstruction. Multiple fixation techniques exist in PCL reconstruction. These include suture endobuttons, interference screws, cortical screws, cross pins, or a combination of these options. Tibial fixation also includes options of distal and proximal fixation or a combination of both. Kitamura et al. compared multiple fixation techniques after 5000 cycles of loading. They found increased displacement in the endobutton group compared to the interference screw and cortical screw interference screw hybrid groups [71]. Another cadaveric study using achilles allograft found that combined distal and proximal tibial fixation resulted in a stronger construct that more closely resembled native PCL kinematics compared to distal fixation alone [72]. In tibial inlay techniques, no difference was found between bioabsorbable interference screws and metallic screws [73]. These findings tend to favor screw fixation and combined tibial fixation as compared to suture endobutton and isolated distal tibia fixation.

25.7 Conclusion

Overall, the posterior cruciate ligament is an important stabilizer of the knee resisting posterior tibial translation and rotational forces. It is made up of two bundles that work together to provide these functions. The anterolateral bundle is noted to be the larger and stronger of the two bundles. Given this, ALB anatomic restoration is the goal of a single bundle PCL reconstruction. Anatomic restorations in general are preferred to isometric in the setting of PCL reconstruction. Double bundle repairs more closely mimic native PCL kinematics and may be particularly helpful when an associated posterolateral corner injury is present. Despite this, there is still a significant role for single bundle repairs. Significant variety exists within repair techniques such as graft choice and fixation methods. Therefore, surgeons have multiple options to biomechanically optimize their construct to best treat a patient's needs. Ultimately through the understanding of PCL anatomy and biomechanics, providers optimally diagnose and treat these injuries.

References

1. Miyasaka KD, DH; Stone ML. The incidence of knee ligament injuries in the general population. Am J Knee Surg. 1991;4:3–8.
2. Fanelli GC. Posterior cruciate ligament injuries in trauma patients. Arthroscopy. 1993;9(3):291–4.
3. Petrigliano FA, McAllister DR. Isolated posterior cruciate ligament injuries of the knee. Sports Med Arthrosc Rev. 2006;14(4):206–12.
4. Schulz MS, Russe K, Weiler A, Eichhorn HJ, Strobel MJ. Epidemiology of posterior cruciate ligament injuries. Arch Orthop Trauma Surg. 2003;123(4):186–91.
5. Makris CA, Georgoulis AD, Papageorgiou CD, Moebius UG, Soucacos PN. Posterior cruciate ligament architecture: evaluation under microsurgical dissection. Arthroscopy. 2000;16(6):627–32.
6. Girgis FG, Marshall JL, Monajem A. The cruciate ligaments of the knee joint. Anatomical, functional and experimental analysis. Clin Orthop Relat Res. 1975;106:216–31.
7. Harner CD, Baek GH, Vogrin TM, Carlin GJ, Kashiwaguchi S, Woo SL. Quantitative analysis of human cruciate ligament insertions. Arthroscopy. 1999;15(7):741–9.
8. Logterman SL, Wydra FB, Frank RM. Posterior cruciate ligament: anatomy and biomechanics. Curr Rev Musculoskelet Med. 2018;11(3):510–4.
9. Edwards A, Bull AM, Amis AA. The attachments of the fiber bundles of the posterior cruciate ligament: an anatomic study. Arthroscopy. 2007;23(3):284–90.
10. Osti M, Tschann P, Künzel KH, Benedetto KP. Anatomic characteristics and radiographic references of the anterolateral and posteromedial bundles of the posterior cruciate ligament. Am J Sports Med. 2012;40(7):1558–63.
11. Anderson CJ, Ziegler CG, Wijdicks CA, Engebretsen L, LaPrade RF. Arthroscopically pertinent anatomy of the anterolateral and posteromedial bundles of the posterior cruciate ligament. JBJS. 2012;94(21):1936–45.
12. Forsythe B, Harner C, Martins CA, Shen W, Lopes OV, Jr., Fu FH. Topography of the femoral attachment of the posterior cruciate ligament. Surgical technique. J Bone Joint Surg Am. 2009;91 Suppl 2:89–100.
13. Morgan CD, Kalman VR, Grawl DM. The anatomic origin of the posterior cruciate ligament: where is it? Reference landmarks for PCL reconstruction. Arthroscopy. 1997;13(3):325–31.
14. Kennedy NI, Wijdicks CA, Goldsmith MT, Michalski MP, Devitt BM, Årøen A, et al. Kinematic analysis of the posterior cruciate ligament, part 1: the individual and collective function of the anterolateral and posteromedial bundles. Am J Sports Med. 2013;41(12):2828–38.
15. LaPrade CM, Civitarese DM, Rasmussen MT, LaPrade RF. Emerging updates on the posterior cruciate ligament: a review of the current literature. Am J Sports Med. 2015;43(12):3077–92.
16. Van Dommelen BA, Fowler PJ. Anatomy of the posterior cruciate ligament. A review. Am J Sports Med. 1989;17(1):24–9.
17. Harner CD, Janaushek MA, Kanamori A, Yagi M, Vogrin TM, Woo SL. Biomechanical analysis of a double-bundle posterior cruciate ligament reconstruction. Am J Sports Med. 2000;28(2):144–51.
18. Kennedy NI, LaPrade RF, Goldsmith MT, Faucett SC, Rasmussen MT, Coatney GA, et al. Posterior cruciate ligament graft fixation angles, part 1: biomechanical evaluation for anatomic single-bundle reconstruction. Am J Sports Med. 2014;42(10):2338–45.
19. Ahmad CS, Cohen ZA, Levine WN, Gardner TR, Ateshian GA, Mow VC. Codominance of the individual posterior cruciate ligament bundles. An analysis of bundle lengths and orientation. Am J Sports Med. 2003;31(2):221–5.
20. Papannagari R, DeFrate LE, Nha KW, Moses JM, Moussa M, Gill TJ, et al. Function of posterior cruciate ligament bundles during in vivo knee flexion. Am J Sports Med. 2007;35(9):1507–12.

21. Trent PS, Walker PS, Wolf B. Ligament length patterns, strength, and rotational axes of the knee joint. Clin Orthop Relat Res. 1976;117:263–70.
22. Marinozzi G, Pappalardo S, Steindler R. Human knee ligaments: mechanical tests and ultrastructural observations. Ital J Orthop Traumatol. 1983;9(2):231–40.
23. Kennedy JC, Hawkins RJ, Willis RB, Danylchuck KD. Tension studies of human knee ligaments. Yield point, ultimate failure, and disruption of the cruciate and tibial collateral ligaments. J Bone Joint Surg Am. 1976;58(3):350–5.
24. Race A, Amis AA. The mechanical properties of the two bundles of the human posterior cruciate ligament. J Biomech. 1994;27(1):13–24.
25. Jeong WS, Yoo YS, Kim DY, Shetty NS, Smolinski P, Logishetty K, et al. An analysis of the posterior cruciate ligament isometric position using an in vivo 3-dimensional computed tomography-based knee joint model. Arthroscopy. 2010;26(10):1333–9.
26. Nakagawa S, Johal P, Pinskerova V, Komatsu T, Sosna A, Williams A, et al. The posterior cruciate ligament during flexion of the normal knee. J Bone Joint Surg Br. 2004;86(3):450–6.
27. Vahey JW, Draganich LF. Tensions in the anterior and posterior cruciate ligaments of the knee during passive loading: predicting ligament loads from in situ measurements. J Orthop Res. 1991;9(4):529–38.
28. DeFrate LE, Gill TJ, Li G. In vivo function of the posterior cruciate ligament during weightbearing knee flexion. Am J Sports Med. 2004;32(8):1923–8.
29. Inderster A, Benedetto KP, Klestil T, Künzel KH, Gaber O. Fiber orientation of posterior cruciate ligament: an experimental morphological and functional study, part 2. Clin Anat. 1995;8(5):315–22.
30. Hosseini Nasab SH, List R, Oberhofer K, Fucentese SF, Snedeker JG, Taylor WR. Loading patterns of the posterior cruciate ligament in the healthy knee: a systematic review. PLoS One. 2016;11(11):e0167106.
31. Wascher DC, Markolf KL, Shapiro MS, Finerman GA. Direct in vitro measurement of forces in the cruciate ligaments. Part I: the effect of multiplane loading in the intact knee. J Bone Joint Surg Am. 1993;75(3):377–86.
32. Miyasaka T, Matsumoto H, Suda Y, Otani T, Toyama Y. Coordination of the anterior and posterior cruciate ligaments in constraining the varus-valgus and internal-external rotatory instability of the knee. J Orthop Sci. 2002;7(3):348–53.
33. Toutoungi DE, Lu TW, Leardini A, Catani F, O'Connor JJ. Cruciate ligament forces in the human knee during rehabilitation exercises. Clin Biomech (Bristol, Avon). 2000;15(3):176–87.
34. Markolf KL, Slauterbeck JL, Armstrong KL, Shapiro MM, Finerman GA. Effects of combined knee loadings on posterior cruciate ligament force generation. J Orthop Res. 1996;14(4):633–8.
35. Markolf KL, O'Neill G, Jackson SR, McAllister DR. Effects of applied quadriceps and hamstrings muscle loads on forces in the anterior and posterior cruciate ligaments. Am J Sports Med. 2004;32(5):1144–9.
36. Kumagai M, Mizuno Y, Mattessich SM, Elias JJ, Cosgarea AJ, Chao EY. Posterior cruciate ligament rupture alters in vitro knee kinematics. Clin Orthop Relat Res. 2002;395:241–8.
37. Li G, Gill TJ, DeFrate LE, Zayontz S, Glatt V, Zarins B. Biomechanical consequences of PCL deficiency in the knee under simulated muscle loads--an in vitro experimental study. J Orthop Res. 2002;20(4):887–92.
38. Pearsall AW, Hollis JM. The effect of posterior cruciate ligament injury and reconstruction on meniscal strain. Am J Sports Med. 2004;32(7):1675–80.
39. Grood ES, Stowers SF, Noyes FR. Limits of movement in the human knee. Effect of sectioning the posterior cruciate ligament and posterolateral structures. J Bone Joint Surg Am. 1988;70(1):88–97.
40. Gill TJ, DeFrate LE, Wang C, Carey CT, Zayontz S, Zarins B, et al. The biomechanical effect of posterior cruciate ligament reconstruction on knee joint function. Kinematic response to simulated muscle loads. Am J Sports Med. 2003;31(4):530–6.
41. Li G, Papannagari R, Li M, Bingham J, Nha KW, Allred D, et al. Effect of posterior cruciate ligament deficiency on in vivo translation and rotation of the knee during weightbearing flexion. Am J Sports Med. 2008;36(3):474–9.
42. MacDonald P, Miniaci A, Fowler P, Marks P, Finlay B. A biomechanical analysis of joint contact forces in the posterior cruciate deficient knee. Knee Surg Sports Traumatol Arthrosc. 1996;3(4):252–5.
43. Skyhar MJ, Warren RF, Ortiz GJ, Schwartz E, Otis JC. The effects of sectioning of the posterior cruciate ligament and the posterolateral complex on the articular contact pressures within the knee. J Bone Joint Surg Am. 1993;75(5):694–9.
44. Boynton MD, Tietjens BR. Long-term followup of the untreated isolated posterior cruciate ligament-deficient knee. Am J Sports Med. 1996;24(3):306–10.
45. Parolie JM, Bergfeld JA. Long-term results of non-operative treatment of isolated posterior cruciate ligament injuries in the athlete. Am J Sports Med. 1986;14(1):35–8.
46. Becker EH, Watson JD, Dreese JC. Investigation of multiligamentous knee injury patterns with associated injuries presenting at a level I trauma center. J Orthop Trauma. 2013;27(4):226–31.
47. Robertson A, Nutton RW, Keating JF. Dislocation of the knee. J Bone Joint Surg Br. 2006;88(6):706–11.
48. Peltola EK, Lindahl J, Hietaranta H, Koskinen SK. Knee dislocation in overweight patients. AJR Am J Roentgenol. 2009;192(1):101–6.
49. Nannaparaju M, Mortada S, Wiik A, Khan W, Alam M. Posterolateral corner injuries: epidemiology, anatomy, biomechanics and diagnosis. Injury. 2018;49(6):1024–31.
50. Shon OJ, Park JW, Kim BJ. Current concepts of posterolateral corner injuries of the knee. Knee Surg Relat Res. 2017;29(4):256–68.
51. Gollehon DL, Torzilli PA, Warren RF. The role of the posterolateral and cruciate ligaments in the stability of

the human knee. A biomechanical study. J Bone Joint Surg Am. 1987;69(2):233–42.
52. Li G, Most E, DeFrate LE, Suggs JF, Gill TJ, Rubash HE. Effect of the posterior cruciate ligament on posterior stability of the knee in high flexion. J Biomech. 2004;37(5):779–83.
53. Castle TH Jr, Noyes FR, Grood ES. Posterior tibial subluxation of the posterior cruciate-deficient knee. Clin Orthop Relat Res. 1992;284:193–202.
54. Logan M, Williams A, Lavelle J, Gedroyc W, Freeman M. The effect of posterior cruciate ligament deficiency on knee kinematics. Am J Sports Med. 2004;32(8):1915–22.
55. Harner CD, Höher J, Vogrin TM, Carlin GJ, Woo SL. The effects of a popliteus muscle load on in situ forces in the posterior cruciate ligament and on knee kinematics. A human cadaveric study. Am J Sports Med. 1998;26(5):669–73.
56. Gill GK, Gwathmey FW. Revision PCL reconstruction review/update. Curr Rev Musculoskelet Med. 2018;11(2):320–4.
57. Nielsen S, Helmig P. Posterior instability of the knee joint. An experimental study. Arch Orthop Trauma Surg. 1986;105(2):121–5.
58. Paterno MV, Hewett TE. Biomechanics of multi-ligament knee injuries (MLKI) and effects on gait. N Am J Sports Phys Ther. 2008;3(4):234–41.
59. Bergfeld JA, McAllister DR, Parker RD, Valdevit AD, Kambic HE. A biomechanical comparison of posterior cruciate ligament reconstruction techniques. Am J Sports Med. 2001;29(2):129–36.
60. Markolf KL, Zemanovic JR, McAllister DR. Cyclic loading of posterior cruciate ligament replacements fixed with tibial tunnel and tibial inlay methods. J Bone Joint Surg Am. 2002;84(4):518–24.
61. Markolf KL, Feeley BT, Jackson SR, McAllister DR. Where should the femoral tunnel of a posterior cruciate ligament reconstruction be placed to best restore anteroposterior laxity and ligament forces? Am J Sports Med. 2006;34(4):604–11.
62. Race A, Amis AA. PCL reconstruction. In vitro biomechanical comparison of 'isometric' versus single and double-bundled 'anatomic' grafts. J Bone Joint Surg Br. 1998;80(1):173–9.
63. Voos JE, Mauro CS, Wente T, Warren RF, Wickiewicz TL. Posterior cruciate ligament: anatomy, biomechanics, and outcomes. Am J Sports Med. 2012;40(1):222–31.
64. McAllister DR, Markolf KL, Oakes DA, Young CR, McWilliams J. A biomechanical comparison of tibial inlay and tibial tunnel posterior cruciate ligament reconstruction techniques: graft pretension and knee laxity. Am J Sports Med. 2002;30(3):312–7.
65. Margheritini F, Mauro CS, Rihn JA, Stabile KJ, Woo SL, Harner CD. Biomechanical comparison of tibial inlay versus transtibial techniques for posterior cruciate ligament reconstruction: analysis of knee kinematics and graft in situ forces. Am J Sports Med. 2004;32(3):587–93.
66. Weimann A, Wolfert A, Zantop T, Eggers AK, Raschke M, Petersen W. Reducing the "killer turn" in posterior cruciate ligament reconstruction by fixation level and smoothing the tibial aperture. Arthroscopy. 2007;23(10):1104–11.
67. Kim YM, Lee CA, Matava MJ. Clinical results of arthroscopic single-bundle transtibial posterior cruciate ligament reconstruction: a systematic review. Am J Sports Med. 2011;39(2):425–34.
68. Chen CH, Chou SW, Chen WJ, Shih CH. Fixation strength of three different graft types used in posterior cruciate ligament reconstruction. Knee Surg Sports Traumatol Arthrosc. 2004;12(5):371–5.
69. Wiley WB, Askew MJ, Melby A 3rd, Noe DA. Kinematics of the posterior cruciate ligament/posterolateral corner-injured knee after reconstruction by single- and double-bundle intra-articular grafts. Am J Sports Med. 2006;34(5):741–8.
70. Whiddon DR, Zehms CT, Miller MD, Quinby JS, Montgomery SL, Sekiya JK. Double compared with single-bundle open inlay posterior cruciate ligament reconstruction in a cadaver model. J Bone Joint Surg Am. 2008;90(9):1820–9.
71. Kitamura N, Yasuda K, Yamanaka M, Tohyama H. Biomechanical comparisons of three posterior cruciate ligament reconstruction procedures with load-controlled and displacement-controlled cyclic tests. Am J Sports Med. 2003;31(6):907–14.
72. Margheritini F, Rihn JA, Mauro CS, Stabile KJ, Woo SL, Harner CD. Biomechanics of initial tibial fixation in posterior cruciate ligament reconstruction. Arthroscopy. 2005;21(10):1164–71.
73. Gupta A, Lattermann C, Busam M, Riff A, Bach BR Jr, Wang VM. Biomechanical evaluation of bioabsorbable versus metallic screws for posterior cruciate ligament inlay graft fixation: a comparative study. Am J Sports Med. 2009;37(4):748–53.

26 Biomechanics of Osteotomies around the Knee

Dominic T. Mathis and Michael T. Hirschmann

26.1 Introduction

Osteotomies around the knee joint have a long tradition and are a well-established and an important part of joint-preserving therapy. Numerous studies have shown that at least 30% of the males and almost 20% of the females in western countries have a constitutional varus deformity of more than 3° [1, 2].

The underlying cause of this varus alignment may vary; however, it subsequently leads to increased pressure loads and peak loading areas in the medial compartment resulting in mechanical abrasion. The patient enters a vicious circle of progressive loading leading to increasing cartilage loss and increased varus alignment, which then leads to even more increased loading [3–6]. Malalignment in varus or valgus direction are therefore unfavourable for joint loading and have a major influence on the development or progression of osteoarthritis (OA).

Biomechanical studies have clearly shown that the correction of the malaligned knee unloads the cartilage and that the extent of the shift of the mechanical weight-bearing line correlates directly with the reduction in cartilage loading [7, 8]. Clinical studies have confirmed the positive influence on the pain level and the resilience of the knee joint [9–11]. Therefore, it is well evidenced that an osteotomy is an effective way of realigning and treating malalignment around the knee. Depending on the site of coronal plane malalignment, a varus or valgus correction osteotomy can be performed in an opening- or closing-wedge manner. These should be always done at the site of malalignment and hence these can be carried out laterally or medially at the distal femur or proximal tibia.

One of the reasons for the recently increasing interest in osteotomies is the improved technique, which allows the procedure to be performed safely without loss of correction. These advances have been made possible by the introduction of internal plate fixators, combined with an improved osteotomy technique (biplanar technique). Angle-stable internal fixators have previously proven its effectiveness in trauma settings [12–14].

26.2 General Aspects of Osteotomies around the Knee

26.2.1 Degree of Osteoarthritis

The malalignment-correcting osteotomy causes a shift of the peak load areas from the painful joint compartment to the intact opposite side. The clin-

D. T. Mathis (✉) · M. T. Hirschmann
University of Basel, Basel, Switzerland

Department of Orthopaedic Surgery and Traumatology, Kantonsspital Baselland (Bruderholz, Liestal, Laufen), Bruderholz, Switzerland
e-mail: dominic.mathis@unibas.ch

J. Koh et al. (eds.), *Orthopaedic Biomechanics in Sports Medicine*,
https://doi.org/10.1007/978-3-030-81549-3_26

ical outcome tends to be more favourable in knees with only moderate OA compared to advanced unicompartmental OA [10]. In the case of a more severe OA, the patient must be informed that a decrease of symptoms and increase of activity, but no complete relief of symptoms and pain can be expected. If there is a considerable extension deficit (over 10°), it must be considered whether an additional removal of intra-articular osteophytes could be helpful [15].

The conventional valgus-producing tibial head osteotomy is not indicated in cases of substantial loss of the outer meniscus and manifest lateral OA (extensive third- or fourth-degree damage, cartilage ulcers). In case of doubt, a stress radiograph should be taken under valgus load. If this results in a loss of height in the lateral joint section, a total knee arthroplasty is more appropriate [15].

26.2.2 Patellofemoral Instability

Valgus deformities can occur in combination with lateral instability of the patella. This problem can be well treated with an osseous and soft tissue combined medial intervention. The biplanar distal femoral varus osteotomy allows a correction of the axis and, if necessary, torsion, while at the same time reconstruction of the medial patellofemoral ligament can be performed using the same approach [16].

26.2.3 Patellofemoral Osteoarthritis

Many patients with unicompartmental OA also show degenerative changes in the patellofemoral joint. The evaluation of these patients is challenging. The medical history and the clinical examination are important factors. The retropatellar changes should not be decisive for the decision against a joint-preserving procedure such as osteotomy. Leg axis correction normalizes the alignment of the extensor mechanism and generally improves the loading conditions in the trochlear groove. If an HTO is indicated, an opening-wedge osteotomy using a biplanar technique with distal tuberosity incision can be selected to avoid distalization of the patella and an increase in patellar pressure [17]. Therefore, patellofemoral degenerations that are clinically mostly asymptomatic do not represent a contraindication for osteotomy around the knee in the case of unicompartmental OA [18].

26.2.4 Imaging

In all patients with possible OA in the knee anteroposterior, lateral radiographs with patella view as well as long-leg full weight-bearing views should be performed. When making the long-leg radiographs, it is important to ensure that both knees are extended maximally and the patellae are pointing forward. MRI scans may add valuable information on cartilage condition, meniscus, ligament and soft tissue damage. Also the location of nerves and vessels relative to the area of deformity correction can be assessed [19]. If a torsion deformity is found at physical examination, a computerized tomography (CT) scan with measurements of axial slides at standardized positions is mandatory. Furthermore, combined single photon emission-computerized tomography and conventional computerized tomography (SPECT/CT) has proved to be helpful in the assessment, pre- and postoperatively, of osteotomy patients [20, 21]. Mucha et al. have shown a significant decrease of bone tracer uptake (BTU) after HTO in the medial joint compartments in patients with medial compartment overloading due to varus malalignment (Fig. 26.1) [21]. The authors concluded that the evaluation of patients before and after HTO using SPECT/CT with regard to the mechanical leg alignment provides the surgeon with helpful additional information about the loading history of the knee joint. SPECT/CT could be further used to identify the optimal individualized correction for each patient and clinical scenario [21]. It is the only imaging modality, which allows a direct visualization of the unloading effect in the relevant compartment after osteotomy.

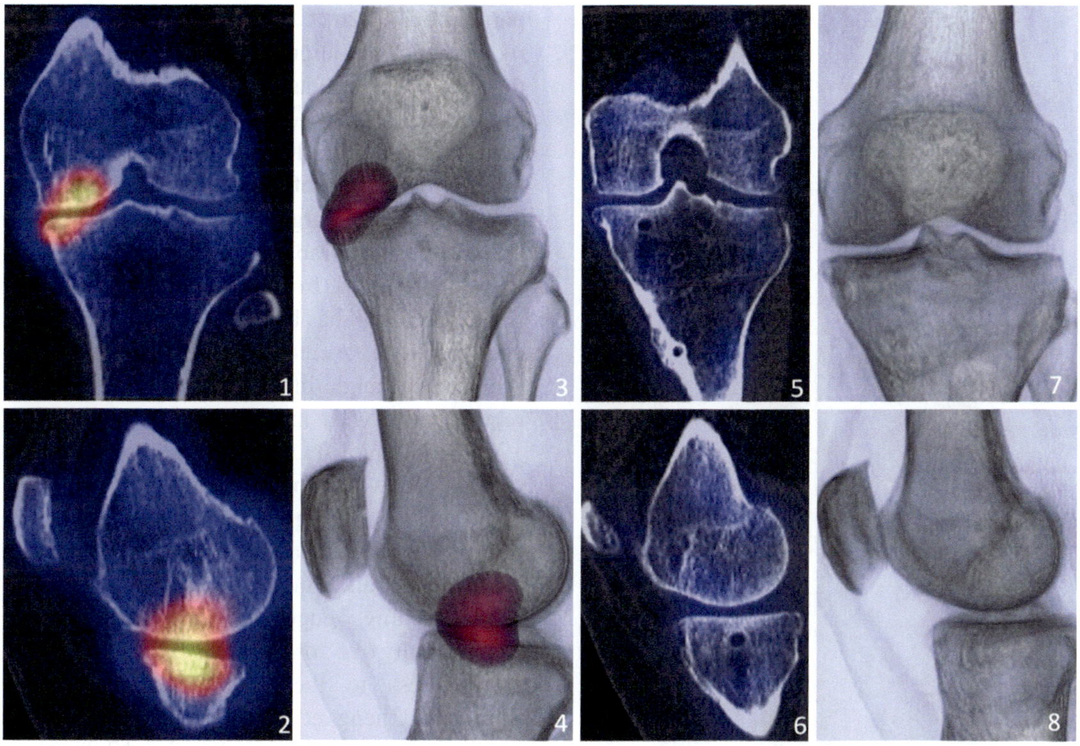

Fig. 26.1 *Left* SPECT/CT images (1, 2) and 3D radiolucent reconstructions (3, 4) of a 52-year-old female patient before HTO showing medial overloading. *Right* SPECT/CT images (5, 6) and 3D reconstructions (7, 8) 16 months postoperatively reveal an unloading effect of the medial joint compartment and a consolidation of the osteotomy gap. Reprinted with permission from Knee Surg Sports Traumatol Arthrosc (2015) 23:2315–2323

26.3 Biomechanical Considerations of High Tibial Osteotomy

Both valgus and varus malalignment are unfavourable for the joint mechanics and have a major influence on the development or progression of OA. A correction of the axis deformity thus results in cartilage decompression; the position of the loading axis in the frontal plane correlates directly with the tibiofemoral cartilage pressure distribution in the knee [7]. The normalization of the mechanical load conditions leads to a positive influence on the homeostasis of the knee joint.

Under normal conditions, the mechanical axes of femur and tibia are colinear (articular surface of the tibia averages 3° varus (medial proximal tibia angle, MPTA) and that of the femur 3° valgus (mechanical lateral distal femoral angle, mLDFA) relative to the mechanical axis) and the mechanical weight-bearing line (WBL) crosses the knee joint in the area of the medial spina (Fig. 26.2) [22]. In a neural aligned knee, 55–70% of the load is transmitted on the medial compartment during the stance phase of gait [23]. A deviation in the varus or valgus direction can be caused by a bony malposition in the femur and/or tibia, by a defect in the knee joint or by ligamentous instability. In a varus aligned knee, a deviation of 1° varus from the neutral alignment will cause an increase of the medial load of 5% [24].

First described by Jackson and Waugh in [25], high tibial osteotomy (HTO) is a well-established procedure for treating medial compartment OA of the varus deformed knee. In HTO, the bone of the proximal tibia is cut, and either the osteotomy gap is opened in a wedge shape (opening-wedge HTO) or a bone wedge is removed (closing-wedge HTO). With correct planning, it can normalize the bony anatomy and therefore create

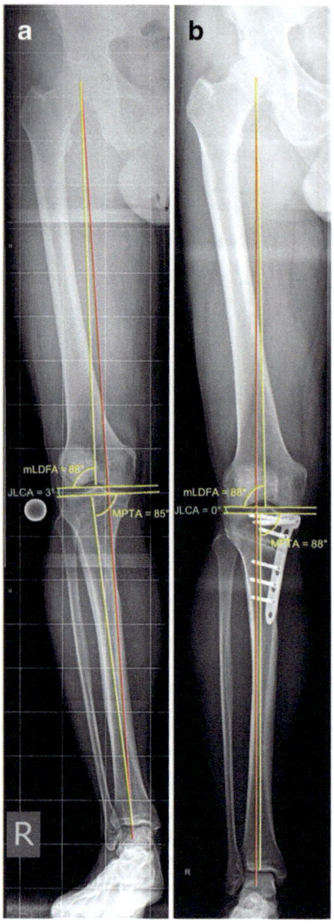

Fig. 26.2 50-year-old male patient before (**a**) and after (**b**) opening-wedge high tibial osteotomy (HTO) due to symptomatic varus alignment. The mechanical weight-bearing line (brown) crosses the knee joint preoperatively (**a**) in the medial compartment and after HTO (**b**) in the area of the lateral spina. The postoperative correction (**b**) of the medial proximal tibial angle (MPTA) is 3°, which results in a horizontal joint line (joint line convergence angle (JLCA), the angle between the tangent to the distal femoral condyles and the tangent to the tibia plateau). The mechanical lateral distal femoral angle (mLDFA) remains unaffected with 88° valgus (**a**, **b**)

physiological loading conditions for the entire leg (Fig. 26.2). The gait pattern is normalized and the dynamic distribution of load becomes physiological.

Regardless of the type of the osteotomy, the biomechanical objective of HTO is to realign the WBL in the coronal plane. The aim is to achieve the shift of the WBL from the arthritic compartment to the opposite tibiofemoral healthy compartment [7, 26]. Fujisawa et al. [27, 28] recommended to align the WBL of HTO through the 65–70% coordinate of the width of the tibial plateau, which has been refined recently to 62.5% towards to the Mikulicz line to restore the kinematic alignment profile [1, 29, 30]. Hence, the influence of the targeted limb alignment after HTO on cartilage repair is under heavy debate in literature [31–35]. In a recent retrospective comparative study, it has been reported that no difference between overcorrected knees with mean femorotibial angle of 165° and moderately corrected knees with mean femorotibial angle of 170° was found [31]. Martay et al. proposed correcting the weight-bearing axis to 55% tibial width (1.7°–1.9° valgus) for the optimal distribution of medial and lateral contact stresses [32]. Nakayama et al. found a large amount of correction in opening-wedge HTO with a resultant joint line obliquity of 5° or more may induce excessive shear stress to the articular cartilage [33]. Similarly, Zheng et al. reported that balanced loading occurred at angles of 4.3° and 2.9° valgus for the femoral and tibial cartilage, respectively [34]. Contradictory, Trad et al. suggested that a balanced stress distribution between two compartments was achieved under a valgus hypercorrection angle of 4.5° [35]. Clinical studies suggest that excessive overcorrection leads to poor functional outcomes and degeneration in the lateral compartment, while undercorrection does not relieve the pain of the medial compartment [27, 36, 37].

To date, it is unclear how articular cartilage repair in the medial compartment is affected by the grade of preoperative degeneration of the articular cartilage. Koshino et al. reported that knees with advanced degeneration of articular cartilage at lateral closing-wedge HTO showed better repair compared with knees with early degeneration [38], whereas Fujisawa et al. reported conflicting arthroscopic findings [28].

As a result, the question remains unsolved whether a "safety corrective range" for HTO in patients with OA exists. The effect of excessive stress on soft tissue wear or repair and the remodelling process after corrective osteotomy is still unknown.

26.3.1 Mediolateral Stability

Of particular importance in HTO patients is the medial collateral ligament (MCL). A medial opening-wedge HTO increases the strain on the superficial distal part of the MCL by spreading the osteotomy gap, whereas a lateral closing-wedge procedure has only a minor effect on the MCL. In this context, Agneskirchner et al. have shown the opening-wedge HTO without MCL release resulted in a significant increase of the pressure medially. Only after a complete release of the MCL a significant decrease of pressure medially was observed after opening-wedge HTO [7]. Conversely, if HTO has to be performed in case of a tibial valgus deformity [2], lateral opening-wedge HTO technique or alternatively, a medial closing-wedge HTO can be performed to correct the valgus leg alignment [39]. However, in the medial closing-wedge HTO, the medial MCL laxity has been found to increase [40, 41]. Hence, it was suggested to perform a surgical reefing procedure at all times to tighten the MCL in these patients [40].

26.3.2 Influence of Tibial Slope Change on Stability

In recent years, it has been shown that the inclination of the tibial plateau in the sagittal plane ("slope") affects the stability of the knee joint [42–44]. Physiologically, the tibial plateau is slightly tilted posteriorly. To describe the posterior inclination of the tibial plateau, the angle of the medial tibial plateau to the right angle to the proximal tibial axis is usually stated in literature. The mean values of the tibial slope reported in literature vary between 5° and 8° with a variance between 0° and 14°. In 19% of the population, there is a posterior slope of more than 10° [45]. An increased tibial slope can accentuate an anterior instability; however, it may also lead to a reduction of the posterior drawer, whereas a decreased tibial slope leads to a reduction of an anterior knee instability [43, 46].

It is well known that all techniques which correct frontal plane misalignment may also change sagittal plane alignment [47, 48]. Posterior tibial slope is considered to be an important factor in knee joint kinematics [42, 49–53]. Schaefer et al. have analysed the frontal and sagittal femorotibial knee alignment after opening- and closing-wedge HTO. Postoperatively, tibial slope had decreased by −0.5° in closing-wedge HTO and increased significantly by +3° in opening-wedge HTO [54].

The combination of symptomatic varus OA with significant knee instability due to overloading of the antero- and posterolateral structures is quite common in the younger group of patients and can be well treated by an HTO [42, 55]. Even a relative loosening of the collateral ligament can be easily eliminated by an opening-wedge HTO [15].

The following paragraphs provide biomechanical principles on how changes in the sagittal and frontal plane of the knee may alter the stability of the joint [56].

26.3.2.1 Coronal Alignment

The lateral joint opening and the tension of the anterior cruciate ligament (ACL) on human knee specimens with neutral mechanical axis and with varus axis have been measured by van de Pol et al. in 2009 [57]. There was no lateral opening of the joint in the neutral axis, but it was increased in the varus axis. The tension in the ACL also increased significantly with increasing varus deformity.

In a biomechanical study, La Prade et al. examined the effect of the varus axis on the posterolateral structures [58]. In this study, a significant increase in varus rotation (ligamentous varus) occurred after transecting the posterolateral structures. However, the opening-wedge osteotomy of the tibia was able to reduce both the varus rotation and external rotation, which were caused by the transection of the posterolateral structures. La Prade et al. also attribute the stabilizing effect of the osteotomy to increased tension in the medial collateral ligament.

A recent meta-analysis has shown that frontal deformities have no influence on the risk of primary ACL ruptures [59]. However, various studies have shown that patients with recurrent instability after ACL reconstruction were significantly more likely to have a varus deformity (>5°) [59–61].

Clinical studies are also available on the influence of frontal alignment on the results of posterior cruciate ligament (PCL) and posterolateral reconstruction [62, 63]. In both studies, varus deformity was considered as a risk factor for a reinjury after PCL and posterolateral reconstruction.

26.3.2.2 Sagittal Alignment

In a biomechanical study, Agneskirchner et al. have changed the tibial slope in human cadaveric knees by flexion osteotomies and then measured the anterior tibial translation of the tibial plateau to the femur [42]. This study demonstrated that an increase of posterior slope intensifies the anterior translation of the tibia. In addition, the tibiofemoral contact area and pressure was shifted anteriorly, resulting in decompression of the posteromedial tibial plateau.

Similarly, Giffin et al. were able to show that under axial compressive load increased slope of the tibia led to an anterior translation of the tibia in relation to the femur [44]. In addition, the in-situ forces in the ACL increased with increasing slope. Shelburne et al. were able to confirm the results of both studies in a computer model [64]. In the computer simulation, an increase in slope led to an increased anterior translation during daily activities like standing, squatting or walking.

Yet, there are three meta-analyses that can show that both the medial tibial slope and the lateral tibial slope are a risk factor for suffering an ACL rupture [59, 65, 66]. In this context, the study of Webb et al. should be highlighted: It was found that the risk of a further ACL injury was increased by factor 5 in patients with a slope of >12° [67]. Significantly fewer studies deal with the influence of the posterior tibial slope on posterior instability. Schatka et al. were able to show that in the uninjured knee, a low posterior slope correlates with an increased posterior translation of the tibia [68]. Bernhardson et al. found that a lower posterior slope is a risk factor for a PCL rupture [69].

26.3.2.3 Valgus HTO in Patients with Anterior Instability

The triad of anterior instability, medial OA and varus deformity [70, 71] as well as an isolated double or triple varus deformity without medial OA [61, 72, 73] are recognised as indications for HTO in patients with anterior instability [56]. Double varus occurs due to tibiofemoral varus alignment and separation of the lateral tibiofemoral compartment due to deficiency of lateral soft tissues (= joint line conversion angle, JLCA) [74]. Triple varus occurs due to deficiency of the posterolateral corner ligament and results in varus with recurvatum. This arises because of varus osseous alignment (primary varus), separation of lateral tibiofemoral compartment (double varus) and increased external rotation and hyperextension caused by posterolateral instability [75].

With regard to postoperative results, all studies on HTO in anterior instabilities show that clinical scores can be improved by HTO alone or by the combined procedure (HTO plus ligament reconstruction) [70–72, 76]. It is irrelevant whether the ligament reconstruction is performed in one or two stages [76]. However, the increased complication rate of 63% must also be pointed out for the combined procedure [70].

26.3.2.4 Slope Correction during Valgus HTO in Patients with Anterior Instability

As described above, tibial slope can also be changed during valgus HTO. Unfortunately, this can happen unintentionally when the surgeon is inexperienced and the slope is not observed or controlled during a tibial head osteotomy (K-wire and lateral image intensifier control). It is therefore inevitable that this potential change is taken into consideration every time an osteotomy is performed on the tibial head. The intentional reduction of the slope can clearly improve anterior instability—in contrast, an increase of the tibial slope can reduce posterior instability. Arun et al. showed that patients after HTO with a posterior slope reduction of more than 5° achieved better functional scores than patients with a slope reduction of less than 5° [77].

Hence, in addition to HTO the intentional and correct reduction of the tibial slope can improve postoperative results in patients with anterior instability. However, in many cases an ACL reconstruction may also be necessary [78].

26.3.2.5 Slope Correction during Valgus HTO in Patients with Posterior Instability

Studies have shown that functional clinical scores and subjective stability can be improved by an isolated valgus medial opening-wedge HTO [72, 79, 80]. Often effectively enough that secondary ligament reconstruction was no longer necessary. Reichwein and Nebelung were able to significantly improve knee function in patients after failed PCL reconstruction with an isolated slope-increasing osteotomy [81].

26.4 Biomechanical Consideration of Distal Femoral Osteotomy

Distal femoral deformities are observed in valgus deformities and also in severe varus deformities. However, there are some biomechanical differences compared to the proximal tibia. The lever arm is longer and the surface at the level of the osteotomy is smaller on the femoral side. There is no "hinge-preserver" such as fibres of the proximal tibiofibular joint in the area of the safe zone. Furthermore, the blood circulation at the distal femur differ fundamentally from the proximal tibia [82, 83]. As a result, DFO is inherently more unstable and considered to be difficult procedures with high potential risk of complications (3.2% non-union and 3.8% delayed union) [84–86]. Distal femoral osteotomies can be performed with lateral opening- or medial closing-wedge osteotomy. However, healing complications and irritation of the iliotibial band by the fixator have been described more frequently for the lateral opening distal femoral osteotomy [87]. For this reason, the medial closing osteotomy of the distal femur has become increasingly popular in recent years [88, 89].

Varus-producing osteotomies of the distal femur are a good surgical option for the purpose of unloading the affected lateral compartment and correcting underlying valgus malalignment in high-demand active patients with symptomatic unicompartmental OA [90, 91]. While clinical studies have demonstrated successful outcomes following distal femoral varus osteotomies (DFVO) in the treatment of lateral compartment OA [86, 89, 92–95], to date there is scarce knowledge on biomechanical effects of the load redistribution produced by the DFVO in orthopaedic literature. In a recent biomechanical cadaveric study, Quirno et al. found progressive unloading of the lateral tibiofemoral compartment with increasing DFVO correction angles (25% decrease in mean contact pressure with 15° osteotomy) [96]. The authors recommended, when performing a DFVO for valgus malalignment, to aim for an overcorrection of 5° to restore near-normal contact pressures and contact areas in the lateral compartment rather than the traditional teaching of correcting to neutral alignment [96]. Conversely, clinical studies are less conclusive with regard to their recommended correction of valgus malalignment with no uniform trend towards any particular correction goal being definitive [92–94, 97].

Based on biomechanical examinations and clinical experience, biplanar osteotomies for the distal femur are recommended [98, 99]. The biplanar technique has geometrical advantages by reducing the volume of the osteotomy, approximating the metaphysis with better bone healing, increasing axial stability, protecting against the potential issue of malrotation, and allowing open reduction in case of a hinge fracture [12, 98, 99].

The biplanar technique, along with angle-stable plate fixators, can be used both laterally for valgus corrections and medially for varus corrections with very good midterm results and patient satisfaction [84, 100, 101].

26.5 Biomechanical Considerations of Intra-Articular Osteotomy

The deviation of the WBL can be caused by a bony deformity of the femur and/or tibia (primary, constitutional deformity) on the one hand, and by a defect in the knee joint itself on the other hand.

For metaphyseal deformities, opening and closing tibial osteotomies can be performed, as developed for the correction of constitutional deformities. If the deformity is located clearly within the joint, an intra-articular osteotomy can be discussed [102–105]. They directly address the incongruent joint surface and can be used for deformities in the sagittal and coronal plane. Indications for an intra-articular osteotomy may be: malunions of the tibial plateau with significant intra-articular depression and/or steps; deviation of Mikulicz line in the overloaded compartment; flexion–/extension deformity with significant restriction of range of motion but also constitutional deformities such as Blount disease, Ellis–van Creveld syndrome and some types of achondroplasia [19, 106].

Posttraumatic intra-articular deformities state the main indications for corrective intra-articular osteotomies. This is explained by the fact that tibial plateau fractures may result in knee incongruity and instability. The incongruity is produced by the mismatch between the tibial and femoral articular surfaces [107]. The lack of containment of the rim of the joint generates instability. The biomechanical aim of the treatment is to restore the rim and its containment and thus stability, as well as a physiological WBL.

26.5.1 Tibial Plateau Widening

Insufficient anatomical reduction of the articular surface may produce secondary depression with angular deformity, widening of the tibial plateau and subluxation of the joint. The goal of correction is to re-establish "normal" relationships in relation to the contralateral side. The widening of more than 5 mm is usually considered to have worse functional outcomes [102, 108]. Johannsen et al. have distinguished residual widening within normal variation from pathological widening and found even a lower threshold with 2.1 mm [109]. Kumar et al. suggested that 4% of extra width relative to femoral articular surface can be considered normal for the tibia plateau [108]. However, pathological widening puts undue stress on surrounding ligaments and capsule but also alters biomechanics which could affect the knee function [108]. An intra-articular closing-wedge osteotomy can be performed to restore the width and height of the tibial plateau and thus joint congruity and stability.

26.5.2 Unicompartmental Angulation

As described by Paley et al., the physiological mechanical proximal tibia angle measures 87 ± 3 degrees [110]. Deviations between the articular surface and the 87° proximal tibia angle are often caused in posttraumatic situations by a malunited split wedge plateau fragment after a tibia plateau fracture. Clinically relevant deviation which requires surgery may include a change of ≥5° in lower limb alignment (varus or valgus), articular surface compression ≥5 mm, and a plateau shift and axial instability ≥5° [102]. A change in posterior slope angle of ≥10° is also considered to be an indication for operation [102].

The correction of unicompartmental angulation is normally performed by an opening-wedge intra-articular osteotomy in the plane of the deformity and with a hinge at its apex (at the level of the tibial spines) [107]. Thereby, the joint line can be elevated in order to restore joint congruity and containment of the tibial plateau rim with respect to the femoral condyle, and thus malalignment will be corrected. Beside relevant articular deviations, surgical indications include joint instability and residual knee pain in daily activities [111].

26.6 Conclusion

The osteotomy around the knee is an evidence-based joint-preserving procedure for the therapy of unicompartmental osteoarthritis with good long-term results. A correction of the axis deformity results in cartilage decompression—the position of the loading axis in the frontal plane correlates directly with the tibiofemoral cartilage pressure distribution in the knee. The normalization of the mechanical load conditions leads to a positive influence on the homeostasis of the knee joint. However, the recommended target for alignment correction is under debate for both the HTO and the DFO with regard to the biomechanical and clinical findings.

Posterior tibial slope is considered to be an important factor in knee joint kinematics. All techniques which correct frontal plane misalignment may also change sagittal plane alignment. It is well known that opening-wedge HTO generally increases and closing-wedge HTO decreases tibial slope. An increased tibial slope can accentuate an anterior instability, however may also lead to a reduction of the posterior drawer, whereas a decreased tibial slope lead to a reduction of an anterior knee instability.

The osteotomy around the knee is a reliable technique with significant biomechanical effects on the entire lower extremity and, if performed correctly, can bring significant benefits to the patient.

Declaration of Competing Interest **No conflicts of interest.**

References

1. Victor JM, Bassens D, Bellemans J, Gursu S, Dhollander AA, Verdonk PC. Constitutional varus does not affect joint line orientation in the coronal plane. Clin Orthop Relat Res. 2014;472(1):98–104.
2. Bellemans J, Colyn W, Vandenneucker H, Victor J. The Chitranjan Ranawat award: is neutral mechanical alignment normal for all patients? The concept of constitutional varus. Clin Orthop Relat Res. 2012;470(1):45–53.
3. Felson DT, Niu J, Gross KD, Englund M, Sharma L, Cooke TD, et al. Valgus malalignment is a risk factor for lateral knee osteoarthritis incidence and progression: findings from the multicenter osteoarthritis study and the osteoarthritis initiative. Arthritis Rheum. 2013;65(2):355–62.
4. Felson DT, Niu J, Yang T, Torner J, Lewis CE, Aliabadi P, et al. Physical activity, alignment and knee osteoarthritis: data from MOST and the OAI. Osteoarthr Cartil. 2013;21(6):789–95.
5. Sharma L, Song J, Dunlop D, Felson D, Lewis CE, Segal N, et al. Varus and valgus alignment and incident and progressive knee osteoarthritis. Ann Rheum Dis. 2010;69(11):1940–5.
6. Sharma L, Song J, Felson DT, Cahue S, Shamiyeh E, Dunlop DD. The role of knee alignment in disease progression and functional decline in knee osteoarthritis. JAMA. 2001;286(2):188–95.
7. Agneskirchner JD, Hurschler C, Wrann CD, Lobenhoffer P. The effects of valgus medial opening wedge high tibial osteotomy on articular cartilage pressure of the knee: a biomechanical study. Arthroscopy. 2007;23(8):852–61.
8. Heller MO, Matziolis G, Konig C, Taylor WR, Hinterwimmer S, Graichen H, et al. Musculoskeletal biomechanics of the knee joint. Principles of preoperative planning for osteotomy and joint replacement. Orthopade. 2007;36(7):628–34.
9. Brouwer RW, van TM R, Bierma-Zeinstra SM, Verhagen AP, Jakma TS, Verhaar JA. Osteotomy for treating knee osteoarthritis. Cochrane Database Syst Rev. 2007;3:CD004019.
10. Floerkemeier S, Staubli AE, Schroeter S, Goldhahn S, Lobenhoffer P. Outcome after high tibial openwedge osteotomy: a retrospective evaluation of 533 patients. Knee Surg Sports Traumatol Arthrosc. 2013;21(1):170–80.
11. Spahn G, Hofmann GO, von Engelhardt LV, Li M, Neubauer H, Klinger HM. The impact of a high tibial valgus osteotomy and unicondylar medial arthroplasty on the treatment for knee osteoarthritis: a meta-analysis. Knee Surg Sports Traumatol Arthrosc. 2013;21(1):96–112.
12. Brinkman JM, Lobenhoffer P, Agneskirchner JD, Staubli AE, Wymenga AB, van Heerwaarden RJ. Osteotomies around the knee: patient selection, stability of fixation and bone healing in high tibial osteotomies. J Bone Joint Surg Br. 2008;90(12):1548–57.
13. Lobenhoffer P, Agneskirchner JD. Improvements in surgical technique of valgus high tibial osteotomy. Knee Surg Sports Traumatol Arthrosc. 2003;11(3):132–8.
14. Luites JW, Brinkman JM, Wymenga AB, van Heerwaarden RJ. Fixation stability of openingversus closing-wedge high tibial osteotomy: a randomised clinical trial using radiostereometry. J Bone Joint Surg Br. 2009;91(11):1459–65.
15. Lobenhoffer P. Importance of osteotomy around to the knee for medial gonarthritis. Indications, technique and results. Orthopade. 2014;43(5):425–31.
16. Hinterwimmer S, Rosenstiel N, Lenich A, Waldt S, Imhoff AB. Femoral osteotomy for patellofemoral instability. Unfallchirurg. 2012;115(5):410–6.
17. Gaasbeek RD, Sonneveld H, van Heerwaarden RJ, Jacobs WC, Wymenga AB. Distal tuberosity osteotomy in open wedge high tibial osteotomy can prevent patella infera: a new technique. Knee. 2004;11(6):457–61.
18. Beard DJ, Pandit H, Gill HS, Hollinghurst D, Dodd CA, Murray DW. The influence of the presence and severity of pre-existing patellofemoral degenerative changes on the outcome of the Oxford medial uni-

compartmental knee replacement. J Bone Joint Surg Br. 2007;89(12):1597–601.
19. Kerkhoffs G, Haddad FS, Hirschmann MT, Karlsson J, Seil R. ESSKA Instructional Course Lecture Book - Glasgow. Ther Ber: Springer; 2018.
20. Mucha A, Dordevic M, Testa EA, Rasch H, Hirschmann MT. Assessment of the loading history of patients after high tibial osteotomy using SPECT/CT--a new diagnostic tool and algorithm. J Orthop Surg Res. 2013;8:46.
21. Mucha A, Dordevic M, Hirschmann A, Rasch H, Amsler F, Arnold MP, et al. Effect of high tibial osteotomy on joint loading in symptomatic patients with varus aligned knees: a study using SPECT/CT. Knee Surg Sports Traumatol Arthrosc. 2015;23(8):2315–23.
22. Paley D, Pfeil J. Principles of deformity correction around the knee. Orthopade. 2000;29(1):18–38.
23. Schipplein OD, Andriacchi TP. Interaction between active and passive knee stabilizers during level walking. J Orthop Res. 1991;9(1):113–9.
24. Halder A, Kutzner I, Graichen F, Heinlein B, Beier A, Bergmann G. Influence of limb alignment on mediolateral loading in total knee replacement: in vivo measurements in five patients. J Bone Joint Surg Am. 2012;94(11):1023–9.
25. Jackson JP, Waugh W. Tibial osteotomy for osteoarthritis of the knee. J Bone Joint Surg Br. 1961;43-B:746–51.
26. Liu X, Chen Z, Gao Y, Zhang J, Jin Z. High Tibial osteotomy: review of techniques and biomechanics. J Healthc Eng. 2019;2019:8363128.
27. Dugdale TW, Noyes FR, Styer D. Preoperative planning for high tibial osteotomy. The effect of lateral tibiofemoral separation and tibiofemoral length Clin Orthop Relat Res. 1992;274:248–64.
28. Fujisawa Y, Masuhara K, Shiomi S. The effect of high tibial osteotomy on osteoarthritis of the knee. An arthroscopic study of 54 knee joints. Orthop Clin North Am. 1979;10(3):585–608.
29. Cooke TD, Pichora D, Siu D, Scudamore RA, Bryant JT. Surgical implications of varus deformity of the knee with obliquity of joint surfaces. J Bone Joint Surg Br. 1989;71(4):560–5.
30. Babis GC, An KN, Chao EY, Rand JA, Sim FH. Double level osteotomy of the knee: a method to retain joint-line obliquity. Clinical results. J Bone Joint Surg Am. 2002;84(8):1380–8.
31. Tsukada S, Wakui M. Is overcorrection preferable for repair of degenerated articular cartilage after open-wedge high tibial osteotomy? Knee Surg Sports Traumatol Arthrosc. 2017;25(3):785–92.
32. Martay JL, Palmer AJ, Bangerter NK, Clare S, Monk AP, Brown CP, et al. A preliminary modeling investigation into the safe correction zone for high tibial osteotomy. Knee. 2018;25(2):286–95.
33. Nakayama H, Schroter S, Yamamoto C, Iseki T, Kanto R, Kurosaka K, et al. Large correction in opening wedge high tibial osteotomy with resultant joint-line obliquity induces excessive shear stress on the articular cartilage. Knee Surg Sports Traumatol Arthrosc. 2018;26(6):1873–8.
34. Zheng K, Scholes CJ, Chen J, Parker D, Li Q. Multiobjective optimization of cartilage stress for non-invasive, patient-specific recommendations of high tibial osteotomy correction angle - a novel method to investigate alignment correction. Med Eng Phys. 2017;42:26–34.
35. Trad Z, Barkaoui A, Chafra M, Tavares JMR. Finite element analysis of the effect of high tibial osteotomy correction angle on articular cartilage loading. Proc Inst Mech Eng H. 2018;232(6):553–64.
36. Coventry MB, Ilstrup DM, Wallrichs SL. Proximal tibial osteotomy. A critical long-term study of eighty-seven cases. J Bone Joint Surg Am. 1993;75(2):196–201.
37. Hernigou P, Medevielle D, Debeyre J, Goutallier D. Proximal tibial osteotomy for osteoarthritis with varus deformity. A ten to thirteen-year follow-up study. J Bone Joint Surg Am. 1987;69(3):332–54.
38. Koshino T, Wada S, Ara Y, Saito T. Regeneration of degenerated articular cartilage after high tibial valgus osteotomy for medial compartmental osteoarthritis of the knee. Knee. 2003;10(3):229–36.
39. Cerciello S, Lustig S, Servien E, Batailler C, Neyret P. Correction of Tibial Valgus deformity. J Knee Surg. 2017;30(5):421–5.
40. van Lieshout WAM, van Ginneken BJT, Kerkhoffs G, van Heerwaarden RJ. Medial closing wedge high tibial osteotomy for valgus tibial deformities: good clinical results and survival with a mean 4.5 years of follow-up in 113 patients. Knee Surg Sports Traumatol Arthrosc. 2019.
41. van Lieshout WAM, Martijn CD, van Ginneken BTJ, van Heerwaarden RJ. Medial collateral ligament laxity in valgus knee deformity before and after medial closing wedge high tibial osteotomy measured with instrumented laxity measurements and patient reported outcome. J Exp Orthop. 2018;5(1):49.
42. Agneskirchner JD, Hurschler C, Stukenborg-Colsman C, Imhoff AB, Lobenhoffer P. Effect of high tibial flexion osteotomy on cartilage pressure and joint kinematics: a biomechanical study in human cadaveric knees. Winner of the AGA-DonJoy award 2004. Arch Orthop Trauma Surg. 2004;124(9):575–84.
43. Brandon ML, Haynes PT, Bonamo JR, Flynn MI, Barrett GR, Sherman MF. The association between posterior-inferior tibial slope and anterior cruciate ligament insufficiency. Arthroscopy. 2006;22(8):894–9.
44. Giffin JR, Stabile KJ, Zantop T, Vogrin TM, Woo SL, Harner CD. Importance of tibial slope for stability of the posterior cruciate ligament deficient knee. Am J Sports Med. 2007;35(9):1443–9.
45. Nunley RM, Nam D, Johnson SR, Barnes CL. Extreme variability in posterior slope of the proximal tibia: measurements on 2395 CT scans of patients undergoing UKA? J Arthroplast. 2014;29(8):1677–80.

46. Feucht MJ, Mauro CS, Brucker PU, Imhoff AB, Hinterwimmer S. The role of the tibial slope in sustaining and treating anterior cruciate ligament injuries. Knee Surg Sports Traumatol Arthrosc. 2013;21(1):134–45.
47. Cullu E, Aydogdu S, Alparslan B, Sur H. Tibial slope changes following dome-type high tibial osteotomy. Knee Surg Sports Traumatol Arthrosc. 2005;13(1):38–43.
48. Marti CB, Gautier E, Wachtl SW, Jakob RP. Accuracy of frontal and sagittal plane correction in open-wedge high tibial osteotomy. Arthroscopy. 2004;20(4):366–72.
49. Fowler JL, Gie GA, Maceachern AG. Upper tibial valgus osteotomy using a dynamic external fixator. J Bone Joint Surg Br. 1991;73(4):690–1.
50. Giffin JR, Vogrin TM, Zantop T, Woo SL, Harner CD. Effects of increasing tibial slope on the biomechanics of the knee. Am J Sports Med. 2004;32(2):376–82.
51. Slocum B, Devine T. Cranial tibial thrust: a primary force in the canine stifle. J Am Vet Med Assoc. 1983;183(4):456–9.
52. Slocum B, Devine T. Cranial tibial wedge osteotomy: a technique for eliminating cranial tibial thrust in cranial cruciate ligament repair. J Am Vet Med Assoc. 1984;184(5):564–9.
53. Slocum B, Slocum TD. Tibial plateau leveling osteotomy for repair of cranial cruciate ligament rupture in the canine. Vet Clin North Am Small Anim Pract. 1993;23(4):777–95.
54. Schaefer TK, Majewski M, Hirschmann MT, Friederich NF. Comparison of sagittal and frontal plane alignment after open- and closed-wedge osteotomy: a matched-pair analysis. J Int Med Res. 2008;36(5):1085–93.
55. Lobenhoffer P, van Heerwaarden R, Staubli A, Jakob RP, Galla M, Agneskirchner J. Osteotomies around the knee. Indications - Planning - Surgical techniques using plate fixators Thieme, Stuttgart; 2013.
56. Petersen W, Hees T, Harrer J. Bony deformity correction and anterior instability: slope and varus thrust. Knie J. 2019;1:7–16.
57. van de Pol GJ, Arnold MP, Verdonschot N, van Kampen A. Varus alignment leads to increased forces in the anterior cruciate ligament. Am J Sports Med. 2009;37(3):481–7.
58. Laprade RF, Engebretsen L, Johansen S, Wentorf FA, Kurtenbach C. The effect of a proximal tibial medial opening wedge osteotomy on posterolateral knee instability: a biomechanical study. Am J Sports Med. 2008;36(5):956–60.
59. Wang Y, Yang T, Zeng C, Wei J, Xie D, Yang Y, et al. Association between tibial plateau slopes and anterior cruciate ligament injury: a meta-analysis. Arthroscopy. 2017;33(6):1248–59.
60. Group M, Wright RW, Huston LJ, Spindler KP, Dunn WR, Haas AK, et al. Descriptive epidemiology of the multicenter ACL revision study (MARS) cohort. Am J Sports Med. 2010;38(10):1979–86.
61. Noyes FR, Barber-Westin SD, Hewett TE. High tibial osteotomy and ligament reconstruction for varus angulated anterior cruciate ligament-deficient knees. Am J Sports Med. 2000;28(3):282–96.
62. Noyes FR, Barber-Westin SD. Posterior cruciate ligament revision reconstruction, part 1: causes of surgical failure in 52 consecutive operations. Am J Sports Med. 2005;33(5):646–54.
63. Noyes FR, Barber-Westin SD, Albright JC. An analysis of the causes of failure in 57 consecutive posterolateral operative procedures. Am J Sports Med. 2006;34(9):1419–30.
64. Shelburne KB, Kim HJ, Sterett WI, Pandy MG. Effect of posterior tibial slope on knee biomechanics during functional activity. J Orthop Res. 2011;29(2):223–31.
65. Wordeman SC, Quatman CE, Kaeding CC, Hewett TE. In vivo evidence for tibial plateau slope as a risk factor for anterior cruciate ligament injury: a systematic review and meta-analysis. Am J Sports Med. 2012;40(7):1673–81.
66. Zeng C, Cheng L, Wei J, Gao SG, Yang TB, Luo W, et al. The influence of the tibial plateau slopes on injury of the anterior cruciate ligament: a meta-analysis. Knee Surg Sports Traumatol Arthrosc. 2014;22(1):53–65.
67. Webb JM, Salmon LJ, Leclerc E, Pinczewski LA, Roe JP. Posterior tibial slope and further anterior cruciate ligament injuries in the anterior cruciate ligament-reconstructed patient. Am J Sports Med. 2013;41(12):2800–4.
68. Schatka I, Weiler A, Jung TM, Walter TC, Gwinner C. High tibial slope correlates with increased posterior tibial translation in healthy knees. Knee Surg Sports Traumatol Arthrosc. 2018;26(9):2697–703.
69. Bernhardson AS, DePhillipo NN, Daney BT, Kennedy MI, Aman ZS, LaPrade RF. Posterior Tibial slope and risk of posterior cruciate ligament injury. Am J Sports Med. 2019;47(2):312–7.
70. Lattermann C, Jakob RP. High tibial osteotomy alone or combined with ligament reconstruction in anterior cruciate ligament-deficient knees. Knee Surg Sports Traumatol Arthrosc. 1996;4(1):32–8.
71. Williams RJ 3rd, Kelly BT, Wickiewicz TL, Altchek DW, Warren RF. The short-term outcome of surgical treatment for painful varus arthritis in association with chronic ACL deficiency. J Knee Surg. 2003;16(1):9–16.
72. Badhe NP, Forster IW. High tibial osteotomy in knee instability: the rationale of treatment and early results. Knee Surg Sports Traumatol Arthrosc. 2002;10(1):38–43.
73. Zaffagnini S, Bonanzinga T, Grassi A, Marcheggiani Muccioli GM, Musiani C, Raggi F, et al. Combined ACL reconstruction and closing-wedge HTO for varus angulated ACL-deficient knees. Knee Surg Sports Traumatol Arthrosc. 2013;21(4):934–41.
74. Markolf KL, Bargar WL, Shoemaker SC, Amstutz HC. The role of joint load in knee stability. J Bone Joint Surg Am. 1981;63(4):570–85.

75. Hughston JC, Jacobson KE. Chronic posterolateral rotatory instability of the knee. J Bone Joint Surg Am. 1985;67(3):351–9.
76. Tischer T, Paul J, Pape D, Hirschmann MT, Imhoff AB, Hinterwimmer S, et al. The impact of osseous malalignment and realignment procedures in knee ligament surgery: a systematic review of the clinical evidence. Orthop J Sports Med. 2017;5(3):2325967117697287.
77. Arun GR, Kumaraswamy V, Rajan D, Vinodh K, Singh AK, Kumar P, et al. Long-term follow up of single-stage anterior cruciate ligament reconstruction and high tibial osteotomy and its relation with posterior tibial slope. Arch Orthop Trauma Surg. 2016;136(4):505–11.
78. Dejour D, Saffarini M, Demey G, Baverel L. Tibial slope correction combined with second revision ACL produces good knee stability and prevents graft rupture. Knee Surg Sports Traumatol Arthrosc. 2015;23(10):2846–52.
79. Arthur A, LaPrade RF, Agel J. Proximal tibial opening wedge osteotomy as the initial treatment for chronic posterolateral corner deficiency in the varus knee: a prospective clinical study. Am J Sports Med. 2007;35(11):1844–50.
80. Naudie DD, Amendola A, Fowler PJ. Opening wedge high tibial osteotomy for symptomatic hyperextension-varus thrust. Am J Sports Med. 2004;32(1):60–70.
81. Reichwein F, Nebelung W. High tibial flexion osteotomy for revision of posterior cruciate ligament instability. Unfallchirurg. 2007;110(7): 597–602.
82. Han SB, Lee DH, Shetty GM, Chae DJ, Song JG, Nha KW. A "safe zone" in medial open-wedge high tibia osteotomy to prevent lateral cortex fracture. Knee Surg Sports Traumatol Arthrosc. 2013;21(1):90–5.
83. Vanadurongwan B, Siripisitsak T, Sudjai N, Harnroongroj T. The anatomical safe zone for medial opening oblique wedge high tibial osteotomy. Singap Med J. 2013;54(2):102–4.
84. Rosso F, Margheritini F. Distal femoral osteotomy. Curr Rev Musculoskelet Med. 2014;7(4):302–11.
85. Saithna A, Kundra R, Modi CS, Getgood A, Spalding T. Distal femoral varus osteotomy for lateral compartment osteoarthritis in the valgus knee. A systematic review of the literature. Open Orthop J. 2012;6:313–9.
86. Wylie JD, Jones DL, Hartley MK, Kapron AL, Krych AJ, Aoki SK, et al. Distal femoral osteotomy for the Valgus knee: medial closing wedge versus lateral opening wedge: a systematic review. Arthroscopy. 2016;32(10):2141–7.
87. Jacobi M, Wahl P, Bouaicha S, Jakob RP, Gautier E. Distal femoral varus osteotomy: problems associated with the lateral open-wedge technique. Arch Orthop Trauma Surg. 2011;131(6):725–8.
88. Freiling D, van Heerwaarden R, Staubli A, Lobenhoffer P. The medial closed-wedge osteotomy of the distal femur for the treatment of unicompartmental lateral osteoarthritis of the knee. Oper Orthop Traumatol. 2010;22(3):317–34.
89. Forkel P, Achtnich A, Metzlaff S, Zantop T, Petersen W. Midterm results following medial closed wedge distal femoral osteotomy stabilized with a locking internal fixation device. Knee Surg Sports Traumatol Arthrosc. 2015;23(7):2061–7.
90. Omidi-Kashani F, Hasankhani IG, Mazlumi M, Ebrahimzadeh MH. Varus distal femoral osteotomy in young adults with valgus knee. J Orthop Surg Res. 2009;4:15.
91. Sprenger TR, Doerzbacher JF. Tibial osteotomy for the treatment of varus gonarthrosis. Survival and failure analysis to twenty-two years. J Bone Joint Surg Am. 2003;85(3):469–74.
92. Backstein D, Morag G, Hanna S, Safir O, Gross A. Long-term follow-up of distal femoral varus osteotomy of the knee. J Arthroplast. 2007;22(4 Suppl 1):2–6.
93. Dewilde TR, Dauw J, Vandenneucker H, Bellemans J. Opening wedge distal femoral varus osteotomy using the Puddu plate and calcium phosphate bone cement. Knee Surg Sports Traumatol Arthrosc. 2013;21(1):249–54.
94. Wang JW, Hsu CC. Distal femoral varus osteotomy for osteoarthritis of the knee. J Bone Joint Surg Am. 2005;87(1):127–33.
95. Chahla J, Mitchell JJ, Liechti DJ, Moatshe G, Menge TJ, Dean CS, et al. Opening- and closing-wedge distal femoral osteotomy: a systematic review of outcomes for isolated lateral compartment osteoarthritis. Orthop J Sports Med. 2016;4(6):2325967116649901.
96. Quirno M, Campbell KA, Singh B, Hasan S, Jazrawi L, Kummer F, et al. Distal femoral varus osteotomy for unloading valgus knee malalignment: a biomechanical analysis. Knee Surg Sports Traumatol Arthrosc. 2017;25(3):863–8.
97. Saithna A, Kundra R, Getgood A, Spalding T. Opening wedge distal femoral varus osteotomy for lateral compartment osteoarthritis in the valgus knee. Knee. 2014;21(1):172–5.
98. van Heerwaarden R, Najfeld M, Brinkman M, Seil R, Madry H, Pape D. Wedge volume and osteotomy surface depend on surgical technique for distal femoral osteotomy. Knee Surg Sports Traumatol Arthrosc. 2013;21(1):206–12.
99. Brinkman JM, Hurschler C, Agneskirchner JD, Freiling D, van Heerwaarden RJ. Axial and torsional stability of supracondylar femur osteotomies: biomechanical comparison of the stability of five different plate and osteotomy configurations. Knee Surg Sports Traumatol Arthrosc. 2011;19(4):579–87.
100. Hofmann S, Lobenhoffer P, Staubli A, Van Heerwaarden R. Osteotomies of the knee joint in patients with monocompartmental arthritis. Orthopade. 2009;38(8):755–69.
101. Bonasia DE, Governale G, Spolaore S, Rossi R, Amendola A. High tibial osteotomy. Curr Rev Musculoskelet Med. 2014;7(4):292–301.

102. Liangjun J, Qiang Z, Zhijun P, Li H. Revision strategy for malunited tibial plateau fracture caused by failure of initial treatment. Acta Orthop Traumatol Turc. 2019;53(6):432–41.
103. Kerkhoffs GM, Rademakers MV, Altena M, Marti RK. Combined intra-articular and varus opening wedge osteotomy for lateral depression and valgus malunion of the proximal part of the tibia. Surgical technique. J Bone Joint Surg Am. 2009;91 Suppl 2:101–15.
104. Kerkhoffs GM, Rademakers MV, Altena M, Marti RK. Combined intra-articular and varus opening wedge osteotomy for lateral depression and valgus malunion of the proximal part of the tibia. J Bone Joint Surg Am. 2008;90(6):1252–7.
105. Krettek C, Hawi N, Jagodzinski M. Intracondylar segment osteotomy: correction of intra-articular malalignment after fracture of the tibial plateau. Unfallchirurg. 2013;116(5):413–26.
106. Lobenhoffer P. Intra-articular osteotomy for malunion of the tibial plateau. Oper Orthop Traumatol. 2020;32(4):367–84.
107. Kfuri M, Schatzker J. Corrective Intra-articular Osteotomies for Tibial Plateau Malunion. J Knee Surg. 2017;30(8):784–92.
108. Kumar A, Passey J, Khan R, Arora R, Kumar S, Chouhan D, et al. Defining the "mediolateral widening of tibial plateau" as a guide for reduction in tibial plateau fractures: An Indian perspective. J Clin Orthop Trauma. 2020;11(Suppl 1):S66–70.
109. Johannsen AM, Cook AM, Gardner MJ, Bishop JA. Defining the width of the normal tibial plateau relative to the distal femur: critical normative data for identifying pathologic widening in tibial plateau fractures. Clin Anat. 2018;31(5):688–92.
110. Paley D. Principles of deformity correction. Berlin: Springer; 2003.
111. Wang Y, Luo C, Hu C, Sun H, Zhan Y. An innovative intra-articular osteotomy in the treatment of posterolateral Tibial plateau fracture Malunion. J Knee Surg. 2017;30(4):329–35.

Meniscus Biomechanics

27

Alberto Grassi, Giacomo Dal Fabbro, Stefano Di Paolo, Gian Andrea Lucidi, Luca Macchiarola, Khalid Al-Khelaifi, and Stefano Zaffagnini

27.1 Introduction

Meniscal tears are common injuries in sports and in active population, and their surgical treatment is one of the most performed procedures in orthopaedic surgery. The comprehension of the normal and pathological biomechanics of meniscus is mandatory to fully address this kind of injury and restore the native behaviour of the knee joint. This chapter provides an overview of the normal meniscus biomechanics, followed by a description of the main mechanical features and consequences of meniscal tears and treatment choices. In particular, in the first part the authors analysed the literature available about the effects of meniscus status on the tibiofemoral contact area and pressure, while in the second part the role of different meniscal status in knee kinematics was described.

A. Grassi · G. Dal Fabbro (✉) · S. Di Paolo
G. A. Lucidi · L. Macchiarola
IRCCS - Istituto Ortopedico Rizzoli - Clinica Ortopedica e Traumatologica II, Bologna, Italy
e-mail: alberto.grassi@ior.it; giacomo.dalfabbro@studio.unibo.it; Stefano.dipaolo@ior.it; gianandrea.lucidi@studio.unibo.it

K. Al-Khelaifi
Orthopaedic Surgery Department, Aspetar Orthopaedic and Sports Medicine Hospital, Doha, Qatar
e-mail: Khalid.Alkhelaifi@aspetar.com

S. Zaffagnini
IRCCS – Rizzoli Orthopaedic Institute – 2nd Orthopaedic and Trauma Unit, University of Bologna, Bologna, Italy
e-mail: Stefano.zaffagnini@unibo.it

27.2 The Normal Meniscus

The biomechanical functions of the meniscus are crucial to chondroprotection and joint stability. In addition to load-sharing, menisci play a key role in joint lubrication, stabilization, congruity and proprioception of the knee.

27.2.1 Morphology

The menisci are two crescent-shaped fibrocartilaginous structures that partially cover the tibiofemoral joint surface. While medial meniscus covers between around 50–54% of the concave medial compartment, the lateral meniscus has been shown to cover around 59–70% of the slightly convex lateral tibial compartment [1, 2]. Both the menisci conform to the tibial and femoral bony geometry: the medial meniscus is larger and more like a capital letter "C" compared with the lateral meniscus, which is smaller and resembles an incomplete "o" shape (Fig. 27.1). In cadaver studies, the wide variation in meniscal size between different individuals was reported [3], underlining the critical importance of

J. Koh et al. (eds.), *Orthopaedic Biomechanics in Sports Medicine*,
https://doi.org/10.1007/978-3-030-81549-3_27

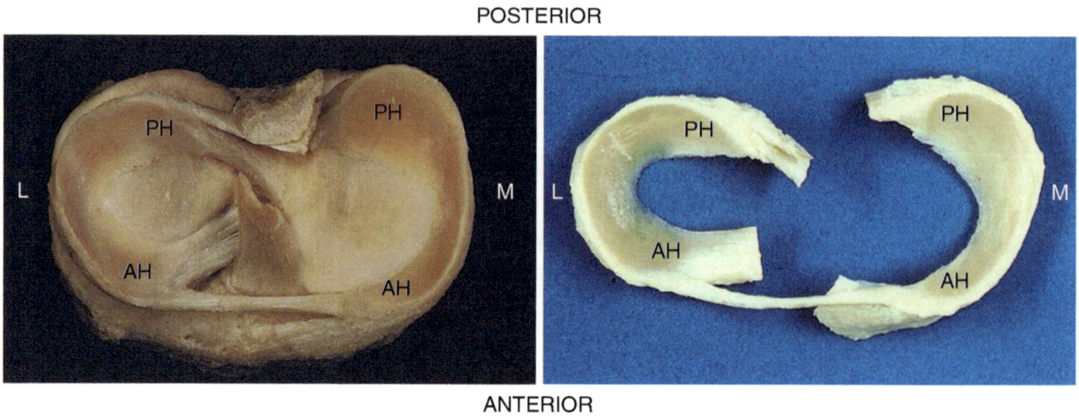

Fig. 27.1 Morphology of menisci. The different anatomical features of the menisci lead to different biomechanical behaviour of the medial and lateral meniscus. L, lateral; M, medial; AH, anterior horn; PH, posterior horn. Courtesy of Pau Golanò

accurate allograft and scaffold sizing. In an in vivo MRI-based study, on the other side, a similar shape between the right and left menisci was founded, suggesting that the contralateral meniscus could be used as a template for three-dimensional meniscus allograft sizing [4]. The joint capsule is attached to the complete periphery of each meniscus but adheres more firmly to the medial meniscus. The lateral meniscus presented an interruption in the attachment of the joint capsule in correspondence to the popliteal hiatus which allows the popliteus tendon to pass through its femoral attachment site. The medial meniscus, otherwise, has not direct muscular connection [5]. The menisci are attached to the surrounding tibial plateau by the meniscal ligaments, which form a functional unit, the meniscus-meniscal ligament construct.

27.2.2 Material Properties

The fine meniscal microstructure, with its circumferentially arranged fibres, defines the meniscal behaviour in both tension and compression.

The **tensile properties** of the human meniscus have been analysed in several studies, in which uniformly shaped specimens were harvested from the whole menisci to perform the tension test, taking it in the radial or circumferential direction. The results showed a higher elastic modulus in the specimen harvested from the anterior area of the meniscus compared with those harvested from the central and posterior portion. Moreover, the circumferential specimens taken resulted 10 times stronger than those taken radially. These findings may partly explain why, in the clinical practice, the circumferential splits are more common than radial tears [6].

The **compressive properties** of the meniscus have been investigated through three different types of compression tests: unconfined compression, confined compression and indentation [7–9]. These tests provided an analysis of the meniscal stiffness and permeability. The results, despite a considerable variation due to different experimental methods and data interpretation, showed that the meniscus is considerably less stiff in transverse compression than it is in circumferential tension. This feature allows menisci to conform to the variable geometry of the articular surface under compression when the knee is moving, particularly during deep knee flexion on the medial side. Moreover, the low meniscal permeability allows menisci to retain their load-bearing capacity during gait by resisting fluid loss under pressure [6].

27.2.3 The Meniscal Ligaments

The menisci are attached to the surrounding tibial plateau by the meniscal ligaments, which form a functional unit, the meniscus-meniscal ligament construct. The functions and biomechanics of menisci are strongly related to their attachments to the surrounding structures.

The menisco-tibial ligaments (Insertional ligaments). The tibial insertional ligaments connect the four horns of the menisci to the bone of the tibial plateau and represent the main restrictors of meniscal extrusion. These ligaments are an extension of the collagen fibres that run circumferentially through the bulk of the menisci and play the hoop stress, which resists the tendency for the meniscus to be forced out from between the joint surfaces during the knee loading. In a controlled laboratory study among 64 human meniscal roots [10], the failure at the same force of the connections was reported, with the exception of the anteromedial root, which was significantly weaker than both the posterior attachments. All the insertion failed at the meniscus-bone interface. In another cadaveric study a range of maximum failure loads of 506–565 N for the four meniscal roots was provided [11]. In a porcine model study in which the biomechanical comparison between two repair techniques, the suture anchor technique, and the transtibial pull-out technique, was performed, none of the repair methods resulted in adequate restoration of the pull-out strength of the meniscal root [12].

The menisco-femoral ligaments (MFL). These ligaments neither appear in all people, although the rate of incidence varies considerably in the literature. They are classified in anterior (the ligament of Humphry) and posterior (the ligament of Wrisberg), and connect the posterior horn of the lateral meniscus to the femur. It has been shown that they play a role in restraining the posterior tibial translation between 15° and 90° of knee flexion and that they contribute to external rotational stability between 60° and 120° of knee flexion [13].

The deep medial collateral ligament (dMCL). It is connected to the outer rim of the medial meniscus and, with the medial meniscus and posterior obliquus ligament, is part of the postero-medial complex of the knee. It provides secondary varus-valgus restraint to the knee joint [14].

The anterior inter-meniscal ligament (Transverse geniculate ligament). It connects the anterior horn of the medial and lateral meniscus and appears to be present in most of the knee joints. Its function is not well understood and does not appear to contribute to tibiofemoral contact characteristics. However, the position and orientation of this ligament suggest that it may be involved in the knee rotational stability [15].

The coronary ligaments. These ligaments connect the outer circumference of the menisci to the proximal tibia. Not a great deal is known about their function. It has been demonstrated that damage to the anterior and posterior portions of the medial coronary ligament does not have a significant effect on tibiofemoral contact pressure [16].

The meniscofibular ligament. This ligament connects the lateral meniscus to the head of the fibula and augments the lateral coronary ligament. It is tense while the knee is extended and the tibia externally rotated, suggesting that it may have a function in controlling tibial external rotation [17].

27.2.4 Functional Biomechanics

During the normal functional activities, the knee joint is subject to the axial compression which leads to the development of high contact stresses in the articular cartilage. The menisci are exposed to compressive, radial tensile, and shear stress: they contribute to the distribution of the load, creating a more congruent articulation and increasing the contact area between the tibia and the femur. This increase of the contact area allows decreasing the contact pressure, thus protecting the articular cartilage. The different anatomical features of the menisci led to different biomechanical behaviour of the medial and lateral meniscus. The 70% and 50% of the load are transferred through the lateral and the medial menisci respectively in their corresponding compartments [18]. On one hand, the lateral meniscus covers a higher articular surface compared with its medial counterpart and is, therefore, more involved in absorption and load transmission. On the other hand, in the medial compartment the load transmission is more evenly distributed between the cartilage surfaces and the medial meniscus [19]. Under an axial load, the menisci tend to extrude radially out of the joint, causing tissue stress which is resisted by the insertional ligaments. The essential mechanical property of the meniscus is the circumferential or hoop strain developed in its circumferential fibres, first described by Krause et al. in [20]. The resulting hoop stress, played by the circumferential fibres, resists the tendency for the meniscus to be forced out from between the joint surfaces during the knee loading, transmitting the femorotibial load [21].

The load-bearing role of the menisci occurs across the whole knee range of motion since the meniscal insertional ligaments allow menisci to move during knee flexion. Indeed, during the knee flexion the menisci translate posteriorly, ensuring congruency between the femoral condyle and tibial plateau. In particular, the lateral meniscus was shown to be more mobile than the medial meniscus, due to both the deep medial collateral ligament attachment and the concave morphology of the medial tibial plateau [22]. Moreover, the knee flexion leads to a shortening of the anterior-posterior diameter of the menisci, which can be related to the positioning and curvature of femoral condyles at the tibiofemoral contact point [23]. The results of an in vivo dynamic-MRI-assessed study [24] confirmed that both menisci moved posteriorly as the knee flexed and that the lateral meniscus appeared more mobile than the medial meniscus. Moreover, the anterior meniscal horns resulted to be more mobile than the posterior horns. The lower posterior translation of the medial meniscus, in particular of its posterior horn, compared with its lateral counterpart, leads to the tibial internal-external rotation (screw-home mechanism) during flexion and extension, respectively [24]. Furthermore, these findings may explain the higher frequency of medial meniscal tears compared to lateral meniscus tears [25, 26], and the observation of medial meniscus tears being located more frequently in the posterior horn [27].

The menisci undergo different load-bearing stresses at different knee flexion angles. In an MRI analysis among normal knees [28], in which the effects of medial meniscus on tibiofemoral contact area were provided, a larger role in load distribution of both medial and lateral meniscus at higher flexion angles was reported, compared with knees in full extension. These results suggest that meniscus undergoes higher loads during a more flexed position than knee extension.

27.3 Meniscus Tears: Mechanism of Injury and Mechanical Consequences

The acute traumatic meniscal tears are common in young and active individuals engaged in level 1 contact sports that comprise frequent pivoting such as soccer, American football and rugby. Moreover, waiting for more than 12 months between anterior cruciate ligament injury and reconstruction was found to be a risk factor for meniscal tears with strong evidence [29]. On the other hand, meniscal tears occur frequently also in the general population during an apparently innocuous activity such as walking or squatting. Advanced age (>60 years old), male gender, work-related kneeling, squatting and climbing stairs were found as risk factors for the degenerative meniscal tears with strong supporting evidence [30]. In both, sports and the general population, the global joint laxity represents a risk factor for meniscal tears [31]. The contact sports have been correlated with an increased risk of meniscal injuries [32]. Moreover, despite their low-contact profile, swimming and running might be also considered as risk factors [30, 33].

Several patterns of meniscal injury mechanisms have been described. The most frequent traumatic mechanism is a twisting or rotating forceful movement of the knee while the leg is bent. A torsional loading or a high compressive force between femoral and tibial surfaces could lead to meniscal damage at different angles of knee flexion [34]. Another typical mechanism of meniscal injury is a sudden transition from knee's flexion to full extension, catching the meniscus trapped between the femur and the tibia [35]. Moreover, varus and valgus stresses could lead to a meniscal tear. A valgus impact associated with tibial external rotation might also cause a triad of injuries involving menisci, the lateral collateral ligament and the anterior cruciate ligament [36]. The most recent reliable and valid classification system for meniscal tears is the ISAKOS classification, which considers the tear's depth, pattern, length and location [37] (Table 27.1).

Longitudinal-vertical, horizontal and radial tears. The most common meniscal tears in young and active people are the longitudinal-vertical tears [29]. The peripheral longitudinal-vertical tears are characterized by a great healing potential, probably due to the higher vascularization of the peripheral zone compared with the other meniscal zones. In a multi-centre study among patients who underwent ACL reconstruction [38], all lateral meniscal tears left untreated that required subsequent surgery measured more than 10 mm in length at six-years-minimum follow-up, while several medial compartment tears that required reoperation measured less than 10 mm. These findings, in line with the results of other clinical studies [39–42], suggested that peripheral longitudinal tears of the lateral meniscus with less than 10 mm in length may be left untreated.

Table 27.1 ISAKOS classification of meniscal tears: description categories

	Category	Descriptors
1	Tear depth	Partial or complete
2	Rim width	Zone 1 (< 3 mm) zone 2 (3 to 5 mm) zone 3 (> 5 mm)
3	Location	Anterior, middle, posterior
4	Central to popliteal hiatus	Yes or no
5	Tear pattern	Longitudinal-vertical, horizontal, radial, vertical/horizontal flap, complex
6	Quality of tissue	Nondegenerative, degenerative or undetermined
7	Length of tear	Length in millimetres
8	Amount of meniscus excised	Percentage of surface area that was excised

Horizontal and longitudinal-vertical circumferential tears may cause pain and lead to mechanical instability of a portion of the meniscus, but they will not disrupt the functional continuity of the circumferential native collagen fibres. Hence, the load-bearing and shock-absorbing functions of the tissue should be largely preserved. The radial tears, on the other hand, cause the discontinuity of the circumferential fibres, preventing the formation of the hoop strains that develop within the menisci as they are loaded, thus effectively compromising the function of the meniscal tissue [21, 43, 44] (Fig. 27.2). In a cadaveric study in which a simulation of the human gait cycle was performed [45], a significant increase in the peak of contact pressure and decrease in contact area was reported with a 90% width radial tear of the body-posterior horn junction of the lateral meniscus, compared with the intact status. In particular, the most pronounced changes were observed in the postero-peripheral quadrant at 45% of the gait cycle [45].

Root tears. The meniscal root tears represent the disruption in the attachment points of the meniscus. The anterior and posterior roots are the only part of menisci with direct insertion into the bone and act as main restrictors of meniscal extrusion. The traumatic meniscal root tears are mainly radial lesions or avulsion, frequently

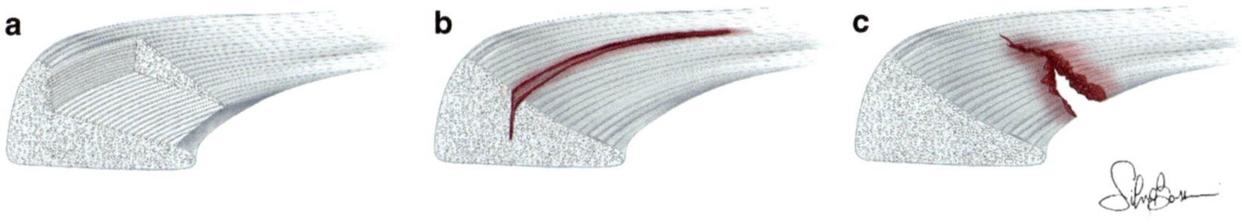

Fig. 27.2 Different biomechanical effects produced by different tear patterns. While longitudinal-vertical tears will not disrupt the functional continuity of the circumferential native collagen fibres, radial tears cause the discontinuity of the circumferential fibres, preventing the formation of the hoop strains that develop within the menisci as they are loaded, and compromising the normal biomechanical function of the meniscal tissue. (**a**) meniscal representation showing the circumferential fibres; (**b**) longitudinal tear; (**c**) radial tear

associated with rupture of the ACL and located in the posterior horn of the lateral meniscus. The lesion of the horn fibres, which warrant the hoop strength resistant, will finally result in extrusion of menisci and loss of their biomechanical properties. A cadaveric study [46] showed that a simulated lateral meniscus posterior root avulsion generated a significant decrease of contact area and an increase in contact pressure at the lateral knee compartment between the knee full extension and 90° flexion, compared with the intact status. These findings are in line with the results reported in another cadaveric study in which the avulsion of the posterior root of the lateral meniscus or an adjacent radial tear resulted in significantly decreased contact area and increased mean and peak contact pressures in the lateral compartment, across all the knee flexion angles except at 0° [47]. Another biomechanical study investigated the changes in tibiofemoral contact pressure and area after a simulated medial meniscus root avulsion and posterior horn radial lesions at varying distances from the root attachment site found that the root avulsion and all the radial tears resulted in a significantly decreased contact area and increased mean contact pressure compared with the intact state [48]. Furthermore, the authors reported that the derangement in medial compartment biomechanics caused by the medial posterior root avulsion was similar to that caused by the medial posterior horn radial tears in load-bearing terms [48]. In a finite element analysis of the effects of the longitudinal tears affecting the anterior and the posterior horns of the menisci during a static stance and a slight flexion simulation [49], the longitudinal tear of the posterior horn of the medial meniscus was found to be the most relevant: the peak compressive and the shear stress of the meniscus, and the load on the subchondral bone and cartilage of the femur and tibia was significantly increased compared with the absence of tear status. Furthermore, the meniscus was significantly more extruded under both simulations when the longitudinal tear occurred at the posterior horn of the medial meniscus, compared to that of the lateral meniscus [49].

Posterior medial menisco-capsular tears. The tears of the peripheral attachment of the posterior horn of the medial meniscus of less than 25 mm in length have been defined as ramp lesions and are frequently associated with injuries to the menisco-tibial ligament [50]. The interest in the postero-medial menisco-capsular lesions increased since they have been identified in up to 30% of patients with ACL injury [51] and they have been associated with significant changes in the kinematics of both ACL-deficient and ACL-reconstructed knee [52]. On the other hand, regarding the effect of the ramp lesion on the medial compartment bony contact forces and the in situ forces in the ACL under 134 N anterior load and 200 N compression, no significant differences were found in a recent cadaveric study between the simulated ramp tear status and the meniscus intact status [50].

Bucket-handle tears. In regard to lesions in which a higher amount of tissue is involved, the bucket-handle tears represent the displacement of the inner fragment of a longitudinal tear in the intercondylar notch, and, since they lead to knee locking or catching, they necessitate surgical treatment [53]. Whenever possible, these lesions should be reduced and repaired, because they represent a big part of the meniscal tissue, and a failure to treat or treatment with a medial meniscectomy can result in the early progression of osteoarthritis. In a cadaveric study [54] in which the effect of different treatment for a medial bucket-handle tear in the setting of ACL reconstruction was investigated, a significant increase of the medial average contact pressure was reported in the bucket tear status compared with the meniscus intact status between 30° and 60° of flexion under 1000-N axial load condition [54].

27.4 Biomechanical Effects of Meniscectomy

The fundamental biomechanical role of menisci is underlined by the well-established degenerative effects of the meniscectomy on the articular surface [43]. Although in the past 30 years the approach to the treatment of meniscal tears has

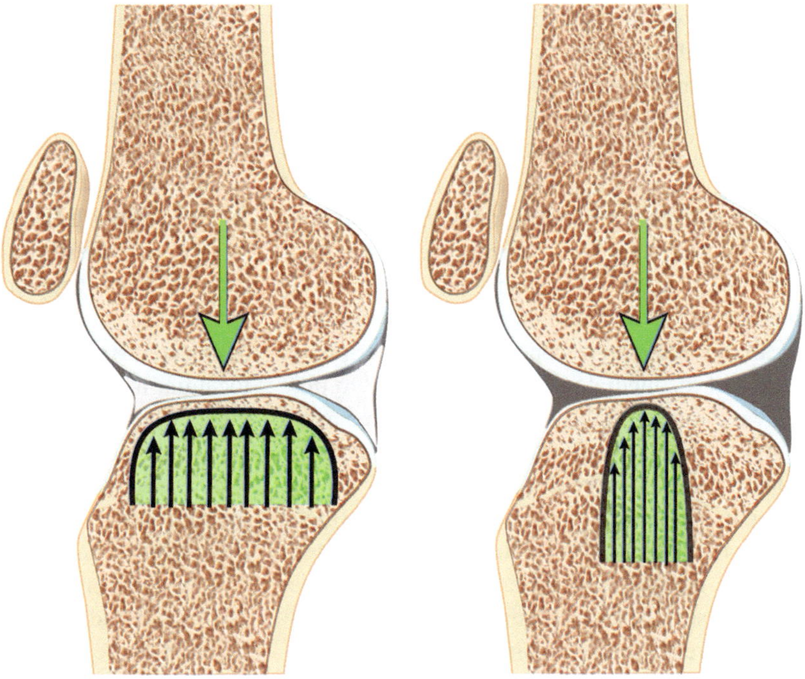

Fig. 27.3 Effect of meniscectomy on bony contact force. The loss of meniscal tissue leads to a decrease of contact area associated with a focal increase of peak contact pressure

shifted to the preservation of the greatest amount of the meniscus possible, meniscectomy is still a commonly performed procedure, because of a quick return to competition time, good-to-excellent short-term results and severity of meniscal tears [55].

Meniscectomy leads to a decrease in the joint contact area, thus significantly increasing the stress acting on the joint surfaces [20, 56] (Fig. 27.3). The increase of the contact stresses after meniscectomy has been associated with overloading of the articular cartilage which produces changes in material properties, such as loss and disaggregation of proteoglycan, increase in the synthesis of proteoglycan and increase in hydration. These impairments lead to failure of articular cartilage, ranging from fibrillation of the surface to necrosis and osteochondral tears with loss of the cartilage layer and significant functional impairment [43]. Regarding radiological assessment, several studies reported degenerative articular changes after meniscectomy [43]. In one of the more comprehensive studies present in literature, 107 of 123 patients were analysed at 21 years follow-up, reporting a relative risk for more advanced osteoarthritic radiological changes of 14.0 (95% CI 3.5 to 121.2) [57]. In an analysis of clinical and radiological results after ACL reconstruction at minimum-20-years follow-up, the concomitant medial meniscectomy resulted in a significant narrowing of medial joint space and presence of radiological osteoarthritis signs [58].

Several cadaveric studies showed a significant decrease of contact area associated with the increase of stress concentration and mean contact pressure in the joint compartment after meniscectomy [59, 60]. Because of the clear anatomical differences between the medial and lateral compartments of the knee and the corresponding meniscus, it is implied that the lateral meniscectomy would present a greater risk of osteoarthritis development than the medial one [22]. In particular, the absence of the lateral meniscus results in an increase in the peak contact pressure and a greater tendency towards point loading compared to the medial compartment, due to the lower degree of congruity between the articular surfaces [43].

Regarding the differences in biomechanical effects between partial and total meniscectomy, a biomechanical study [61] in which an axial load of 1800 N was applied with the knee flexed at 0°, 30° and 60°, showed that a progressive medial meniscectomy resulted in a significant decrease of the contact area and in a significant increase of both mean and peak contact stress in comparison with the intact knee. However, the amount of incremental changes in the contact area and contact stresses suggested the presence of a nonlinear relationship between the degree of meniscectomy and the alteration of the contact area and the mean contact stress: peripheral region of the medial meniscus was found to play a greater role in modifying contact area and mean contact stress. By contrast, the medial peak of contact stress increased proportionally to the amount of meniscus removed, demonstrating a linear relationship [61]. The linear correlation between the increase of peak stress on the tibial articular surface and the amount of meniscal tissue removed was also reported in previous cadaveric studies [60, 62]. In a clinical and radiological evaluation at mean 10.3 years follow-up, no significant correlation was found between the amount of tissue resected and the subjective, clinical and radiological score. Furthermore, the patient who underwent lateral meniscectomy presented a high incidence of degenerative changes coupled with a high rate of reoperation and a relatively low functional outcome score [44].

Several biomechanical studies reported the potential of medial meniscectomy to develop degenerative osteoarthritis in both the medial and lateral compartment. In a three-dimensional finite element model analysis, the peak contact pressure in the lateral meniscus was found to be increased significantly after total medial meniscectomy [63]. In a cadaveric study with a setting of ACL-reconstructed knee, the medial meniscectomy for a bucket-handle tear resulted in a significantly increase in mean and peak pressure of both medial and lateral compartments across all tested flexion angles, compared with meniscal repair [54]. These findings emphasized the negative effect of a medial meniscectomy on the whole joint, and not just on the medial compartment.

Considering the alteration of knee articular contact mechanics and the degenerative consequences of meniscectomy, this procedure, whenever possible, should be avoided. By contrast, meniscus-sparing or replacing treatment strategies should be performed in order to preserve the native tibiofemoral biomechanics of the knee.

27.5 Biomechanical Effects of Meniscal Repair and Replacing Strategies

Meniscus Repair. Depending on its size and location, a meniscal injury that does not require meniscectomy should be repaired in order to avoid the biomechanical effects of the meniscal lesions such as the increase of contact pressure, the meniscal extrusion and the kinematics deleterious consequences.

In a cadaveric study, in which the effect of a 90%-width radial tear of the body-posterior horn junction of the lateral meniscus was investigated, the inside-out repair resulted in a significant reduction of the contact pressure compared to the radial tear status, while no significant difference was found in the contact area between repair status and the radial tear status [45].

Different biomechanical studies investigated the effects of the repair treatment for the meniscal root tears, with regard to the restoration of the meniscal anchor point and the native articular load distribution. In a porcine model, the nonanatomic positioning of the root attachment resulted in a significant effect on meniscus hoop tension [64]; furthermore, the lower levels of meniscus hoop tension caused increased local stress and an increased cartilage deformation [64]. In a cadaveric study [48], a significant increase in medial contact area resulted after an in situ pull-out repairs for root avulsion and radial tear of the posterior horn of the medial meniscus compared with the corresponding tear condition. In the same study, the authors reported no significant differences between the average contact pressure of the in situ pull-out repair of the root avulsion and radial tears versus the intact meniscus status [48]. A significant improvement of the joint contact pressure after in situ pull-out repair was also found in a study investigating the effect of posterior horn and root avulsion of the lateral meniscus [47]. In another biomechanical study, in which the effect of transosseus repair of a lateral meniscus posterior root was investigated, the contact area was partially recovered and the mean contact pressure significantly decreases after the repair at all flexion angles compared with the injured and intact status [46]. Moreover, the authors reported that the subsequent total lateral meniscectomy produced a significant reduction of the contact area in the injured compartment at 90° of flexion and a significant increase of mean pressure at 0° of flexion compared with the torn status, showing that the meniscectomy leads to a higher alteration in contact mechanics compared to the root avulsion [46]. These findings suggested that, while the meniscectomy causes greater disorders than the avulsion left in situ, the repair of meniscal root tears restores the mechanical alterations, in particular between 0° and 60° of knee flexion.

Regarding the effect of ramp lesions repair on knee mechanics, clear evidence still lacks. In a cadaveric study, no significant differences with regard to medial compartment bony contact forces and in situ force in ACL were reported between the repair status and the injured and normal knee status. The authors of the latter study, considering the negligible effects also on knee kinematics of the ACL-intact knee after a simulated ramp tear and its subsequent repair, suggested that the indication for the repair of this kind of lesion may be limited [50].

In a cutting study, in which the effect of the medial bucket-handle tear in the setting of ACL reconstruction was investigated, no significant differences were found between the intact and the bucket-handle repair states at any knee flexion angle with regard to the medial contact pressure. Moreover, the authors of the study showed that the bucket-handle repaired status presented a significantly decreased medial contact pressure compared with the meniscectomy status between 30° and 60° of knee flexion [54].

However, despite recent advances, a large proportion of meniscal tears remain irreparable, and partial or sub-total meniscectomy is often necessary, regardless of the recognized consequences. Complex lesions and large defects that require partial meniscectomy should be addressed using replacing strategies such as scaffold or allograft, with the purpose to avoid the long-term sequelae of meniscal injury and meniscectomy.

Meniscus scaffold. The concept of meniscal scaffold was introduced in the early nineties to prevent or delay the deleterious effects of meniscal deficiency. The rationale behind the use of a meniscal scaffold was to replace meniscal deficiency with a three-dimensional structure capable of supporting the production of a meniscus-like fibrocartilaginous tissue. In vitro and animal studies of collagen and synthetic-based meniscal scaffold showed healing of the implant with regenerated tissue at the host meniscus rim, meniscus-like tissue ingrowth [65, 66]. In clinical studies good-to-excellent results have been reported, ranging from 70 to 90% [67], and the scaffolds have been reported to undergo substitution with a meniscus-like tissue with the potential of chondroprotection. Despite the clinical benefits of this scaffold, further long-term studies are needed to confirm outcome over time and the protective effects on articular cartilage [68].

Meniscus allograft transplantation (MAT). MAT has been reported to be effective in the treatment of meniscus injury, and to partially restore the biomechanical function of the knee after meniscectomy [69]. An in vitro study [59] demonstrated that joint contact pressure in the knee with the previous medial meniscectomy was significantly higher compared with the MAT status, especially at 30° and 60° of knee flexion. In line with these findings, another in vitro study [70] showed that joint contact pressure after MAT was close to the ones in the intact knee, after being significantly risen in knees with meniscectomy. These results confirmed the potential chondroprotective effect of MAT in knee osteoarthritis.

27.6 Meniscus Role in Knee Kinematics

The menisci play a synergic role with the ACL, resulting key structures in joint stability and kinematics. Indeed, while the ACL represents the main component to control antero-posterior and rotatory knee laxity, the medial and lateral menisci contribute to knee stability, acting as secondary restraints for anterior tibial displacement and rotations, respectively.

Several biomechanical studies investigated the role of the menisci in the kinematics of the knee in different conditions with regard to ACL status, and, recently, increased the interests in kinematics consequences of the typical meniscal tears associated with ACL-injuries as menisco-capsular and posterior horn root lesions.

27.6.1 ACL-Intact Knee

The medial translation of the tibia after complete radial tears in the lateral meniscus has been demonstrated in porcine knees [71]. This result was in line with the findings reported in a cadaveric study in which, in response to coupled internal and valgus tibial torque, one-third partial lateral meniscectomy of the posterior horn significantly increased the medial tibial translation in the ACL-intact knee [72]. In the same study, the total lateral meniscectomy led to a further significant increase in medial translation of the tibia compared with the intact status [72]. In an in vivo preliminary kinematics evaluation of the biomechanical role of meniscus allograft transplantation (MAT), in a patient with intact ACL and previous sub-total lateral meniscectomy, the authors reported a decrease of knee laxity after lateral MAT was performed, compared with the meniscus-deficient status, and in particular of anterior tibial translation (ATT) at 30° (from 4.5 mm to 2.5 mm) and 90° of knee flexion (from 3 mm to 1,5 mm), internal-external rotation at 30° (from 21° to 13°) and 90° (from 23.5° to 17.5°), varus-valgus stress at 0° (from 2° to 1°) and 30° (from 3° to 1.8°) [73]. These results suggest that lateral meniscus plays a not neglectable role in the stability of the intact-ACL knee and that surgeons should always consider repairing or replacing strategies for the meniscus, such as allograft transplantation, with the aim to fully restore the native knee kinematics.

On the other hand, in a cadaveric evaluation of the effect of a medial meniscus posterior menisco-capsular tear of less than 25 mm (ramp tear) in ACL-intact knee kinematics, in response to 90-N anterior load and coupled rotation torques and axial load, no significant differences were found with respect to knee kinematics between meniscus intact status and ramp tear status [50].

27.6.2 ACL-Deficient Knee

It has been shown that meniscus plays a greater role in contributing to the stability in an ACL-deficient knee than in an ACL-intact knee [74, 75]. Several cadaveric studies investigated the effects on knee laxity of meniscectomy and meniscus tears in the ACL-deficient knee setting (Tables 27.2 and 27.3).

A recent systematic review reported that medial and lateral meniscus act differently in providing secondary stability in the ACL-deficient knee: while the medial meniscus is more important than the lateral meniscus in restraining uniplanar anterior load of the tibia, the lateral meniscus represents a critical secondary stabilizer of the knee under combined rotatory loads [76]. Different cadaveric studies reported that the progressive medial meniscal resection led to an increase of the ATT, with the most significant effect when the posterior horn was involved [77–79]. On the other hand, the medial meniscal resection resulted to be not associated with changes in internal and external rotation, with the exception of the posterior menisco-capsular tears such as the ramp lesions: in two cadaveric studies a significant increase of internal and external tibial rotation after a simulated ramp tear was reported, with a significant decrease of rotatory laxity after that the repair of the simulated lesions was performed [52, 80]. These findings are in line with the results of a recent in vivo pre-operative evaluation of 275 patients with an ACL injury, in which a higher prevalence of grade III pivot shift test was founded in the patient with an associated ramp tear compared with isolated ACL injury patients [81].

Regarding the lateral meniscus, while no significant differences were reported in ATT of

Table 27.2 Cadaveric studies about medial meniscus role in the ACL-deficient knee kinematics

Study	Year	Specimen number	Type of tear	Test	Force	Kinematics evaluation system
Levy et al. [19]	1982	8 knees	TSM	att 0-90°	125 N ap	Dynamic knee-testing servohydraulc testing machine
Shoemaker et al. [20]	1986	11 knees	removed BH tear; TSM; DM	att 20°	(0.320, 925 N al) + 200, 100, 50 N ap	Dynamic knee-testing servohydraulc testing machine
Bonnin et al. [21]	1996	12 knees	PRL; TSM	att 5-60°	300 N al	3D Lateral knee x-ray based model
Allen et al. [22]	2000	10 knees	TSM	att 0-90°; ier 0-90°	134-N ap	Robotic/universal force moment sensor testing system
Seon et al. [23]	2009	8 knees	TSM	att 0, 15, 30, 60, 90°	130 N ap; 400N ql	Robotic/universal force moment sensor testing system
Musahl et al. [24]	2010	8 knees	TSM	att 30°; ps 0-30°	68-N ap; mechanized ps	Surgical navigation system
Ahn et al. [25]	2011	10 knees	PH tear; TSM	att 0-90°; ier 0-90°; ps 15-30°	(200-N al) + 134 N ap; 10Nm vs +5Nm irt	Instron testing machine with two optical encoders
McCulloch et al. [26]	2013	6 knees	PH tear; TSM; PRL	att 30,60°; ps 30, 60°	134 N ap; 10 Nm vs +5 Nm irt	Surgical navigation system
Lorbach et al. [27]	2015	12 knees	PH tear	att 30,90°, ps 0-30°	134 N ap; 10 Nm vs + 4Nm irt	Robotic/universal force moment sensor testing system
Lorbach et al. [28]	2015	18 knees	BH tear; removed BH tear	att 30,90°, ps 0-30°	134 N ap; 10 Nm vs + 4 Nm irt	Robotic/universal force moment sensor testing system
Peltier et al. [16]	2015	10 knees	menisco-capsular (ramp) tear	att 0-90°, ier 0-90°	134 N ap; 5 N iert	Surgical navigation system
Stephen et al. [11]	2016	9 knees	menisco-capsular (ramp) tear	att 0-100°, ier 0-100°, att + er 0-100°	90 N ap; 5Nm iert; 90 N ap + 5Nm ert	Surgical navigation system
DePhillipo et al. [12]	2018	12 knees	menisco-capsular (ramp) tear	att 30, 90°, ier 0-90°, ps 15, 30°	(10 N al) + 88N ap; 5 Nm ier; 10 N vs + 5Nm irt	Robotic/universal force moment sensor testing system

TSM total/sub-total meniscectomy, *BH* bucket handle, *DM* double meniscectomy, *PH* posterior horn, *PRL* posterior root lesion, *att* anterior tibial translation, *ier* internal-external rotation, *ir* internal rotation, *ps* pivot shift, *ap* anterior tibial load, *irt* internal rotation torque, *ert* external rotation torque, *vs* valgo stress, *vv* varo-valgo stress, *al* axial load, *ql* quadriceps load. Modified by Grassi et al. [76]

ACL-deficient knee specimens after posterior horn tears or lateral meniscectomy, a critical role was showed in control of the rotational and dynamic knee laxity. The lateral meniscus posterior root lesion resulted in a significant increase of anterior tibial translation under a simulated pivot shift test [82, 83] and in a significant increase of internal tibial rotation [84]. The significant increase of pivot shift was reported also in a cadaveric study in which the analysis of the effect of a total lateral meniscectomy in an ACL-deficient knee specimen was provided [85].

The different role of medial and lateral meniscus in control of the ACL-deficient knee laxities was confirmed in an in vivo study in which analysis of the knee kinematics was performed with a surgical navigation system (Fig. 27.4) in patients who underwent ACL reconstruction: the authors found significantly higher ATT at 30° and 90° of knee flexion before the ACL reconstruction in the

Table 27.3 Cadaveric studies about lateral meniscus role in the ACL-deficient knee kinematics

Study	Year	Specimen number	Type of tear	Test	force	Kinematics evaluation system
Levy et al. [30]	1989	11	TSM; BM	att 0-90°	125 N ap	Dynamic knee-testing servohydraulc testing machine
Musahl et al. [24]	2010	8	TSM; DM	att 30°; ps 0-30°	68-N ap; mechanized ps	Surgical navigation system
Shybut et al. [29]	2015	8	Posterior root tear	att 15-90°; ps 15-90°	90 N ap; vs 5,7N + irt 1,2,3 Nm + IT 50-175N	3D CT scan-based model of computer controlled loading system
Lording et al. [13]	2017	16	Posterior root tear	att 0-90°; ir 0-90°	5N irt	Surgical navigation system
Frank et al. [15]	2017	19	Posterior root tear	att 0-90°; ir 0-90°; ps 0-90°	(10 N al) + 88N ap; 5N irt; 10N vs +5Nm irt	Robotic/universal force moment sensor testing system
Forkel et al. [14]	2018	8	Posterior root tear	att 20°; ir 0°-90°	(20N al) + 50N ap; 5N irt	Robotic/universal force moment sensor testing system

TSM total/sub-total meniscectomy, *BH* bucket handle, *DM* double meniscectomy, *PH* posterior horn, *PRL* posterior root lesion, *att* anterior tibial translation, *ier* internal-external rotation, *ir* internal rotation, *ps* pivot shift, *ap* anterior tibial load, *irt* internal rotation torque, *ert* external rotation torque, *vs* valgo stress, *vv* varo-valgo stress, *al* axial load, *ql* quadriceps load. Modified by Grassi et al. [76]

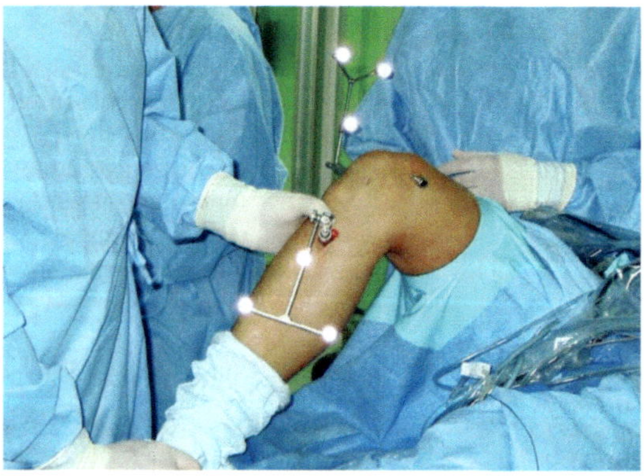

Fig. 27.4 Intraoperative assessment of knee kinematics. Surgical navigation represents the gold standard for knee laxity assessment

patients who underwent medial meniscectomy compared with the intact menisci status; the dynamic laxity evaluated through the pivot shift, on the other hand, significantly increased in the presence of lateral meniscectomy [86].

27.6.3 ACL-Reconstructed Knee

In a cadaveric study investigating the role of medial meniscectomy on knee kinematics in the setting of ACL reconstruction, the ATT in knee specimens with sub-total medial meniscectomy after ACL reconstruction was higher than the intact knee [87]. Furthermore, among knee specimens of another study, no significant differences in knee kinematics were found between ACL-reconstructed knee with medial meniscus tear repair and the intact status [88]. The significant role of medial meniscus in the ACL-reconstructed knee was confirmed in an in vivo kinematic evaluation in which a residual ATT at 90° of flexion was reported in patients with concomitant partial medial meniscectomy with respect to those with an intact meniscus [86].

A biomechanical study among thirteen knee specimens reported that, in the ACL-reconstructed knee, a root tear of the lateral meniscus significantly increased knee laxity under anterior loading. Furthermore, the authors showed that the repair of the root lesion improved knee stability under anterior tibial and simulated pivot shift loading [89].

In conclusion, the critical role of menisci in contributing to knee stability, in particular in ACL-deficient and ACL-reconstructed knee, should suggest that the preservation of the greatest amount of meniscal tissue possible is mandatory in order to better restore knee kinematics and prevent possible ACL graft overload and failure in the setting of ACL reconstruction [83, 87].

27.7 Conclusion

Medial and lateral meniscus played a critical role in the biomechanics of the knee, in particular in load distribution and knee kinematics, protecting the joint from degenerative process.

Meniscal injuries are common in sports patients, and the tear and loss of meniscal tissue leads to cartilage damage, impairment in knee kinematics and onset of osteoarthritis. The approach to the treatment of meniscal tears and pathologies must be aimed to preservation of the greatest amount of meniscus possible in order to fully restore knee biomechanics and prevent early degenerative changes.

Acknowledgments Authors would like to express special thanks to Ms. Silvia Bassini for the pictures and graphic support.

References

1. Bloecker K, Englund M, Wirth W, Hudelmaier M, Burgkart R, Frobell RB, et al. Revision 1 size and position of the healthy meniscus, and its correlation with sex, height, weight, and bone area- a cross-sectional study. BMC Musculoskelet Disord. 2011;12:248.
2. Clark CR, Ogden JA. Development of the menisci of the human knee joint. Morphological changes and their potential role in childhood meniscal injury. J Bone Joint Surg Am. 1983;65(4):538–47.
3. McDermott ID, Sharifi F, Bull AM, Gupte CM, Thomas RW, Amis AA. An anatomical study of meniscal allograft sizing. Knee Surg Sports Traumatol Arthrosc. 2004;12(2):130–5.
4. Beeler S, Vlachopoulos L, Jud L, Sutter R, Fürnstahl P, Fucentese SF. Contralateral MRI scan can be used reliably for three-dimensional meniscus sizing - retrospective analysis of 160 healthy menisci. Knee. 2019;26(5):954–61.
5. Proffen BL, McElfresh M, Fleming BC, Murray MM. A comparative anatomical study of the human knee and six animal species. Knee. 2012;19(4):493–9.
6. Halewood C, Masouros S, Amis AA. Structure and function of the menisci. In: Getgood A, Spalding T, Cole B, Gersoff W, Verdonk P, editors. Meniscal allograft transplantation a comprehensive review. Guilford: DJO publications; 2015. p. 19–33.
7. Moyer JT, Abraham AC, Haut Donahue TL. Nanoindentation of human meniscal surfaces. J Biomech. 2012;45(13):2230–5.
8. Joshi MD, Suh JK, Marui T, Woo SL. Interspecies variation of compressive biomechanical properties of the meniscus. J Biomed Mater Res. 1995;29(7):823–8.
9. Sweigart MA, Zhu CF, Burt DM, DeHoll PD, Agrawal CM, Clanton TO, et al. Intraspecies and interspecies comparison of the compressive properties of the medial meniscus. Ann Biomed Eng. 2004;32(11):1569–79.
10. Kopf S, Colvin AC, Muriuki M, Zhang X, Harner CD. Meniscal root suturing techniques: implications for root fixation. Am J Sports Med. 2011;39(10):2141–6.
11. Ellman MB, LaPrade CM, Smith SD, Rasmussen MT, Engebretsen L, Wijdicks CA, et al. Structural properties of the meniscal roots. Am J Sports Med. 2014;42(8):1881–7.
12. Feucht MJ, Grande E, Brunhuber J, Burgkart R, Imhoff AB, Braun S. Biomechanical evaluation of different suture techniques for arthroscopic transtibial pull-out repair of posterior medial meniscus root tears. Am J Sports Med. 2013;41(12):2784–90.
13. Gupte CM, Bull AM, Thomas RD, Amis AA. The meniscofemoral ligaments: secondary restraints to the posterior drawer. Analysis of anteroposterior and rotary laxity in the intact and posterior-cruciate-deficient knee. J Bone Joint Surg Br. 2003;85(5):765–73.
14. Robinson JR, Bull AM, Thomas RR, Amis AA. The role of the medial collateral ligament and posteromedial capsule in controlling knee laxity. Am J Sports Med. 2006;34(11):1815–23.
15. Poh SY, Yew KS, Wong PL, Koh SB, Chia SL, Fook-Chong S, et al. Role of the anterior intermeniscal ligament in tibiofemoral contact mechanics during axial joint loading. Knee. 2012;19(2):135–9.
16. Martens TA, Hull ML, Howell SM. An in vitro osteotomy method to expose the medial compartment of the human knee. J Biomech Eng. 1997;119(4):379–85.
17. Natsis K, Paraskevas G, Anastasopoulos N, Papamitsou T, Sioga A. Meniscofibular ligament: morphology and functional significance of a relatively unknown anatomical structure. Anat Res Int. 2012;2012:214784.
18. Bourne RB, Finlay JB, Papadopoulos P, Andreae P. The effect of medial meniscectomy on strain distribution in the proximal part of the tibia. J Bone Joint Surg Am. 1984;66(9):1431–7.
19. Walker PS, Hajek JV. The load-bearing area in the knee joint. J Biomech. 1972;5(6):581–9.
20. Krause WR, Pope MH, Johnson RJ, Wilder DG. Mechanical changes in the knee after meniscectomy. J Bone Joint Surg Am. 1976;58(5):599–604.
21. Jones RS, Keene GC, Learmonth DJ, Bickerstaff D, Nawana NS, Costi JJ, et al. Direct measurement of hoop strains in the intact and torn human medial meniscus. Clin Biomech (Bristol, Avon). 1996;11(5):295–300.
22. Halewood C, Amis AA. Physiology: biomechanics. In: Hulet C, Pereira H, Peretti G, Denti M, editors. Surgery of the meniscus. Berlin: Springer; 2016. p. 35–45.
23. Kawahara Y, Uetani M, Fuchi K, Eguchi H, Hayashi K. MR assessment of movement and morphologic

change in the menisci during knee flexion. Acta Radiol. 1999;40(6):610–4.
24. Vedi V, Williams A, Tennant SJ, Spouse E, Hunt DM, Gedroyc WM. Meniscal movement. An in-vivo study using dynamic MRI. J Bone Joint Surg Br. 1999;81(1):37–41.
25. Masouros SD, McDermott ID, Amis AA, Bull AM. Biomechanics of the meniscus-meniscal ligament construct of the knee. Knee Surg Sports Traumatol Arthrosc. 2008;16(12):1121–32.
26. Campbell SE, Sanders TG, Morrison WB. MR imaging of meniscal cysts: incidence, location, and clinical significance. AJR Am J Roentgenol. 2001;177(2):409–13.
27. Drosos GI, Pozo JL. The causes and mechanisms of meniscal injuries in the sporting and non-sporting environment in an unselected population. Knee. 2004;11(2):143–9.
28. Yao J, Lancianese SL, Hovinga KR, Lee J, Lerner AL. Magnetic resonance image analysis of meniscal translation and tibio-menisco-femoral contact in deep knee flexion. J Orthop Res. 2008;26(5):673–84.
29. Doral MN, Bilge O, Huri G, Turhan E, Verdonk R. Modern treatment of meniscal tears. EFORT Open Rev. 2018;3(5):260–8.
30. Snoeker BA, Bakker EW, Kegel CA, Lucas C. Risk factors for meniscal tears: a systematic review including meta-analysis. J Orthop Sports Phys Ther. 2013;43(6):352–67.
31. Baker BE, Peckham AC, Pupparo F, Sanborn JC. Review of meniscal injury and associated sports. Am J Sports Med. 1985;13(1):1–4.
32. Makris EA, Hadidi P, Athanasiou KA. The knee meniscus: structure-function, pathophysiology, current repair techniques, and prospects for regeneration. Biomaterials. 2011;32(30):7411–31.
33. Baker P, Reading I, Cooper C, Coggon D. Knee disorders in the general population and their relation to occupation. Occup Environ Med. 2003;60(10): 794–7.
34. Frizziero A, Ferrari R, Giannotti E, Ferroni C, Poli P, Masiero S. The meniscus tear. State of the art of rehabilitation protocols related to surgical procedures. Muscles Ligaments Tendons J. 2012;2(4):295–301.
35. Pereira H, Varatojo R, Sevivas N, Serratosa L, Ripoll P, Oliveira J, et al. Physiopathology of the Meniscal Lesions. In: Hulet C., Pereira H, Peretti G, Denti M. Surgery of the Meniscus. Berlin: Springer; 2016., editor. Surgery of the Meniscus: Springer; 2016. p. 47–61.
36. Shelbourne KD, Nitz PA. The O'Donoghue triad revisited. Combined knee injuries involving anterior cruciate and medial collateral ligament tears. Am J Sports Med. 1991;19(5):474–7.
37. Anderson AF, Irrgang JJ, Dunn W, Beaufils P, Cohen M, Cole BJ, et al. Interobserver reliability of the International Society of Arthroscopy, knee surgery and Orthopaedic sports medicine (ISAKOS) classification of meniscal tears. Am J Sports Med. 2011;39(5):926–32.
38. Duchman KR, Westermann RW, Spindler KP, Reinke EK, Huston LJ, Amendola A, et al. The fate of meniscus tears left in situ at the time of anterior cruciate ligament reconstruction: a 6-year follow-up study from the MOON cohort. Am J Sports Med. 2015;43(11):2688–95.
39. Shelbourne KD, Heinrich J. The long-term evaluation of lateral meniscus tears left in situ at the time of anterior cruciate ligament reconstruction. Arthroscopy. 2004;20(4):346–51.
40. Yagishita K, Muneta T, Ogiuchi T, Sekiya I, Shinomiya K. Healing potential of meniscal tears without repair in knees with anterior cruciate ligament reconstruction. Am J Sports Med. 2004;32(8):1953–61.
41. Weiss CB, Lundberg M, Hamberg P, DeHaven KE, Gillquist J. Non-operative treatment of meniscal tears. J Bone Joint Surg Am. 1989;71(6):811–22.
42. Fitzgibbons RE, Shelbourne KD. "aggressive" non-treatment of lateral meniscal tears seen during anterior cruciate ligament reconstruction. Am J Sports Med. 1995;23(2):156–9.
43. McDermott ID, Amis AA. The consequences of meniscectomy. J Bone Joint Surg Br. 2006;88(12):1549–56.
44. Hoser C, Fink C, Brown C, Reichkendler M, Hackl W, Bartlett J. Long-term results of arthroscopic partial lateral meniscectomy in knees without associated damage. J Bone Joint Surg Br. 2001;83(4):513–6.
45. Bedi A, Kelly N, Baad M, Fox AJ, Ma Y, Warren RF, et al. Dynamic contact mechanics of radial tears of the lateral meniscus: implications for treatment. Arthroscopy. 2012;28(3):372–81.
46. Perez-Blanca A, Espejo-Baena A, Amat Trujillo D, Prado Nóvoa M, Espejo-Reina A, Quintero López C, et al. Comparative biomechanical study on contact alterations after lateral meniscus posterior root avulsion, Transosseous reinsertion, and Total meniscectomy. Arthroscopy. 2016;32(4):624–33.
47. LaPrade CM, Jansson KS, Dornan G, Smith SD, Wijdicks CA, LaPrade RF. Altered tibiofemoral contact mechanics due to lateral meniscus posterior horn root avulsions and radial tears can be restored with in situ pull-out suture repairs. J Bone Joint Surg Am. 2014;96(6):471–9.
48. Padalecki JR, Jansson KS, Smith SD, Dornan GJ, Pierce CM, Wijdicks CA, et al. Biomechanical consequences of a complete radial tear adjacent to the medial meniscus posterior root attachment site: in situ pull-out repair restores derangement of joint mechanics. Am J Sports Med. 2014;42(3):699–707.
49. Zhang K, Li L, Yang L, Shi J, Zhu L, Liang H, et al. Effect of degenerative and radial tears of the meniscus and resultant meniscectomy on the knee joint: a finite element analysis. J Orthop Translat. 2019;18:20–31.
50. Naendrup JH, Pfeiffer TR, Chan C, Nagai K, Novaretti JV, Sheean AJ, et al. Effect of meniscal ramp lesion repair on knee kinematics, bony contact forces, and in situ forces in the anterior cruciate ligament. Am J Sports Med. 2019;47(13):3195–202.
51. Liu X, Feng H, Zhang H, Hong L, Wang XS, Zhang J. Arthroscopic prevalence of ramp lesion in 868

patients with anterior cruciate ligament injury. Am J Sports Med. 2011;39(4):832–7.
52. Stephen JM, Halewood C, Kittl C, Bollen SR, Williams A, Amis AA. Posteromedial Meniscocapsular lesions increase tibiofemoral joint laxity with anterior cruciate ligament deficiency, and their repair reduces laxity. Am J Sports Med. 2016;44(2):400–8.
53. Denti M, Espregueira-Mendes J, Pereira H, Raoulis V, Hantes M. Traumatic meniscal lesions. In: Hulet C, Pereira H, Peretti G, Denti M, editors. Surgery of the meniscus. Berlin: Springer; 2016. p. 67–78.
54. Logan CA, Aman ZS, Kemler BR, Storaci HW, Dornan GJ, LaPrade RF. Influence of medial meniscus bucket-handle repair in setting of anterior cruciate ligament reconstruction on tibiofemoral contact mechanics: a biomechanical study. Arthroscopy. 2019;35(8):2412–20.
55. Abram SGF, Judge A, Beard DJ, Wilson HA, Price AJ. Temporal trends and regional variation in the rate of arthroscopic knee surgery in England: analysis of over 1.7 million procedures between 1997 and 2017. Has practice changed in response to new evidence? Br J Sports Med. 2019;53(24):1533–8.
56. Baratz ME, Fu FH, Mengato R. Meniscal tears: the effect of meniscectomy and of repair on intraarticular contact areas and stress in the human knee. A preliminary report. Am J Sports Med. 1986;14(4):270–5.
57. Roos H, Laurén M, Adalberth T, Roos EM, Jonsson K, Lohmander LS. Knee osteoarthritis after meniscectomy: prevalence of radiographic changes after twenty-one years, compared with matched controls. Arthritis Rheum. 1998;41(4):687–93.
58. Zaffagnini S, Marcheggiani Muccioli GM, Grassi A. Roberti di Sarsina T, Raggi F, Signorelli C, et al. over-the-top ACL reconstruction plus extra-articular lateral Tenodesis with hamstring tendon grafts: prospective evaluation with 20-year minimum follow-up. Am J Sports Med. 2017;45(14):3233–42.
59. Kim JG, Lee YS, Bae TS, Ha JK, Lee DH, Kim YJ, et al. Tibiofemoral contact mechanics following posterior root of medial meniscus tear, repair, meniscectomy, and allograft transplantation. Knee Surg Sports Traumatol Arthrosc. 2013;21(9):2121–5.
60. Ihn JC, Kim SJ, Park IH. In vitro study of contact area and pressure distribution in the human knee after partial and total meniscectomy. Int Orthop. 1993;17(4):214–8.
61. Lee SJ, Aadalen KJ, Malaviya P, Lorenz EP, Hayden JK, Farr J, et al. Tibiofemoral contact mechanics after serial medial meniscectomies in the human cadaveric knee. Am J Sports Med. 2006;34(8):1334–44.
62. Burke DL AA, Miller J A biomechanical study of partial and total medial meniscectomy of the knee. Trans OrthopRes Soc. 1978;91(3).
63. Bae JY, Park KS, Seon JK, Kwak DS, Jeon I, Song EK. Biomechanical analysis of the effects of medial meniscectomy on degenerative osteoarthritis. Med Biol Eng Comput. 2012;50(1):53–60.
64. Stärke C, Kopf S, Gröbel KH, Becker R. The effect of a nonanatomic repair of the meniscal horn attachment on meniscal tension: a biomechanical study. Arthroscopy. 2010;26(3):358–65.
65. Stone KR, Rodkey WG, Webber R, McKinney L, Steadman JR. Meniscal regeneration with copolymeric collagen scaffolds. In vitro and in vivo studies evaluated clinically, histologically, and biochemically. Am J Sports Med. 1992;20(2):104–11.
66. Maher SA, Rodeo SA, Doty SB, Brophy R, Potter H, Foo LF, et al. Evaluation of a porous polyurethane scaffold in a partial meniscal defect ovine model. Arthroscopy. 2010;26(11):1510–9.
67. Zaffagnini S, Marcheggiani Muccioli GM, Lopomo N, Bruni D, Giordano G, Ravazzolo G, et al. Prospective long-term outcomes of the medial collagen meniscus implant versus partial medial meniscectomy: a minimum 10-year follow-up study. Am J Sports Med. 2011;39(5):977–85.
68. Verdonk R, Verdonk P, Huysse W, Forsyth R, Heinrichs EL. Tissue ingrowth after implantation of a novel, biodegradable polyurethane scaffold for treatment of partial meniscal lesions. Am J Sports Med. 2011;39(4):774–82.
69. Seitz AM, Dürselen L. Biomechanical considerations are crucial for the success of tendon and meniscus allograft integration-a systematic review. Knee Surg Sports Traumatol Arthrosc. 2019;27(6):1708–16.
70. McDermott ID, Lie DT, Edwards A, Bull AM, Amis AA. The effects of lateral meniscal allograft transplantation techniques on tibio-femoral contact pressures. Knee Surg Sports Traumatol Arthrosc. 2008;16(6):553–60.
71. Tachibana Y, Mae T, Fujie H, Shino K, Ohori T, Yoshikawa H, et al. Effect of radial meniscal tear on in situ forces of meniscus and tibiofemoral relationship. Knee Surg Sports Traumatol Arthrosc. 2017;25(2):355–61.
72. Novaretti JV, Lian J, Patel NK, Chan CK, Cohen M, Musahl V, et al. Partial lateral meniscectomy affects knee stability even in anterior cruciate ligament-intact knees. J Bone Joint Surg Am. 2020;102(7):567–73.
73. Zaffagnini S, Di Paolo S, Stefanelli F, Dal Fabbro G, Macchiarola L, Lucidi GA, et al. The biomechanical role of meniscal allograft transplantation and preliminary in-vivo kinematic evaluation. J Exp Orthop. 2019;6(1):27.
74. Allen CR, Wong EK, Livesay GA, Sakane M, Fu FH, Woo SL. Importance of the medial meniscus in the anterior cruciate ligament-deficient knee. J Orthop Res. 2000;18(1):109–15.
75. Levy IM, Torzilli PA, Warren RF. The effect of medial meniscectomy on anterior-posterior motion of the knee. J Bone Joint Surg Am. 1982;64(6):883–8.
76. Grassi A, Dal Fabbro G, Di Paolo S, Stefanelli F, Macchiarola L, Lucidi GA, et al. Medial and lateral meniscus have a different role in kinematics of the ACL-deficient knee: a systematic review. Journal of ISAKOS: Joint Disorders & Orthopaedic Sports Medicine. 2019;4(5):233–41.
77. Ahn JH, Bae TS, Kang KS, Kang SY, Lee SH. Longitudinal tear of the medial meniscus poste-

rior horn in the anterior cruciate ligament-deficient knee significantly influences anterior stability. Am J Sports Med. 2011;39(10):2187–93.
78. McCulloch PC, Shybut TB, Isamaily SK, Durrani S, Gold JE, Noble PC, et al. The effect of progressive degrees of medial meniscal loss on stability after anterior cruciate ligament reconstruction. J Knee Surg. 2013;26(5):363–9.
79. Lorbach O, Kieb M, Herbort M, Weyers I, Raschke M, Engelhardt M. The influence of the medial meniscus in different conditions on anterior tibial translation in the anterior cruciate deficient knee. Int Orthop. 2015;39(4):681–7.
80. DePhillipo NN, Moatshe G, Brady A, Chahla J, Aman ZS, Dornan GJ, et al. Effect of Meniscocapsular and Meniscotibial lesions in ACL-deficient and ACL-reconstructed knees: a biomechanical study. Am J Sports Med. 2018;46(10):2422–31.
81. Mouton C, Magosch A, Pape D, Hoffmann A, Nührenbörger C, Seil R. Ramp lesions of the medial meniscus are associated with a higher grade of dynamic rotatory laxity in ACL-injured patients in comparison to patients with an isolated injury. Knee Surg Sports Traumatol Arthrosc. 2020;28(4):1023–8.
82. Shybut TB, Vega CE, Haddad J, Alexander JW, Gold JE, Noble PC, et al. Effect of lateral meniscal root tear on the stability of the anterior cruciate ligament–deficient knee. Am J Sports Med. 2015;43(4):905–11.
83. Frank JM, Moatshe G, Brady AW, Dornan GJ, Coggins A, Muckenhirn KJ, et al. Lateral meniscus posterior root and Meniscofemoral ligaments as stabilizing structures in the ACL-deficient knee: a biomechanical study. Orthop J Sports Med. 2017;5(6):2325967117695756.
84. Forkel P, von Deimling C, Lacheta L, Imhoff FB, Foehr P, Willinger L, et al. Repair of the lateral posterior meniscal root improves stability in an ACL-deficient knee. Knee Surg Sports Traumatol Arthrosc. 2018;26(8):2302–9.
85. Musahl V, Citak M, O'Loughlin PF, Choi D, Bedi A, Pearle AD. The effect of medial versus lateral meniscectomy on the stability of the anterior cruciate ligament-deficient knee. Am J Sports Med. 2010;38(8):1591–7.
86. Grassi A, Di Paolo S, Lucidi GA, Macchiarola L, Raggi F, Zaffagnini S. The contribution of partial meniscectomy to preoperative laxity and laxity after anatomic single-bundle anterior cruciate ligament reconstruction: in vivo kinematics with navigation. Am J Sports Med. 2019;47(13):3203–11.
87. Seon JK, Gadikota HR, Kozanek M, Oh LS, Gill TJ, Li G. The effect of anterior cruciate ligament reconstruction on kinematics of the knee with combined anterior cruciate ligament injury and subtotal medial meniscectomy: an in vitro robotic investigation. Arthroscopy. 2009;25(2):123–30.
88. Lorbach O, Kieb M, Domnick C, Herbort M, Weyers I, Raschke M, et al. Biomechanical evaluation of knee kinematics after anatomic single- and anatomic double-bundle ACL reconstructions with medial meniscal repair. Knee Surg Sports Traumatol Arthrosc. 2015;23(9):2734–41.
89. Tang X, Marshall B, Wang JH, Zhu J, Li J, Smolinski P, et al. Lateral meniscal posterior root repair with anterior cruciate ligament reconstruction better restores knee stability. Am J Sports Med. 2019;47(1):59–6.

Patellofemoral Biomechanics 28

John J. Elias and S. Cyrus Rezvanifar

The patella is a sesamoid bone located within the quadriceps extensor muscle tendon (Fig. 28.1) [1], which covers and articulates with the anterior surface of the distal femur, known as the intercondylar or trochlear groove, forming the patellofemoral joint. The patella facilitates knee extension by increasing the moment arm of the extensor mechanism [2–4] throughout the range of motion. The axial cross-section of a patella typically resembles a triangle, with a vertical ridge dividing the articular surface of the patella into a medial and a lateral facet. Wiberg [5] conducted one of the earliest studies investigating morphology of the patella, introducing a three-level classification based on the proportion of the width of the lateral to medial facet. Later studies further investigated and compared morphology in normal and pathologic patellofemoral joints [6, 7], laying the groundwork for additional efforts that reported an elevated ratio of lateral facet to medial facet width for pathologic knees [8, 9]. The kinematics of the patellofemoral joint primarily depends upon the forces that are applied on the patella through the quadriceps muscles and patellar tendon, medial and lateral ligaments and retinacular structures, as well as the geometry of the trochlear groove on which the patellofemoral articulation occurs.

The quadriceps muscle group is mainly comprised of the vastus medialis obliquus (VMO), vastus lateralis (VL), vastus intermedius (VI), rectus femoris (RF), and vastus medialis longus (VML). Based on muscle fiber orientations and physiological cross-sectional area (PCSA) reported in the literature [10, 11], VMO has an orientation approximately 47° medial to a proximal-distal axis and applies nearly 10% of the total quadriceps force. In contrast, VL is oriented approximately 25° lateral to a proximal-distal axis and applies nearly 40% of the total quadriceps force. The remaining 50% of the total quadriceps force is applied through the combination of the RF, VI, and VML, with an orientation approximately 5° lateral to a proximal-distal axis. Activation of the quadriceps creates passive loading within the patellar tendon. With the knee extended and a primarily superior-inferior force acting on the patella, the patellar tendon force is similar to the total quadriceps force. As the patella engages with the trochlear groove and the patellar tendon balances the flexion moment applied by the quadriceps muscles, the patellar tendon to quadriceps tendon force ratio increases initially before decreasing to approximately 0.7 by 90° of flexion [12].

J. J. Elias (✉)
Department of Research, Cleveland Clinic Akron General, Akron, OH, USA
e-mail: eliasj@ccf.org

S. C. Rezvanifar
Department of Rehabilitation Medicine, Medical School, University of Minnesota, Minneapolis, MN, USA
e-mail: rezva006@umn.edu

J. Koh et al. (eds.), *Orthopaedic Biomechanics in Sports Medicine*, https://doi.org/10.1007/978-3-030-81549-3_28

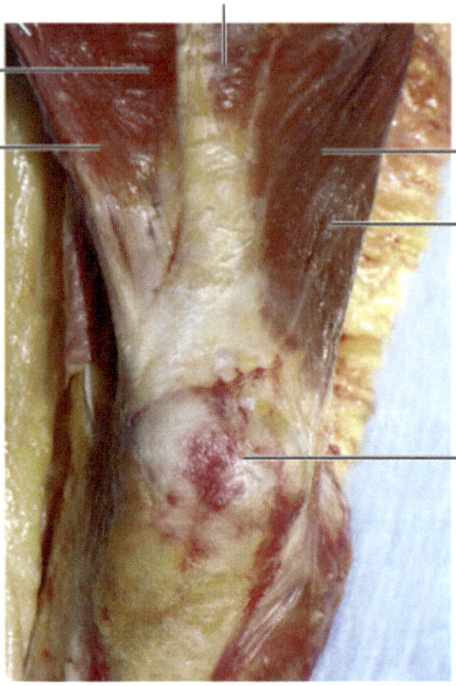

Fig. 28.1 Anatomy of the patella and the surrounding quadriceps muscles. Reprinted with permission from Noyes' Knee Disorders: Surgery, Rehabilitation, Clinical Outcomes, Elsevier Books

Previous studies have also described the characteristics of patellofemoral articular contact through the knee range of motion [13–15]. Based on in vitro simulation of closed-chain knee flexion with physiological quadriceps orientation and forces, normal patellofemoral contact area has been reported to be nearly 2.6 cm^2 (20.5% of total articular area) at 20° of knee flexion angle, increasing to a maximum average of 4.1 cm^2 (32.2% of total articular area) at 90° flexion, with average patellofemoral contact pressure ranging from 2.0 MPa to a maximum of 4.4 MPa within the same range of knee flexion. The steepest increase in patellofemoral contact pressure has been observed to occur from 30° to 60° of flexion, increasing from an average value of 2.4–4.1 MPa [13]. As the knee flexes from 20° to 90° of flexion, contact area on the patellar articular surface moves from the distal third to the proximal half of the cartilage, reaching the proximal margin of the patellar cartilage by 120° of knee flexion [13] (Fig. 28.2). Normal knees with a physiological quadriceps orientation have indicated similar maximum and mean contact pressures for the medial and lateral facets, with a larger contact area for the lateral facet [13, 15–17]. However, pathologic conditions can result in different contact pressure magnitudes and distributions between the two facets. For instance, an increased lateral orientation of the quadriceps or a lateralized tibial tuberosity has been observed to increase contact pressures on the lateral facet and decrease the pressure on the medial facet [15, 18]. In such cases, improving the forces applied by the VMO muscle can enhance the force balance and increase the medial facet contact pressure toward a more uniform pressure distribution [17].

The physiological response to the contact pressure distribution is a cartilage thickness distribution to most effectively dampen pressures where needed on both patellar and femoral counterparts. Many studies have used high-resolution magnetic resonance (MR) images to quantify patellofemoral cartilage thickness [19–23]. Patellar cartilage is reported to have an average thickness of approximately 7 mm in the central part of the medial and lateral facets where the

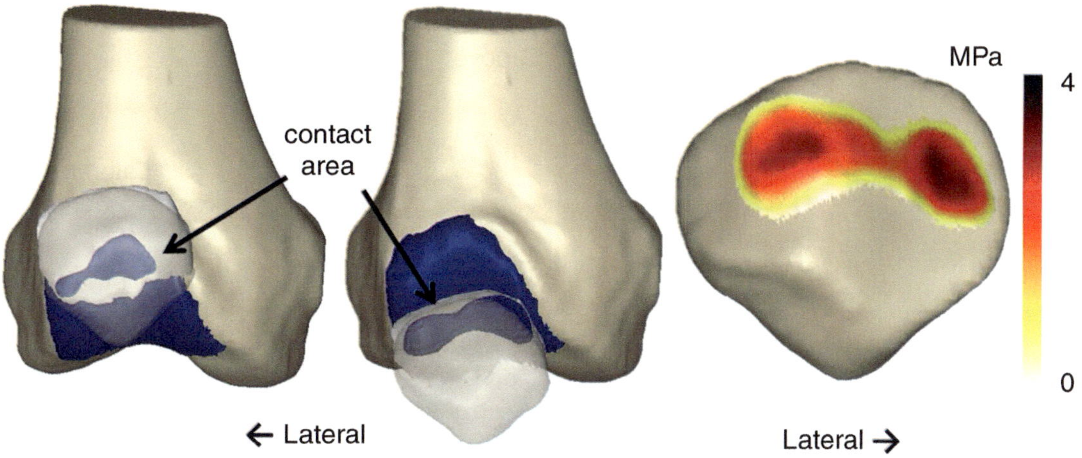

Fig. 28.2 Representation of patellar tracking based on computational reconstruction of in vivo function [101]. Viewing through the semi-transparent patella shows the area of contact as the patella enters the trochlear groove (left) and in deeper flexion (center). The patellofemoral contact pressure distribution is also shown for the deeper flexion angle (right)

main articular contact occurs, tapering down to a thickness of nearly 1 mm in the borders [19, 23]. Cartilage thickness on the femoral side is roughly 3 mm throughout the trochlear groove, increasing to an approximate thickness of 4.5 mm within the deepest part of the groove [23].

28.1 Patellar Tracking

Patellofemoral disorders are typically related to patellar malalignment and maltracking. Pathologic patellofemoral tracking is commonly observed within young, active patients, with a predominant occurrence among females [24]. The normal valgus orientation of the knee causes a lateral orientation in the resultant force acting on the patella from the quadriceps muscles and patellar tendon. This lateral force component can be elevated by conditions such as genu valgum, femoral anteversion, and tibial torsion. External rotation of the tibia related to the screw-home mechanism, as well as internal rotation of the femur associated with weak hip external rotators can also contribute to an elevated lateral force acting on the patella. Hence, active and passive anatomical components including articular constraints, soft tissues, and muscle forces are required to maintain a stable mediolateral force balance and ensure nonpathologic patellar tracking.

A sufficiently deep trochlear groove provides significant bony containment for normal patellar tracking. The patella engages with the trochlear groove as the knee flexes, typically becoming fully engaged by 30° of knee flexion. When the patella is engaged with the trochlear groove, the osteochondral constraint is the primary stabilizer of the patellofemoral joint. For knees with normal anatomy, the mean depth of the trochlea is 4.0 mm, with the average value slightly greater in males (4.2 mm) compared to females (3.4 mm) [25]. Depth of the trochlear groove can also be expressed in terms of the orientation of the lateral ridge of the groove with respect to the posterior condylar axis, also known as lateral trochlear inclination (Fig. 28.3). The lateral trochlear inclination is typically measured from MRI at the proximal aspect of the groove, with a value less than 11° considered to represent a shallow groove [26].

Patellar lateral maltracking is clinically expressed with respect to the trochlear groove. Bisect offset index (portion of patellar width lateral to the deepest point of the trochlear groove) and patellar lateral tilt (angle between the

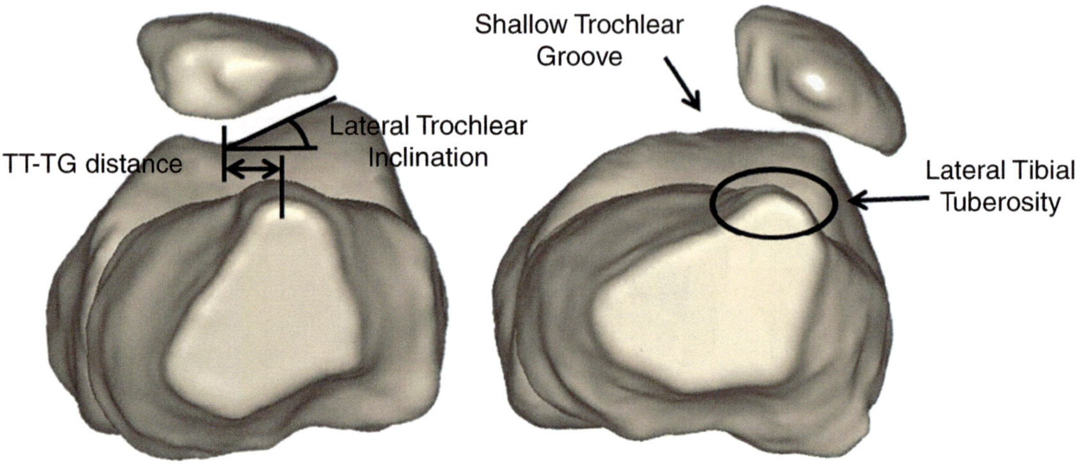

Fig. 28.3 Patellar tracking for a knee with a normal position of the tibial tuberosity and normal trochlear groove (left) and for a knee with a lateral position of the tibial tuberosity and shallow trochlear groove (right). The tibial tuberosity to trochlear groove (TT-TG) distance and the lateral trochlear inclination are also shown

mediolateral axis of the patella and the femoral posterior condylar axis) are common measures to quantify patellar tracking with respect to the femur (Fig. 28.4). Patellar tracking is considered to be normal with a bisect offset index and patellar lateral tilt up to 0.75 [27] and 15° [28], respectively.

The medial patellofemoral ligament (MPFL) is a fan-shaped ligament connecting the medial facet of the patella to the medial femoral condyle (Fig. 28.5). Based on cadaveric studies applying an isolated lateral force to the patella, the MPFL provides nearly 60% of the medial soft tissue restraint against excessive patellar lateral translation. Thus, the MPFL is considered the primary passive stabilizer against patellar lateral maltracking [29–31]. Additional fibers extend to the quadriceps tendon, and the combination of soft tissues is commonly referred to as the medial patellofemoral complex (MPFC). The femoral origin of the MPFL is located approximately 10 mm distal and 10 mm anterior to the adductor tubercle [32]. The MPFL merges with the attachment of the vastus medialis obliquus (VMO) on the patella and extends to the medial border of the patella and vastus intermedius tendon. The average width at the patella insertion is 30 mm. The fibers attach to the patella and the quadriceps tendon, with a slight majority attached to the bone [33]. Tension

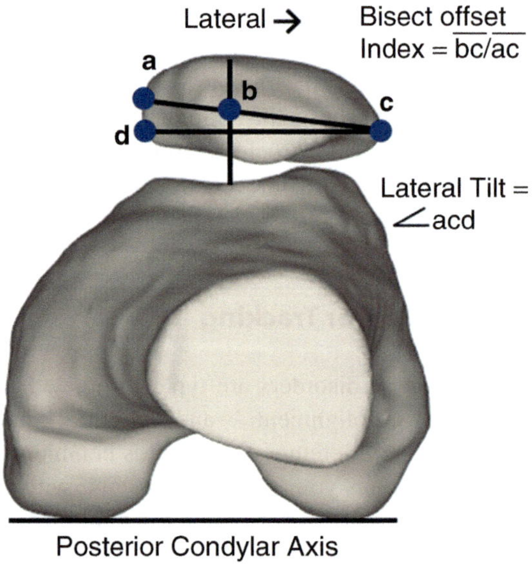

Fig. 28.4 Bisect offset index and patellar lateral tilt used to characterize patellar tracking. Bisect offset index is the length of the line between points b and c divided by the length of the line between points a and c. Lateral tilt is the angle between the line connecting points a and c and the line connecting points d and c

within the MPFL produces a force with primary components in a medial and posterior direction. The MPFL is primarily under tension when the patella is proximal to the trochlear groove at low flexion angles. The path length from the femoral

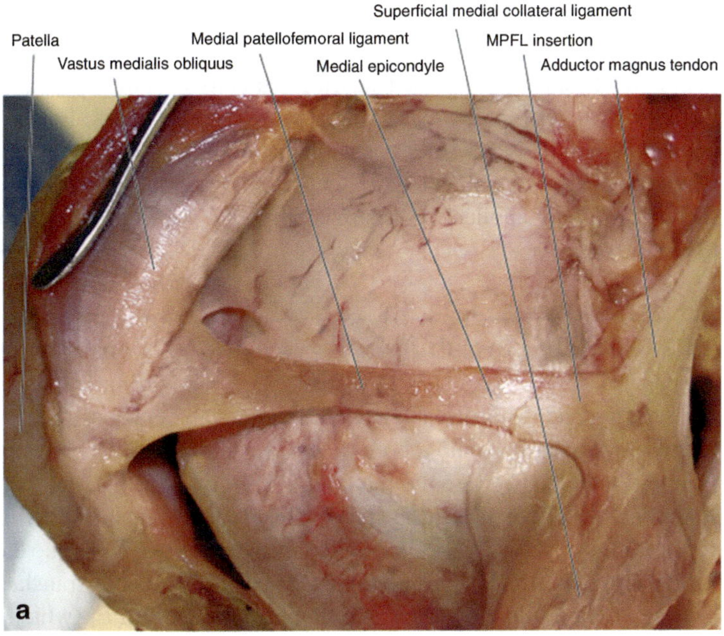

Fig. 28.5 Medial aspect of the patella, highlighting the medial patellofemoral ligament and vastus medialis obliquus. Reprinted with permission from Noyes' Knee Disorders: Surgery, Rehabilitation, Clinical Outcomes, Elsevier Books

attachment to the patellar attachment has been shown to peak at low flexion angles and decrease as the patella becomes engaged with the trochlear groove [34–37]. The average ultimate strength of the MPFL has been measured to be 178 N [38]. The medial patellotibial ligament (MPTL) and medial patellomeniscal ligament (MPML) also act as passive stabilizers, attaching to the distal inferomedial patella. These two ligaments are primarily under tension when the MPFL is slack with the knee flexed [39].

The primary active component of the quadriceps that resists excessive lateral patellar translation is the vastus medialis obliquus (VMO). Previous studies have provided evidence on the role of the vastus medialis muscle in resisting excessive patellar lateral tracking and reducing the pressure applied to the lateral cartilage [17, 40–44]. The VMO attaches on the proximal-medial surface of the patella, with a medial orientation that provides resistance to lateral patellar translation. The VMO also includes a posterior component that provides resistance to lateral tilting of the patella. Both elevated lateral shift and tilt increase compression of the cartilage on the lateral facet of the patella with the cartilage on the lateral ridge of the trochlear groove.

In vitro studies have indicated weakness of the VMO increases the maximum pressure applied to cartilage on the lateral facet of the patella and decreases pressure applied to cartilage on the medial facet at multiple flexion angles [17], with the increased pressure attributed to increased lateral shift and tilt of the patella. Administering a motor branch block to the VMO has been shown to increase lateral patellar shift during knee extension as measured by dynamic cine phase-contrast MRI [45]. Prospective studies have identified delayed activation of the VMO with respect to the vastus lateralis as a source contributing to patellofemoral pain [46]. Lateral patellar translation and tilt have been correlated with a delay in the activation of the VMO in patients with pain [46].

Overloading of the patellofemoral cartilage can be a source of pain at the pain receptors of the subchondral bone [47]. Continued overloading during function can also contribute to accumulating cartilage degradation and development of patellofemoral osteoarthritis (OA). Patellofemoral pain has been shown to be correlated with patellofemoral OA, with an elevated risk of OA due to adolescent anterior knee pain noted to be 7.5 [48]. Overloading of patellofemoral cartilage can lead to degradation that accelerates continued degeneration. As cartilage is overloaded and begins to degenerate, the load bearing capacity decreases, thereby reducing the area absorbing contact forces. In vitro work has shown that areas of cartilage degradation on the lateral cartilage increase pressure applied to the surrounding cartilage [49].

28.2 Patellar Instability

Subluxation and dislocation episodes are characterized as patellar instability, both of which cause pain and apprehension within the knee and limit functional activities. Patellar instability accounts for more than 10% of the office visits to musculoskeletal specialists [50]. A lateral patellar dislocation is a traumatic event that commonly requires medical care for reduction of the joint and accounts for approximately 3% of all sports-related knee injuries [51]. The incidence rate is highest in young individuals aged 14–18 and 19–25 years at 148 and 54 per 100,000 persons, respectively [52]. Dislocation episodes also induce cartilage degradation, with cartilage lesions commonly noted for up to 95% of patients treated for patellar instability [53]. Initial cartilage degradation can also develop into osteoarthritis, as a dislocation episode has been shown to increase the risk of developing patellofemoral osteoarthritis by a factor of 8 [52].

Several anatomical factors are related to patellar instability. These anatomical factors are linked with the lateral forces applied to the patella, as well as the articular constraints and passive soft tissue constraints that stabilize the patella. The primary anatomical parameter associated with the active lateral force vector applied to the patella is the lateral position of the tibial tuberosity. The lateral position of the tibial tuberosity, expressed with respect to another landmark on the femur or tibia, characterizes the lateral force vector applied to the patella by the quadriceps muscles as translated through the patellar tendon. The tibial tuberosity is typically positioned in the most lateral position with the knee extended due to the normal external rotation of the tibia with terminal knee extension (Fig. 28.5) [54, 55]. Two conditions related to the articular constraint are commonly associated with patellar instability: patella alta and trochlear dysplasia. Patella alta is a condition in which the superior position, or height, of the patella is larger than normal (Fig. 28.6). The position of the patella delays engagement with the trochlear groove as the knee begins to flex, decreasing the articular constraint to resist lateral translation at low flexion angles. Trochlear dysplasia refers to pathologic anatomy primarily represented by a shallow proximal entry of the trochlear groove (Fig. 28.3). A shallow trochlear groove decreases engagement of the patella, thereby decreasing the articular resistance to lateral translation. Following a dislocation event, the passive resistance to patellar lateral translation is limited by injury to the MPFL, which is typically overloaded as the patella dislocates from the groove [56].

All the conditions related to patellar instability are primarily expressed at low flexion angles. Patellar instability is primarily a condition noted at low flexion angles, where reduced engagement of the patella within the trochlear groove, limited passive resistance due to injury of the MPFL, and the largest lateral orientation of the patellar tendon make the patella susceptible to lateral maltracking. Patellar dislocations primarily occur at low flexion angles as the patella is pushed lateral to the trochlear groove or fails to be captured by the trochlear groove as the knee flexes. One mechanism of patellar dislocation is a direct lateral blow to the patella, but non-contact mechanisms are also common. A twisting motion away from a planted foot is associated with patellar instability as external rotation of the tibia increases the lateral force vector acting on the patella [24]. A landing or knee flexion activity with the knee going into a valgus orientation, thereby increasing the lateral force on the patella is also associated with patellar instability [57].

The primary tracking pattern associated with patellar instability is a J-sign during knee extension, with the patella constrained by the trochlear groove in more flexed positions, but making a dramatic lateral shift in the form of an inverted "J" as the patella moves to the proximal trochlear

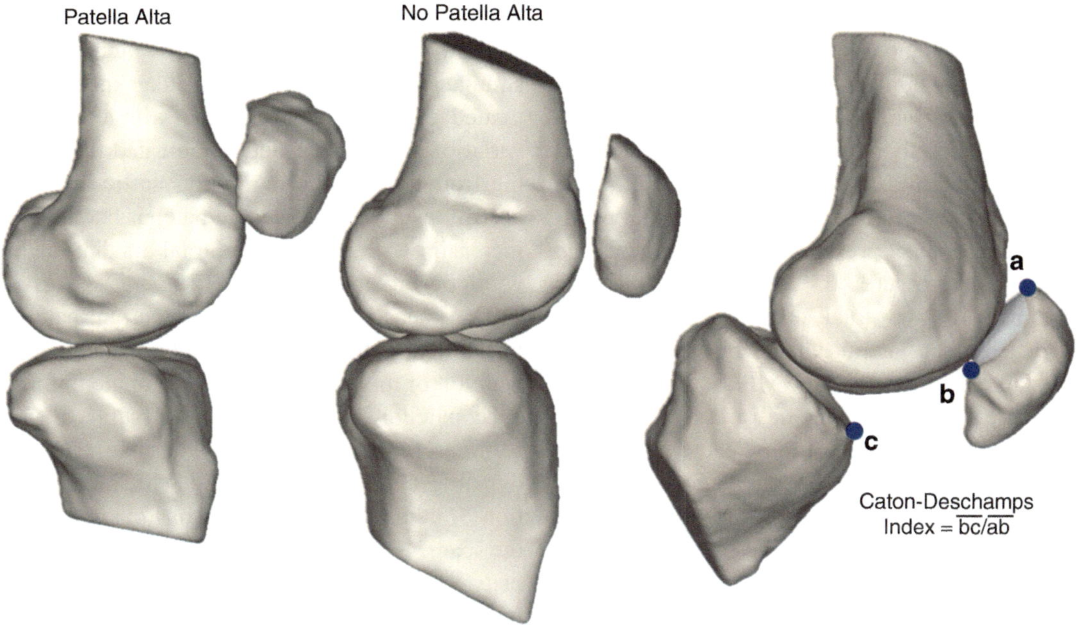

Fig. 28.6 Representation of a knee with (left) and without (center) patella alta. The Caton-Deschamps index to quantify patella alta is also shown

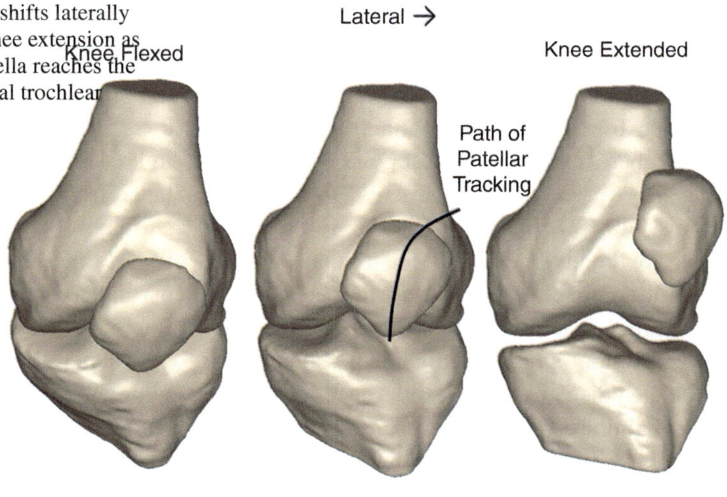

Fig. 28.7 Representation of patellar tracking producing a J-sign. The patella shifts laterally with knee extension as the patella reaches the proximal trochlear groove

groove during knee extension (Fig. 28.7). A J-sign is noted for more than 80% of knees being treated for recurrent instability, with various degrees ranging from a maximum bisect offset index of 0.75 to more than 1.25 [27]. Other forms of tracking are also noted less regularly, such as an inverted J-sign, lateral maltracking throughout the range of knee flexion, and nearly normal tracking. These tracking patterns are indicative of the wide variety of anatomical conditions and loading mechanisms associated with patellar instability.

As the primary passive stabilizer to lateral patellar translation, a dislocation event typically injures the MPFL. Injury to the MPFL further compromises patellar stability. Several in vitro studies performed with cadaveric knees tested on a loading frame have evaluated the influence of the native MPFL on patellar tracking. Sectioning the MPFL has increased patellar lateral shift and tilt by up to 9 mm and 8°, respectively, primarily near full extension [39, 58–62]. Applying a lateral force directly to the patella increased the lateral shift and tilt [60, 63, 64]. Cutting the MPFL has been shown to reduce the peak medial patellofemoral contact pressure with no influence on the lateral pressure [65]. Similar results were obtained for the total contact pressure [66]. Elevated mean lateral contact pressure has also been found after cutting the MPFL, with a similar decrease in the mean medial pressure [62, 63]. The primary limitation of these cadaveric studies is lack of representation of the pathologic conditions associated with patellar instability that can influence patellar tracking and the contact pressure distribution.

In the absence of the MPFL following an injury, the influence of knee pathology on patellar tracking increases. The lateral position of the tibial tuberosity has particularly been associated with lateral patellar tilt and shift in patients being treated for recurrent patellar instability [54, 67]. The lateral position of the tibial tuberosity is typically characterized clinically in terms of the distance from the most prominent point of the tibial tuberosity to the deepest point of the trochlear groove or tibial tuberosity to trochlear groove (TT-TG) distance (Fig. 28.3), along the posterior condylar axis of the femur. Another clinical approach to measuring the lateral position of the tibial tuberosity is the tibial tuberosity to posterior cruciate ligament attachment (TT-PCL) distance along a reference axis at the proximal tibia. This measurement eliminates the variation with respect to knee flexion angle associated with external rotation of the tibia with knee extension.

Another anatomical factor influencing lateral patellar maltracking related to patellar instability is trochlear dysplasia. Trochlear dysplasia is typically quantified based on the lateral trochlear inclination, or the orientation of the lateral ridge of the trochlear groove with respect to the posterior condylar axis of the femur. Multiple biomechanical studies have shown elevated lateral maltracking as the depth of the trochlear groove decreases from a normal value to one representative of trochlear dysplasia [68–70]. Computational simulation of knee function based on knees being treated for patellar instability, with controlled variation of trochlear depth, indicated that varying from a normal to highly dysplastic trochlear groove results in the average patellar lateral shift increasing by 15% of the total medial-lateral width during knee extension [71].

For knees being treated for recurrent patellar instability, both trochlear dysplasia and lateral position of the tibial tuberosity combine to influence patellar tracking. Based on subjects performing functional activities during MRI and dynamic CT scans, combined with reconstruction of computational models of the knees from the scans, both lateral position of the tibial tuberosity and trochlear depth influence bisect offset index and patellar lateral tilt [72, 73]. For the combination of the two parameters, high correlations were measured at low flexion angles, with r^2 values greater than 0.7 [73].

Patella alta is another anatomical factor commonly associated with patellar instability, although the correlation with lateral patellar maltracking is not as well established. Patella alta is commonly expressed by the Caton-Deschamps index (Fig. 28.6), defined as the ratio of the distance from the most distal point on the articular surface to the anterior superior position of the tibia to the length of the articular surface of the cartilage on the patella, with the knee at 30° of flexion. Patella alta has been associated with elevated lateral shift and tilt of the patella in subjects with patellar pain [28], although this has not yet been identified in a population of patients with patellar instability.

28.3 Surgical Treatment for Patellar Instability

An initial patellar dislocation is typically treated conservatively, with a treatment protocol including activity modification, bracing, and physical therapy. Physical therapy includes components such as general quadriceps and core muscle strengthening and stretching of soft tissues, as well as components that improve the medial-lateral force balance acting on the patella. Physical therapy routines can include exercises to strengthen the VMO and improve activation timing with respect to the vastus lateralis. Exercises to improve the strength of the hip external rotators are also commonly prescribed to lateralize the trochlear groove to constraint the patella. Following conservative treatment, recurrent patellar instability has commonly been noted for more than 35% of patients [74–77], leading to consideration of surgical patellar stabilization.

Currently the most common surgical approach for patellar stabilization is reconstruction of the MPFL. The MPFL is typically reconstructed with a hamstring tendon autograft or allograft. The graft is typically positioned to approximate the anatomical attachment points of the native MPFL on the femur and the patella. An MPFL graft is designed to act as a checkrein to prevent patellar subluxation, particularly at low flexion angles when the patella is not fully constrained by the trochlear groove. Care must be taken to avoid over constraining the patella, however, as it can limit knee extension and overload cartilage on the medial facet of the patella when the patella is constrained within the trochlear groove. Overloading medial cartilage is a particular concern due to the high rate of developing medial cartilage lesions following a lateral patellar dislocation [78, 79]. The formation of these lesions has been attributed to the impact of the medial facet against the lateral femoral condyle and subsequent abrasion to relocate the patella.

Because MPFL reconstruction acts to address the passive resistance to patellar subluxation without addressing elevated lateral forces acting on the patella or reduced articular constraints, the approach is typically recommended for patients with mild levels of pathologic anatomy. Other approaches are typically recommended for severe forms of pathology related to lateral position of the tibial tuberosity, patella alta, and trochlear dysplasia. An alternative to MPFL reconstruction for a lateral position of the tibial tuberosity is osteotomy and medialization of the tuberosity for a TT-TG distance exceeding 20 mm [80, 81]. An alternative to MPFL reconstruction for patella alta is osteotomy and distalization of the tibial tuberosity for a Caton-Deschamps exceeding 1.4 [80, 82]. An alternative to MPFL reconstruction for trochlear dysplasia is a trochleoplasty to deepen the trochlear groove, although the criteria for consideration are not as well established due to the complexity of the surgical technique. MPFL reconstruction can also be performed in combination with all of the other surgical techniques.

Several computational simulation as well as in vitro experimental studies showed that MPFL reconstruction decreases lateral patellar maltracking [63, 64, 66, 83, 84]. Several studies were performed with computational models based on knees being treated for recurrent patellar instability, including the anatomical pathology and maltracking associated with patellar instability [85–89]. These studies simulated dynamic knee squatting and showed that MPFL reconstruction primarily decreases lateral patellar maltracking with the knee extended. As the knee flexes, grafts were shown to guide the patella into the trochlear groove, so that articular constraints resist lateral patellar maltracking even as the graft becomes slack and unloaded with the knee flexed. MPFL reconstruction generally decreases the peak bisect offset index from conditions that would be considered maltracking (>0.75) [27] to one that would be considered normal tracking [86]. The peak value of patellar tilt has been shown to decrease by 3°, in comparison to measures of elevated patellar tilt for patients with patellar instability of approximately 15° [90, 91], although a difference of 3° in patellar tilt can be beneficial. A 3° difference was identified between initial instability patients who went on to a second dislocation within 2 years compared to those who did not [92]. The results of MPFL reconstruction are not uniform, however, as the postoperative bisect offset index and lateral tilt were found to increase as the lateral position of the tibial tuberosity increased and the slope of the lateral ridge of the trochlear groove decreased [88]. For the squatting motions simulated, graft tension values with the knee extended were less than 35 N. In vitro testing has identified failure loads of greater than 100 N for MPFL grafts [93,

94], although higher graft forces could be produced for activities that induce a larger lateral force acting on the patella and more likely to create a dislocation episode.

A primary concern with MPFL reconstruction is over constraint of the patella. In vitro studies have indicated that graft forces of 10 N or larger over constrain the patella [16, 39, 62, 63, 95] and elevate the force applied to medial cartilage. MPFL grafts have generally been shown to have little influence on the pressure applied to lateral cartilage [16, 85, 89]. The graft tension associated with elevated medial cartilage pressures is primarily associated with a poorly positioned attachment point on the femur or intra-operative over-tensioning. Errors in either parameter could occur intra-operatively [96, 97]. Previous dynamic simulation studies indicate that constraining the patella so there is no allowable glide within the trochlear groove as well as femoral fixation that over tensions the graft can elevate pressure applied to medial patellofemoral cartilage [87, 89]. The general conclusions related to these studies are not uniformly applicable, however, as factors such as patella alta and initial lateral patella tracking likely influence the fixation conditions that over constrain the patella.

Tibial tuberosity medialization is reserved for patients with a lateralized tibial tuberosity and persistent lateral maltracking. In vitro biomechanical studies have shown that tibial tuberosity medialization decreases lateral patellar tracking [15, 61, 49]. In patients treated for recurrent instability, computational reconstruction of function performed within a dynamic CT scanner indicated that surgical stabilization of the patella including tibial tuberosity medialization reduces patellar lateral shift and tilt in patients with recurrent instability [98]. The changes due to realignment were greatest near full extension, corresponding to the largest values of patellar lateral shift and tilt. A dynamic simulation comparison between tuberosity medialization and MPFL reconstruction for patients being treated for patellar instability indicated that tuberosity medialization produces the greater decrease in patellar lateral tracking [85]. The results are dependent on the surgical parameters, however. The previous study focused on tuberosity medialization of 10 mm combined with a 5 mm anteriorization component to minimize increases in patellofemoral contact pressure and MPFL reconstruction that allowed freedom of patellar lateral glide of one-eighth of the patellar width when setting the graft length. Tibial tuberosity medialization has also been shown to increase external rotation of the tibia due to the altered force vector of the patellar tendon [98, 99].

Tibial tuberosity medialization generally has a greater influence on patellofemoral contact pressures than MPFL reconstruction. In vitro studies have indicated that tuberosity medialization and anteromedialization decrease peak lateral contact pressures [15, 49, 100], but this phenomenon has not been identified when simulating knee function with knees displaying the pathology related to patellar instability [85, 101]. In vitro studies have also indicated that tuberosity medialization and anteromedialization shift contact pressure medially and increase peak medial contact pressures [65, 49, 100]. This trend has also been noted during dynamic simulation of knee function [85, 101].

Tuberosity distalization for patella alta and trochleoplasty for trochlear dysplasia have not been studied to the same extent as MPFL reconstruction and tibial tuberosity medialization. Previous computational simulation studies showed that patella alta increases the maximum patellofemoral contact pressure at 30° of flexion [102], while tuberosity distalization decreases the mean contact pressure at 45° [103], without publications specifically focused on the influence of tuberosity distalization on patellar tracking published to date. Trochlear dysplasia has primarily been addressed with respect to patellar tracking based on the studies noted previously. The simulation study that represented knees being treated for recurrent instability with varying trochlear depth indicated that changing the depth from a highly dysplastic value (lateral trochlear inclination = 6°) to a normal value (lateral trochlear inclination = 24°) would represent a change in patellar tracking from a high risk of patellar instability to still maltracking, but with a lower risk of recurrent instability [27, 71].

28.4 Conclusions

Proper knee function requires interaction of the patella with the trochlear groove that varies dramatically over the full range of knee flexion. Patellar tracking and distribution of contact pressures depend on a complex interaction of muscle forces, passive soft tissue restraints, and articular constraints. All of these factors can be influenced by pathological conditions that make the joint susceptible to disorders related to pain and instability. Appropriate care of patellofemoral disorders requires patient-specific diagnosis and treatment to address the variety of pathological contributors.

References

1. Samuels ME, Regnault S, Hutchinson JR. Evolution of the patellar sesamoid bone in mammals. PeerJ. 2017;5:e3103. https://doi.org/10.7717/peerj.3103.
2. Kaufer H. Mechanical function of the Patella. J Bone Jt Surg. 1971;53:1551–60. https://doi.org/10.2106/00004623-197153080-00007.
3. Ahmed AM, Burke DL, Hyder A. Force analysis of the patellar mechanism. J Orthop Res. 1987;5:69–85. https://doi.org/10.1002/jor.1100050110.
4. Fox A, Wanivenhaus F, Rodeo S. The basic science of the Patella: structure, composition, and function. J Knee Surg. 2012;25:127–42. https://doi.org/10.1055/s-0032-1313741.
5. Wiberg G. Roentgenographs and anatomic studies on the Femoropatellar joint: with special reference to chondromalacia patellae. Acta Orthop Scand. 1941;12:319–410. https://doi.org/10.3109/17453674108988818.
6. Grelsamer RP, Proctor CS, Bazos AN. Evaluation of patellar shape in the sagittal plane: a clinical analysis. Am J Sports Med. 1994; https://doi.org/10.1177/036354659402200111.
7. Stäubli H-U, Dürrenmatt U, Porcellini B, Rauschning W. Anatomy and surface geometry of the patellofemoral joint in the axial plane. J Bone Jt Surg. 1999;81:452–8. https://doi.org/10.1302/0301-620X.81B3.8758.
8. Pfirrmann CWA, Zanetti M, Romero J, Hodler J. Femoral trochlear dysplasia: MR findings. Radiology. 2000;216:858–64. https://doi.org/10.1148/radiology.216.3.r00se38858.
9. Fucentese SF, von Roll A, Koch PP, Epari DR, Fuchs B, Schottle PB. The patella morphology in trochlear dysplasia — a comparative MRI study. Knee. 2006;13:145–50. https://doi.org/10.1016/j.knee.2005.12.005.
10. Farahmand F, Sejiavongse W, Amis AA. Quantitative study of the quadriceps muscles and trochlear groove geometry related to instability of the patellofemoral joint. J Orthop Res. 1998;16:136–43. https://doi.org/10.1002/jor.1100160123.
11. Kwak SD, Ahmad CS, Gardner TR, Grelsamer RP, Henry JH, Blankevoort L, Ateshian GA, Mow VC. Hamstrings and iliotibial band forces affect knee kinematics and contact pattern. J Orthop Res. 2000;18:101–8. https://doi.org/10.1002/jor.1100180115.
12. Huberti HH, Hayes WC, Stone JL, Shybut GT. Force ratios in the quadriceps tendon and ligamentum patellae. J Orthop Res. 1984;2:49–54. https://doi.org/10.1002/jor.1100020108.
13. Huberti HH, Hayes WC. Patellofemoral contact pressures. The influence of Q-angle and tendofemoral contact. J Bone Jt Surg- Ser A. 1984;66:715–24. https://doi.org/10.2106/00004623-198466050-00010.
14. Li G, DeFrate LE, Zayontz S, Park SE, Gill TJ. The effect of tibiofemoral joint kinematics on patellofemoral contact pressures under simulated muscle loads. J Orthop Res. 2004;22:801–6. https://doi.org/10.1016/j.orthres.2003.11.011.
15. Ramappa AJ, Apreleva M, Harrold FR, Fitzgibbons PG, Wilson DR, Gill TJ. The effects of medialization and anteromedialization of the tibial tubercle on patellofemoral mechanics and kinematics. Am J Sport Med. 2006;34:749–56. https://doi.org/10.1177/0363546505283460.
16. Stephen JM, Kaider D, Lumpaopong P, Deehan DJ, Amis A a (2014) The effect of femoral tunnel position and graft tension on patellar contact mechanics and kinematics after medial patellofemoral ligament reconstruction. Am J Sports Med 42:364–372. doi:https://doi.org/10.1177/0363546513509230.
17. Elias JJ, Kilambi S, Goerke DR, Cosgarea AJ. Improving vastus medialis obliquus function reduces pressure applied to lateral patellofemoral cartilage. J Orthop Res. 2009;27:578–83. https://doi.org/10.1002/jor.20791.
18. Elias JJ, Wilson DR, Adamson R, Cosgarea AJ. Evaluation of a computational model used to predict the patellofemoral contact pressure distribution. J Biomech. 2004;37:295–302. https://doi.org/10.1016/S0021-9290(03)00306-3.
19. Sittek H, Eckstein F, Gavazzeni A, Milz S, Kiefer B, Schulte E, Reiser M. Assessment of normal patellar cartilage volume and thickness using MRI: an analysis of currently available pulse sequences. Skelet Radiol. 1996;25:55–62. https://doi.org/10.1007/s002560050032.
20. Eckstein F, Adam C, Sittek H, Becker C, Milz S, Schulte E, Reiser M, Putz R. Non-invasive determination of cartilage thickness throughout joint surfaces using magnetic resonance imaging. J Biomech. 1997;30:285–9. https://doi.org/10.1016/S0021-9290(97)81146-3.
21. Cohen ZA, McCarthy DM, Kwak SD, Legrand P, Fogarasi F, Ciaccio EJ, Ateshian GA. Knee cartilage

topography, thickness, and contact areas from MRI: in- vitro calibration and in-vivo measurements. Osteoarthr Cartil. 1999;7:95–109. https://doi.org/10.1053/joca.1998.0165.
22. Cohen ZA, Mow VC, Henry JH, Levine WN, Ateshian GA. Templates of the cartilage layers of the patellofemoral joint and their use in the assessment of osteoarthritic cartilage damage. Osteoarthr Cartil. 2003;11:569–79. https://doi.org/10.1016/S1063-4584(03)00091-8.
23. Draper CE, Besier TF, Gold GE, Fredericson M, Fiene A, Beaupre GS, Delp SL. Is cartilage thickness different in young subjects with and without patellofemoral pain? Osteoarthr Cartil. 2006;14:931–7. https://doi.org/10.1016/j.joca.2006.03.006.
24. Smith TO, Donell ST, Chester R, Clark A, Stephenson R. What activities do patients with patellar instability perceive makes their patella unstable? Knee. 2011;18:333–9. https://doi.org/10.1016/j.knee.2010.07.003.
25. Hasler RM, Gal I, Biedert RM. Landmarks of the normal adult human trochlea based on axial MRI measurements: a cross-sectional study. Knee Surgery, Sport Traumatol Arthrosc. 2014;22:2372–6. https://doi.org/10.1007/s00167-014-3152-9.
26. Batailler C, Neyret P. Trochlear dysplasia: imaging and treatment options. EFORT Open Rev. 2018;3:240–7. https://doi.org/10.1302/2058-5241.3.170058.
27. Tanaka MJ, Elias JJ, Williams AA, Demehri S, Cosgarea AJ. Characterization of patellar maltracking using dynamic kinematic CT imaging in patients with patellar instability. Knee Surgery, Sport Traumatol Arthrosc. 2016;24:3634–41. https://doi.org/10.1007/s00167-016-4216-9.
28. Pal S, Besier TF, Beaupre GS, Fredericson M, Delp SL, Gold GE. Patellar maltracking is prevalent among patellofemoral pain subjects with patella Alta: An upright, weightbearing MRI study. J Orthop Res. 2013;31:448–57. https://doi.org/10.1002/jor.22256.
29. Conlan T, Garth WP, Lemons JE. Evaluation of the medial soft-tissue restraints of the extensor mechanism of the knee. J Bone Joint Surg Am. 1993;75:682–93. https://doi.org/10.2106/00004623-199305000-00007.
30. Desio SM, Burks RT, Bachus KN. Soft tissue restraints to lateral patellar translation in the human knee. Am J Sports Med. 1998;26:59–65. https://doi.org/10.1177/03635465980260012701.
31. Smith TO, Walker J, Russell N. Outcomes of medial patellofemoral ligament reconstruction for patellar instability: a systematic review. Knee Surg Sport Traumatol Arthrosc. 2007;15:1301–14. https://doi.org/10.1007/s00167-007-0390-0.
32. Placella G, Tei MM, Sebastiani E, Criscenti G, Speziali A, Mazzola C, Georgoulis A, Cerulli G. Shape and size of the medial patellofemoral ligament for the best surgical reconstruction: a human cadaveric study. Knee Surgery, Sport Traumatol Arthrosc. 2014;22:2327–33. https://doi.org/10.1007/s00167-014-3207-y.
33. Tanaka MJ. Variability in the patellar attachment of the medial patellofemoral ligament. Arthrosc - J Arthrosc Relat Surg. 2016; https://doi.org/10.1016/j.arthro.2016.01.046.
34. Smirk C, Morris H. The anatomy and reconstruction of the medial patellofemoral ligament. Knee. 2003;10:221–7. https://doi.org/10.1016/S0968-0160(03)00038-3.
35. Gobbi RG, Pereira CAM, Sadigursky D, Demange MK, Tírico LEP, Pécora JR, Camanho GL. Evaluation of the isometry of different points of the patella and femur for medial patellofemoral ligament reconstruction. Clin Biomech. 2016;38:8–12. https://doi.org/10.1016/j.clinbiomech.2016.08.002.
36. Redler LH, Meyers KN, Brady JM, Dennis ER, Nguyen JT, Shubin Stein BE. Anisometry of medial patellofemoral ligament reconstruction in the setting of increased Tibial tubercle–trochlear groove distance and Patella Alta. Arthrosc J Arthrosc Relat Surg. 2018;34:502–10. https://doi.org/10.1016/j.arthro.2017.08.256.
37. Oka S, Matsushita T, Kubo S, Matsumoto T, Tajimi H, Kurosaka M, Kuroda R. Simulation of the optimal femoral insertion site in medial patellofemoral ligament reconstruction. Knee Surgery, Sport Traumatol Arthrosc. 2014;22:2364–71. https://doi.org/10.1007/s00167-014-3192-1.
38. LaPrade MD, Kallenbach SL, Aman ZS, Moatshe G, Storaci HW, Turnbull TL, Arendt EA, Chahla J, LaPrade RF. Biomechanical evaluation of the medial stabilizers of the Patella. Am J Sports Med. 2018;46:1575–82. https://doi.org/10.1177/0363546518758654.
39. Philippot R, Boyer B, Testa R, Farizon F, Moyen B. Study of patellar kinematics after reconstruction of the medial patellofemoral ligament. Clin Biomech. 2012;27:22–6. https://doi.org/10.1016/j.clinbiomech.2011.08.001.
40. Goh J, Lee P, Bose K. A cadaver study of the function of the oblique part of vastus medialis. J Bone Joint Surg Br. 1995;77-B:225–31. https://doi.org/10.1302/0301-620X.77B2.7706335.
41. Sakai N, Luo Z-P, Rand JA, An K-N. The influence of weakness in the vastus medialis oblique muscle on the patellofemoral joint: an in vitro biomechanical study. Clin Biomech. 2000;15:335–9. https://doi.org/10.1016/S0268-0033(99)00089-3.
42. Csintalan RP, Schulz MM, Woo J, McMahon PJ, Lee TQ. Gender differences in patellofemoral joint biomechanics. Clin Orthop Relat Res. 2002;402:260–9. https://doi.org/10.1097/00003086-200209000-00026.
43. Dhaher YY, Kahn LE. The effect of vastus Medialis forces on Patello-femoral contact: a model-based study. J Biomech Eng. 2002;124:758–67. https://doi.org/10.1115/1.1516196.
44. Lee TQ, Sandusky MD, Adeli A, McMahon PJ. Effects of simulated vastus medialis strength variation on patellofemoral joint biomechanics in human cadaver knees. J Rehabil Res Dev. 2002;39:429–38.

45. Sheehan FT, Borotikar BS, Behnam AJ, Alter KE. Alterations in in vivo knee joint kinematics following a femoral nerve branch block of the vastus medialis: implications for patellofemoral pain syndrome. Clin Biomech. 2012;27:525–31. https://doi.org/10.1016/j.clinbiomech.2011.12.012.
46. Pal S, Draper CE, Fredericson M, Gold GE, Delp SL, Beaupre GS, Besier TF. Patellar maltracking correlates with vastus medialis activation delay in patellofemoral pain patients. Am J Sports Med. 2011;39:590–8. https://doi.org/10.1177/0363546510384233.
47. Biedert RM, Sanchis-Alfonso V. Sources of anterior knee pain. Clin Sports Med. 2002;21:335–47. https://doi.org/10.1016/S0278-5919(02)00026-1.
48. Conchie H, Clark D, Metcalfe A, Eldridge J, Whitehouse M. Adolescent knee pain and patellar dislocations are associated with patellofemoral osteoarthritis in adulthood: a case control study. Knee. 2016;23:708–11. https://doi.org/10.1016/j.knee.2016.04.009.
49. Saranathan A, Kirkpatrick MS, Mani S, Smith LG, Cosgarea AJ, Tan JS, Elias JJ. The effect of tibial tuberosity realignment procedures on the patellofemoral pressure distribution. Knee Surg Sports Traumatol Arthrosc. 2012;20:2054–61. https://doi.org/10.1007/s00167-011-1802-8.
50. Redziniak D, Diduch D, Mihalko W, Fulkerson J. Patellar instability. J Bone Jt Surg. 2009;91:2264–75.
51. Majewski M, Susanne H, Klaus S. Epidemiology of athletic knee injuries: a 10-year study. Knee. 2006;13:184–8. https://doi.org/10.1016/j.knee.2006.01.005.
52. Sanders TL, Pareek A, Hewett TE, Stuart MJ, Dahm DL, Krych AJ. High rate of recurrent patellar dislocation in skeletally immature patients: a long-term population-based study. Knee Surgery, Sport Traumatol Arthrosc. 2018;26:1037–43. https://doi.org/10.1007/s00167-017-4505-y.
53. Nomura E, Inoue M, Kurimura M. Chondral and osteochondral injuries associated with acute patellar dislocation. Arthrosc J Arthrosc Relat Surg. 2003;19:717–21. https://doi.org/10.1016/S0749-8063(03)00401-8.
54. Williams AA, Elias JJ, Tanaka MJ, Thawait GK, Demehri S, Carrino JA, Cosgarea AJ. The relationship between Tibial tuberosity–trochlear groove distance and abnormal patellar tracking in patients with unilateral patellar instability. Arthrosc J Arthrosc Relat Surg. 2016;32:55–61. https://doi.org/10.1016/j.arthro.2015.06.037.
55. Izadpanah K, Weitzel E, Vicari M, Hennig J, Weigel M, Südkamp NP, Niemeyer P. Influence of knee flexion angle and weight bearing on the Tibial tuberosity-trochlear groove (TTTG) distance for evaluation of patellofemoral alignment. Knee Surgery, Sport Traumatol Arthrosc. 2014;22:2655–61. https://doi.org/10.1007/s00167-013-2537-5.
56. Balcarek P, Ammon J, Frosch S, Walde TA, Schüttrumpf JP, Ferlemann KG, Lill H, Stürmer KM, Frosch K-H. Magnetic resonance imaging characteristics of the medial patellofemoral ligament lesion in acute lateral patellar dislocations considering trochlear dysplasia, Patella Alta, and Tibial tuberosity–trochlear groove distance. Arthrosc J Arthrosc Relat Surg. 2010;26:926–35. https://doi.org/10.1016/j.arthro.2009.11.004.
57. Sillanpää P, Mattila VM, Iivonen T, Visuri T, Pihlajamäki H. Incidence and risk factors of acute traumatic primary patellar dislocation. Med Sci Sports Exerc. 2008; https://doi.org/10.1249/MSS.0b013e318160740f.
58. Ostermeier S, Stukenborg-Colsman C, Hurschler C, Wirth CJ. In vitro investigation of the effect of medial patellofemoral ligament reconstruction and medial tibial tuberosity transfer on lateral patellar stability. Arthrosc - J Arthrosc Relat Surg. 2006;22:308–19. https://doi.org/10.1016/j.arthro.2005.09.024.
59. Ostermeier S, Holst M, Bohnsack M, Hurschler C, Stukenborg-Colsman C, Wirth CJ. In vitro measurement of patellar kinematics following reconstruction of the medial patellofemoral ligament. Knee Surgery, Sport Traumatol Arthrosc. 2007;15:276–85. https://doi.org/10.1007/s00167-006-0200-0.
60. Zaffagnini S, Colle F, Lopomo N, Sharma B, Bignozzi S, Dejour D, Marcacci M. The influence of medial patellofemoral ligament on patellofemoral joint kinematics and patellar stability. Knee Surgery, Sport Traumatol Arthrosc. 2013;21:2164–71. https://doi.org/10.1007/s00167-012-2307-9.
61. Stephen JM, Dodds AL, Lumpaopong P, Kader D, Williams A, Amis AA. The ability of medial patellofemoral ligament reconstruction to correct patellar kinematics and contact mechanics in the presence of a lateralized Tibial tubercle. Am J Sports Med. 2015;43:2198–207. https://doi.org/10.1177/0363546515597906.
62. Stephen JM, Kittl C, Williams A, Zaffagnini S, Marcheggiani Muccioli GM, Fink C, Amis AA. Effect of medial patellofemoral ligament reconstruction method on patellofemoral contact pressures and kinematics. Am J Sports Med. 2016;44:1186–94. https://doi.org/10.1177/0363546516631736.
63. Beck P, Brown NAT, Greis PE, Burks RT. Patellofemoral contact pressures and lateral patellar translation after medial patellofemoral ligament reconstruction. Am J Sports Med. 2007;35:1557–63. https://doi.org/10.1177/0363546507300872.
64. Sandmeier RH, Burks RT, Bachus KN, Billings A. The effect of reconstruction of the medial patellofemoral ligament on patellar tracking. Am J Sports Med. 2000;28:345–9. https://doi.org/10.1177/03635465000280031001.
65. Beck PR, Thomas AL, Farr J, Lewis PB, Cole BJ. Trochlear contact pressures after Anteromedialization of the Tibial tubercle. Am J Sports Med. 2005;33:1710–5. https://doi.org/10.1177/0363546505278300.
66. Lorbach O, Haupert A, Efe T, Pizanis A, Weyers I, Kohn D, Kieb M. Biomechanical evaluation of

MPFL reconstructions: differences in dynamic contact pressure between gracilis and fascia lata graft. Knee Surgery, Sport Traumatol Arthrosc. 2017; https://doi.org/10.1007/s00167-016-4005-5.
67. Tanaka MJ, Elias JJ, Williams AA, Carrino JA, Cosgarea AJ. Correlation between changes in Tibial tuberosity–trochlear groove distance and patellar position during active knee extension on dynamic kinematic computed tomographic imaging. Arthrosc J Arthrosc Relat Surg. 2015;31:1748–55. https://doi.org/10.1016/j.arthro.2015.03.015.
68. Amis AA, Oguz C, Bull AMJ, Senavongse W, Dejour D. The effect of trochleoplasty on patellar stability and kinematics. J Bone Joint Surg Br. 2008;90-B:864–9. https://doi.org/10.1302/0301-620X.90B7.20447.
69. Van Haver A, De Roo K, De Beule M, Labey L, De Baets P, Dejour D, Claessens T, Verdonk P. The effect of trochlear dysplasia on patellofemoral biomechanics. Am J Sports Med. 2015;43:1354–61. https://doi.org/10.1177/0363546515572143.
70. Fitzpatrick CK, Steensen RN, Tumuluri A, Trinh T, Bentley J, Rullkoetter PJ. Computational analysis of factors contributing to patellar dislocation. J Orthop Res. 2016;34:444–53. https://doi.org/10.1002/jor.23041.
71. Rezvanifar SC, Flesher BL, Jones KC, Elias JJ. Lateral patellar maltracking due to trochlear dysplasia: a computational study. Knee. 2019;26:1234–42. https://doi.org/10.1016/j.knee.2019.11.006.
72. Biyani R, Elias JJ, Saranathan A, Feng H, Guseila LM, Morscher MA, Jones KC. Anatomical factors influencing patellar tracking in the unstable patellofemoral joint. Knee Surgery, Sport Traumatol Arthrosc. 2014;22:2334–41. https://doi.org/10.1007/s00167-014-3195-y.
73. Elias JJ, Soehnlen NT, Guseila LM, Cosgarea AJ. Dynamic tracking influenced by anatomy in patellar instability. Knee. 2016;23:450–5. https://doi.org/10.1016/j.knee.2016.01.021.
74. Sanders TL, Pareek A, Johnson NR, Stuart MJ, Dahm DL, Krych AJ. Patellofemoral arthritis after lateral patellar dislocation: a matched population-based analysis. Am J Sports Med. 2017;45:1012–7. https://doi.org/10.1177/0363546516680604.
75. Christensen TC, Sanders TL, Pareek A, Mohan R, Dahm DL, Krych AJ. Risk factors and time to recurrent ipsilateral and contralateral patellar dislocations. Am J Sports Med. 2017;45:2105–10. https://doi.org/10.1177/0363546517704178.
76. Arendt EA, Askenberger M, Agel J, Tompkins MA. Risk of Redislocation after primary patellar dislocation: a clinical prediction model based on magnetic resonance imaging variables. Am J Sports Med. 2018;46:3385–90. https://doi.org/10.1177/0363546518803936.
77. Ying ZG, Yu DH, Miao LE, Zheng L, Wu BZ, Shi H, Jing FF, Guo D. Incidence of second-time lateral patellar dislocation is associated with anatomic factors, age and injury patterns of medial patellofemoral ligament in first-time lateral patellar dislocation: a prospective magnetic resonance imaging study with 5-year fol. Knee Surgery, Sport Traumatol Arthrosc. 2019;27:197–205. https://doi.org/10.1007/s00167-018-5062-8.
78. Salonen EE, Magga T, Sillanpää PJ, Kiekara T, Mäenpää H, Mattila VM. Traumatic patellar dislocation and cartilage injury: a follow-up study of long-term cartilage deterioration. Am J Sports Med. 2017;45:1376–82. https://doi.org/10.1177/0363546516687549.
79. Nomura E, Inoue M. Cartilage lesions of the Patella in recurrent patellar dislocation. Am J Sports Med. 2004;32:498–502. https://doi.org/10.1177/0095399703258677.
80. Weber AE, Nathani A, Dines JS, Allen AA, Shubin-Stein BE, Arendt EA, Bedi A. An algorithmic approach to the Management of Recurrent Lateral Patellar Dislocation. J Bone Jt Surg. 2016;98:417–27. https://doi.org/10.2106/JBJS.O.00354.
81. Post WR, Fithian DC. Patellofemoral instability: a consensus statement from the AOSSM/PFF patellofemoral instability workshop. Orthop J Sport Med. 2018;6:232596711775035. https://doi.org/10.1177/2325967117750352.
82. Bartsch A, Lubberts B, Mumme M, Egloff C, Pagenstert G. Does patella Alta lead to worse clinical outcome in patients who undergo isolated medial patellofemoral ligament reconstruction? A systematic review. Arch Orthop Trauma Surg. 2018;138:1563–73. https://doi.org/10.1007/s00402-018-2971-4.
83. Melegari TM, Parks BG, Matthews LS. Patellofemoral contact area and pressure after medial patellofemoral ligament reconstruction. Am J Sports Med. 2008;36:747–52. https://doi.org/10.1177/0363546508314410.
84. Alvarez O, Steensen RN, Rullkoetter PJ, Fitzpatrick CK. Computational approach to correcting joint instability in patients with recurrent patellar dislocation. J Orthop Res. 2020; https://doi.org/10.1002/jor.24526.
85. Elias JJ, Tanaka MJ, Jones KC, Cosgarea AJ. Tibial tuberosity anteriomedialization vs. medial patellofemoral ligament reconstruction for treatment of patellar instability related to malalignment: computational simulation. Clin Biomech. 2020;74:111–7. https://doi.org/10.1016/j.clinbiomech.2020.01.019.
86. Tanaka MJ, Cosgarea AJ, Forman JM, Elias JJ. Factors influencing graft function following MPFL reconstruction: a dynamic simulation study. J Knee Surg. 2020; https://doi.org/10.1055/s-0040-1702185.
87. Elias JJ, Kelly MJ, Smith KE, Gall KA, Farr J. Dynamic simulation of the effects of graft fixation errors during medial patellofemoral ligament reconstruction. Orthop J Sport Med. 2016;4:232596711666508. https://doi.org/10.1177/2325967116665080.
88. Elias JJ, Jones KC, Cyrus Rezvanifar S, Gabra JN, Morscher MA, Cosgarea AJ. Dynamic track-

ing influenced by anatomy following medial patellofemoral ligament reconstruction: computational simulation. Knee. 2018;25:262–70. https://doi.org/10.1016/j.knee.2018.02.002.
89. Elias JJ, Jones KC, Lalonde MK, Gabra JN, Rezvanifar SC, Cosgarea AJ. Allowing one quadrant of patellar lateral translation during medial patellofemoral ligament reconstruction successfully limits maltracking without overconstraining the patella. Knee Surgery, Sport Traumatol Arthrosc. 2018;26:2883–90. https://doi.org/10.1007/s00167-017-4799-9.
90. Charles MD, Haloman S, Chen L, Ward SR, Fithian D, Afra R. Magnetic resonance imaging–based topographical differences between control and recurrent patellofemoral instability patients. Am J Sports Med. 2013;41:374–84. https://doi.org/10.1177/0363546512472441.
91. Askenberger M, Janarv PM, Finnbogason T, Arendt EA. Morphology and Anatomic Patellar Instability Risk Factors in First-Time Traumatic Lateral Patellar Dislocations: A Prospective Magnetic Resonance Imaging Study in Skeletally Immature Children. Am J Sports Med. 2017;45:50–58. https://doi.org/10.1177/0363546516663498.
92. Balcarek P, Oberthür S, Hopfensitz S, Frosch S, Walde TA, Wachowski MM, Schüttrumpf JP, Stürmer KM. Which patellae are likely to redislocate? Knee Surgery, Sport Traumatol Arthrosc. 2014;22:2308–14. https://doi.org/10.1007/s00167-013-2650-5.
93. Saper MG, Meijer K, Winnier S, Popovich J, Andrews JR, Roth C. Biomechanical evaluation of classic solid and all-soft suture anchors for medial patellofemoral ligament reconstruction. Am J Sports Med. 2017;45:1622–6. https://doi.org/10.1177/0363546517691951.
94. Joyner PW, Bruce J, Roth TS, et al. Biomechanical tensile strength analysis for medial patellofemoral ligament reconstruction. Knee. 2017;24:965–76. https://doi.org/10.1016/j.knee.2017.04.013.
95. Mehl J, Otto A, Comer B, Kia C, Liska F, Obopilwe E, Beitzel K, Imhoff AB, Fulkerson JP, Imhoff FB. Repair of the medial patellofemoral ligament with suture tape augmentation leads to similar primary contact pressures and joint kinematics like reconstruction with a tendon graft: a biomechanical comparison. Knee Surgery, Sport Traumatol Arthrosc. 2020;28:478–88. https://doi.org/10.1007/s00167-019-05668-z.
96. Nelitz M, Williams RS, Lippacher S, Reichel H, Dornacher D. Analysis of failure and clinical outcome after unsuccessful medial patellofemoral ligament reconstruction in young patients. Int Orthop. 2014;38:2265–72. https://doi.org/10.1007/s00264-014-2437-4.
97. Servien E, Fritsch B, Lustig S, Demey G, Debarge R, Lapra C, Neyret P. In vivo positioning analysis of medial patellofemoral ligament reconstruction. Am J Sports Med. 2011;39:134–9. https://doi.org/10.1177/0363546510381362.
98. Elias JJ, Carrino JA, Saranathan A, Guseila LM, Tanaka MJ, Cosgarea AJ. Variations in kinematics and function following patellar stabilization including tibial tuberosity realignment. Knee Surg Sports Traumatol Arthrosc. 2014;22:2350–6. https://doi.org/10.1007/s00167-014-2905-9.
99. Mani S, Kirkpatrick MS, Saranathan A, Smith LG, Cosgarea AJ, Elias JJ. Tibial tuberosity osteotomy for patellofemoral realignment alters tibiofemoral kinematics. Am J Sport Med. 2011;39:1024–31. https://doi.org/10.1177/0363546510390188.
100. Stephen JM, Lumpaopong P, Dodds AL, Williams A, Amis A a (2015) The effect of Tibial tuberosity medialization and lateralization on patellofemoral joint kinematics, contact mechanics, and stability. Am J Sports Med 43:186–194. doi:https://doi.org/10.1177/0363546514554553.
101. Elias JJ, Jones KC, Copa AJ, Cosgarea AJ. Computational simulation of medial versus anteromedial tibial tuberosity transfer for patellar instability. J Orthop Res. 2018; https://doi.org/10.1002/jor.24108.
102. Watson NA, Duchman KR, Grosland NM, Bollier MJ. Finite element analysis of Patella Alta: a patellofemoral instability model. Iowa Orthop J. 2017;37:101–8.
103. Yin L, Liao T-CC, Yang L, Powers CM. Does Patella tendon Tenodesis improve Tibial tubercle Distalization in treating Patella Alta? A computational study. Clin Orthop Relat Res. 2016;474:2451–61. https://doi.org/10.1007/s11999-016-5027-5.

Biomechanics of Cruciate Retaining and Posterior Stabilised Total Knee Arthroplasty and Return to Sports

29

A. Kropelnicki and D. A. Parker

29.1 Introduction

Total knee arthroplasty (TKA) has shown to be an effective and reliable treatment of end stage arthritis of the knee [1–5], once medical management has become ineffective [6]. Some 1.1 million knee joint replacements have been performed since the start of the National Joint Registry of the United Kingdom in 2003 [7], with this demand set to increase [4]. Knee replacement procedures produce obvious benefits in relieving arthritic pain, correcting deformity and improving ambulation, thus allowing an improvement in physical and mental health [1, 8–10]. Clinical studies on total knee arthroplasty demonstrate considerable variation in kinematics and functional performance [11].

The knee itself is a complex, incongruent joint [12] with little inherent stability [13] and correspondingly complex kinematics [14]. It is reinforced by the surrounding capsule, ligaments, menisci and muscles, all of which contribute to knee stability and movement [8, 15]. The normal anatomy of the knee varies widely, and pathological changes increase its variability further [14]. For example, approximately 20% of the population have varus knees of 3° or greater, such that a mechanically neutral-aligned TKA is a significant anatomical adjustment and would require considerable soft tissue re-balancing [14].

The movement of the knee is considered a complex hinge with femoral roll-back and the 'screw home' mechanism, which are unique to the knee [8]. These have been discussed in detail in earlier chapters. Restoration of the functional anatomy of the knee, including alignment, soft tissue balancing, and restoration of the joint line are integral to improving function [8]. The complexity of movement as described in previous chapters needs to be understood and reproduced as much as possible when performing a TKA, in order to achieve the best outcome.

The average age at which patients seek TKA surgery is decreasing. Comparing the cumulative TKA data from the 19th NJR report against that of 5 years ago, the mean age has dropped from 70 to 69 years [7, 16]. Year-by-year data is likely to reflect an ongoing trend. For example, in Canada the number of 45- to 64-year olds diagnosed with arthritis is set to double between 1991 and 2031 to 2.1 million [17]. In general, with decreasing age comes increasing expectations, and patients' functional demands and expectations of a TKR have evolved, with many wishing to resume some form of sporting activity after surgery [6, 18].

A. Kropelnicki
Sports Medicine Institute, Miranda, NSW, Australia
e-mail: anna@krop.co.uk, anna@headsafe.com

D. A. Parker (✉)
Sydney Orthopaedic Research Institute, Chatswood, NSW, Australia

University of Sydney, Sydney, Australia
e-mail: dparker@sydneyortho.com.au

J. Koh et al. (eds.), *Orthopaedic Biomechanics in Sports Medicine*,
https://doi.org/10.1007/978-3-030-81549-3_29

Furthermore, it is well understood that meeting patients' expectations is highly correlated with satisfaction following TKA [19]. It is well-established that regular physical activity has significant health benefits, yet sports activity after TKA has not been particularly well studied [2], and, to our knowledge, no prospective controlled trials exist assessing sporting outcomes with different TKA designs or techniques.

29.2 TKA Design

There are many designs in current use for total knee replacement. The Australian National Joint Registry lists 68 common total knee replacement designs with another 194 forming an 'Other' category [20].

Although knee replacement design has been based on native knee anatomy, it is not a pure anatomic reproduction of the native knee [1, 2, 14], particularly given the wide variation in individual anatomy, which clearly presents challenges and limitations when trying to reproduce normal knee function. The majority of total knee arthroplasty designs can be primarily divided into whether they are cruciate retaining (CR; sacrificing the ACL, but preserving the PCL) or cruciate sacrificing (PS; removing both cruciate ligaments) (Fig. 29.1). Both of these have good success rates over time with 4.9% revised at 10 years for the CR, and 6% for the PS [20]. At 15 years, these rates remain similar for the CR compared with the PS with revision rates of 7% and 8%, respectively [20]. However, it is well recognised that despite these relatively low revision rates, up to 20% of patients with a total knee replacement are dissatisfied with their outcome [14].

Forces transmitted across the articulating surfaces are determined by a combination of the alignment, movement and integrity of anatomical structures within the knee [8]. More constrained designs offer greater global stability and rely less on soft tissues, but increase the stresses on the bone–implant interface [21], whereas less constrained designs rely more on the soft tissues for stability. It is important to remember that the sta-

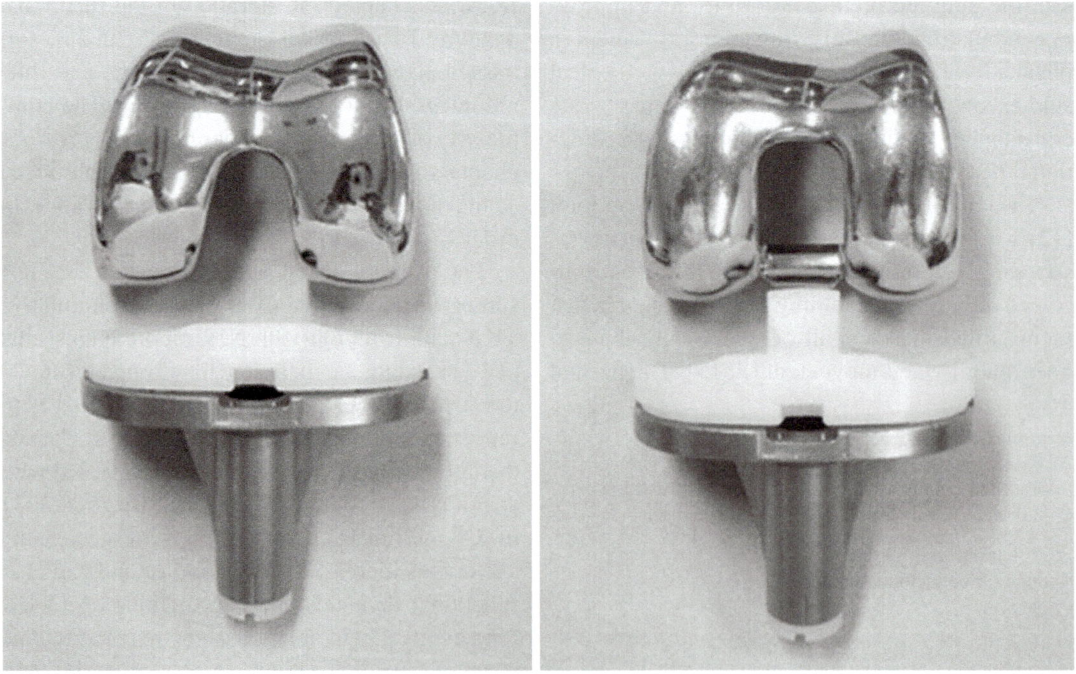

Fig. 29.1 Cruciate retaining TKA (left) and posterior stabilised TKA (right). (Reprinted with permission from Yagashita K, et al., J Arthroplasty 2012)

bility of the TKA is highly dependent on not only the conforming design of the prosthesis but also the surgical techniques employed whilst implanting the prosthesis [21]. Soft tissue releases may help with varus/valgus balance, but can introduce both antero-posterior and rotational instability [22].

29.2.1 Anatomy

The role of the cruciates is debated in TKA with most prosthetic designs requiring complete excision of the ACL [15]. This is frequently justified by the fact that in end stage arthritis the ACL is often compromised or ruptured, and also that patients in this demographic are unlikely to experience instability when not competing in more vigorous twisting and pivoting activities. Many current TKA designs can result in anterior tibial subluxation in full extension [15], which could potentially compromise certain movements involved in sports, such as jumping and pivoting. There have been several attempts to design an ACL preserving TKA, but with limited success and ongoing investigation.

29.2.1.1 Soft Tissues

Accurate soft tissue balancing with preservation of normal anatomy is an important component of success with TKR, influencing range of motion, proprioception and kinematics [8, 15]. Balance of the flexion–extension gaps is required to produce stability throughout the arc of motion [1]. Navigated or computer-assisted surgery enables the surgeon to more accurately achieve alignment of the implants in the three planes and provides feedback on soft tissue balancing [12]. Carefully planned and accurately executed surgical technique, therefore, contributes to achieving optimal range of motion with good stability [12].

Restoration of the native joint line has been shown to be important for both tibiofemoral and patellofemoral kinematics, and achieving optimal soft tissue balance [1]. Modelling also showed that the most influential factor was restoration of the joint line and coronal plane alignment [11]. Recent debate has centred around the concept of 'kinematic' or 'constitutional' alignment, which aims to recreate a patient's presumed pre-arthritic anatomy, with early reports from selected centres reporting improved functional outcomes when compared to more traditional mechanical alignment strategies [23]. At this point in time, the debate continues, but these more anatomical strategies do seem to show some promise in improving outcomes and restoring more normal function for TKA recipients. Longer term outcomes studies will be necessary to demonstrate that these early promising results are sustained and that there is no compromise to longer term survival.

29.2.2 Design Considerations

29.2.2.1 Femoral Component

Design features influence tibiofemoral contact mechanics and kinematics [11]. There are many different femoral designs on the market; however, substantial differences in patient outcomes has not been proven [13].

The trochlea grooves of many modern designs has been deepened, laterally flared and flattened to improve patella tracking and congruity throughout the range of motion [13].

The single axis total knee replacement was designed on the concept of a single flexion extension axis of the knee. The single radius design, however, lengthens the quadriceps moment arm, with a resulting decrease in quadriceps force [8]. Consistent with this, biomechanical studies have shown less eccentric knee extensor muscle activation and greater mediolateral stability during the stand-to-sit movement with this single-radius design. Multiple radius designs of the femoral component have been shown to develop several compensatory mechanisms, such as increased hamstring co-activation in order to increase joint stability [8]. Overall, the increased power output shows better functional recovery for the patient [8]. However, the multi-radius knee has been associated with mid-range instability despite rectangular flexion and extension gaps at 0 and 90° [14].

29.2.2.2 High Flexion Implants

Range of motion is an important component of patient satisfaction after TKA. In an effort to improve flexion post TKA, designers have produced femoral prostheses with both increased posterior femoral condylar offset and shortened posterior condyles to improve 'clearance' in deep flexion and reduce pressure over the extensor mechanism. Such designs also aim to achieve a larger contact area during high flexion, thus theoretically reducing contact pressures and wear, whilst a deepened trochlea avoids impingement of the extensor mechanism [13]. Reviews have however failed to show sufficient evidence of consistently improved function or range of motion [8].

29.2.2.3 Mobile Bearing Versus Fixed Bearing

Conformity of the femoral implant with the polyethylene insert has a conflicting impact on native knee kinematics and contact stresses. A highly conforming design in a fixed bearing prosthesis decreases contact stress and hence polythene wear but adversely affecting normal knee kinematics [13]. A low conforming design, however, results in high contact stresses increasing wear and the prospect of failure whilst allowing more normal knee kinematics [13]. Reproducing the natural movement of the knee requires a design that allows movement in all three planes [8], and mobile bearing prosthesis was designed with the intention of allowing both antero-posterior and axial rotation to be accommodated along with greater congruency, theoretically providing improved kinematics and reduced polyethylene wear [8]. Yet, clinical and cadaver studies have failed to show any significant benefit of these prostheses, for either kinematics or polyethylene wear [8]. Furthermore, there are concerns that the mobility of the bearing increases the risk of polyethylene wear and osteolysis.

29.2.3 Cruciate Retaining

The CR prosthesis was developed based on the concept that motion is guided by the soft tissue, with the ligaments helping to drive the kinematics of the knee [13]. Preservation of the PCL has potential advantages of bone preservation, improved proprioception, femoral rollback, kinematics closer to that of the native knee, and improved stability [15]. In this design, the PCL alone provides antero-posterior stability and rollback of the femoral condyles as long as its structure is preserved by careful surgical technique, and normal function preserved by adequately balancing the flexion and extension gaps [1, 8]. The tibial surface must be relatively flat in order to allow this rollback and prevent excessive tension in the PCL [8], yet have a degree of concavity to contribute towards the antero-posterior stability of the prosthesis [21]. Previous kinematic studies have, however, demonstrated limited success with the PCL alone consistently producing roll-back, and do not reproduce the screw-home mechanism [1].

29.2.4 Cruciate Sacrificing

When the posterior cruciate ligament is sacrificed, the posterior stabilised (PS) design uses a cam-post mechanism to create an artificially produced roll-back [1, 15, 21]. This implant is based more on a functional philosophy than an anatomical one with the implant driving the kinematics rather than the soft tissues [13]. The PS implant produces more reliable rollback, and therefore usually results in improved flexion, but does not reproduce the screw-home mechanism [1, 13, 15]. The centre post may also contribute to antero-posterior stability in the case of extensor mechanism weakness [8]. The PS however may have decreased mid-range stability [1] and have greater lift-off in the gait cycle that may be detrimental to the polyethylene bearing [1].

29.2.5 Kinematics of CR Compared with PS

The decision to retain or sacrifice the posterior cruciate ligament has been a longstanding debate within orthopaedics [24, 25]. Both prostheses came about at a similar time, and the PS is obviously required in the event of PCL insufficiency. However, the selection of the sub-type of TKA is often driven by the surgeon's training and experience rather than biomechanical or kinematic evidence [25]. As discussed in previous chapters, there are arguments that the CR implant retains more natural knee kinematics, better proprioception and thus greater stability [24], whereas those arguing for the PS total knee replacement suggest this implant produces more predictable kinematics, reduces tibiofemoral loads and is simpler to balance [24]. Overall, the PS is also considered to allow a greater range of movement post-operatively [24]. In order to address this, many using the CR implant increase the posterior slope as part of their surgical technique [26]. However, fluoroscopic studies show that all total knee arthroplasties reduce the range of motion compared with native knees regardless of subtype [27]. On average, TKA does not gain flexion above 120° [28]. Despite the above differences, there is no good evidence to support functional superiority between these two implant sub-types [29]. Singleton et al. [29] did note less stiffness at 1 year post-operatively in the PS knee, but in comparisons at 5 and 10 years showed there was no difference between the outcomes of the CR and PS knees. Whilst efforts to reproduce anatomy and function more closely to the native knee continue, it would certainly be universally agreed that, although TKR usually results in good outcomes, it does not reliably reproduce normal knee kinematics, irrespective of the design.

29.2.5.1 Gait

Gait in patients following a TKA does not normalise, rather frequently returns to pre-surgery joint loading patterns and abnormal biomechanics or adaptive changes to compensate to the effects of a TKA [38]. Modelling shows that there is a complex interaction of patient, surgical and implant design factors which vary in their predominance throughout the gait cycle [11]. Therefore, it is extremely difficult to tease out the sole influence of the biomechanics of the TKA prostheses. Gait analysis has been criticised for its diversity in reporting due to comparing different patient characteristics, different prosthetics and different methodologies of analysis [5]. A systematic review by McClelland et al. [5] found that patients do not demonstrate a normal walking gait after TKA, the most consistent findings being walking with a reduced range of motion, specifically less flexion, compared with control subjects, as well as reduced loading in the stance phase. These findings would certainly be consistent with reduced ability to play a pivoting or twisting sport, but should allow a reasonable level of leisure activity such as golf or hiking.

Comparing the gait achieved with CR or PS TKA designs, Andriacchi found that the CR implant produced statistically normal ranges of motion compared to the PS implant [39]. This was attributed to the lack of dynamic interaction between the PCL, tibial rollback with flexion, and the changing of the lever arm of the quadriceps during flexion. Patients with the PS implant leant forward in order to compensate for a lack of flexion, despite those with a PS implant having an identical range of motion to those with a CR TKA, which was then interpreted to reflect either weakness or avoidance of use of the quadriceps [39]. More recently, as prosthetic designs have improved, studies have failed to find a significant difference in gait when comparing CR and PS prostheses [5]. It is noted that quadricep avoidance is a phenomenon also seen in the ACL-deficient gait [39]. It is therefore possible that the ACL is more critical than previously thought for stability, and its absence an explanation for these gait changes seen post-operatively, regardless of the type of TKA.

29.3 Sports After TKA

As discussed above, no TKA restores normal knee biomechanics. The abnormal kinematics are likely to be detrimental to the performance of the TKA [30], and it is this that necessitates activity restrictions that are discussed below. Since TKA does not restore normal knee function, the challenging movement patterns and speed required to perform sports could be significantly impacted. More than half of patients report some degree of limitation in their activities following TKA, compared with only one fifth in age-matched subjects with no previous knee disorders [31]. Further, up to one quarter of patients have reported dissatisfaction following TKA [15]. Therefore, clinicians are increasingly forced to question how much activity a TKA patient can perform and what sports are acceptable [10].

Patients are increasingly involved in the shared decision-making process, thus the clinician must be able to counsel and inform them according to the available evidence [10]. However, to the best of our knowledge, there are no studies that specifically compare CR and PS implants with regard to sporting activity, as studies tend to group all TKA designs together. Furthermore, there appear to be no trials looking specifically at biomechanics and the challenges of pivoting sports. Much of the evidence in this field of return to sports and TKA is sparse and appears to be low level retrospective studies predominantly using questionnaires on self-reported outcomes. They are, therefore, heavily flawed with recall bias.

Return to sports is a complex and multifactorial outcome, and of course there will always be examples of patients who get back to relatively high-level sports, largely related to preoperative fitness, skill, experience, and motivation. There will also be others who are at the other extreme and fail to return to any level of sport. The reason for this variation goes well beyond the design of the TKA implant. Healy et al. [32] suggest the ability to return to sports participation after TKA is dependent on several factors: (1) pre-operative athletic ability, (2) pre-operative (p)rehabilitation, (3) surgical reconstruction, (4) implant failure, (5) implant fixation, (6) wear of the bearing surface and (7) trauma [1]. Further, others show that sports participation is dependent on ageing and motivation [31].

Functional results in athletes are primarily linked to the flexion of the knee and strength of the quadriceps [12]. Flexion is multifactorial concerning the patient, the surgical technique, the implant design and post-operative rehabilitation [12]. These factors influencing return to sport could simply be divided into the implant design, surgical technique and patient variables [11]. It might therefore be reasonable to question how critical a factor the biomechanics of the TKA implant actually is in a patient's ability to return to sport.

29.3.1 What Is Sport?

Most surgeons recommend low impact, low demand, low duration sports following TKA [1, 33]. Flecher et al. [12] state that the majority of activities can be resumed excluding team sports, ball sports and jogging. However, there is little evidence to support these suggestions, which are largely based on first principles and left to surgeons' discretion [33]. Guidelines published by Healy et al. [34] after surveying the members of the Knee Society (1999) [34] set out their recommendations for sports considered appropriate after TKA (see Table 29.1). Furthermore, these recommendations appear based on opinion rather than evidence, thus it is not entirely clear what should and should not be recommended for each individual after TKA. It is, of course, important to define what may be considered 'sport' versus 'leisure activities' in order to clearly counsel patients appropriately and ensure their expectations are addressed accurately.

29.3.2 Patient Concerns

Patients in general are not as likely to return to sports after a TKA compared with those having a total hip replacement [6], and tend to decrease their participation in, and intensity of, athletic activities [4]. Recommendations for participation in sports after TKA appear to be based predominantly on opinions rather than evidence [1]. However, as discussed above, if the average age of the TKA patient is decreasing and the demand and expectations are rising, we must consider sporting options for the more active patient.

Kawamura showed that preoperative flexion (positively) and varus/valgus deformity (nega-

Table 29.1 1999 Knee Society Survey Recommendations for Activity after TKA

Recommended	Allowed with prior experience	Not recommended	No conclusion
Walking	Road cycling	Squash	Fencing
Swimming	Doubles tennis	Soccer	Downhill skiing
Low-impact aerobics	Rowing	Jogging	Rollerblading
Stationary bike	Weight machines	Singles tennis	Weightlifting
Hiking	Cross-country skiing	Volleyball	
Ballroom dancing		Hockey	
Bowling		Basketball	
Croquet		Handball	
Golf		Rock climbing	

Modified from [4, 34]

tively) affects post-operative flexion [12], supporting the importance of good physiotherapy prior to surgery. Silva et al. note that quadriceps strength decreases by around 30% compared with the contralateral side following a TKA [12]. Lamb et al. state that the two most predictive factors of post-operative strength are BMI and pre-operative strength, again highlighting the increasingly recognised importance of 'pre-habilitation'.

Regarding the impact of body weight, biomechanical modelling has shown a significant effect of vertical hip load (related to body weight) on compressive load to the knee and the contact areas of both the tibio-femoral and patellofemoral joints, demonstrating the potentially deleterious effect of obesity on outcomes [11].

29.3.3 Implant Concerns

Typically the concern over sports participation relates to early component failure through excessive polyethylene wear, and excessive stress leading to loosening or trauma [1, 32]. Accumulating data suggests that prosthetic wear is not simply a function of time but one of use [3, 32]. The forces encountered by the TKA vary from 3 to 8 times body weight when ascending and descending stairs [1]. Jogging exceeds 7 times body weight and isokinetic exercise can exceed 12 times body weight [1]. Even whilst hiking, flexion between 40 and 60° loads the joint by 5–10 times body weight [10]. Thus, even the most basic of anticipated activities could be significantly contributing towards early wear of the prosthesis. This raises concerns about implant longevity for those wishing to return to sports [12, 33], and surgeons should educate patients regarding these risks. Table 29.2 illustrates the joint reaction forces through the knee during various activities.

Table 29.2 Knee joint reaction forces during different sports

Knee joint reaction forces	Body weight
Walking	3
Cycling	1.2
Stair ascent	5
Stair descent	6
Isokinetic knee extension	9
Jogging	8–9
Running	14
Skiing	3.5–10

Adapted from Hartford et al. [1]

29.3.3.1 Longevity

Studies taken from the Norwegian Arthroplasty Registry find a higher risk of revision of TKA in males under 65 year of age [2]. To our knowledge, there are no long-term studies on TKA longevity as related to sporting activities, but it would seem to be a reasonable assumption that younger age is a surrogate for higher activity levels, which in turn contributes to the higher revision rate.

Trauma caused by athletic participation remains a primary concern in advising on return to sports. Component dislocation, particularly in the PS TKA, and periprosthetic fracture are two severe complications that may occur during sports participation [1].

29.3.4 Sports/Leisure Activities Following TKA

Several retrospective studies have suggested athletic activity decreases after TKA, which is applicable to all ages [32]. In a survey of patients with TKA, Dahm et al. [3] found the mean activity level reported according to the UCLA activity scale was 'regularly participating in active events' (level 7). However, what one interprets as an 'active event' is open to interpretation as the only example given is cycling. About 16% of patients reported participating in heavy lifting, strenuous tasks or 'not recommended' sports. Baumann et al. [17] found a similar mean level of activity as assessed by the UCLA scale (mean of 6—moderate activity) as reported by Dahm et al. [3], but with no patients involved in any impact activity.

Bock et al. also reported a reduction in higher impact-type sports such as climbing, soccer and tennis, with no one returning to these sports, but an increase in participation of low-impact activities such as walking and swimming [10]. Bradbury et al. [6] also found that no one could return to high impact/pivoting sports and noted a decrease in participation in most low-level activities with the exception of cycling, in which participation increased. In a retrospective survey in Korean patients by Chang et al. [35] participation in a number of low-impact activities increased following TKA, but more challenging and higher impact sports showed a decrease in participation. Similarly, Bradbury et al. [6] showed an overall decrease in all activities following TKA with only 77% returning to any level of activity. There was an increase in cycling, but a complete avoidance of any high impact/contact sports such as running, skiing, football (soccer) and rugby.

Diduch et al. [36] found that only 24% of TKA patients reached level 5 on the Tegner activity scale. Bonnin et al. [18] used a retrospective self-reported questionnaire to TKA patients, with an average length of time from surgery to study of 44 months and a low respondent rate. More than a quarter reported they were less active than prior to the operation, and more than half considered their activities to be limited by their knee. Only 10% of patients participated in what was considered 'strenuous sports', defined as skiing, tennis and running over 500 m. However, a closer look at the data showed only 1 of the 347 respondents actually participated in tennis and only 3 could run more than 500 m. Furthermore, the level of activity was only rated in terms of level of participation rather than the level of performance. This study highlights the importance of appropriately counselling prospective TKA patients and managing their expectations.

Hepperger et al. [37] reported that sports activity is maintained or increased following TKA for 78% of their cohort. Their prospective study was based on standardised outcome scores over 2 years and showed statistical improvements, although the median Tegner score, for example, did not change from level 3 at any time point from pre-operative to 24 months post-operative. This is a lower level than reported by Diduch et al. [36]. The sports chosen in this cohort were not discussed in detail but was predominantly cycling, swimming, hiking and skiing, and did not include any more vigorous pivoting sports. Bradbury reported that the 75% of patients who performed a sport (walking, bowling, golf) pre-operatively could continue post-operatively, but only 20% of those who played tennis could return [12, 33]. Clearly the baseline level of activity within any cohort will influence patients' expectations for activity post TKA and the potential for improvement in activity level.

The difficulty with interpreting survey data is that it is always prone to selection bias, in that only those who want to participate respond, usually selecting those with the more positive outlook and results. Further caution must always be placed on measurement of level of participation in any sport. There is a clear difference between performing at a competitive level compared with social level, and self-reporting is not always reliable or objectively accurate. Furthermore, the retrospective nature of these surveys is additionally subject to recall bias, and bias based on the relationship between the patient and surgeon. No data exists with regard to TKA kinematics during running, turning, cutting or pivoting sports [1], only modelling and cadaver studies. Return to sport must be gentle and progressive with moderate activities limited to short sessions. Some TKA patients may be able to return to moderate activities proving they have prior experience and an adequate level of technique for that sport [2].

29.3.4.1 Specific Sports

Golf

Golf is a sport that patients would usually expect to be able to return to after TKA, and indeed an inability to play golf or walk the course is a common catalyst for patients to seek surgery. The reported results around patients returning to golf after TKA have been somewhat variable. Limitations affecting patients' ability to return to golf largely relate to the ability to walk comfortably over relatively long distances on an uneven soft surface (requiring good proprioception) and a tolerance for the pivoting movement involved with the golf swing (Fig. 29.2).

Mallon and Callaghan studied 83 active golfers who had undergone TKA. All returned following surgery, although 87% used a cart whilst playing and 36% reported experiencing a mild ache in the operated knee after playing. Naal et al. [40] reported an increase in golf participation from 5 to 8 within their cohort. However, other authors have found a significant reduction (19 to 9) in those continuing to play golf following TKA [33]. The total of those leaving the sport represented a reduction of 44% (109 to 61), suggesting golf may be a realistic sport to anticipate returning to post-operatively, but very unlikely as a new sport to undertake.

Tennis

Tennis demands a higher level of impact than golf and requires more rapid movements including pivoting and lunging. In a small survey of 33 highly experienced tennis players (mean experience of 35 years, range 15–70 years) who had undergone TKA, Mont et al. [42] found a 100% return to tennis, both singles and doubles play. These players returned to the same level as pre-operatively, but at a relatively low level, and all players reported a significant loss of court speed [42].

These findings are not replicated when looking at non-selected patient groups with all of these studies reporting a significant drop in par-

Fig. 29.2 Demonstration of knee movement during the golf swing. Note the valgus stress and pivot stability required

ticipation. Table 29.3 shows a number of studies citing rates of participation in tennis before and after TKA. Totalling the numbers from these studies, the expected percentage returning to tennis is considerably lower than thought, at 18% (15 from 84 pre-operative tennis players). Thus, the experience, motivation and conditioning of the patient prior to surgery would appear to be a significant factor in return to tennis rather than specific TKA implant.

Tennis demands knee stability in antero-posterior and medial-lateral directions in both extension and flexion extension and flexion (Fig. 29.3).

Table 29.3 Return to tennis following total knee arthroplasty

Study	No. playing tennis pre-operatively	No. playing tennis post TKA	Percentage drop (%)
Bock et al. [10]	2	0	100
Bradbury et al. [6]	30	6	80
Chatterji et al. [33]	14	2	86
Huch et al. [43]	6	2	67
Naal et al. [40]	21	3	86
Walton et al. [41]	11	2	88

29.4 Summary

The understanding of native knee kinematics has been, and continues to be, used by implant manufacturers to innovate and improve TKA outcomes and longevity, as well as accommodate patient demands with a goal of allowing patients to return to their chosen activities, including sports, postoperatively and increase patient satisfaction. The result is a wide spectrum of TKR implant designs with varied biomechanical philosophies. Following this review, it would seem that being able to reliably restore a patient's ability to return to sports is affected by multiple factors, of which implant design is only one. Only with all of these factors including patient characteristics, surgical technique and implant factors considered, can the surgeon offer the patient realistic and evidence-based information regarding the likely outcomes of TKA and create valid expectations.

The other concern is the actual definition of 'sport'. In the case of the vast majority of the studies reviewed above, one might question if the activities described can be fairly described as sport or should be more accurately labelled as' leisure activities'. Cycling, swimming and hiking once or twice a week at a Tegner-defined level 3 is, at best, recreational. It is unfair to give the impression to patients that they might return to sports quoting upwards of 77% when in fact

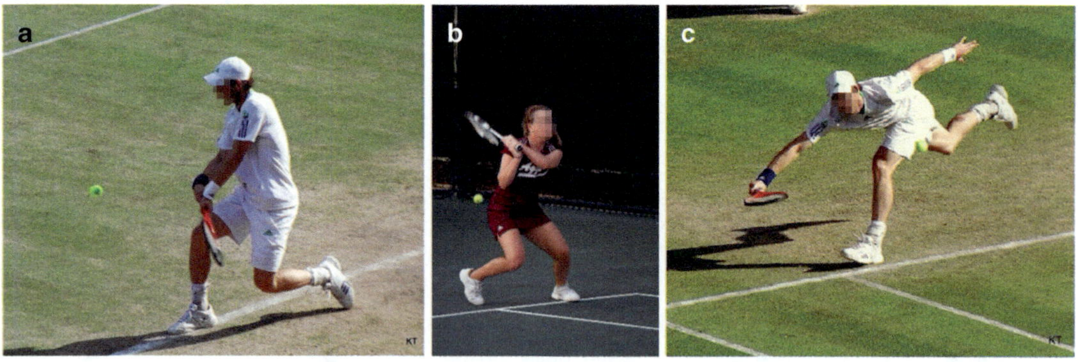

Fig. 29.3 (**a**) Knee flexion in tennis. (CC Search (creativecommons.org). "Andy Murray" by Carine06 is licensed with CC BY-SA 2.0. To view a copy of this license, visit https://creativecommons.org/licenses/by-sa/2.0/. Andy Murray | Andy Murray during his semi-final with Rafael … | Flickr). (**b**) Pivoting on knee. ("Aggie Women's Tennis—58" by StuSeeger is licensed with CC BY 2.0. To view a copy of this license, visit https://creativecommons.org/licenses/by/2.0/)). (**c**) High loads are experienced by the knee. (CC Search (creativecommons.org). "Andy Murray" by Carine06 is licensed with CC BY-SA 2.0. To view a copy of this license, visit https://creativecommons.org/licenses/by-sa/2.0/. Andy Murray | Andy Murray during his semi-final with Rafael … | Flickr)

the numbers of patients involved in pivoting sports is less than 5%, and all of these had significant prior experience within that sport. Patients' expectations should be fully explored on an individual basis and a good understanding of the actual activities one is able to participate in should be realistically discussed. This discussion should clearly include advice around which sports the patients are realistically likely to be capable of returning to, as well as which sports are less advisable due to concerns for implant longevity.

In summary, participation in leisure activities is deemed to be acceptable following a TKA, but this is based on consensus statements, surveys of orthopaedic surgeons and retrospective studies [17]. Recommendations regarding appropriate activities should be made on a case-by-case basis with a full understanding of the patient's expectations, pre-operative activity level and sporting experience.

Life-long and ex-professional surfer following a right total knee replacement

References

1. Hartford JM. Sports after arthroplasty of the knee. Sports Med Arthrosc. 2003;11:149–54.
2. Dagneaux L, Degeorge B. Return to sport after total or unicompartmental knee arthroplasty: an informative guide for residents to patients. EFORT Open Rev. 2017;2(December):496–501.
3. Dahm DL, Barnes SA, Harrington JR, Sayeed SA, Berry DJ. Patient-reported activity level after total knee arthroplasty. J Arthroplasty. 2008;23(3): 401–7.
4. Healy WL, R Iorio, M J Lemos. Athletic activity after joint replacement. Am J Sports Med. 2001;29(3): 377–88.
5. McClelland JA, Webster KE, Feller JA. Gait analysis of patients following total knee replacement: a systematic review. Knee. 2007;14:253–63.
6. Bradbury N, Borton D, Spoo G, Cross MJ. Participation in sports after total knee replacement. Am J Sports Med. 1998;26(4):530–5.
7. Brittain R, Young E, Mccormack V, Swanson M. 16th Annual report. 16th Annual report 2019 National Joint Registry England, Wales, North Ireland and Isle of Man, December 2018; 2019.
8. Shenoy R, Pastides PS, Nathwani D. Biomechanics of the knee and TKR. Orthop Trauma. 2013;27(6):364–71. https://doi.org/10.1016/j.mporth.2013.10.003.
9. Collins M, Lavigne M, Girard J, Vendittoli P. Joint perception after hip or knee replacement surgery. Orthop Traumatol Surg Res. 2012;98(3):275–80. https://doi.org/10.1016/j.otsr.2011.08.021.
10. Witjes S, Gouttebarge V, Kuijer PPFM, Van Geenen RCI, Poolman RW, Kerkhoffs GMMJ. Return to sports and physical activity after total and unicondylar knee arthroplasty: a systematic review and meta-analysis. Sports Med. 2016;46(2):269–92.

11. Fitzpatrick CK, Clary CW, Rullkoetter PJ. The role of patient, surgical, and implant design variation in total knee replacement performance. J Biomech. 2012;45(12):2092–102. https://doi.org/10.1016/j.jbiomech.2012.05.035.
12. Flecher X, Argenson JN, Aubaniac JM. Prothèse de hanche, du genou et sport Hip and knee replacement and sport. Rehabil Phys Med. 2004;47:382–8.
13. Aweid O, Melton J. Biomechanics of the knee. Orthop Trauma. 2019;33(4):224–30. https://doi.org/10.1016/j.mporth.2019.05.004.
14. Vendittoli P, Blakeney W. Redefining knee replacement. Orthop Traumatol Surg Res. 2017; https://doi.org/10.1016/j.otsr.2017.09.003.
15. Muckenhirn K, Chahla J, Laprade RF. Anatomy and biomechanics of the native knee and its relevance for total knee replacement. Chapter 1, Anatomy and biomechanics of the native knee and its relevance for total knee replacement, ISAKOS. 2017;3–15.
16. Wishart N, Beaumont R, Young E, Mccormack V, Swanson M. The NJR 11th Annual Report 2014 (December 2013). https://doi.org/10.13140/2.1.1191.2481.
17. Bauman S, Williams D, Petruccelli D, Elliott W, De Beer J. Physical activity after total joint replacement: a cross-sectional survey. Clin J Sport Med. 2007;17(2):104–8.
18. Bonnin M, Parratte JRLS, Bissery FZRBA, Bissery A. Can patients really do sport after TKA? Knee Surg Traumatol Arthrosc. 2010;18:853–62.
19. Scott CEH, Howie CR, Macdonald D, Biant LC. Predicting dissatisfaction following total knee replacement. A prospective study of 1217 patients. J Bone Joint Surg Br. 2010;92(9):1253–8.
20. Australian Orthopaedic Association National Joint Replacement Registry (AOANJRR). Hip, Knee & Shoulder Arthroplasty: 2019 Annual Report. Adelaide: AOA, 2019.
21. Athwal KK, Hunt NC, Davies AJ, Deehan DJ, Amis AA. Clinical biomechanics of instability related to total knee arthroplasty. Clin Biomech. 2014;29(2):119–28. https://doi.org/10.1016/j.clinbiomech.2013.11.004.
22. Whiteside LA. Soft tissue balancing: the knee. J Arthroplasty. 2002;17(4 Suppl 1):23–7.
23. Howell SM, Hull ML. Kinematic alignment in total knee arthroplasty. Chapter 121, Knee Arthroplasty 2011, p. 1255–69.
24. Broberg JS, Ndoja S, Macdonald SJ, Lanting BA, Teeter MG. Comparison of contact kinematics in posterior-stabilized and cruciate-retaining total knee arthroplasty at long-term follow-up. J Arthroplasty. 2020;35(1):272–7. https://doi.org/10.1016/j.arth.2019.07.046.
25. Song SJ, Park CH, Bae DK. What to know for selecting cruciate-retaining or posterior-stabilized total knee arthroplasty. Clin Orthop Surg. 2019;11:142–50.
26. Kang K, Koh Y, Son J, Kwon O, Lee J, Kwon SK. Comparison of kinematics in cruciate retaining and posterior stabilized for fixed and rotating platform mobile-bearing total knee arthroplasty with respect to different posterior tibial slope. Biomed Res Int. 2018; 5139074.
27. Cates HE, Komistek RD, Mahfouz MR, Schmidt MA, Anderle M. In vivo comparison of knee kinematics for subjects having either a posterior stabilized or cruciate retaining highflexion total knee arthroplasty. J Arthroplasty. 2008;23(7):1057–67.
28. Li G, Kernkamp WA, Rubash HE. In vitro and in vivo kinematics of total knee arthroplasty — a review of the research at the Orthopaedic Bioengineering Laboratory of the Massachusetts General Hospital (MGH). Ann Joint. 2016;1:20.
29. Singleton N, Nicholas B, Gormack N, Stokes A. Differences in outcome after cruciate retaining and posterior stabilized total knee arthroplasty. J Orthop Surg. 2019;27(2):1–8.
30. Stiehl J, Komistek R, Dennis D, Paxson R, Hoff W. Fluoroscopic analysis of kinematics after posterior-cruciate-retaining knee arthroplasty. J Bone Joint Surg Br. 1995;77:884–9.
31. Noble PC, Gordon MJ, Weiss JM, Reddix RN, Conditt MA, Mathis KB. Does total knee replacement restore normal knee function? Clin Orthop Relat Res. 2005;431:157–65.
32. Healy BWL, Sharma S, Schwartz B, Iorio R. Athletic activity after total joint arthroplasty. J Bone Joint Surg. 2008;90:2245–52.
33. Chatterji U, Ashworth M, Lewis P, Dobson P. Effect of total knee arthroplasty on recreational and sporting activity. ANZ J Surg. 2005;75:405–8.
34. Healy WL, Iorio R, Lemos MJ. Athletic activity after total knee arthroplasty. Clin Orthop Relat Res. 2000;380:65–71.
35. Chang MJ, Kim SH, Kang YG, Chang CB, Kim TK. Activity levels and participation in physical activities by Korean patients following total knee arthroplasty. BMC Musculoskelet Disord. 2014;15(1):2–7.
36. Diduch D, Insall J, Scott W, Scuderi G, Font-Rodriguez D. Total knee replacement in young, active patients: long-term follow-up and functional outcome. J Bone Joint Surg. 1997;79(4):575–82.
37. Hepperger C, Gföller P, Christian EA, Ulmer H, Herbst E, Fink C. Sports activity is maintained or increased following total knee arthroplasty. Knee Surg Sport Traumatol Arthrosc. 2018;26(5):1515–23.
38. Levinger P, Menz HB, Morrow AD, Pod M, Feller JA, Bartlett JR, et al. Lower limb biomechanics in individuals with knee osteoarthritis before and after total knee arthroplasty surgery. J Arthroplasty. 2013;28(6):994–9. https://doi.org/10.1016/j.arth.2012.10.018.
39. Andriacchi TP. Functional analysis of pre and post-knee surgery: total knee arthroplasty and ACL reconstruction. J Biomech Eng. 1993;115:575–81.
40. Naal FD, Impellizzeri FM, Leunig M. Which is the best activity rating scale for patients undergo-

ing total joint arthroplasty? Clin Orthop Relat Res. 2009;467(4):958–65.
41. Walton NP, Tr F, Lewis PL, Dobson PJ, Angel KR, Campbell DG. Patient-perceived outcomes and return to sport and work: TKA versus mini-incision unicompartmental knee arthroplasty. J Knee Surg. 2006;19(2):112–6.
42. Mont MA, Rajadhyaksha AD, Marxen JL, Silberstein CE, Hungerford DS. Tennis after total knee arthroplasty. Am J Sports Med. 2002;30(2):163–6.
43. Huch K, Muller K, Sturmer T, Brenner H, Puhl W, Gunther K-P. Sports activities 5 years after total knee or hip arthroplasty: the Ulm Osteoarthritis Study. Ann Rheum Dis. 2005;64:1715–20.

30 Biomechanics of Unicompartmental Knee Replacement

Johanna Elliott and Myles Coolican

30.1 Introduction

Osteoarthritis of the knee is a complex disorder that may involve any one of the three compartments of the knee alone or in combination. Proponents of unicompartmental knee arthroplasty (UKA) point to the clinical benefits that UKA offers knee arthritis patients compared to total knee arthroplasty (TKA) when managing isolated medial or lateral wear. In summary, they feel a well-functioning UKA is better than a well-functioning TKA with patients describing a more naturally behaving knee. Tailoring the treatment to the involved compartment avoids replacing non-worn parts as can be the case with many patients undergoing TKA. Given that unicompartmental knee arthritis is not necessarily the beginning of a progressive global knee arthritis, preference for patient-tailored solutions is growing, but it has been a journey with some twist and turns.

J. Elliott · M. Coolican (✉)
Sydney Orthopaedic Research Institute, Chatswood, NSW, Australia
e-mail: myles@mylescoolican.com.au

30.2 Historical Perspective on Implant Design and Survivorship

The modern era of UKA began with the St George Sled (1969) and Marmor (1972) prostheses, the designs of which were based on a polycentric metal femoral condyle, articulating with a flat and non-congruent tibial polyethylene component designed to minimize constraint, thus reducing shear forces at the implant bone interface. In 1974, the Oxford mobile bearing UKA combined femoral and polyethylene congruency and reduced shear forces by introducing a gliding minimally constrained mobile bearing articulation with a spherical femoral component [1].

Evaluation of early failures demonstrated that altered knee kinematics frequently contributed to poor outcomes. Overstuffing, mobile bearing dislocation, polyethylene wear and malalignment following overcorrection of limb deformity resulted in failures with malalignment causing symptoms and progressive wear in the opposite non-resurfaced compartment. Whilst stress shielding and polyethylene wear were recognized as technical and design problems, there is little doubt that the quality of polyethylene from that period and its sterilization in air rather than an inert gas contributed to the higher failure rate and that further improvements such as cross-linking polyethylene with radiation have helped lower revision rates.

J. Koh et al. (eds.), *Orthopaedic Biomechanics in Sports Medicine*,
https://doi.org/10.1007/978-3-030-81549-3_30

Polyradial femoral designs demonstrated increased joint compression forces in flexion, leading to excessive polyethylene wear particularly posteriorly. Coronal axis over-correction shifted load to the uninvolved compartment and was associated with progressive symptoms in the remaining compartment as well as stress shielding of the medial tibia. Ligament balancing as an essential part of the surgical technique to avoid polyethylene dislocation and joint overstuffing gained appreciation. However, few if any studies showed long-term survivorship of UKA to match that of TKA. More recently, data from joint replacement registries has consistently shown UKA revision rates to be two to three times that of TKA. As a consequence, UKA as a proportion of all knee arthroplasty has fallen in the past 10–15 years. It has become clear that UKA is a less forgiving procedure than TKA in terms of imperfect implant positioning. It must also be acknowledged that the threshold to revise a well fixed but painful UKR is lower than for a well fixed but painful TKR. A recent resurgence has been seen driven in part by new technology to improve implant positioning, in particular computer navigated and robotic surgery.

30.3 Biomechanics of the Native Knee and Limitations of FEA to Predict Outcome in UKA

Radiostereometric analysis (RSA) and MRI studies have shown that the tibia rotates internally with increasing flexion. Whilst this rotation—known as roll-back—is greater on the lateral side, some medial femoral rollback does occur, creating shear forces at the interface for a medial UKA.

Force transmission during the gait cycle is impacted by the mechanical properties of the materials used. Many finite element analysis (FEA) studies have shown that implanting a UKA in the anatomical position without altering lower limb alignment increases load transmission to the remaining un-resurfaced compartment. This is in part because metal-backed tibial trays demonstrate a relative strain shielding [2]. The aim of surgery should be to balance load transmission across medial and lateral compartments to limit wear of polyethylene and prevent progressive arthritis in the remaining compartment. It is important for the surgeon to recognize that load transmission characteristics do not necessarily correlate with lower limb alignment once the material properties of the relative compartments are altered by the UKA implant.

Heyse et al. in cadaver studies [3] demonstrated that UKA replicates well the biomechanics of the native knee, and vivo studies [4] support these findings. However, a limitation is that many FEA studies have chosen as their computer model a healthy adult knee, despite the fact that the kinematics of an arthritic knee may be altered.

Clinical studies have shown residual varus does not correlate with a poor clinical outcome [5], and the results are best when the implant minimally adjusts the biomechanics of the knee, without attempting to restore "normal" alignment. An FEA study by Guo et al. [6] supports the concept of leaving the limb in some residual varus to prevent early progressive lateral compartment osteoarthritis. In effect, the implant functions best as a joint resurfacing of the involved compartment.

Bearing in mind the limitations of direct transference of in vitro results to the clinical situation, a summary of the current knowledge and research in this field is presented below.

30.4 Surgical Considerations

There are three matters for the surgeon to consider when performing UKA, preserving the original joint line, resecting the correct amount of bone on both sides of the joint and creating an articulation with balanced soft tissues. Although there is little evidence that subtle modification of the joint line has much clinical impact on the on the patellofemoral joint in UKA, it does alter the distribution of forces to the opposite compartment particularly if associated with an alteration in alignment.

Determining optimal tibial resection remains one of the most challenging aspects of this procedure. Advice on planning tibial resection varies between prosthetic designs. The Oxford UKA which is indicated for anteromedial osteoarthritis acknowledges tibial bone loss results in a pseudolaxity of the medial collateral ligament. A stated goal of implanting an Oxford UKR is to restore the MCL to its physiological length. Other implant designs are not as specific in their advice. Many designs suggest a minimal tibial resection of 4 mm in order to accommodate the implant and avoid overstuffing. The surgeon must choose between either under resection with subsequent over stuffing the replaced compartment or excessive tibial resection removing the harder bone under the plateau, thereby increasing the risk of both tibial subsidence and fracture.

30.5 Joint Line Preservation

Kwon et al. [7] in 2017 with a fixed bearing design and Kang et al. [8] in 2018 with a mobile bearing design performed FEA studies on the effect of proximalizing and distalizing the joint line over a range of 6 mm above and below neutral. They observed the contact stress on the polyethylene in the replaced medial compartment as well as the stress on the articular cartilage of the lateral compartment. They described a reciprocal relationship for stress distribution, being that when stress was increased on the polyethylene, it was reduced on the remaining (lateral) compartment articular cartilage. They noted that the effects were proportionately greater for proximalizing the joint line than for distalizing it and concluded that preservation of the original joint line would provide the best conditions for implant survival.

30.6 Tibial Component Positioning

30.6.1 Limb Alignment

Clinical studies demonstrating improved outcomes for under-corrected lower limb alignment led to a general acceptance of residual varus as the goal of medial UKA. A figure up to 3° mechanical varus has been cited in many publications [9–11].

30.6.2 Tibial Cut Orientation

Several biomechanical studies describe the effect of variations in frontal and coronal plane orientation of the tibial component on polyethylene and on the remaining (lateral) joint articular cartilage throughout the gait cycle. Innocenti et al. [12] performed a finite element analysis to quantify bone stress, load distribution, ligament strain and polyethylene stress distribution over a range of angles from 6° varus to 6° valgus. Their results agreed with the current consensus that stress is better balanced across the proximal tibia, and the strain in ligaments is optimal when the coronal axis of the tibial cut remains within three degrees varus.

In their finite element analysis, Inoue et al. [13] examined tibial tray angles from 6° of varus to 6° of valgus combined with sagittal extension of the posterior tibial cut. They concluded that increasing valgus cuts and greater loss of bone volume medially increase the risk of medial tibial condylar fracture.

Guo et al. [14] performed a finite element analysis of an Oxford medial UKA with a mobile bearing design and focused on the stress transfer to medial tibial bone over a range of 4° varus to 4° valgus for the tibial implant. They found increased compressive strain on the medial tibia with increasing valgus positioning of the tibial component and theorized that this could clinically lead to pain. Varus above 4° resulted in high compressive forces at the keel slot, potentially causing component migration. They cautioned that the tibial implant should not exceed these parameters. However, the findings of this FEA study are not supported in the earlier published clinical findings of Clarius et al.

The effect of posterior tibial slope (PTS) on the outcome of UKA has received attention. Most authors report reduced contact stress with a PTS of 7°. A recent FEA by Kang et al. [15] demonstrated the effects of PTS on contact stress of the polyethylene insert as well as the remaining compartment. They determined that the relationship between polyethylene stress and stress on the remaining (lateral) compartment was reciprocal, under the influence of PTS. With increasing slope, polyethylene stress decreased, whilst opposite compartment stress increased. The authors recommended aiming for a slope within 2° of the native slope and generally agreed with a consensus slope of 7°.

30.6.3 Medial Tibial Condylar Fracture Risk

As proximal medial tibial fracture is one of the most devastating complications of medial UKA, the biomechanical factors that may contribute to the risk of fracture have extensively been investigated.

Scott et al. [2] used composite tibial saw bones to assess the proximal tibial strain transmission under all-polyethylene and metal-backed medial tibial components. The measured strain transmission demonstrated strain shielding with metal backed implants, for both fixed and mobile bearings. The authors caution surgeons against the use of all polyethylene tibial components in patients whose activity level is likely to impose higher medial strains on their knees.

Pegg et al. [16] use a finite element analysis to apply a probabilistic approach to predict the aetiology of medial tibial condylar fracture after unicompartmental knee replacement. In their analysis, they considered several factors:

- Excessive depth of surgical cuts made for the tray, tray keel, or pegs
- Multiple cortical holes for alignment guide placement
- Perforation of the tibial cortex
- Under-sizing of the tibial tray
- Use of excessive force when impacting the implant
- Excessive bone removal

They concluded that the parameters most likely to contribute to medial tibial condylar fracture were resection depth (excessive removal of tibial bone) and distal extension of the posterior part of the sagittal plane cut.

30.7 Femoral Component Positioning

Given correct restoration of the joint line with the correct depth of cut, the femoral component can vary in position in the sagittal and coronal planes and can vary its rotation in the coronal, sagittal and axial planes. From the surgeon's perspective, positional options include anteroposterior and mediolateral placement with rotational options as varus valgus, flexion extension and internal or external rotation.

Limb alignment is usually established in extension by most systems with the tibial cut first. Following this a measured distal femoral resection occurs, producing a range of varus–valgus alignments that varies with the amount of bone removed and the implant thickness. Many systems use ligament balancing to determine combined implant thickness. Femoral component positioning is critical for its impact on the medial collateral ligament and mobile bearing inserts.

Many biomechanical studies based on FEA have been performed, considering varying configurations of femoral component placement. Outcomes measured include strain on the remaining compartment, the polyethylene insert, as well as ligamentous structures. A further variable in these analyses is bearing mobility. Given the limitations of comparing the outcomes of these varied studies, most authors conclude that restoration of the rotational centre represents the biomechanical ideal.

Pegg et al. [17] demonstrated increased tibial strain with medialization of the femoral component of more than 3 mm in their FEA of a virtually implanted mobile bearing medial Oxford UKA. These findings were supported by the FEA study of Kang et al. [18] using a fixed bearing model where the effect of mediolateral translation of the femoral component was studied. The authors describe the effect on contact stress with varying the position of the femoral component had over five locations from medial to lateral, including the centre and 3 and 5 mm medial and lateral, respectively. Again, an inverse relationship was determined, whereby medializing the femoral component reduced the contact stress on the polyethylene but increased the stress on the remaining (lateral) articular cartilage. This pattern was maintained throughout the gait cycle. The authors concluded that the optimum position to be the centre of the femoral condyle, hypothesizing that medialization of the femoral component could be a basis for persistent medial pain clinically.

Inui et al. [19] in their clinical study looking at femoral component positioning described an association between lateralization of the femoral component and mobile bearing tilt. Again, a central location of the femoral component was recommended.

Kang et al. performed a similarly validated FEA method in 2018 [20] to observe the distribution of contact stresses between the polyethylene and the remaining articular cartilage, and the forces on collateral constraints across a range of femoral coronal plane alignments from 9° of varus to 9° of valgus. Under the conditions of a preserved neutral lower limb mechanical alignment and femoral and tibial cuts, isolated varus–valgus positioning of the femoral component resulted in an inverse relationship of increasing stress on the polyethylene, resulting in decreased contact stress on the opposite compartment. Valgus alignment resulted in increased medial edge loading of the polyethylene, whilst varus femoral component alignment led to increased contact stress on the lateral aspect of the polyethylene and increased stress on the lateral articular cartilage. Overall, the effect of valgus malalignment of the femoral component was more pronounced than varus malalignment over the range considered with respect to redistribution of loading forces over the polyethylene and lateral cartilage.

The same authors showed that valgus malalignment produced increased strain in the medial collateral ligament. Again, this was more pronounced than the increased strain observed in the lateral collateral ligament by varus malalignment. The popliteo-fibular and the anterolateral ligament demonstrated no change in strain with respect to the malpositioning of femoral component. The authors concluded that valgus malpositioning of the femoral component should be avoided to minimize the risk of MCL-related complications.

A further study by Park et al. [21] used a similarly validated technique of FEA to study the effect of variations in sagittal position of the femoral component over a range of 10° of flexion through to 10° of extension. Virtual implantation of a fixed bearing medial UKA showed that particularly flexion of the femoral component resulted in increased contact stress on both the polyethylene and the un-resurfaced (lateral) compartment. Extension of the femoral component appeared to be more forgiving, showing only slight increase in contact stress on polyethylene and no increase in stress on the lateral articular cartilage. Extension of the femoral component had the effect of increasing the strain in the medial collateral ligament, whilst reducing strain in the lateral collateral, patellofemoral and anterolateral ligaments. The opposite effect was noted with flexion of the femoral component.

Clinical validation of the extensive laboratory analyses discussed above remains to be published, and in all likelihood, we will never see published results of a human trial where implants are inserted in markedly non anatomical positions. However, a review of outcomes where implant position is closely measured may confirm laboratory findings. Clarius et al. [22] reported their medium-term results in 59 patients, where only 29% of components were placed centrally on the medial femoral condyle. At that point in follow-up, there was no evidence of poorer clinical outcome. Varus–valgus alignment, in their series, however, was much more closely controlled with 96% of cases falling within the recommended values. The authors concluded that there is room for variation in implant positioning, given the biomechanical advantages specific to the mobile bearing Oxford UKA, provided the surgeon adheres to the principles of restoring the mechanical axis and preserving ligament balancing.

30.8 Ligament Balancing

Heyse et al. [3] performed a cadaver study looking at the impact of overstuffing the medial compartment in a medial UKR. They observed that a subjective assessment of a 'balanced' knee resulted in valgus malalignment of the limb and significantly increased strain in the superficial medial collateral ligament. Results most comparable to the native knee kinematics could be achieved by understuffing the medial compartment.

30.9 Fixed Bearing Versus Mobile Bearing

Brockett et al. [23] looked at wear patterns in fixed and mobile bearing UKRs. Despite theoretically reduced shear forces with the Oxford mobile bearing inlay, the guide railing on the lateral side of the tibial tray can block rotation between the femoral condyle and the mobile bearing, resulting in multidirectional motion and shear forces which will precipitate wear [23].

Ettinger et al. [24] performed a cadaver study comparing fixed and mobile bearing UKA designs assessing the effect of quadriceps extension force on tibiofemoral contact pressures. They determined a mechanical advantage for mobile bearing designs with respect to quadriceps extension force. Pressure distribution analysis showed higher concentrated peak pressures for fixed bearing designs, but lower pressure introduction in deep flexion when compared with mobile designs.

30.10 Clinical Reports

Gulati et al. [25] in their 4-year clinical and radiological follow-up of a cohort of medial UKAs concluded that due to the spherical nature of the Oxford femoral implant, component malalignment was relatively well tolerated. Most cases (98%) lay within a range of femoral component varus–valgus alignment of up to 10°, as well as flexion–extension alignment up to 10°. This variation of femoral alignment was well tolerated, with no significant difference in radiographic evidence of loosening, nor significant difference in Oxford Knee Score (OKS) between groups. Positioning of the tibial components was also variable. Most patients (92%) lay within a range between 5° of varus to 5° of valgus, and the same percentage were within 5° superior/inferior tilt (neutral tilt being 7°), with no significant differences between categories for radiolucency or OKS.

In their clinical and radiological report on the results of the Oxford medial UKA, Clarius et al. [22] reported their results for 59 patients at a minimum follow-up of 4 years. They considered similar radiographic parameters of femoral alignment (10° varus to 10° valgus, 5° extension to 5° flexion) and tibial alignment (10° varus to 5° valgus, slope +7° to −5°) but also included mediolateral positioning of the femoral component, posterior overhang of the femoral component and tibial overhang including anterior (+10 mm to −12 mm), posterior (−12 mm to +5 mm) and medial (+5 mm to −5 mm). They observed no correlation between implant positioning and radiological or clinical outcome, although they stressed the importance of slope as this is 'key for normal kinematics of the joint with intact ligaments'.

30.11 Lateral UKA

Whilst the biomechanics of the lateral compartment and its variations in the arthritic state has been studied, there is scarce literature when compared to that available for the medial compartment. This is not surprising given that isolated lateral compartment osteoarthritis is less common and lateral UKR is performed less frequently.

The biomechanics of the lateral compartment differ significantly from the medial compartment. The medial compartment is closer to a ball and socket through motion whilst with increasing knee flexion, the lateral femoral condyle translates posteriorly. This is known as femoral rollback and results in internal tibial rotation. As a consequence, the observed arthritic changes isolated to the lateral compartment tend to be located posteriorly [26]. Shear forces tend to be greater in the lateral compartment in flexion during the gait cycle, whilst load transmission in the stance phase is more in the medial compartment. This consideration has implications for alignment correction. Khamaisy et al. [27] demonstrated that despite robotic-assisted surgery, there was a tendency to over-correct lower limb alignment with lateral UKR into varus, thereby increasing the risk of progressive medial compartment arthritis.

The posterior slope of the lateral tibial can vary dramatically with that of the medial tibia within the same knee. Preventing lateral lift-off in flexion is an acknowledged challenge and persisting lift-off is a reason for the relative poor performance of mobile bearing lateral UKA designs when compared to medial UKR [28]. Experienced centres warn against the risk of elevating the lateral joint line in a mobile bearing lateral UKR [29] to minimize the chance of mobile bearing dislocation.

In their cadaver study, Ali et al. [30] found that excessive (>5°) PTS resulted in a significant increase in mean major principal strain in the posterior tibial zone. They concluded that small changes in implant orientation could significantly affect tibial cortical strain and cautioned against excessive PTS in lateral UKA.

A Swedish Registry study [31] demonstrated a higher revision rate than for medial UKR, and it is the authors' opinion that lateral UKR should be performed in centres with experience in this procedure.

30.12 Summary

Whilst it is commonly acknowledged that function in a well-executed UKA is generally better than a well-done TKR, the procedure is highly dependent on good surgical technique. Surgical decisions including variation in the plane and depth of bone resection for both the tibial and femoral implants affects the load transmission in the resurfaced and also the un-resurfaced compartments as well as loads on the collateral ligaments and this has a marked bearing on the outcome. It is likely that further refinements in surgical accuracy will enhance both function and longevity of implants.

References

1. Price AJ, O'Connor JJ, Murray DW, Dodd CAF, Goodfellow JW. A history of Oxford unicompartmental knee arthroplasty. Orthopedics. 2007;30(5 Suppl):7–10.
2. Scott CE, Eaton MJ, Nutton RW, Wade FA, Pankaj P, Evans SL. Proximal tibial strain in medial unicompartmental knee replacements: a biomechanical study of implant design. Bone Joint J. 2013;95-B(10):1339–47.
3. Heyse TJ, El-Zayat BF, De Corte R, Chevalier Y, Scheys L, Innocenti B, et al. UKA closely preserves natural knee kinematics in vitro. Knee Surg Sports Traumatol Arthrosc. 2014;22(8):1902–10.
4. Pegg EC, Bare J, Gill HS, Pandit HG, O'Connor JJ, Murray DW, et al. Influence of consciousness, muscle action and activity on medial condyle translation after Oxford unicompartmental knee replacement. Knee. 2015;22(6):646–52.
5. Gulati A, Pandit H, Jenkins C, Chau R, Dodd CA, Murray DW. The effect of leg alignment on the outcome of unicompartmental knee replacement. J Bone Joint Surg Br. 2009;91(4):469–74.
6. Wen PF, Guo WS, Gao FQ, Zhang QD, Yue JA, Cheng LM, et al. Effects of lower limb alignment and tibial component inclination on the biomechanics of lateral compartment in unicompartmental knee arthroplasty. Chin Med J (Engl). 2017;130(21):2563–8.
7. Kwon OR, Kang KT, Son J, Suh DS, Baek C, Koh YG. Importance of joint line preservation in unicompartmental knee arthroplasty: finite element analysis. J Orthop Res. 2017;35(2):347–52.
8. Kang KT, Kwon OR, Son J, Suh DS, Kwon SK, Koh YG. Effect of joint line preservation on mobile-type bearing unicompartmental knee arthroplasty: finite element analysis. Australas Phys Eng Sci Med. 2018;41(1):201–8.
9. van der List JP, Chawla H, Villa JC, Pearle AD. Different optimal alignment but equivalent functional outcomes in medial and lateral unicompartmental knee arthroplasty. Knee. 2016;23(6):987–95.
10. Petterson, SC, Blood, TD and Plancher KD Role of alignment in successful clinical outcomesfollowing medial unicompartmental knee arthroplasty:current concepts. Journal of ISAKOS: Joint Disorders and Orthopaedic Sports Medicine. 2020;5:224–228.
11. Vasso M, Del Regno C, D'Amelio A, Viggiano D, Corona K, Schiavone PA. Minor varus alignment provides better results than neutral alignment in medial UKA. Knee. 2015;22(2):117–21.

12. Innocenti B, Pianigiani S, Ramundo G, Thienpont E. Biomechanical effects of different varus and valgus alignments in medial unicompartmental knee arthroplasty. J Arthroplasty. 2016;31(12):2685–91.
13. Inoue S, Akagi M, Asada S, Mori S, Zaima H, Hashida M. The valgus inclination of the tibial component increases the risk of medial tibial condylar fractures in unicompartmental knee arthroplasty. J Arthroplasty. 2016;31(9):2025–30.
14. Zhu GD, Guo WS, Zhang QD, Liu ZH, Cheng LM. Finite element analysis of mobile-bearing unicompartmental knee arthroplasty: the influence of tibial component coronal alignment. Chin Med J (Engl). 2015;128(21):2873–8.
15. Kang KT, Park JH, Koh YG, Shin J, Park KK. Biomechanical effects of posterior tibial slope on unicompartmental knee arthroplasty using finite element analysis. Biomed Mater Eng. 2019;30(2):133–44.
16. Pegg EC, Walter J, D'Lima DD, Fregly BJ, Gill HS, Murray DW. Minimising tibial fracture after unicompartmental knee replacement: a probabilistic finite element study. Clin Biomech (Bristol, Avon). 2019;73:46–54.
17. Pegg EC, Walter J, Mellon SJ, Pandit HG, Murray DW, D'Lima DD, et al. Evaluation of factors affecting tibial bone strain after unicompartmental knee replacement. J Orthop Res. 2013;31(5):821–8.
18. Kang KT, Son J, Koh YG, Kwon OR, Kwon SK, Lee YJ, et al. Effect of femoral component position on biomechanical outcomes of unicompartmental knee arthroplasty. Knee. 2018;25(3):491–8.
19. Inui H, Taketomi S, Yamagami R, Sanada T, Shirakawa N, Tanaka S. Impingement of the mobile bearing on the lateral wall of the tibial tray in unicompartmental knee arthroplasty. J Arthroplasty. 2016;31(7):1459–64.
20. Kang KT, Son J, Baek C, Kwon OR, Koh YG. Femoral component alignment in unicompartmental knee arthroplasty leads to biomechanical change in contact stress and collateral ligament force in knee joint. Arch Orthop Trauma Surg. 2018;138(4):563–72.
21. Park KK, Koh YG, Park KM, Park JH, Kang KT. Biomechanical effect with respect to the sagittal positioning of the femoral component in unicompartmental knee arthroplasty. Biomed Mater Eng. 2019;30(2):171–82.
22. Clarius M, Hauck C, Seeger JB, Pritsch M, Merle C, Aldinger PR. Correlation of positioning and clinical results in Oxford UKA. Int Orthop. 2010;34(8):1145–51.
23. Brockett CL, Jennings LM, Fisher J. The wear of fixed and mobile bearing unicompartmental knee replacements. Proc Inst Mech Eng H J Eng Med. 2011;225(5):511–9.
24. Ettinger M, Zoch JM, Becher C, Hurschler C, Stukenborg-Colsman C, Claassen L, et al. In vitro kinematics of fixed versus mobile bearing in unicondylar knee arthroplasty. Arch Orthop Trauma Surg. 2015;135(6):871–7.
25. Gulati A, Chau R, Simpson DJ, Dodd CA, Gill HS, Murray DW. Influence of component alignment on outcome for unicompartmental knee replacement. Knee. 2009;16(3):196–9.
26. Heyse TJ, Tibesku CO. Lateral unicompartmental knee arthroplasty: a review. Arch Orthop Trauma Surg. 2010;130:1539–48.
27. Khamaisy S, Gladnick BP, Nam D, Reinhardt KR, Heyse TJ, Pearle AD. Lower limb alignment control: is it more challenging in lateral compared to medial unicondylar knee arthroplasty? Knee. 2015;22(4):347–50.
28. Pandit H, Jenkins C, Beard DJ, Price AJ, Gill HS, Dodd CA, et al. Mobile bearing dislocation in lateral unicompartmental knee replacement. Knee. 2010;17(6):392–7.
29. Streit MR, Walker T, Bruckner T, Merle C, Kretzer JP, Clarius M, et al. Mobile-bearing lateral unicompartmental knee replacement with the Oxford domed tibial component: an independent series. J Bone Joint Surg Br. 2012;94(10):1356–61.
30. Ali AM, Newman SDS, Hooper PA, Davies CM, Cobb JP. The effect of implant position on bone strain following lateral unicompartmental knee arthroplasty: a biomechanical model using digital image correlation. Bone Joint Res. 2017;6(8):522–9.
31. Lewold S, O Robertson K, Knutson K, Lidgren L. Revision of unicompartmental knee arthroplasty: outcome in 1,135 cases from the Swedish Knee Arthroplasty study. Acta Orthop Scand. 1998;69:469–74.

Part VII

Ankle and Foot Biomechanics

Biomechanics of the Ankle Joint

31

Krishna C. Ravella, Jamal Ahmad, and Farid Amirouche

31.1 Introduction

The ankle joint is one of the most complex articulations in the human body. It provides highly stable and efficient interactions between the body and ground during stance, gait, and other daily activities. The ankle must rapidly adapt to variable ground surfaces, absorb, and translate forces placed upon it, and provide rigid propulsion during stance. The ankle joint must also be durable enough to withstand considerable compressive, shear, and rotatory force quantities from body weight during activities of daily life (ADLs). As a result, it is less susceptible than other joints such as the hip and knee to primary degenerative joint disease (DJD) or osteoarthritis (OA) [1, 2]. Rather, the ankle is more commonly affected by DJD secondary to bony and/or cartilaginous injury and inflammatory joint disease such as rheumatoid arthritis (RA).

The purpose of this chapter is to discuss normal anatomy and biomechanics of the ankle joint complex. DJD has a significant impact upon normal ankle anatomy and biomechanics, which will be explored in this text. This will include discussion of current surgical treatment options for ankle arthritis, such as ankle arthrodesis and arthroplasty, and their effect upon biomechanics of the ankle joint.

31.2 Anatomy and Kinematics

The ankle is a constrained mortise-and-tenon joint that consists of the distal tibia and fibula articulating with the dome of the talus [3, 4]. The distal tibial articular surface or plafond has a quadrilateral shape to properly conform to the talar dome [5]. The primary articulations of the ankle joint complex are the tibiotalar or talocrural, talofibular, and distal tibiofibular joints. An important anatomical aspect of the distal tibia and fibula is that the former's medial projection or malleolus is anterior and superior to the latter's lateral malleolus. This contributes to the ankle having a slight oblique axis of motion. Specifically, the ankle's axis is 10° and 6° from the horizontal medial–lateral axis in the coronal and transverse planes, respectively.

While the foot is beyond the scope of this chapter, some discussion of its anatomy is warranted since the talus is an integral part of both the ankle and foot. Including the talus, the foot accounts for 28 bones and nearly 40 joints (see Fig. 31.1). The foot can be divided into three anatomical units

K. C. Ravella
Department of Orthopaedics, University of Illinois at Chicago – College of Medicine, Chicago, IL, USA

J. Ahmad · F. Amirouche (✉)
Department of Orthopaedics, University of Illinois at Chicago – College of Medicine, Chicago, IL, USA

NorthShore Orthopaedic Institute, NorthShore University Health System, Evanston, IL, USA
e-mail: amirouch@uic.edu

J. Koh et al. (eds.), *Orthopaedic Biomechanics in Sports Medicine*,
https://doi.org/10.1007/978-3-030-81549-3_31

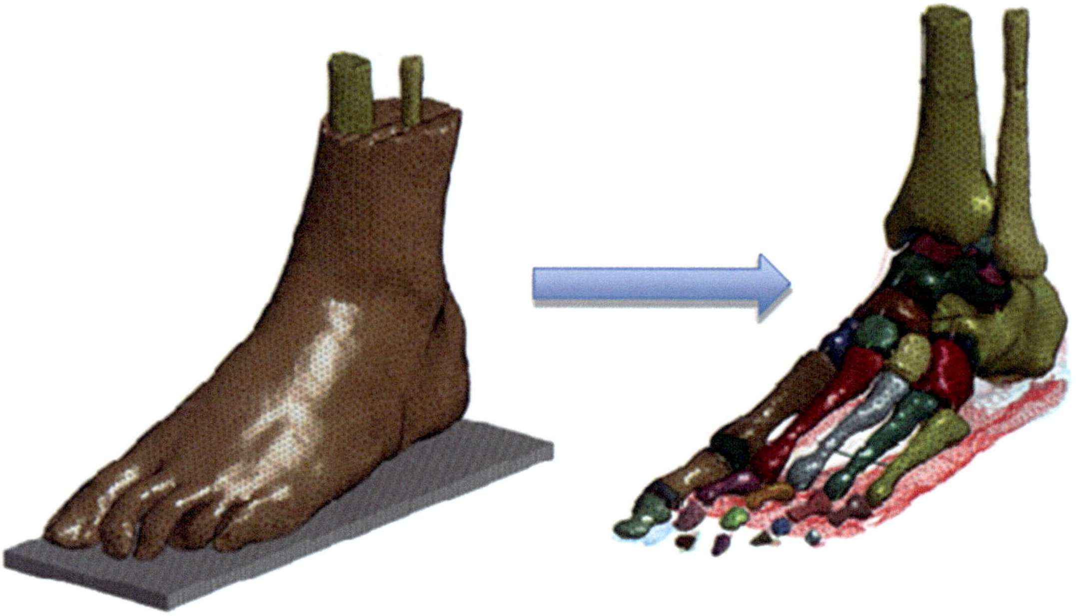

Fig. 31.1 Anatomy of the foot–ankle based on CT imaging and CAD reconstruction techniques. (DOI: https://doi.org/10.1016/j.clinbiomech.2012.05.005)

from distal to proximal: the forefoot, midfoot, and hindfoot. The forefoot includes the phalanges of all five toes, all five metatarsals, and the sesamoids of the plantar first metatarsal head with inter-phalangeal (IP) and metatarso-phalangeal (MTP) joints. The midfoot consists of the navicular, cuboid, and all three cuneiforms with naviculocuneiform (NC), metatarso-cuneiform (MTC), tarsometatarsal (TMT), and cubo-metatarsal joints. The hindfoot consists of the talus and calcaneus with the talocalcaneal or subtalar (ST), talonavicular (TN), and calcaneocuboid (CC) joints.

31.2.1 Kinematics

Ankle motion or kinematics is highly complex, involves particular movements around three axes, and involves the hindfoot's ST joint to achieve a full range of motion (ROM). Ankle flexion and extension are better known as plantarflexion and dorsiflexion, respectively. This occurs at the tibiotalar joint around a slightly oblique medial to lateral axis in the sagittal plane. Normal goniometric measurements of these motions are 10–20° of dorsiflexion and 40–55° of plantarflexion. Ankle inversion and eversion occur around an anterior to posterior axis in the coronal plane, but primarily at the ST rather than the ankle joint. Ankle abduction and adduction occur around a vertical axis in the axial plane, but mainly at the ST rather than the ankle joint as well. Of note, these movements are quite limited.

When describing ankle kinematics completely, it is important to understand that the ankle and ST joints have additional axes of motion that are oblique to the standard orthogonal axes discussed earlier. Ankle supination and pronation occur around oblique axes that are composed of simultaneous ankle and ST motion in the three orthogonal axes. Supination is triplanar with components of plantarflexion, adduction, and inversion. Pronation is also triplanar with elements of dorsiflexion, abduction, and eversion.

31.2.2 Tibiotalar and Talofibular Joints

The talus articulates with the distal tibia and fibula simultaneously. Specifically, the talar dome and body sit in a groove formed by the distal tibia's medial malleolus and plafond, and distal fibula's lateral malleolus. This bony anatomy unique to the ankle joint provides it with stability.

The primary weight-bearing surface of the ankle is the tibiotalar joint. The tibial plafond is the weight-bearing surface of the distal tibia and concave to articulate appropriately with the talar dome or trochlea underneath. While the distal tibia's medial malleolus does not provide a weight-bearing articular surface at the tibiotalar joint, it provides a medial restraint to the talus within the ankle mortise. At the dome and body, the talus is shaped like a truncated cone. Inman et al. found the dome and body of the talus to be 4.2 mm wider anteriorly than posteriorly, which offers greater stability with the ankle in dorsiflexion versus plantarflexion [2, 6]. In dorsiflexion, the anterior talus is tightly compressed between the distal tibia and fibula. These authors also measured the ankle mortise's arc of curvature to be within 1 mm of the talus laterally and within 1–3.2 mm of the talus medially [2, 7].

Forward progression during locomotion occurs at the tibiotalar joint, where the ankle's instant center of rotation varies slightly within the talus. Sammarco et al. measured ankle joint surface motion and found joint distraction and posterior talar gliding with initial dorsiflexion. It was only with complete dorsiflexion that the talus was fully compressed by the distal tibia and fibula.

Compared to dorsiflexion and plantarflexion, the talus is much more constrained within the vertical and anterior–posterior (AP) axes which causes limited transverse and coronal plane motion. Such stability allows torque from transverse plane forces at the leg to be transmitted across the ankle to the ST joint. This allows the ST joint to provide inversion, eversion, abduction, and adduction. In addition to its articulation with the calcaneus, the talus also interacts with the navicular to form the TN joint at the hindfoot.

In contrast to the tibiotalar joint, the talofibular joint is primarily non-weight-bearing at the ankle. The specific portion of the distal fibula that articulates with the lateral talar dome, body, and tubercle is called the incisura fibularis, which is shaped as a concavity to allow for optimal conformity. While the fibula can rotate and move sagitally within the incisura during gait, the amount of ankle rotation (<5°) and weight-bearing through this joint is minimal. Akin to the medial malleolus providing medial constraint, the lateral malleolus at the talofibular joint contains the talus in the ankle mortise laterally.

31.2.3 Distal Tibiofibular Joint

The distal tibiofibular joint is not truly a joint lined by articular cartilage, but rather a syndesmosis or ligamentous connection between the distal tibia and fibula just superior to the talar dome at an area called the sigmoid notch. This syndesmosis consists of four ligaments, which are the anterior inferior tibiofibular ligament (AITFL), posterior inferior tibiofibular ligament (PITFL), transverse tibiofibular ligament (TTL), and interosseous membrane. These ligaments provide stability between the distal tibia and fibula, allowing both bones to move as one unit with ankle motion. This stability is conferred to the ankle joint as it provides additional lateral constraint to the talus in the ankle mortise laterally. Ogilvie-Harris et al. found the AITFL, PITFL, TTL, and interosseous membrane to provide 35%, 40%, 22%, and less than 10% resistance to lateral talar translation within the mortise, respectively. While the fibula is able to move in the sagittal plane against the distal tibia at the sigmoid notch, such motion is minimal as 1 mm and 2° of external rotation. However, this small amount of distal fibular migration is enough to deepen the ankle mortise in dorsiflexion for additional ankle stability. As the distal tibiofibular joint contributes much to ankle stability, many consider it to be part of the tibiotalar and talofibular joints rather than a separate entity [1, 6, 8].

31.2.4 Subtalar (ST) Joint

The ST joint has one degree of freedom at an oblique angle directed upwards at 42° from the horizontal and 16° medially from the midline. This split between the AP and vertical axes create equal amounts of inversion/eversion and abduction/adduction with ST motion. In contrast to the ankle, the ST joint has very little dorsiflexion and plantarflexion. While non-weight-bearing (NWB), the talus is static and the calcaneus is mobile. With weight-bearing and locomotion, the talus is more mobile as the calcaneus is stationary at the ST joint. This allows the lower leg to rotate in the transverse and coronal planes without requiring movement from the foot. Combined with the ankle joint's function, the limb is able to progress forwards, balance, change direction, and adjust to variable ground surfaces.

The ST joint consists of anterior, middle, and posterior facets where the plantar talus articulates with the superior calcaneus. The largest of these is the posterior facet, which accounts for 70% of the ST joint's articular surface. These facets are concave at the talus and convex at the calcaneus to allow for ankle and foot inversion and eversion [1, 6]. These facets also resemble a spiral or screw, where the calcaneus translates posterior to anterior along the ST joint's axis as it rotates counter-clockwise to clockwise. Akin to the threads of a screw, the ST joint has a single degree of freedom. Specifically, the calcaneus moves constrained at the ST joint across all three planes in eversion/inversion and abduction/adduction. Eversion/inversion are easier to measure than abduction/adduction and are approximately 22–23°/12–13°, respectively. Subtalar joint motion is also related to the rotation of the tibia. Subtalar joint inversion and eversion occur with tibial external and internal rotation, respectively [5].

Subtalar joint stability is primarily capsular and ligamentous. The posterior and combined anterior and middle facets have their own joint capsules. The interosseous and cervical anterior ligaments are between the posterior and anterior/middle facets to provide maximal ST joint stability. Other ST ligaments include the weaker posterior, lateral, and medial talocalcaneal ligaments. Changes to ST joint stability and reactive forces can increase or decrease contact stresses at the subtalar joint and lead to abnormal ankle and foot kinematics.

31.2.5 Muscles and Tendons

Ankle motion is highly complex, where each muscle–tendon unit that crosses the joint has a specific function and role to provide full range of mobility. These muscles are separated by fascia into four groups or compartments: the anterior, lateral, superficial posterior, and deep posterior compartments. Anterior compartment muscles are primarily the tibialis anterior (TA), extensor hallucis longus (EHL), and extensor digitorum longus (EDL) which provide ankle dorsiflexion, first toe extension, and lesser (second through fifth) toe extension, respectively. Some individuals have an additional anterior muscle called the peroneus tertius, which aids in ankle dorsiflexion and eversion. Lateral compartment muscles are the peroneus longus (PL) and peroneus brevis (PB), which provide ankle and foot eversion and weak plantarflexion. Superficial posterior compartment muscles are the medial and lateral gastrocnemius and soleus, which are also known as the triceps surae. These three muscles conjoin distally to form the Achilles tendon, which inserts upon the posterior calcaneus to provide ankle plantarflexion. Some individuals have an additional superficial posterior muscle called the plantaris, which aids in ankle plantarflexion. Deep posterior compartment muscles are the posterior tibial tendon (PTT), flexor hallucis longus (FHL), and flexor digitorum longus (FDL), which provide ankle inversion, first toe flexion, and lesser (second through fifth) toe flexion, respectively [1, 5].

31.2.6 Ligaments

The ankle has several ligaments that traverse the joint which are classified as complexes based on anatomic location. The lateral ankle ligament complex includes the anterior and posterior talofibular ligaments (ATFL and PTFL), and the calcaneo-fibular ligament (CFL) (see Fig. 31.2). These ligaments resist inversion and varus joint stresses. The ATFL is the primary restriction of excess talar internal rotation, but has the greatest strain and lowest load to failure within the lateral

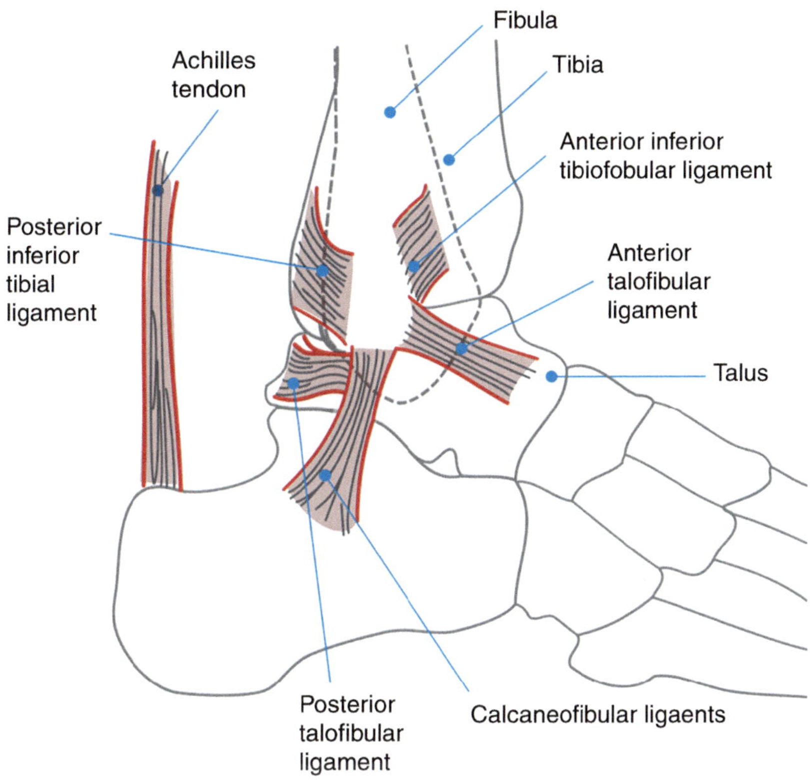

Fig. 31.2 Ankle ligaments

ankle ligament complex. Additional functions of the ATFL are resistance to anterior talar and internal rotational displacement within the mortise. The CFL provides stability to the ankle and subtalar joint by restricting excessive adduction of both joints. Of note, the ATFL and CFL are positioned perpendicular to one another to form an angle of 90–105°. This anatomy allows the ATFL to limit inversion during plantarflexion while the CFL constrains inversion during dorsiflexion [2]. The PTFL is the strongest ligament within the lateral ankle ligament complex with the lowest strain and highest load to failure to rupture less frequently than the ATFL and CFL [9]. The PTFL is under maximal strain in dorsiflexion and limits posterior translation and external rotation within the mortise.

The medial ankle ligaments are collectively called the deltoid ligament complex as they resemble the Greek letter, delta, in appearance. The superficial deltoid ligaments originate at the medial malleolus' anterior and posterior colliculi and insert upon the medial calcaneus, talar neck, and navicular bones as the tibiocalcaneal, posterior tibiotalar, and tibionavicular ligaments, respectively. The deep deltoid ligaments begin at the medial malleolus' anterior colliculi as the anterior tibiotalar ligaments, but have three fascicles that attach to three distinct regions of the medial talar body [2, 7, 10]. Both the superficial and deep portions of the deltoid ligament complex are extra synovial and intra-articular. This ligamentous complex helps to resist eversion and valgus joint stresses upon the ankle joint. The deep deltoid provides greater resistance against lateral ankle translation with eversion than the superficial deltoid [9]. Studies have shown no lateral displacement or valgus tilt of the talus within the ankle mortise when both superficial and deep deltoid ligaments are intact. When either the superficial or deep ligaments were transected individually, ankle joint stability remained. Rather, ankle instability in valgus occurs when both the superficial and deep portions are sectioned or ruptured [2, 11, 12].

Other ligaments at the ankle joint include the distal tibiofibular syndesmotic and subtalar joint ligaments that were discussed earlier. These ligaments cooperate with the medial and lateral ankle ligament complexes to optimize ankle joint stability. The syndesmosis and subtalar joint's interosseous, anterior, posterior, and medial talocalcaneal ligaments work with the deltoid ligaments to provide resistance against abduction, eversion, and external rotation forces applied to the ankle joint. The interosseous, anterior, posterior, and lateral talocalcaneal ligaments offer constraint from adduction, inversion, and internal rotation at the ankle and subtalar joint alongside the CFL at the lateral ankle [5, 9].

Combined with bony congruency, the ankle's ligaments provide joint stability. Under physiologic loads of weight-bearing, bony congruency provides 100% resistance to inversion/eversion and 30% resistance to internal/external rotation. During weight-bearing, ligaments provide the additional rotational stability needed to resist joint translation and excessive motion. This has been demonstrated in cadaveric models where absent lateral ankle ligamentous constraint causes as much as 20° varus talar tilt under the tibial plafond.

31.3 Normal Ankle Joint Biomechanics

31.3.1 Gait

Gait or locomotion is a complex process in which the ankle plays a critical role. A full cycle of gait between heel strikes to the ground is known as stride. Each stride consists of a stance and swing phase. Stance occupies nearly 2/3 of stride length, which progresses from heel strike to toe-off from the ground. Swing phase accounts for the remainder of stride time, which lasts from toe-off to heel strike with the ground [5]. To provide efficient gait, the ankle joint is critical in providing forward progression while adapting to individual body weight distributions, external walking environments, and any alterations that can occur with either.

During stride, the ankle joint employs the three primary cardinal planes of motion discussed earlier. Dorsiflexion and plantarflexion occur in the sagittal plane (see Fig. 31.3). Inversion and eversion occur in the frontal plane. Adduction and abduction occur in the transverse plane. As states earlier, the ankle joint mechanical axes are not perpendicular to these cardinal planes. This allows the ankle to achieve complex movements involving any and/or all of its three cardinal planes of motion during stride, such as supination and pronation [13].

During stride, stance encompasses more complex interactions between the ankle and ground than during swing. Stance specifically progresses from heel strike to the foot flat to mid-stance to the heel off from the ground to end with the toes off from the ground. In terms of limb support, stance advances from double-limb to single-limb to revert back to double-limb support by the end of stance [5]. Most individuals have mean walking parameters of 82 m and 58 heel strikes per minute. Running is speed at which double stance disappears into a float phase where both limbs are off the ground at mean velocities of 201 meters/minute (m/min).

Ankle motion during gait has been greatly studied. At heel strike, the leg internally rotates and the ankle is in slight plantarflexion. At this time, the ST joint everts as the calcaneus' point of contact to the ground is lateral to the ankle joint's center. Ankle plantarflexion continues until the foot is flat, but then rapidly reverses to dorsiflexion at mid-stance. Simultaneously, the leg externally rotates and the ST joint inverts. In turn, this increases the foot's rigidity to allow for propulsion and the ankle resumes plantarflexion to allow the limb clear the ground for toe-off. The ankle begins dorsiflexion from early to mid-swing and back to slight plantarflexion for another heel strike. In walking, mean ankle dorsiflexion and plantarflexion is 10° and 14°, respectively. Maximum ankle dorsiflexion and plantarflexion occur at 2/3 through stance and toe-off, respectively. ST pronation occurs from heel strike to foot flat. ST supination occurs at 1/3 through gait to approach a maximum of 6° of supination just before toe-off.

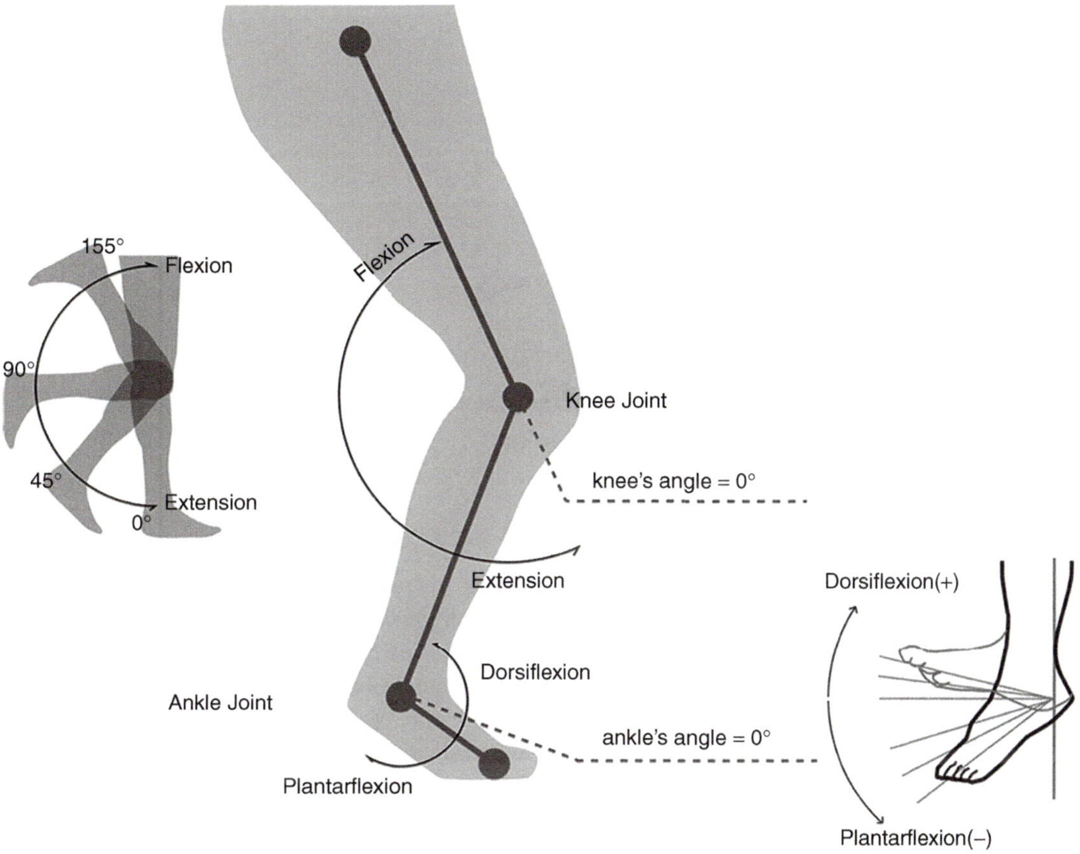

Fig. 31.3 Foot–ankle kinematics with plantarflexion and dorsiflexion angles in a sagittal plane and flexion–extension at the knee joint

Daniels et al. examined gait within the context of likening the ankle and foot to a three-rocker steady-state mechanism (see Fig. 31.4). The three rockers are the calcaneus, ankle joint, and forefoot. The calcaneus is the first rocker through heel strike and the first third of stance. The ankle joint is the second rocker throughout the mid-stance phase, where it bears much of the load of body weight before this is transferred to the forefoot. Finally, the forefoot is the third rocker as it supports body weight at the end of stance [14].

Dubin et al. offer another look at gait analysis focusing on optimizing energy expenditure in relation to the stance and swing phase in gait. This is obviously not a trivial problem, however if we examine the basic relation between force, height and energy in its simplified form, we see that: Body mass × gravity × change in height ($mg\Delta h$) = potential energy during ambulation [15]. This formula dictates that decreasing joint forces, such as those transmitted to the ankle joint, allows for less energy spent in locomotion. Optimizing energy used by muscles at the ankle translates into less work and more efficient gait on the part of the ankle. A particular adaptation of the ankle to sustain loading forces is that the joint's normal soft tissue and bony anatomy allows for eccentric loading during stride. This allows for optimal distribution of forces upon the ankle with less energy and potential for injury imparted to the joint. An important clinical implication from this is that normal ankle anatomy, without pathologies such as bone and/or joint disease, is critical to an efficient gait that minimizes energy and work spent while ambulating.

The muscles of the ankle act during stride to allow for efficient transfer of muscle force to the floor and smooth limb and body movement along

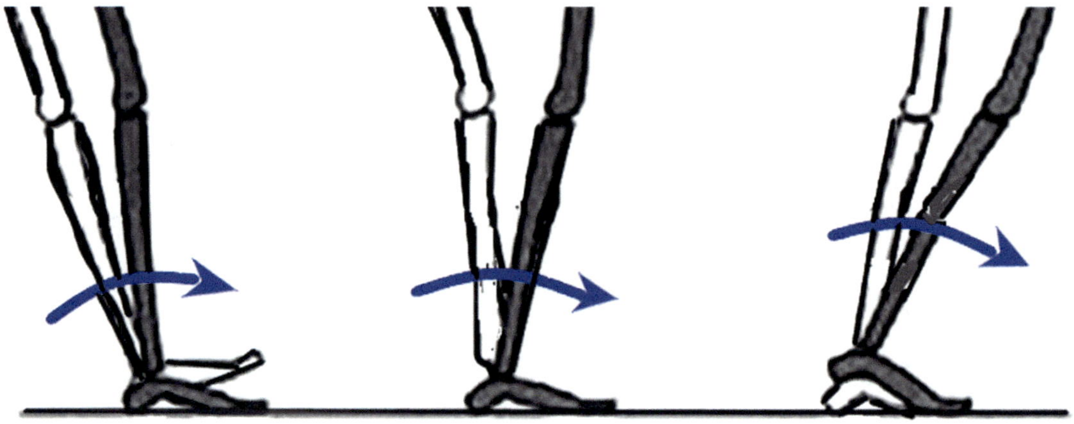

Fig. 31.4 Ankle–foot three-rocker steady-state mechanism

the axis of progression. In stance, the gastrocnemius and soleus provide plantarflexion to decelerate the leg over the foot. At the same point, the TA eccentrically fires to prevent the foot from slapping the ground. Furthermore, the PTT inverts the ST joint during stance to lock the hindfoot and make the foot rigid enough by the point of toe-off. In swing, the TA allows the foot to clear the floor.

In a static analysis of the ankle joint, forces produced by contraction of the gastrocnemius and soleus muscles are directed along the Achilles to the calcaneus. Ankle joint reactive forces are applied from the talar dome. Studies have found the Achilles tendon and joint reactive forces to be 1.2 and 2.1 times body weight, respectively. More loading forces are applied through the tibiotalar rather than the talofibular joint. Wang et al. measured that 17% and 83% of body weight forces are transmitted to the fibula and tibia, respectively. Fibular loading forces decrease with ankle varus and/or plantarflexion but increase with valgus and/or dorsiflexion.

In a dynamic analysis of the ankle joint, forces produced by contraction of the gastrocnemius and soleus muscles and ankle reactive forces are much increased than in the static phase. At late stance, compressive forces from the gastrocnemius and soleus muscles are 4–5 times body weight. At faster walking speeds, ankle joint reactive forces are 3–5 times body weight with peaks at early and late stance. With running, ankle joint reactive forces can be as high as 13 times body weight.

The ankle has a load bearing surface area of 11–13 cm^2, where load distribution is partially determined by ankle position. With weight-bearing, 77–90% of loading forces are transmitted to the central talar dome. With inversion, the medial talar dome is loaded more. During eversion, the lateral talar dome is loaded more. With dorsiflexion, the central contact area at the talar dome moves anterior which is greatest at this point.

31.4 Ankle Degenerative Joint Disease and Its Effect on Biomechanics

31.4.1 Degenerative Joint Disease

Degenerative joint disease (DJD) is more commonly known as arthritis. This condition is characterized by diffuse breakdown of articular hyaline cartilage and the underlying or subchondral bone at the affected joint. At times, inflammation of the joint capsule and lining or synovium can occur. On gross examination, DJD displays joint space-narrowing, sclerosis or altered mineralization of subchondral bone, and presence of bone spurs or osteophytes. These spurs occur at the margins of the joint and are believed to form as the human body attempts to restore joint congruence in the absence of cartilage.

When the ankle is affected, the joint can become significantly altered from its normal state. DJD commonly causes ankle pain, stiffness, weakness, joint incongruence, abnormal gait, functional disability, and ultimately a poor quality of life. Most ankle arthritis is posttraumatic [16]. Snedeker et al. reported that nearly 70% of cases of ankle DJD result from prior trauma, such as ankle fractures or chronic ligamentous injuries. Other causes of arthritis include infection, talar osteonecrosis, and Charcot neuroarthropathy [3, 4, 17]. The ankle joint appears less susceptible to primary osteoarthritis (OA) than other joints such as the hip and knee. This may stem from a number of factors, including different joint loading configurations which could have less dramatic effects on articular cartilage [9]. DJD can eventually lead to varus or valgus deformity at the ankle which contributes to limitations in joint range of motion [5]. As DJD worsens at the ankle joint, normal ankle kinematics and biomechanics become effected and altered. This leads to altered ankle joint function, which will be discussed below.

31.4.2 Effect of DJD upon Ankle Biomechanics

DJD has a dramatic effect upon the normal ankle joint and its biomechanics. As articular cartilage degenerates and DJD develops, the ankle becomes painful. Daniels et al. discuss the pathway in which a painful ankle joint can lead to inefficient biomechanics, with a shortened stance and antalgic gait in order to avoid ankle pain [14]. As arthritis progresses, the joint becomes increasingly stiff with significantly decreased range of motion (ROM) in all planes of motion [18–20]. This leads to increased contact forces at the ankle joint, which alters gait. Snedeker et al. further examined forces at the ankle joint in arthritic patients and recorded diminished peak ground reactive forces and power in these individuals. They also found these altered forces clinically correlated to a decreased ankle joint ROM in all cardinal planes [9].

The particular pattern of gait that occurs in the arthritic ankle is known as an antalgic or vaulting gait [21, 22]. Specifically, the arthritic patient will maintain his/her knee in greater extension than flexion during stride to decrease forces, pressure, and pain at the ankle joint. It has also been suggested that the ankle's muscles have to work harder during gait to minimize forces, pressure, and pain at the ankle. Valderbano et al. examined gait parameters and found a mean 9.1% decrease in cadence, 16.2% decrease in walking speed, 37.5% reduction in total plantarflexion movement, 26.7% decrease in maximal medial ground reactive forces, and 54.8% decrease in joint power among patients with ankle arthritis compared to those without [20]. In Daniels' model of the ankle and foot being three rockers, these changes shorten stance phase and limits the functions of the second and third rocker functions during gait. All of these above alterations involving the ankle joint lead to increased energy expenditure and decreased efficiency. These results demonstrate the extent to which ankle arthritis can affect multiple areas of ankle joint biomechanics that include range of motion, gait cycle, and joint forces. Increased energy expenditure, painful gait, and decreased joint motion and function lead to the conversation of methods to improve such developments.

31.4.3 Treatment of Ankle Joint Arthritis and Its Effect on Biomechanics

31.4.3.1 Ankle Joint Fusion or Arthrodesis

Ankle joint fusion or arthrodesis is the traditional method of treating severe ankle DJD that has failed nonsurgical treatment options [23–26]. This procedure involves removal or resection of the diseased articular cartilage and connecting or fusing the subchondral bone of the distal tibia and talar body at both sides at the tibiotalar joint. The distal fibula at the ankle joint can either be morcellized for use as bone graft or left intact for added bulk at the tibiotalar fusion so long as its talofibular joint is free of degenerated cartilage. The means used to achieve a rigid ankle arthrodesis with complete bony healing or fusion across the tibiotalar joint include screws and/or plate-and-screw constructs.

The goal of ankle arthrodesis is to construct a plantigrade foot that will allow for pain-free weight-bearing and walking. The position and alignment of the lower limb and ankle joint for fusion is critical. The optimal position for the ankle joint to be fused in is neutral in the sagittal plane with 0–5° of valgus and 0–10° of external rotation [3, 4, 23, 27, 28]. This position results in a plantigrade foot and allows for maximal foot function (see Fig. 31.5). Malpositioning of the ankle arthrodesis can lead to persistent pain and/or unwanted stresses on the collateral ligaments of neighboring joints such as the knee [23, 29]. From numerous outcome studies, ankle arthrodesis has been shown to offer arthritic patients decreased pain, improved function, and high rates of satisfaction [23, 30].

It is important to recognize that while ankle arthrodesis can treat DJD, the procedure does so without restoring or improving ankle joint biomechanics. By its very nature, ankle fusion leads to decreased joint ROM in all planes [30–34]. Thomas et al. found that patients from ankle arthrodesis demonstrated significantly lower sagittal, coronal, and transverse plane ROM, with the greatest decrease seen in the sagittal plane. Specifically, the authors measured mean hindfoot and forefoot sagittal ROM at 10.25° and 14.99° in fused ankles, compared to 17.09° and 21.45°, respectively, in normal ankles [30]. Such biomechanical alterations at the ankle due to its fusion can lead to consequences within the foot in the intermediate to long-term. Following ankle arthrodesis, the ST joint and the midfoot' s MTC and TMT joint compensate with hyper-extension and hyper-flexion to provide more sagittal plane motion than in the normal state. Such altered ST and midfoot motion can increase joint forces that can lead to future DJD at any or all of those particular joints [21, 30, 35].

As expected, gait is greatly affected from ankle joint fusion. Beyaert et al. assessed the stride in patients after ankle arthrodesis and observed a reduction in walking speed from 1.28 to 1.09 (m/s), a reduction in stride length from 152 to 135 (% leg length), and earlier heel-off and knee extension compared to cohort of patients with normal ankle function [36]. Other authors have corroborated these findings, but also shown improvements compared to those with untreated ankle arthritis [30, 31, 33]. Barton et al. studied gait in patients with ankle arthritis that was untreated and treated with arthrodesis [34]. They found mean walking speed, cadence, and stride length improved with arthrodesis compared to those without surgical treatment. These

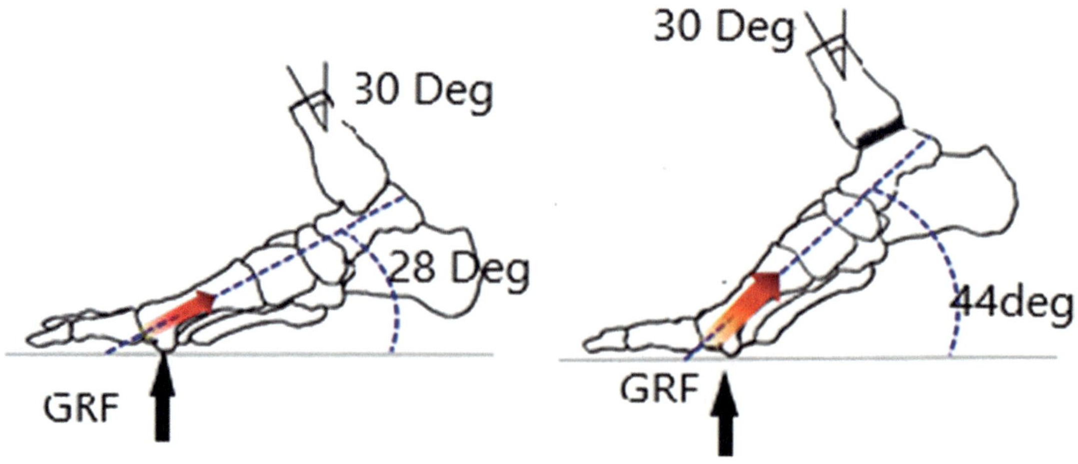

Fig. 31.5 Range of motion for normal and ankle arthrodesis

authors suggest that while patients truly have less pain after ankle arthrodesis, some abnormal gait persists as patients walk with more caution and less speed to avoid movements or ankle joint positions that caused discomfort prior to surgery.

Ground reaction force and joint power are also affected from ankle arthrodesis [31–34]. As discussed earlier, ground reaction forces and power at the ankle are diminished in arthritic patients. With increased ankle joint stiffness from arthrodesis, these parameters remain decreased rather than normalize. Wu et al. examined gait analysis in patients following ankle arthrodesis and found significantly decreased ground reaction forces in the sagittal, coronal, and transverse planes compared to normal controls [33]. Similar studies support these findings and have recorded significantly higher ankle joint power at the contralateral limb [31, 32]. Such an effect on the contralateral ankle is important to recognize as this increases stresses upon that ankle. A clinical implication is that an ankle arthrodesis may lead to initial or worsen preexisting DJD at the contralateral ankle. However, this has not yet been shown in the orthopedic literature.

31.4.3.2 Total Ankle Arthroplasty (TAA)

Total ankle arthroplasty (TAA) is an alternative to arthrodesis for treating severe ankle DJD that has failed nonsurgical treatment [23]. Akin to arthrodesis, TAA involves removal or resection of the diseased articular cartilage of the distal tibia and talar dome at both sides at the tibiotalar joint. Unlike arthrodesis, porous-coated metal implants that are shaped to recreate the normal anatomy of the tibial plafond and talar dome are placed under and above the distal tibial and talar body's subchondral bone respectively to allow for bony ingrowth of these implants. To recreate ankle motion, a polyethylene spacer is placed between the tibial and talar implants that the talar implant can move or glide against. Some TAAs, known as third generation TAAs, are designed to allow the polyethylene to move against both the tibial and talar implants. For most current TAAs, the distal fibula and talofibular joint at the ankle joint is left untouched with neither resection nor removal.

Similar to arthrodesis, the goal of TAA is to provide a plantigrade foot with some ankle motion that will allow for pain-free weight-bearing and walking. Based on design, the TAA is intended to be placed in neutral in the sagittal plane with 0° of valgus and 0–10° of external rotation within the ankle joint [3, 4, 23, 27, 28]. This position results in a plantigrade foot and allows for maximal post-surgical foot and ankle function with TAA stability. TAA malposition within the ankle joint can cause continued pain and/or dysfunction with increased edge loading of the TAA to cause early implant wear. From numerous outcome studies, TAA has been shown to offer arthritic patients decreased pain, improved function, and high rates of satisfaction.

As clinical symptoms improve, ankle biomechanical parameters are affected positively after TAA. Valderrabano et al. assessed ROM before and after TAA to find mean increases in ankle plantarflexion, inversion, and adduction of 4.3°, 1.1°, and 1.6°, respectively. Similar to these investigators, others have found the greatest gains in ankle ROM from arthroplasty to be in sagittal plane motion. Valderrabano et al. examined gait before and after TAA to find improvements in stride length, cadence, walking speed, and support time by 5 cm, 5.5 cm/s, 12 cm/s, and 0.04 s, respectively, after this surgery. Brodsky et al. confirmed these benefits from TAA and reported increases in stride length, cadence, and walking speed by 17 cm, 12.9 steps/min, and cm/s after this surgery. While ankle ROM and gait clearly improve from before to after TAA, the surgery does not restore ankle biomechanics to normal.

Several studies have compared ankle biomechanics after arthrodesis and TAA [21, 37]. Unlike arthrodesis, TAA improves upon preoperative arthritic ankle ROM. Preserved ankle ROM from TAA provides less stress to the ST joint and midfoot than ankle arthrodesis. However, current TAA designs do not allow a return to normal ankle ROM and biomechanics. During stride, TAA is able to significantly improve walking speed and gait symmetry compared to ankle arthrodesis. However, these gains from TAA remain just below normal parameters.

Currently, neither arthrodesis nor TAA are able to increase ankle joint power measures during stance and gait to normal values [37].

31.5 Conclusion

The ankle is one of the most complex joints in the human body. In normal conditions, the ankle provides stable and efficient interactions between the body and ground during stance, gait, and other daily activities. The ankle simultaneously adapts to variable ground, absorbs, and translates forces placed upon it and provides rigid propulsion during stride. The ankle joint is also durable enough to withstand considerable compressive, shear, and rotatory force quantities from body weight during daily activities. Knowledge of ankle joint biomechanics allows for a better understanding of ankle function in normal and abnormal states.

One of the most common joint diseases to involve the ankle is DJD, which has a dramatic negative effect upon the joint. As DJD affects the ankle, ROM decreases and gait becomes inefficient and dysfunctional with greater energy expenditures. Currently, the primary surgical treatment options are ankle joint arthrodesis or arthroplasty. Both alleviate pain, but neither restores normal ankle motion, biomechanics, and gait. It is important to recognize that arthroplasty preserves more ankle motion than arthrodesis. This feature of joint arthroplasty improves gait parameters such as gait symmetry, walking speed, and hindfoot and midfoot motion in ways that an ankle arthrodesis cannot. In time, arthroplasty design may improve to the point where such future implants can offer restored normal ankle biomechanics.

References

1. Brockett CL, Chapman GJ. Biomechanics of the ankle. Orthop Trauma. 2016;30:232–8.
2. Castro MD. Ankle biomechanics. Foot Ankle Clin. 2002;7:679–93.
3. Ahmad J. Total ankle arthroplasty as current treatment for ankle arthritis. Semin Arthroplast. 2010;21:247–52.
4. Ahmad J, Raikin SM. Ankle arthrodesis: the simple and the complex. Foot Ankle Clin. 2008;13:381–400.
5. Oliver J. Miller's Review of Orthopaedics, 8th Ed. Miller MD, Thompson SR. J Chem Inf Model. 2013; https://doi.org/10.1017/CBO9781107415324.004.
6. Medina McKeon JM, Hoch MC. The ankle-joint complex: a kinesiologic approach to lateral ankle sprains. J Athl Train. 2019;54:589–602.
7. Isman RE, Inman VT. Anthropometric studies of the human foot and ankle. Foot Ankle. 1969;11:97–129.
8. Kundert H-P. Mann's surgery of the foot and ankle. Fuß Sprunggelenk. 2015;13:138–9.
9. Snedeker JG, Wirth SH, Espinosa N. Biomechanics of the normal and arthritic ankle joint. Foot Ankle Clin. 2012;17:517–28.
10. Close JR. Some applications of the functional anatomy of the ankle joint. J Bone Joint Surg Am. 1956;38(A):761–81.
11. Michelson JD, Clarke HJ, Jinnah RH. The effect of loading on tibiotalar alignment in cadaver ankles. Foot Ankle Int. 1990;10:280–4.
12. Harper MC. The deltoid ligament. Clin Orthop Relat Res. 1988;226:156–68.
13. CHAN CW, RUDINS A. Foot biomechanics during walking and running. Mayo Clin Proc. 1994;69:448–61.
14. Daniels T, Thomas R. Etiology and biomechanics of ankle arthritis. Foot Ankle Clin. 2008;13:341–52.
15. Dubin A. Gait. The role of the ankle and foot in walking. Med Clin North Am. 2014;98:205–11.
16. Chou LB, Coughlin MT, Hansen S, Haskell A, Lundeen G, Saltzman CL, Mann RA. Osteoarthritis of the ankle: the role of arthroplasty. J Am Acad Orthop Surg. 2008;16:249–59.
17. Thomas RH, Daniels TR. Ankle arthritis. J Bone Joint Surg Am. 2003;85:923–36.
18. Khazzam M, Long JT, Marks RM, Harris GF. Preoperative gait characterization of patients with ankle arthrosis. Gait Posture. 2006;24:85–93.
19. Stauffer RN, Chao EYS, Brewster RC. Force and motion analysis of the normal, diseased, and prosthetic ankle joint. Clin Orthop Relat Res. 1977;127: 189–96.
20. Valderrabano V, Nigg BM, von Tscharner V, Stefanyshyn DJ, Goepfert B, Hintermann B. Gait analysis in ankle osteoarthritis and total ankle replacement. Clin Biomech. 2007;22:894–904.
21. Coester LM, Saltzman CL, Leupold J, Pontarelli W. Long-term results following ankle arthrodesis for post-traumatic arthritis. J Bone Joint Surg Am. 2001;83:219–28.
22. Muehleman C, Bareither D, Huch K, Cole AA, Kuettner KE. Prevalence of degenerative morphological changes in the joints of the lower extremity. Osteoarthr Cartil. 1997;5:23–37.
23. Nihal A, Gellman RE, Embil JM, Trepman E. Ankle arthrodesis. Foot Ankle Surg. 2008;14:1–10.
24. Scranton PE. An overview of ankle arthrodesis. Clin Orthop Relat Res. 1991;(268):96–101.

25. Katcherian DA. Treatment of ankle arthrosis. Clin Orthop Relat Res. 1998;(349):48–57.
26. Mann RA, Rongstad KM. Arthrodesis of the ankle: a critical analysis. Foot Ankle Int. 1998;19:3–9.
27. King HA, Watkins TB, Samuelson KM. Analysis of foot position in ankle arthrodesis and its influence on gait. Foot Ankle Int. 1980;1:44–9.
28. Hefti FL, Baumann JU, Morscher EW. Ankle joint fusion - determination of optimal position by gait analysis. Arch Orthop Trauma Surg. 1980;96:187–95.
29. Wang Y, Li Z, Wong DWC, Zhang M. Effects of ankle arthrodesis on biomechanical performance of the entire foot. PLoS One. 2015;10:e0134340. https://doi.org/10.1371/journal.pone.0134340.
30. Thomas R, Daniels TR, Parker K. Gait analysis and functional outcomes following ankle arthrodesis for isolated ankle arthritis. J Bone Joint Surg Am. 2006;88:526–35.
31. Kerkhoff YRA, van Boxtel W, Louwerens JWK, Keijsers NLW. Asymmetries in gait and balance control after ankle arthrodesis. J Foot Ankle Surg. 2018;57:899–903.
32. Tenenbaum S, Coleman SC, Brodsky JW. Improvement in gait following combined ankle and subtalar arthrodesis. J Bone Joint Surg Am. 2014;96:1863–9.
33. Wu WL, Su FC, Cheng YM, Huang PJ, Chou YL, Chou CK. Gait analysis after ankle arthrodesis. Gait Posture. 2000;11:54–61.
34. Barton T, Lintz F, Winson I. Biomechanical changes associated with the osteoarthritic, arthrodesed, and prosthetic ankle joint. Foot Ankle Surg. 2011;17:52–7.
35. Fuchs S, Sandmann C, Skwara A, Chylarecki C. Quality of life 20 years after arthrodesis of the ankle. A study of adjacent joints. J Bone Joint Surg Br. 2003;85:994–8.
36. Beyaert C, Sirveaux F, Paysant J, Molé D, André JM. The effect of tibio-talar arthrodesis on foot kinematics and ground reaction force progression during walking. Gait Posture. 2004;20:84–91.
37. Singer S, Klejman S, Pinsker E, Houck J, Daniels T. Ankle arthroplasty and ankle arthrodesis: gait analysis compared with normal controls. J Bone Joint Surg Am. 2013;95:e191. https://doi.org/10.2106/JBJS.L.00465.

32 Biomechanics of the Ankle Joint in Relation to Ankle Ligament Injuries

Marshall Haden, Jamal Ahmad, and Farid Amirouche

32.1 Introduction

The ankle is one of the most complex joints in the human body. It provides limb balance and efficient interactions between the body and ground during stance, locomotion, and other daily activities. The ankle must be stable enough to adapt to ground, absorb and translate forces placed upon it, and provide movement while simultaneously providing limb and body support. Much ankle stability comes from ligaments that traverse its joint. These ligaments are classified as lateral, medial, and syndesmotic complexes based on their anatomic location. Injuries to any or all of these ligaments can greatly affect ankle biomechanics.

The purpose of this chapter is to discuss biomechanics of the ankle within the context of the ligaments that provide joint stability. This will include an explanation of normal ankle ligament function, to be followed by consideration as to how ligamentous injuries affect ankle stability and biomechanics. This will include discussion of current surgical treatment options for ankle ligamentous injury and their effect upon biomechanics of the ankle joint.

32.2 Anatomy

As discussed in an earlier chapter, the ankle joint is a constrained mortise-and-tenon that consists of the distal tibia and fibula articulating with the dome of the talus [1, 2]. The main articulations of the ankle joint complex are the tibiotalar or talocrural, talofibular, and distal tibiofibular joints. An important anatomical aspect of the distal tibia and fibula is that the former's medial projection or malleolus is anterior and superior to the latter's lateral malleolus. These features contribute to ankle ligamentous anatomy.

While the foot is beyond the scope of this chapter, some discussion of its anatomy is warranted due to bony areas where many of the ankle's ligaments insert. The foot can be divided into three anatomical units from proximal to distal: the forefoot, midfoot, and hindfoot. The hindfoot consists of the talus and calcaneus with the talocalcaneal or subtalar (ST), talonavicular (TN), and calcaneocuboid (CC) joints. The midfoot consists of the navicular, cuboid, and all three cuneiforms with naviculocuneiform (NC), metatarso-cuneiform (MTC), tarso-metatarsal (TMT), and cubo-metatarsal joints. The forefoot

M. Haden
Department of Orthopaedics, University of Illinois at Chicago – College of Medicine, Chicago, IL, USA

J. Ahmad · F. Amirouche (✉)
Department of Orthopaedics, University of Illinois at Chicago – College of Medicine, Chicago, IL, USA

Orthopaedic and Spine Institute, Northshore University HealthSystem, Evanston, IL, USA
e-mail: amirouch@uic.edu

J. Koh et al. (eds.), *Orthopaedic Biomechanics in Sports Medicine*,
https://doi.org/10.1007/978-3-030-81549-3_32

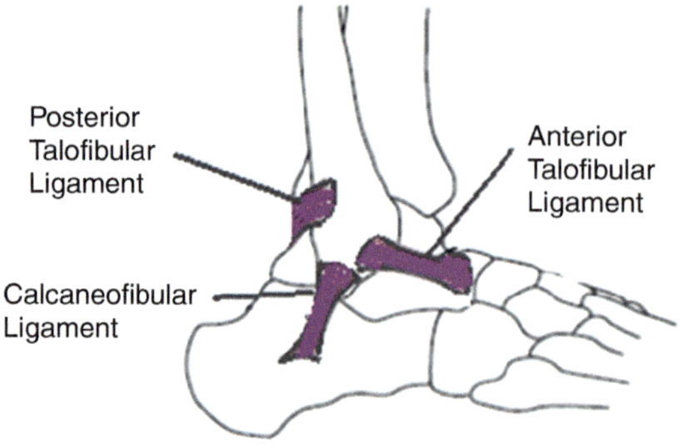

Fig. 32.1 Lateral view of the ankle joint ligaments

includes the phalanges of all five toes, all five metatarsals, and the sesamoids of the plantar first metatarsal head with inter-phalangeal (IP) and metatarso-phalangeal (MTP) joints.

32.2.1 Lateral Ankle Ligament Anatomy

The lateral ankle ligament complex includes the anterior and posterior talofibular ligaments (ATFL and PTFL), and the calcaneofibular ligament (CFL) (see Fig. 32.1). These ligaments resist inversion and varus joint stresses. The ATFL spans the antero-lateral malleolus to the lateral talar body. It is the primary restriction of excess talar internal rotation, but has the greatest strain and lowest load to failure within the lateral ankle ligament complex. Additional functions of the ATFL are resistance to anterior talar and internal rotational displacement within the mortise. The CFL runs from the inferior lateral malleolus to the lateral calcaneal wall. It provides stability to the ankle and ST joint by restricting excessive adduction of both joints. Of note, the ATFL and CFL are positioned perpendicular to each other to form an angle of 90°–105°. This anatomy allows the ATFL to limit inversion during plantarflexion while the CFL constrains inversion during dorsiflexion [3]. The PTFL originates at the posterior lateral malleolus and inserts at the posterolateral talus. It is the strongest ligament within the lateral ankle ligament complex with the lowest strain and highest load to failure. Subsequently, the PTFL ruptures less frequently than the ATFL and CFL [4]. The PTFL is under maximal strain in dorsiflexion and limits posterior translation and external rotation within the mortise.

32.2.2 Medial Ankle Ligament Anatomy

The medial ankle ligaments are collectively called the deltoid ligament complex as they resemble the Greek letter, delta, in appearance. The superficial deltoid ligaments originate at the medial malleolus' anterior and posterior colliculi and insert upon the medial calcaneus, talar neck, and navicular bones as the tibiocalcaneal, posterior tibiotalar, and tibionavicular ligament, respectively (see Fig. 32.2). The deep deltoid ligaments begin at the medial malleolus' anterior colliculi as the anterior tibiotalar ligaments, but have three fascicles that attach to three distinct regions of the medial talar body [3, 5, 6]. Both the superficial and deep portions of the deltoid ligament complex are extrasynovial and intra-articular. This ligamentous complex helps to resist eversion and valgus joint stresses upon the ankle joint. Each portion of the superficial deltoid contributes to medial ankle instability at different ankle positions. The tibiocalcaneal ligament provides stability in ankle pronation, where it is tight at greater than and equal to 0° of dorsiflexion. The posterior tibiotalar ligament provides stability in ankle dorsiflexion, where it is tight at greater than and equal to 0° of dorsiflexion. The tibionavicular ligament provides stability in

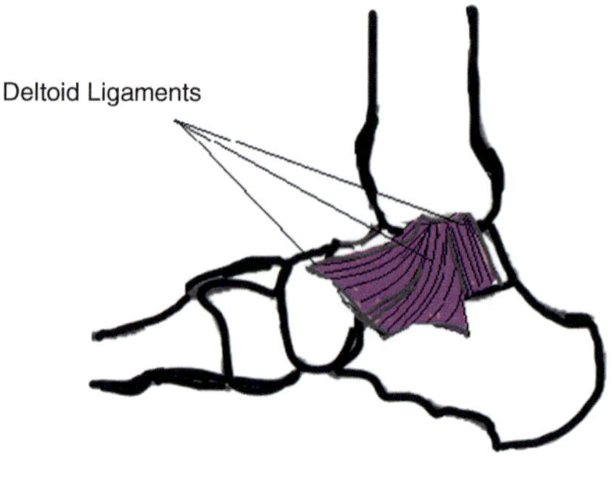

Fig. 32.2 Medial view of the ankle joint deltoid ligaments

ankle plantarflexion, where it is tight at greater than and equal to 10° of plantarflexion. The deep deltoid has a higher load to failure and provides greater resistance against lateral ankle translation with eversion than the superficial deltoid [4, 7].

32.2.3 Syndesmotic Ankle Ligament Anatomy

The syndesmotic ligaments connect the distal tibia and fibula to each other just superior to the talar dome at an area called the sigmoid notch. This syndesmosis consists of four ligaments, which are the anterior inferior tibiofibular ligament (AITFL), posterior inferior tibiofibular ligament (PITFL), transverse tibiofibular ligament (TTL), and interosseous membrane. These ligaments provide stability between the distal tibia and fibula, allowing both bones to move as one unit with ankle motion. This stability is conferred to the ankle joint as it provides additional lateral constraint to the talus in the ankle mortise laterally. Ogilvie-Harris et al. found the AITFL, PITFL, TTL, and interosseous membrane to provide 35%, 40%, 22%, and less than 10% resistance to lateral talar translation within the mortise respectively [8]. While the fibula is able to move in the sagittal plane against the distal tibia at the sigmoid notch, such motion is minimal as 2°–5° of external rotation and 1–3 mm of posterior and/or medial translation [9]. However, this small amount of distal fibular migration is enough to deepen the ankle mortise in dorsiflexion for additional ankle stability. An additional feature of the syndesmotic ligaments is that they work with the deltoid ligaments to provide resistance against abduction, eversion, and external rotation forces applied to the ankle joint.

32.3 Lateral Ankle Ligament Injuries

32.3.1 Introduction

Ligamentous injuries of the ankle are one of the most commonly seen orthopedic conditions by general and specialized physicians in an active patient population, particularly among athletes [10]. Studies of United States Emergency Department visits have shown that ankle ligament injuries or sprains occur in the general population at a rate of 2.15–3.29 sprains/1000 person years with roughly half occurring during athletic activity [11, 12]. The vast majority of these injuries are to the lateral ankle ligaments, which typically occur with excessive inversion and/or internal rotation stress applied to the ankle (see Fig. 32.3).

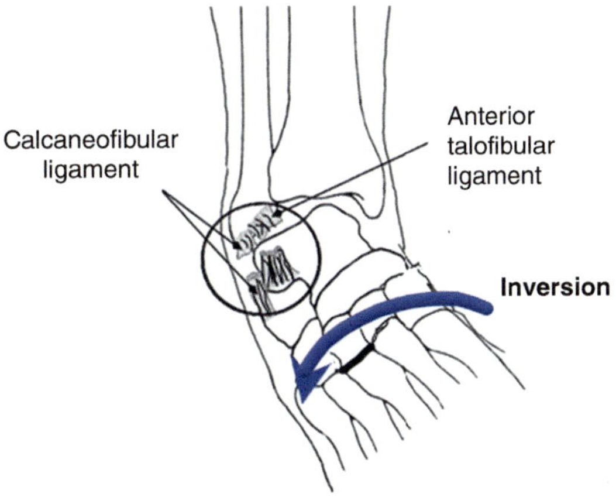

Fig. 32.3 Ankle joint ligament injury

32.3.2 Biomechanical Effects of Lateral Ankle Ligament Injuries

It is important to recognize that ligament injuries do not occur in a loaded ankle, but rather in the process of loading the ankle such as landing or cutting.

As discussed in an earlier chapter, both ligamentous structures and articular joint anatomy are primary determinants in ankle stability. Under an axial load, the ankle's articular congruency is shown to be the primary restraint to rotation and the sole constraint to translation and inversion or eversion [13, 14].

In lateral ankle ligament injuries, any or all can be affected. Of the three lateral ligaments, the most commonly injured is the ATFL. This is seen clinically and confirmed biomechanically where the ATFL has lower loads to failure than the CFL and PTFL [15]. The ATFL also experiences the highest strain to failure, which is typically with ankle plantarflexion and inversion. In comparison, the CFL is the second most common lateral ligament to be injured. It has a lower strain to failure than the ATFL, where the CFL's peak strain occurs with ankle dorsiflexion and inversion. This shows synergy between the ATFL and CFL for providing ankle stability against inversion during physiologic range of motion (ROM). The PTFL is the least likely lateral ligament to sprain or tear. It has a lower strain to failure than the CFL, where the PTFL's peak strain occurs with ankle dorsiflexion and external rotation. This shows that the PTFL has a lesser role in providing ankle stability to inversion and internal rotation than the ATFL and CFL [16]. While the PTFL is subject to injury during ankle external rotation, isolated injuries to it are rare. This is likely due to its low strain to failure. In biomechanical studies, external rotation stresses have been shown to produce higher strain in either the deltoid or syndesmotic ligaments than the PTFL [17].

Without an axial load, the lateral ankle ligaments are the primary restraint to anterior translation of the talus from the tibial plafond [14]. The ATFL is the primary restraint to internal rotation in plantarflexion, with less contribution as the ankle progresses to dorsiflexion. The CFL is the primary restraint to inversion in the sagittal plane with decreased contribution from the ATFL in plantarflexion [13]. With ATFL tear or rupture, there is increased anterior translation and internal rotation of the ankle, with minimal to mild increased inversion. With CFL injury, there is significantly increased talar inversion, with absent to minimal change in internal rotation or anterior translation [18–21]. These biomechanical findings correlate with clinical findings in patients with ATFL and CFL injuries. With ATFL tear, patients will exhibit an anterior drawer where the talus can subluxate anteriorly from the distal tibial plafond. With CFL injury, patients can demonstrate abnormal talar tilt on both physical examination and stress radiographs, where the talus is inverted under the distal tibial plafond.

Whether any or all of the lateral ankle ligaments are sprained, partially torn, or completely ruptured, the initial management for these injuries is nonsurgical. This includes providing rest, ice, compression, and elevation (RICE) to the limb. Those individuals with ligament sprains that are typically not very painful can be mobilized immediately with a functional ankle brace. More severe injuries will often have symptoms of ankle instability, pain with ankle motion, and/or pain with weight-bearing. Such patients may benefit from a brief period of ankle immobilization, whether in a cast or controlled ankle motion (CAM) boot. While there is no current consensus regarding how long patients should be immobilized for treatment of severe lateral ankle ligament injuries, most agree that such problems can heal well nonsurgically. In collegiate athletes with a variety of lateral ligament sprains that ranged from mild to severe, Roos et al. found that 44% and 96% of patients were able to return to play within 24 h and 3 weeks, respectively [10]. In patients with more severe ligament injuries and more persistent symptoms, proprioceptive physical therapy is included as acute treatment. Several investigators have shown proprioceptive therapy to improve patient reported outcomes and decrease risk of recurrent lateral ankle ligament injury [22, 23].

While many patients fully heal after acute lateral ankle ligament injuries, some individuals do not and subsequently develop chronic ankle instability (CAI) with lasting alterations in ankle biomechanics. Brown et al. studied gait parameters in patients with painful CAI, painless CAI, and uninjured ankle. As expected, they found that those with CAI had increased mean inversion laxity compared to the control population [24]. However, the authors and others observed patients with painful CAI to have increased inversion during swing phase and heel strike in stride than those with painless CAI [25]. Moisan et al. found that CAI patients have increased inversion, plantarflexion, and forces through the lateral foot during stance with increased magnitude and duration of peroneal activation [26]. Of note, some investigators have recorded proximal limb adaptations with CAI. Such altered biomechanics include increased hip flexion and adduction, decreased gluteal strength, and increased knee flexion with gait, jumping, and cutting. However, these studies are heterogeneous in quality and methodology [27]. Further investigation is warranted to clearly characterize the specific nature and significance of proximal limb adaptations with CAI.

32.3.3 Biomechanical Effects of Treatment of Lateral Ankle Ligament Injuries

The initial management for patients with unhealed lateral ankle ligaments with CAI is nonsurgical. This typically involves brace immobilization to the ankle, where options include over-the-counter braces and custom molded ankle foot orthoses (AFO). Other nonsurgical treatment modalities include physical therapy that focus primarily on proprioception and peroneal muscular activity. In this situation, the goal of increasing peroneal recruitment is to transform these muscle–tendon units into alternate lateral ankle stabilizers rather than promote lateral ankle ligament healing. Specifically, intensifying peroneal activation and the resultant ankle eversion is intended to overcome the increased ankle inversion laxity that occurs with CAI. Surgery upon the lateral ankle ligaments becomes a viable treatment option should patients do poorly with these nonsurgical management options.

The traditional surgical treatment for patients with unhealed lateral ankle ligaments with CAI is a Gould modification of a Brostrom lateral ankle ligament reconstruction. This involves a direct repair of the lateral ankle ligaments and imbrication of the lateral ankle joint capsule or inferior extensor retinaculum (IER) to the inferior lateral malleolus. This procedure is highly successful in providing lateral ankle ligament healing and improving lateral ankle instability. An important factor of this procedure is that it not only repairs but also tightens the lateral ankle ligament complex. This has a variable effect on lateral ankle stability. Bahr et al. studied biomechanics in ankle specimens with normal lateral ankle ligaments and those with transected lateral ligaments before and after the Brostrom procedure [28]. They found that the Brostrom provided specimens with decreased ankle joint laxity and more normalized ankle ROM and force patterns. Boey et al. performed a similar study in ankle specimens with normal lateral ankle ligaments and those with transected lateral ligaments before and after the Brostrom procedure [29]. Their study differs from Bahr et al.'s, where some of their specimens received a thick strand or tape of fiberwire suture (Internal Brace, Arthrex, Naples, FL) connecting the inferior lateral malleolus to the lateral talar body for augmenting the Brostrom. They found that not all specimens that received the Brostrom were able to achieve normal ankle ROM. They observed a greater percentage of specimens attained normal ankle parameters with suture tape augmenting the Brostrom rather than without, but this was not statistically significant. Further study is needed to determine the extent to which the Brostrom procedure with or without modifications is able to normalize ankle biomechanics.

32.4 Medial or Deltoid Ankle Ligament Injuries

32.4.1 Introduction

Harm to the medial or deltoid ligaments account for 5–10% of all ankle ligament injuries [30–32]. Most deltoid sprains and/or tears occur in a younger, athletic population, where the most common sports involved are football, gymnastics, and soccer. The most common mechanism of ligament injury is forceful or high-energy contact with another individual in which a large external rotation stress or moment is applied to the ankle [33]. This may occur with the ankle in supination, pronation, or neutral. Due to the robust nature of the deltoid ligament complex, sprains are not commonly seen in low-energy injuries. Rather, deltoid ligament damage implies a high-energy injury to the ankle where it is not uncommon for other structures at the ankle to be harmed simultaneously. Many deltoid ligament injuries can occur with lateral ankle ligament rupture and/or bony fracture to the distal fibula's lateral and/or distal tibia's posterior malleoli [30].

32.4.2 Biomechanical Effect of Deltoid Ligament Injuries

The deltoid ligaments' contributions to ankle stability vary based on loading and position of the ankle. With axial loads, cadaveric biomechanical studies demonstrate that the ankle's bone and joint articulations provide 100% of inversion, eversion, and medial/lateral translational stability. In addition, this particular congruity provides 70–80% anterior translational, 50–80% posterior translational, and 20–74% rotational stability. Rather it is when the ankle is unloaded and plantarflexed that the deltoid ligaments provide the primary restraint to lateral translation and eversion. Investigators have also shown that the deltoid ligaments work cooperatively with the lateral ligament complex to provide rotational stability, where the ankle exhibits increased internal and external rotation with disruption of either ligament complex [13, 14].

The effect that deltoid ligament injury has upon ankle biomechanics is well studied. As stated earlier, the deltoid provides little ankle stability during locomotion on a flat surface. Tochigi reconciled this in suggesting the deltoid ligaments likely provide ankle stability in maximal plantarflexion and/or dorsiflexion during stride [34].

Both the superficial and deep deltoid ligaments provide ankle stability, but in different ways. Simulating isolated injuries to the superficial deltoid ligament causes increased talar external rotation within the mortise with little eversion. Earll et al. found that the condition also led to decreased ankle joint contact areas by 43% and increased peak joint pressures by 30% [35]. Such altered biomechanics have clinical implications where patients can develop ankle degenerative joint disease (DJD) as a result. In addition, Ziai et al. found that the superficial deltoid has some role in resisting ankle inversion stresses. In simulated injuries to the lateral ankle and superficial deltoid ligaments, they measured increased talar inversion within the mortise when both structures are injured compared to isolated damage to the lateral ankle ligaments [36]. Others have shown that subsequent transection of the deep deltoid with the superficial ligaments cause increased talar eversion and lateral translation with external rotation within the mortise [37].

Deltoid ligament injuries can also significantly affect ankle biomechanics with respect to joint congruency. Years ago, Ramsey et al. showed up to 42% reductions in ankle joint contact area with every 1 mm of lateral talar translation within the mortise [38]. As stated earlier, this finding has clinical implications where patients can develop secondary ankle DJD. Harper subsequently demonstrated that complete superficial deltoid ligament rupture caused no significant talar translation in the mortise [39]. Rather, he found that a combined deep deltoid ligament rupture and distal fibular fracture led to lateral talar translation. As previously discussed, Earl et al. achieved different results and found that superficial deltoid ligament damage caused mean decreased ankle joint contact areas by 43%, increased peak joint pressures by 30%, and lateralized peak ankle joint pressures by 4 mm [35]. Such altered biomechanics have clinical implications where patients can develop ankle degenerative joint disease (DJD). While the particular contributions between the superficial and deep deltoid ligaments require further study, it is agreed that complete deltoid ligament injury can significantly alter the ankle's articular contact pressures, which can contribute to development of ankle DJD.

As long as the deltoid ligaments are the only structure harmed at the ankle, the initial management for these injuries is nonsurgical whether these ligaments are sprained, partially torn, or completely ruptured [40]. This includes providing rest, ice, compression, and elevation (RICE) to the limb. Those individuals with ligament strains that are not very painful can be mobilized immediately with a functional ankle brace. Other persons with more acute ankle instability, pain with ankle motion, and/or pain with weight-bearing may benefit from a brief period of ankle immobilization, whether in a cast or CAM boot. While there is no current consensus regarding how long patients should be immobilized for treatment of severe isolated deltoid ligament injuries, most agree that such problems can heal well nonsurgically. However, these conditions often take longer to heal than injuries to the lateral ankle ligaments. To the authors' knowledge, there are no studies addressing the effect of nonoperative treatment of isolated deltoid ligament injury on ankle biomechanics.

Of note, isolated complete deltoid ligament ruptures are rare. Rather, these more commonly occur with other injuries such proximal or distal fibular fracture, posterior malleolar fracture, and/or syndesmotic ligament injury which itself will be discussed later in this chapter. To date, there is limited literature regarding acute deltoid ligament repair. While associated injuries such as fracture may necessitate surgical intervention, there is no consensus regarding treatment of the deltoid ligaments in the setting of concomitant fracture. Given the rarity of isolated deltoid rupture, there is scant literature regarding acute ligament repair in this situation.

32.4.3 Biomechanical Effects of Treatment of Deltoid Ligament Injuries

While many patients heal after acute deltoid ligament injuries, some individuals do not and can subsequently develop CAI with lasting alterations in ankle biomechanics. As rare as isolated deltoid ligament injuries are, those that lead to CAI are even less frequent. However, existing literature shows that the talus can become malaligned in valgus and varying amounts of external rotation in the ankle mortise with longstanding CAI of the deltoid ligaments.

Currently, there are a variety of surgical techniques for treatment of unhealed deltoid ligaments with CAI. Most of these surgeries can be divided into deltoid ligament repair versus reconstruction. In cases where there is adequate deltoid ligament tissue available or preserved, the ligament can be primarily repaired with suture. In cases of deltoid ligament avulsion from the medial malleolus or talar body, suture anchor fixation can be used between the ligament and the affected bone. When there is inadequate deltoid tissue where the ligament may be greatly attenuated, the ligament requires a reconstruction utilizing tendon autograft or allograft [41].

Most biomechanical studies regarding surgical treatment of deltoid ligament instability focus upon ligament reconstruction. Haddad et al. described a technique of deltoid ligament reconstruction with an anterior tibial tendon graft secured through bone tunnels at the talar and calcaneal insertions of the ligament [42]. The remaining loop of graft is passed through a bone tunnel at the medial malleolus. These investigators recorded medial ankle stability in eversion and external rotation that was equivalent to the native deltoid ligaments.

Xu et al. compared four different types of deltoid ligament reconstruction [43]. Deland, Kitaoka, Hintermann, and Wiltberger previously described those particular techniques separately [44–46]. Deland's method involves passing a peroneus longus tendon graft through the talus from lateral to medial and through another bone tunnel in the medial malleolus with fixation to the tibia. Kitaoka's technique utilized a free extensor digitorum longus graft through bone tunnels in the medial malleolus and medial cuneiform. Hintermann described suture imbrication of the existing deltoid ligament and augmentation with plantaris tendon autograft through bone tunnels in the medial malleolus and navicular tuberosity. Finally, Wiltberger utilizes a split posterior tibial tendon graft from the navicular through a bone tunnel in the medial malleolus. While these four procedures were all originally described in the context of deltoid reconstruction for certain cases of flatfoot deformity with deltoid instability, Xu et al. compared them at the ankle through finite element analysis. The investigators demonstrated improved medial ankle stability in external rotation and eversion for all four techniques when compared to a deltoid deficient ankle. They clarified that none of these methods of deltoid reconstruction restored normal external rotation stability, but the Kitaoka procedure was closest among all four to normalized deltoid ligament biomechanics. With regard to restoration of normal eversion stability, Xu et al. found that the Kitaoka's and Deland's techniques for deltoid reconstruction allowed for near-normal and above normal stability, respectively, compared to the native deltoid ligaments. Further study is needed to determine if other types of deltoid ligament reconstruction can achieve normal or above normal medial ankle stability.

32.5 Syndesmotic Ankle Ligament Injuries

32.5.1 Introduction

Harm to the syndesmotic ligaments account for 1–11% of all ankle ligament injuries. Akin to deltoid ligament injuries, most syndesmotic strains and/or tears occur in a younger, athletic population, where the most common sports involved are football and hockey. The most common mechanism of syndesmotic injury is a high-energy external rotation stress or moment applied to the ankle joint as the ankle is supinated or pronated. Due to the robust nature of the syndesmotic ligament complex, sprains are not commonly seen in low-energy injuries. Rather, syndesmotic ligament damage implies a high-energy injury to the ankle where it is not uncommon for other structures at the ankle to be harmed simultaneously. Many syndesmotic ligament injuries can occur with deltoid ligament rupture and/or bony fracture to the proximal fibula and distal fibula's lateral and/or distal tibia's posterior and/or medial malleoli.

32.5.2 Biomechanical Effect of Syndesmotic Ligament Injuries

The effect that syndesmotic ligament injury has upon ankle biomechanics is the subject of much recent research. Clanton et al. simulated isolated injuries to the syndesmosis and found abnormal distal fibular motion at the sigmoid notch of greater than 5° of external rotation and 3 mm of anterior–posterior translation [47]. Complete syndesmotic ligament transection significantly decreased resistance to external and internal ankle rotation. With regard to contributions from individual ligaments, these investigators found the AITFL provided the most constraint against external rotation. However, no one syndesmotic ligament provided more resistance to internal ankle rotation than another.

Syndesmotic ligament injuries can also significantly affect ankle biomechanics with respect to joint congruency. Teramoto et al. simulated syndesmotic damage and applied abduction and external rotation torques to the ankle joint. With syndesmotic transection and abduction, tibiofibular diastasis and talar tilt in the ankle mortise increased significantly by 2 mm and 5°, respectively [48]. Such abnormal biomechanics may have clinical implications where patients may develop abnormal ankle joint contact areas due to talar translation, which can progress to secondary ankle DJD. However, this particular consideration requires further study.

As long as the syndesmotic ligaments are the only structure harmed at the ankle, the initial management for these injuries is nonsurgical if these ligaments are sprained or partially torn. This includes providing rest, ice, compression, and elevation (RICE) to the limb. Those individuals with ligament strains that are not very painful can be mobilized immediately with a functional ankle brace. Other persons may present with more acute ankle instability, pain with ankle motion, and/or pain with weightbearing. These patients benefit from a brief period of ankle immobilization, whether in a cast or CAM boot. While there is no current consensus regarding how long patients should be immobilized for treatment of isolated syndesmotic ligament injuries, most agree that strains and partial tears can heal well nonsurgically.

Some complete tears of the syndesmosis can be managed nonsurgically in the same above manner so long as distal tibio-fibular stability is preserved. Radiographically, this means that there is neither gapping nor diastasis between the distal tibia and fibula at the sigmoid notch.

Furthermore, the talus must remain well aligned in the ankle mortise for complete syndesmotic tears to be deemed stable and appropriate for nonsurgical treatment. To the authors' knowledge, there are no studies regarding the effect of nonoperative treatment of isolated stable syndesmotic injury on ankle biomechanics whether the ligaments are sprained, partially torn, or completely torn.

32.5.3 Biomechanical Effects of Treatment of Syndesmotic Ligament Injuries

Surgical treatment is indicated for complete syndesmotic ligament rupture with distal tibio-fibular and/or ankle instability. This is demonstrated with distal tibio-fibular diastasis at the sigmoid notch and/or talus malalignment in the ankle mortise.

The traditional surgical management for complete syndesmotic ligament rupture with distal tibio-fibular and/or ankle instability involves compression of the distal fibula to the distal tibia at and/or just above the sigmoid notch with steel screws. Several investigators have studied the biomechanical effects of this procedure upon the ankle joint. Clanton et al. found that syndesmotic screw fixation restored normal posterior fibular translation in the sagittal plane while restricting abnormal fibular motion to cause syndesmotic and/or ankle instability [49].

In contrast, Goetz et al. found detrimental effects upon the ankle joint from this means of syndesmotic ligament fixation [50]. In a cadaveric model, this group observed 6–32% decreased mean peak contact pressure on the talar dome and 2.4-to 6.6-fold increased mean peak contact pressure on the medial and lateral ankle joint gutters with steel screw fixation of the syndesmosis. While such results may have clinical implications such as progression to secondary ankle DJD, this has yet to be shown in the orthopedic literature. There are important aspects of this means of syndesmotic ligament fixation to consider. One is that the use of screws may be too rigid to allow the ankle joint to resume normal motion parameters after ligamentous healing. The other is that the screws do not directly repair the syndesmosis. Rather, they serve to compress the distal fibula to the distal tibia at and/or just above the sigmoid notch to allow for indirect ligamentous healing.

Those above shortcomings have influenced many to consider other means of surgically treating complete syndesmotic ligament rupture with distal tibio-fibular and/or ankle instability. One particular implant that has been utilized and studied as an alternative to steel screw syndesmotic fixation is the use of a suture endo-button construct known commercially as the Tight-Rope (Tight-Rope, Arthrex, Naples, FL). Similar to steel screws, this implant is placed through the distal fibula to the distal tibia at and/or just above the sigmoid notch to provide compression between the two bones. While this device also provides an indirect rather than a direct syndesmotic ligament repair, its theoretical benefit is that it is not as rigid as steel screws. This potentially avoids any negative effects on ankle biomechanics. While this implant performs well in clinical trials, it has mixed success in biomechanical studies. Pang et al. found more normalized mean contact pressures, peak pressures, contact area reduction, center of pressure translation, and relative talar and fibular motion with Tight-Rope over steel screw fixation of the syndesmosis [51]. However, others have observed suboptimal ankle biomechanics with this particular implant at the syndesmosis. Shoji et al. recorded continued abnormal tibiofibular diastasis and fibular rotation at the ankle with the Tight-Rope in a cadaveric model of syndesmotic instability [52]. Rather, it was the addition of weaving suture tape between the distal fibula and tibia at the AITFL that restored normal tibiofibular diastasis and fibular rotation. In two separate studies, two different investigators found persistent abnormal sagittal fibular translation at the sigmoid notch with the Tight-Rope with or without ankle rotational moments [49, 53]. Further study may be needed to clarify the biomechanical advantages and/or disadvantages of using a suture endo-button over steel screw usage to address complete syndesmotic ligament rupture with distal tibio-fibular and/or ankle instability.

An additional alternate means for treating complete syndesmotic ligament rupture with distal tibio-fibular and/or ankle instability is to formally repair the affected ligaments by weaving a graft of tendon between the distal fibula and tibia at its rupture site. This procedure is more demanding and technically challenging than compression of the distal fibula to the distal tibia at the sigmoid notch with steel screws or suture endo-buttons. In a comparative biomechanical study of syndesmotic instability, Che et al. recorded absent posterior fibular displacement on axial load and external rotation with an AITFL/PITFL/interosseous ligament reconstruction with a split peroneal brevis transfer [54]. In contrast, those specimens that received steel screw syndesmotic fixation performed worse than those with ligament reconstruction and experienced persistent abnormal fibular displacement at the sigmoid notch. While this technique of syndesmotic ligament reconstruction holds promise as it replaces the injured ligaments, further study is needed to confirm these positive results.

32.6 Future Directions in Foot and Ankle Soft Tissue Modeling Techniques

The biomechanics of soft tissue injuries at the foot and ankle joints are still a subject of research. Ligament injuries cause instability and imbalance of force transmission through the ankle joint. Delayed or non-healed ligaments lead to early osteoarthritis, which continues to be the most common joint disorder. The process of healing can be studied through gait analysis, X-rays, CTs, and MRIs. The future of foot and ankle must rely on modeling techniques (see Fig. 32.4) that combine imaging, deep understanding of the mechanics of the ankle joint and the role ligaments play in locomotion.

Despite advances in therapeutics, many ligaments do not regain their normal tensile strength. To address the different activities that the patient must undergo while in motion, testing might help shed the light on how ligament strength can be regained over time.

The biomechanical properties of ligament healing can be investigated by examining the foot and ankle kinematics to determine the effectiveness of the treatments. A combination of clinical studies and modeling using FEA and gait analysis may provide guidance to improve recovery time and reduce ligament injuries.

Once the FE models are created and validated, the analysis and potential outcomes can be very helpful to clinicians studying different foot and ankle pathologies in sport injuries. These models when used in combination with gait data and real-time imaging can set the stage for future clinical tools that are patient specific and deterministic in terms of selecting the right procedures with expected and measurable outcomes.

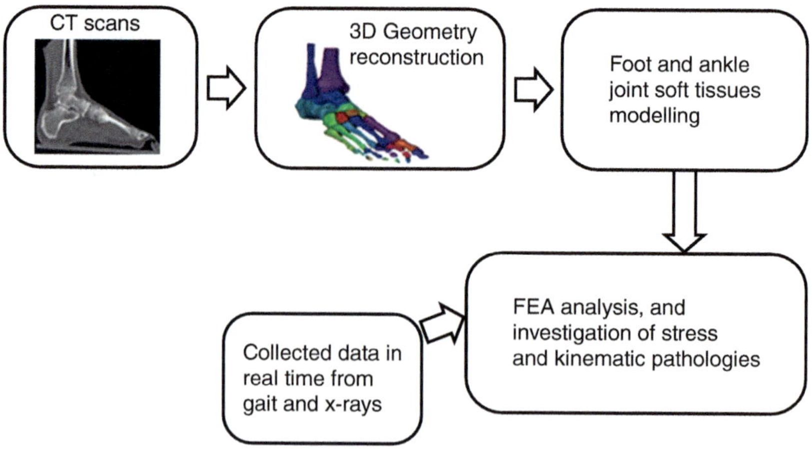

Fig. 32.4 Future direction in 3D finite element analysis of foot ankle pathologies during gait

32.7 Conclusion

The lateral, medial, and syndesmotic ankle ligaments contribute much stability to the ankle joint to allow for normal ankle biomechanics. Most acute injuries to these ligaments, particularly to the lateral and medial ankle, heal well nonsurgically without lasting detrimental effects to the ankle. Acute surgical treatment is indicated for complete tears of the syndesmosis in the presence of distal tibio-fibular diastasis. Surgery is also recommended for chronic and unhealed lateral and medial ankle ligament injuries that lead to CAI. If left untreated, ligament instability alters talar alignment within the ankle mortise, ROM, and other biomechanical parameters. Given enough time, such instability can cause ankle DJD by chronically altering ankle joint contact pressures. Most ligament repairs or reconstructions indeed allow for improved ankle stability, ROM, and biomechanics than before surgery. However, most surgical techniques do not completely normalize ankle ROM and biomechanics. A comprehensive understanding of ankle biomechanics in the context of ligamentous anatomy and injury may allow for future surgeries to restore the ankle joint to its pre-injury state.

References

1. Ahmad J, Raikin SM. Ankle arthrodesis: the simple and the complex. Foot Ankle Clin. 2008;13(3):381–400., viii. https://doi.org/10.1016/j.fcl.2008.04.007.
2. Manganaro D, Alsayouri K. Anatomy, bony pelvis and lower limb, ankle joint. StatPearls [Internet]. Treasure Island: StatPearls Publishing; 2020.
3. Golanó P, Vega J, de Leeuw PA, Malagelada F, Manzanares MC, Götzens V, et al. Anatomy of the ankle ligaments: a pictorial essay. Knee Surg Sports Traumatol Arthrosc. 2016;24(4):944–56. https://doi.org/10.1007/s00167-016-4059-4.
4. Petersen W, Rembitzki IV, Koppenburg AG, Ellerman A, Liebau C, Bruggemann GP, et al. Treatment of acute ankle ligament injuries: a systematic review. Arch Orthop Trauma Surg. 2013 Aug;133(8):1129–41. https://doi.org/10.1007/s00402-013-1742-5.
5. Campbell KJ, Michalski MP, Wilson KJ, Goldsmith MT, Wijdicks CA, LaPrade RF, et al. The ligament anatomy of the deltoid complex of the ankle: a qualitative and quantitative anatomical study. J Bone Joint Surg Am. 2014;96(8):e62. https://doi.org/10.2106/JBJS.M.00870.
6. Won HJ, Koh IJ, Won HS. Morphological variations of the deltoid ligament of the medial ankle. Clin Anat. 2016;29(8):1059–65. https://doi.org/10.1002/ca.22793.
7. Attarian DE, McCrackin HJ, DeVito DP, McElhaney JH, Garrett WE Jr. Biomechanical characteristics of human ankle ligaments. Foot Ankle. 1985;6(2):54–8. https://doi.org/10.1177/107110078500600202.
8. Ogilvie-Harris DJ, Reed SC, Hedman TP. Disruption of the ankle syndesmosis: biomechanical study of the ligamentous restraints. Arthroscopy. 1994;10(5):558–60. https://doi.org/10.1016/s0749-8063(05)80014-3.
9. Beumer A, Valstar ER, Garling EH, Niesing R, Ranstam J, Löfvenberg R, et al. Kinematics of the distal tibiofibular syndesmosis: radiostereometry in 11 normal ankles. Acta Orthop Scand. 2003;74(3):337–43. https://doi.org/10.1080/00016470310014283.
10. Roos KG, Kerr ZY, Mauntel TC, Djoko A, Dompier TP, Wikstrom EA. The epidemiology of lateral ligament complex ankle sprains in National Collegiate Athletic Association sports. Am J Sports Med. 2017;45(1):201–9. https://doi.org/10.1177/0363546516660980.
11. Waterman BR, Owens BD, Davey S, Zacchilli MA, Belmont PJ Jr. The epidemiology of ankle sprains in the United States. J Bone Joint Surg Am. 2010;92(13):2279–84. https://doi.org/10.2106/JBJS.I.01537.
12. Shah S, Thomas AC, Noone JM, Blanchette CM, Wikstrom EA. Incidence and cost of ankle sprains in United States Emergency Departments. Sports Health. 2016;8(6):547–52. https://doi.org/10.1177/1941738116659639.
13. Stormont DM, Morrey BF, An KN, Cass JR. Stability of the loaded ankle. Relation between articular restraint and primary and secondary static restraints. Am J Sports Med. 1985;13(5):295–300. https://doi.org/10.1177/036354658501300502.
14. Watanabe K, Kitaoka HB, Berglund LJ, Zhao KD, Kaufman KR, An KN. The role of ankle ligaments and articular geometry in stabilizing the ankle. Clin Biomech (Bristol, Avon). 2012;27(2):189–95. https://doi.org/10.1016/j.clinbiomech.2011.08.015.
15. Attarian DE, McCrackin HJ, Devito DP, McElhaney JH, Garrett WE Jr. A biomechanical study of human lateral ankle ligaments and autogenous reconstructive grafts. Am J Sports Med. 1985;13(6):377–81. https://doi.org/10.1177/036354658501300602.
16. Colville MR, Marder RA, Boyle JJ, Zarins B. Strain measurement in lateral ankle ligaments. Am J Sports Med. 1990;18(2):196–200. https://doi.org/10.1177/036354659001800214.
17. Wei F, Braman JE, Weaver BT, Haut RC. Determination of dynamic ankle ligament strains from a computational model driven by motion analysis based kinematic data. J Biomech. 2011;44(15):2636–41. https://doi.org/10.1016/j.jbiomech.2011.08.010.
18. Fujii T, Kitaoka HB, Watanabe K, Luo ZP, An KN. Ankle stability in simulated lateral ankle liga-

ment injuries. Foot Ankle Int. 2010;31(6):531–7. https://doi.org/10.3113/FAI.2010.0531.
19. Akhbari B, Dickinson MH, Louie EG, Shalhoub S, Maletsky LP. Characterization of ankle kinematics and constraint following ligament rupture in a cadaveric model [published online ahead of print, 2019 Jul 17]. J Biomech Eng. 2019;141(11):1110121–8. https://doi.org/10.1115/1.4044234.
20. Hollis JM, Blasier RD, Flahiff CM. Simulated lateral ankle ligamentous injury. Change in ankle stability. Am J Sports Med. 1995;23(6):672–7. https://doi.org/10.1177/036354659502300606.
21. Ringleb SI, Dhakal A, Anderson CD, Bawab S, Paranjape R. Effects of lateral ligament sectioning on the stability of the ankle and subtalar joint. J Orthop Res. 2011;29(10):1459–64. https://doi.org/10.1002/jor.21407.
22. Doherty C, Bleakley C, Delahunt E, Holden S. Treatment and prevention of acute and recurrent ankle sprain: an overview of systematic reviews with meta-analysis. Br J Sports Med. 2017;51(2):113–25. https://doi.org/10.1136/bjsports-2016-096178.
23. Herzog MM, Kerr ZY, Marshall SW, Wikstrom EA. Epidemiology of ankle sprains and chronic ankle instability. J Athl Train. 2019;54(6):603–10. https://doi.org/10.4085/1062-6050-447-17.
24. Brown CN, Rosen AB, Ko J. Ankle ligament laxity and stiffness in chronic ankle instability. Foot Ankle Int. 2015;36(5):565–72. https://doi.org/10.1177/1071100714561057.
25. Koldenhoven RM, Hart J, Saliba S, Abel MF, Hertel J. Gait kinematics & kinetics at three walking speeds in individuals with chronic ankle instability and ankle sprain copers. Gait Posture. 2019;74:169–75. https://doi.org/10.1016/j.gaitpost.2019.09.010.
26. Moisan G, Descarreaux M, Cantin V. Effects of chronic ankle instability on kinetics, kinematics and muscle activity during walking and running: a systematic review. Gait Posture. 2017;52:381–99. https://doi.org/10.1016/j.gaitpost.2016.11.037.
27. Dejong AF, Koldenhoven RM, Hertel J. Proximal adaptations in chronic ankle instability: systematic review and meta-analysis. Med Sci Sports Exerc. 2020;52(7):1563–75. https://doi.org/10.1249/MSS.0000000000002282.
28. Bahr R, Pena F, Shine J, Lew WD, Tyrdal S, Engebretsen L. Biomechanics of ankle ligament reconstruction. An in vitro comparison of the Broström repair, Watson-Jones reconstruction, and a new anatomic reconstruction technique. Am J Sports Med. 1997;25(4):424–32. https://doi.org/10.1177/036354659702500402.
29. Boey H, Verfaillie S, Natsakis T, Vander Sloten J, Jonkers I. Augmented ligament reconstruction partially restores hindfoot and midfoot kinematics after lateral ligament ruptures. Am J Sports Med. 2019;47(8):1921–30. https://doi.org/10.1177/0363546519848421.
30. Swenson DM, Collins CL, Fields SK, Comstock RD. Epidemiology of U.S. high school sports-related ligamentous ankle injuries, 2005/06-2010/11. Clin J Sport Med. 2013;23(3):190–6. https://doi.org/10.1097/JSM.0b013e31827d21fe.
31. Tummala SV, Hartigan DE, Makovicka JL, Patel KA, Chhabra A. 10-Year epidemiology of ankle injuries in men's and women's collegiate basketball [published correction appears in Orthop J Sports Med. 2019 Mar 29;7(3):2325967119837985]. Orthop J Sports Med. 2018;6(11):2325967118805400.
32. Gulbrandsen M, Hartigan DE, Patel KA, Makovicka JL, Tummala SV, Chhabra A. Ten-year epidemiology of ankle injuries in men's and women's collegiate soccer players. J Athl Train. 2019;54(8):881–8. https://doi.org/10.4085/1062-6050-144-18.
33. Kopec TJ, Hibberd EE, Roos KG, Djoko A, Dompier TP, Kerr ZY. The epidemiology of deltoid ligament sprains in 25 National Collegiate Athletic Association sports, 2009-2010 through 2014-2015 academic years. J Athl Train. 2017;52(4):350–9. https://doi.org/10.4085/1062.6050-52.2.01.
34. Tochigi Y, Rudert MJ, Amendola A, Brown TD, Saltzman CL. Tensile engagement of the peri-ankle ligaments in stance phase. Foot Ankle Int. 2005;26(12):1067–73. https://doi.org/10.1177/107110070502601212.
35. Earll M, Wayne J, Brodrick C, Vokshoor A, Adelaar R. Contribution of the deltoid ligament to ankle joint contact characteristics: a cadaver study. Foot Ankle Int. 1996;17(6):317–24. https://doi.org/10.1177/107110079601700604.
36. Ziai P, Benca E, Skrbensky GV, Wenzel F, Auffarth A, Krpo S, et al. The role of the medial ligaments in lateral stabilization of the ankle joint: an in vitro study. Knee Surg Sports Traumatol Arthrosc. 2015;23(7):1900–6. https://doi.org/10.1007/s00167-013-2708-4.
37. Rasmussen O, Kromann-Andersen C, Boe S. Deltoid ligament. Functional analysis of the medial collateral ligamentous apparatus of the ankle joint. Acta Orthop Scand. 1983;54(1):36–44. https://doi.org/10.3109/17453678308992867.
38. Ramsey PL, Hamilton W. Changes in tibiotalar area of contact caused by lateral talar shift. J Bone Joint Surg Am. 1976;58(3):356–7.
39. Harper MC. Deltoid ligament: an anatomical evaluation of function. Foot Ankle. 1987;8(1):19–22. https://doi.org/10.1177/107110078700800104.
40. McCollum GA, van den Bekerom MP, Kerkhoffs GM, Calder JD, van Dijk CN. Syndesmosis and deltoid ligament injuries in the athlete. Knee Surg Sports Traumatol Arthrosc. 2013;21(6):1328–37. https://doi.org/10.1007/s00167-012-2205-1.
41. Savage-Elliott I, Murawski CD, Smyth NA, Golanó P, Kennedy JG. The deltoid ligament: an in-depth review of anatomy, function, and treatment strategies. Knee Surg Sports Traumatol Arthrosc. 2013;21(6):1316–27. https://doi.org/10.1007/s00167-012-2159-3.
42. Haddad SL, Dedhia S, Ren Y, Rotstein J, Zhang LQ. Deltoid ligament reconstruction: a novel technique with biomechanical analysis. Foot Ankle Int. 2010;31(7):639–51. https://doi.org/10.3113/FAI.2010.0639.

43. Xu C, Zhang MY, Lei GH, Zhang C, Gao SG, Ting W, et al. Biomechanical evaluation of tenodesis reconstruction in ankle with deltoid ligament deficiency: a finite element analysis. Knee Surg Sports Traumatol Arthrosc. 2012;20(9):1854–62. https://doi.org/10.1007/s00167-011-1762-z.
44. Deland JT, de Asla RJ, Segal A. Reconstruction of the chronically failed deltoid ligament: a new technique. Foot Ankle Int. 2004;25(11):795–9. https://doi.org/10.1177/107110070402501107.
45. Kitaoka HB, Luo ZP, An KN. Reconstruction operations for acquired flatfoot: biomechanical evaluation. Foot Ankle Int. 1998;19(4):203–7. https://doi.org/10.1177/107110079801900403.
46. Hintermann B, Valderrabano V, Kundert HP. Lengthening of the lateral column and reconstruction of the medial soft tissue for treatment of acquired flatfoot deformity associated with insufficiency of the posterior tibial tendon. Foot Ankle Int. 1999;20(10):622–9. https://doi.org/10.1177/107110079902001002.
47. Clanton TO, Williams BT, Backus JD, Dornan GJ, Liechti DJ, Whitlow SR, et al. Biomechanical analysis of the individual ligament contributions to syndesmotic stability. Foot Ankle Int. 2017;38(1):66–75. https://doi.org/10.1177/1071100716666277.
48. Teramoto A, Kura H, Uchiyama E, Suzuki D, Yamashita T. Three-dimensional analysis of ankle instability after tibiofibular syndesmosis injuries: a biomechanical experimental study. Am J Sports Med. 2008;36(2):348–52. https://doi.org/10.1177/0363546507308235.
49. Clanton TO, Whitlow SR, Williams BT, Liechti DJ, Backus JD, Dornan GJ, et al. Biomechanical comparison of 3 current ankle syndesmosis repair techniques. Foot Ankle Int. 2017;38(2):200–7. https://doi.org/10.1177/1071100716666278.
50. Goetz JE, Rungprai C, Rudert MJ, Warth LC, Phisitkul P. Screw fixation of the syndesmosis alters joint contact characteristics in an axially loaded cadaveric model. Foot Ankle Surg. 2019;25(5):594–600. https://doi.org/10.1016/j.fas.2018.05.003.
51. Pang EQ, Bedigrew K, Palanca A, Behn AW, Hunt KJ, Chou L. Ankle joint contact loads and displacement in syndesmosis injuries repaired with tightropes compared to screw fixation in a static model. Injury. 2019;50(11):1901–7. https://doi.org/10.1016/j.injury.2019.09.012.
52. Shoji H, Teramoto A, Suzuki D, Okada Y, Sakakibara Y, Matsumara T, et al. Suture-button fixation and anterior inferior tibiofibular ligament augmentation with suture-tape for syndesmosis injury: a biomechanical cadaveric study. Clin Biomech (Bristol, Avon). 2018;60:121–6. https://doi.org/10.1016/j.clinbiomech.2018.10.014.
53. LaMothe JM, Baxter JR, Murphy C, Gilbert S, DeSandis B, Drakos MC. Three-dimensional analysis of fibular motion after fixation of syndesmotic injuries with a screw or suture-cutton construct. Foot Ankle Int. 2016;37(12):1350–6. https://doi.org/10.1177/1071100716666865.
54. Che J, Li C, Gao Z, Qi W, Ji B, Liu Y, et al. Novel anatomical reconstruction of distal tibiofibular ligaments restores syndesmotic biomechanics. Knee Surg Sports Traumatol Arthrosc. 2017;25(6):1866–72. https://doi.org/10.1007/s00167-017-4485-y.

Part VIII

Biomechanics of Tissue Repair Techniques

Biomechanics of Cartilage Repair 33

Tomoya Iseki, Benjamin B. Rothrauff,
Kazunori Shimomura, and Norimasa Nakamura

33.1 Introduction

From routine movements such as standing and walking to fast and complex movements encountered in sport, articular cartilage of the synovial joint functions to permit nearly frictionless interaction of the articular bones. The complex mechanical loads experienced by cartilage facilitate oxygen and nutrient diffusion to the endogenous chondrocytes while also directing the balanced synthesis and degradation of the surrounding cartilage extracellular matrix (ECM). However, traumatic injury or age-associated changes in joint homeostasis can compromise the ability of articular cartilage to efficiently distribute the forces encountered during joint motion, potentially contributing to degenerative joint disease (i.e., osteoarthritis). In order to prevent joint degeneration and/or facilitate cartilage repair with an emerging armamentarium of biological-based treatment strategies, a detailed understanding of the molecular mechanisms governing the effect of mechanical loads on cartilage (i.e., cartilage mechanobiology) is needed. This knowledge may then be applied as part of the treatment algorithm to promote cartilage healing or regeneration. This chapter will review how joint loading influences healthy cartilage structure and, more specifically, how mechanical loading mediates cellular responses that are either chondroprotective or catabolic, contributing to the initiation and progression of osteoarthritis. Examples of the application of mechanical loads to improve cartilage repair will also be highlighted.

33.2 Biomechanical Characteristics of Healthy Articular Cartilage

Synovial joints contain hyaline cartilage, which is a smooth, compressible connective tissue that is largely absent of blood vessels, nerves, and lymphatics. Articular cartilage function is attributed to its unique ECM structure and biochemical composition, comprised of water (65–80% wet weight, ww), collagen (10–15% ww), proteoglycan (10–15% ww), and endogenous

T. Iseki (✉)
Department of Orthpaedic Surgery, Hyogo College of Medicine, Nishinomiya, Hyogo, Japan
e-mail: iseki@hyo-med.ac.jp

B. B. Rothrauff
Department of Orthopaedic Surgery, Center for Cellular and Molecular Engineering, University of Pittsburgh School of Medicine, Pittsburgh, PA, USA

K. Shimomura
Department of Orthopaedic Surgery, Osaka University Graduate School of Medicine, Osaka, Japan

N. Nakamura
Osaka Health Science University| OHSU, Osaka, Japan

Global Centre for Medical Engineering and Informatics, Osaka University, Osaka, Japan
e-mail: norimasa.nakamura@ohsu.ac.jp

J. Koh et al. (eds.), *Orthopaedic Biomechanics in Sports Medicine*,
https://doi.org/10.1007/978-3-030-81549-3_33

chondrocytes, which constitute approximately 5% of the tissue volume. The chondrocytes receive nutrients through diffusion of the synovial fluid, which is a protein-rich joint lubricant produced by synoviocytes lining the inner surface of the joint capsule and positioned in close proximity to blood vessels [1]. The ECM more specifically consists of type II collagen and aggrecan, which serves to maintain the high water content of cartilage ECM due to its abundance of hydrophilic, negatively charged sulfated glycosaminoglycans [2].

Type II collagen in the form of cross-linked microfibrils is the predominant structural protein in hyaline cartilage, with other collagen subtypes, such as types IX and XI collagen, playing important roles in organizing ECM ultrastructure. While these other collagen subtypes constitute only a small proportion of the ECM (in terms of weight and volume), they play a crucial role in the biomechanical behavior of the whole structure [3]. Proteoglycans, mainly aggrecan, are also important molecules in the structure and function of cartilage ECM. Aggrecan, with negatively charged side chains of chondroitin sulfate, is highly hydrophilic and the resulting osmotic pressure resists tissue deformity with compressive loading. Other molecules, such as small-leucine rich proteoglycans biglycan and decorin, organize cartilage ECM structure through interactions with collagen type II fibrils, modulate cell signaling through transforming growth factor (TGF-β) binding, and regulate water concentrations by maintaining charge density [4].

Past studies have explored the relationships between mechanical loading and biomechanical properties of the native cartilage, with methodology including kinematic analyses, in vivo imaging techniques, and finite element models (FEM). Similar comparative studies have been performed with disease cartilage. As an example, the effect of dynamic loading of knee cartilage, using active and passive orthoses during running, on serum levels of cartilage oligomeric matrix protein (COMP), a biomarker of cartilage metabolism, have been performed [5]. COMP levels were found to increase immediately after running, with changes in COMP concentration after physical activity highly influenced by the baseline COMP level [5]. Several recent kinematic studies have been used to identify the effects of knee OA on activities of daily living, particularly gait [6–11]. Using an in vivo gait model, Harkey et al. investigated the association between habitual walking speed and resting femoral cartilage thickness and deformation, as determined by ultrasound, after 30 min of walking [12]. Ultrasound images were acquired by having subjects flex their dominant legs to 140° and imaging the femoral condyles above the superior region of the patella. This study found that while habitual walking speed was not associated with resting cartilage thickness, it was significantly associated with greater medial femoral cartilage deformation in the region superior to the patella [12]. Similarly, Van Rossom et al. used 3D modeling to investigate the relationships between knee cartilage thickness, loading during gait, and cartilage composition measured by T1rho and T2 mapping. Thicker cartilage was associated with higher condylar loading during walking while T1rho and T2 relaxation times correlated with increased compressive forces and pressures. Based on this study, the authors suggested that the femoral condyle cartilage adapts to localized loading during gait (Fig. 33.1) [13, 14]. Likewise, Nakayama and Higa et al. used 3D FEM to analyze the relationship between tibiofemoral alignment and shear stress in the most superficial 2 mm of the articular surface. It was found that the maximum shear stress increased from 1.6 MPa for the normal knee to 3.3, 5.2, and 7.2 MPa with increasing joint line obliquity of 5°, 7.5°, and 10°, respectively, concluding that joint obliquity should be avoided when performing knee osteotomy [15]. Yin et al. analyzed in vivo tibiofemoral articular cartilage contact biomechanics during a dynamic step-up motion, using MRI and dual fluoroscopic imaging, and found that both the medial and lateral compartments of the knee experienced convex (femur) to convex (tibia) contact in the sagittal plane upon fitting a circle to the curvature of the femoral condyles and tibial plateaus [16]. Together, these studies provide important baseline information regarding cartilage composition and morphology during various activities, which

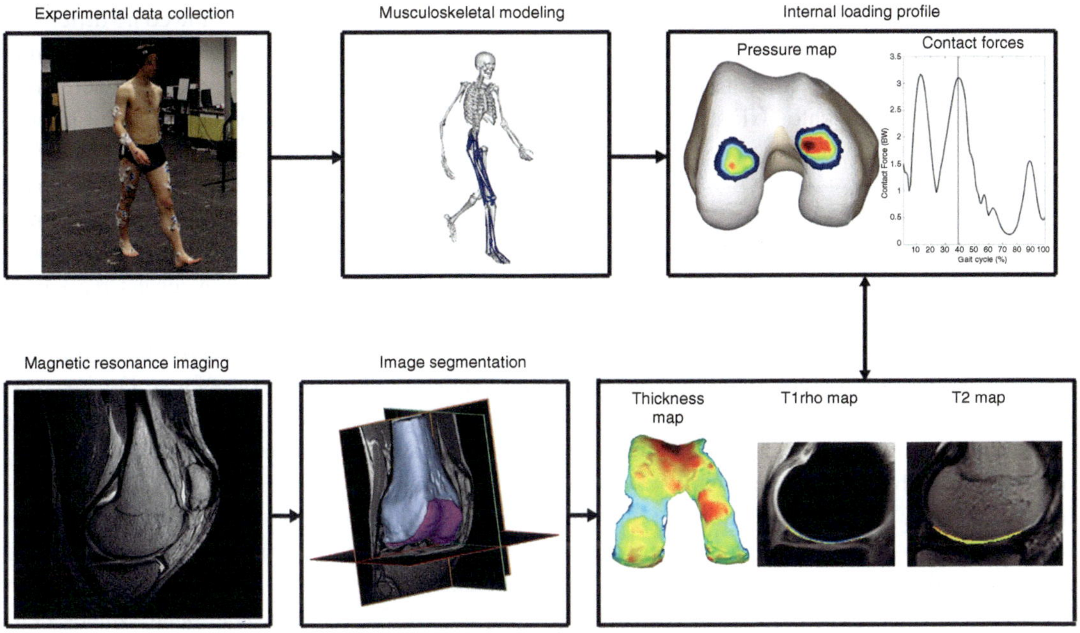

Fig. 33.1 Schematic overview of 3D modeling to investigate the relationships between knee cartilage thickness, loading during gait, and cartilage composition measured by T1rho and T2 mapping. Thicker cartilage was correlated with higher condylar loading during walking, suggesting that cartilage thickness is increased in those areas experiencing higher loading during a cyclic activity

can serve as a point of comparison for identifying the development and progression of diseased cartilage.

33.3 Mechanobiology in OA

Biochemical factors are known to influence the development and progression of OA. As acute trauma to articular cartilage provides a definite time point from which biochemical and biomechanical changes in the cartilage can be studied, it is commonly used experimental model [17, 18]. Early injury to cartilage, as encountered through a focal chondral defect, subsequently leads to proteolytic breakdown of the ECM. Chondrocytes may attempt to synthesize new matrix by upregulating anabolic processes to offset the increased catabolic cytokines and matrix-degrading enzymes associated with injury [19]. Subsequently, beginning at the cartilage surface, there is proteoglycan loss and breakdown of type II collagen, depending on a circumstance of excessive mechanical loading. There is often an associated increase in water concentration. ECM proteolysis in turn leads to critical reductions of the tensile strength of the ECM and to aberrant perturbation of the mechanosensors involved in normal articular cartilage homeostasis [20]. With abnormal articular cartilage homeostasis in lasting an unsuitable mechanical loading, there is continued degradation of ECM and development of synovial inflammation.

Matrix metalloproteinases (MMPs) are central to cartilage ECM homeostasis. Previous studies have shown that tissue-specific MMPs 1, 8, and 13 (collagenases I, II, and III, respectively) are involved in OA. The subsequent step is aggrecan degradation produced by aggrecanases 1 and 2, which are family of the ADAMs (a desintegrin and metalloproteinases) attached to type 1 thrombospondin (TS1) [21]. Growth of bone spurs thereby progresses during disease development, which will further lead to articular inflammation and joint pain [22]. In this state, the presence of interleukin 1 (IL-1) and tumor necrosis factor α

(TNF-α) induces the synthesis of other inflammatory factors like cyclooxygenases (COX-1, COX-2), mitogen-activated protein kinase (MAPK), as well nuclear factor-kappa B (NF-kB) [23].

33.4 Mechanical Loading Influences Cartilage Repair

If articular cartilage is focally damaged, several treatments of the defect site exist, including microfracturing, autologous chondrocyte implantation (ACI), and osteochondral autograft transplantation (OAT). While short- to medium-term outcomes have been good, long-term results are limited by either the initial formation of fibrocartilage or the progressive deterioration of tissue properties at the repair site. In an effort to improve the formation and stability of hyaline cartilage at the repair site, there has been increasing attention paid to the mechanical environment in which the neocartilage forms. It is hoped that targeted rehabilitation may be able to provide the appropriate mechanotransductive cues to achieve better long-term repair outcomes.

In vitro studies have suggested that shear forces mimicking early passive range of motion (ROM) may stimulate proteoglycan metabolism in the native articular surface [24–26] and potentially guide the cellular response in an implanted graft, resulting in neomatrix production [25]. While translation of in vitro findings to clinical recommendations is not adequately supported in the literature, some clinicians have implemented these findings by recommending 8 weeks of continuous passive motion (CPM) following certain cartilage repair procedures [27]. In support of this clinical implementation, Nugent-Derfus et al. reported that CPM of explanted bovine knee joints induced proteoglycan 4 (*PRG 4*) expression after 24 h of continuous stimulation (Fig. 33.2) [28]. However, Iseki et al., using fluid shear stress on an in vitro model of microfracture repair showed that a marked increase in *MMP 3* and *ADAMTS 5* gene expression was observed, which is associated with a catabolic response or with attempts at remodeling by the differentiating cells. As for longer term effects of CPM in a clinical study, Marder et al. showed no difference in clinical outcomes at a mean of 4.2 years after microfracture between patients who were non-weightbearing with CPM versus patients allowed to bear weight as tolerated without CPM [29].

In terms of compressive loading parameters, a number of different culturing and loading conditions are described to promote engineered cartilage maturation [30]. Varying results have been reported, depending on the specific combination of bioreactor, scaffolds, growth factor supplementation, and loading parameters. For instance, Thorpe et al. reported that compressive loading inhibited chondrogenesis of MSCs [31]. Stolberg et al. also, using the impact model of compressive loading to mimic the post traumatic arthritis, showed that compressive strains resulted in significantly reduced chondrocyte viability (Fig. 33.3) [32]. On the other hand, Mauck et al. demonstrated enhanced chondrogenesis after compressive mechanical activation [33]. Iseki et al. also showed that dynamic compressive loading enhanced neotissue chondrogenesis and maturation in a simulated in vitro model of microfracture, with generation of more hyaline-like cartilage and improved integration with the surrounding tissue (Fig. 33.4) [34]. Due to these conclusions of the effectiveness of the compressive loading on the cartilage repair, an optimal compressive load produced in the course of post-operative rehabilitation must be existed and be beneficial in promoting the facilitative cartilage repair in the future.

As a potential mechanism explaining the chondrogenic effect of compressive loading, Fahy et al. found consistent upregulation of TGF-b expression in mechanically activated MSCs, presumably serving as an autocrine/paracrine factor to enhance chondrogenesis [35]. Thorpe et al. also reported that TGF-b, when added as a medium supplement, further enhanced the effect of compressive loading in upregulating GAG content, as compared to unloaded constructs [31]. Taken together, these results involving the suggest that in vitro experiments mimicking the in vivo forces encountered following cartilage repair may also be able to guide rehabilitation protocols.

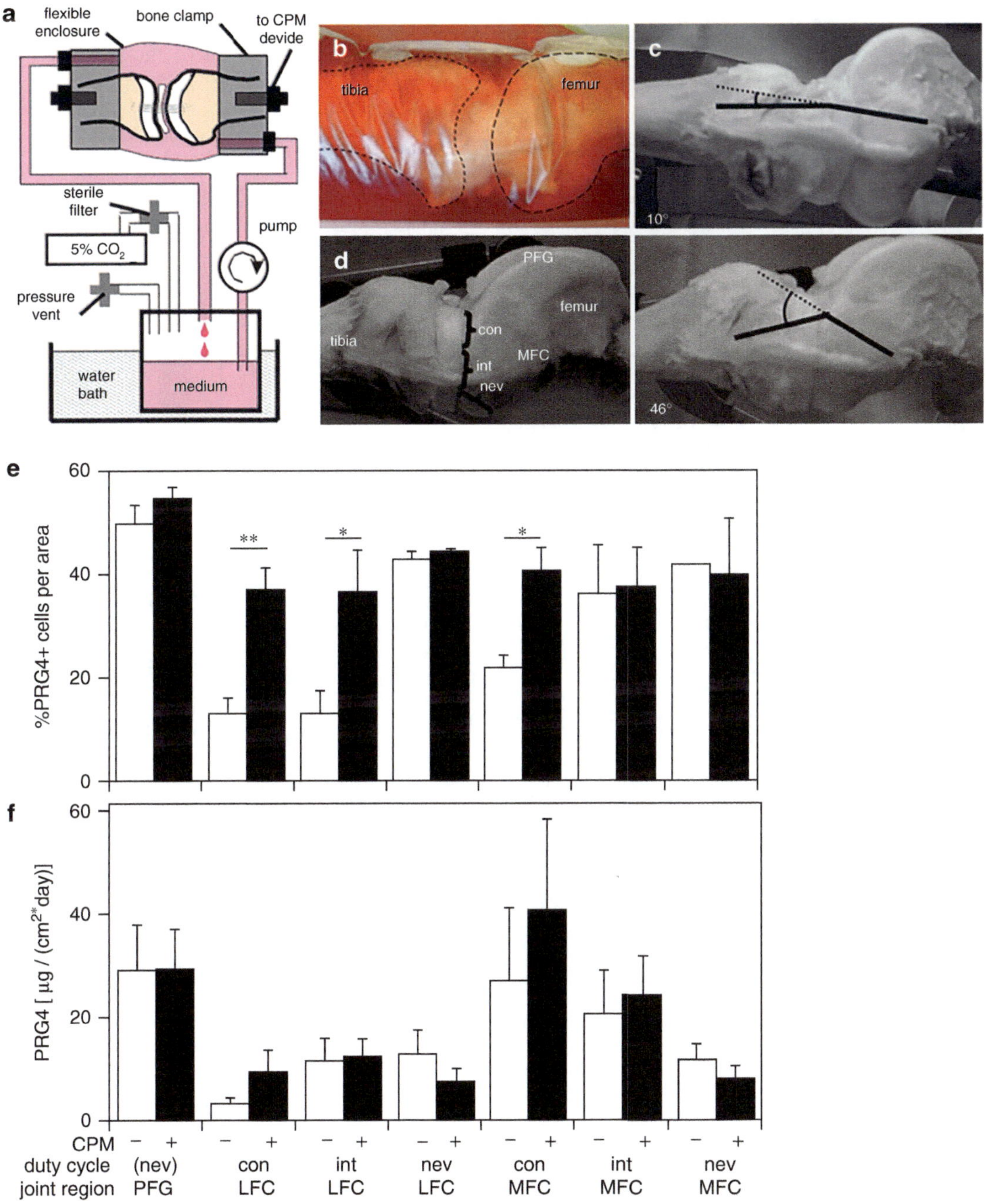

Fig. 33.2 (**a**) Schematic of CPM bioreactor system for maintaining tissue-culture conditions to regulate articular cartilage metabolism of chondrocyte proteoglycan 4 (*PRG4*). (**b**) Actual joint in bioreactor. (**c**, **d**) CPM stimulation for a bovine joint consisted of 24 h of continuous motion, with the joint oscillating. (**e**, **f**) The percent of the chondrocyte population expressing PRG4 was regulated by CPM exercise

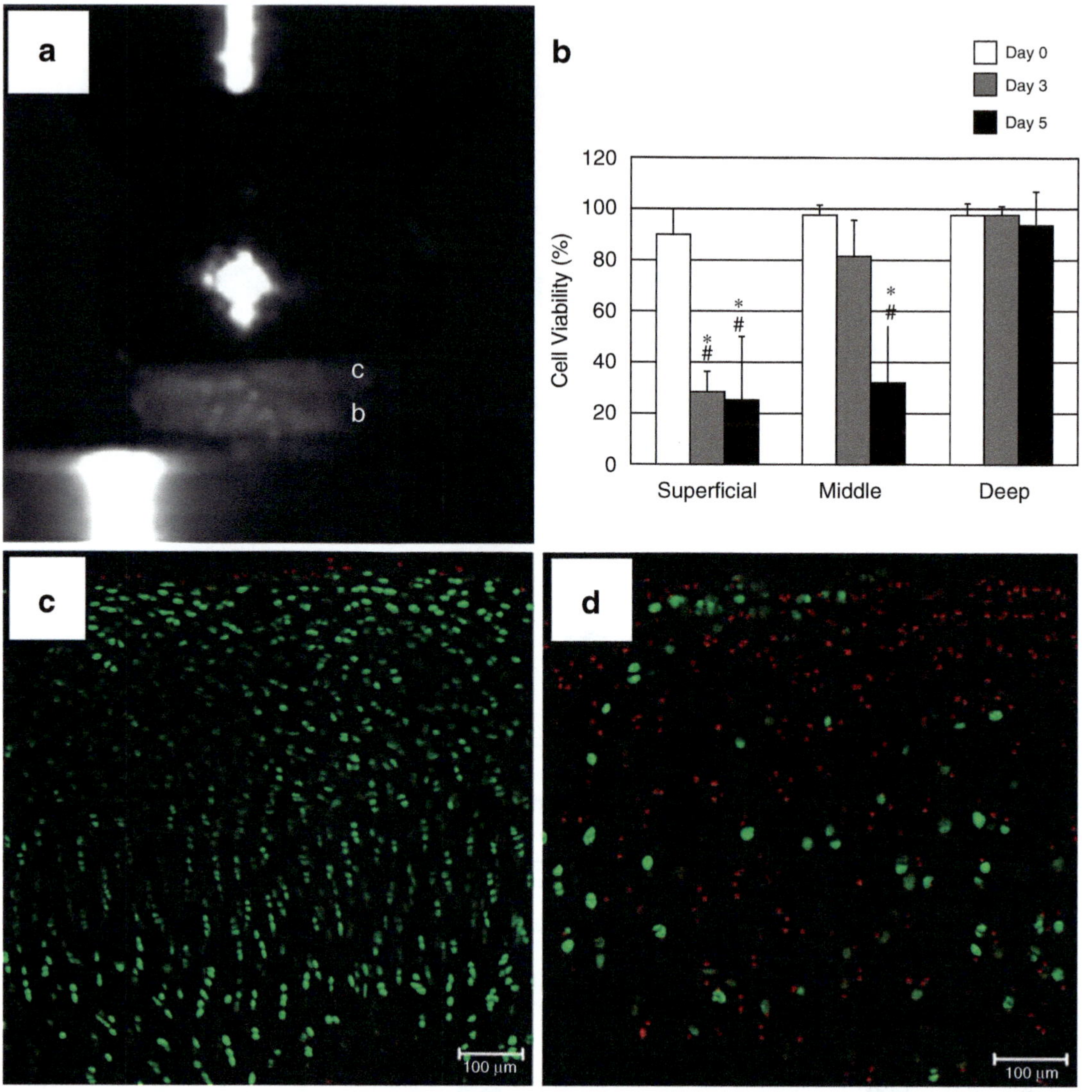

Fig. 33.3 (**a**) Impact model of osteochondral cores that can apply surface-to-surface compressive strain. (**b**) Chondrocyte viability by cartilage zone. A significant decrease of viability over time could only be detected after 70% strain in the superficial and middle zone. (**c**) Live/dead assay confocal microscope images. Seventy percent strain blunt compressive impact caused a significant loss of viability over time from day 0 to (**d**) day 5

33.5 Conclusion

In this chapter, the biochemical and biomechanical properties of native cartilage have been summarized, with review of how these properties change with disease or injury, and brief consideration with how mechanical loads may serve as injurious or therapeutic stimuli. In order to better understand the relationships between mechanical loading and cartilage injury or repair, in order to prevent OA development and progression, in vitro models of the forces encountered during in vivo loading have recently been developed. That said, mapping of cellular responses of cells in vitro studies to their in vivo response remains the target of future research. In addition to mechanical loading, concurrent supplementation of repaired cartilage with biochemical drivers of chondrogenesis and ECM synthesis, such as exogenously delivered or endogenously pro-

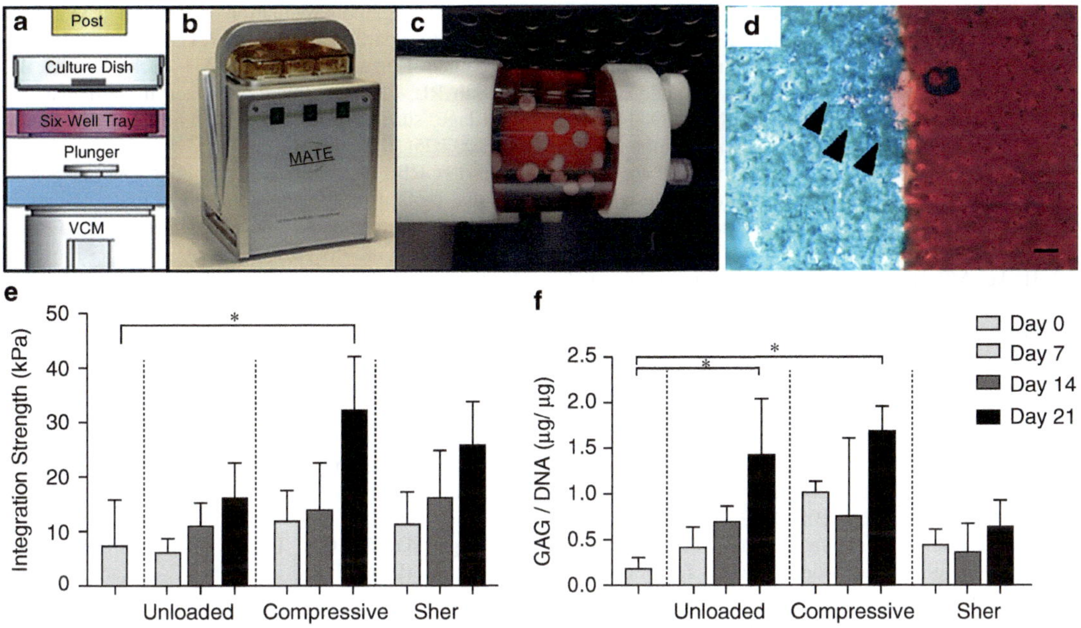

Fig. 33.4 (**a**, **b**) Dynamic compressive loading was provided using the bioreactor system which allows the application of cyclic compression to tissues with controlled force and displacement within an incubator. (**c**) Rotatory cell culture system in the incubator to provide continuous shear loading. (**d**, **e**) Safranin O/Fast Green histological staining of the interface between outer hyaline chondral ring and central insert of progenitor cell in the compressive loading condition. (**f**) Glycosaminoglycan production (GAG/DNA) was significantly higher in compressively loaded constructs compared with day 0 controls

moted TGF-β, may serve to further enhance the formation and stability of hyaline cartilage matching that of native articular cartilage. Therefore, future studies on cartilage repair should investigate the role and therapeutic utility of mechanical loading to improve present of future regenerative cell- or tissue-based therapies.

References

Uncategorized References

1. Rieppo J, Toyras J, Nieminen MT, Kovanen V, Hyttinen MM, Korhonen RK, et al. Structure-function relationships in enzymatically modified articular cartilage. Cells Tissues Organs. 2003;175(3):121–32.
2. Sun DD, Guo XE, Likhitpanichkul M, Lai WM, Mow VC. The influence of the fixed negative charges on mechanical and electrical behaviors of articular cartilage under unconfined compression. J Biomech Eng. 2004;126(1):6–16.
3. Mayne R. Cartilage collagens. What is their function, and are they involved in articular disease? Arthritis Rheum. 1989;32(3):241–6.
4. Hardingham TE, Fosang AJ. Proteoglycans: many forms and many functions. FASEB J. 1992;6(3):861–70.
5. Firner S, Willwacher S, de Marees M, Bleuel J, Zaucke F, Bruggemann GP, et al. Effect of increased mechanical knee joint loading during running on the serum concentration of cartilage oligomeric matrix protein (COMP). J Orthop Res. 2018;36(7):1937–46.
6. Foucher KC. Sex-specific hip osteoarthritis-associated gait abnormalities: alterations in dynamic hip abductor function differ in men and women. Clin Biomech. 2017;48:24–9.
7. Tanimoto K, Takahashi M, Tokuda K, Sawada T, Anan M, Shinkoda K. Lower limb kinematics during the swing phase in patients with knee osteoarthritis measured using an inertial sensor. Gait Posture. 2017;57:236–40.
8. Astephen Wilson JL, Stanish WD, Hubley-Kozey CL. Asymptomatic and symptomatic individuals with the same radiographic evidence of knee osteoarthritis walk with different knee moments and muscle activity. J Orthop Res. 2017;35(8):1661–70.
9. Ogaya S, Kubota R, Chujo Y, Hirooka E, Kwang-Ho K, Hase K. Muscle contributions to knee extension in

the early stance phase in patients with knee osteoarthritis. Gait Posture. 2017;58:88–93.
10. Na A, Piva SR, Buchanan TS. Influences of knee osteoarthritis and walking difficulty on knee kinematics and kinetics. Gait Posture. 2018;61:439–44.
11. Kumar D, Wyatt C, Lee S, Okazaki N, Chiba K, Link TM, et al. Sagittal plane walking patterns are related to MRI changes over 18-months in people with and without mild-moderate hip osteoarthritis. J Orthop Res. 2018;36(5):1472–7.
12. Harkey MS, Blackburn JT, Davis H, Sierra-Arevalo L, Nissman D, Pietrosimone B. The association between habitual walking speed and medial femoral cartilage deformation following 30minutes of walking. Gait Posture. 2018;59:128–33.
13. Van Rossom S, Smith CR, Zevenbergen L, Thelen DG, Vanwanseele B, Van Assche D, et al. Knee cartilage thickness, T1rho and T2 relaxation time are related to articular cartilage loading in healthy adults. PLoS One. 2017;12(1):e0170002.
14. Van Rossom S, Wesseling M, Van Assche D, Jonkers I. Topographical variation of human femoral articular cartilage thickness, T1rho and T2 relaxation times is related to local loading during walking. Cartilage. 2019;10(2):229–37.
15. Nakayama H, Schroter S, Yamamoto C, Iseki T, Kanto R, Kurosaka K, et al. Large correction in opening wedge high tibial osteotomy with resultant joint-line obliquity induces excessive shear stress on the articular cartilage. Knee Surg Sports Traumatol Arthrosc. 2018;26(6):1873–8.
16. Yin P, Li JS, Kernkamp WA, Tsai TY, Baek SH, Hosseini A, et al. Analysis of in-vivo articular cartilage contact surface of the knee during a step-up motion. Clin Biomech. 2017;49:101–6.
17. Ganz R, Leunig M, Leunig-Ganz K, Harris WH. The etiology of osteoarthritis of the hip: an integrated mechanical concept. Clin Orthop Relat Res. 2008;466(2):264–72.
18. Saxby DJ, Lloyd DG. Osteoarthritis year in review 2016: mechanics. Osteoarthr Cartil. 2017;25(2):190–8.
19. Martinez-Moreno D, Jimenez G, Galvez-Martin P, Rus G, Marchal JA. Cartilage biomechanics: a key factor for osteoarthritis regenerative medicine. Biochim Biophys Acta Mol basis Dis. 2019;1865(6):1067–75.
20. Radin EL, Burr DB, Caterson B, Fyhrie D, Brown TD, Boyd RD. Mechanical determinants of osteoarthrosis. Semin Arthritis Rheum. 1991;21(3 Suppl 2):12–21.
21. Abbaszade I, Liu RQ, Yang F, Rosenfeld SA, Ross OH, Link JR, et al. Cloning and characterization of ADAMTS11, an aggrecanase from the ADAMTS family. J Biol Chem. 1999;274(33):23443–50.
22. Scanzello CR, Goldring SR. The role of synovitis in osteoarthritis pathogenesis. Bone. 2012;51(2):249–57.
23. Goldring MB, Berenbaum F. Human chondrocyte culture models for studying cyclooxygenase expression and prostaglandin regulation of collagen gene expression. Osteoarthr Cartil. 1999;7(4):386–8.
24. Nugent-Derfus GE, Chan AH, Schumacher BL, Sah RL. PRG4 exchange between the articular cartilage surface and synovial fluid. J Orthop Res. 2007;25(10):1269–76.
25. Salter RB. The biologic concept of continuous passive motion of synovial joints. The first 18 years of basic research and its clinical application. Clin Orthop Relat Res. 1989;(242):12–25.
26. Ikenoue T, Trindade MC, Lee MS, Lin EY, Schurman DJ, Goodman SB, et al. Mechanoregulation of human articular chondrocyte aggrecan and type II collagen expression by intermittent hydrostatic pressure in vitro. J Orthop Res. 2003;21(1):110–6.
27. Mithoefer K, Hambly K, Logerstedt D, Ricci M, Silvers H, Della VS. Current concepts for rehabilitation and return to sport after knee articular cartilage repair in the athlete. J Orthop Sports Phys Ther. 2012;42(3):254–73.
28. Nugent-Derfus GE, Takara T, O'Neill JK, Cahill SB, Gortz S, Pong T, et al. Continuous passive motion applied to whole joints stimulates chondrocyte biosynthesis of PRG4. Osteoarthr Cartil. 2007;15(5):566–74.
29. Marder RA, Hopkins G Jr, Timmerman LA. Arthroscopic microfracture of chondral defects of the knee: a comparison of two postoperative treatments. Arthroscopy. 2005;21(2):152–8.
30. Filardo G, Kon E, Roffi A, Di Martino A, Marcacci M. Scaffold-based repair for cartilage healing: a systematic review and technical note. Arthroscopy. 2013;29(1):174–86.
31. Thorpe SD, Buckley CT, Vinardell T, O'Brien FJ, Campbell VA, Kelly DJ. Dynamic compression can inhibit chondrogenesis of mesenchymal stem cells. Biochem Biophys Res Commun. 2008;377(2):458–62.
32. Stolberg-Stolberg JA, Furman BD, Garrigues NW, Lee J, Pisetsky DS, Stearns NA, et al. Effects of cartilage impact with and without fracture on chondrocyte viability and the release of inflammatory markers. J Orthop Res. 2013;31(8):1283–92.
33. Mauck RL, Byers BA, Yuan X, Tuan RS. Regulation of cartilaginous ECM gene transcription by chondrocytes and MSCs in 3D culture in response to dynamic loading. Biomech Model Mechanobiol. 2007;6(1–2):113–25.
34. Iseki T, Rothrauff BB, Kihara S, Sasaki H, Yoshiya S, Fu FH, et al. Dynamic compressive loading improves cartilage repair in an in vitro model of microfracture: comparison of 2 mechanical loading regimens on simulated microfracture based on fibrin gel scaffolds encapsulating connective tissue progenitor cells. Am J Sports Med. 2019;47(9):2188–99.
35. Fahy N, Alini M, Stoddart MJ. Mechanical stimulation of mesenchymal stem cells: implications for cartilage tissue engineering. J Orthop Res. 2018;36(1):52–63.

GPSR Compliance

The European Union's (EU) General Product Safety Regulation (GPSR) is a set of rules that requires consumer products to be safe and our obligations to ensure this.

If you have any concerns about our products, you can contact us on ProductSafety@springernature.com

In case Publisher is established outside the EU, the EU authorized representative is:

Springer Nature Customer Service Center GmbH
Europaplatz 3
69115 Heidelberg, Germany

Batch number: 10732799

Printed by Printforce, the Netherlands